Plates are referred to in Chapter 10, pages 255 and 265

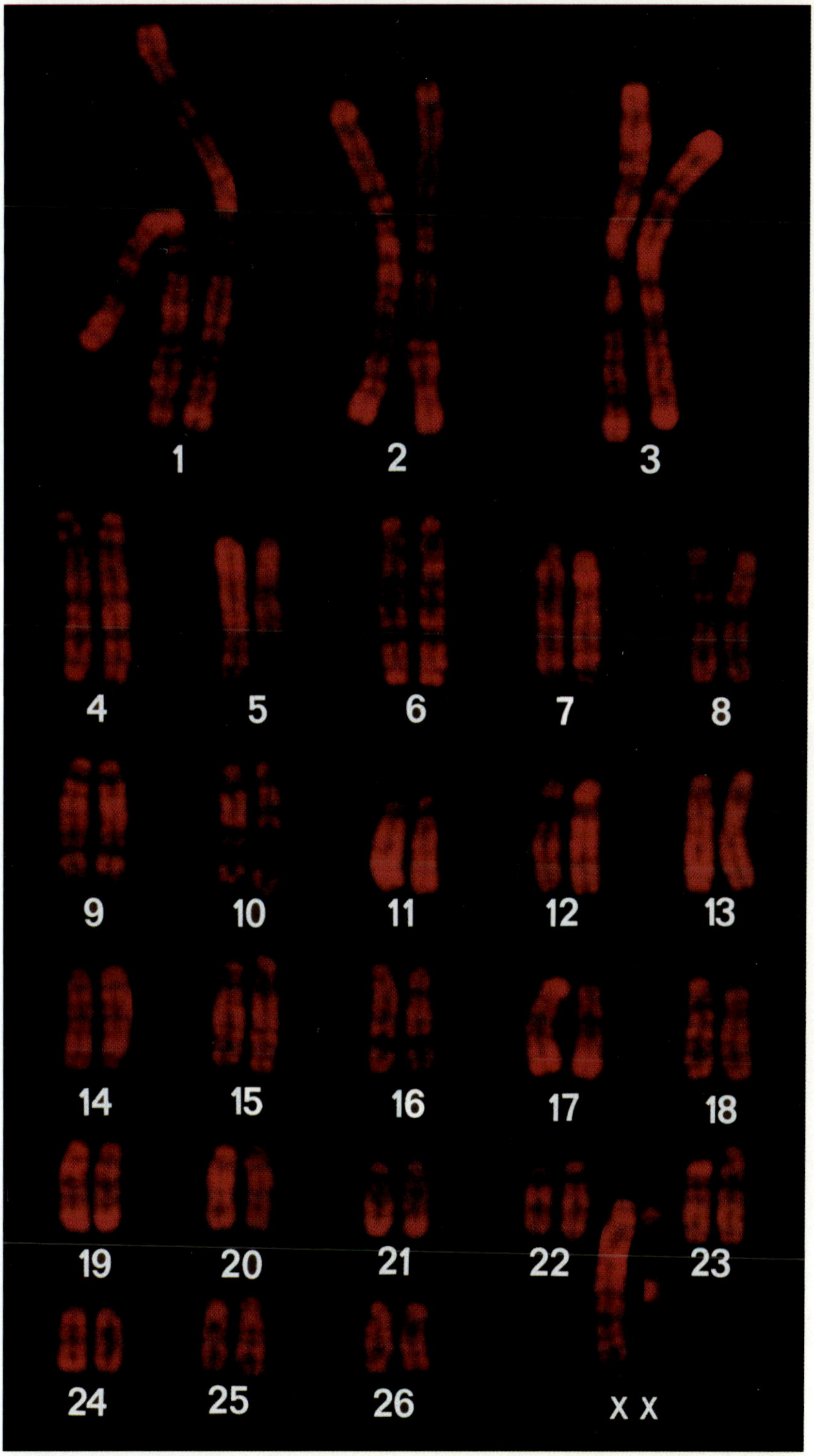

Plate 1. A RBP-banded sheep karyotype (prepared by Dr Helene Hayes).

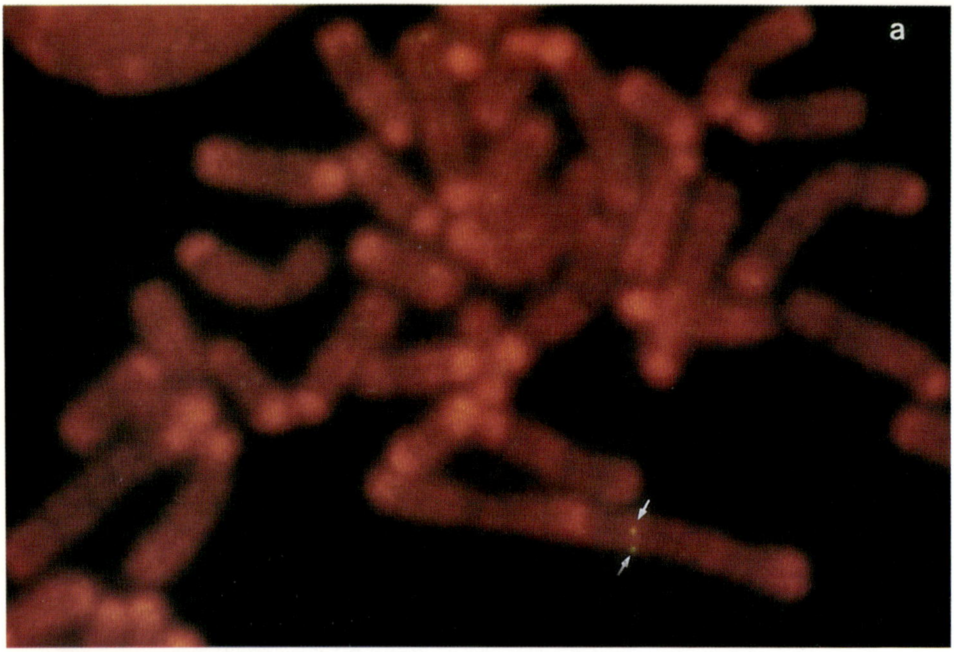

Plate 2a. Fluorescence *in situ* hybridization of *alpha lactalbumin (LALBA)* to sheep chromosome 3q21. The probe, which was a genomic DNA fragment of 11.5 kb from goat, was labelled with biotin by nick-translation (from Hayes *et al.*, 1993; with permission from Springer-Verlag GmbH and Company, Berlin).

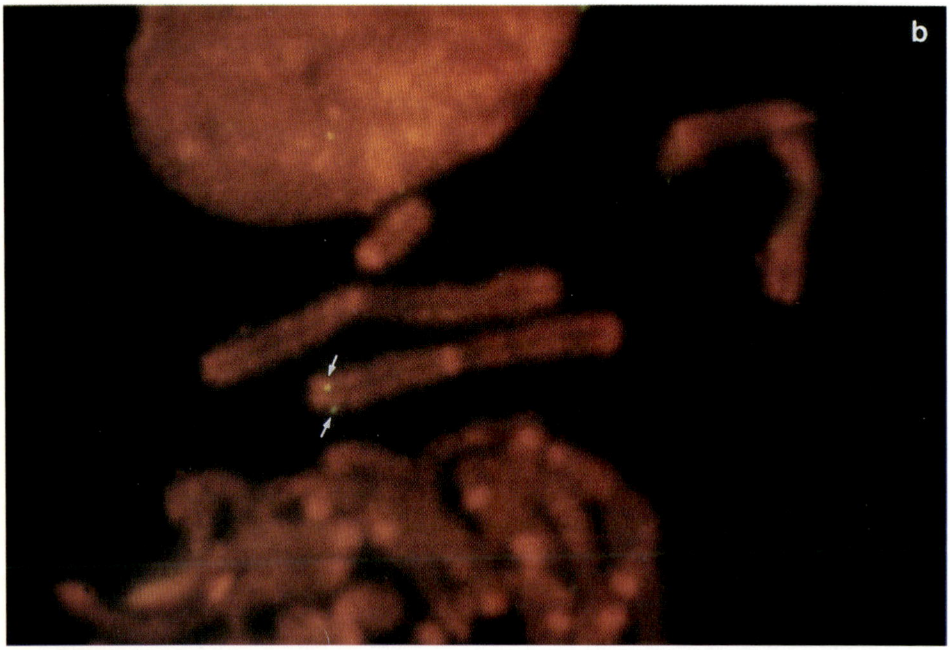

Plate 2b. Fluorescence *in situ* hybridization of *glycoprotein alpha galactosyltransferase (GGTA1)* to sheep chromosome 3p26. The probe, which was a bovine cDNA fragment of 0.94 kb, was labelled with biotin by PCR (from Hayes *et al.*, 1996; with permission from S. Karger AG, Basel).

THE GENETICS
of Sheep

THE GENETICS *of Sheep*

Edited by

LAURIE PIPER

CRC for Premium Quality Wool,
University of New England, Armidale 2351,
NSW, Australia

and

ANATOLY RUVINSKY

Department of Animal Science,
University of New England, Armidale 2351,
NSW, Australia

CAB INTERNATIONAL

CAB INTERNATIONAL
Wallingford
Oxon OX10 8DE
UK

CAB INTERNATIONAL
198 Madison Avenue
New York, NY 10016–4314
USA

Tel: +44 (0)1491 832111
Fax: +44 (0)1491 833508
E-mail: cabi@cabi.org

Tel: +1 212 726 6490
Fax: +1 212 686 7993
E-mail: cabi-nao@cabi.org

A catalogue record for this book is available from the British Library, London, UK.

Library of Congress Cataloging-in-Publication Data
The genetics of sheep / edited by Laurie Piper, Anatoly Ruvinsky.
p. cm.
Includes index.
ISBN 0–85199–200–5 (alk. paper)
1. Sheep–Breeding. 2. Sheep–Genetics. I. Piper, Laurie.
II. Ruvinsky. Anatoly. III. C.A.B. International.
SF376.2.G475 1997 97–14632
636.3 '0821–dc21 CIP

ISBN 0 85199 200 5

Typeset in Garamond by Solidus (Bristol) Ltd
Printed and bound in the UK at the University Press, Cambridge

Contents

Preface
L. Piper and A. Ruvinsky vii

1. Systematics and Phylogeny of the Sheep 1
   I.R. Franklin
2. Genetic Aspects of Domestication, Common Breeds and Their Origin 13
   K. Maijala
3. Genetics of Colour and Hair Texture 51
   D.P. Sponenberg
4. Genetics of Morphological Traits and Inherited Disorders 87
   F.W. Nicholas
5. Biochemical Genetics 133
   L. Di Stasio
6. Molecular Genetics of Sheep 149
   B.C. Powell
7. Molecular Genetics of Immune Molecules 183
   W.R. Hein
8. Genetics of Disease Resistance and Vaccine Response 199
   H.W. Raadsma, G.D. Gray and R.R. Woolaston
9. Molecular Biology and Genetics of Scrapie in Sheep 225
   N. Hunter
10. Cytogenetics: Physical Chromosome Maps 241
    T.E. Broad, H. Hayes and S.E. Long
11. The Sheep Linkage Map 297
    G.W. Montgomery and A.M. Crawford
12. Genetics of Behaviour 353
    G.N. Hinch
13. Biology and Genetics of Reproduction 375
    I.W. Purvis and M. Hillard

14. Modern Reproduction Technologies and Transgenics 395
I. Wilmut, K.H.S. Campbell and L. Young
15. Developmental Genetics 413
N.E. Cockett
16. Genetic Resources and Conservation 437
R.W. Ponzoni
17. Genetic Improvement of Wool Production 471
K.D. Atkins
18. Genetics of Lamb and Meat Production 505
R.G. Banks
19. Genetics of Meat Quality 523
J.M. Thompson and A.J. Ball
20. Genetics of Milk Production 539
F. Barillet
21. Genetic Improvement of Sheep 565
B.P. Kinghorn
22. Standardized Genetic Nomenclature for Sheep 593
C.H.S. Dolling

Index 603

Preface

It seems very likely that sheep were among the earlier domesticated species and played a unique role in the development of several civilizations. Perhaps none of the domesticated species has been used for such a variety of purposes as sheep. The total number of sheep worldwide was around 1087 million in 1994.

Wool has been regarded as a very valuable textile fibre throughout history and remains the main source of natural fibre of animal origin. The Merino and related breeds are of paramount importance in apparel production. The Karakul breed is a source of high quality pelt, produced widely in central Asia, South Africa and other regions. Sheep producing carpet wool are common breeds in many countries. The total annual world production of sheep wool was around 1,450,579 tonnes in 1994 (cleaned basis).

Meat is also an important commodity produced by sheep. On a world scale, lamb and mutton together represent about 4% of all meat eaten. However, in many countries the proportion is much higher. It is noteworthy that there are no religious or cultural restrictions on the use of sheep meat. The utilization of sheep for milk and cheese production is also common, especially in Mediterranean countries.

The hardiness of sheep is well recognized. They are ruminants designed to utilize low-quality pastures and can thrive in very adverse climatic conditions. They are found from the cold dry deserts of central Asia to the high Andes of South America and in a wide variety of other environments including highly improved pasture and intensive, housed production systems. They have an important impact on the economy and ecology of large regions of the civilized world.

The last two decades have drastically changed the landscape of mammalian genetics. The new technologies of molecular and cytogenetics have greatly enlarged the amount and quality of genetic information. Until recently, only the genome of the house mouse and human was mapped satisfactorily. Genetic data for other mammals were very limited, with only a small number of genes being known. Since the new analytical methods appeared, progress achieved in genome mapping has been dramatic. New approaches to the genetic improvement of domestic livestock, which combine marker-assisted selection (MAS), utilization of loci of major effect (quantitative trait loci (QTL)) and modern reproductive technologies with traditional quantitative genetics methodologies, create optimistic expectations for future applications of genetic knowledge in breeding and selection.

These theoretical developments were a stimulus for significant progress in sheep genome mapping. In 1990, only about 30 gene loci were mapped or assigned to several ovine chromosomes. Six years later, the number of mapped loci and molecular markers approaches 600 and continues to grow rapidly. Our purpose in the production of this book has been to recognize the major changes that have occurred in the past decade and to present a comprehensive and updated description of the genetics of the sheep.

The first two chapters provide information about evolution, taxonomy and domestication and descriptions of the common breeds of sheep. Chapters 3 and 4 are devoted to the genetics of colour and texture variation, other morphological traits and inherited disorders. The next two chapters cover molecular and biochemical genetics. Chapters 7, 8 and 9 deal with immunogenetics, immune response and disease resistance. Cytogenetics and gene mapping are covered in Chapters 10 and 11. Chapter 12 is concerned with genetic aspects of the behaviour of sheep. The biology and genetics of reproduction, as well as the utilization of modern reproductive technologies, are discussed in Chapters 13 and 14. Developmental genetics of sheep is covered in Chapter 15. Genetic resources and their conservation are the focus of Chapter 16. Quantitative genetics, selection and sheep breeding are covered in Chapters 17–21. The last chapter is devoted to the genetic nomenclature of sheep.

This book is written for a broad audience, which includes students, farmers, researchers and lecturers. *The Genetics of Sheep* is the first publication in a series of monographs devoted to genetics of mammalian species published by CAB INTERNATIONAL. The next books in the series, *The Genetics of the Pig* and *The Genetics of Cattle*, are currently in preparation.

The authors have made every attempt to highlight the most recent and important references but errors and omissions are unfortunately inevitable in such publications. During the writing of this book several changes in genetic nomenclature were approved by the nomenclature commitees. There was no opportunity to implement all these changes in the book, but they are discussed in Chapter 22. The editors apologize for that.

The book is a result of truly international efforts and we are happy to express our gratitude to the authors who contributed to the book and made this publication possible. We greatly appreciate the efforts of numerous scientists who helped to review all materials. Our particular thanks are due to S.A. Barwick, J.N. Clarke, B.J. Crook, G. Echard, H.-U. Grasser, J.P. Hanrahan, F. Hughes, J.W. James, A.B. Lawrence, R.S. Lundie, A.E. McClintock, J. Maddox, A.L. Rae, R.M. Ridley, L. Ruvinsky, K.L. White and C.W. Wong. The assistance of the publisher, CAB INTERNATIONAL, is acknowledged and appreciated. It is our hope that this text will serve as a useful resource for those people throughout the world who study or work with sheep.

Laurie Piper
Anatoly Ruvinsky
Armidale, New South Wales, Australia
May 1997

Systematics and Phylogeny of the Sheep

1

I.R. FRANKLIN
CSIRO, Division of Animal Production, Locked Bag 1, Delivery Centre, Blacktown, NSW 2148, Australia

Introduction	1
A discursion into systematics	2
The Phylogenetic Position of the Genus *Ovis*	3
The order Artiodactyla	4
The family Bovidae	5
The subfamily Caprinae	7
The genus *Ovis*	8
Discussion	10
References	11

Introduction

The domesticated sheep, *Ovis aries*, is classified as a member of the family Bovidae of the order Artiodactyla. The artiodactyls are one of the most successful of the mammalian orders, comprising ten extant families and approximately 200 species; the exact number is a matter of choice rather than fact. The genus *Ovis* is a good example of the arbitrary nature of classification at the species level. Some taxonomists recognize as few as four species and others as many as eight, but, since all appear capable of interbreeding with each other and with domesticated sheep, the number of species, subspecies and races will always remain contentious.

Clearly, sheep are closely related to goats, and members of the genera *Ovis* and *Capra* are generally grouped together, with a number of sheep–goat-like species, such as the Barbary sheep (*Ammotragus*) and the 'goat-antelopes', into the subfamily Caprinae.

Before discussing the evolutionary position of the genus *Ovis* in detail, it is appropriate to discuss briefly the current techniques in systematics and to define a few terms that will be used later.

A discursion into systematics

Systematics is the science of constructing an evolutionary history, or phylogeny, of a group of species. There are, essentially, four approaches to systematics: classical evolutionary taxonomy, phenetics, cladistics and molecular taxonomy. Classical systematics is still the most widely used method for reconstructing evolutionary history, one that depends heavily on the experience and judgement of the systematist. It is a procedure in which a group is examined for as many 'independent' characteristics as possible, classifying individuals according to differences or similarities, assuming that similarities reflect true phylogenetic history. The greatest problem faced by systematists is distinguishing between characteristics that are similar because of shared evolutionary descent (homologous) and those that are similar but have arisen independently in different lineages because of parallel or convergent evolutionary pressures (analogous).

Phenetics, which has its roots in numerical taxonomy, uses multivariate statistical methods in an attempt to remove the subjectivity of classical systematics by considering a large number of characteristics simultaneously and grouping taxa according to the number of characteristics in common. It is assumed that, using this approach, the number of homologous characteristics will far outweigh those that are analogous. Cladistics, on the other hand, seeks greater objectivity by focusing only on shared derived characters, or synapomorphies – traits that are shared by a group and define their common evolutionary history. Cladistics defines a set of rules to construct a lineage of ascent along a cladogram, in which the breakpoints are defined by the acquisition of the shared derived character that delineates the group.

Because cladistic analysis places great weight on characteristics that define a lineage and that separate this lineage from others, considerable care is taken in choosing these characteristics. For example, almost all ruminants have horns, but this characteristic cannot be given much weight; it is a trait under relatively simple genetic control and may disappear or reappear through a lineage. In contrast, complex multifactorial characteristics, such as the development of a multichambered stomach and the adaptations associated with rumination, are likely to be shared by all members of a lineage and unlikely to appear in the same form in unrelated species.

Of increasing importance to phylogenetic studies is the karyotype, especially as methods such as fluorescent *in situ* hybridizsation (FISH) and chromosome painting have become available. These techniques, which combine the use of specific molecular probes and classical cytogenetic methodology, allow rapid and more accurate assessment of chromosomal homologies and rearrangements that accumulate during the evolution of a lineage. However, little weight can be placed on simple measures such as chromosome number. Both cattle and goats, for example, have 60 chromosomes, whereas sheep have 54. This does not mean that goats and cattle are more closely related than goats and sheep. Chromosome number can be easily changed by centric fusion (Robertsonian translocations); such fusions are commonly observed in natural populations and it is possible to develop sheep strains with chromosome numbers as few as 44 chromosomes.

However, some chromosomal rearrangements are essentially irreversible; apart from centric fusions and fissions, the probability that a chromosome will fragment at exactly the same place in different lineages is, essentially, zero. If such breakpoints can be defined accurately, these changes in karyotype are ideal for cladistic analysis, and much of the current effort in comparative physical and linkage mapping is directed at identifying these breakpoints.

Molecular taxonomy began in the 1960s with comparisons between the composition of proteins, such as the globin families, over a range of taxonomic groups, and the assumption that the degree of divergence in composition reflected the time since divergence of the taxa. Further, with the hypothesis that the rate of molecular evolution for a particular protein is approximately linear in time (as it would be for neutral substitutions), the time of divergence of the two taxa could be estimated, something hitherto impossible without a rich fossil record. Attention has now been redirected to deoxyribonucleic acid (DNA) sequences, which are easier to obtain and allow stronger inferences. Much effort has been devoted to mitochondrial DNA (mtDNA) sequences, which appear to evolve rapidly in comparison with many nuclear genes. It is also possible to compute genetic distances, using simply inherited genetic polymorphisms, such as blood groups, allozymes or microsatellites; these are particularly useful when the groups are closely related, as in different races of the same species.

In fact, molecular taxonomy suffers from many of the difficulties faced by classical systematics; because of gene duplication, the difficulties in establishing homologies may remain, and it is clear that homologous genes do not evolve at a constant rate throughout a lineage. One example is the apparent rapid evolutionary change early in bovid lineages of growth hormone and placental lactogen (Wallis, 1993, 1994); another is the divergence of the lysozyme gene family, which has arisen, presumably, as an adaptation to rumination (Irwin, 1995). An additional complication is that, as coalescent theory demonstrates, molecular divergence does not necessarily coincide with speciation.

While there are purists, phylogenetic reconstruction in practice often draws upon all of the above approaches. Accordingly, I present a classical systematic description of the phylogeny of the genus *Ovis*, descending from the order Artiodactyla, but use interpretations from cladistic and molecular taxonomic analysis where appropriate.

The Phylogenetic Position of the Genus *Ovis*

In discussing the formal taxonomy of the sheep, I shall discuss briefly some of the anatomical, physiological and behavioural characteristics of the species, topics covered in more detail elsewhere in this volume. However, I confine discussion only to those traits that delineate the taxon from other related taxonomic groups.

The order Artiodactyla

Members of the order Artiodactyla (Greek: *artios*, 'even'; *dactylos*, 'toed') first appear in the fossil record during the Eocene, approximately 50–60 million years ago. The first fossils indicate that the early artiodactyls were small, omnivorous, forest dwelling mammals. The commonly described nine families and the time of the first appearance in the fossil record are shown in Table 1.1. In recent classifications, a tenth family, the Moschidae (musk deer), is added within the suborder Ruminantia.

A distinguishing feature of artiodactyls is that the main axis of each limb passes through the third and fourth toes (in contrast to, for example, the Perissodactyla, or odd-toed ungulates, whose axis passes through the middle toes). The outer (second and fifth) toes are diminished or absent, and the toes of all recent artiodactyls are enclosed in hooves. Most artiodactyls, particularly males, possess one or, rarely, two pairs of horns; these may be permanent or shed periodically. Artiodactyls are typically terrestrial, herbivorous (pigs are omnivores) and gregarious. They are native to all continents except Australia and Antarctica.

The suborder Ruminantia is the most widespread and varied of the artiodactyls and is the dominant herbivore group of the world. The name derives from the ruminating habit, associated with the division of the stomach into three or four connected chambers. In general, they lack upper incisors and the first premolar. The lower molars have grinding surfaces of the seledont (crescent shape) type.

Table 1.1. Extant families of the order Artiodactyla.

Taxon	Members	Approximate time of first appearance (MYBP)
Suborder Suiformes		
Family Suidae	Pigs	Early Miocene (25–30)
Family Tayasuidae	Peccaries	Early Miocene (25–30)
Family Hippopotamidae	Hippopotamuses	Miocene (20–25)
Suborder Tylopoda		
Family Camelidae	Camels, llamas, vicuna	Eocene (40–45)
Suborder Ruminantia		
Family Tragulidae	Chevrotains	Middle Eocene (45–50)
Family Moschidae	Musk deer	Late Eocene
Family Giraffidae	Giraffes, okapi	Early Miocene (25)
Family Cervidae	Deer	Miocene (20)
Family Antilocapridae	Pronghorn antelope	Early Miocene (25)
Family Bovidae	Cattle, sheep, antelope	Miocene (20–25)

MYBP, million years before present.

Many have pedal glands between their toes, which produce an odoriferous secretion, and some have another (preorbital) below the eye, which secretes during the rutting season. The preorbital gland is particularly evident in the Cervidae (deer), but is one of those characteristics that distinguish sheep from goats.

The most primitive, and the smallest, of the recent ruminants are the chevrotains (or mouse deer) of the family Tragulidae; they possess, like camels, a three chambered stomach. The musk deer are considered to be closely related to the tragulids. These two families are usually set apart from the remainder of the Ruminantia, which are often classified into a separate infraorder, Pecora, comprising the true ruminants.

The pecorans include all of the remaining ruminant genera. This group has been subjected to extensive cladistic and molecular analysis. Mitochondrial DNA (Miyamoto *et al.*, 1993) resolves poorly the relationships among the pecoran families; this has been attributed (Kraus and Miyamoto, 1991) to a rapid radiation of these lineages over a relatively short time span (of the order of 5 million years) approximately 23–28 million years ago.

The family Bovidae

The family Bovidae includes 45 genera grouped into nine subfamilies (Table 1.2), and has been considered to be among the most difficult groups of mammals to systematize. There are over 100 species, including many of our important commercial species, such as sheep, goats, buffalo and cattle, but also including gazelles, wildebeest and many other species native to Africa. They are characterized by the presence of unbranched, hollow, continuously growing horns.

This family comprises the most diverse group of large mammals originating since the Miocene, and has an abundant fossil record. All of the major subfamilies are present in the fossil record 7 or more million years ago. While the group is generally considered monophyletic, cladistic analysis (e.g. Scott and

Table 1.2. Subfamilies of the family Bovidae.

Family Bovidae	
Subfamily Aepycerotinae	Impala
Subfamily Alcelaphinae	Wildebeest, hartebeest
Subfamily Antilopinae	Gazelles, saiga
Subfamily Bovinae	Bison, buffalo, wild and domestic cattle
Subfamily Caprinae	Sheep, goats
Subfamily Cephalophinae	Duikers
Subfamily Hippotraginae	Oryx, addax, blesbok
Subfamily Peleinae	Rhebok
Subfamily Reduncinae	Kob, waterbuck, puku

Janis, 1993) concludes that the split between the bovids, cervoids and giraffoids is poorly defined. The scarcity of suitable synapomorphic traits is presumed to be a consequence of rapid evolutionary divergence within the Ruminantia, and the possibility of paraphyletic origins of the major families cannot be ruled out. Similarly, the presumed rapid divergence among the subfamilies Bovinae and Caprinae obscures their phylogenetic relationship, as deduced from either cladistic or molecular analysis.

Figure 1.1 shows a phylogenetic analysis by Allard *et al.* (1992) of these subfamilies, using a 2.7-kilobase region of mtDNA – a region that includes the complete 12S and 16S ribosomal ribonucleic acid (rRNA) genes and three transfer RNA (tRNA) genes. The tree shown is based on transversions only and is far from robust; if all base substitutions are used, the tree is altered substantially.

This study and other similar analyses of divergence among mitochondrial genomes support a general conclusion that the bovid subfamily lineages diverged rapidly around 17–25 million years ago, shortly after the divergence of the major pecoran families.

Studies of chromosomal evolution in the pecoran families may shed light on their phylogeny. As indicated earlier, both cattle and goats have a superficially similar karyotype, namely a chromosome number of $2n = 60$, comprising 29 pairs of acrocentric autosomes and two sex chromosomes. In fact, most of the Bovidae have either 58 or 56 autosomal chromosome arms, and Gallagher and Womack (1994) propose that $2n = 60$ is the primitive arrangement and that the

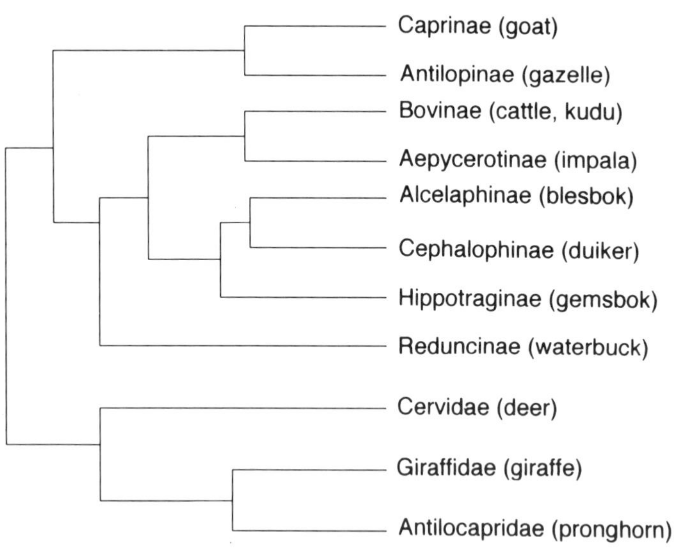

Fig. 1.1. Phylogenetic relationships among eight bovid subfamilies, with three pecoran families as outgroups. Data are based on 2.7 kilobases of mitochrondrial DNA, using transversions only. (Adapted from Allard *et al.*, 1992.)

chromosome numbers of all members of the group are related through a series of centric fusions.

This view contrasts with Todd's (1975) earlier supposition that the giraffe, with 30 pairs, primarily metacentrics, possesses the more primitive karyotype and that the karyotypes observed in the extant species are the result, largely, of centric fission. Gallagher and Womack (1992, 1994) base their conclusion on their observations that conservation of chromosome arms is extensive, but there is little conservation of whole (biarmed) chromosomes.

Certainly, it appears that Robertsonian translocations are common in most of the bovid lineages: a recent study by Vassart *et al.* (1995) of chromosomal evolution in gazelles showed chromosome numbers varying from 30 to 58. Other rearrangements comprise inversions within or between arms or translocations between chromosome arms, but these are much less common. However, chromosome banding has limited power to detect small rearrangements. As mapping data, both physical and linkage, become available, we shall be able to describe, in much finer detail, the evolutionarily conserved syntenic relationships among species and hence make much stronger inferences about phylogenetic relationships. Comparisons between genetic maps of sheep and cattle have already begun to define these evolutionary breakpoints (see Chapters 10 and 11).

The subfamily Caprinae

Table 1.3 shows the 10 genera usually included in the subfamily Caprinae. Two of these, the musk ox (*Ovibos*) and takin (*Budorcas*), are similar to each other, and probably represent a subgroup that split early from other members of the group. Pasitschniak-Arts *et al.* (1994) have studied the karyotype of these two species and, despite the differences in chromosome number (48 for the musk ox, 52 for the takin), have concluded that the karyotypes of these two genera are quite similar and clearly different from sheep. They propose that sheep and musk ox–takin have evolved independently from a goat-like ancestor.

Table 1.3. Genera of subfamily Caprinae.

Ammotragus	Aoudad or Barbary sheep
Budorcas	Takin
Capra	Goats
Hemitragus	Tahr
Nemorhedus	Goral
Oreamnos	Rocky Mountain goat
Ovibos	Musk ox
Ovis	Sheep
Pseudois	Bharal
Rupicapra	Chamois

Many taxonomists believe that both sheep and goats are descended from the goat-antelopes, the Rupicaprini, an ancient group whose modern representatives include chamois (*Rupicapra*), goral (*Nemorhedus*) and the North American mountain goat (*Oreamnos*). However, evidence from globin gene sequences (Li and Gojobori, 1983), mtDNA (Upholt and Dawid, 1977) and electrophoretic data (Hartl *et al.*, 1990; Randi *et al.*, 1991) suggest that *Ovis*, *Capra* and *Rupicapra* diverged contemporaneously around 5 million years ago.

Sheep are phenotypically similar to goats and, at the margins, appear to merge almost imperceptibly. They differ by the presence of scent glands in the face and hind feet and in the absence of beards in the males. Sheep typically possess transversely ribbed horns, which tend to curl into spirals. Like goats, they have a narrow hairy muzzle and narrow upper molar teeth.

The term 'sheep' includes a number of genera considered to be intermediate forms between sheep and goats. These include *Pseudois* (the bharal or 'blue sheep' of Tibet and western China), *Hemitragus* (the tahr, which seems to occupy an intermediate position between mountain goats and American sheep) and the Barbary sheep, *Ammotragus*. Only the last appears to be truly a sheep physiologically, behaviourally and anatomically, even though it has been shown occasionally to hybridize with goats but not with sheep. Indeed, Geist (1971) considers the Barbary sheep to be 'a near perfect ancestral form for sheep' and places it at one end of a cline between the extremes of Eurasian 'wild' sheep – urials and argalis – the latter representing, in his view, a more recently evolved form.

The genus Ovis

The earliest fossil record of sheep dates to about 2.5 million years ago, but, as we have indicated, sheep and goats probably separated at least 5 million years ago. The fossil record is poor, probably because the normal habitat is unfavourable for fossil formation. Sheep first appear in the Villafranchian and from then on appear sporadically in the Pleistocene fossil record. Like many of the mammalian fauna of the ice ages, these were large animals and were later replaced by smaller mouflon-like animals.

As sheep increased their range and habitat, ranging from Arctic mountains to hot desert regions, they evolved a large number of races, varying in size from the Siberian argalis, which may weigh up to 200 kg, down to the smaller (35 kg) urials from Cyprus. They may have long fleeces with heavy manes or, like the desert bighorns, be short-haired. The Asiatic sheep may have long legs with medium bones and muscling or be short-legged and heavily muscled. The coat colour ranges from white to black. Sheep are found on all continents except Australia and Antarctica, and because of their habitat range must be considered one of the most successful of the ruminant groups.

The taxonomy of the genus *Ovis* is controversial and, indeed, since it appears that all the 'species' can interbreed, it could be argued that they are all races of the same species. However, most taxonomists agree that the 30–40 recognized races fall into two main groups – the Eurasian group, comprising the mouflons, the urials and the argalis, and the American wild sheep, comprising snow sheep, thinhorns and bighorns.

As we might expect, the origins of the domesticated sheep, *O. aries*, are obscure, although most authors believe that domestication occurred 9000–11,000 years ago in western Asia (see Chapter 2). Apart from the dog, sheep were one of the first of the domesticated animals, and are thought to have been derived from the urial, whose distribution is from present day Pakistan to Iraq (the small urials of Cyprus were probably introduced by humans). However, contributions from the mouflon and the argali cannot be ruled out (Ryder and Stephenson, 1968).

While the coat of wild sheep is made up of a hairy outer coat (composed of kemps) and a woolly undercoat, all of which moults each spring, domestication has been accompanied by a reduced tendency to moult and an increased development of wool at the expense of hair or kemp. There are around 800–1000 breeds of domesticated sheep, descriptions of which can be found in Epstein (1969), Ponting (1980), Acharya (1982) and a series of compendia by Mason (1967, 1980, 1996). The more important of these breeds are discussed in Chapter 2.

Sheep breeds are often classified into groups recognized by simple phenotypic characteristics, such as tail type (thin, fat or short) or coat type (long- versus short-woolled, fine- versus strong-woolled, hair or fur), or by other characteristics, such as fat rump or lop-eared breeds. By far the most important commercially are the wool breeds, particularly the Merino and its descendants, followed by the short-woolled European breeds (such as the Southdown, Dorset, Suffolk and Île-de-France) and the British longwools (such as Leicester, Lincoln and Romney). Other major groups include the fat-tailed breeds of western Asia, Africa and the former USSR, whose long wool is used for carpets and which are also kept for meat and milk. Another major group is the haired sheep, found primarily in tropical regions and used primarily for meat.

There has been little attempt to classify this wide diversity of domestic breeds, apart from the speculation by Ryder and Stephenson (1968) on the phylogenetic relationships among the British breeds. Initial attempts using protein data (Manwell and Baker, 1977) produced unreasonable estimates of genetic distance, but recently Buchanan *et al.* (1994) and Forbes *et al.* (1995) have used microsatellites, which are highly polymorphic short DNA repeats, to characterize variation between and within a number of domesticated and wild sheep strains. Forbes *et al.* (1995) found that domesticated strains were much more variable than wild (bighorn) races, and both studies showed that microsatellites could be used to compute genetic distances that appear to accord with biogeographical and historical data. Very large numbers of microsatellites are now available for sheep and goats, and it is likely that these DNA markers will provide a means of establishing phylogenetic relationships among races and strains within species.

Discussion

Sheep are typical members of one of the most recent and successful of grazing hoofed mammals, the infraorder Pecora of the order Artiodactyla. The available evidence suggests that they diverged from their nearest taxonomic relatives, the *Capra* and *Rupicapra*, approximately 5 million years ago, and from the bovids approximately 20 million years ago in the late Miocene, following the explosive diversification of ruminant families that accompanied the first appearance of grasslands. The taxonomy of the pecoran lineages is continually undergoing revision, and much remains to be done to resolve the evolutionary history of all of the ruminant families. In particular, apart from some very recent work on karyotypic differences, little has been done to sort out the relationships among the subfamily Caprinae, a group that comprises a number of sheep and goat-like genera, such as the tahr, goral, bharal and musk ox. Presumably, as molecular data are collected from representatives of these genera, the interrelationships among this subfamily will become much clearer.

Within the genus *Ovis* itself, molecular and karyotypic analysis should illuminate the interrelationships among the wild species and domesticated strains and, perhaps, examine some of the current beliefs about the origins of our commercially important sheep breeds. Recent studies using DNA typing of polymorphic markers, such as microsatellites, suggest that these may be used profitably to define genetic distances between domesticated breeds.

What impact will this have on animal breeders, nutritionists and physiologists who are concerned to improve both production efficiency and product quality in our most important domesticated species? Very little, I suspect. Already, the most important of the wool-producing breeds, the Merino, is well outside the range in wool production and quality of any of the other races, both wild and domesticated. However, our understanding of wool biology and wool follicle development may benefit from comparative studies and from gene mapping in wide crosses, just as our understanding of sheep behaviour has been improved by studies of their wild relatives.

Similarly, the specialized meat breeds, such as the Suffolk and Texel, have been subject to intense selection for production efficiency and meat quality, and there are probably few genes that we could incorporate by introgression from wild or distantly related domesticated sheep. In addition, DNA studies have indicated that there is ample genetic variation within each of our domesticated breeds to ensure continued response to genetic selection. One area that might be pursued profitably is parasite resistance, where there may be simply inherited characteristics that could be incorporated to improve the resistance of our important commercial strains to internal and external parasites. As far as I am aware, however, little is known about either tolerance or resistance to these parasites in the wild relatives.

Technically, however, if single genes were to be identified that might be usefully introgressed into our commercial strains, there should be no difficulty. The available evidence indicates that all of the species of the genus *Ovis* can intercross freely, and the sheep genome is now sufficiently well marked for a small chromosome segment to be transferred in a relatively straightforward manner.

References

Acharya, R.M. (1982) *Sheep and Goat Breeds of India.* Animal Health Paper No. 30, FAO, Rome.

Allard, M.W., Miyamoto, M.M., Jarecki, L., Kraus, F. and Tennant, M.R. (1992) DNA systematics and evolution of the artiodactyl family Bovidae. *Proceedings of the National Academy of Sciences of the USA* 89, 3972–3976.

Buchanan, F.C., Adams, L.J., Littlejohn, R.P., Maddox, J.F. and Crawford, A.M. (1994) Determination of evolutionary relationships among sheep breeds using microsatellites. *Genomics* 22, 397–403.

Epstein, H. (1969) *Domestic Animals of China.* Commonwealth Agricultural Bureaux, Farnham Royal.

Forbes, S.H., Hogg, J.T., Buchanan, F.C., Crawford, A.M and Allendorf, F.W. (1995) Microsatellite evolution in congeneric mammals: domestic and bighorn sheep. *Molecular Biology and Evolution* 12, 1106–1113.

Gallagher, D.S and Womack, J.E. (1992) Chromosome conservation in the Bovidae. *Journal of Heredity* 83, 287–298.

Gallagher, D.S and Womack, J.E. (1994) Chromosome conservation among the advanced pecorans and determination of the primitive bovid karyotype. *Journal of Heredity* 85, 204–210.

Geist, V. (1971) *Mountain Sheep.* University of Chicago Press, Chicago, 383 pp.

Hartl, G.B., Burger, H. Willing, R. and Suchentrunk, F. (1990) On the biochemical systematics of the Caprini and Rupicaprini. *Biochemical Systematics and Ecology* 18, 175–182.

Irwin, D.M. (1995) Evolution of the bovine lysozyme gene family: changes in gene expression and reversion of function. *Journal of Molecular Evolution* 41, 299–312.

Kraus, F. and Miyamoto, M.M. (1991) Rapid cladogenesis among the pecoran ruminants: evidence from mitochondrial DNA sequences. *Systematic Zoology* 40, 117–130.

Li, W.-H. and Gojobori, T. (1983) Rapid evolution of sheep and goat globin genes following gene duplication. *Molecular Biology and Evolution* 1, 94–108.

Manwell, C. and Baker, C.M.A. (1977) Genetic distance between the Australian Merino and the Poll Dorset Sheep. *Genetical Research, Cambridge* 29, 239–253.

Mason, I.L. (1967) *The Sheep Breeds of the Mediterranean.* FAO, Rome, and Commonwealth Agricultural Bureaux, Farnham Royal.

Mason, I.L. (1980) *Prolific Tropical Sheep.* Animal Production and Health Paper No. 17, FAO, Rome.

Mason, I.L. (1996) *A World Dictionary of Livestock Breeds, Types and Varieties.* 4th edn. CAB International, Wallingford.

Miyamoto, M.M., Kraus, F., Laipis, P.J., Tanhauser, S.M. and Webb, S.D. (1993) Mitochondrial DNA phylogenies within Artiodactyla. In: Szalay, F.S., Novacek, M.J. and McKenna, M.C. (eds) *Mammal Phylogeny: Placentals.* Springer-Verlag, New York, pp. 268–281.

Pasitschniak-Arts, M., Flood, P.F., Schmutz, S.M. and Seidel, B. (1994) A comparison of G-band patterns of the muskox and takin and their evolutionary relationship to sheep. *Journal of Heredity* 85, 143–147.

Ponting, K. (1980) *Sheep of the World*. Blandford Press, Poole, Dorset, 155 pp.
Randi, E., Fusco, G., Lorenzini, R., Toso, S. and Tosi, G. (1991) Allozyme divergence and phylogenetic relationships among *Capra*, *Ovis* and *Rupicapra* (Artiodactyla, Bovidae). *Heredity* 67, 281–286.
Ryder, M.L. and Stephenson, S.K. (1968) *Wool Growth*. Academic Press, London, 805 pp.
Scott, K.M. and Janis, C.M. (1993) Relationship of the Ruminantia (Artiodactyla) and an analysis of the characters used in ruminant taxonomy. In: Szalay, F.S., Novacek, M.J. and McKenna, M.C. (eds) *Mammal Phylogeny: Placentals*. Springer-Verlag, New York, pp. 282–302.
Todd, N.B. (1975) Chromosomal mechanisms in the evolution of the artiodactyls. *Paleobiology* 1, 175–188.
Upholt, W.B. and Dawid, I.B. (1977) Mapping of mitochondrial DNA of individual sheep and goats: rapid evolution in the D loop region. *Cell* 11, 571–583.
Vassart, M., Séguéla, A. and Hayes, H. (1995) Chromosomal evolution in gazelles. *Journal of Heredity* 86, 217–227.
Wallis, M. (1993) Remarkably high rate of molecular evolution of ruminant placental lactogens. *Journal of Molecular Evolution* 37, 86–88.
Wallis, M. (1994) Variable evolutionary rates in the molecular evolution of mammalian growth hormones. *Journal of Molecular Evolution* 38, 619–627.

Genetic Aspects of Domestication, Common Breeds and their Origin

2

K. MAIJALA
Haapatie 13 D, 00780 Helsinki, Finland

Introduction 14
Centres of Wild-sheep Domestication 14
Preconditions for domestication 14
The wild ancestors of sheep 15
Places and dates of domestication 16
Motives for domesticating sheep 17
Effects of Domestication on Genetic and Physiological Mechanisms 18
Genetic variation 18
Selection 19
Inbreeding 20
Outbreeding 21
Changes of Sheep after Domestication 21
Horns 22
Coat and fleece 22
Moulting 22
Colour 23
Behaviour 23
Skeleton 24
Tail length 24
Other features 24
Production traits 24
Common Breeds of Sheep, Their Origin and Characteristics 25
Definition of breed 25
Classification of sheep breeds 26
Common breeds 27
Conclusions 45
References 47

Introduction

Knowing the history of a certain area of activity is often an important precondition for sound and safe visions of what should be done for continued future development in the area, even though overemphasis of history could also prevent creative courage. In this book, there are two chapters looking mainly at the past: the evolution and taxonomic position of sheep are described by Franklin (Chapter 1) and the present chapter aims at describing the historical aspects of sheep breeding and breed formation from the beginning of domestication until the application of modern breeding technology. Other authors discuss present knowledge about the genetics of various traits and the techniques available for future development.

Centres of Wild-sheep Domestication

There are several valuable reviews of the history of domestication of animals, including sheep (e.g. Zeuner, 1963; Epstein, 1971; Bökönyi, 1974; Clutton-Brock, 1981; Ryder, 1983, 1984, 1991; Reed, 1984). Based on findings of archaeological bone remains, there appears to be good agreement that sheep and goats were domesticated in south-western Asia about 9000 BC, 14 million years after the development of hominids and two millennia after the domestication of the dog and before that of other agricultural animals. It is not clear which of these two species were domesticated first, because the nature of the skeletal remains is fragmentary and it is difficult to distinguish sheep and goat bones (Ryder, 1984).

Preconditions for domestication

The region and time of sheep domestication were largely determined by the development of various preconditions for it. To these belong, among others, the following steps and factors (Reed, 1984).

1. Contacts of humans with sheep. Humans started to hunt sheep for meat and skins. The two species learned features of each other's social behaviour. Each social group learned to tolerate members of the other species as a part of a larger social group. This led to close contacts and symbiotic relationships between humans and animals, a mutual understanding of each other's behaviour and the taming of newborn lambs and using them as pets.

2. The worldwide change in environment that accompanied and followed the end of the last glacial period in the twelfth to ninth millennia BC. South-western Asia had not been covered by glaciers, so that people could live there through the ice age. Wild sheep had adapted to high latitudes and cold climates by delaying the breeding season or by extending the gestation length. Growth of wild grasses (ancestors of wheat and barley) in the hills of south-western Asia, the gathering and preserving of their grains as food by people and the

development of cultivation of these grasses, of little villages and of a sedentary lifestyle were important factors. The crops and their leftovers attracted sheep, bringing them close to humans.
3. Increased birth rates and lowered mortality of humans, increased sizes of settlements and diminishing supply of large game, resulting from overhunting.
4. Physiological ability of ruminants, developed during millions of years, to eat, digest and convert cellulose-rich grasses and straws to products useful for humans. As a ruminant, the sheep was able to recycle its urea and thus to survive on protein-poor diets.

These preconditions prevailed especially in south-western Asia, thus explaining why that region became the centre of domestication of many ruminant species, including sheep. The behaviour and intelligence of these animals could have made their domestication possible at any time during the preceding several million years (Reed, 1984).

The wild ancestors of sheep

The sheep is an exceptional domestic animal, since its wild ancestors are still living in large numbers and can thus be studied. Their main present home is in the mountain ranges of central Asia. They spread in the Pleistocene period westwards into Europe and eastwards into America (Reed, 1984). This indicates the success of the group, due mainly to behavioural and reproductive adaptations (Geist, 1971, quoted in Ryder, 1984).

The Latin name *Ovis aries* is used for the domestic sheep to distinguish it from the wild type, which by some authors is divided into seven species (Ryder, 1984) (diploid chromosome number at the end).

1. *Ovis ammon* (argali) in the mountains of central Asia (Altai mountains, Pamirs, Tien Shan, Himalayas, Tibet) (56).
2. *Ovis vignei* (urial) west of the argali (mountain areas from north-eastern Iran to Afghanistan and north-western India) (58).
3. *Ovis orientalis* (Asian mouflon) west of the urial (probably the ancestor of all domestic sheep and also of the European mouflon) (54).
4. *Ovis musimon* (European mouflon) extending into Europe, mainly to the islands of Corsica and Sardinia (now considered to be feral domesticates, but appear no different from truly wild sheep and so may have been introduced by humans at a very early date) (54).
5. *Ovis canadensis* (bighorn) in North America (54).
6. *Ovis nivicola* (snow sheep) of Siberia (52).
7. *Ovis dalli* (thinhorn) of Alaska (54).

The last three 'species' have never been domesticated. The Latin names are no longer used in practice, since the types represent varieties or breeds and not different species (Ryder, 1984). Since domestic sheep first appeared in the urial area, it was thought that the urial was its main ancestor, while the mouflon was assumed to have contributed to European breeds and the argali to those of Asia.

When the number of chromosomes in the mouflon was found to be the same as in domestic sheep, it was thought that all sheep must have the same number and that the mouflon might be the only ancestor of domestic sheep. Newer techniques and studies on several wild types showed, however, that the urial, argali and Siberian snow sheep have a differing number of chromosomes, while the thinhorn and bighorn sheep had the same as the mouflon (Clutton-Brock, 1981). It has also been found that the argali and urial will hybridize with domestic sheep, the hybrids having intermediate numbers. However, no modern breeds with such numbers have been found. This can be explained by observations on hybrids between the argali and the mouflon: the hybrid ewes had 55 chromosomes, but they produced ova with 27 chromosomes, suggesting prezygotic selection towards a lower chromosome number. Thus, the 54 chromosomes of modern domestic sheep could also have arisen by hybridization between the mouflon and another wild type, followed by selection for a reduced chromosome number. Still, the Asian mouflon is generally considered the ancestor of all domestic sheep and of the European mouflon. This is considered to be a relic of the first domestic sheep taken to Europe around 7000 BC (Clutton-Brock, 1981). Another example of ancient sheep of this type is the Soay from the islands of St Kilda in Scotland (Ryder, 1984).

The size range among the four Asian wild types (Asian and European mouflon, argali, urial) is from 36 kg in the Cyprus mouflon to 200 kg in the Siberian argali. Their colour is generally brown, but the North American sheep include all-white and all-black varieties.

Places and dates of domestication

It is assumed that the cultivation of plants and the domestication of herd animals began on the grassy, open-forested hills of the crescent-shaped fertile area extending from Palestine northwards to Lebanon and southern Turkey, and then southwards through the foothills of the Zagros mountains on the Iraq–Iran border. The three neighbouring areas are especially associated with sheep domestication: the Kermanshah plateau and Luristan, which descend from there to the plain in Iran, and the western foothills of the Zagros mountains in Iraq (Ryder, 1984).

The importance of the region as a centre of domestication was largely explained in the section on preconditions. The time of domesticating sheep was the prehistoric period at the end of the middle Stone Age, by the primitive peoples settled there and cultivating plants (Ryder, 1984). The first evidence for sheep domestication comes from bone remains dated *c.* 9000 BC at Zawi Chemi Shanidar in Iraq, where 50% of the bones were from sheep (> 40% of them immature). The high proportion of young animals is considered a sign of a systematic influence of humans on the deaths of sheep and of economic development. Other sites and dates of remains showing early domestication are Tepe Sarab in Iran (8000 BC), Tebe Sabz and Jarmo in Iran and Afghanistan (7000 BC), Turkey, Baluchistan and the Indus valley in India (6500 BC) and Syria,

Crete, Greece and the Quetta valley in Pakistan (6000 BC). By 5500 BC domestic sheep had evidently reached Corsica and by 4000 BC the Atlantic coast, Britain and Scandinavia.

Relatively few bone remains are adequate to show age and sex ratios or the relative importance of sheep as compared with other livestock. In the Zawi Chemi Shanidar excavations, 50 and 42% were from sheep and goats, respectively. At Tepe Sabz the goat predominated until *c.* 5500 BC, when both species became more common and only 40% reached the age of 3 years. Thereafter, skin-working tools decreased and spindle whorls used in wool spinning increased. After 4000 BC, sheep predominated. The excess of young males on many early sites can be interpreted as differential mortality rather than killing, but it might have created a surplus of ewe's milk, leading to the origin of milking. The knowledge obtained from the sheep remains can be supplemented, for example, by rock paintings in Tassili, Sahara, which show horned sheep about 4000 BC, with a smooth coat very similar to that of the wild type (Ryder, 1984).

Motives for domesticating sheep

Only assumptions are available about the reasons for domesticating sheep, since it happened long before any records were kept (Ryder, 1984). The first domesticators may not have known what they were doing, but they had known animals as living creatures to be hunted and killed or to be avoided in the case of large predators. The change from aggressive hunting to protective keeping was primarily a change in attitude of humans. This change left no direct traces for the archaeologist to discover, but only indirect results of changes in human behaviour following modifications in human thinking. The process was probably gradual, beginning long before there is evidence for it. It may have been preceded by a 'game-management' phase, and was an unconscious, unplanned and gradual strengthening of an association between humans and sheep, both of which were social species, preadapted by their evolutions to be of benefit to each other (Ryder, 1984). The process can be understood on the basis of ethnographic and animal behaviour studies. Men may have caught newborn lambs of killed ewes as hunting decoys or as pets and brought them to their camps, and their wives then suckled them and raised them with the children. The young lamb might then have become attached to the foster mother by 'imprinting'. The lamb had to be handled and fed at a personal and individual level, to avoid its running wild. Those lambs that inherited genetic combinations for continuing wildness either escaped or were killed, so their genes did not persist in the population undergoing domestication. Thus, the women may have had a decisive role (Ryder, 1984).

It is often easier to obtain food by hunting than by farming; for example, North American Indians did not domesticate the bighorn. Thus, the motives may have been largely ecological; for example, the desiccation caused by the retreating ice-cap forced humans to share food sources with the decreasing numbers of wild animals and at the same time to search for alternative sources. Another motive was supplied by the extinction of the Pleistocene megafauna, causing a shift to the hunting of smaller animals for food and thus a closer association of humans with sheep (Ryder, 1984).

The hunter-gatherers might have seen the uses for meat, bones and skins. After long experience and the development of a more sedentary lifestyle by humans and the accumulation of random mutations in sheep, their secondary uses (milk, wool) could have been realized. After blood and milk were developed as sources of food, the animals became more valuable alive, so that meat consumption probably declined after domestication. However, it was not possible to foresee the value of milk before domestication. The same is true of wool, which in the wild animals is obscured by the hairy outer coat, demanding to be made finer before one can obtain fleece suitable for textile use.

Zeuner (1963) considered confinement and breeding in captivity, with separation from the wild type, an important stage of domestication, since it allowed the emergence of a distinct domestic type. Then came selective breeding for certain features, with occasional mating to wild forms. This was already practised in the Bronze Age, since in the Middle East distinct 'breeds' can be recognized by 3000 BC. This was followed by gradual intensification of the development of different breeds with desired economic characters and elimination of wild and primitive domestic types.

Effects of Domestication on Genetic and Physiological Mechanisms

The mechanisms of evolution that led to the development of, for example, wild sheep were mutations and chromosomal rearrangements, selection, migration, geographical isolation and fixation of diversity by inbreeding and random genetic drift in small populations (Dobzhansky, 1951). Domestication increased the speed or the size of the effects of many factors (e.g. inbreeding, outbreeding, genetic drift) and added artificial selection and assortive mating to the forces changing gene frequencies (Lush, 1945).

Genetic variation

The change to domesticated conditions did not create any new inheritance, but the environmental changes following domestication allowed many genetic differences to show themselves more clearly than before and thus to be more readily and accurately selected. Examples that can be mentioned include growth rate, fattening ability and milk yield, in which good feeding increases differences. The change permitted a greater range of variation to survive through reduced natural selection (Lush, 1945).

Belyaev (1979) considered the enormous increase in the rate and range of variability to be the main result of domestication. He developed an interesting theory of 'destabilizing selection' on the basis of his experiments on domesticating silver foxes. He observed that selection for tame behaviour also led to earlier onset of reproductive activity within the breeding season, to longer moulting time and to changes in some dominantly inherited morphological traits. He interpreted these changes as being caused by changes in the timing and amount of gene expression, due to changes in the systems of neuroendocrine control of ontogenesis, caused in turn by stressful factors in connection with domestication. In fact, the level of serotonin and one of its metabolites turned out to be higher in tame than in wild females. Belyaev thought that a number of 'dormant' genes could be expressed under the influence of an altered equilibrium. Ruvinsky (1988) thought that 'dominant and semi-dominant mutation may be involved in the morpho-functional reorganization of species' and that the use of this reserve 'can produce a drastic increase in hereditary variability and evolutionary novelties'.

Means for increasing the mutation rate have been invented during the twentieth century, but their utilization has been restricted mainly to microbes and plants. Still more recently, gene technology has made it possible to recognize individual genes and to transfer these to animals of other breeds with the aid of introgressive crossing. The best-known example is the major gene affecting ovulation rate in the Booroola Merino line, selected for prolificacy (Piper and Bindon, 1982), which has been introduced to several breeds in different parts of the world. Similarly, single genes for polledness, colours and wool qualities have been transferred to other breeds. The problem in this kind of introgression is that the undesired genes follow the desired ones, so that it takes a long time to get rid of them. Hopes were raised in the 1980s that gene transfer would make it possible to bring in only the desired gene, even from another species. However, there appeared to be many problems and, since the conventional selection methods with modern statistical techniques give good rates of progress, molecular-genetic techniques have not yet led to new breeds in sheep (reviewed by Maijala and Terrill, 1991).

Selection

Selection means differences in reproductive rates in a population, increasing the relative number of offspring of animals with desired characteristics and thus making the number of their genes more abundant in the population (Lush, 1945). The introduction of artificial selection in addition to natural selection was certainly an important factor. Artificial selection differs from natural selection in the kind or degree of characteristics that are favoured, often acting in the opposite direction. It can also be more intense, decreasing the role of chance or accidental circumstances in decision-making. Different conditions and the demands of humans favoured new traits and opposed some of those favoured by nature. Natural selection still continued to work: more of the weak and sickly than of the strong and vigorous are still doomed to die before breeding age, independently of the breeder's will to select, for example, a type that is weak or susceptible to disease. In domesticated animals, natural selection is merely supplemented by artificial selection. In deciding which animals should have many offspring, humans often strongly emphasize traits that were almost useless in a state of nature. Other qualities useful in the wild state lost their value when the animals were protected against their enemies, cold and starvation by humans.

An important aid in selection was obviously the preferential killing of young rams, for which the earliest bone remains give indications. Neolithic horn cores, intermediate in size between rams and ewes, also give evidence of castration. Thus, elimination of undesirable animals could have been the origin of conscious selective breeding. Culling of aggressive animals might have led to submissiveness, and the animals breeding best in captivity might have contributed most to the gene pool (Reed, 1969).

Inbreeding

Extra inbreeding happened because domesticated animals were narrowly restricted (tethered, herded, fenced) to grow up, stay and reproduce in their birth region, soon becoming related to each other. Even primitive peoples intentionally tried to avoid the very closest inbreeding, but they knew the pedigrees only for one or two generations or only in the female line. Thus, attempts to avoid inbreeding only decreased its intensity and increased the time for producing the same amount of fixation and uniformity. That gave more opportunity for selection to discard undesired results of the inbreeding. Even at times of difficult relations among tribes, some introduction of outside genes went on. The exchanges often involved stock from only the closest neighbouring tribes. As a result of previous exchanges, such animals would already be more closely related to those among which they were introduced than with animals chosen randomly from the whole species. A large amount of such interchange would have been needed to prevent this inbreeding from leading to a situation where most tribes would have their own type of each animal species. The intensity of this inbreeding obviously varied between regions, according to geographical barriers and social customs preventing extensive trade with neighbouring tribes and to beliefs about disadvantages of inbreeding and the desire to get sires from unrelated or remotely related stocks.

The extra inbreeding caused by restricted movements of animals and by the limited number of breeding males might have been one of the most important factors leading to the production of diverse breeds in domestic animals. Its past importance is impossible to measure accurately, since so little is known about early breeding customs. It is probable that the many artificial mating barriers humans have added under domestication are much more effective than the geographical and physiological isolation, which in nature divides many species into small subgroups between which crosses occur only at rare intervals, although the processes are similar in principle (Lush, 1945).

Outbreeding

Outbreeding has sometimes increased as a result of domestication. Humans have transported breeding animals far beyond their birth or domestication area. The increased migrations of people, with the development of transport aids, such as horses and ships, increased these possibilities. As examples can be mentioned the Phoenician sailors (1400–800 BC), the military expeditions of Alexander the Great, the existence of the Roman Empire, the raids of Huns into Europe, the crusades and the age of discovery, not to mention the era of trains, cars and aeroplanes for transportation and of the development of sciences and information systems. Such cases known from documented history make it reasonable to suppose that similar exchange by migrations or wars had occurred almost since the beginnings of domestication. As in the case of geographical isolation, there have been more of such exchanges in some parts of the world than in others. Thus, animals could be crossed with races more diverse than they would have encountered if they had remained in the wild state. The spread of Merino from Spain to many lands and other examples show how this process went on in historical times. Presumably, it has gone on more rapidly in the last three or four centuries than earlier. The interchange of people and animals between continents has increased possibilities of outbreeding at an accelerating rate. Alternations of moderate inbreeding with wide outbreeding have helped decisively in producing many distinct families and new breeds. The combined effect of all these factors was that sheep changed more following domestication and evolved faster than under natural selection (Lush, 1945).

Changes of Sheep after Domestication

Unconscious selection caused by cherishing desirable individuals probably led to some genetic changes, e.g. in temperament, but the life of primitive humans close to their animals may also have given them an anatomical instinct and observation of simple truths, such as 'like begets like', long before any theories developed (Ryder, 1991). On the basis of present knowledge about genetics, the traits of sheep could be divided into those easily observable by humans and usually determined by relatively few genes (morphological traits) and the economically important genes affecting complicated physiological and quantitative productive performances, determined by many genes. Most of the major changes had taken place by the time that illustrations and records first appeared *c.* 3000 BC. These changes concerned traits determined by relatively few genes, e.g. type and colour. According to Ryder (1984), the major morphological changes following domestication were reduction in the length of the legs, followed more recently by a thickening of the limb bones, lengthening of the tail and reduction in the size and shape of horns.

Horns

In the wild ancestor, both sexes were horned, and such breeds also occur in domesticated sheep. In addition, breeds developed in which the ewes or both sexes were hornless. It is not known whether the lack of horns in some mouflon and urial ewes is due to a mutation in these wild types or to crossbreeding with polled domestic ewes. There is evidence of hornless ewes in neolithic times (7000 BC) in Iran and on the continent of Europe, but not of hornless rams until after the Middle Ages. One of the first changes was reduction in horn size (Ryder, 1984).

Coat and fleece

The coat of the first domestic sheep was very probably the same as that of the wild ancestor. Remains of neolithic wool have not been found, but modern 'hair' sheep with a coat of kemp and wool can be considered survivals of this stage of evolution. These are found in hot areas and it has been thought that the simple coat has been an adaptation to a hot environment. However, since there is no motive for breeding a woolly fleece for clothing in a hot environment, the sheep can be considered to represent the first animals that reached these areas (Ryder, 1991). Since domestication, the hairy outer coat of the wild sheep has become finer and the annual spring moult has almost disappeared. Ryder (1983) proposed a probable order of the evolutionary changes from the archaeological remains in a diagram. According to this, the first change was an increasing fineness of the coarse, outer-coat kemps of the wild sheep to produce the hairy medium fleece of the hairy Soay. Artificial selection further reduced the diameter of these fine kemps, changing them into wool fibres of medium diameter, as in the fleece of the woolly Soay.

Selection for less hairy fleeces may have started when skins were still used in clothing. The earliest signs of change are found in wool from textile remains from the European Bronze Age (*c.* 1500 BC), after several millennia of breeding. In the textile remains there are both woolly and hairy fleeces, but the latter are much less hairy than the wild type. The similarity of this wool and that of the small, brown Soay sheep of St Kilda (north-west Scotland) suggests that the Soay is a survival of Bronze Age sheep. The Soay shared other primitive features with the wild ancestor, such as short tail, white belly and an annual moult, the wool being harvested by plucking (Ryder, 1991).

Moulting

A moulting fleece is not desired by the sheep-keeper, since it is difficult to pluck a sheep before the moult has started, but the wool can be lost if plucking is left too late. Therefore, sheep have been selected for continuous growth of wool, but probably only after the invention of iron shears for removing the fleece in the Middle East about 1000 BC. The loss of moulting was evidently associated with the development of heterotype hairs. This took place on an evolutionary line in which the kemps of the hairy medium fleece changed to long hairs. Hairs are usually long and are found in the true hairy fleeces used now in carpets. This type was rare until the Middle Ages. They are intermediate between kemp and wool: coarse and kemp-like in summer, finer in winter when they lose their internal medulla and are like true wool fibres. The kemps cease to grow before moulting, but hairs only thin down and continue to grow (Ryder, 1991).

Colour

There has been a loss of colour. Wool in the textile remains is brown, and in wild sheep only the belly is white, while the upper parts have brown agouti kemps and all-black and all-white patches and the underwool is grey by dilution. The range of colour increased in the Iron Age (Ryder, 1991). Records and illustrations show white sheep in the Middle East during the Bronze Age *c.* 3000 BC. In Europe, cloth remains and surviving sheep from the Iron Age show three colour categories: all coloured (black or brown), some coloured and some white (e.g. grey) and none coloured. The surviving breeds have black, grey and white individuals in addition to the brown of the Soay. The white belly of wild sheep is rare. Most modern breeds have a white fleece. According to observations on mutations by Adalsteinsson (1983) in Icelandic sheep this colour range evolved from the brown, wild-pattern Soay as follows: the white-bellied gene mutated to all-white, this in turn to self-colour (giving all-black or all-brown animals) and this to a gene giving a mixture of coloured and white fibres (e.g. grey). In the Soay breed, white sheep are rare, but some self-coloured animals (with a coloured belly) occur. Grey sheep have not been reported in the Soay but they were common in the Iron Age type. Details of the genetics of colours are given by Sponenberg (Chapter 3).

The motive for breeding sheep with white wool may have come after the development of dyes. The breeding could have been carried out either by selecting white-spotted sheep with greater areas of white wool or by selection of all-white mutants. The great variation in the extent of white markings in piebald Soays indicates the feasibility of the first approach, but this would have resulted in recessive white sheep, which does not occur. In general, white is dominant to coloured, suggesting the second approach. In fact, the two approaches were probably carried out at the same time, because coloured sheep in modern breeds frequently have white markings (Ryder, 1991).

Behaviour

The changes in behaviour have often been neglected when the emphasis has been on the search for morphological changes after domestication. However, modification of behaviour was obviously important in adapting wild sheep to new environments. The ability to change behaviour has certainly been an important factor facilitating the domestication of sheep (Ryder, 1991).

Skeleton

Archaeological findings show an early reduction in body size, especially the height, as a result of shortening of the leg bones. Also, wild sheep have slender bones, and this trait occurred also in domestic sheep until after the Middle Ages. Wild sheep have large forequarters, and this difference appears exaggerated in rams, because of their mane and throat fringe. The full development of the hindquarters is probably a recent development for meat. Bökönyi (1974) used other limb bones to calculate changes in the wither height of sheep in Hungary, with the following results: Neolithic 60 cm, Bronze Age 70 cm, Iron Age 60 cm, Roman period 70 cm, Middle Ages 56 cm, modern 80 cm. Measurements of Soay sheep by Ryder (1991) gave a wither height of 52 cm in ewes and 56 cm in rams, and the results are supported by the similarity of bone remains to the skeleton of the Soay.

Tail length

Ryder (1984) considered it unlikely that it will become possible to follow changes in tail length except in surviving breeds. At birth, the Soay (representing the Bronze Age) has a mean tail length of 60 mm, compared with 90 mm in the Shetland (the Iron Age), while in a selection of modern British breeds it has a range of 120–190 mm. Many of the latter also have a short tail.

Other features

Ryder (1991) reported that domestic sheep have a smaller heart and a smaller eye socket than wild sheep, and the brain is about 20% smaller. There is also a reduction in brain size of c. 8% between primitive domestic and modern breeds.

Production traits

Ryder (1991) considered that the adoption of selective breeding for economic characters such as milk yield, controlled by many genes and showing continuous variation, has been a gradual process. In the last two centuries, the development has taken place at an accelerated rate, by using modern theories and techniques. Inclusion of wild sheep or very old sheep breeds, like the Soay, in breed comparisons with modern breeds in similar environments would be possible, but has not been made. However, it is apparent, without such experiments, that specialized breeds differ decisively from the wild-type breeds in wool yield and quality, growth rate, muscularity, milk yield, etc. Examples can be found in the next section.

Common Breeds of Sheep, Their Origin and Characteristics

Definition of breed

The concept of breed was first used in animal breeding in the sixteenth century, but its meaning has undergone many changes with time and is still different in the minds of different people. Among others, the following definitions have recently been used.

1. Many encyclopaedias: a breed is a population or a group of populations which can be distinguished from other populations of the same species on the basis of different allele frequencies, chromosomal changes or genetically determined phenotypical characteristics.
2. Turton (1974) used two alternative definitions:
(a) a homogeneous, subspecific group of domestic livestock with definable and identifiable external characteristics that enable it to be separated by visual appraisal from other similarly defined groups within the same species;
(b) a homogeneous group for which geographical separation from phenotypically similar groups has led to general acceptance of its separate identity.
3. Carter and Cox (1982): a breed is a subgroup of a species possessing certain recognizable characteristics and maintained as a closed breeding population, historically in a single geographical area, after which it is frequently named. In discussing sheep breeds, they defined breeds 'as those subgroups which have been recognized through formation of a breed society or registration of stock in an official flockbook. Strains will denote subdivisions of a breed not represented by separate breed associations.'

The Food and Agriculture Organization of the United Nations (FAO) has adopted Turton's definitions in its programme on animal genetic resources, especially in developing countries, which do not have breeding organizations (Loftus and Scherf, 1993). It has also realized that breeds have been developed according to geographical and cultural differences and to meet human food and agricultural requirements. It considers that breed is more a cultural than a strict scientific term. Much of the development of animal breeds took place in Britain in the eighteenth and nineteenth centuries. Ryder (1964) considered that

> breed differences have resulted partly from man's attempts to select, transport and crossbreed for obtaining desired animals and partly from natural process of selection and genetic drift. New breeds originated as strains of existing breeds. A too heavy emphasis on 'breed purity' is not well-founded, since today's crossbreds often can become pure breeds of tomorrow.

The term has obviously received too much attention in animal breeding, where the concept of 'breed purity' was central for a couple of centuries from the middle of the eighteenth century and was based largely on morphological traits that are now known to be determined by only a few of the *c.* 100,000 genes. With the development of animal genetics and breeding as sciences, the term has lost some of its value. For the purposes of conservation and use of breeds, it would be important to measure their mutual relationships. Several local studies on the genetic distances based on gene frequencies in biochemical polymorphisms have already been made, but it is important to include those based on deoxyribonucleic acid (DNA). Buchanan *et al.* (1994) used microsatellite frequencies in five sheep breeds, and more are expected in the near future from large international projects.

Classification of sheep breeds

The sheep is the most versatile farm animal: it can be kept in many kinds of environments and gives many kinds of products. Hence, the number of sheep breeds in a given region is usually larger than that of other species. In Mason's dictionary (1996), 144 pages were devoted to sheep breeds, while 108, 40, 39 and 29 pages were used for cattle, horses, pigs and goats, respectively. In the FAO Data Bank, the numbers of breeds were 863, 783, 357, 263 and 313, respectively (Loftus and Scherf, 1993). Some breeds with the same name in different countries may differ from each other more than breeds with different names. Similarly, the present population of a known breed may represent essentially different qualities from those 50 years ago.

Many breeds are native ones, developed over a long time for local, rather static environments and needs, without planned efforts by humans. They may have developed different kinds of adaptations to climates, feeds and disease environments, which may also be important in the future. The rate of changes in production conditions and market demands has been accelerating during the last two centuries. Hence, breeds have been moved from their native areas and used for upgrading and changing other breeds. The arguments for developing new breeds, in order to combine the desirable traits of existing breeds, are very similar to the economic–biological reasons for conserving existing variation for possible future needs.

Several systems have been used for classifying sheep breeds. As long as wool was the main product, its length and fineness were used as grouping bases. Other alternatives have been horns, colours, the length and fatness of tail and the origin and the main use of the breed. They could also be grouped according to continents or regions, or whether they are indigenous, locally important improved breeds, or international. Mason (1991) arranged the breeds in groups according to similarity in appearance and origin and geographical proximity, and in further subgroups according to coat type and the length, width and fatness of tail. Dolling and Jefferies (1991) grouped wool breeds on the basis of fineness and staple length of fleece. After the Second World War, grouping based on the main use has increased in importance and would be helpful for the users of sheep. However, the versatility of sheep and its utilization render a clear classification on this basis difficult: the preferential uses of the same breed vary between countries or regions and it should thus be included in several groups. Hence, 'common' breeds in the following description are presented in an alphabetical order according to their English names used by Mason (1996). Many of them have an English origin, since England was the forerunner of breed development in the AD 1700–1800s and many English people with their animals emigrated to other continents during the expansion of the British Empire in the eighteenth and nineteenth centuries.

Common breeds

In choosing 'common' breeds, the main attention was paid to their present numbers in the world and their international importance in production or in the development of new breeds. A population size of *c.* 5 million or presence in at least three countries was used as a crude rule. Thus, several famous breeds that have lost their importance are not included. Instead, some local, indigenous breeds, used in wide areas or utilized in creating new breeds, are included. Some breeds have been included in order to represent countries with large sheep populations (e.g. China, India, former USSR, some African countries), even though they are not used in other countries or do not have breed societies or herd-books. Of the many new breeds developed in the nineteenth century, only those few are included which have found use in several countries. Information on origin, population sizes, some typical features published by various authors and suitabilities for different uses and environments and some estimates of measurable traits will be given.

Arab

Arab is a group of similar breeds in at least seven African countries (among others, the Sudan desert sheep in Sudan, Maure in West Africa, Somali Arab in Somalia, Tuareg in Sahara). They are mainly used for meat and some of them also for milk. Their fleece is of the hairy type, and the fibre diameter is 28–31 μm. It is used for blankets, carpets and tent cloth. The animals are mostly white, but some types may have faces with shades of red-brown. White-faced sheep are made light-sensitive by *Hypericum* plants. The ewes are polled, but Maure and Tuareg rams have spiral horns. They have fat tails and lop ears. In good conditions some ewes may lamb twice a year. They are hardy (Ryder, 1983).

Awassi

Awassi is a breed of the Near East fat-tailed type, originating from the fat-tail of antiquity (Ryder, 1983). It is the main breed in Iraq, Syria, Israel, Jordan and Lebanon, and occurs also in Turkey, Cyprus and Saudi Arabia. It is used for meat, milk and carpet wool. It can have different colours: white with brown head and legs, sometimes with black, white, grey or spotted face, but there are also all-brown or all-black animals. The nose is convex and the ears are long and drooping. Rams are horned and most ewes are polled (Mason, 1996). The average height of adult ewes is 65 cm, live weight 46 kg, fleece weight 2.1 kg, clean wool yield 63%, staple length 14 cm and fibre diameter 34 μm. There are 3–6% medullated fibres. The birth rate is 80–85% and twinning rate 6–10%. The milk yield is 90–155 kg in 170–200 days, with 6.1–7.0% fat (Yalcin, 1986). Awassi is able to use dry pastures and to withstand intense heat and heavy winter rainfalls (Epstein, 1982).

Baluchi

Baluchi is a fat-tailed breed kept in Pakistan, Iran and Afghanistan (Ryder, 1983). It is used for carpet wool, milk and meat. It is white with black or brown spots on head and legs, rams are horned or polled, ewes polled (Mason, 1996). The average height is 60 cm, the weight of rams/ewes 37/33 kg. The age at first breeding is *c.* 18 months, and 55 single lambs are obtained per 100 ewes per year. Fleece weight is 2.4 kg year^{-1}, 90% is white, 8% grey and 2% black. Staple length is 7.3 cm and clean yield 50%. There are 69% true fibres, 24% heterotypical, 6% medullated and 1% kemp (Hasnain, 1985).

Barbary

Barbary is a native breed typical of Libya and Tunisia. There have probably been fat-tailed sheep in Tunisia since Carthaginian times and they may have been brought from Syria by the Phoenicians *c.* 400 BC. About 700 years later they were replaced by another type of sheep, but after an additional 500 years the fat-tailed sheep were reintroduced by Arabs in *c.* AD 900. The fat-tailed sheep came to Libya from Egypt *c.* 1000 BC (Mason, 1967). The Barbary sheep were probably introduced to Tunisia and Libya simultaneously, originating from the Asian steppes. In both countries the breed has its local name. 'Barbary sheep' is the European name stemming from the ancient inhabitants of North Africa (Berbers) (Khaldi, 1989). The breed is also found in Algeria, Egypt (Barki) and Italy (Campanian and Sicilian B). Its main uses vary from country to country, but meat and carpet wool are the common uses. Wool varies from coarse and kempy to medium-fine and wavy (Mason, 1967). Milk production is important in Libya, Egypt and Italy. White is the main colour, but there are also pied, black or brown animals. The faces and legs are often red or black or have dark spots. The ewes are polled, the rams usually horned but may be polled in some types. Some types may also have lop-ears. The Tunisian and Italian types have had herd-books for *c.* 50 years (Mason, 1996).

Bleu du Maine

Bleu du Maine is a French long-woolled breed, which originated in the Maine region from crosses of Leicester Longwool and Wensleydale with a local marsh breed. In the first part of the twentieth century, a blue line was developed, for which a breed society was established in 1927. There are herd-books in Germany, the UK and South Africa, and animals in Belgium, the Czech Republic, The Netherlands and Switzerland. It is used for meat and wool production (Mason, 1996). It is mainly white with blue head and legs, but there can be blue also in the body. It is polled, matures early and has a litter size of 2.0 and a short breeding season. The fibre diameter is 37 μm and medullation 9.6% (Steinhagen *et al.*, 1986). The average height of adult rams/ewes is 84/74 cm and weight 123/83 kg. It surpassed the reference breed in litter size and earliness of sexual maturity and was equal in daily gain, milk yield and carcass leanness (Simon and Buchenauer, 1993).

Booroola Merino
Booroola Merino is a strain of medium non Peppin Australian Merino carrying a gene (*Fec*B) for high ovulation rate. It is often considered a separate breed. The selection of ewes for prolificacy was started by a private breeder in 1945 at a farm named Booroola and was continued on both males and females by Commonwealth Scientific and Industrial Research Organisation (CSIRO) since 1958. Its rams are horned and ewes usually polled (Mason, 1996). It has spread to New Zealand, the USA, Canada, Algeria, Egypt, Israel, ten European countries, six Central or South American countries and South Africa for improving prolificacy (Fahmy, 1996a). It does not differ significantly from its ancestor Merino strain in aspects other than ovulation rate (5.7) and litter size (2.6). The fleece weight is 3.8 kg, clean yield 68%, fibre diameter 19.4 μm and staple length 10.3 cm (Piper and Bindon, 1996).

Border Leicester
Border Leicester was developed in the border region of Scotland and England at the turn of the nineteenth century from the Leicester Longwool and the Cheviot. It was recognized as a breed in 1869 (herd-book in 1898). It is long-woolled and was also used with the Cheviot to produce the Scottish Halfbred ewe for crosses with meat breeds for lamb production. There are herdbooks in the USA, Canada and New Zealand and animals in Ireland, Australia, South Africa, Colombia and China. It is white and polled. In UK, the height of rams/ewes is 81/77 cm, the live weight 135/90 kg. In the European Association for Animal Production (EAAP) Data Bank it was used as standard, so there are no relative performances (Simon and Buchenauer, 1993).

Charollais
Charollais originated in Nièvre, France, from crossing the Leicester Longwool with a local breed since 1825. It was named in 1964 and has herd-books in France, the UK and Switzerland and animals in ten other European countries, Israel, Canada and China. It is used for producing meat and wool. The animals are white, but some may have a pinkish or woolless head or brown face and legs. It is polled, and ewes average 1.9 lambs per litter. The average wither height of rams/ewes is 80/71 cm and live weight 118/80 kg. The litter size is at the level of the reference breeds, muscularity equal or above, daily gain, carcass leanness and age at sexual maturity equal, wool yield and length of mating season below (Simon and Buchenauer, 1993).

Cheviot
Cheviot originated at the border of England and Scotland at the end of the eighteenth century from a local breed and Leicester Longwool. It was named in 1791. A breed society was founded in 1891 and later also in the USA and New Zealand. There are herdbooks in Australia and Canada (Mason, 1996) and animals in Colombia, Chile, Ireland, Norway, South Africa and Taiwan. It is used for meat and wool production. The animals are white, rams horned and ewes polled. The height of rams/ewes is 75/65 cm and the weight 100/55 kg. They are considered hardy and of a good conformation. Early sexual maturity is one of its relative advantages, while its muscularity, carcass leanness, milk yield and lambing interval are equal with Suffolk and daily gain, litter size and wool yield poorer (Simon and Buchenauer, 1993). It has a well-known variety, North Country Cheviot, developed since 1792 (breed society in 1912), which has herd-books in the USA and Canada.

Chios
Chios is a long-fat-tailed breed, developed on the Chios island in Greece. It may have originated from a cross between the Greek Zackel and the Turkish fat-tailed Karaman, or between the Turkish Kivircik and Daglic breeds. Mason (1967) preferred the latter alternative. In the 1960s the number of Chios sheep decreased, but, because of its reproductive characteristics (Zervas *et al.*, 1988) and milking ability, the number increased in the 1970s. It found demand on the mainland of Greece and on other Aegean islands, so that the number increased manyfold (Brooke and Ryder, 1978). It became popular in Cyprus and is also kept in western Turkey (as Sakiz) and in nine other Near Eastern and two other Balkan countries (Galal, 1996; Hatziminaoglou *et al.*, 1996). It is white, but has black or brown spots on face, belly and legs; sometimes the whole face is black. The ewes are usually polled, but some have small horns, and the rams have spiral horns. It is used for milk, meat and carpet wool. The height of rams/ewes is 81/73 cm and weight 72/50 kg. Ewe lambs can be bred at 8–9 months of age. The breeding season is long. The average litter size is 2.0. Average fleece weight of rams/ewes is 2.0/1.4 kg. The average milk yield is *c.* 190 kg in 180 days (Lysandrides, 1981). The staple length is 10 cm and mean fibre diameter 42 μm and there are 6–7% medullated fibres (Brooke and Ryder, 1978).

Churra
Churra (Churro) (Ch.) is a group of coarse-woolled autochthonous breeds of Spain, Portugal and their former possessions in South America. In Spain their share of sheep population is about one-fifth. The Andalusian Ch. is the largest variety of Spanish Ch. and may have some Celtic origin (Ryder, 1983). Because of crossbreeding the breed is endangered. Tensina Ch. has a bigger population and is not endangered. In Portugal, there are three main varieties: autochthonous Mondegueira, Andalusian Ch.-based Algarve Ch. and Mondegueira-based Ch. da Torre Quenta. They represent > 20% of the Portuguese sheep population. In South America, there are Ch. sheep in Peru, Chile, Argentina and Uruguay (since 1550). The main use of three varieties is for meat and two are

also used for milk. Wool is the third use in two varieties. The mean heights and weights of the types vary depending on feeding conditions. The average height of rams/ewes of the five types is 87/78 kg and weight 74/54 kg (Simon and Buchenauer, 1993). All have white fleeces, but most of them have black or brown spots around the eyes and on the legs. The rams are usually horned and ewes polled, but in Tensina Ch. there are also horned ewes. The Andalusian Ch. is claimed to be resistant to foot-rot and fascioliasis and adapted to damp, the Tensina Ch. to both cold and humid and hot and arid climates. All the types have been considered to be above reference breeds in litter size, most of them also in fibre thickness and two in milk yield. They are close to average in age at sexual maturity, lambing interval and length of mating season, but below it in daily gain, muscularity and carcass leanness. In wool yield and dry matter content of milk one type was average and one inferior (Simon and Buchenauer, 1993). The fleece weight of rams/ewes is 2.5/1.5 kg, fibre thickness 30–40 μm and milk yield 50–60 kg lactation^{-1} (Brooke and Ryder, 1978).

Columbia
Columbia originated from crossing Lincoln with American Rambouillet (similar to Corriedale), started in 1912 in Idaho, USA. The first herd-book appeared in 1942. It occurs also in Canada (Shrestha and Vesely, 1986). It is polled and mainly used for producing wool and meat. The weight of rams/ewes is 102–136/57–91 kg. They are robust, but have long legs and a rangy body. The face is open. There is very little wool below the knees and hocks, although there may be occasional black hairs. The carcass conformation has suffered from aiming at sheep that could walk readily and stand rigorous range conditions. A typical fleece weight in ewes is 5.4 kg, of which 50% is clean wool. Its fibre thickness is 28–33 μm (Briggs, 1969).

Coopworth
Coopworth originated in New Zealand in 1956–1968 from a cross between Border Leicester and Romney sheep. A breed society was established in 1968, later in Australia and USA (Mason, 1996). It has the third largest ewe population in New Zealand. It is used mainly for meat and wool production (Carter and Cox, 1982), as also in Canada (Shrestha and Vesely, 1986).

Cormo
Cormo is a new breed, originating in Australia since 1960 from crossing Polwarth with Corriedale × Tasmanian Merino ewes (Mason, 1996). It has breed societies in Australia and the USA. In Argentine, there is a similar new breed, Argentine Cormo, based on Corriedale × Merino crosses. The breed is used for medium-wool and meat production.

Corriedale
Corriedale originated in New Zealand in 1880–1910 from crossing Lincoln or Leicester Longwool rams with Merino ewes. A breed society was established in 1910. In 1977 it comprised 8.8% of the New Zealand ewe population (Carter

and Cox, 1982). Australia and South Africa also have breed societies and even South American countries have altogether over 10 million Corriedales (Ryder, 1983). They were first imported into the USA in 1914 and spread rapidly to most states producing sheep and also to Canada. In Asia, they comprise 90% of the sheep of Japan and the breed is also represented in Korea, Taiwan, Papua, China and India. It is a dual-purpose breed, producing wool and mutton. The rams weigh 84–114 kg and ewes 57–84 kg. They have open, white faces, but black nostrils. They mature relatively early and have good mutton conformation (Briggs, 1969). Using Suffolk as reference breed, Corriedales have in experiments obtained relative values > 100 for number and weight of lambs weaned per ewe exposed to ram, but < 100 for daily gain, weaning weight and carcass leanness. Their carcasses have contained more fat than those of Suffolk (Terrill and Maijala, 1991).

Criollo

Criollo originated in Spanish America in 1548–1812 mainly from Spanish Churro, possibly also partly from Spanish Merino. The animals are white, black or pied (Mason, 1996). It is kept especially on highlands, comprises a substantial part of the sheep populations in Chile, Bolivia and Colombia and is also kept in seven other South American countries and in the West Indies. It is used for meat and carpet wool. The animals are small: the weight of ewes is *c.* 22 kg and the height 45 cm (Ryder, 1983).

Deccani

Deccani originated naturally through adaptation to the arid and semiarid conditions of the Deccan plateau of India. It comprises practically all sheep in this region and is also kept in the adjoining regions. It is one of the most important of the 40 sheep breeds of India and is used for meat and carpet wool production. It is a hairy, rather heterogeneous breed, which has black, white, brown and pied animals. The animals are hardy and long-legged and have a narrow chest, long thin neck, Roman nose, short muzzle and ears of varying shape. Rams are usually horned, ewes polled. The average height of rams/ewes is 73/66 cm and weight 42/26 kg. The annual fleece yield from two clips for rams/ewes is 1.35/1 kg. The wool is not of high quality, being a mixture of hair and fine fibres. The average staple length is 8.6 cm and fibre diameter 52.4 μm. The average fertility is 82% and litter size 1.0. They lamb for the first time at 3 years of age and the average lambing interval is 433 days (Nimbkar, 1993).

Dorset Down

Dorset Down originated in South England in the first part of the nineteenth century from Berkshire, Southdown, Hampshire and Wiltshire. A breed society was founded in 1906 (Mason, 1996). It was established permanently in New Zealand in 1947, and the herd-book was refounded in 1950 (Carter and Cox, 1982). There are herd-books in France and Australia (Mason, 1996) and animals in India and Ireland. The animals are polled and white with brown face, ears, legs and feet. It is used for meat (sire breed) and short fine wool. It is close to

Suffolk in daily weight gain, in age at sexual maturity, milk yield and muscularity, but its carcasses have a little more fat and less bone (Terrill and Maijala, 1991). The average height of rams/ewes is 68/64 cm and weight 135/100 kg (Simon and Buchenauer, 1993).

Dorset Horn

Dorset Horn (DH) probably originated from the Portland and the Merino in southern England in the nineteenth century. It was recognized in 1862 and a breed society was established in 1891. There are societies/herd-books in the USA, Australia, Canada, South Africa and New Zealand (Mason, 1996) and animals in Ireland, Colombia, South Africa, Malaysia and Taiwan. Both sexes are horned, but a polled variety (Poll D) was developed in Australia by crossing DH with Corriedales or Ryelands and then repeatedly backcrossing with DH while selecting against horns. A separate breed society was founded in 1954. In New Zealand, there were 20 times as many Poll D ewes as DH ewes in 1977 (Carter and Cox, 1982). There are also Poll D animals in the UK. The breed is used for meat, wool and milk. Its special advantage in meat production is the ability to lamb twice a year and at most times of the year. This was observed as early as the 1600s and has (besides early sexual maturity) recently made it popular as a sire of dams in systematic crossing or as a component of new dam breeds (Maijala and Terrill, 1991). The fleece weight of DH is *c.* 2.8 kg. The wool is free from kemp, the fibre diameter is 26–29 μm and the staple length 9 cm. The adult weight of rams/ewes is 105/93 kg. The breed is close to Suffolk in muscularity, daily gain, carcass leanness, litter size and milk yield (Simon and Buchenauer, 1993).

East Friesian

East Friesian is based on an autochthonous marsh sheep of northern Holland and Germany. It may include some inheritance from Leicester rams imported in the nineteenth century. There are herd-books in Germany, The Netherlands and the UK, and it has spread to at least eight other European and four South American countries, Canada, Israel, Cyprus, New Zealand and Japan, for experimentation and formation of new dam breeds. It is used mainly for milk production, in which it is considered the world's most highly developed breed. The average yield of recorded ewes is around 600 kg with 6–7% fat in a 200-day lactation. Because of its high milk yield, early sexual maturity and high prolificacy (litter size 1.7), it has been used as a component of at least five new dam breeds (Maijala and Terrill, 1991). Earlier it was used for crossing to improve mutton production in south and south-east European countries. The average height of rams/ewes is 86/80 cm and weight 110/80 kg (Simon and Buchenauer, 1993). It is mostly white, but some black and brown animals occur. It is polled and has a thin woolless rat-tail. Its head and legs are also woolless. The average fleece weight is 4 kg, staple length 15 cm and fibre diameter 40 μm (Farid and Fahmy, 1996). It is adapted to intensive management in small herds, but not always to extensive management in big flocks. It is considered a 'smallholders' cow' (Alderson, 1976).

Finnsheep

Finnsheep (F) is an autochthonous breed of the northern short-tailed type, other members of which are found in Iceland, Norway, Sweden, the UK and Russia (see Romanov) (Mason, 1996). The basic animals may have arrived from the east two millennia ago, but it has common features with the mouflon sheep and may have received genes from the west and south. A breed society was founded in 1918 (Maijala, 1988). Since 1962, more than 40 countries from each continent have imported F for experiments on prolificacy and for creating new dam breeds (Fahmy, 1996a). There are societies/herd-books in the USA, Canada and France. Pure F and its crosses have been compared with many breeds for many traits. It has been used as a component of 23 new prolific breeds (Maijala,1984; Maijala and Terrill,1991). Finnsheep was used mainly for serving domestic needs for wool and fur, but since the 1950s meat has been the main product. Multiutilization has recently been considered important. About 90% are white, 8% black, 1% brown and 1% grey. The head and legs are woolless. Both sexes are usually polled. The weight of recorded rams/ewes in 1992 was 95/69 kg, i.e. 85% of that of three foreign breeds, which agrees well with results from 27 foreign experiments. These showed the F to differ from most other breeds in several fertility traits. The average conception rate is 97–98%, litter size 2.8. The relative value for daily gain was 90%, dressing percentage 100%, carcass leanness 94%, eye muscle area 81%, carcass fatness 102%, kidney fatness 169%, bone content 93%, daily milk yield 92%, fleece weight 74%, clean yield 104%, staple length 133% and fibre diameter 77%. The mean fat content of the milk is 5.9%. Many animals are suitable for producing luxury furs; the pure white colour is favoured and allows many kinds of dyeing. Adaptability to cold temperature is poorer than in some Scottish breeds and to hot temperature poorer than in Egyptian breeds (reviewed in Maijala, 1996).

Fulani

Fulani is a native hairy breed of the Sahel type in Central Africa from Senegal to Cameroon. Three of its varieties (Toronké, Uda and Bornu) are found in at least two of the countries and another three in one country each. They are used for meat, are white, pied or coloured and have lop-ears. Rams are horned, ewes horned or polled (Mason, 1996). Ewes lamb the first time at 18–20 months and twins are rare. A weight of adult rams/ewes of 49/32 kg has been recorded (Wilson, 1984).

German Mutton Merino

German Mutton Merino (GMM) originated in Germany in 1904 from Précoce, imported 30 years earlier. There are also herd-books in Bulgaria, Spain and South Africa (Mason, 1996) and animals in Israel. It is used for meat and fine-wool production. The animals are white and polled. The height of rams/ewes is 81/77 cm and adult weight 100/65 kg. They have a good carcass quality and high twinning rate (60%). They have a larger wool yield, larger litters and thicker wool fibres than German Merinos. In daily gain, muscularity, carcass leanness and age at sexual maturity they are equal. There is another GMM strain based on importations from Spain, France and UK. It has a separate herd-book and a better muscularity and finer wool fibres than the main strain. It also has the ability for off-season breeding and does well under dry climatic conditions (Simon and Buchenauer, 1993).

Hair sheep
Hair sheep are a group of sheep with fleece similar to wild sheep. They comprise nine hairy varieties in Africa, 12 in India and one subgroup in tropical America. Almost all are used for meat, but one in Sudan is mainly used for milk. The ewes are polled, but in two varieties rams have horns. In most strains white is the usual colour, but some also have tan or pied animals or black heads or spots. Lop-ears are typical of some varieties, and one variety is often earless. Most strains have long tails, some are fat-tailed, but some have short or thin tails. Rams of the American variety usually have a mane and throat ruff. The West African dwarf sheep is trypanotolerant (Mason, 1996).

Hampshire
Hampshire (Down) originated in the first half of the nineteenth century in England from crossing Southdown rams with Wiltshire Horn and Berkshire Knot ewes. It was recognized in 1859. Breed societies were founded in England and the USA in 1889. There are herd-books in Australia, Canada, New Zealand, South Africa and France (Mason, 1996). The breed has also spread to Chile, Urugay, Colombia and Brazil. It has been used for forming many other breeds. It was used mainly for short-wool production, but has become a popular sire breed for meat lambs. It is white with a black-brown face, ears and legs. Its forehead is wool-covered and polled (Mason, 1996). The adult rams/ewes weigh 100/70 kg. It has larger wool yields (2.5 kg) and thicker fibres (26–28 μm) than Suffolk, but a somewhat lower daily gain, milk yield and litter size (1.4). In muscularity, carcass leanness and age at sexual maturity it is equal with Suffolk (Simon and Buchenauer, 1993). Staple length is 7 cm. The legs and body are short (Alderson, 1976).

Han
Han originated from Mongolian sheep and is kept in the semihumid agricultural areas of China (Henan, Hebei, Shandong, Anhui and Jiangsu provinces). It has several common features with other Chinese breeds (% medullated fibres, adaptation to unfavourable environments, ability to deposit fat in the body and low wool production). There are two types of Han, which can be considered two different breeds: largetail (Lt) and smalltail (St). Largetail is polled and has a long, broad, fat tail (length in rams/ewes 49/33 cm, width 33/21 cm, weight in rams can be 25 kg). Because of the heavy tail, the Lt type is only adaptable to the plains. Smalltail rams have large horns and the ewes small horns. The tails are much shorter (23/18 cm) and narrower (19/13 cm) than in Lt. This type is adaptable to hilly lands for grazing. It is used for production of medium-fine and carpet wool and of meat. The height of Lt rams/ewes is 75/66 cm, of St 76/69 cm, and the weights 60/42 and 58/38 kg, respectively. Han sheep are

precocious and prolific. Largetail ewes give 163 and St ewes 229 lambs per 100 ewes lambing. The average fleece weight of Lt rams/ewes is 0.9/0.4 kg and of St 2.0/1.3 kg. In the wool of Lt, there is 97% true wool (fibre diameter 14.5 μm), 1.4% heterotype wool (34.6 μm), 1.4% hair (57.4 μm) and 0% kemp. Dressing % of carcasses has been 51.3 in Lt and 50.7 in St (Cheng, 1984).

Île-de-France

Île-de-France originated in France as Dishley Merino from a Leicester Longwool × Merino cross in 1833–1900. It also has breed societies or herd-books in the UK and Spain, and has spread to Canada, Ireland, Turkey, The Netherlands, South Africa, Brazil and Morocco. It is used for the production of meat (sire breed) and medium-fine wool. It is white and polled (Mason, 1996). The height of rams/ewes is 80/69 cm and the adult weight 130/87 kg. Fleece weight is *c.* 4 kg, staple length 13 cm. It equals Suffolk and Texel in muscularity, daily gain and age at sexual maturity, excels Suffolk and equals Texel in litter size and equals Suffolk and is below Texel in carcass leanness and milk yield. It has thicker wool fibres, longer lambing intervals and lower wool yields than Texel (Simon and Buchenauer, 1993).

Karakul

Karakul (K) is a fat-tailed desert sheep, on whose origin there are varying views. One theory assumes it originated from crossing long-tailed Danadar sheep with native sheep called Arabi and with fat-rumped sheep (Briggs, 1969). The development may have taken place in the eighth century in Buchara, West Turkestan, where there is a village called Kara-Kul. Another theory assumes that K emerged through centuries-long selection of lambs on the basis of the curl in the birth coat. According to a third theory, Karakul sheep appeared many centuries BC by mutations for fat tail and curly lamb coat and in the eighth century in Turkestan (where it is also known as Bukhara), when the territory was conquered by the Arabs. In Uzbekistan, K sheep are called Arabi, suggesting that K are related to Arab sheep. Present-day K may not be the pure fur sheep brought by the Arabs, but crossed with fat-rumped sheep. Constitution and productive traits of K were shaped by the year-long extensive husbandry and a dry, hot climate. They are characterized by endurance and adaptability to life in deserts. They resist dry and hot weather and frequent drought, and utilize efficiently the poor and salty grazing lands of the northern plain. Karakul is the most classical and important fur breed but is also used for milk production after slaughtering the lambs after birth (Dmitriev and Ernst, 1989). There are 12 million K animals in six republics of the former USSR. They are kept in Asia (Afghanistan, China, Iran, Turkey, Uzbekistan), Africa (Botswana, Namibia, Mali, Rwanda, South Africa), Europe (Germany, Czechoslovakia, Austria, Romania, Spain), South America (Brazil) and North America (USA) (Ryder, 1983; Wilson,1991). Karakul is widely used for improving other fur breeds. The black variety (Arabi) accounts for more than 80%. The rest are grey, brown, pink-roan or white. Adult K sheep are grey, varying from ash grey to white. The head, ears and legs have hair of the same colour as the newborn lamb. The rams are horned. The skin is thin. The wool generally consists of true wool and intermediate fibres with some guard hair (Dmitriev and Ernst, 1989). The mature weight of rams/ewes is 70/48 kg, daily gain until 120 days 0.22 kg. The milk yield during 2–3 months is 400–500 g day^{-1} with 9% fat. The annual wool yield is *c.* 2 kg of carpet-type wool (Nel, 1991). Karakul ewes lamb the first time at 15–22 months, the conception rate is 84%, lambing interval 8.5 months and litter size born 1.01 (Wilson, 1991). Because of continuous matings, 150 lambs per ewe can be obtained. Many details of the traits of lamb pelts are given by Nel (1991) and Wilson (1991) and of body measurements and performances by Yalcin (1986).

Kazakh Finewool

Kazakh Finewool was bred in the Alma-Ata region of Kazakhstan in 1931–1946, based on Kazakh fat-rumped ewes mated to Précoce and American Rambouillet rams. In China, there is a native, fat-rumped, carpet-wool breed called Kazakh, which is obviously the female ancestor of Kazakh Finewool. The breed is used for production of meat and fine wool. It is polled and skin folds or wrinkles are rare. It has a large body, strong constitution and skeleton, good conformation for meat and high carcass weight. The adult weight of rams/ewes is 112/63 kg, of which 71/86% is reached by 18 months of age. The wool is similar to that of Précoce, white and strong, fibre diameter 22–25 μm. The fleece weight of rams/ewes is 10.7/4.9 kg (50% clean) and staple length 8.5/7.5 cm. The litter size at birth is 1.05–1.35. The breed is suitable for transhumance management systems (Dmitriev and Ernst, 1989).

Kirgiz Finewool

Kirgiz Finewool was developed in 1932–1956 at the high altitudes (3000 m) and continental climate of Kirgizia, based on ewes of coarse-wooled Kirgiz fat-rumped breed and rams of six different fine-wooled Merino-based breeds. It was recognized as a breed in 1946 and is raised in most regions of Kirgizia and parts of Tadzikistan. It is used for meat and fine wool and is polled. The animals rarely have skin folds. They are white, but some animals have brown spots on the ear tips, around the eyes and on the legs. The adult weight of rams/ewes is 105/60 kg, of which 86% is reached by 18 months of age. The average fleece weight of rams/ewes is 12.0/4.3 kg, clean yield 54% and fibre diameter 24–27/22–25 μm. The staple length is 7–8 cm. The head is wool-covered up to the eyes and the limbs down to the knees and hocks. The litter size at birth is 1.3. They are able to walk long distances and to survive on mountain and valley pastures (Dmitriev and Ernst, 1989).

Leicester Longwool

Leicester Longwool was developed by the forerunner of animal breeding, R. Bakewell, in England in 1755–1790 from Old Leicester by introducing additional genetic variation from Lincoln and Ryeland. The breed has contributed to the creation of many English long-wool breeds. There are also breed societies/herd-books in Australia, Canada and New Zealand (Mason, 1996) and animals in Ireland, the USA, Uruguay, China and Taiwan (Ryder, 1983). It is used for production of coarse wool and meat, and is white and polled. The weight of ewes is 82 kg. The fleece weighs 5.5 kg and fibre diameter is 36 μm (Alderson, 1976).

Lincoln Longwool

Lincoln Longwool originates from crossing Leicester Longwool with Old Lincoln in England. Breed societies in the USA and UK were founded in 1891–1892 and a herd-book in New Zealand in 1912. It has also been exported to Argentina, Australia and several Eastern European countries, but has decreased in numbers in many countries. It was one of the six breeds used in the foundation of the Corriedale. It is used for meat and wool. It is white with a blue or white face, and polled. The height of rams/ewes is 90/80 cm and the weight 113/91 kg. The breed is considered the heaviest in the world and to have the longest fleece fibres (staple length 35 cm). The fibre diameter is 39 μm (Alderson, 1976). It is the coarsest and most lustrous of the English long-wool breeds. The fleece weight is 6.8 kg. It is claimed to excel Suffolk in wool yield and earliness of sexual maturity and to be equal in daily gain, carcass leanness, milk yield and lambing interval, but it is inferior in muscularity and litter size (Simon and Buchenauer, 1993).

Marwari

Marwari is one of India's major breeds, being kept in many districts. It is used for carpet wool and meat. It is white with a black face, is polled and the ears are small. The tail is thin and 18 cm long. The average height of adult rams/ewes is 62/59 cm, weight 31/26 kg and body length 72/69 cm. The age at first breeding of rams/ewes is 12/18 months. The lambing % is 87 and litter size 1.0. The average fleece weight in two annual shearings is 1.8 kg, staple length 6.6 cm, fibre diameter 37 μm and medullation 65% (Acharya, 1982).

Merino

Merino is one of the oldest known sheep breeds. Its exact origin is not known, but sheep in Spain were known for the fineness of their wool almost two millennia ago (Briggs, 1969). They have been assumed to have come from Phoenicia and Carthage. The Moors developed a prosperous wool industry in Spain in *c.* AD 700–1200 and the Spanish textile products became appreciated in Europe and Africa. The high prices obtained for the textiles increased the interest in improving the quality of wool, and so the Spanish Merino became the world's best wool-producing sheep. It originated in Spain in the thirteenth century and was named in the fifteenth century. Its first exports to other countries were to Sweden in 1723, Germany in 1765 (see Merinolandschaf) and France in 1767–1786 (see Rambouillet, Précoce). Before 1800 it was also exported to Austria, England, South Africa, the USA and Australia. Its arrival in Australia in 1793 was decisive for the role of Merinos, since Australia developed into the most important wool-producing country in the world. The history and expansion of Merinos are described by, among others, Ryder (1983).

Even in 1993 there were more sheep in Australia than in Europe. There are at least 45 different strains of Merino or Merino-based breeds in at least 27 countries in all continents. One strain, differing from Australian Merino with regard to a single gene, is often considered a separate breed (see Booroola Merino). A similar example is the Poll Merino, selected for absence of horns (Carter and Cox, 1982). Although Merinos were originally developed as wool breeds, more and more Merino-based breeds are used as meat breeds. Merino types comprise more than one-third of the world population of sheep (Flamant, 1991).

The population of the Spanish Merino, from which the development started, is approaching endangered level, but it has 14 derivatives in Spain, which together with their ancestor comprised almost one-quarter of the Spanish sheep population in 1978. The average height of rams/ewes is 82/77 cm, the weight 83/52 kg. The animals are white. Rams have spiral horns, but ewes are polled. It is used for meat, wool and fur. It is adapted to extreme temperatures. It has a higher wool yield and thinner fibres than Précoce, is equal in litter size and length of mating season, is inferior in daily gain and has a higher age at sexual maturity and a longer lambing interval (Simon and Buchenauer, 1993). Information on the wool types is given by, among others, Dolling and Jefferies (1991).

Merinolandschaf

Merinolandschaf originated in Germany from crossing Merino with Württemberg Land sheep. It was recognized in 1887–1906 and finally named in 1950. It has been imported by South Africa and Spain. It is used for meat, wool and vegetation management. The animals are white and polled (Mason, 1996). The height of rams/ewes is 95/80 cm and weight 130/80 kg. They are long-distance walkers and have aseasonal breeding (Simon and Buchenauer, 1993).

Mongolian

Mongolian is a fat-tailed breed in Mongolia and in the pastoral areas of northern China. It is used for production of carpet wool, meat and milk. It is usually white with a black or brown head, pied or self-coloured. Rams have spiral-shaped horns, ewes are polled. There are several varieties in both countries. The breed is the origin of many breeds in China (Mason, 1996). It is adapted to unfavourable environments, can deposit fat in the tail and has a low wool production. The height of rams/ewes is 67/62 cm and weight 45/39 kg. The fleece weight of rams/ewes is 1.5/1.2 kg, fibre length 7–8 cm, true wool 76%, heterotype wool 8%, hair 5% and kemp 11%. Dressing % of males/females was 51.4/46.5 (Cheng, 1984).

Oxford Down

Oxford Down originated in England in *c.* 1830 from crosses of Cotswold with Hampshire and Southdown. It was recognized in 1851, and a breed society was founded in 1888. There are also societies/herd-books in the USA and Canada and animals in New Zealand, Ireland, Switzerland, Norway and Finland

(Mason, 1996). It is used for meat (terminal sire) and wool production (short wool). It is white with a wool-covered dark brown face and legs, and it is polled. The height of rams/ewes is 75/65 cm, weight 120/90 kg. It equals Suffolk in daily gain, carcass leanness, litter size and milk yield, and is inferior in muscularity (Simon and Buchenauer, 1993). Its carcasses have a high fat content (Terrill and Maijala, 1991).

Perendale

Perendale originated in 1938–1960 in New Zealand from a cross of Cheviot rams with New Zealand Romney ewes, primarily to improve production in difficult hill country. A breed society was established in 1960 and a herd-book in 1961 (Mason, 1996). Its contribution over time to the New Zealand sheep industry has been about the same as that of the Coopworth. In 1977 it had the second largest registered ewe population after Romney in the country (Carter and Cox, 1982). It has a breed society in Australia and has also spread to other countries. It is polled and used for meat and medium wool production (Mason, 1996).

Polwarth

Polwarth originated in *c.* 1880 in Australia from a backcross of Merino to Lincoln Longwool, using then at least five generations of *inter se* breeding; it is 75% Merino. It comprises 2% of the Australian sheep population and is also used in New Zealand, Uruguay, Argentina, Brazil and Korea for production of medium-fine wool and meat and for formation of a new wool breed in India. There are horned and polled varieties (Mason, 1996).

Précoce

Précoce originated in France from Spanish Merino (imported in 1799–1811), with special emphasis on early sexual maturity. It has herd-books in Portugal and Spain and occurs in Morocco (Mason, 1996). It has been used for the formation of, among others, Arles Merino in France, Polish Merino, GMM and North Caucasus Semifinewool. It is used for meat and wool. It is white, the rams horned, ewes polled. The height of rams/ewes is 80/73 cm and the weight 199/65 kg. It surpasses Rambouillet in muscularity, carcass leanness, daily gain, litter size and milk yield and is equal in age at sexual maturity, lambing interval and fibre thickness (Simon and Buchenauer, 1993).

Rambouillet

Rambouillet is another strain originating from Spanish Merino, which was imported to Rambouillet state farm near Paris (1786, 1799–1803) and bred exclusively there. It has been considered a breed in the USA since its importation from France (1840–1860) and Germany (1882–1900). A society was founded in the USA in 1889. Some decades ago, *c.* 50% of all sheep in the USA carried some Rambouillet breeding, and sheep was equated with Rambouillet in many areas in the USA. There are a herd-book in Canada and animals in Argentina, Colombia and Uruguay. It is used for wool and meat. It is white, the rams are horned, ewes polled. In the USA, there is also a totally polled variety. It is said to be thermotolerant (Briggs, 1969). The height of rams/ewes is 72/62 cm, the weight 80/50 kg. It has a higher age at sexual maturity and better daily gain than F but is equal in milk yield and has a lower wool yield and finer fibres (Simon and Buchenauer, 1993).

Romanov

Romanov originated from a local stock of the northern short-tailed sheep, which have been raised from time immemorial in northern, north-western and central regions of Russia and throughout the extreme north of Europe. It developed in the Yaroslavl region in Russia at the end of the seventeenth century. Its name (first mentioned in literature in 1802) stems from a small town of Romanov in the Volga valley, north-east of Moscow. It is widely distributed in north-western and central Russia (Dmitriev and Ernst, 1989). Since 1963 it has spread to France, nine other European and three North American countries, Cuba, Venezuela, Israel, Algeria and Egypt (Fahmy, 1996a). It was used in Russia mainly for producing furs and coarse wool, but the foreign interest is based on its prolificacy, needed in lamb-meat production. It is usually grey, with a black head and legs and a white face stripe. It can be horned or polled. Its physical, reproductive and productive characteristics have been reviewed by Fahmy (1996b). The average weight of rams/ewes is 90/55 kg. The legs are long and fine, covered with short hair. Lambs are born with a pelt-like cover, but the coat gradually turns grey, when two types of wool fibres are developed: an inner coat of white fibres and an outer coat of black hairy ones (ratio *c.* 1 : 5). Both sexes can mate at 4–6 months. An average breeding season contains eight to nine oestrus periods, and the conception rate of mature ewes in a normal season was 93% and the litter size 2.8. Two annual lambings have been recorded on 30–40% of ewes. The average daily milk yield is 1.1–1.6 kg in a 100-day period, with 6.4% fat. The average fleece weight of mature ewes is 2.1 kg/year, of which 60–70% is clean. The mixture of the fine, white fibres of the inner coat (19–21 μm) and the coarse, black guard hair (61–72 μm) together give the grey appearance. The great length of the down fibres above the guard hair, their resistance to interweaving, their good thermal insulation and the fineness of the wool make the skins of 0.5-year-old lambs valuable as pelts (Young and Dickerson, 1987). The animals tolerate cold weather and temperature fluctuations. In France, Romanovs were estimated to be better than Limousin in litter size and milk yield, equal in carcass leanness and wool yield and inferior in muscularity and daily gain (Simon and Buchenauer, 1993).

Romney

Romney is a long-wool breed that originated in England from Old Romney Marsh sheep. A breed society was established in 1895. Societies/herd-books are also found in the USA, Australia, New Zealand (50% of sheep) and Canada (Mason, 1996). Romney has also spread to seven South American countries and to Asia (Taiwan, Papua, China) (Ryder, 1983). It is used for meat and wool. The fibre diameter is 30–33 μm, fleece weight 4 kg and staple length 15 cm. It is white with a black nose and feet, and polled. Lambing rate is *c.* 125% (Alderson, 1976). The height of rams/ewes in the UK is 82/80 cm, weight 90/70 kg. According to the information in the European Data Bank, it equals Suffolk in muscularity, daily gain, carcass leanness, litter size and milk yield and has a lower age at sexual maturity (Simon and Buchenauer, 1993).

Scottish Blackface

Scottish Blackface originated from mountain breeds of the English Pennine hills, from where it spread throughout Scotland in the seventeenth and eighteenth centuries. It was probably influenced by the tan-faced mountain breed and Cheviot while moving simultaneously northwards (Alderson, 1976). A breed society was founded in the UK in *c.* 1890 and in the USA in 1907. The population is larger in Ireland than in Scotland. It also occurs in Colombia and was imported to New Zealand, but did not establish. It is used for production of meat and carpet wool. Its body is white, but the face and legs are mottled black and white and woolless. Both sexes are horned, and the face is strong, with a Roman nose. The height of Irish rams/ewes is 55/50 cm and weight 80/60 kg (Simon and Buchenauer, 1993). The average fleece weight is 2.5 kg, staple length 30–33 cm and fibre diameter > 40 μm. Alderson (1976) reported a litter size of 1.5.

Somali

Somali is an autochthonous breed developed in the dry tropics of Somalia. It also occurs in Ethiopia and Kenya and is the origin of Blackhead Persian, kept in South Africa, and partly of Brazilian Somali. It is raised for meat. It is a white hairy breed with a black head, and polled (Mason, 1996). The adult body weight of ewes is 52 kg and lamb marking percentage 93 (Turner, 1977).

Southdown

Southdown originated in 1780–1829 in England by selection from local heath-breed Sussex. A breed society was established in Britain in 1892 and in the USA in 1882. There are also herd-books in Australia, New Zealand, Canada and France and animals in Uruguay and Taiwan. It has played an important role in the formation of, among others, Dorset Down, Hampshire, Shropshire and Suffolk. It was the sire breed in the South Suffolk and the South Dorset Down, established in New Zealand in the last 30 years. It is used for meat (as sire breed) and short-wool production. It is white with a grey-brown face and legs, and polled (Mason, 1996). Height of rams/ewes is 67/60 cm, weight 82/60 kg. The body is barrel-shaped and the legs are short. It has finer fibres and a lower age at sexual maturity than Suffolk. The fleece weight is *c.* 2.8 kg, fibre diameter 26 μm and staple length 6 cm. It almost equals Suffolk and Texel in muscularity, Suffolk in daily gain, litter size, milk yield and % fat and Texel in age at sexual maturity. It is inferior to both in carcass leanness, has a finer wool fibre and earlier sexual maturity than Suffolk and lower wool and milk yields and daily gain than Texel (Simon and Buchenauer, 1993).

Soviet Merino

Soviet Merino originated in North Caucasus in Russia, mainly in 1925–1946, from crosses of two local coarse-wool types with fine-wool rams of American Rambouillet and several Merino- or Rambouillet-based Russian breeds. It was named in 1938. It has several varieties, differing greatly from each other in productivity and constitution. It is Russia's most numerous and widespread fine-wool breed, raised in North Caucasus, the Volga area, the Urals and the central regions of Russia and Kazakhstan. It has been used for the formation of four new breeds. It is used for wool and meat production. It is white and polled. Well-developed neck wrinkles and body folds are common. The weight of rams/ ewes is 110/50 kg. The fleece weight of rams/ewes is 11.5/6.2 kg (64's quality (fibre thickness 22 μm)) and staple length 8.7/8.0 cm. Litter size is 1.3–1.4 (Dmitriev and Ernst, 1989).

Suffolk

Suffolk originated in England in the early nineteenth century from a cross of Southdown and Norfolk Horn. It was recognized as a breed in 1810 and named in 1859. A society was founded in 1886 and in the USA in 1892. There are societies/herd-books in Australia, New Zealand, South Africa, Canada and France. The breed occurs in many other countries (among others, Ireland, Germany, Chile, Brazil, Norway and Israel). It is mainly used for meat production (sire breed) and also for short wool. It is white with a black head and legs, and polled (Mason, 1996). The average fleece weight is 3 kg, staple length 6.5 cm and fibre diameter 28 μm. The average lambing % is 150 (Alderson, 1976). Average height of adult rams/ewes in Europe is 76/73 cm, weight 120/84 kg. Typical of Suffolk is the high percentage meat and minimum fat cover in carcasses. It has a clearly higher daily gain and age at sexual maturity than Texel, a little higher milk yield and litter size and about equal carcass leanness, muscularity, length of breeding season and lambing interval, but a lower wool yield (Simon and Buchenauer, 1993). A summary of breed comparisons made in different countries showed it superior to 100% of compared breeds in daily gain, to 87% in weaning weight of lambs, to 60% in weight weaned/ewes exposed to ram and in percentage bone in carcass, to 50% in percentage lean and to 33% in percentage fat in the carcass (Terrill and Maijala, 1991).

Texel

Texel originated at the turn of the twentieth century on the isle of Texel in The Netherlands from crosses of Leicester Longwool, Lincoln and some other breeds with the local breed. The first herd-book appeared in 1909. It comprises nearly 100% of the Dutch sheep population. There are societies/herd-books in Ireland, the UK, France, Germany, Luxemburg and Finland, and it has spread recently to Norway, New Zealand, the USA, Brazil and South Africa. It is used for meat (sire breed) and for long wool. It is usually white, sometimes blond, occasionally with black spots on nose and legs. Both sexes are polled. The average height of rams/ewes is 83/74 cm and weight 114/84 kg. The recent international interest is based on its good conformation, muscularity and carcass leanness, in which it surpasses Suffolk. It equals the Suffolk in daily gain, milk yield, length of breeding season and lambing interval and in some countries also in age at sexual maturity and litter size (Simon and Buchenauer, 1993). The average fleece weight is 4.5 kg, clean yield 60%, staple length 15–20 cm, fibre diameter 30 μm, litter size 1.8 and milk yield 300 kg per lactation (Alderson, 1976).

Tibetan

Tibetan is an autochthonous breed developed in the Qinghai–Tibet mountain plateau (now in China), at more than 3000 m of altitude. It is almost the only breed of Tibet and is also kept in northern Nepal and three regions of India. It is a carpet-wool sheep with a small tail. It is adapted to unfavourable environments and can deposit fat in the viscera around the stomach and kidneys and in the mesentery. The fleece weight of rams/ewes is 1.4/0.7 kg, staple length 9.5 cm and fibre diameter 20.6 μm. There are 80% of true fibres, 5% of heterotype fibres, 15% hair and no kemp. The dressing % in males/females is 48.3/54.0. The average height of rams/ewes is 66/42 cm and weight 45/34 kg, and lambing rate 103% (Cheng, 1984).

Tsigai

Tsigai is an autochthonous breed, which may have originated from a cross between Zackel and a fine-woolled sheep, e.g. the Kivircik of Turkey. It occurs in many Balkan countries (Bulgaria, Greece, Turkey, former Yugoslavia) and in some neighbouring countries (Hungary, Slovakia, Romania). In 1914 it was imported into Russia, where it has become popular in several southern regions. It even occurs in China (Ryder, 1983). It is used for production of medium to coarse wool, milk and meat, the emphasis varying among regions. Some of these breed wool–mutton and some mutton–wool types. All types have much in common in constitution and productivity. There are dirty-white and black varieties. The white variety may have a white, red or black face. The rams are horned, ewes polled. The legs are strong and woolled up to the knees and hocks, the face down to the eyes. The live weight of the wool–mutton-type rams/ewes in Russia is 90/47 kg, of the mutton–wool type 105/57 kg. The fleece weights are, respectively, 7.0/3.8 kg and 8.0/4.3 kg, the clean wool yield 57%. The pelts are used for manufacturing fur articles. The ewes are good milk-producers (Dmitriev and Ernst, 1989). In Balkan and neighbouring countries, three of five varieties have milk production as the main use, one as the second use. Wool is first in two cases. The average height of rams/ewes is 67/61 cm and weight 79/48 kg. Most varieties are white with a black or brown face and legs. The rams are horned in three of the five types and the ewes in one. Four of five types excelled their reference breed in milk yield, three had a higher age at sexual maturity and two had a higher daily gain or fat content of milk. Three are inferior in wool yield or muscularity, two in carcass leanness or litter size (Simon and Buchenauer, 1993).

Vendéen

Vendéen originated in the Bas-Poitou region of France in the mid-1800s from a cross between Southdown and local sheep. A breed society was established in 1967 (Farid and Fahmy, 1996). Since 1981 it has spread to the UK, Belgium, Ireland, Italy and Spain (Veress, 1996). It is used for meat and wool. It is white with a grey head and legs, and polled. The legs are longer than in Southdown. The average weight of rams/ewes is 100/65 kg, fleece weight 3.8/2.8 kg and staple length 7 cm, with medium-fine wool. It adapts well to both extensive and intensive pasture conditions. The animals can be mated at 10–13 months and have a long breeding season. The average litter size of adult ewes is 1.8 (Farid and Fahmy, 1996). It is claimed to surpass Texel in litter size, earliness of sexual maturity and shortness of lambing interval, to equal it in muscularity and to be inferior in wool yield, daily gain, carcass leanness and milk yield (Simon and Buchenauer, 1993).

Zackel

Zackel is a group of autochthonous breeds developed in south-eastern Europe. There are many breeds with local names. They comprise the majority of sheep in Albania, Bulgaria (ten Zackel breeds), former Czechoslovakia, Greece, Romania and former Yugoslavia and are represented in Hungary, Italy, Macedonia, Poland and Russia. Some types occur in several Balkan countries (Ryder, 1983). The Pramenka in Yugoslavia is composed of *c.* 20 'breeds', commonly called by that name (Brook and Ryder, 1978). At the European Data Bank (Simon and Buchenauer, 1993), information was obtained on ten Zackel breeds from seven countries. The numbers used firstly for milk, meat and wool were 5/4/1, secondly 3/7/0 and thirdly 2/1/6, respectively. The ranges and averages for mean height of rams were 53–78/57 cm, of ewes 50–69/59 cm, and for mean weights 35–80/60 kg and 30–60/42 kg, respectively, so there are large size differences. The same concerns colours: four are white, two white with dark spots and four self-coloured black, brown or white animals or combinations. In seven breeds rams are horned and ewes polled, in two both are horned and in one both are polled. The rams of two breeds have spiral horns and in Hungarian Racka both sexes have corkscrew horns. Three breeds are marked for high quality cheese. Compared with reference breeds they were more often superior than inferior in carcass leanness and percentage fat and protein in milk, and vice versa in daily gain, litter size and milk yield. In muscularity, wool yield and fibre thickness they were equal. The age at sexual maturity and lambing interval are longer and breeding season shorter. Brooke and Ryder (1978) estimated the average staple length of Pramenka to be 17.3 cm and mean fibre diameter 42.1 μm. There were 13.2% medullated fibres. The fleeces were of the hairy type.

Conclusions

The ideas about the domestication of sheep are based on archaeological bone remains, which indicate that sheep were domesticated in south-western Asia about 9000 BC. The region and time of this depended largely on the development of various preconditions, e.g. social behaviour of and close contacts between sheep and humans, worldwide changes in environment, adaptation of wild sheep to high latitudes and cold climates, growth and cultivation of cellulose-rich green grasses and the ability of ruminants to digest these and to recycle urea and thus to live on protein-rich diets. The prevalence of these in the region explains its role as the centre of domestication of ruminants. Seven wild types of sheep are known, some of which still survive. The Asian mouflon is considered the ancestor of domestic sheep. Changes in the attitudes of humans, the need for sharing reduced food resources and for finding new ones, the extinction of megafauna and the usefulness of sheep meat, bones and skins may have been important motives. Utilization of blood, milk and wool came after domestication and gave additional motives to keep the animals alive.

The mechanisms of genetic evolution did not change essentially after domestication, but their speed and size increased. The environmental change allowed the genetic differences to appear and different kinds of individuals to survive. The role of chance decreased, and selection for traits unfavourable in nature became possible. Preferential killing and castration of young males were important factors. The restriction of animals to limited areas increased inbreeding and fixation of genes, leading to production of diverse 'breeds'. The artificial mating barriers were more effective than geographical and physiological isolation in nature. Transportation of animals by humans facilitated outbreeding with different stocks.

Many traits obviously changed before any illustrations were conserved, i.e. by 3000 BC: behaviour, legs, bones, tail and horns. The invention of shearing led to replacement of hairs and that of dyes to the development of white-woolled sheep. Changes in production traits were gradual, but occurred at an accelerated rate.

The concept of breed, introduced to animal breeding in the sixteenth century, has been interpreted in different ways. In the developed countries, establishment of a herd-book has been presupposed, but in many developing countries there are none. Hence, FAO has applied wider definitions, based on homogeneity of external characteristics or on observable identity of animals of a geographically separated group. In 1993, there were 863 sheep breeds in the FAO Data Bank, but it did not contain all breeds. There are large differences among breeds in many respects, so they could be classified in many ways. No system appears ideal, since the same breed can have different uses, yields and qualities of different products and be adapted to different kinds of environment in different countries. Origins, uses and characteristics of 53 breeds, of which about a third are from countries without herd-books, were described and partly compared with other breeds.

References

Acharya, R.M. (1982) *Sheep and Goat Breeds of India.* FAO Animal Production Health Paper 30, FAO, Rome. 190 pp.

Adalsteinsson, S. (1983) Inheritance of colours, fur characteristics and skin quality in North European sheep breeds. *Livestock Production Science* 10, 555–567.

Alderson, L. (1976) *The Observer's Book of Farm Animals.* Frederick Warne, London, 191 pp.

Belyaev, D.K. (1979) Destabilizing selection as a factor in domestication. *Journal of Heredity* 70, 301–308.

Bökönyi, S. (1974) *History of Domestic Mammals in Central and Eastern Europe.* Akademy Kiadó, Budapest, 596 pp.

Briggs, H.M. (1969) *Modern Breeds of Livestock*, 3rd edn. Macmillan, London, 714 pp.

Brooke, C.H. and Ryder, M.L. (1978) *Declining Breeds of Mediterranean Sheep.* FAO Animal Production Health Paper 8, FAO, Rome. 78 pp.

Buchanan, F.C., Adams, L.J., Littlejohn, R.P., Maddox, J.F. and Crawford, A.M. (1994) Determination of evolutionary relationships among sheep breeds using microsatellites. *Genomics* 22, 397–403.

Carter, A.H. and Cox, E.H. (1982) Sheep breeds in New Zealand. In: Wickham, G.A. and McDonald, M.F. (eds) *Sheep Production*, Vol. 1. *Breeding and Reproduction.* Ray Richards, Auckland, pp. 11–38.

Cheng, P. (1984) *Livestock Breeds of China.* FAO Animal Production Health Paper 46, FAO, Rome. 217 pp.

Clutton-Brock, J. (1981) *Domesticated Animals from Early Times.* Heinemann/British Museum, London, 208 pp.

Dmitriev, N.G. and Ernst, L.K. (1989) *Animal Genetic Resources of the USSR.* FAO Animal Production Health Paper 65, FAO, Rome. 517 pp.

Dobzhansky, T. (1951) *Genetics and the Origin of Species*, 3rd edn. Columbia University Press, New York, 364 pp.

Dolling, C.H.S. and Jefferies, B.C. (1991) Genetic resources for wool production. In: Maijala, K. (ed.) *World Animal Science*, B 8. *Genetic Resources of Pig, Sheep and Goat.* Elsevier, Amsterdam, pp. 277–290.

Epstein, H. (1971) *The Origin of Domestic Animals of Africa*, Vol. 1. African Publication Corporation, Munich, xi + 573 pp.

Epstein, H. (1982) Awassi sheep. *World Animal Review* 44, 9–18.

Fahmy, M.H. (1996a) Preface. In: Fahmy, M.H. (ed.) *Prolific Sheep.* CAB International, Wallingford, pp. ix–xi.

Fahmy, M.H. (1996b) The Romanov. In: Fahmy, M.H. (ed.) *Prolific Sheep.* CAB International, Wallingford, pp. 47–72.

Farid, A.-H. and Fahmy, M.H. (1996) The East Friesian and other European breeds. In: Fahmy, M.H. (ed.) *Prolific Sheep.* CAB International, Wallingford, pp. 93–108.

Flamant, J.C. (1991) International exchange of genetic material. In: Maijala, K. (ed.) *World Animal Science*, B 8. *Genetic Resources in Pig, Sheep and Goat.* Elsevier, Amsterdam, pp. 357–363.

Galal, E.S.E. (1996) The tropics and subtropics. In: Fahmy, M.H. (ed.) *Prolific Sheep.* CAB International, Wallingford, pp. 236–241.

Hasnain, H.U. (1985) *Sheep and Goats in Pakistan.* FAO Animal Production Health Paper 56, FAO, Rome, 135 pp.

Hatziminaoglou, I., Georgoudis, A., Zervas, N. and Boyazoglu, J. (1996) Prolific breeds in Greece. In: Fahmy, M.H. (ed.) *Prolific Sheep*. CAB International, Wallingford, pp. 73–92.

Khaldi, G. (1989) *The Barbary Sheep*. FAO Animal Production Health Paper 74, Rome. pp. 96–135.

Loftus, R. and Scherf, B. (eds.) (1993) *World Watch List for Domestic Animal Diversity*. FAO, Rome, 376 pp.

Lush, J.L. (1945) *Animal Breeding Plans*, 3rd edn. Iowa State University Press, Ames, Iowa, 443 pp.

Lysandrides, P. (1981) The Chios sheep in Cyprus. *World Animal Review* 39, 12–16.

Maijala, K. (1984) Review of experiences about the use of Finnsheep in improving fertility. In: Hofmeyr, J.H. and Meyer, E.H.H. (eds) *Proceedings 2nd World Congress Sheep, Beef, Cattle Breed.* South African Stud Book and Livestock Improvement Association, Pretoria, South Africa, pp. 519–531.

Maijala, K. (1988) History, recent development and uses of Finnsheep. *Journal Agricultural Sciences Finland* 60, 449–454.

Maijala, K. (1996) The Finnsheep. In: Fahmy, M.H. (ed.) *Prolific Sheep*. CAB International, Wallingford, pp. 10–48.

Maijala, K. and Terrill, C.E. (1991) Breed structure, dynamics and new breed development in sheep. In: Maijala, K. (ed.) *World Animal Science*, B 8. *Genetic Resources of Pig, Sheep and Goat*. Elsevier, Amsterdam, pp. 305–326.

Mason, I.L. (1967) *Sheep Breeds in Mediterranean*. Commonwealth Agricultural Bureaux, Farnham Royal, UK, 215 pp.

Mason, I.L. (1991) Classification and distribution of sheep breeds. In: Maijala, K. (ed.) *World Animal Science*, B 8. *Genetic Resources of Pig, Sheep and Goat*. Elsevier, Amsterdam, pp. 179–193.

Mason, I.L. (1996) *World Dictionary of Livestock Breeds, Types and Varieties*, 4th edn. CAB International, Wallingford, 348 pp.

Nel, J.A. (1991) Fur breeds of sheep. A. Karakul. In: Maijala, K. (ed.) *World Animal Science*, B 8. *Genetic Resources of Pig, Sheep and Goat*. Elsevier, Amsterdam, pp. 291–303.

Nimbkar, C. (1993) The goats and sheep of the Deccan Plateau in the Maharastra State of India. *Animal Genetics Resource Information* 12, 81–91.

Piper, L.R. and Bindon, B.M. (1982) Genetic segregation for fecundity in Booroola Merino sheep. *Proceedings of the World Congress on Sheep Beef Cattle Breeding*, New Zealand, Vol. 1. Technical, pp. 395–400.

Piper, L.R. and Bindon, B.M. (1996) The Booroola Merino. In: Fahmy, M.H. (ed.) *Prolific Sheep*. CAB International, Wallingford, pp. 152–160.

Reed, C.A. (1969) The pattern of animal domestication in the prehistoric Near East. In: Ucko, P.J. and Dimbleby, G.W. (eds) *Domestication and Exploitation of Plants and Animals*. Duckworth, London, pp. 361–380.

Reed, C.A. (1984) The beginnings of animal domestication. In: Mason, I.L. (ed.) *Evolution of Domesticated Animals*. Longman, London, pp. 1–6.

Ruvinsky, A.O. (1988) Inheritance of dominant genes with variable penetrance: an evolutionary aspect. *Journal of Animal Breeding Genetics* 105, 103–111.

Ryder, M.L. (1964) The history of sheep breeds in Britain. *Agricultural History Review* 12, 1–12.

Ryder, M.L. (1983) *Sheep and Man*. Duckworth, London, 846 pp.

Ryder, M.L. (1984) Sheep. In: Mason, I.L. (ed.) *Evolution of Domestic Animals*. Longman, London, pp. 63–85.

Ryder, M.L. (1991) Domestication, history and breed evaluation in sheep. In: Maijala, K.

(ed.) *World Animal Science*, B 8. *Genetic Resources of Pig, Sheep and Goat*. Elsevier, Amsterdam, pp. 157–177.

Shrestha, J.N.B. and Vesely, J.A. (1986) Evaluation of established breeds of sheep in Canada for daily gain and body weights. *Canadian Journal of Animal Science* 66, 897–904.

Simon, D.L. and Buchenauer, D. (1993) *Genetic Diversity of European Livestock Breeds*. EAAP Publication 66, Wageningen Press, Wageningen, Netherlands. 581 pp.

Steinhagen, O., Dreyer, J.H. and Hofmeyr, J. (1986) Histological differences in the skin and fibre characteristics of ten white-woolled sheep breeds. *South African Journal of Animal Science* 18, 90–94.

Terrill, C.E. and Maijala, K. (1991) Breed comparisons for meat production in sheep. In: Maijala, K. (ed.) *World Animal Science*, B 8. *Genetic Resources of Pig, Sheep and Goat*. Elsevier, Amsterdam, pp. 257–276.

Turner, H.N. (1977) *Some Aspects of Sheep Breeding in the Tropics*. FAO Animal Production Health Paper 1, FAO, Rome, pp. 115–121.

Turton, J.D. (1974) The collection, storage and dissemination of information on breeds of livestock. In: *Proceeding 1st World Congress Genetics and Applied Livestock Production*, Vol. II. Madrid, pp. 61–74.

Veress, L. (1996) Europe. In: Fahmy, M.H. (ed.) *Prolific Sheep*. CAB International, Wallingford, pp. 218–227.

Wilson, R.T. (1984) Livestock production in central Mali: sheep husbandry in the traditional sector. *World Animal Review* 53, 8–14.

Wilson, R.T. (1991) *Small Ruminant Production and the Small Ruminant Genetic Resource in Tropical Africa*. FAO Animal Production Health Paper 88, FAO, Rome, 231 pp.

Yalcin, B.C. (1986) *Sheep and Goats in Turkey*. Animal Production Health Paper 60, FAO, Rome, 168 pp.

Young, L.D. and Dickerson, G.E. (1987) *Importation of Romanov Sheep*. ARS Sheep Research Program Progress Report 3, USDA, Fort Collins, Colorado, USA, p. 8.

Zervas, N.P., Hatjiminaoglou, J., Georgoudis, A, and Boyazoglu, J.G. (1988) Characterization and experiences of Chios breed. *Journal of Agricultural Science Finland* 60, 576–584.

Zeuner, F.E. (1963) *A History of Domestic Animals*. Harper and Row, New York, 560 pp.

Genetics of Colour and Hair Texture 3

D.P. SPONENBERG
Virginia–Maryland Regional College of Veterinary Medicine, Virginia Polytechnic Institute and State University, Blacksburg, VA 24061, USA

Introduction 52
General Considerations of Sheep Pigmentation 52
Factors Affecting Colour of Sheep 56
  The *Agouti* locus 56
  The *Extension* locus 64
  The *Albino* locus 65
  Expression and modification of phaeomelanin 66
  Expression and modification of eumelanin 67
  Colour of black- or dark-faced breeds 69
  Sur 70
  The *Australian Piebald* locus 70
Loci Controlling White Spotting 71
  The *Spotting* locus 71
  The *Pigmented Head* locus 73
  The *Roan* locus 75
  Other loci affecting white spotting 76
  The *Ticking* locus 76
Strategies for Selection of White Wool 76
Strategies for Selection of Coloured Wools 78
Single-gene Effects on Hair and Wool Structure 80
  Fibre Loss 80
  Lustrous Wool 81
  Silky Wool 81
  Carpet wool 81
References 82

Introduction

Sheep-colour variations were among the very first characters of any organism to be studied during the infancy of the science of genetics (Davenport, 1905; Vacher, 1905; Wood, 1905; Adametz, 1917). The early investigations of sheep colour genetics were undertaken from the viewpoint that departures from white sheep were the phenomenon of interest. The main purpose of the early investigations was to assist breeders in getting rid of coloured sheep and coloured fibres.

The emphasis on white sheep drove most investigations until the 1960s, when a comparative approach began to be used, at which time investigations began to concentrate more on the general biology of pigmentation of sheep (Rendel, 1957; Adalsteinsson, 1970; Lauvergne, 1975, 1976a, b, 1979). The comparative approach yielded many results that shed light on the basic biology and genetics of sheep colour, especially as these relate to the specific selection either for white wool or for coloured wool (Sponenberg *et al.*, 1988; Adalsteinsson, 1991).

This chapter focuses first on the general biology of sheep colour. It then examines the consequences of the biology of pigmentation for selection in sheep either for or against pigment. A clear understanding of the basic genetic phenomena is essential for those interested in the production of white wool, as well as for those interested in the production of coloured wool. A complete historical review of the world literature dealing with the loci and alleles affecting colour in sheep has recently been completed by the Committee on Genetic Nomenclature of Sheep and Goats (COGNOSAG, 1989, 1993). The COGNOSAG recommendations on nomenclature are used in this chapter, which will focus on the action and interaction of the various loci that control pigmentation in sheep and the biological processes that are involved. Single-locus effects on wool texture will also be discussed.

General Considerations of Sheep Pigmentation

Colour in mammals depends on the presence of melanins in skin and hair (Searle, 1968). Melanins reside in cellular organelles called melanosomes, which are produced in melanocytes. Melanosomes reside in the melanocyte cytoplasm, and are also deposited into epidermis and hair by a process of exocytosis, which transfers the melanosomes from the dendritic processes of the melanocytes into these other (generally epidermal) cells. Melanocytes are specialized cells that migrate from the neural crest during embryological development, and melanocytes have close embryological ties to cells of the neurological system. Pigmentation of skin, hair and wool depends on melanocytes being present and on their relative level of melanogenic and exocytotic activity. The two main mechanisms by which pigmentation is reduced or absent are relative inactivity or absence of melanocytes. Decrease or elimination of pigment by either of these basic mechanisms can be regional or can involve entire animals.

Melanins are large polymers formed of varying amounts of tyrosine and cysteine (Prota, 1992; Jackson, 1994). Melanins occur as two types: eumelanin and phaeomelanin. Eumelanin is generally black or chocolate-brown and is predominantly made of tyrosine. Phaeomelanin is generally reddish-brown or yellowish-tan and has varying amounts of cysteine in its polymers. The two types of melanin are important visually and genetically and are usually considered as two discrete classes of pigment. Some overlap occurs between the two, but this overlap is usually trivial to an understanding of the genetic control of colour in sheep. Overlap can also occur visually, with very dark phaeomelanin being similar to lighter forms of eumelanin, but usually the two classes are distinct (Sponenberg *et al.*, 1988; Sumner *et al.*, 1994). The differences between eumelanin and phaeomelanin have been demonstrated by electron magnetic resonance (Vsesolodov *et al.*, 1981) as well as by analysis of chemical degradation products (Sponenberg *et al.*, 1988). Other techniques to demonstrate the difference between the two melanins include histological evaluation of melanosomes (Alexieva *et al.*, 1985; Renieri *et al.*, 1995) and the use of colour charts (Sumner *et al.*, 1994).

Melanins are formed from tyrosine and cysteine by a process of catalysis, and the main enzyme involved in this process is tyrosinase (Jackson, 1994). Tyrosinase activity is essential for melanin synthesis. Two other proteins, tyrosinase-related protein 1 (TRP-1) and TRP-2, are also important for melanogenesis. The exact function of TRP-1 is uncertain, although it is well documented to have an important role in the synthesis of eumelanin and has little or no role in the synthesis of phaeomelanin. Tyrosinase-related protein 2 has dopachrome tautomerase activity.

Melanocytes are capable of forming both eumelanin and phaeomelanin, although they usually only produce one or the other at any one time. The dedication of melanocytes for eumelanin production depends on the presence of α-melanocyte-stimulating hormone (α-MSH), which is a pituitary hormone (Jackson, 1994). Melanocytes have surface receptors that bind this hormone. When α-MSH binds to the surface receptors, the result is a cascade of events that activates adenylate cyclase. This activation in turn stimulates the melanocyte to produce eumelanin. In the absence of this signal, which is dependent on α-MSH as well as the surface receptors, melanocytes produce phaeomelanin. The switch between eumelanogenesis and phaeomelanogensis depends on the function of the receptors to α-MSH.

The control of melanocyte function is intricate, and many loci have mutants that affect different components of the melanogenic control mechanisms (Jackson, 1994). Some loci affect cell differentiation or migration from the neural crest. Other loci affect the morphology of the melanocyte or its ability to deposit melanosomes in hair and epidermis. Other loci directly affect the enzymes and related proteins that are responsible for melanogenesis. A few loci have mutants that affect the interaction of α-MSH with the target melanocytes. All of these loci interact to give the final colour phenotype of sheep. Table 3.1 lists loci and alleles that affect sheep colour.

White is obviously a very important colour type in sheep, to the extent that

Table 3.1. Loci and alleles affecting colour in sheep.

Locus	Symbol	Allele	Symbol
Agouti	*A*	*white or tan*	A^{Wt}
		wild	A^{+}
		grey and tan	A^{gt}
		light grey	A^{kg}
		light badgerface	A^{lb}
		badgerface	A^{b}
		light blue	A^{lbl}
		blue	A^{bl}
		grey	A^{g}
		gotland grey	A^{gg}
		black and tan	A^{t}
		swiss marked	A^{s}
		lateral stripes	A^{ls}
		pale cheek/eye ring	A^{pc}
		eye patch	A^{ep}
		sooty	–
		non-agouti	A^{a}
Albino	*C*	*wild*	C^{+}
		albino	C^{a}
		albino marrabel	C^{mar}
Australian Piebald	*AsP*	*wild*	AsP^{+}
		piebald	AsP^{p}
Brown	*B*	*wild*	B^{+}
		brown	B^{b}
Extension	*E*	*dominant black*	E^{D}
		wild	E^{+}
Pigmented Head	*Ph*	*afghan lethal*	Ph^{afl}
		turkish	Ph^{T}
		persian	Ph^{P}
Roan	*Rn*	*lethal roan*	Rn^{Rn}
		wild	Rn^{+}
Spotting	*S*	*wild*	S^{+}
		spotted	S^{s}
		bizet spotting	S^{b}
Sur Bukhara	*SuB*	*wild*	SuB^{+}
		sur bukhara	SuB^{s}
Sur Surkhandarya	*Sus*	*wild*	SuS^{+}
		sur surkhandarya	SuS^{s}
Ticking	*Ti*	*ticked*	Ti^{Ti}
		wild	Ti^{+}

coloured sheep are very much a minority of sheep in most countries and in most breeds of sheep. The expression of whiteness in mammalian hair can come about through a variety of different genetic mechanisms (Searle, 1968). Some of these mechanisms appear to be widespread throughout domesticated sheep populations, but this is no guarantee that other mechanisms are not important to some specific populations.

One general mechanism for whiteness in mammals is due to white spotting. White spotting in mammals occurs in regions where skin or hair follicles lack melanocytes, since these regions are incapable of forming melanins. Absence of melanocytes can occur from a variety of mechanisms, including melanoblasts failing to differentiate from the neural crest, failing to migrate to skin regions or failing to survive once migration has taken place (Searle, 1968). The variety of mechanisms responsible for white spotting are affected by several different loci, each of which controls a different aspect of the process of populating skin and hair follicles with melanocytes.

A second general mechanism for whiteness is achieved through dilution of pigment, which is accomplished by decreased effectiveness of melanin production by melanocytes. Melanocytes are present, but are ineffective at forming melanins, placing them in the wool and hair fibres, or both. Their ineffectiveness at melanosome production or transfer results in unpigmented or weakly pigmented wool, hair and skin, even though melanocytes are present. The effects of both white spotting and dilution are important in sheep, and separating these two concepts is important to understanding pigmentation in sheep.

The expression of colour in sheep is affected not only by genes that directly control distribution and activity of melanocytes, but also by the textural qualities of the hair coat (Aliev and Rachkovsky, 1987a, b). Coarse hairy fibres and very fine wool fibres express colour genes differently, and this phenomenon is extremely important in sheep since most sheep are valued for specific fibre types. A general trend is that fibres from primary follicles (coarse hairy fibres in many breeds) have the potential to be darkly pigmented. Fibres from secondary follicles (fine fibres in double-coated fleece types) have less potential for full pigmentation. Secondary fibres seem to become pigmented with eumelanin much more easily than they are with phaeomelanin. The result of this phenomenon is that phaeomelanic colours are rare among sheep. This is especially true among the types which have fine or medium wool rather than hair or very coarse wool (Sponenberg *et al.*, 1988).

The wild-type colour in sheep must be considered as the standard type if the genetics of sheep colour is to make sense. The wild ancestor of the domesticated sheep is uncertain, although the Asiatic mouflon, urial and argali are the most likely progenitors (Ryder, 1984). The relative contributions of these three may vary with different breed groups. The colour of the various species of wild sheep has subtle differences one from the other, but similarities are striking (Novak, 1991). Wild sheep are generally tan-bodied (phaeomelanin) with pale bellies (dilute phaeomelanin, rather than non-pigmented white). Black areas (eumelanin) are present as a dorsal stripe and striping patterns on the heads and legs. Some sexual dimorphism is present, with rams darker than ewes due to more extensive eumelanic patterns on the head and shoulders. The eumelanic areas, especially on males, become more extensive with age, and mature dominant rams are therefore darker than younger ones. The interplay of eumelanic areas with phaeomelanic areas is critical to the overall appearance of the colour pattern, as is the presence of intense, dark phaeomelanic areas along with dilute, pale phaeomelanic areas.

The study of sheep colour has been complicated by the fact that wild-type colour in domesticated sheep is very rare. As a result, most genetic studies have taken wholly white animals as the standard or reference type, which has obscured understanding of the genetic control of colour in sheep. This can be avoided by considering the wild type as the standard with which to compare genetic variation.

Factors Affecting Colour of Sheep

The Agouti *locus*

One of the main determiners of sheep colour is the *Agouti* locus. This locus has many allelic variants, making it one of the more complex loci governing colour in sheep (Adalsteinsson, 1970; Lundie, 1984). Variation at the *Agouti* locus is widespread among sheep breeds, such that *Agouti* locus phenomena account for the majority of colour variants in most sheep breeds. The concept of multiple alleles is essential to understanding the *Agouti* locus. Roberts (1928) first noted that multiple patterns (his badgerface, reversed badgerface, black and white) segregated as alleles at a single locus. Rendel (1957) first suggested the *Agouti* locus as the site for these patterns. The four patterns suggested by Roberts were confirmed as multiple alleles at a single locus by Brooker and Dolling (1969a). Alleles documented by these researchers, as well as several newly described ones, were ascribed by Adalsteinsson (1970) to the *Agouti* locus, and it is this work which put the entire study of sheep colour on a sound comparative basis since this work was widely used as a base by other researchers. Demonstrating the character of the *Agouti* locus patterns in sheep was the breakthrough that finally provided for more rapid advances in the understanding of the genetic control of sheep colour, since it put the science on a sound foundation of comparative genetics. Following this, sheep phenomena could be more readily coupled to homologous phenomena of other species.

The manifestations of the *Agouti* locus in sheep are relatively complex, although studies in mice indicate that the locus itself is relatively simple in its biology (Jackson, 1994). Many different alleles occur within the *Agouti* locus of sheep (Brooker and Dolling, 1969a; Adalsteinsson, 1970; Lauvergne and Adalsteinsson, 1976; Lundie, 1984); however, all of the ovine *Agouti*-locus patterns are characterized by symmetrical arrangements of eumelanic and phaeomelanic areas. In sheep the eumelanic areas are usually black, although they can be brown when the genotype at other loci modifies black to brown.

Phaeomelanic areas on *Agouti*-locus patterns vary in colour, depending on fleece type and the specific *Agouti*-locus allele involved. In most wool sheep, any phaeomelanic areas are diluted to cream or white. The paleness of the phaeomelanic areas on wool sheep results in most *Agouti*-locus patterns appearing as black, grey and white areas in different symmetrical arrangements. In some coarse-wool types and in hair sheep, it is possible for phaeomelanic areas to be tan to reddish-brown. The *Agouti*-locus patterns therefore take on a different appearance in hair sheep from that in wool sheep, due to the more intense phaeomelanic colours. The distribution and colour of eumelanic areas are constant, though, whether the sheep has wool or hair.

In general, primary follicles produce fibres that are more capable of being pigmented with phaeomelanin than do secondary follicles (Aliev and Rachkovsky, 1987a, b). Some evidence exists that the birth coat is more subject to phaeomelanic pigmentation than the later coat (Adalsteinsson, 1970), so that early classification of lamb colour patterns is more accurate than delayed classification. While the usual interpretation of the differences in intensity of phaeomelanic pigmentation relates these to the differences between the birth coat and the subsequent coat, it is also possible that some *Agouti* alleles directly control the intensity of phaeomelanin to an extent.

Mechanism of action

The *Agouti* locus is responsible for the formation of a protein that acts to nullify the action of α-MSH on melanocytes (Jackson, 1994). In regions in which this protein is present the melanocytes fail to respond to α-MSH and therefore form phaeomelanin and not eumelanin. Melanocytes in regions lacking this protein have full capability for stimulation by α-MSH and therefore form eumelanin (Jackson, 1994). In regions with a pulsatile formation of the agouti protein, the result is banded hairs. These are important in mouse phenotypes but, while present in some sheep patterns, they are less important as a cause of colour variation than they are in mice.

In mice, the *Agouti* locus has been extensively studied, and implications of homology suggest that its biology is similar in other species, including the sheep. The series of murine *Agouti* alleles is a consistent array as it progresses from the most recessive to the most dominant allele (Jackson, 1994). The most recessive murine *Agouti* allele allows for no agouti protein production in the skin, and therefore results in a completely eumelanic phenotype. The more recessive alleles, above the one coding for a completely eumelanic phenotype, allow for agouti protein in ventral body regions, resulting in these being phaeomelanic. Successive alleles then add body regions to those of lower alleles, such that phaeomelanic areas of the coat are added successively as the series of alleles progresses from more recessive to more dominant. In a few of the intermediate alleles, some regions have pulsatile agouti-protein production, resulting in the banded hairs (on the dorsum) typical of the wild pattern in mice. These hairs have a eumelanic tip and base and a phaeomelanic middle.

The murine *Agouti*-locus patterns exhibit no reversal of pigment type among the alleles, so that regions which have become phaeomelanic from the action of more recessive alleles retain this pigment type in all patterns determined by the more dominant alleles. In this regard, the murine *Agouti* locus acts as though the various alleles were simply shifting melanin production in an orderly stepwise progression from one extreme to the other, adding phaeomelanic areas body region by body region. The more dominant alleles result in phenotypes that are entirely phaeomelanic, while the most recessive alleles have phenotypes that are entirely eumelanic. The stepwise progress of *Agouti*-locus patterns gave rise to early hypotheses that the locus was a complex of small loci that each controlled pigment type in only a specific body region. Recent studies have shown that the complex hypotheses are mistaken and that the locus is indeed single and simple rather than an array of miniloci (Jackson, 1994).

The most dominant murine *Agouti* allele is *lethal yellow* (Jackson, 1994). This allele is lethal to homozygotes, and the mechanism of lethality appears to be that the very high levels of expression of the agouti protein disrupt fetal differentiation and development. The heterozygotes for this allele also have somatic manifestations of it, such as obesity, reduced fertility and increased incidences of certain neoplasms.

The action of *Agouti*-locus alleles in sheep is not as neat and orderly as in the murine series. It is fairly common for the intermediate agouti patterns in sheep to have reversals of pigmentation type, so that areas that are eumelanic in one pattern are phaeomelanic in another, and vice versa. This reversal of pigmentation type results in some ovine patterns that appear to be opposites of one another, in contrast to the murine situation, in which the *Agouti*-locus patterns are characterized by a stepwise increase in phaeomelanic areas. The reversal of pigmentation patterns in ovine phenotypes is most remarkable in the 'badgerface' and 'black-and-tan' patterns, which are nearly perfect opposites of one another in pigment distribution (Figs 3.1 and 3.2).

The action of ovine *Agouti* alleles is to consistently express phaeomelanic areas on a eumelanic background (Adalsteinsson, 1991). The result of this (in wool sheep) is that the white or pale areas are dominant to the dark areas. This is a key concept for the understanding of the function of the *Agouti* locus. Some mild degree of intermediate dominance is present in some animals of some breeds with certain combinations of *Agouti* alleles, but as a general rule it holds true that the phaeomelanic areas are consistently expressed, whether heterozygous or homozygous. Heterozygotes for intermediate alleles therefore express the phaeomelanic areas for each of the alleles, and this results in phenotypes with combinations of both phaeomelanic patterns. Some heterozygotes for intermediate alleles and the entirely eumelanic *non-agouti* (A^a) allele are darker than homozygotes for the same intermediate alleles, although this is variable. This incomplete dominance has been noted for the *badgerface* allele, and it has also been found that overlapping phenotypes are more characteristic of females than of males (Brooker and Dolling, 1969a).

Fig. 3.1. Welsh Torddu sheep with typical badgerface phenotype, which is pale phaeomelanic above and dark eumelanic below (photograph © Evelyn Simak).

The allele white or tan

The most dominant allele at the ovine *Agouti* locus is *white or tan* (A^{Wt}). This is logical, since this controls the most phaeomelanic phenotype of all the alleles (Brooker and Dolling, 1969a; Adalsteinsson, 1970). The A^{Wt} allele is the cause of white phenotypes in most breeds of sheep in which colour inheritance has been investigated. While other mechanisms for white sheep may well turn up, the A^{Wt} allele is by far the most common, and selection has fixed this allele in most breeds of sheep in which white is an important colour. The allele A^{Wt} is completely dominant for wool and hair colour, although heterozygotes for A^{a} can have increased pigment in hooves and skin when compared with homozygotes. The colour scores of sheep homozygous for the A^{Wt} allele overlap with the scores of those heterozygous for the allele, to the extent that visual inspection is not accurate in separating the two genotype classes (Brooker and Dolling, 1965; Fleet *et al.*, 1989).

The A^{Wt} allele has pleiotropic effects, in keeping with its status as the top dominant allele at the *Agouti* locus. Ewes with the A^{Wt} allele are on average less fecund than those without it. In Icelandic sheep, in which this has been best investigated, sheep bearing the A^{Wt} allele are on average 15% less fecund than those lacking this allele (Adalsteinsson, 1975). In addition, sheep bearing the A^{Wt} allele are more strongly seasonal in reproduction than those lacking it

Fig. 3.2. North Country Cheviot with black-and-tan pattern. This pattern has a pigment distribution very nearly opposite to the badgerface pattern. (Photograph © Evelyn Simak).

(Dyrmundsson and Adalsteinsson, 1980). Obviously the breed genotype other than that at the *Agouti* locus is equally important in the overall fecundity and seasonality of any sheep, but the effects of this single allele are dramatic in breeds in which variants at the *Agouti* locus are segregating.

At least some of the whiteness of sheep with the A^{Wt} allele is due to delayed or reduced migration and proliferation of melanoblasts from the neural crest. This occurs early in fetal development (Adalsteinsson, 1970; Fleet *et al.*, 1993; Fleet, 1994). Interactions between A^{Wt} and the *spotted* allele at the *Spotting* locus are also important, and are further considered under that locus.

Sheep that are white from the A^{Wt} allele are subject to some incursion of black or grey spotting, either in skin or wool. This spotting especially occurs on older sheep, and is related to shearing and to the effects of sunlight on recently shorn skin (Kelly and Shaw, 1942; Fleet and Forrest, 1984).

The allele non-agouti

The allele A^a is completely recessive, since it results in a completely eumelanic phenotype (Adalsteinsson, 1970; Brooker and Dolling, 1965). In sheep of most breeds this results in a completely black sheep at birth. The final colour does depend on modifiers at independent loci. Some of these modifiers lighten the black to blue-grey, others to brown. Most black sheep do indeed fade somewhat to grey, so that truly black sheep are somewhat unusual in most breeds. The A^a

allele is very widespread, and is common among coloured sheep that segregate from otherwise white breeds. Ryder *et al.* (1974) reported that the homozygous A^a genotype is less viable than others (usually *black and tan*) in the Soay sheep. Soays are feral, and whether this trend holds true for sheep in less demanding environments is not known.

Agouti-*locus alleles causing patterns*
Between the extremes of completely white sheep and completely black sheep are many other patterns caused by alleles at the *Agouti* locus (reviewed in Lundie, 1984; COGNOSAG, 1986, 1993). Some of these alleles produce patterns that are relatively dark due to preponderance of eumelanic areas. Others produce patterns that are relatively pale due to preponderance of phaeomelanic areas. A few patterns are intermediate and consist of relatively even mixtures of the two pigment types. The main phenomenon that reflects the residence of these alleles at the *Agouti* locus is that the dark and pale areas are arranged in symmetrical patterns. *Agouti*-locus patterns are numerous, and the differences between some of them are subtle. A very large number of alleles are ultimately likely to be documented as residing at this locus in sheep, some with minor but consistent differences in phenotype from the ones already documented. The patterns can be placed into groups based on predominance of pigment types.

Agouti-locus patterns that are predominantly phaeomelanic The *Agouti*-locus patterns that consist primarily of phaeomelanic areas include badgerface, light badgerface, wild-type, blue and light blue. Most of these patterns are phaeomelanic on the majority of the body, with striking patterns of eumelanic areas on the belly and legs, along the spine and on the head.

The *wild*-type allele causes a very typical pattern in hair sheep, although this pattern is rarely recognized in wool sheep. Lydekker (1912) describes this colour pattern on wild sheep (specifically the mouflon):

> The body is reddish fawn to reddish grey with a yellowish to white saddle. A black line is present down the back, and a black stripe separates the body from the pale belly. The shoulder and neck are dark, black stripes are present on the front of the legs, and black marks occur on the head. The muzzle is light to white, as are the eye rings, the inner surfaces of the ears, and the lower legs and belly. Some fibres are banded different colours.

The *wild*-type allele in this case is responsible not only for the distribution of eumelanin and phaeomelanin but also for the variations in intensity of the phaeomelanic areas, so that some are tan (body) while others are nearly white (belly).

The *badgerface* allele causes phenotypes similar to the wild pattern, but with more extensive eumelanic areas (Roberts, 1924; Brooker and Dolling, 1969a; Adalsteinsson, 1970). The badgerface pattern has a phaeomelanic body with the following eumelanic areas: belly, entire lower legs, inside of legs to underside of tail, stripe down the back, on shoulders and up the middle of the underside of the neck, covering the lower jaw, and inside the ears. Other eumelanic areas are variable, including areas on the head and ill-defined dark areas on the rear flanks. A light spot beneath the eyes is consistent. The *badgerface* allele is the cause of the distinctive colour pattern of Barbados Blackbelly hair sheep, on which the phaeomelanic areas fully express the reddish-tan pigment (Maule, 1977). On wool sheep the badgerface pattern is usually expressed as an interplay of nearly white and nearly black areas, which can resemble the facial pattern of the badger. The dramatic interplay is less like a badger in hair sheep with intense phaeomelanin, and this pattern is usually called blackbelly instead of badgerface in these breeds.

Two *Agouti*-locus patterns are very similar to badgerface but have darker, greyer bodies than do badgerface sheep. These include the patterns resulting from the *blue* and the *light blue* alleles (Hoogschagen *et al.*, 1978; Lundie, 1984). The blue pattern is especially common in long-wool breeds and is the main *Agouti*-locus allele that segregates in these breeds. The fleece from the bodies of blue and light blue sheep is usually grey (darker than that of badgerface sheep), with the typical darker areas of the badgerface pattern. Two other alleles, *light badgerface* and *grey badgerface*, are identical in origin and appearance (Adalsteinsson, 1983; Lundie, 1984; Sumner, 1980) and should be considered the same and not as two distinct patterns. This allele results in a typical badgerface distribution of phaeomelanic and eumelanic areas that is even paler than the badgerface pattern, since the eumelanic areas are grey rather than black.

Agouti-locus patterns that are predominantly eumelanic Several patterns caused by *Agouti*-locus alleles in sheep are mainly eumelanic, and therefore tend to have rather dark phenotypes. Within this group are many distinct patterns, each the result of a separate allele. These patterns have eumelanic bodies with minor phaeomelanic regions on the belly and head. Some are approximate reversals of specific *Agouti*-locus patterns that are mainly phaeomelanic.

The black-and-tan *Agouti*-locus pattern can be loosely thought of as opposite to the badgerface pattern, and in some references this pattern is referred to as 'reverse badgerface' (Roberts and White, 1930a; Brooker and Dolling, 1969a). The black-and-tan pattern has also been called 'mouflon', due to the pale belly reminiscent of wild sheep. The eumelanic body of black-and-tan sheep is distinct from the phaeomelanic body of the wild type, and therefore the term mouflon only contributes to the confusion surrounding this allele. It is for this reason that COGNOSAG (1986) suggested the use of *black and tan*, since this name describes a similar *Agouti*-locus pattern in several other species. The *black and tan* allele causes a phaeomelanic belly, perineum, throat, inside the ears and stripes above the eyes. This pattern is similar to the black-and-tan pattern in many other mammals, such as dogs, rabbits and mice, and referring to it by the name 'black and tan' in sheep helps to reflect the homology between the ovine allele and that of the other species. In most wool sheep the phaeomelanic areas are modified to be very pale or white, and so the pattern is not truly 'black and tan' but is more likely to be 'black and white', as are most of the *Agouti*-locus patterns in wool sheep.

The lateral-stripes pattern is darker than the black-and-tan pattern (Lundie, 1984). It consists of a pale band at the border of the belly and the body (both of which are eumelanic), a pale scrotum and pale inside the ears. The outside of the ear is dark. A pale moustache is present, which extends to the medial border of the eye and up over the nose. The chin varies from dark to light. Swiss marked has slightly more extensive phaeomelanic areas than does lateral stripes (Lundie, 1984). The pale areas are the muzzle, stripes to above the eyes, pale under the jaw and pale lower legs with a dark stripe down the front of each leg.

The *eyepatch* allele (Lauvergne and Adalsteinsson, 1976; Lauvergne *et al.*, 1981) produces one of the darker *Agouti*-locus patterns in sheep. Eyepatch consists of phaeomelanic areas around the eyes and covering the cheeks on an otherwise eumelanic coat. Light red cheek is similar to eyepatch but lighter and more extensively marked with phaeomelanin (Lundie, 1984).

Intermediate *Agouti*-locus patterns The intermediate *Agouti*-locus alleles all produce patterns that are mixtures of eumelanic and phaeomelanic fibres throughout the coat. The resulting patterns are therefore distinct from the regional contrasts typical of the other *Agouti*-locus patterns (Adalsteinsson, 1970), since they are less obviously symmetrical and instead are characterized by more uniform mixtures of phaeomelanic and eumelanic fibres.

The grey pattern consists of a white undercoat (secondary fibres) and a black outer coat (primary fibres) throughout the fleece. Lambs may be born dark enough to be misclassed as non-agouti, but quickly develop white hairs near the eyes and also have light hairs on the nose, inside the ears, the flank and the scrotum and on the feet above the hoofs. Grey phenotypes are variable, and in general the homozygotes are paler than heterozygotes that also carry A^a. The *grey* allele causes gametes to have a decreased fertilization rate, resulting in under-representation of this genotype from that which is expected (Adalsteinsson, 1970). This is not due to lethality, but to some disruption of fertilization.

The Gotland grey pattern is darker than grey, and may be due to a separate *Agouti*-locus allele, or alternatively may be the result of long selection for a narrow range of dark shades of grey (Adalsteinsson *et al.*, 1978). This is the colour typical of the Gotland breed of sheep used for sheep pelts, and a narrow range of colour variation is essential for the matching of the pelts used in the final products. The pelt colour score for Gotland sheep has a heritability of 0.8 (Skårman, 1977), which is evidence for modification of relative shade independent of the *Agouti* locus. A light grey pattern has also been documented (Adalsteinsson *et al.*, 1978), and is lighter than the standard grey pattern and caused by a separate allele (Lundie, 1984). Sooty is another pattern that is a uniform mixture of eumelanic and phaeomelanic hairs, and in this regard is somewhat similar to the grey patterns (Lauvergne and Adalsteinsson, 1976), although it is darker and more eumelanic overall.

The grey-and-tan pattern also fits into this group, and appears phenotypically as if the grey pattern had been combined with the black and tan pattern (Adalsteinsson, 1970). The eumelanic regions of the black-and-tan pattern are grey instead of black, resulting in a much paler pattern. This may be very similar or identical to the wild-type pattern, only on a wool sheep instead of a hair sheep. Patterns such as grey and tan appear to be the result of combining two distinct *Agouti*-locus patterns, and are the phenomenon that made attractive the hypothesis that the *Agouti* locus is a complex of miniloci. While the exact organization of the ovine *Agouti* locus is uncertain, it is unlikely to be very different from that of the murine locus, which has been shown to be simple rather than complex (Jackson, 1994). The ovine *Agouti* alleles do have differences from those of the murine locus, though, especially in the relatively high number of patterns that have reversals of pigment-type distribution.

The Extension *locus*

Mechanism of action

The *Extension* locus encodes one member of a group of seven α-MSH receptor proteins (Jackson, 1994). This protein, when present, results in eumelanin formation by melanocytes. The α-MSH receptor, when activated, results in increased cyclic adenosine monophosphate (cAMP) levels and increased protein kinase activity. The *wild*-type *Extension*-locus allele in most species allows for control of eumelanin and phaeomelanin production to be governed by the *Agouti* locus, as determined by the regional distribution of the agouti protein, which inhibits eumelanin formation.

Many species have dominant alleles at the *Extension* locus, which act to increase adenylate cyclase either spontaneously or in response to α-MSH. The specific mechanisms are different for different alleles in the mouse, and involve either responsiveness to α-MSH or autonomous adenylate cyclase production (Jackson, 1994). Either mechanism results in an entirely eumelanic phenotype, and these alleles are usually considered as *dominant black* (E^D), regardless of the underlying mechanism used to achieve this phenotype. Recessive alleles at the *Extension* locus consistently result in an inactive receptor that is unresponsive to stimulation by α-MSH. The inactive receptor results in a completely phaeomelanic phenotype. In this regard the *Extension* locus acts opposite to the *Agouti* locus, since more dominant phenotypes are completely eumelanic and more recessive ones are completely phaeomelanic. *Extension*-locus alleles tend to affect the entire animal, so that this locus does not produce patterns like those typical of the *Agouti* alleles, with interplay of the two pigment types. The *Extension* locus is usually responsible for completely eumelanic or completely phaeomelanic phenotypes, although exceptions do occur in some other species (Searle, 1968).

Alleles

The *Extension* locus has important interactions with the *Agouti* locus in determining the colour of sheep. The *wild*-type allele at the *Extension* locus allows the expression of the *Agouti* locus. It is therefore the most common *Extension* allele in most European breeds of sheep, in which segregation at the

Agouti locus accounts for the majority of colour variation. The dominant allele at the *Extension* locus is usually called *dominant black* (Roberts, 1926; Vasin, 1928) and is symbolized E^D (COGNOSAG, 1986). This allele causes a uniformly eumelanic coat, which in most breeds of sheep is black. The allele E^D has a very widespread occurrence worldwide, although it is obviously rare in areas where white wool is desired, because selection against black quickly removes this allele from any such population. The E^D allele accounts for the black of many coloured breeds, including Black Welsh Mountain (Roberts, 1924), Jacob (black and white) (Roberts, 1926), Persian and Somali Blackhead and Karakul (Vasin, 1928). It also accounted for a portion of the black sheep in multicoloured breeds, such as Shetland (Roberts and White, 1930a). In some of the originally multicoloured breeds, it might now be extinct, due to genetic drift or recent selection in favour of white sheep.

In some uniformly black breeds, the similarity of the phenotypes resulting from E^D at the *Extension* locus and A^a at the *Agouti* locus has caused both alleles to be present (Vasin, 1928; Roberts and White, 1930a). Selection for eumelanic phenotypes alone cannot distinguish between these two biological mechanisms that achieve similar phenotypes. The persistence of both mechanisms can result in some strange segregation of colour unless the observer is aware that both the E^D of the *Extension* locus and the recessive A^a of the *Agouti* locus can be present in some breeds.

Other alleles have been postulated at the *Extension* locus, and these will be discussed below with the colour of sheep that have white wool and dark points. The only alleles that are well documented at the ovine *Extension* locus are *wild-type* and E^D. It is unusual that a recessive phaeomelanic allele is not documented in sheep, as is the case for many other mammalian species (Searle, 1968). Such a recessive allele, if it were found in sheep, would have an advantage over A^{Wt} as a source of phaeomelanic phenotypes (potentially white) in sheep, since coloured sheep would be unable to segregate out of white flocks whose whiteness resulted from a recessive mechanism for whiteness.

The Albino *locus*

Mechanism of action

The *Albino* locus codes for the tyrosinase enzyme that is essential for melanogenesis. Recessive alleles at this locus are responsible for the production of abnormal forms of tyrosinase that have either reduced or nearly completely absent activity (Jackson, 1994). As a result, melanocytes are present but are incapable of melanogenesis.

Alleles

The *Albino* locus in sheep has been documented as having a few recessive alleles. The *albino* mutant (Adalsteinsson, 1977, 1978a) completely inhibits both phaeomelanin and eumelanin formation in wool, hair, eyes and skin. Eyes are bluish-pink, the blue cast coming from Tyndall scattering of light, as the iris is thicker than in smaller mammals whose albinos have thinner irises and therefore the more classic pink eyes expected of albino animals. The *albino marrabel* allele (Rowlett and Fleet, 1993) has been documented in Suffolk sheep in Australia. The *albino marrabel* allele does allow for some tan or light yellow pigment in the hoofs and in leg hairs, and some pigment at the pupillary border of the iris. Lambs homozygous for *albino marrabel* have transient photophobia, but adults usually tolerate sunlight well. Albino phenotypes are very rare among sheep, even though they are of at least some usefulness in removing pigment from wool.

Expression and modification of phaeomelanin

Phaeomelanic colours are subject to various modifications, some of which are well understood and many others of which are not. Many of these modifications do appear to have a genetic basis. Phaeomelanic colours are modified by independent loci controlling colour directly, and are also modified by the physical structure of the hair or wool coat. It is difficult to separate the results of these two phenomena.

In general, the birth coat of lambs is more intensely pigmented with phaeomelanin than is the subsequent coat (Adalsteinsson, 1970). Coarser fibres are also more intensely pigmented than finer fibres, which usually correlates to the origin of fibres as from primary rather than secondary follicles (Adalsteinsson, 1970; Aliev and Rachkovsky, 1987a, b). The dark extremes of phaeomelanin expression occur on hair sheep, in which intense red-brown phaeomelanic pigmentation is possible. Dark red-brown phaeomelanin colours are rare or non-existent on wool sheep. At least some of this difference is related more to coat structure than genetic background for colour, since certain breed types, such as the American Tunis and the Solognot, are born intensely phaeomelanic and then fade quickly to ivory or white in the woolled portions of the coat while retaining phaeomelanin in the haired portions. Occasional red lambs are born in the Arles Merino breed, and these fade quickly (Lauvergne *et al.*, 1981). Such fading red sheep also occur in other breeds (Fleet *et al.*, 1991a).

Aliev and Rachkovsky (1987a, b) determined that genes responsible for the change from a hairy coat to a woolly coat are also related to the fading of phaeomelanin from the wool coat. Whether this change is related only to hair structure or results from a direct effect on melanocytes is uncertain. Whatever the mechanism, this phenomenon is of importance to the potential colours of wool, since intensely phaeomelanic fine wools are counter to the basic biology of pigmentation in sheep.

Adalsteinsson (1976) found that the score for tan areas on A^{Wt} Icelandic sheep has a heritability of 0.46 ± 0.05. Icelandic sheep generally have fairly coarse primary fibres. Merino sheep have a heritability of 0.79 ± 0.20 (Fleet *et al.*, 1991b) for such tan fibres. These numbers indicate some genetic control of the level of phaeomelanin production for the expression of phaeomelanin in genotypes capable of forming this pigment.

Expression and modification of eumelanin

Eumelanic areas on sheep are subject to modification by mechanisms distinct from those affecting phaeomelanic colours. One fairly common modification is the replacement of black eumelanin with brown eumelanin.

The Brown *locus*

Mechanism of action   The *Brown* locus is the major source of brown colours in sheep. This locus controls TRP-1, which has important but poorly documented action within melanocytes (Jackson, 1994). The recessive *brown* allele (B^b) reduces activity of TRP-1, and the result is that all regions that are eumelanic have the black form replaced by a brown form. Brown melanosomes are structurally different from black ones (Alexieva *et al.*, 1985). In the brown phenotypes the catalase normally involved in melanogenesis fails to protect the eumelanin from the action of hydrogen peroxide, which is produced during melanogenesis.

The murine *Brown* locus has a few dominant alleles in addition to the recessive ones. These usually have apical pigmentation and basal pallor on hair shafts, and similar mutants have not been documented in sheep. The banding on hairs that is caused by these murine alleles is due to the build-up of toxic metabolites within the melanocytes, which increases as the hair cycle progresses. The result is that the apical portions are pigmented, since melanocytes are functional, but as the hair grows the melanocytes undergo toxicosis and pigmentation is therefore diminished as the base of the hair is produced.

Alleles   Brown eumelanin in sheep results from the action of the recessive B^b at the *Brown* locus (Adalsteinsson, 1970). Brown eumelanin results in a range of colours which have become known by the name 'moorit' when they occur on sheep (Adalsteinsson, 1979). Moorit is a Shetland word for 'red as the moor', and it or similar words are used throughout the North Atlantic region to describe the fleece colours caused by B^b. Moorit sheep vary from dark chocolate to very pale, and some of the shades are reddish. The term moorit is specifically for the eumelanic browns, and this avoids confusion with the phaeomelanic red browns of some sheep (Karakul lambs and hair sheep). Brown phenotypes are relatively rare in most breeds of sheep unless selection has specifically favoured these colours (some Shetland sheep, for example, are valued for moorit wool). Brown eumelanin can replace black eumelanin in any of the *Agouti*-locus patterns, as well as in sheep with the *dominant black* allele of the *Extension* locus (E). The brown pigmented version of the various *Agouti*-locus patterns is usually more subtle than the black one, since the contrast of the pale areas with the brown is less than that of pale areas with black.

Brown eumelanin is somewhat weaker than black eumelanin and fades more readily with age and sunlight. As a result, moorit sheep are generally paler than their black counterparts. Very dark brown moorits are fairly rare, and brown wools are usually relatively pale. However, selection in some specific strains of sheep for wool for handicraft use has succeeded in obtaining a relatively dark range of moorit shades of colour from B^b.

The B^b allele has pleiotrophic effects in addition to its action on colour. Ryder *et al.* (1974) found that Soay sheep homozygous for B^b were less viable than sheep not homozygous for this allele. Homozygous sheep were also smaller-bodied than those of other genotypes. This effect was most noticed in $A^aA^aB^bB^b$ sheep. This work concerned the feral Soay sheep, in which the B^b allele is relatively common. A similar effect on non-feral sheep has not been documented.

Brown phenotypes occur as rare variants in a number of sheep breeds worldwide. Brown phenotypes are rare in any breed that is selected for a black or a blue face. Black faces result from black hair, and any replacement of black by brown would be quickly noted and culled and the allele causing the brown faces would tend to be removed from the population by that means. Blue faced breeds result from black-pigmented skin overlain by white hair. The brown phenotypes in such breeds lack the bluishness of black phenotypes, since the skin colour is not as dark. This has resulted in negative selection for any expression of brown phenotypes in blue-faced breeds. Many of these are long-wool breeds, and moorit brown is therefore rare in these. A few reports, such as Dry (1924), suggest that perhaps brown did segregate in breeds such as the Wensleydale as a very rare variant, at least earlier in the century. If the 'dirty brownish-grey' lamb of Dry (1924) was moorit, the allele appears to have been lost to the Wensleydale breed since then. This is probably the result of genetic drift due to this breed's becoming rare in this century, with rare variants drifting to extinction. Duck (1921, 1922) presents data that suggest that moorit may also have occurred in the Leicester earlier in this century. Other more common long-wool breeds, such as the Romney and Border Leicester (if these are accepted as long-wools), still do segregate out occasional brown lambs. Breeds selected specifically for a range of coloured wools have used brown phenotypes to good advantage, and selection has increased the relative number of moorit sheep in these breeds. Shetland and Icelandic sheep are good examples. Recent efforts to raise coloured sheep of usually white breeds, such as Romney, Border Leicester and Merino, also tend to use the B^b allele to generate multiple colours of wool.

Brown phenotypes (moorit as well as others) segregate relatively more commonly in sheep breeds that are expected to be stark white than in those requiring black or blue faces. In Merinos, for example, the moorit colours (among coloured sheep produced in this usually white breed) seem relatively more common than they do in the long-wool breeds. This might be due in part to the fact that any phenomenon leading to paleness in these breeds is likely to be the subject of subtle positive selection, because the breeders have so heavily penalized any obviously pigmented wool fibres for so many years. Isolated pigmented fibres in white sheep capable of forming black pigment (B^+) are likely to result in the culling of the sheep bearing these. Isolated pigmented spots of B^bB^b phenotypes are more likely to be missed, since some of these are very pale. The result, over centuries of selection, is that the brown-pigmented types are relatively more common in the fine-wools than they are in the longwools, since they can more easily escape the high levels of negative selection than can the black-pigmented types.

Alleles other than B^b have not yet been documented as residing at the *Brown* locus, although the coffee-brown of Nel (1967) may be a candidate. Nel (1967) ascribes this variant to the *E* locus, which has contributed to confusion in interpreting his work. A *dark brown* and a *light brown* have both been documented in goats, and each is dominant to the wild type (Adalsteinsson *et al.*, 1994; Sponenberg and La Marsh, 1996). Such alleles have yet to be documented in sheep.

Dark brown

A second modification of black eumelanin has been called 'dark brown' (Lundie, 1984). This results in a colour intermediate between moorit and black. This colour has been poorly investigated, although early results suggest that it is due to a dominant allele and that the homozygotes are much lighter (pale brown) than are the heterozygotes (dark brown). The locus at which this resides is not certain, and this mechanism for brown eumelanin appears to be more rare than that afforded by the B^b allele. Dark brown lambs are born nearly black and then fade to a colour intermediate to black and moorit. The paler brown homozygotes fade rapidly to a grey-brown colour, which is somewhat darker along the spine than it is on the rest of the body.

Colour of black- or dark-faced breeds

Many breeds of sheep have white wool but dark-haired portions of the body, such as heads and legs. The dark portions (or 'points') can be solid-coloured black, brown or tan, or can be speckled with one of these colours and white. These sheep have adult coats that have dark or speckled faces and legs (hair-bearing skin) and white wool. Most of these sheep are born with the dark hairy portions, and some also have dark wool or mottled and spotted wool on the wool-bearing areas of skin. These wool-bearing areas then fade to the adult white wool.

Aliev and Rachkovsky (1987a, b) propose that a series of alleles at the *E* locus account for the colours that have dark points in Asian breeds, such as the Tajik and Hissar. These alleles are *blackish*, *brownish* and *yellowish*. These breeds segregate for these colours, although the wool they produce is consistently white or nearly so, despite the variation in point colour. These alleles are dominant to the *E*-locus *wild* allele, with the order of dominance E^D, *blackish*, *brownish*, *yellowish* and then *wild*. Another *E* allele, *japanese brindling*, is proposed by Mahler and Denis (1990) to account for the colour of the Down breeds and breeds derived from them. These proposed *E*-locus alleles are reputed to interact with A^{Wt} to provide for the various colour phenotypes in breeds of sheep with dark points and light wool.

Sur

The Sur modification of colours is important only to breeders of the Karakul and other breeds from which neonatal lamb pelts are used as the main commercial product (Aliev and Rachkovsky, 1987a, b). The Sur variants occur on most, if not all, background colours of sheep, and result in pale apical segments on the birth-coat fibres. These pale terminal segments are white or yellow. Some evidence suggests that the Sur variants are due to single genes, although they may equally be due to polygenic modifications of the base colour. These variants appear to modify all base colours, even though some evidence suggests that E^D phenotypes escape this modification. Two types of Sur are documented, the Surkhandarya type and the Bokhara type. These are different in origin but produce identical phenotypes. The two types originated in different strains of Karakul sheep; hence the separate designations. Both are reputed to be recessive, although most data sets include an annoyingly low percentage of non-Sur lambs from Sur to Sur matings. The preponderance of the evidence, though, suggests that these phenotypes are due to recessive genes. Tests for allelism suggest that the two different types of Sur might be at independent loci. An alternative interpretation of the data associated with the Sur colours are that they are caused by polygenes and are a quantitative trait rather than a Mendelian trait. The two types are assigned to two loci by COGNOSAG (1993): *Sur Surkhandarya* and *Sur Bukhara*.

The Australian Piebald *locus*

Australian piebald is a modification of pigmentation. The *Australian Piebald* gene was originally documented as the source of occasional non-symmetrical black spots on otherwise white sheep (Brooker and Dolling, 1965, 1969b) (Fig. 3.3). Since its original documentation as an incompletely penetrant recessive gene, it has undergone further investigation, which has revealed details of the biology of this phenomenon. *Australian Piebald* is indeed the source of isolated black spots, but it works on any *Agouti*-locus genotype and not only on white. The intermediate *Agouti*-locus alleles are also modified to have isolated, usually single, black spots on body regions that are expected to be phaeomelanic. The result is that badgerface sheep with the appropriate genotype can also have isolated black spots in otherwise pale (phaeomelanic) regions of the colour pattern. Obviously some patterns, such as the black-and-tan patterns, would be difficult backgrounds on which to detect this effect, since they are predominantly black to begin with. Australian piebald has features consistent with somatic reversion from higher *Agouti* alleles to the A^a allele. No evidence of germinal reversion has been noted, though, and so this phenomenon might not be due to reversion. The Australian piebald phenotype has thus far only been documented in Australian Merino sheep.

Fig. 3.3. Australian Merino wether exhibiting the Australian Piebald phenotype. (Photograph by C.H.S. Dolling).

Loci Controlling White Spotting

Patterns of white spotting are added independently to any coloured background in sheep. Multiple spotting patterns occur, and some are important in the breeds that are commercially important throughout the world. Others have special importance to the pelt breeds.

The Spotting *locus*

Mechanism of action

In mice the *Piebald* (*S*) locus has been documented to act on the differentiation of melanocytes at the neural crest, as well as on their migration from the neural crest to the rest of the body (Jackson, 1994). The result is an array of white regions on otherwise pigmented mice. The specific colour of the pigmented regions is governed by the other coat-colour loci, so that white spotting can be superimposed over any colour or pattern.

Mice with extremely white manifestations of spotting are afflicted with aganglionic megacolon, which is a lethal condition that occurs relatively early in life. This condition results from failure of neurons to migrate from the neural crest to the colon, much as the white spotting results from failure of the migration of melanocytes from the neural crest to the skin. The alleles for spotting at the *S* locus are recessive in most species, and result in unpigmented regions of skin and hair.

Alleles

The spotting that is due to the *spotted* allele at the *Spotting* locus is relatively common in sheep (Roberts, 1926; Adalsteinsson, 1970) (Fig. 3.4). The white spotting is usually distal on the limbs and head, and is also roughly symmetrical side to side on the animal. The *spotted* allele is likely to be recessive on eumelanic backgrounds, although 50% of *spotted* heterozygotes on black-colour phenotypes have minor white spotting on head or legs. In A^{Wt} sheep nearly all heterozygotes are spotted (Lauvergne, 1969). More extensive spotting in A^{Wt} phenotypes than in other *Agouti*-locus phenotypes has also been demonstrated by Adalsteinsson (1970). The interaction of *spotted* with A^{Wt} is important, since this interaction appears to provide not only for extensive white spotting but also for a decrease in expression of phaeomelanin (Adalsteinsson, 1970; Lauvergne, 1975; Fleet *et al.*, 1991b; Fleet, 1994). This interaction is crucial to the production of white fleeces in many breeds, because it removes pigment both by white spotting and by dilution.

The extent of spotting conditioned by the *spotted* allele varies from animal to animal as well as among breeds in which spotting occurs. It is unclear if such variation is due to different *Spotting*-locus alleles, or whether it is due to

Fig. 3.4. Jacob sheep with a dominant black phenotype modified by spotting. Three of the animals are ticked, with varying degrees of expression.

modification at other loci. Adalsteinsson (1978b) showed that selection could easily increase the proportion of white regions in spotted phenotypes of sheep. His work eventually produced starkly white sheep which were fully viable, in contrast to the non-viability of markedly white-spotted mice. Adalsteinsson (1978b) demonstrated a heritability of 0.8 in the extent of white spotting. Various studies involving the Karakul breed have also shown that selection in favour of extensive whiteness in spotted phenotypes can be effective in producing white or nearly white sheep (Panfilova, 1939; Cook, 1950; Nel, 1967). In the light of these results, the importance of multiple alleles at the *Spotted* locus is difficult to assess. It is still unclear if all spotting is due to this one allele, or if there are multiple alleles in a series with overlapping phenotypes (Adalsteinsson, 1970).

One specific allele, *bizet spotting*, has been proposed as a cause of minor white marks on head and legs of sheep, and as an allele separate from *spotted*. The *bizet spotting* allele may be incompletely dominant to the *wild* type rather than recessive to it (Lauvergne, 1975). About 50% of E^D Bizet sheep with this allele in the heterozygous condition are spotted. In the phaeomelanic Solognot breed the spotting is nearly 100% in sheep that are heterozygotes for the *bizet spotting* allele. Homozygous *bizet spotting* sheep in the Solognot breed are nearly all white, due to the extensiveness of the spotting (Lauvergne, 1975). This allele may be the same as the *spotted* allele, only with low grades of expression.

The *Spotting* locus is important in the determination of skin pigmentation of white wool-bearing sheep, which is frequently an important aspect of breed character. The sheep of those breeds for which the standard calls for pigmented skin generally lack any spotting. As a result, the occasional coloured sheep that segregates from white sheep of breeds having pigmented skin tends to lack spotting and is whole-coloured. Breeds characterized by pink skin and hooves usually accomplish this by white spotting, since even minor grades of spotting affect the nose and distal limbs. Therefore the coloured sheep that segregate out of breeds with pink noses and hooves tend to have white spotting. This spotting can be extensive and very noticeable, or may be minor white spotting limited to the head and legs. The spotting is accomplishing the pinkness of the skin and hooves, since spotting removes melanocytes from these regions. This phenomenon, striking in coloured sheep, is frequently overlooked in white sheep.

The *spotted* allele appears to directly affect fleece texture in some breeds. In Jacob sheep Ryder and Adalsteinsson (1987) documented that the black areas have longer and coarser wool than the white areas.

The Pigmented Head *locus*

Other types of white spotting patterns occur in sheep, but most of these occur only in a limited array of breeds. Many of these spotting patterns are associated with Asiatic breeds and rarely, if ever, occur in the European breeds that are so common internationally (Vasin, 1928; Aliev and Rachkovsky, 1986). At the *Pigmented Head* locus there are several alleles of importance to certain breeds.

One of these is the *afghan lethal*. This is a dominant allele, and the homozygotes are white and die soon after birth, due to gastrointestinal abnormalities similar to those of excessively spotted mice. Heterozygotes have pigmented hair and skin on the nose, small eye patches and ear tips and small spots on the knees and hocks. Were it not for the lethality of the homozygous condition, this allele would be a very satisfactory mechanism for whiteness in the fleece-bearing portions of sheep. This allele is almost certainly the same as the lethal dominant white described by Nel (1967).

A second allele at the *Pigmented Head* locus is also dominant, and is called *turkish* (Lauvergne, 1976b; Aliev and Rachkovsky, 1986) This has been called Akaraman spotting in the past. Homozygotes have pigmented noses, eye patches, ears and distal legs. They are similar to the *afghan lethal* heterozygotes, only more extensively pigmented. Heterozygotes are usually classed as piebald, although the specific pattern is usually not characterized in detail, so it is uncertain how it differs from *Spotting*-locus patterns. The *turkish* allele has been used as a mechanism to produce white fleeces in some Asian breeds. The Akaraman breed is among these. The effectiveness of this mechanism for white wool depends very much on a lack of pigmentation in the woolled portions of the body. The desire for uniformly white wool should drive selection for modification to the whiter extremes of spotting, at which point some lethality may well occur.

The *persian* allele at the *Pigmented Head* locus results in viable homozygotes with a pigmented head but a white body (Aliev and Rachkovsky, 1986) (Fig. 3.5). The pigment can extend to the neck and shoulder on extensively pigmented animals. Heterozygotes are described as piebald, although this is likely to be in patterns that are distinct from those of the *Spotting*-locus alleles. The *spotted* allele, even in low grades of expression, results in facial white, and this is in contrast to the solid-coloured heads associated with the patterns caused by the *persian* allele. The *persian* allele is responsible for the colour pattern of Persian and Somali Blackheaded sheep, as well as the Dorper breed. In these breeds the black-pigmented areas are usually due to E^D, over which the white spotting is superimposed. The black areas on these sheep are extensive enough to affect the fleece-bearing portions of the coat, and consequently this allele is of limited use as a means of producing white wool uncontaminated with coloured fibres. This is especially so since *persian* heterozygotes are more extensively pigmented than homozygotes, so that any crossbreeding with breeds whose whiteness comes from the *persian* allele would result in extensively pigmented crossbred progeny.

The alleles at the *Pigmented Head* locus have a consistent effect in that the more dominant alleles cause more extensive spotting than the more recessive ones. The whitest of these fail to thrive, and this is due to gastrointestinal problems caused by lack of innervation of portions of the intestine (Nel, 1967; Aliev and Rachkovsky, 1986). This phenomenon is fairly common across species with certain spotting patterns, and is due to faulty innervation of the gut as is similar to the situation with excessively spotted mice. Pigment cells and nerve cells both migrate from the neural crest and are closely related embryologically, and disruption of differentiation or migration of neural-crest cells can easily result in manifestations in both the pigmentary system (usually white spotting) and the neurological system.

The Roan *locus*

Another type of white spotting is the roan pattern, which is due to the dominant *lethal roan* allele (Vasin, 1928; Nel, 1967). This allele causes intermixture of white hairs into the coat regardless of background colour. Heterozygotes usually have a fairly even mixture of dark and white hairs. Homozygotes are very pale, and die due to abnormalities of gastrointestinal innervation (Constantinescu, 1932a, b, 1933). The colour classes of heterozygotes and homozygotes are somewhat variable, and some of the darkest homozygotes do survive and reproduce (Contescu, 1938; Constantinescu, 1939; Contescu and Epureanu, 1939). The lethality of the homozygote seems to be specifically related to the whiteness and not the genetic complement *per se*. This allele is the source of important colours in the Karakul and other pelt sheep, but of limited importance in other sheep. *Roan* on a black phenotype results in a grey colour, caused by intermixture of black and white hairs, and it is important not to

Fig. 3.5. Somali sheep with the *persian*-allele spotting pattern on a dominant black (left and right) and tan (middle) colour phenotype. (Photograph © Evelyn Simak).

confuse this colour with the grey caused by the *Agouti* locus. Some evidence suggests that *Roan* may be allelic to the *Pigmented Head* alleles (Nel, 1967).

Other loci affecting white spotting

A fairly rare pattern in coloured sheep consists of small flecks of white on an otherwise coloured sheep. These spots are usually lateral on the body in lower grades of expression. A study on Merinos (C.H.S. Dolling, South Australia, 1995, personal communication) suggests that the trait is due to a dominant allele. In the population studied, the pattern goes by the name 'Patel' after the original ram with this spotting pattern. This pattern is visually similar or identical to the pattern that distinguishes the Florida Sevillana goat breed. In sheep this pattern is most easily appreciated following shearing, since long fleece obscures the subtle character of the small flecks of white.

The Ticking *locus*

Small spots of pigment grow into the fleece or hair coat of some sheep that are spotted by virtue of the *spotted* allele (Adalsteinsson, 1970, 1978b; Dolling, 1989) (Fig. 3.4). This is called ticking, and the small spots are not present at birth but grow into the white areas later. This phenomenon is only important in certain breeds in which spotting is desired, such as the Jacob sheep. These dark spots are sometimes called 'freckles', especially in the Jacob breed. Ticking is dominant to its absence and varies considerably in extent. The mechanism that allows melanocytes to populate regions that are embryologically devoid of them has yet to be elucidated. The *Ticking* gene is analogous to ticking in dogs, on which species it is extremely variable in extent and in size of the tick marks (Little, 1957).

Strategies for Selection of White Wool

The elimination of all pigmented fibres from white wool is a daunting task, as evidenced by the continued effort this endeavour requires (Fleet, 1994). Part of the reason for this is that the usual mechanism for the whiteness of sheep and wool uses the *Agouti* allele A^{Wt}. This allele has the potential for pigment production in various portions of the coat. Observations on goats suggest that the analogous caprine allele can vary in phenotype from stark white to very dark red-brown, and that it even includes occasional phenotypes with minor regions of eumelanic pigmentation. The importance of these observations is that the usual genotype of white sheep is one that at least potentially can provide for pigmented fibre. Selection has succeeded in eliminating most of the propensity for this genotype to express pigment, but the problem remains that white-sheep production involves attempting to eliminate pigment completely by using complex modification of a potentially pigmented genotype.

Dilution of pigment is one mechanism that can be useful in elimination of pigment from A^{Wt} phenotypes. Dilution of the phenotype caused by A^{Wt} appears to have been widely used as a strategy in the breeding of white sheep. This phenomenon is so widespread that intensely phaeomelanic sheep are noteworthy due to their rarity. Even the breeds selected for relatively intensely phaeomelanic-haired portions (American Tunis, Solognot, Fuchschaf) are noteworthy for having wool that is white or nearly so. The paleness of the wool on these breeds is probably due in part to the structural modifications that are important in generating wool instead of hair. These modifications decrease the propensity for phaeomelanin to pigment the fibres. Since dilution is such an important aspect of the production of white wool, it is probably accurate to place at least some negative selection pressure on sheep that express any intensely phaeomelanic fibres, since these sheep are reversing the usual biological directive for dilution of this pigment.

Spotting is another mechanism that is important in the generation of white wool. Spotting reduces the availability of melanocytes to varying degrees. Spotting, from whatever allele, seems to have been used very rarely as the sole source of whiteness in sheep. The few breeds, such as the Akaraman, that are rendered white by virtue of spotting are noteworthy for their rarity. Spotting of various types has been used in concert with the A^{Wt} allele to take advantage of the synergism of the *spotted* allele with the A^{Wt} allele. Spotting of various sorts interacts with A^{Wt} to provide not only for extensive grades of spotting (therefore more white) but also for dilution of remnant phaeomelanic areas. The interaction of these two alleles is important to many breeds that produce white wool. Since this interaction is important, selection does need to penalize sheep that do not exhibit extensive spotting in breeds in which this mechanism is important. This especially includes breeds with non-pigmented skin and hoofs. The breeds for which the standard requires pigmented skin do not have white spotting and therefore cannot take advantage of the synergism of white spotting with phaeomelanic phenotypes.

Age-related effects on white-wool sheep are also important. Some lambs, for example, have transitory phaeomelanic pigmentation of fibres, which then fade to white with age. This process is usually complete by a year or so of age. The fading of these fibres suggests that older sheep are less likely to contain fleeces contaminated by such fibres. Unfortunately, an opposite phenomenon is also typical of many white wool-bearing sheep, and the incidence of pigmented skin spots and isolated pigmented (usually eumelanic) fibres also increases with age. Some of the development of these spots is directly related to exposure to sun post-shearing. Whatever the cause, the result is that older sheep are more likely than younger sheep to have such fibres. The combination of the two age-related effects on isolated pigmentation suggests that no single strategy based on maintaining specific ages of sheep is going to be entirely successful in eliminating the minor but important contamination of white wool with isolated pigmented fibres.

The biological mechanisms that provide for white wool are themselves the main limitation to the success of selection for white wools. Since the usual mechanism is the presence of the A^{Wt} allele, most white-wool sheep do have the potential for producing at least isolated pigmented fibres. The use of mechanisms other than A^{Wt}, with or without *spotted*, to generate white wool could be very successful. Candidate mechanisms include the extreme dilution provided by the *albino* genotypes or the use of extensive white spotting, such as the *turkish* allele. These mechanism have the potential advantage that white wool resulting from these is likely to be entirely free of pigmentation faults. The genetic variants for these are rare to non-existent in most breeds, though, and so are unlikely to ever gain widespread use in the production of white wool.

Strategies for Selection of Coloured Wools

The breeder who favours pigmented wools has a selection problem distinct from that faced by breeders of white wool. The potential range of wool colours is determined by the biological pigments available in sheep, and these are eumelanin and phaeomelanin.

Eumelanic colours in wools include many that are based on black or brown. Selection for eumelanic sheep first needs to involve the expression of eumelanin and then the specific colour of eumelanin that is desired. The A^a and E^D alleles are useful for breeders desiring eumelanic wools. The intermediate *Agouti*-locus alleles are generally less useful since they provide for phenotypes with variable shades and colours in the fleece. In some situations (especially the hand-spinning market), this variability is desired because individual fleeces can then provide a range of colour all on wool of relatively uniform grade and texture.

The E^D allele can be readily used to provide for the introduction of eumelanic phenotypes into a flock, but this allele does have the disadvantage that white sheep can segregate as a recessive phenomenon. The *Agouti*-locus alleles are more difficult to introduce, since they are recessive to A^{Wt}, but once they are introduced they are more likely to consistently produce eumelanic lambs since white lambs cannot segregate from such a flock. The phenotypic similarity of E^D and A^a can cause problems for flock owners selecting for eumelanic colours in situations where both alleles are present in a single flock.

Eumelanic colours can be based on black or brown. Black-based colours vary from truly black through shades of grey, and finally to very pale greys that are nearly white. Selection within the eumelanic phenotypes can either narrow or broaden the range of shades available, depending on breeder goals. As a general rule, most eumelanic phenotypes fade somewhat with age, so that having various ages of sheep in a flock will help to generate a variety of shades of wool.

Brown eumelanic colours are also variable, from dark chocolate-brown through a large range of intermediate fawn colours. These shades are subject to the same modifications as the black-based colours. Most of the shades of colour afforded by brown eumelanin lack much redness, although a few do have moderate redness. Selection is very important for breeders trying to secure the darker range of the brown eumelanic shades, since brown eumelanin is generally weaker than black eumelanin and more subject to the effects of weather and age-related changes.

Selection for phaeomelanic-coloured wools is tantalizing, since phaeomelanin offers the potential of reddish-brown colours in addition to rich tan colours. Unfortunately for breeders interested in such colours, millennia of selection for white wool in sheep have resulted in very white wool on most breeds of wool sheep. The most common mechanism used in this process has been to select for the A^{Wt} allele. This results in sheep at least theoretically capable of forming phaeomelanin throughout the coat. Selection, either directly for whiteness or indirectly through fibre characteristics, has removed most or all of the propensity for phaeomelanin to be deposited in the fibres. Selection in favour of phaeomelanin phenotypes is therefore going counter to millennia of selection in the opposite direction. Selection for intense phaeomelanin is unlikely to be simple, since the basic genotypes used for white wool and those used for phaeomelanic colours are the same.

A few breeds stand in stark contrast to the general rule that phaeomelanic sheep are pale. Hair sheep are one such group, and intensely phaeomelanic sheep do occur in these breeds. In wool breeds, intense phaeomelanic colour types are rare and usually fade rapidly from a dark birth coat. Red (usually called brown in the industry, but no doubt phaeomelanic and not eumelanic) Karakul sheep are valued for their birth coat, which can be intensely red brown (Nel, 1967). These red Karakuls do fade, though, so that the adult fleeces are generally very pale. In some red Karakul adults the fibres from primary follicles do maintain red pigment, but such sheep are rare even in the Karakul, which has been selected for centuries for a heavily pigmented birth coat (Sponenberg *et al.*, 1988). Karakuls of eumelanic types also fade considerably from the birth coat, so absolute statements about the pigmentation type and its relationship to fading are complicated in this breed.

A few wool breeds, such as the Portland and the American Tunis, retain a phaeomelanic birth coat. In these breeds the adult fleece is generally ivory or white, although the phaeomelanic colour is retained on the face, legs and ears, where the coat is hair instead of wool. For whatever reason, intensely phaeomelanic wool is a very rare phenomenon. This is in stark contrast to phaeomelanic colours of mohair from angora goats, in which selection rapidly produces darkly pigmented phaeomelanic colours (Sponenberg *et al.*, 1988).

Early work by Duck (1922) suggested that red in Karakuls is dominant to white and, while detailed interpretations may be overstepping the data, it is likely that the modifiers that change phaeomelanin from red to pale (white) have recessive tendencies. This could account for the paucity of red phenotypes in sheep, since selection has long ago removed the dominant intense modifiers from most wool-sheep gene pools.

Spotting patterns are of some potential use in the production of coloured wools, although these act to introduce white fibres or areas into coloured sheep. In some situations, the white spots may be desired since they provide for variability of colours in a single fleece. One problem with the use of the *spotted* allele, at least in the Jacob breed, is that the white areas and the coloured areas tend to grow wool with different characteristics. This negates the potential advantage of spotted sheep growing white and coloured wool of identical grade and texture, and most breeders of coloured sheep will therefore select away from spotting rather than toward it.

Single-gene Effects on Hair and Wool Structure

The hair coat of sheep has been drastically altered by selection following domestication. Many of the changes that modify hair to wool are genetically complex and are not due to single genes or combinations of a few genes. The genetic mechanisms used to derive many of the more useful coat types in sheep are poorly understood. In stark contrast to these are several documented single genes that have profound effects on the coat of sheep (Table 3.2). It is these that are discussed in this section.

Fibre Loss

One autosomal recessive trait is *Fibre Loss* (Dolling and Brooker, 1966). This has only been reported in Poll Dorset sheep in Australia. The homozygotes shed fibres from the hair-bearing skin *in utero* and then from the wool-bearing portions of skin throughout life.

Table 3.2. Loci and alleles affecting fibre quality and character.

Locus	Symbol	Allele	Symbol
Fibre Loss	*Fl*	*wild*	Fl^{+}
		loss	Fl^{l}
Halo Hair 1	*HH1*	*high halo hair T*	$HH1^{T}$
		high halo hair J	$HH1^{J}$
		high halo hair N	$HH1^{N}$
		wild	$HH1^{+}$
Halo Hair 2	*HH2*	*wild*	$HH2^{+}$
		high halo hair	$HH2^{nr}$
Lustrous Wool	*LW*	*lustrous*	LW^{L}
		wild	LW^{+}
Silky Wool	*SW*	*wild*	SW^{+}
		silky	SW^{s}

Lustrous Wool

A dominant gene at the *Lustrous Wool* locus causes Merino fleeces to lack crimp, have great lustre and be subject to felting (Short, 1958). This is a dramatic fleece defect, and should not be confused with the lustre normally associated with the wool of many long-wool breeds as a highly desired characteristic.

Silky Wool

An autosomal recessive allele at the *Silky Wool* locus has an action similar to the *lustrous* allele (Warwick *et al.*, 1960). This allele is found in Delaine Merino sheep. The primary follicles in homozygotes are degenerate.

Carpet wool

Genetic investigations surrounding the development of carpet-wool breeds from the Romney have revealed simple genetic mechanisms for changing the usual uniform fibre type of Romneys into the carpet-wool type fleece (Dry *et al.*, 1940). The carpet-wool fleece exhibits an increase of halo hairs in the lamb birth coat, and these loci go by the designation of *Halo Hair 1* (*HH1*) and *Halo Hair 2* (*HH2*). At the *HH1* locus are three dominant alleles, in addition to the *wild*-type allele.

The allele $HH1^N$ is incompletely dominant over the *wild* type (Dry, 1955a; Wickham and Rae, 1977). Homozygotes have more complete halo-hair coverage over the shoulder region compared with heterozygotes. The homozygotes are also heavily horned in both sexes, as compared with heterozygotes, in which only the rams are heavily horned. This effect is noted on a Romney-breed genetic background, which breed is uniformly polled in both sexes when lacking this allele.

The allele $HH1^T$ is completely dominant (Wickham and Rae, 1977). Both sexes are strongly horned, whether homozygous or heterozygous. The birth coat of heterozygotes and homozygotes is likewise indistinguishable and resembles that of $HH1^N$ homozygotes. The allele $HH1^J$ is very similar to the allele $HH1^T$.

At the *HH2* locus resides a recessive allele which has an effect similar to the dominant alleles at *HH1* (Dry, 1955b). Homozygotes for this allele are intermediate in type between animals heterozygous and homozygous for the allele $HH1^N$. Homozygous ewes are polled and most, but not all, rams are horned.

References

Adalsteinsson, S. (1970) Colour inheritance in Icelandic sheep and relation between colour, fertility, and fertilization. *Journal of Agricultural Research in Iceland* 2, 3–135.

Adalsteinsson, S. (1975) Depressed fertility in Icelandic sheep caused by a single colour gene. *Annales de Génétique et Sélection Animal* 7, 445–447.

Adalsteinsson, S. (1976) Occurrence and inheritance of tan colour in Icelandic sheep. *Journal of Agricultural Research in Iceland* 7, 55–62.

Adalsteinsson, S. (1977) Albinism in Icelandic sheep. *Journal of Heredity* 68, 347–349.

Adalsteinsson, S. (1978a) A note on an Icelandic albino ram. *Journal of Heredity* 69, 430.

Adalsteinsson, S. (1978b) Ræktun á dropóttu fé (Breeeding of piebald sheep with undercoat drops). *Radunautafundur* 6–10 February, 209–214.

Adalsteinsson, S. (1979) Genetics of breeding coloured sheep in Iceland and Scandinavia. In: *Breeding Coloured Sheep and Using Coloured Wool.* Peacock Publications, Hyde Park, South Australia, pp. 17–25.

Adalsteinsson, S. (1983) Inheritance of colours, fur characteristics and skin quality traits in North European sheep breeds: a review. *Livestock Production Science* 10, 555–567.

Adalsteinsson, S. (1991) Gene identification in sheep. In: Maijala, K. (ed.) *Genetic Resources of Pig, Sheep, and Goat.* Elsevier, Amsterdam, pp. 365–378.

Adalsteinsson, S., Lauvergne, J.J., Boyazoglu, J.G. and Ryder, M.L. (1978) A possible genetic interpretation of the colour variants in the fleece of the Gotland and Goth sheep. *Annales de Génétique et Sélection Animal* 10, 329–342.

Adalsteinsson, S., Sponenberg, D.P., Aliexeva, S, and Russel, A.J.F. (1994) Inheritance of goat colours. *Journal of Heredity* 85, 267–272.

Ademetz, L. (1917) Studies of mendelian heredity of main breed characteristics of Karakul sheep in purebreeding and in crossing with the Rambouillet. *Bibliotheca Genetica* 1, 1–258.

Alexieva, S.A., Césarini, J.P. and Lauvergne, J.J. (1985) Histologic study of melanogenesis in sheep with different variants of hairy cover pigmentation. In: *Technological and Breeding Requirements in Both Sheep and Cattle Rearing.* School for Specialists, Graders and Breeders, Varna, Agricultural Academy of Bulgaria, Sofia, pp. 85–87.

Aliev, G.A. and Rachkovsky, M.L. (1986) Interaction of A, E, Sb, and Ss gene alleles in formation of Sur colour in sheep. *Soviet Agricultural Sciences* 7, 43–47.

Aliev, G.A. and Rachkovsky, M.L. (1987a) *Genetic Studies of Pigmentation in Sheep.* Academy of Science of the Tajik SSR, Dushanbe, USSR, 200 pp.

Aliev, G.A. and Rachkovsky, M.L. (1987b) Genetic control of variation in the grey colour in the fleece of sheep. *Sel'skokhozyaistvennaya Biologiya* 3, 102–108.

Brooker, M.G. and Dolling, C.H.S. (1965) Pigmentation of sheep. I. Inheritance of pigmented wool in the Merino. *Australian Journal of Agricultural Research* 16, 219–228.

Brooker, M.G. and Dolling, C.H.S. (1969a) Pigmentation of sheep. II. The inheritance of coloured patterns in black Merino. *Australian Journal of Agricultural Research* 20, 387–394.

Brooker, M.G. and Dolling, C.H.S. (1969b) Pigmentation of sheep. III. Piebald patterns in Merinos. *Australian Journal of Agricultural Research* 20, 523–532.

COGNOSAG (1986) *Proceedings of COGNOSAG Workshop 1986.* Bureau des

Ressources Génétiques, Paris, Committee on Genetic Nomenclature of Sheep and Goats, Clamart, France, 112 pp.

COGNOSAG (1989) *Loci for Coat Colour of Sheep and Goats. Proceedings, COGNOSAG 1988 and 1989 Workshop, Manosque, France, and Eugene, USA.* Committee on Genetic Nomenclature of Sheep and Goats, Clamart, France, 84 pp.

COGNOSAG (1993) Unpublished recommendations from the 1993 COGNOSAG workshop, South Australia.

Constantinescu, G.K. (1932a) One lethal factor in sheep. In: *Proceedings of the International Congress on Genetics, Ithaca,* Vol. 2, p. 31.

Constantinescu, G.K. (1932b) New research on the heredity of grey colour in Turkana sheep. *Annals of the Institute of Natural Zootechnology in Rumania* 1, 15–29.

Constantinescu, G.K. (1933) Research on heredity of grey colour in Turkana sheep. *Bibliotheca Genetica* 8.

Constantinescu, G.K. (1939) The inheritance of black colour in the Rumanian sheep: the Turkana and Tsigai breeds. *Annals of the Institute of Natural Zootechnology in Rumania* 7, 9–28.

Contescu, D. (1938) Studies on the inheritance of white and black colour in Turkana breed. *Zeitschrift für Tierzuchtung und Züchtungsbiologie* 42, 124–127.

Contescu, D. and Epureanu (1939) On the inheritance of grey colour in Turkana sheep. *Zeitschrift für Tierzuchtung und Züchtungsbiologie* 44, 211–218.

Cook, O.M. (1950) Idaho man puts white Karakuls on the map. *Fur Farming Journal* 7, 6–7.

Davenport, C.B. (1905) The origin of black sheep in the flock. *Science* 22, 674.

Dolling, C.H.S. (1989) Sheep colour and genes revisited. *South Australian Coloured Sheep Owners' Society Newsletter* 75, 10–12.

Dolling, C.H.S. and Brooker, M.G. (1966) A viable hypotrichosis in Poll Dorset sheep. *Journal of Heredity* 57, 86–90.

Dry, F.W. (1924) The genetics of the Wensleydale breed of sheep. I. The occurrence of black lambs – an examination of flock records. *Journal of Genetics* 14, 203–218.

Dry, F.W. (1955a) The dominant N-gene in New Zealand Romney sheep. *Australian Journal of Agricultural Research* 6, 725–769.

Dry, F.W. (1955b) The recessive N-gene in New Zealand Romney sheep. *Australian Journal of Agricultural Research* 6, 833–862.

Dry, F.W., McMahon, P.R. and Sutherland, J.A. (1940) A mendelian situation in the birthcoat of the New Zealand lamb. *Nature* 145, 390–391.

Duck, R.W. (1921) Mendelism in fur sheep crosses. *Journal of Heredity* 12, 410–413.

Duck, R.W. (1922) Mendelism in fur sheep crosses II. The zygotic cause of red lambs when fur sheep are crossed on longwools or their grade offspring. *Journal of Heredity* 13, 63–68.

Dyrmundsson, O.R. and Adalsteinsson, S. (1980) Coat-colour gene suppresses sexual activity in Icelandic sheep. *Journal of Heredity* 71, 363–364.

Fleet, M.R. (1994) Control of pigmentation in white sheep – commercial relevance and mechanisms involved. In: Alderson, G.L.H. (ed.) *Proceedings of the Fourth World Congress on Coloured Sheep: Coloured Sheep and Coloured Fibres of Animal Origin.* Rare Breeds Survival Trust, Stoneleigh, pp. 25–36.

Fleet, M.R. and Forrest, J.W. (1984) The occurrence of spots of pigmented skin and pigmented wool fibres in adult Merino sheep. *Wool Technology and Sheep Breeding* 32(2), 83–90.

Fleet, M.R., Smith, D.H. and Stafford, J.E. (1989) Comparison between Merino sheep homozygous or heterozygous for white fleece on the incidence of isolated

pigmented wool fibres and types of non-fleece pigmentation. *Australian Journal of Agricultural Research* 41, 155–166.
Fleet, M.R., Smith, D.H. and Pourbeik, T. (1991a) Age-related changes in pigmentation traits in Merino sheep. *Wool Technology and Sheep Breeding* 39(1), 26–36.
Fleet, M.R., Pourbeik, T., Mortimer, S.I., Ponzoni, R.W. and Rogan, I.M. (1991b) Heritabilities and phenotypic correlations for pigmentation traits in Peppin Merino sheep. *Proceeding of the Australian Association of Animal Breeding and Genetics* 9, 400–403.
Fleet, M.R., Lush, B.J. and Hynd, P.I. (1993) Age-related changes in white wool-bearing skin melanocytes. *Wool Technology and Sheep Breeding* 41(2), 83–90.
Hoogschagen P., Adalsteinsson, S. and Lauvergne, J.J. (1978) A badger-face-like colour variant in Texel and in Dutch sheep in the Netherlands. *Annales de Génétique et Sélection Animal* 10, 517–523.
Jackson, I.J. (1994) Molecular and developmental genetics of mouse coat colour. *Annual Review of Genetics* 28, 189–217.
Kelly, R.B. and Shaw, H.E.B (1942) A note on the occurrence and inheritance of pigmented wool. *Journal of the Council of Science and Industrial Research in Australia* 15, 1.
Lauvergne, J.J. (1969) Inheritance of white colour of Berrichon sheep crossed with Solognot. *Annales de Génétique et Sélection Animal* 1, 219–226.
Lauvergne, J.J. (1975) Genetics of fleece colour in the French sheep breeds Sologne, Bizet, and Berrichon. *Annales de Génétique et Sélection Animal* 7, 263–276.
Lauvergne, J.J. (1976a) Compared fertility of E^+E^+ and E^DE^D ewes. *Annales de Génétique et Sélection Animal* 8, 47–49.
Lauvergne, J.J. (1976b) Fleece colour gene formulae for the German Mutton Merino and White Karaman. *Annales de Génétique et Sélection Animal* 8, 141–151.
Lauvergne, J.J. (1979) Homology of alleles at the Agouti locus in ruminants. *Annales de Dermatologie et Vénérologie* 106, 411.
Lauvergne, J.J. and Adalsteinsson, S. (1976) Genes for the colour of the fleece in the Corsican ewe. *Annales de Génétique et Sélection Animal* 8, 153–172.
Lauvergne, J.J., Danzart, M. and Burrill, M.J. (1981) Colour variants in Arles Merino breed II. Isolable factors in the progeny, their frequency in the breed. *Annales de Génétique et Sélection Animal* 13, 219–243.
Little, C.C. (1957) *The Inheritance of Coat Colour in Dogs.* Cornell University Press, Ithaca, New York.
Lundie, R.S. (1984) Colour patterns in New Zealand Sheep. In: Blair, H.T. (ed.) *Coloured Sheep and Their Products.* New Zealand Black and Coloured Sheep Breeders Association. Masterton, New Zealand, pp. 128–138.
Lydekker, R. (1912) *The Sheep and Its Cousins.* London.
Mahler, X. and Denis, B. (1990) Inheritance of coloured face and legs in Vendeen sheep. *Recueil de Médecine Vétérinaire* 166, 501–504.
Maule, J.P. (1977) Barbados Blackbelly sheep. *World Review of Animal Production* 24, 19–23.
Nel, J.A. (1967) Genetic studies in karakul sheep. *Annals of the University of Stellenbosch* 42(3), 153–317.
Novak, R.M. (1991) *Walker's Mammals of the World*, Vol. II, 5th edn. Johns Hopkins University Press, Baltimore, London.
Panfilova, E.P. (1939) Inheritance of piebaldness in black Karakul sheep. *Trudy Insitute of Genetics, Moscow* 13, 331–338.
Prota, G. (1992) *Melanins and Melanogenesis.* Academic Press, London.

Rendel, J. (1957) Heredity of colour and white designs in farm animals. *Kunglishe Hogskole Landbruksadamie Tidskrift* 96, 207–263.

Renieri, C., Lauvergne, J.J. and Antonini, M. (1995) Quantitative determination of melanins in black, brown, grey and red phenotypes of sheep, goat, and alpaca. In: *Ninth International Wool Textile Research Conference, Biella, Italy.*

Roberts, J.A.F. (1924) Colour inheritance in sheep. I. Black colour and badger-face pattern in Welsh Mountain sheep. *Journal of Genetics* 14, 367–374.

Roberts, J.A.F. (1926) Colour inheritance in sheep. II. The piebald pattern of the piebald breed. *Journal of Genetics* 17, 77–83.

Roberts, J.A.F. (1928) Colour inheritance in sheep. III. Face and leg colour. *Journal of Genetics* 19, 261–268.

Roberts, J.A.F. and White, R.G. (1930a) Colour inheritance in sheep. IV. White colour, recessive black colour, recessive brown colour, badger-face pattern and reversed badger-face pattern. *Journal of Genetics* 22, 165–180.

Roberts, J.A.F. and White, R.G. (1930b) Colour inheritance in sheep. V. Dominant black. *Journal of Genetics* 22, 181–190.

Rowlett, M.A. and Fleet, M.R. (1993) Albinism in a Suffolk sheep. *Journal of Heredity* 84, 67–69.

Ryder, M.L. (1984) Sheep. In: Mason, I.L. (ed.) *Evolution of Domesticated Animals.* Longman, London.

Ryder, M.L. and Adalsteinsson, S. (1987) A note on the differences in coat structure between the black and white areas of piebald lambs. *Journal of Agricultural Science, United Kingdom* 108, 379–382.

Ryder, M.L., Land, R.B. and Ditchburn, R. (1974) Coat colour inheritance in Soay, Orkney, and Shetland sheep. *Journal of Zoology* 173, 477–485.

Searle, A.G. (1968) *Comparative Genetics of Coat Colour in Mammals.* Logos Press, Academic Press, London.

Short, B.F. (1958) A dominant felting lustre mutant fleece type in the Australian Merino sheep. *Nature* 181, 1414.

Skårman, S. (1977) Coat colour changes in fur sheep. *Fårskötsel* 57(7), 6–9.

Sponenberg, D.P. and La Marsh, C. (1996). Dominant and recessive brown in goats. *Genetics, Selection, Evolution* 28, 117–120.

Sponenberg, D.P., Ito, S., Wakamatsu, K. and Eng, L. (1988) Pigment types in sheep, goats, and llamas. *Pigment Cell Research* 1, 414–418.

Sumner, R.M.W. (1980) Sheep breeding for coloured wool. In: *Farming Production Practices*, 373. Media Services, Wellington, New Zealand, 4 pp.

Sumner, R.M.W., Ito, S. and Wakamatsu, K. (1994) Using objective measurements to understand the histology of wool colour. In: Alderson, G.L.H. (ed.) *Proceedings: The Fourth World Congress on Coloured Sheep. Coloured Sheep and Coloured Fibres of Animal Origin.* Rare Breeds Survival Trust, Stoneleigh, pp. 47–56.

Vacher, M. (1905) Transmission of colour in animals. *Bulletin of the National Agricultural Society of France* 65, 76–83.

Vasin, B. (1928) *The Genetics of the Sheep. I. Inheritance of Colour and Pattern.* Publications of the Minister of Agriculture, Centre for Studies of Genetics in Domesticated Animals, Moscow, USSR.

Vsesolodov, E.B., Latypov, I.R., Martynov, Y.F. and Razozanev, K.M. (1981) Detection of sheep heterozygous for dominant white by electron paramagnetic resonance spectrometry. *Trudy Instituta Eksperimental'noi Biologii, Akademiya Nauk Kazakhskoi, SSR* 15, 61–73.

Warwick, B.L., Davis, S.P., Berry, R.O. and Patterson, R.E. (1960) 'Silky', a dominant

mutation in sheep. *Journal of Heredity* 51, 39–42.
Wickham, G.L. and Rae, A.L. (1977) Investigations of carpet wool genes in sheep. *Proceedings of the New Zealand Society of Animal Production* 37, 213.
Wood, T.B. (1905) Note on the inheritance of horns and face colour in sheep. *Journal of Agricultural Science* 1, 364–365.

Genetics of Morphological Traits and Inherited Disorders 4

F.W. NICHOLAS
Department of Animal Science, University of Sydney, NSW 2006, Australia

Introduction	87
The Range of Possibilities	88
Previous Reviews	89
Current Sources of Information	89
An Overview	90
Inherited Disorders	90
Conclusion	91
Acknowledgements	91
References	118

Introduction

Most of the morphological traits and inherited disorders that occur in sheep have not yet been documented. While this is a sobering thought, it also represents a major challenge for sheep breeders and researchers.

Among the few morphological traits that have been well documented in sheep, there are some that have proved to be extremely important in practical sheep production; some have even given rise to new breeds and new sheep industries.

Inherited disorders are also important. The need for knowledge of inherited disorders in sheep and other domesticated animal species is continually increasing, mainly because of the ever-increasing need for large-animal models of human inherited disorders. Although mice and other laboratory rodents have provided, and continue to provide, invaluable models of human inherited disorders, the small size of these animals imposes severe restrictions on their utility for clinical research. Species such as sheep do not have these restrictions. In addition to their utility as animal models, inherited disorders in sheep sometimes create a substantial challenge for individual breeders. Studs that have been closed for many decades, for example, sometimes encounter a frustratingly high incidence of an inherited disorder. More often than not, current knowledge about such disorders is so incomplete as to be practically useless; all that can be done is to provide breeders with general advice about how to decrease the incidence of inherited disorders (e.g. Nicholas, 1996; Chapter 11).

The major aim of this chapter is to provide a summary of current knowledge about morphological traits and inherited disorders in sheep. With the molecular revolution now in full swing and, in particular, with the development of deoxyribonucleic acid (DNA) markers covering all regions of all sheep chromosomes (see Chapter 11), knowledge of such traits and disorders in sheep will increase rapidly in the decades ahead. Since regularly updated information is available on the Internet, the other purpose of this chapter is to show readers where to find the latest information in an electronic form.

The Range of Possibilities

The spectrum of morphological traits and inherited disorders ranges from those that are definitely due to the action of just one gene to those that are due to the combined action of many genes and many non-genetic (environmental) factors. In between these two extremes are many traits and disorders which appear to run in families, but for which there is insufficient information to enable a conclusion to be drawn about whether one or more genes are involved. Unfortunately, the literature abounds with examples of traits and disorders that have been claimed to be due to just one gene, despite the data being so sparse that such a claim cannot be justified. Similar problems exist with claims of inheritance being recessive or dominant; in most cases, there is insufficient information to justify the claims that have been made. In the fullness of time, of course, additional data might support the initial claims. But we must be careful not to jump the gun.

This scarcity of reliable data on the inheritance of traits and disorders poses a challenge to those who are asked to compile lists of such traits – as required for this chapter. As will be seen below, the present author's solution has been to follow McKusick's (1997) well-established practice of providing a full list of all traits and disorders for which claims of inheritance have been made, using an asterisk to indicate those for which reasonable evidence of single-locus inheritance has been published.

During the 1980s, there have been considerable advances made in identifying the molecular basis of inherited disorders in domestic animal species. Unfortunately, at the time of writing, there are no sheep traits or disorders which have been documented at the DNA level, i.e. there are no sheep traits or disorders for which the nucleotide-sequence differences between normal and mutant alleles have been described. This situation is bound to change in the future.

Previous Reviews

There have been several comprehensive reviews of inherited traits in sheep published over the years. The first major summary of knowledge was by Roberts and Crew (1925). Three decades later, Rae (1956) provided a comprehensive review of all aspects of sheep genetics. The most recent reviews were published as several chapters in Dennis (1993). A full list of reviews is presented in the Reviews section of the References at the end of this chapter. It should be noted that many of these reviews are concerned with congenital traits and disorders, i.e. traits and disorders that are present at birth. Not all such traits and disorders are inherited.

Current Sources of Information

While a list of reviews is useful, it is even more useful to have a single catalogue of morphological traits and inherited disorders that is regularly updated, and which is made available both electronically (on the Internet) and in hard-cover (book) form on a regular basis. Human geneticists have long had access to such a resource – McKusick's *Mendelian Inheritance in Man* (MIM). Now in its twelfth edition as a book (McKusick, 1997) and accessible on the Internet as Online Mendelian Inheritance in Man (OMIM) at http://www3.ncbi.nlm.nih.gov:80/omim/, this catalogue contains a wealth of information on thousands of morphological traits and inherited disorders in humans. It also contains a surprising quantity of information on sheep, because McKusick has always been interested in potential animal models of human disorders. Indeed, when Montgomery *et al.* (1993) reported that the *Fecundity Booroola* locus is located in a region of the sheep genome that is homologous with human chromosome 4q, McKusick created an entry in the human catalogue entitled FECUNDITY GENE, BOOROOLA, OF SHEEP, HOMOLOG OF, to highlight the likelihood that a human homologue of this gene exists and will be discovered, sooner or later.

As described in Chapter 18 of the present book, the Committee on Genetic Nomenclature of Sheep and Goats (COGNOSAG) commenced assembling a catalogue of genes and inherited traits in sheep as soon as the committee came into existence in 1984. As also explained in that same chapter, COGNOSAG's catalogue exists in four categories, of which category 2 (loci for visible traits other than coat colour) is the most relevant to the present chapter. This catalogue was first published by COGNOSAG (1989) and an updated version is included in COGNOSAG (1996). An electronic version of this catalogue is accessible on the Internet at http://www. angis.su.oz.au. Starting in 1996, this catalogue is being updated after each annual COGNOSAG meeting. At the time of writing, category 2 of the COGNOSAG catalogue comprises 46 inherited traits. In a useful attempt to come to terms with the scarcity of knowledge available for so many of these traits, COGNOSAG has categorized the knowledge status of each trait according to whether the evidence for single-locus inheritance is 'beyond reasonable doubt' (status 1), 'not unequivocal' (status 2) or 'little or no evidence' (status 3). In their list of 46 traits, only ten have status 1, 19 have status 2, 14 have status 3 and three remain unclassified. Thus, considerable uncertainty exists for more than three-quarters of the traits in the COGNOSAG list.

In 1978, the present author commenced compiling a catalogue of inherited traits and disorders in a wide range of animal species. Being modelled on, and complementary to, McKusick's catalogue, this catalogue is called Mendelian Inheritance in Animals (MIA). It is accessible on the Internet as Online Mendelian Inheritance in Animals (OMIA) at http://morgan.angis su.oz au in the same format as OMIM, and at http://tetra.gig.usda.gov in a different format.

In OMIA are included entries for all inherited disorders in sheep, together with other sheep traits for which single-locus inheritance has been claimed, however dubiously. Each entry consists of a list of references arranged chronologically, so as to present a convenient history of knowledge about each disorder or trait. For some entries, there is additional information on inheritance or molecular genetics. If the disorder or trait has a human homologue, the relevant MIM numbers are included, providing a direct hyperlink to the relevant entry in McKusick's on-line catalogue OMIM.

An Overview

Table 4.1 (pp. 92–116) provides a list of sheep traits and disorders – a total of 139 – together with the earliest report plus the most recent reference, extracted from OMIA. A complete set of up-to-date references for each entry can be obtained by accessing OMIA on the Internet. Those traits and disorders for which there is reasonable evidence of single-locus inheritance (49 in total) are indicated by an asterisk.

Many of the traits and disorders are of only minor interest, having been the subject of one or a few poorly documented reports. Some have been discussed in detail in other chapters of the present book, e.g. coat colour and halo hair (carpet wool) in Chapter 3, blood groups in Chapter 5, spongiform encephalopathy (scrapie) in Chapter 9 and fecundity in Chapter 13.

Inherited Disorders

None of the inherited disorders that have been documented in sheep has caused major economic problems on the same scale as inherited disorders in, for example, the dairy-cattle industry. However, some disorders have caused considerable anguish in particular breeds.

Probably the most notable is spider-lamb syndrome or bent-limb syndrome, more formally known as chondrodysplasia, which occurs in Suffolks and Hampshires. The main distinguishing feature is long, spindly and distorted limbs, due to abnormal cartilage growth (Nakano *et al.*, 1994). Initially occurring only in the USA, the mutant was imported into Australia in the late 1980s (Phillips *et al.*, 1993).

By far the greatest interest in inherited sheep disorders arises from their potential as models for human disorders. Table 4.2 lists the substantial number of sheep traits that fit into this category. Of particular interest to medical research (both veterinary and human) are those disorders for which the faulty peptide is known (Table 4.3). Since most of the corresponding genes have been cloned in at least one species of mammal, it is only a matter of time before the molecular basis of many of the sheep disorders listed in Table 4.3 is worked out.

Sheep are attractive as animal models because they are sufficiently large to be amenable to surgery (cf. mice) and yet are not so large as to involve prohibitive expense (cf. cattle). Because of these attractions, Tebbutt *et al.* (1995) have embarked on a novel search for a sheep model for cystic fibrosis. The disorder has not been documented in sheep, but these workers have studied the sheep homologue of one of the main cystic fibrosis genes in humans, and are now busy screening flocks of sheep, searching for mutant alleles at this locus. Once such alleles are found, their effects will then be studied in detail by observing the results of planned matings. This novel approach is quite likely to bear fruit, especially in a species like sheep, where the occasional abnormal animal in a commercial flock is rarely noticed.

Conclusion

The list of inherited morphological traits and disorders presented in this chapter provides an indication of the range of such traits and disorders that have been observed and studied in sheep. The molecular and gene-mapping revolutions now under way will lead to an explosion of knowledge in this area in the years ahead. To exploit fully the genetic variation that does occur, breeders and researchers need to be continually on the lookout for unusual animals, saving them where possible. If DNA can be sampled from several generations of a family in which a particular trait or disorder occurs and if careful records on the occurrence of the trait or disorder in that family have been kept, it will be an increasingly straightforward matter to identify the gene responsible.

Acknowledgements

The author is grateful to Philippa Beard for her very helpful contribution to the entries on hyperbilirubinaemia and mesangiocapillary glomerulonephritis, to Chris Moran for valuable comments on the first draft of this chapter and to Phil Sponenberg for going through the whole chapter with a fine-tooth comb.

Table 4.1. A list of inherited traits and disorders in sheep (from Online Mendelian Inheritance in Animals). An asterisk indicates reasonable evidence for single-locus inheritance. (Clinical descriptions drawn substantially from Blood and Studdert, 1988.)

Inherited trait or disorder	Brief description	Earliest reference; latest reference (if different from earliest reference)
Adactyly	Congenital absence of claws (cleats)	Leipold *et al.* (1972)
Agnathia	Congenital absence of the lower jaw	McFarland and Deniz (1964); Dennis and Leipold (1972a)
Agnathia (type-3 otocephaly)	Congenital absence of the lower jaw, with the ears united below the face	Glahn-Luft *et al.* (1978)
Alopecia	Lack of wool	Lazovskii (1983)
Anencephaly	Congenital absence of the cranial vault, with the cerebral hemispheres completely missing or reduced to small masses	Dennis and Leipold (1972d)
Aprosopia	Partial or complete absence of the face	Dennis and Leipold (1972b)
*Argininaemia	Deficiency of the enzyme arginase in red blood cells. The enzyme's normal task is to convert arginine to ornithine and urea. Homozygotes for the deficiency have zero levels of the enzyme; heterozygotes appear to have somewhat lower levels than homozygote normals (Wright *et al.*, 1977). In contrast to humans, deficient sheep show no clinical signs, despite having elevated levels of arginine in red blood cells	Tucker *et al.* (1977); Wright *et al.* (1977)
Arthrogryposis	Persistent flexion of a joint	Roberts (1929); Parsonson *et al.* (1977)
Ataxia	Failure of muscle coordination	Binns *et al.* (1972); Hill *et al.* (1993)
Ataxia, progressive, with degenerative thoracic myelopathy	Progressive failure of muscle coordination, associated with degeneration of the thoracic region of the spinal cord	Harper *et al.* (1991)
Atresia ani	Congenital absence of the anus, causing a build-up of faeces and consequent distension of the abdomen	Fischer and Adinata (1957); Dennis and Leipold (1972c)

Table 4.1. Continued

Inherited trait or disorder	Brief description	Earliest reference; latest reference (if different from earliest reference)
Atresia ani vaginalis	In females, congenital absence of the anus is often accompanied by a rectal–vaginal fistula (tube-like passage between the rectum and vagina), allowing urine and faeces to be voided through the vulvar orifice	Swartz *et al.* (1985)
Atresia ilei	Congenital absence of the ileum (the distal portion of the small intestine, extending from the jejunum to the caecum)	Dennis and Seddon (1971)
Atresia jejuni	Congenital absence of the jejunum (that part of the small intestine extending from the duodenum to the ileum)	Leipold and Dennis (1973)
Birth coat	Adult production of a fleece typical of newborn lambs (consisting of hairy (medullated) fibres produced almost entirely by primary follicles; in contrast to normal adult wool, which has no medulla, and which is produced by secondary follicles and modified primary follicles)	Schinckel (1951); Schinckel (1955)
Blindness, congenital	Congenital blindness	Zwiep (1958)
Blood-group system, general	Each blood-group system consists of a set of blood types, each of which corresponds to a particular antigen (usually a glycoprotein) on the surface of red blood cells. The different types within a system are the result of the action of different alleles at a locus that usually encodes an enzyme that catalyses the creation of the unique feature of the glycoprotein	Evans and Phillipson (1958); Nguyen *et al.* (1992)
*Blood-group system A	This locus has at least three alleles	Rasmusen *et al.* (1960); Casati and Carroll (1982)
*Blood-group system B	This locus has at least 52 alleles. Appears to be analogous to the B system in cattle	Rasmusen (1960); Casati and Carroll (1982)
*Blood-group system C	This locus has at least four alleles. Appears to be analogous to the C system in cattle	Rasmusen *et al.* (1960); Nguyen (1985)

continued overleaf

Table 4.1. Continued

Inherited trait or disorder	Brief description	Earliest reference; latest reference (if different from earliest reference)
*Blood-group system D	This locus has at least two alleles	Rasmusen *et al.* (1960); Casati and Caroll (1982)
*Blood-group system F30	This locus has at least two alleles	Nguyen and Bunch (1980); Nguyen *et al.* (1992)
*Blood-group system F41	This locus has at least two alleles	Nguyen and Bunch (1980)
*Blood-group system I	This locus has at least two alleles, which control the expression of the R blood-group system (see below). Also called the secretor locus	Rasmusen (1966); Nguyen (1985)
*Blood-group system M	This locus has at least four alleles. The Mb factor of this system (previously called the L factor) appears to be an inhibitor of potassium transport; in fact, this locus may be identical with the potassium-transport locus (see below)	Rasmusen *et al.* (1960); Casati and Caroll (1982)
*Blood-group system R	This locus has at least two alleles. Gene expression at this locus is controlled by alleles at the locus determining blood-group system I (see above)	Nguyen and Bunch (1980); Casati and Caroll (1982)
*Blood-group system X	This locus has at least two alleles	Rasmusen (1958)
Brachygnathia	Congenital abnormal shortness of the mandible (lower jaw), resulting in protrusion of the maxilla (upper jaw). Also called overshot jaw or parrot mouth	Hamori (1959); Singh *et al.* (1970)
Brachygnathia superior	Congenital abnormal shortness of the maxilla (upper jaw), resulting in protrusion of the mandible (lower jaw). Also called monkey mouth	Luft (1972a); Luft (1972b)
Cataract	Opacity of the lens of the eye	Brooks *et al.* (1982)
Cerebellar cortical atrophy	Wasting (decrease) of the outer layer of the grey matter of the cerebellum (the part of the brain situated behind the two hemispheres and above the pons and medulla oblongata)	Innes *et al.* (1949); Terlecki *et al.* (1978)
*Ceroid lipofuscinosis, general	A disorder characterized by blindness, mental dullness and abnormal behaviour, caused by abnormal storage	Jolly and West (1976); Westlake *et al.* (1995)

Table 4.1. Continued

Inherited trait or disorder	Brief description	Earliest reference; latest reference (if different from earliest reference)
	(build-up) of lipofuscins, which are fatty pigments (lipopigments or ceroid bodies) formed by the solution of a pigment in fat. The brain atrophies, showing signs of extensive accumulation of ceroid bodies in nerve tissue and later in viscera. The locus for this disorder in sheep has not yet been identified. Possibly a good animal model for Batten's disease in humans	
*Ceroid lipofuscinosis, juvenile onset	A juvenile form of ceroid lipofuscinosis, which differs from all other forms reported in sheep, in that the build-up of lipofuscins is confined to nervous tissue. Reported in a flock of Rambouillet sheep in the USA by Edwards *et al.* (1994).	Edwards *et al.* (1994); Jolly (1995)
*Chondrodysplasia	Abnormal growth of cartilage, leading to disproportionate dwarfism. Also called spider syndrome	Hidiroglou *et al.* (1978); West *et al.* (1995)
Cleft palate	Congenital fissure (split) that involves the hard or soft palate (roof of the mouth). Also called palatoschisis	Binns *et al.* (1972)
*Coat colour, general	Determined by several loci, which are discussed in detail in Chapter 3 of the present book	Roberts (1926); Fleet *et al.* (1995)
*Coat colour, albinism	Congenital lack of pigment in all parts of the body. Due to a non-functional form of the enzyme tyrosinase	Adalsteinsson (1977); Rowett and Fleet (1993)
Cryptorchidism	Failure of one or both testes to descend	Wendt *et al.* (1960); Keel *et al.* (1994)
Cystic renal dysplasia	Congenital occurrence of small, misshapen kidneys	Jones *et al.* (1990)
Dandy–Walker syndrome	Congenital hydrocephalus (enlargement of the cranium due to accumulation of fluid; 'water on the brain') due to obstruction of the foramina of Magendie and Luschka (openings within the brain)	Linklater (1994); Pritchard *et al.* (1994)

continued overleaf

Table 4.1. Continued

Inherited trait or disorder	Brief description	Earliest reference; latest reference (if different from earliest reference)
Doggy wool	Short, coarse, glossy wool of nondescript type	Balevska and Tanev (1964); Tanev (1964)
Dwarfism, general	Abnormally small animals	Wray *et al.* (1971)
*Dwarfism, ancon	The first recorded case of this type of dwarfirm was a ram lamb born on a farm in Massachusetts in 1791 (Humphreys, 1813). An entertaining history of the mutation (on which the following account is largely based) has been presented by Schwartz and Vogel (1994). The farmer, Seth Wight (or Wright or Wite), kept the ram and, by mating him to related ewes, produced several more dwarfs, eventually giving rise to a new breed of short-legged sheep, which were quite popular in New England, because they were unable to jump over the low stone fences. The results of the initial matings and the true-breeding nature of the breed suggested that this form of dwarfism was an autosomal recessive trait. The main distinguishing characteristics of the mutant were short and crooked limbs and loose articulation. When the animals walked, their crooked forelimbs looked like elbows, which led a Dr Shattuck, who dissected a specimen in Boston, to suggest the name 'Ancon', from the Greek word for elbow (Humphreys, 1813). Humphreys suggested the name 'Otter', because of the 'real or imaginary resemblance to that animal . . . by some supposed to have been caused by an unnatural intercourse' (i.e. by a mating between an otter and a sheep!). Darwin (1859, 1868) was impressed with the way in which offspring of ancons mated to non-dwarf breeds 'perfectly resemble either parent' rather than 'being intermediate in character', citing this as evidence against blending inheritance. Despite its supposed advantages, the	Humphreys (1813); Schwartz and Vogel (1994)

Table 4.1. Continued

Inherited trait or disorder	Brief description	Earliest reference; latest reference (if different from earliest reference)
	breed appears to have become extinct by the end of the nineteenth century. A similar mutation occurred in 1919 in a Cheviot flock in Norway (Wriedt, 1925). Descendants of this mutant were exported to the Storrs Agricultural Experiment Station at the University of Connecticut, where they were studied by Landauer and Chang (1949) and Landauer (1950). Yet another similar mutation occurred in a merino flock at the Livestock and Forage Research Center at McGregor, Texas, in 1962 (Shelton, 1968). Unfortunately, both of these mutants also became extinct	
Ears, short	Short ears	Djorbineva *et al.* (1985)
Ectrodactyly	Congenital absence of all or part of a claw (cleat)	Ramadan (1993)
*Ehlers–Danlos syndrome	Congenital syndrome characterized by joint hyperextensibility, hyperelasticity and fragility of the skin and poor healing of wounds. Also called dermatosparaxis or cutaneous asthenia	Mauch *et al.* (1988)
*Ehlers–Danlos syndrome, type VII	Although evidence is incomplete, it appears that this disorder in sheep is due to a mutation in the gene for the enzyme procollagen I amino proteinase, this being the enzyme responsible for removing 'surplus' amino acids from the N-terminal end of procollagen-I molecules	Helle and Nes (1972); Lapiere and Nusgens (1993)
Entropion	Inversion (turning inward) of the margin of the eyelid	Littlejohn (1954); Sakul *et al.* (1996)
*Epidermolysis bullosa, dystrophic	Blisters on the outer layer of the skin, especially (in sheep) in the mouth and on skin where no wool grows. Bruckner-Tuderman *et al.* (1991) showed that the occurrence of an autosomal recessive form of this disorder in a flock of Weiss	Bruckner-Tuderman *et al.* (1991)

continued overleaf

Table 4.1. Continued

Inherited trait or disorder	Brief description	Earliest reference; latest reference (if different from earliest reference)
	Alpenschaf sheep from Switzerland was due to a deficiency of collagen VII.	
Epidermolysis bullosa, junctionalis	Blisters in and around the mouth and on the limbs, leading in the latter case to separation of the hoof from the corium (dermis), resulting in inability to stand. Also called red-foot disease	Alley *et al.* (1974); McTaggart *et al.* (1974)
Epididymal aplasia	Total or partial failure of development of the epididymis	Ladds *et al.* (1990)
*Epitheliogenesis imperfecta	Congenital absence of the skin. Also called aplasia cutis	Munday (1970); Alani *et al.* (1995)
Fecundity, general	Ability to produce offspring in relatively large numbers. Also called prolificacy	Rendel (1956); Fahmy (1996)
*Fecundity, Booroola	On their farm called 'Booroola' near Cooma in New South Wales, Australia, the Seears brothers developed a prolific strain of merino, with an average litter size of more than two, in contrast to the usual litter size of merinos, which is less than one. Sheep from this flock were studied by the Commonwealth Scientific and Industrial Research Organisation (CSIRO) (Helen Turner, Laurie Piper, Bernie Bindon and colleagues), and it was eventually shown that the prolificacy of the Booroola strain was due to a single gene of large effect, which acts additively for both ovulation rate and litter size. The difference between homozygotes is approximately 2.7–3.0 in ovulation rate and 1.1–1.7 in litter size. The genetic and physiological basis of the Booroola's prolificacy was also studied extensively by New Zealand researchers, who provided one of the first examples of the practical application of gene mapping in sheep, by mapping the Booroola gene to chromosome 6 (Montgomery *et al.*, 1993, 1994)	Turner *et al.* (1962); Heath *et al.* (1996)
*Fecundity, Cambridge	A segregation analysis of data from a flock of 549 Cambridge ewes maintained	Owen *et al.* (1990); Fahmy (1996)

Table 4.1. Continued

Inherited trait or disorder	Brief description	Earliest reference; latest reference (if different from earliest reference)
	at the University of Wales, Bangor, indicated the presence of a gene of major effect on ovulation rate (number of ova shed), with a difference in performance between homozygotes ranging from 2.14 to 3.44 (Owen *et al.*, 1990)	
*Fecundity, Iceland	A segregation analysis of Icelandic sheep by Jonmundsson and Adalsteinsson (1985) provided substantial evidence for a gene of large effect on litter size. These authors named the gene Thoka, after the ewe from whom the gene appears to have originated. Heterozygotes give birth to an extra 0.64 lambs, on average	Jonmundsson and Adalsteinsson (1985); Fahmy (1996)
*Fecundity, Inverdale	The only X-linked fecundity gene was discovered in offspring from a prolific Romney ewe that had been screened from a commercial flock as part of the foundation of a prolific experimental flock in New Zealand in 1979. Heterozygous ewes ovulate approximately one extra egg and give birth to approximately 0.6 extra lambs. Homozygous ewes have 'streak' ovaries: underdeveloped, non-functional, flattened structures about one-eighth the size of normal ovaries. Males with this gene have normal fertility	Davis *et al.* (1991); Fahmy (1996)
*Fecundity, Java	Analysis of reproductive data from three breeds of Indonesian sheep (Javanese Thin Tail, Javanese Fat Tail, Semarang) produced substantial evidence for the presence of a gene of large effect on ovulation rate and litter size in all three breeds (Bradford *et al.*, 1986). The difference between homozygotes was estimated to be 2.6 for ovulation rate and 1.8–2.0 for litter size. Since all three breeds are thought to have intermated	Bradford *et al.* (1986); Fahmy (1996)

continued overleaf

Table 4.1. Continued

Inherited trait or disorder	Brief description	Earliest reference; latest reference (if different from earliest reference)
	from time to time, it was not considered surprising that the same gene apparently occurs in all three breeds	
*Fragile site	A site on a chromosome that does not stain, at which a break in the chromosome often occurs. In cultured cells, chromosomal breakage at fragile sites can be induced by the addition of caffeine, aphidicolon or bromodeoxyuridine to the culture medium. In humans, fragile sites are sometimes associated with tandem repeats of three nucleotides (triplet microsatellites), which, if the number of repeats increases, can cause inherited disorders. To date, no examples of inherited disorders associated with fragile sites have been documented in domesticated animals	Matejka *et al.* (1990)
*Gangliosidosis, GM1	A lysosomal storage disease in which there is a build-up (storage) of GM1 gangliosides (a type of glycolipid) in various tissues, due to the lack of the enzyme β-galactosidase, whose task is to break down the GM1 ganglioside into its constituents. Characterized by progressive neuromuscular dysfunction and impaired growth from an early age	Ahern-Rindell *et al.* (1986); Skelly *et al.* (1995)
*Gaucher disease, type I	A lysosomal storage disease in which there is a build-up (storage) of glucocerebroside in various tissues, due to the lack of the enzyme β-glucocerebrosidase, whose task is to break down glucocerebroside into its constituent molecules. The build-up is associated with large foamy macrophages (Gaucher's cells)	Laws and Saal (1968)
*Glycogen storage disease II	A lysosomal storage disease in which there is a build-up (storage) of glycogen, due to the lack of the enzyme α-glucosidase, whose task is to break	Manktelow and Hartley (1975); Reuser (1993)

Table 4.1. Continued

Inherited trait or disorder	Brief description	Earliest reference; latest reference (if different from earliest reference)
	down glycogen into its constituent glucose molecules. Characterized by poor growth, incoordination, muscle weakness and eventual recumbency. Also called Pompe's disease	
*Goitre, familial	Enlargement of the thyroid gland, causing a swelling in the front of the neck. Mayo and Mulhearn (1969) provided extensive evidence supporting autosomal recessive inheritance of this disorder in South Australian Merinos. Normal thyroglobulin is absent from affected sheep, but seems to be replaced by thyroglobulin-like, albumin-like and immunoglobulin-like iodoproteins, with the latter two most probably arising from outside the gland (Dolling and Good, 1976). The following details are taken from Mayo and Mulhearn (1969). Both lobes of the thyroid gland are enlarged and ovoid, usually in the range 1" to 3" (2.5 cm to 7.6 cm) along their longest axes. Other abnormalities often occur: the area surrounding the nostrils is flattened dorsoventrally, and the transverse folds of skin ('nose wrinkle') over the nasal bones and immediately behind the nostrils are very pronounced and harsh; ears are oedematous and tend to flop, thereby giving greater prominence to the poll; carpal joints are swollen, and legs – particularly the forelegs – are bowed either inwards or outwards, resulting in oblique plantar surfaces of the hooves; some animals are low at the withers. Affected lambs are generally very weak and are often highly sensitive to the effects of harsher environments. Mortality is very high in the first year of life. Wool seems to be normal, except for	Watson *et al.* (1962); Nasseri and Prasad (1990)

continued overleaf

Table 4.1. Continued

Inherited trait or disorder	Brief description	Earliest reference; latest reference (if different from earliest reference)
	birth coats, which tend to be 'silky' (lustrous). At post-mortem, the weight of the thyroid gland varied from greater than 5 g up to 222 g, compared with a maximum of less than 5 g in normal sheep. In cattle and goats, familial goitre has been shown to be due to nonsense mutations in the thyroglobulin gene	
Golden-ram effect	In a large-scale investigation of the quantitative genetics of resistance and resilience to *Haemonchus contortus* in merino sheep, Albers *et al.* (1987) observed one 'outlier' sire whose progeny exhibited extremely high resistance. The evidence justified a tentative conclusion that this sire was carrying a major gene for resistance. Unfortunately, such a gene failed to materialize in subsequent studies	Albers *et al.* (1987)
Gonadal hypoplasia	Incomplete development or underdevelopment of the gonads (testes or ovaries)	Wright *et al.* (1996)
Haemolytic anaemia	Anaemia caused by the increased destruction of red blood cells (erythrocytes)	Hotchkiss (1994)
*Haemophilia A	Impaired coagulability of the blood, with a consequent strong tendency to bleed, due to a deficiency of the clotting factor VIII. Since the gene for this factor is located on the X chromosome, the disorder is X-linked. Also called classic haemophilia	Neuenschwander and Pliska (1990); Neuenschwander and Pliska (1994)
*Halo hair 1	Normal wool fibres are solid. In contrast, hair has a core (medulla) down the centre, filled with air. Sheep that are bred for wool have almost zero medullated (hairy) fibres. In 1931, a purebred Romney Marsh ram lamb with an exceptionally hairy fleece was born in New Zealand, on a property belonging to a Mr Nielson. It was given to Professor	Dry *et al.* (1940); Sides and Banks (1987)

Table 4.1. Continued

Inherited trait or disorder	Brief description	Earliest reference; latest reference (if different from earliest reference)
	Dry at what is now Massey University, where a set of experimental matings showed that the extreme hairiness is due to an incompletely dominant allele at an autosomal locus (Dry *et al.*, 1940; Dry, 1955a, Dry, 1955b). Homozygotes have approximately 65% by weight of medullated fibres. This is no use for traditional wool production, but it is ideal for carpet manufacture. In the following decades, the mutant allele was multiplied within a Romney background, giving rise to a new 'carpet-wool' breed: the Drysdale. Other carpet-wool alleles have appeared in New Zealand and in Australia, also eventually giving rise to new breeds: the Tukidale and the Carpetmaster in New Zealand and the Elliotdale in Tasmania. The New Zealand mutants appear to be alleles at a single locus, but there is some evidence that the Tasmanian mutation is at a different locus (Sides and Banks, 1987). The most comprehensive review of carpet-wool genetics is provided by Wickham (1978).	
Heart defect, congenital	A range of congenital heart defects have been reported in sheep. They include ventricular septal defect, tetralogy of Fallot and the emergence of both the cranial and caudal vena cava and pulmonary veins from the right atrium	Dennis and Leipold (1968b); Anderson *et al.* (1990)
Hemimelia	Congenital absence of all or part of the distal part of a limb	Ramadan (1993)
Hepatic encephalopathy	A degenerative disease of the brain induced by a failure of the liver to metabolize ammonia. Clinical signs include excitability, tremor, compulsive walking, head pressing and apparent blindness, followed by coma and convulsions	Clark and Spence (1994)

continued overleaf

Table 4.1. Continued

Inherited trait or disorder	Brief description	Earliest reference; latest reference (if different from earliest reference)
Hernia, general	Protrusion of part of an organ or tissue through the structures normally containing it	Dennis and Leipold (1968a)
Hernia, inguinal	Protrusion of part of the intestine through the abdominal wall	Carr (1972)
Hernia, umbilical	Protrusion of part of the intestine or other abdominal organs through the abdominal wall at the umbilicus, the protrusion being covered with skin and subcutaneous tissue	Adamu *et al.* (1991)
*Horns	There is substantial variation in the extent of horn growth, making classification often difficult. However, in general, the presence or absence of horns can be attributed to the action of two alleles at an autosomal locus, with the polled condition being dominant to horned. A comprehensive review has been provided by Dolling (1970), while Tibau i Font *et al.* (1982) have prepared a bibliography. The actual gene involved in this trait in sheep has not yet been identified, but a recent discovery in goats provides a clue as to its location: Vaiman *et al.* (1996) have discovered that the horns locus in goats is located in the distal region of goat chromosome 1. Since this chromosome of goats is homologous to one of the arms of sheep chromosome 1 (Hansen, 1973), it is possible that the horns gene in sheep is located on sheep chromosome 1. Interestingly, the horns gene in cattle is located on cattle chromosome 1 (Georges *et al.*, 1993; Brenneman *et al.*, 1996), which is homologous to goat chromosome 1 (Hansen, 1973). However, the locations of the bovine and caprine horns genes are not homologous (Vaiman *et al.*, 1996)	Dolling (1956); Tibau i Font *et al.* (1982)

Table 4.1. Continued

Inherited trait or disorder	Brief description	Earliest reference; latest reference (if different from earliest reference)
Hydrocephalus	Enlargement of the cranium caused by accumulation of fluid. Also called 'water on the brain'	Miller (1993)
Hydrops foetalis	Accumulation of fluid in the whole body of the newborn	Dassonville (1989)
*Hyperbilirubinaemia I	Bilirubin is an orange bile pigment produced by the breakdown of haem and reduction of biliverdinin. An excess of bilirubin in the blood is called hyperbilirubinaemia I or Gilbert's syndrome. The disorder, which results from a defect in uptake of bilirubin by the liver, has been reported in Southdown sheep in New Zealand and in California. The primary clinical sign is photosensitive facial dermatitis, due to an associated defect in liver uptake of phylloerythrin, a metabolite of chlorophyll. If not protected from sunlight, affected lambs die within 2–3 weeks after they begin to eat grass. If the lambs are protected from sunlight, they eventually die within 5 years from renal failure. Inheritance is autosomal recessive in sheep. In humans this disorder is due to a deficiency of the enzyme bilirubin glucuronosyltransferase, but such a deficiency has not yet been demonstrated in sheep	Hancock (1950); Cuppage *et al.* (1979)
*Hyperbilirubinaemia II	Bilirubin often occurs conjugated with glucuronic acid. An excess of conjugated bilirubin in the blood is called hyperbilirubinaemia II or Dublin–Johnson syndrome. This disorder is autosomal recessive in Corriedale sheep. As with hyperbilirubinaemia I, there is also an excess of phylloerythrin in the blood. Clinical signs include photodynamic dermatitis, predominantly around the eyes and on the ears, due to the	Arias *et al.* (1964); Kitamura *et al.* (1992)

continued overleaf

Table 4.1. Continued

Inherited trait or disorder	Brief description	Earliest reference; latest reference (if different from earliest reference)
	increased phylloerythrin concentration. Progressive anorexia and emaciation eventually lead to death within 6 months. At post-mortem, the livers are dark brown to black. On histopathology, brown-black granules are evident in hepatocytes and Kupffer's cells. The cause of this disorder has not yet been determined	
Hypomyelinogenesis, congenital	Congenital deficiency of myelin, especially in the cerebellum and brain stem; includes failure of formation of myelin, plus incomplete and delayed myelination of axons. Clinical signs include inability to rise and severe muscle tremor with periods of spasticity	Markson *et al.* (1959); Patterson and Sweasey (1970)
Hypoplasia of mammary gland, congenital	Incomplete development or underdevelopment of the mammary gland	Minchev (1976)
Hypospadias	Developmental abnormality in which the urethra of males opens on the underside of the penis or on the perineum	Gilanpour (1971)
Hypothyroidism	Deficiency of thyroid-gland activity	Nasseri and Prasad (1990)
Hypotrichosis	Absence of all or part of the normal wool coat	Nel (1964); Mackie and McIntyre (1992)
Infertility	Inability to conceive	Jakobiec (1952); Ayalon (1964)
Intersex	A mixture of male and female characteristics	Brandsch (1964); Fayrerhosken *et al.* (1992)
Iris mottling	Mottling of the iris	Vasiliev and Aripov (1983)
Knobbed acrosome	The presence of a protrusion (knob) on the acrosome (cap) of a sperm. Associated with sterility. No evidence for inheritance in sheep, but a similar disorder appears to be inherited in cattle and pigs	Savage (1984)

Table 4.1. Continued

Inherited trait or disorder	Brief description	Earliest reference; latest reference (if different from earliest reference)
*Krabbe disease	A lysosomal storage disease in which there is a build-up (storage) of myelin, due to the lack of the enzyme galactosylceramidase (galactocerebrosidase), whose task is to break down myelin into its constituents. Characterized by weakness progressing to paralysis. Also called globoid-cell leucodystrophy	Pritchard *et al.* (1980)
Lethal, general	Inherited lethality	Groenewald (1993a); Groenewald (1993b)
Lethal grey	Lethality associated with grey fleece	Nel and Louw (1953); Vasiliev and Aripov (1983)
Light tips, length of	Length of light-coloured region of the tip of the staple	Sagdullaev (1964)
*Lustrous wool	Lustrous, light yellow wool, lacking crimp, occurring in merinos. This trait is characteristic of breeds such as Lincoln and Leicester, but is not normally present in merinos. Short (1958) presented convincing evidence of the autosomal dominant inheritance of this trait in merinos. Several independent occurrences of very similar mutations in Border Leicester–Romney crosses and in Romney flocks in New Zealand were reported by Blair (1990). Lustrous wool is very similar to, and could be the same as, silky wool (see below)	Short (1958); Blair (1990)
*Mannosidosis, alpha	A lysosomal storage disease in which there is a build-up (storage) of mannose-rich compounds, due to the lack of the enzyme α-mannosidase, whose task is to cleave mannose from such compounds. Clinical signs include ataxia, head tremor, aggression and finally paralysis and death	Stegelmeier *et al.* (1995)
Megaoesophagus	Dilatation and atony of the body of the oesophagus, usually associated with	Braun *et al.* (1990)

continued overleaf

Table 4.1. Continued

Inherited trait or disorder	Brief description	Earliest reference; latest reference (if different from earliest reference)
	asynchronous function of the oesophagus and the caudal oesophageal sphincter. Causes regurgitation and aspiration pneumonia	
Mesangiocapillary glomerulonephritis, type I	A chronic, slowly progressive inflammation of the capillary loops in the glomeruli of the kidney, in which the glomeruli are enlarged as a result of proliferation of mesangial cells and irregular thickening of the capillary walls, which narrows the capillary lumina. Abbreviated to MCGN. The type I variant of this disorder (MCGN I) is characterized by a double-contour appearance of the capillary walls. The disorder becomes clinically apparent when lambs are 1–3 months old. Affected animals die after a brief clinical episode, which includes anorexia and signs of central nervous system disturbance. At post-mortem, the kidneys are dramatically enlarged, with pale spotted cortices. Glomeruli show a thickening of the capillary walls, proliferation of mesangial cells and matrix and formation of epithelial crescents. There may be focal haemorrhage and oedema evident on histopathological examination of the brain. The damage to the kidneys and brain are thought to be due to deposition of immune complexes (immunoglobulin A (IgA), IgM, IgG) and complement. Affected lambs are born with a severe deficiency of the third component of complement, but the relevance of this is not known. MCGM I is strongly familial in Finnish Landrace purebred and crossbred flocks, but a single-locus pattern of inheritance has not yet been determined. This ovine disorder appears to provide a good model for the same	Thomson *et al.* (1973); Frelier *et al.* (1990)

Table 4.1. Continued

Inherited trait or disorder	Brief description	Earliest reference; latest reference (if different from earliest reference)
	disorder in humans, for which X-linkage has been claimed but not proved	
Microcephaly	Small size of the head in relation to the rest of the body	Hartley *et al.* (1974); Hartley and Haughey (1974)
Microphthalmia	Abnormal smallness in all dimensions of one or both eyes	Hanset (1961); Jackson (1990)
*Muscular dystrophy	Degenerative muscle disorder, leading to muscle weakness, liberation of myoglobin into the circulation from skeletal muscle and subsequent wasting and possible contracture	Morley (1954); Nakada *et al.* (1990)
Muscular hyperplasia	Abnormal increase in muscular tissue caused by the formation and growth of new, normal muscle cells (cf. muscular hypertrophy)	Dennis (1972a)
*Muscular hypertrophy	Abnormal increase in muscular tissue caused entirely by enlargement of existing cells (cf. muscular hyperplasia). Muscular hypertrophy was first reported in sheep only recently, by Cockett *et al.* (1994), who showed that the locus at which this potentially advantageous mutation occurred is on chromosome 18. In ways that are not yet understood, the mutant allele leads to increased protein synthesis and decreased protein degradation. The practical effects include higher dressing percentage, reduced fat thickness and reduced marbling, while carcass weight is not affected (Koohmaraie *et al.*, 1995, 1996; Carpenter *et al.*, 1996). Also called callipyge	Cockett *et al.* (1994); Koohmaraie *et al.* (1996)
Myofibrillar hypoplasia	Congenital defect of the hindlimbs that prevents standing, because the hindlimbs are splayed. Also called splay-leg	Swatland (1974)
Nephropathy	Disease of the kidneys	Angus (1990)

continued overleaf

Table 4.1. Continued

Inherited trait or disorder	Brief description	Earliest reference; latest reference (if different from earliest reference)
Neuroaxonal dystrophy	Progressive ataxia, hypermetria, proprioceptive deficits, head incoordination and tremors	Cordy *et al.* (1967); Harper and Morton (1991)
Neurogenic atrophy	Wasting of a muscle due to damage to its peripheral nerve supply	Thornton and MacColl (1985)
Neuropathy, thalamic–cerebellar	A chronic locomotor dysfunction characterized by mild to moderately severe pelvic limb paresis, thoracic-limb hypermetria, fine mild head tremor, wide-based stance, uncoordinated gait, muscle asynergy, disturbed equilibrium and a consequent tendency to fall over (Bourke *et al.*, 1993). Post-mortem reveals nerve-cell lesions in many thalamic neurons and many cerebellum Purkinje cells, together with fibre-tract lesions in the thalamus, cerebellar peduncles, cerebellar folia, pons, medula oblongata (ventral tracts) and spinal cord (ventral and lateral columns at all levels). On the combined basis of clinical signs and neuropathology, this disorder can be distinguished from segmental axonopathy, cerebellar axonopathy, neuroaxonal dystrophy and degenerative thoracic myelopathy. Indirect evidence suggests a genetic aetiology	Bourke *et al.* (1993)
*Nucleoside transport defect	Defect in the transport of nucleosides (purine or pyrimidine base attached to a ribose or deoxyribose sugar) across erythrocyte (red blood cell) membranes. In sheep, defective nucleoside transport is autosomal recessive (Jarvis and Young, 1978). No clinical signs have been reported in defective sheep	McManus and Lambe (1972); Young (1983)
Ocular squamous carcinoma	Malignant new growth of squamous epithelial cells in the eye. Also called cancer eye	Walde *et al.* (1991)
Ophthalmia	Severe inflammation of the eye or of the conjunctiva or deeper structures of the eye	Teodoreanu (1958); Teodoreanu *et al.* (1964)

Table 4.1. Continued

Inherited trait or disorder	Brief description	Earliest reference; latest reference (if different from earliest reference)
Osteochondrosis	Abnormal differentiation of growth cartilage. Also called dyschondroplasia	Doherty *et al.* (1996)
Osteogenesis imperfecta	Abnormally brittle bones that fracture easily	Kater *et al.* (1963); Arthur *et al.* (1992)
Otognathia	Congenital presence of a rudimentary accessory mouth at the base of the aural pinna (projecting part of the ear)	Dennis (1970)
Patent urachus	The urachus is a canal present in the fetus, connecting the bladder with the allantois via the navel. It normally disappears after birth. A patent urachus is one that persists after birth, allowing urine to drip out of the bladder through the umbilicus	Dennis (1969)
Polyceraty	Having more horns than the usual two; multihorned. Henson (1981) provided evidence suggesting that the presence of four horns is an autosomal recessive trait	Atmadilaga and Asikin (1962); Henson (1981)
*Potassium transport	An autosomal polymorphism of high potassium and low sodium (HK) versus low potassium and high sodium (LK) in red blood cells has been long recognized in sheep, with LK being dominant or incompletely dominant to HK (Evans and King, 1955). Despite many searches in many populations, there is no convincing evidence of associations between this polymorphism and productivity (Ellory and Tucker, 1983). There is, however, quite strong evidence that this polymorphism is actually the M blood-group (see above), with the L factor (now called Mb) at this locus being an inhibitor of potassium transport into the cell (Bell, 1983)	Evans and King (1955); Xu *et al.* (1993)
Progressive retinal atrophy	Progressive, non-inflammatory degeneration or dysplasia of rods and cones or just of cones, leading first to night-blindness and later to blindness during the day. Also called bright-blindness	Watson *et al.* (1972)

continued overleaf

Table 4.1. Continued

Inherited trait or disorder	Brief description	Earliest reference; latest reference (if different from earliest reference)
*Reduced glutathione deficiency due to GCS deficiency	Glutathione is a small peptide consisting of just three amino acids: glutamic acid, cysteine and glycine. It is a widely distributed molecule, serving as a reducing agent in many different metabolic processes. One of its most notable roles is in protecting red blood cells from oxidation and haemolysis. Glutathione normally exists in its reduced sulphydryl form (GSH) (often called reduced glutathione). Its oxidized form consists of two GSH molecules whose cysteines are joined by a disulphide bridge (symbolized GSSG). Unlike most peptides, glutathione is not the product of a gene; instead, it is manufactured from its three constituent amino acids in a two-step process, catalysed by γ-glutamylcysteine synthetase (GCS) (which creates a dipeptide of glutamic acid and cysteine) and glutathione synthetase (GSHS) (which adds glycine). Oxidation of GSH to GSSG is catalysed by glutathione peroxide (GPX); reduction back to GSH is catalysed by glutathione reductase (GR). In merino sheep, a deficiency of GCS in red blood cells results in a deficiency of erythrocyte GSH but (unlike the situation in humans) no detectable haemolytic anaemia or any other clinical signs, either chronically or after exposure to oxidant drugs. The deficiency is due to an allele at an autosomal locus, but there are conflicting data on whether the allele is recessive or dominant (Board and Agar, 1983)	Tucker and Kilgour (1972); Bayon *et al.* (1994)
*Reduced glutathione deficiency due to an amino acid transport defect	In Finn and Awassi sheep, a different form of erythrocyte GSH deficiency is inherited in an autosomal codominant manner (Young, 1983). In this disorder, all aspects of glutathione metabolism function normally. But a defect in the	Tucker and Kilgour (1970); Young (1983)

Table 4.1. Continued

Inherited trait or disorder	Brief description	Earliest reference; latest reference (if different from earliest reference)
	transport of cysteine across the erythrocyte membrane results in a deficiency of this vital component of GSH and hence in a deficiency of GSH	
Renal dysplasia and bladder aplasia–hypoplasia	Congenital occurrence of small, misshapen kidneys and bladder	O'Toole *et al.* (1993)
*Response to noradrenaline	Approximately one-half of the heat produced in response to exposure to low temperatures (cold exposure) arises from non-shivering thermogenesis, in which exposure to cold stimulates production of endogenous noradrenaline (NA), which in turn stimulates brown adipose tissue to produce heat from the oxidation of free fatty acids derived from the triglycerides in fatty tissue. While investigating the genetic basis of resistance to cold in lambs, Simpson and Slee (1988) measured the extent of non-shivering thermogenesis in newborn lambs injected subcutaneously with a standard dose of NA. Offspring of one sire failed to show any non-shivering thermogenesis. Subsequent investigations of planned matings showed that this lack of response to NA is due to a single gene, which is most probably dominant	Simpson and Slee (1988)
Roll curls, length of	Analysis of the results of matings of karakul sheep suggested that length of the roll curl is partly under genetic control (Djjackov and Cehirinadze, 1965).	Djjackov and Cehirinadze (1965)
Schistosomus reflexus	A fetus in which the edges of the somatic disc in the developing embryo have curved upwards instead of downwards, resulting in the viscera floating free in the amnion. The head and the tail are curved up towards each other. Dystocia is the usual result	Dennis and Meyer (1965); Dennis (1972c)

continued overleaf

Table 4.1. Continued

Inherited trait or disorder	Brief description	Earliest reference; latest reference (if different from earliest reference)
*Silky wool	An autosomal dominant mutation that occurred in a Delaine Merino flock in west Texas in the early 1930s. Newborn lambs lacked the usual halo hairs, had smooth wavy wool and no skin folds, giving a slick, shiny appearance (hence the term 'silky'). Adult fleeces lacked the usual crimp, having instead locks resembling flat locks of mohair with large bold waves. Adult sheep showed a marked tendency to partly shed their wool, resulting in extreme felting of many fibres before shearing. The wool was extremely greasy. Viability of the mutant animals was so poor that the mutation had become extinct by 1960. Silky wool is very similar to, and could be the same as, lustrous wool (see above)	Warwick *et al.* (1960)
Spina bifida	A congenital defect characterized by defective closure of the two halves of the vertebral arch, through which the spinal cord and meninges may or may not protrude	Adalsteinsson and Basrur (1984); Davies (1993)
Split eyelid	Crosses among various rare breeds of sheep (mainly Jacob's sheep and Hebridean) suggested that the presence of a split in the eyelid appears to be associated with the presence of four horns (Henson, 1981). See also polyceraty	Henson (1981)
*Spongiform encephalopathy	Spongiform encephalopathies are a class of fatal neurological diseases. The ovine disease, scrapie, has been known for centuries. Clinical signs are characteristic of a progressive degeneration of the central nervous system; they include pruritus, abnormalities of gait, recumbency and scraping against fence posts (hence the name). Death is inevitable. At post-mortem, brain histology shows a characteristic spongy appearance. The	Parry (1960); Hosie and Dawson (1996)

Table 4.1. Continued

Inherited trait or disorder	Brief description	Earliest reference; latest reference (if different from earliest reference)
	infectious agent is a modified form of a protein encoded by a gene in the host. The name given to this infectious particle is prion. The host gene is called the prion protein (PrP) gene, which is a normal part of the genome of mammals and chickens. Its polypeptide product, called PrP^{c}, is a naturally occurring protein attached to the outer surface of neurons and some other cells. The protein PrP^{c} appears to play a role in maintaining the Purkinje cells of the cerebellum, which are essential for balance and muscular function. The infectious agent, called PrP^{Sc}, is a modified form of PrP^{c}, where the modifications involve glycosylation and the creation of intrastrand disulphide bonds, i.e. there is no change in amino acid sequence. When PrP^{Sc} molecules enter a previously uninfected host, they convert the naturally occurring PrP^{c} molecules, produced by the host gene, into infectious PrP^{Sc} particles, which ultimately cause clinical signs in that animal and which can spread to other animals, both horizontally (by infection) and vertically (by maternal transmission). In sheep, the PrP gene is about 20 kb long, consisting of three exons encoding a peptide comprising approximately 250 amino acids. Extensive polymorphism has been discovered in codons 136, 154 and 171, giving rise to amino acid substitutions at these sites. There is substantial evidence that at least some of this polymorphism is strongly associated with susceptibility/resistance to scrapie. The most likely explanation for this association is that some amino acid sequences are able to form themselves into PrP^{Sc}, without the	

continued overleaf

Table 4.1. Continued

Inherited trait or disorder	Brief description	Earliest reference; latest reference (if different from earliest reference)
	intervention of other PrP^{Sc} particles. Although having been studied for decades, scrapie has been the centre of intense scientific and political attention since 1988, when bovine spongiform encephalopathy (BSE or Mad Cow disease) was recognized as a new disease, apparently having arisen from the practice of feeding sheep offal to cattle.	
Swayback	Incoordination progressing to paralysis, blindness and death, associated with a deficiency of copper, which causes dysmyelination of the nerve cells in the brain and spinal cord. There is substantial genetic variation in ability to withstand a copper deficiency and hence in resistance to swayback (Wiener, 1966)	Wiener (1966); Kavanagh *et al.* (1972)
Syndactyly	Fusion of the claws (cleats)	Dennis and Leipold (1970); Binns *et al.* (1972)
Tail abnormalities	A range of congenital tail defects have been reported, including anury (absence of the tail), wrytail (twisted or kinked tail) and polyury (multiple tails)	Dennis (1972b)
Tail, short	Also called brachyury. There is hope that this trait may provide a form of natural resistance to the sheep blowfly	James *et al.* (1991)
Teat number	Alexander Graeme Bell (1898), of telephone fame, developed by selection a strain of sheep in which most ewes had two to four additional teats	Bell (1898)
Thread defect	A crimp defect characterized by tight spiral curling at the ends of the staple, associated with apparent fineness of the fibres, reported in Polish Merinos by Radomska and Kozanecki (1964)	Radomska and Kozanecki (1964)
Twinning, conjoined	Congenital disorder in which twins are partly joined; 'Siamese' twins	Dennis (1972c); Hiraga and Dennis (1993)

Table 4.1. Continued

Inherited trait or disorder	Brief description	Earliest reference; latest reference (if different from earliest reference)
Urolithiasis	Formation of calculi ('stones') in the urinary tract, or the condition associated with the presence of such calculi	Cornelius (1963); Braun *et al.* (1992)
Ventricular septal defect	A congenital heart defect characterized by persistent patency (openness) of the ventricular septum, permitting flow of blood directly between ventricles, bypassing the pulmonary circulation and resulting in various degrees of cyanosis (blue discoloration of the skin) due to oxygen deficiency. Clinical signs include systolic murmur and a palpable thrill on both sides of the chest, dyspnoea and poor tolerance of exercise	Boucek *et al.* (1989)
Wattles	Appendages suspended from the head. They are fusiform (spindle-shaped), fleshy masses covered with normal skin and suspended usually from the mandibular area. Also called tassels	Lancaster and Medwell (1991)
Wilms' tumour	A rapidly developing malignant mixed tumour of the kidneys. The disorder has not actually been reported in sheep, but the gene responsible for the disorder has been identified in sheep. The Wilms Tumour (WT) gene in human and mouse was cloned in the early 1990s. Since its peptide is involved in the development of the urinary–genital tract and the gonads, the gene is of considerable interest in relation to the genetic basis of sexual development. In 1995, Vaiman *et al.* designed conserved primers from mouse and human sequence data, and successfully amplified a 150-bp fragment of the homologous gene in sheep, cattle and pig, which includes a dinucleotide (TG) microsatellite. Exploiting the polymorphism of the microsatellite, they were able to map the WT gene to cattle chromosome 15, between the genes for follicle stimulating factor, β-polypeptide	Vaiman *et al.* (1995)

continued overleaf

Table 4.1. Continued

Inherited trait or disorder	Brief description	Earliest reference; latest reference (if different from earliest reference)
	(FSHB) and neutral cell adhesion molecule (NCAM), and possibly very near the gene for haemoglobin β (HBB). Using a somatic-cell hybrid panel, they also showed that the WT gene in sheep is syntenic with sheep chromosome 15. Future studies of this gene should shed interesting light on sexual development, and hence on the occurrence of abnormalities of sexual development, in sheep and other domestic animal species	
Wilson disease	A disorder of copper metabolism, due to a deficiency of ceruloplasmin, which forms a complex with copper. The excess copper is deposited in the brain (causing mental retardation) or the liver (causing jaundice and cirrhosis)	Wiener (1987)
*Yellow fat	The presence of three xanthophylls (lutein, lutein-5,6-epoxide and flavoxanthin) in sheep fat, presumably due to the inability to oxidize xanthophylls. Inherited as a single-locus, autosomal recessive trait	Hill (1962); Baker *et al.* (1985)

References

Adalsteinsson, S. (1977) Albinism in Icelandic sheep. *Journal of Heredity* 68, 347–349.

Adalsteinsson, S. and Basrur, P.K. (1984) Inheritance of spina bifida in Icelandic lambs. *Journal of Heredity* 75, 378–382.

Adamu, S.S., Mohammed, A. and Sivachelvan, M.N. (1991) Umbilical hernia with adhesion and rupture of the abomasum in a lamb. *Veterinary Record* 128, 187.

Ahern-Rindell, A.J., Prieur, D.J. and Murnane, R.D. (1986) An inherited lysosomal storage disease of sheep associated with deficiencies of beta-galactosidase and alpha-neuraminidase (abstract). *American Journal of Human Genetics* 39, A3.

Alani, F.K., Aldarraji, A.M. and Hamad, R.M. (1995) Epitheliogenesis imperfecta in a flock of Awassi sheep in Iraq. *Small Ruminant Research* 17, 283–285.

Albers, G.A.A., Gray, G.D., Piper, L.R., Barker, J.S.F., Le Jambre, L.F. and Barger, I.A. (1987) The genetics of resistance and resilience to *Haemonchus contortus* infection in young Merino sheep. *International Journal for Parasitology* 17, 1355–1363.

Table 4.2. Potential sheep models of human disorders.

	MIM number
Agnathia	202650
Alopecia	104110, 104130, 147770, 104100, 104000, 203600, 203650, 203655
Arginaemia	207800
Ataxia	108500, 108600, 108650, 208750, 301835
Cataract	116200, 212600, 212700
Ceroid lipofuscinosis	204200, 204300, 204500, 204600
Ceroid lipofuscinosis, juvenile onset	204200, 204300, 204500, 204600
Cleft palate	119530, 201180, 258320, 129810, 225000, 303400, 119540, 129400, 260150, 119580, 106260, 119550, 120433, 119300, 129900, 119500, 179400, 119570, 129830, 216100, 216300, 218090, 164220, 231060, 241850, 242840, 244300, 255995, 114300, 181180, 261800, 268300, 277170, 301815, 106250, 600331, 600460
Coat colour, albinism	203100
Cryptorchidism	100100, 219050, 314300
Ehlers–Danlos syndrome	130000, 130010, 130020, 130050, 130060, 130070, 130080, 130090, 225320, 128180, 225350, 225360, 305200, 225400, 153454, 120160, 225410, 120150, 130080, 225310, 147900, 229200
Ehlers–Danlos syndrome, type VII	225410
Epidermolysis bullosa, dystrophic	226600, 120120
Epitheliogenesis imperfecta	107600, 107601, 136500, 207700, 207731, 302803, 600268, 600360
Fecundity, Booroola	134720
Gangliosidosis, GM1	230500, 230600, 230650
Gaucher disease, type I	230800
Glycogen storage disease II	232300
Goitre, familial	188450, 138790, 138800, 274600, 274500
Haemophilia A	306700
Hernia, inguinal	142350, 245550
Hydrops foetalis	215140, 236750
Hyperbilirubinaemia I	143500, 191740, 218800
Hyperbilirubinaemia II	237500

Table 4.2. Continued

	MIM number
Hypospadias	145410, 146450, 164220, 176305, 179250, 201181, 139710, 241750, 241760, 264600
Hypothyroidism	225050, 225250, 241850, 243800, 275120
Hypotrichosis	146520, 146550, 250460, 211370, 146530, 183849, 246500, 241900, 278200, 600077
Krabbe disease	245200
Mannosidosis, alpha	248500
Mesangiocapillary glomerulonephritis, type I	305800
Osteogenesis imperfecta	120150, 120160, 166260, 259770, 259420, 259440, 166240, 259400, 259410, 166210, 166200, 259450, 166220, 166230, 166260
Reduced gluthathionine deficiency	230450
Spongiform encephalopathy	176640
Teat number	163700
Wilms' tumour	194070, 194071, 194072, 194080, 194090
Wilson disease	277900

Table 4.3. Sheep disorders with a known peptide.

Arginaemia	Arginase
Coat colour, albinism	Tyrosinase
Ehlers–Danlos syndrome	Procollagen
Ehlers–Danlos syndrome, type VII	Procollagen I amino proteinase
Epidermolysis bullosa, dystrophic	Collagen, type VII
Gangliosidosis, GMI	Galactosidase, beta
Gaucher disease, type I	Glucocerebrosidase
Glycogen storage disease II	Glucosidase, alpha-1, 4
Goitre, familial	Thyroglobulin
Haemophilia A	Factor VIII
Krabbe disease	Galactosylceramidase
Mannosidosis, alpha	Mannosidase, alpha
Reduced glutathione deficiency due to GCS deficiency	Gamma-glutamylcysteine synthetase
Spongiform encephalopathy	Prion protein

Alley, M.R., O'Hara, P.J. and Middelberg, A. (1974) An epidermolysis bullosa of sheep. *New Zealand Veterinary Journal* 22, 55–59.

Anderson, W.I., George, C., Delahunta, A., King, J.M. and Smith, M.C. (1990) Pulmonary hamartoma and multiple cardiac anomalies in a Finn–Dorset cross lamb. *New Zealand Veterinary Journal* 38, 32–33.

Angus, K.W. (1990) Nephropathy in young lambs. *Veterinary Record* 126, 525–528.

Arias, I., Bernstein, L., Toffler, R., Cornelius, C., Novikoff, A.B. and Essner, E. (1964) Black liver disease in Corriedale sheep : a new mutation affecting hepatic excretory function. *Journal of Clinical Investigation* 43, 1249–1250.

Arthur, D.G., Thompson, K.G. and Swarbrick, P. (1992) Lethal osteogenesis imperfecta and skin fragility in newborn New-Zealand Romney lambs. *New Zealand Veterinary Journal* 40, 112–116.

Atmadilaga, D. and Asikin (1962) Some observations on the inheritance of polyceraty in the Priangan sheep. *Communicationes Veterinariae* 6, 63–68.

Ayalon, N. (1964) 'Sterilitas sine materia' – sterility not clinically diagnosable. In: *Proceedings of the Fifth International Congress on Animal Reproduction and Artificial Insemination, Trento* Vol. V. pp. 47–80.

Baker, R.L., Steine, T., Vabeno, A.W. and Breines, D. (1985) The inheritance and incidence of yellow fat in Norwegian sheep. *Acta Agriculturae Scandinavica* 35, 389–397.

Balevska, R.K. and Tanev, D.N. (1964) [Changes with age in the wool characters of lambs with various amounts of doggy wool at birth]. *Nauchni Trudove. Vissh Selskostopanski Institut 'Georgi Dimitrov'. Zootekhnicheski Fakultet* 14, 73–86.

Bayon, Y., Arranz, J.J. and Sanprimitivo, F. (1994) Red cell reduced glutathione concentrations in Spanish Churra sheep. *Animal Genetics* 25, 277–279.

Bell, A.G. (1898) On the development by selection of supernumerary mammae in sheep. *Science* 9, 6337–6339.

Bell, K. (1983) The blood groups of domestic mammals. In: Agar, N.S. and Board, P.G. (eds) *Red Blood Cells of Domestic Animals*. Elsevier, Amsterdam, pp. 133–164.

Binns, W., Keller, R.F. and Balls, L.D. (1972) Congenital deformities in lambs, calves and goats resulting from maternal ingestion of *Veratrum californicum*: hare lip, cleft palate, ataxia, and hypoplasia of metacarpal and metatarsal bones. *Clinical Toxicology* 5, 253–269.

Blair, H.T. (1990) Inheritance of a major gene for excessively lustrous wool in sheep. *Journal of Heredity* 81, 220–222.

Blood, D.C. and Studdert, V.P. (1988) *Baillière's Comprehensive Veterinary Dictionary*. Baillière Tindall, London, 1124 pp.

Board, P.G. and Agar, N.S. (1983) Glutathione metabolism in erythrocytes. In: Agar, N.S. and Board, P.G. (eds) *Red Blood Cells of Domestic Animals*. Elsevier, Amsterdam, pp. 253–270.

Boucek, M.M., Chang, R. and Synhorst, D.P. (1989) Renin–angiotensin-II response to the hemodynamic pathology of ovines with ventricular septal defect. *Circulation Research* 64, 524–531.

Bourke, C.A., Carrigan, M.J. and Dent, C.H.R. (1993) Chronic locomotor dysfunction, associated with a thalamic–cerebellar neuropathy, in Australian Merino sheep. *Australian Veterinary Journal* 70, 232–233.

Bradford, G.E., Quirke, J.F., Sitorus, P., Inounu, I., Tiesnamurti, B., Bell, F.L., Fletcher, I.C. and Torell, D.T. (1986) Reproduction in Javanese sheep: evidence for a gene with large effect on ovulation rate and litter size. *Journal of Animal Science* 63, 418–431.

Brandsch, H. (1964) [Comparative investigations on the inheritance of cryptorchidism and intersexuality in domestic animals]. *Kuhn-Archiv. Arbeiten aus der Landwirtschaftlichen Fakultat der Martin Luther-Universitat Halle-Wittenberg* 77, 323–425.

Braun, U., Steiger, R., Fluckiger, M., Bearth, G. and Guscetti, F. (1990) Regurgitation due to megaesophagus in a ram. *Canadian Veterinary Journal – Revue Vétérinaire Canadienne* 31, 391–392.

Braun, U., Schefer, U. and Fohn, J. (1992) Urinary tract ultrasonography in normal rams and in rams with obstructive urolithiasis. *Canadian Veterinary Journal – Revue Vétérinaire Canadienne* 33, 654–659.

Brenneman, R.A., Davis, S.K., Sanders, J.O., Burns, B.M., Wheeler, T.C., Turner, J.W. and Taylor, J.F. (1996) The polled locus maps to BTA1 in a *Bos indicus* × *Bos taurus* cross. *Journal of Heredity* 87, 156–161.

Brooks, H.V., Jolly, R.D., West, D.M. and Bruere, A.N. (1982) An inherited cataract in New Zealand Romney sheep. *New Zealand Veterinary Journal* 30, 113–114.

Bruckner-Tuderman, L., Guscetti, F. and Ehrensperger, F. (1991) Animal model for dermolytic mechanobullous disease – sheep with recessive dystrophic epidermolysis bullosa lack collagen-VII. *Journal of Investigative Dermatology* 96, 452–458.

Carpenter, E., Rice, O.D., Cockett, N.E. and Snowder, G.D. (1996) Histology and composition of muscles from normal and callipyge lambs. *Journal of Animal Science* 74, 388–393.

Carr, P.M. (1972) An apparently inherited inguinal hernia in the Merino ram. *Australian Veterinary Journal* 48, 126–127.

Casati, M.Z. and Caroll, A. (1982) [Immunogenetic characterisation of a population of Varese sheep]. *Atti della Societa Italiana delle Scienze Veterinarie* 35, 571–572.

Clark, A.M. and Spence, J.A. (1994) Possible familial hepatic encephalopathy in a Suffolk ewe. *Veterinary Record* 135, 463.

Cockett, N.E., Jackson, S.P., Shay, T.L., Nielsen, D., Moore, S.S., Steele, M.R., Barendse, W., Green, R.D. and Georges, M. (1994) Chromosomal localization of the callipyge gene in sheep (*Ovis aries*) using bovine DNA markers. *Proceedings of the National Academy of Sciences of the USA* 91, 3019–3023.

COGNOSAG (1989) *Standardised Genetic Nomenclature for Sheep and Goats 1987: Loci for Visible Traits other than Colour and Blood and Milk Polymorphisms.* (Proceedings of the COGNOSAG Workshop, 1987.) Lavoisier, Paris.

COGNOSAG (1996) *Mendelian Inheritance in Sheep.* University of Camerino, Matalica, Italy.

Cordy, D.R., Richards, W.P.C. and Bradford, G.E. (1967) Systemic neuroaxonal dystrophy in Suffolk sheep. *Acta Neuropathologia* 8, 133–140.

Cornelius, C.E. (1963) Studies on ovine urinery biocolloids and phosphatic calculosis. *Annals of the New York Academy of Sciences* 104, 638–657.

Cuppage, F.E., McGavin, M.D. and Strafuss, A.C. (1979) Animal model: Gilbert's syndrome in Southdown sheep. *American Journal of Pathology* 97, 207–210.

Darwin, C.R. (1859) *On the Origin of Species by Means of Natural Selection.* John Murray, London.

Darwin, C.R. (1868) *The Variation of Animals and Plants under Domestication.* John Murray, London.

Dassonville, J.A. (1989) Hydrops-foetalis in a mutton Merino ewe. *Journal of the South African Veterinary Association – Tydskrif van die Suid-Afrikaanse Veterinere Vereniging* 60, 174–175.

Davies, I.H. (1993) Spina bifida in lambs. *Veterinary Record* 132, 90–91.

Davis, G.H., McEwan, J.C., Fennessy, P.F., Dodds, K.G. and Farquhar, P.A. (1991)

Evidence for the presence of a major gene influencing ovulation rate on the X-chromosome of sheep. *Biology of Reproduction* 44, 620–624.
Dennis, S.M. (1969) Patent urachus in a neonatal lamb. *Cornell Veterinarian* 59, 581–584.
Dennis, S.M. (1970) Otognathia in a neonatal lamb. *American Journal of Veterinary Research* 31, 203–204.
Dennis, S.M. (1972a) Congenital muscular hyperplasia in a lamb. *Cornell Veterinarian* 62, 263–268.
Dennis, S.M. (1972b) Congenital tail defects in lambs. *Cornell Veterinarian* 62, 568–572.
Dennis, S.M. (1972c) Schistosomus reflexus in cojoined twin lambs. *Veterinary Record* 90, 509–510.
Dennis, S.M. and Leipold, H.W. (1968a) Congenital hernias in sheep. *Journal of the American Veterinary Medical Association* 152, 999–1003.
Dennis, S.M. and Leipold, H.W. (1968b) Congenital cardiac defects in lambs. *American Journal of Veterinary Research* 29, 2337–2340.
Dennis, S.M. and Leipold, H.W. (1970) Syndactylism in a neonatal lamb. *Cornell Veterinarian* 60, 23–27.
Dennis, S.M. and Leipold, H.W. (1972a) Agnathia in sheep: external observations. *American Journal of Veterinary Research* 33, 339–347.
Dennis, S.M. and Leipold, H.W. (1972b) Aprosopia (facelessness) in lambs. *Veterinary Record* 90, 365–367.
Dennis, S.M. and Leipold, H.W. (1972c) Atresia ani in sheep. *Veterinary Record* 91, 219–222.
Dennis, S.M. and Leipold, H.W. (1972d) Anencephaly in sheep. *Cornell Veterinarian* 62, 273–281.
Dennis, S.M. and Meyer, E.P. (1965) Schistosomus reflexus in a sheep. *Veterinary Record* 77, 1386–1388.
Dennis, S.M. and Seddon, H.D. (1971) Atresia ilei in a lamb. *Veterinary Record* 89, 254.
Djjackov, I.N. and Cehirinadze, T.B. (1965) [The inheritance of the length of roll curls in Karakul sheep]. *Ovtsevodstvo* 11, 14–16.
Djorbineva, M.K., Aleksieva, S.A. and Lauvergne, J.J. (1985) [Hereditary shortening of the external ear in Karakachan sheep of Bulgaria. Brief report (1)]. *Recueil de Médecine Vétérinaire* 161, 57–58.
Doherty, M.L., McAllister, H., Rackard, S., Skelly, C. and Bassett, H.F. (1996) Osteochondrosis in a pedigree Suffolk ram. *Veterinary Record* 138, 137–138.
Dolling, C.E. and Good, B.F. (1976). Congenital goitre in sheep: isolation of the iodoproteins which replace thyroglobulin. *Journal of Endocrinology* 71, 179–192.
Dolling, C.H.S. (1956) The inheritance of horns in sheep. *Proceedings of the Australian Society of Animal Production* 1, 161–164.
Dolling, C.H.S. (1970) *Breeding Merinos.* Rigby, Adelaide.
Dry, F.W. (1955a) Multifactorial inheritance of halo hair abundance in New Zealand Romney sheep. *Australian Journal of Agricultural Research* 6, 608–623.
Dry, F.W. (1955b) The dominant N gene in New Zealand Romney sheep. *Australian Journal of Agricultural Research* 6, 725–769.
Dry, F.W., McMahon, P.R. and Sutherland, J.A. (1940) A Mendelian situation in the birthcoat of the New Zealand Romney lamb. *Nature* 145, 390–391.
Edwards, J.F., Storts, R.W., Joyce, J.R., Shelton, J.M. and Menzies, C.S. (1994) Juvenile-onset neuronal ceroid-lipofuscinosis in Rambouillet sheep. *Veterinary Pathology* 31, 48–54.
Ellory, J.C. and Tucker, E.M. (1983) Cation transport in red blood cells. In: Agar, N.S.

and Board, P.G. (eds) *Red Blood Cells of Domestic Animals*. Elsevier, Amsterdam, pp. 291–314.

Evans, J.V. and King, J.W.B. (1955) Genetic control of sodium and potassium concentrations in the red blood cells of sheep. *Nature* 176, 171.

Evans, J.V. and Phillipson, A.T. (1958) Heritable characters of the red blood cells of sheep. *New Zealand Veterinary Journal* 6, 12–14.

Fahmy, M.H. (ed.) (1996) *Prolific Sheep*, 1st edn. CAB International, Wallingford.

Fayrerhosken, R.A., Huber, T.L., Miller, D.M., Thompson, F.N. and Caudle, A.B. (1992) Infertility in a ewe as a result of ovotestis. *Journal of the American Veterinary Medical Association* 200, 1528–1529.

Fischer, H. and Adinata, M.M. (1957) Atresia ani – a new lethal factor in sheep. *Hemera Zoa* 64, 98–103.

Fleet, M.R., Foulds, R.A., Pourbeik, T., McInnes, C.B., Smith, D.H. and Burbidge, A. (1995) Pigmentation relationships among young Merino sheep and their processed wool. *Australian Journal of Experimental Agriculture* 35, 343–351.

Frelier, P.F., Armstrong, D.L. and Pritchard, J. (1990) Ovine mesangiocapillary glomerulonephritis type-I and crescent formation. *Veterinary Pathology* 27, 26–34.

Georges, M., Drinkwater, R., King, T., Mishra, A., Moore, S.S., Nielsen, D., Sargeant, L.S., Sorensen, A., Steele, M.R., Zhao, X.Y., Womack, J.E. and Hetzel, J. (1993) Microsatellite mapping of a gene affecting horn development in *Bos taurus*. *Nature Genetics* 4, 206–210.

Gilanpour, H. (1971) Perineal hypospadias in a ram. *Veterinary Record* 88, 657–658.

Glahn-Luft, B., Schneider, H., Schneider, J. and Wassmuth, R. (1978) [Agnathia in sheep with chromosome aberrations and Hb deficiency]. *Deutsche Tierarztliche Wochenschrift* 85, 472–474.

Groenewald, H.B. (1993a) Ultrastructure of the myenteric ganglia in the rumen, reticulum, omasum and abomasum of grey, white and black Karakul lambs. *Onderstepoort Journal of Veterinary Research* 60, 189–195.

Groenewald, H.B. (1993b) Ultrastructure of the epithelium of the rumen, reticulum and omasum of grey, white and black Karakul lambs. *Onderstepoort Journal of Veterinary Research* 60, 197–204.

Hamori, D. (1959) [Genetic pathological investigations. II. Inheritance of brachygnathia in sheep]. *Magyar Allatorvosok Lapja* 14, 341–343.

Hancock, J. (1950) Congenital photosensitivity in Southdown sheep: a new sublethal factor in sheep. *New Zealand Journal of Science and Technology* 32A, 16–24.

Hansen, K.M. (1973) The karyotype of the domestic sheep (*Ovis aries*) and the relation between goat and bovine Q-bands. *Hereditas* 75, 119–130.

Hanset, R. (1961) Hereditary microphthalmia in Texel sheep. *Annales de Médecine Vétérinaire* 105, 443–449.

Harper, P. and Morton, A.G. (1991) Neuroaxonal dystrophy in Merino sheep. *Australian Veterinary Journal* 68, 152–153.

Harper, P.A.W., Plant, J.W., Walker, K.H. and Timmins, K.G. (1991) Progressive ataxia associated with degenerative thoracic myelopathy in Merino sheep. *Australian Veterinary Journal* 68, 357–358.

Hartley, W.J. and Haughey, K.G. (1974) Attempted transmission of micrencephaly in newborn lambs. *Australian Veterinary Journal* 50, 323–324.

Hartley, W.J., Alexander, G. and Edwards, M.J. (1974) Brain cavitation and micrencephaly in lambs exposed to prenatal hyperthermia. *Teratology* 9, 299–304.

Heath, D.A., Caldani, M. and McNatty, K.P. (1996) Relationships between the number of immunostaining gonadotropes and the plasma concentrations of gonadotrophins in ewes with and without the FecB(B) gene. *Journal of Reproduction and Fertility* 106, 73–78.

Helle, O. and Nes, N.N. (1972) A hereditary skin defect in sheep. *Acta Veterinaria Scandinavica* 13, 443–445.

Henson, E. (1981) A study of the congenital defect 'split eyelid' in the multihorned breeds of British sheep. *Ark* 8, 84–90.

Hidiroglou, M., Dukes, T.W., Ho, S.K. and Heaney, D.P. (1978) Bent-limb syndrome in lambs raised in total confinement. *Journal of the American Veterinary Medicine Association* 173, 1571–1574.

Hill, B.D., O'Dempsey, N.D. and Carlisle, C.H. (1993) Cervicothoracic vertebral subluxation causing ataxia in sheep. *Australian Veterinary Journal* 70, 156–157.

Hill, F. (1962) Yellow fat in sheep. *Irish Journal of Agricultural Research* 1, 83–89.

Hiraga, T. and Dennis, S.M. (1993) Congenital duplication. *Veterinary Clinics of North America – Food Animal Practice* 9, 145–161.

Hosie, B.D. and Dawson, M. (1996) Scrapie genotyping for Suffolk sheep. *Veterinary Record* 138, 215–216.

Hotchkiss, C.E. (1994) Hemolytic anemia in several sheep. *Laboratory Animal Science* 44, 513–516.

Humphreys, D. (1813) On a new variety in the breeds of sheep. *Philosophical Transactions of the Royal Society of London* 1, 88–95.

Innes, J.R.M., Rowlands, W.T. and Parry, H.B. (1949) An inherited form of cortical cerebellar atrophy in (daft) lambs in Great Britain. *Veterinary Record* 61, 225–228.

Jackson, E.K.M. (1990) Microphthalmia in sheep. *Veterinary Record* 126, 650.

Jakobiec, J. (1952) Inheritance of fertility in sheep. *Roczniki Nauk Rolniczych* 62, 177–198.

James, P.J., Gare, D.R., Singh, A.W., Clark, J.P., Ponzoni, R.W. and Ancell, P.M. (1991) Studies of the potential for breeding short tail Merinos. *Wool Technology and Sheep Breeding* 38, 106–111.

Jarvis, S.M and Young, J.D. (1978) Genetic control of nucleoside transport in sheep erythrocytes. *Biochemical Genetics* 16, 1035–1043.

Jolly, R.D. (1995) Comparative biology of the neuronal ceroid-lipofuscinoses (NCL): an overview. *American Journal of Medical Genetics* 57, 307–311.

Jolly, R.D. and West, D.M. (1976) Blindness in South Hampshire sheep: a neuronal ceroid lipofuscinosis. *New Zealand Veterinary Journal* 24, 123.

Jones, T.O., Clegg, F.G., Morgan, G. and Wijeratne, W.V.S. (1990) A vertically transmitted cystic renal dysplasia of lambs. *Veterinary Record* 127, 421–424.

Jonmundsson, J.V. and Adalsteinsson, S. (1985) Single genes for fecundity in Icelandic sheep. In: Land, R.B. and Robinson, D.W. (eds) *Genetics of Reproduction in Sheep*. Butterworths, London, pp. 159–168.

Kater, J.C., Hartley, W.J., Dysart, T.H. and Campbell, A.R. (1963) Osteogenesis imperfecta and bone resorption: two unusual skeletal abnormalities in young lambs. *New Zealand Veterinary Journal* 11, 41–46.

Kavanagh, P.J., Purcell, D.A. and Thompson, R.H. (1972) Congenital and delayed swayback in lambs in Northern Ireland. *Veterinary Record* 90, 538–540.

Keel, B.A., Zalesky, D.D., Sohaili, I., Schanbacher, B.D. and Grotjan, H.E. (1994) Heterogeneity of gonadotropins and levels of uncombined luteinizing hormone subunits in pituitaries of cryptorchid rams. *Journal of Andrology* 15, 29–35.

Kitamura, T., Alroy, J., Gatmaitan, Z., Inoue, M., Mikami, T., Jansen, P. and Arias, I.M. (1992) Defective biliary excretion of epinephrine metabolites in mutant (TR-) rats – relation to the pathogenesis of black liver in the Dubin–Johnson syndrome and

Corriedale sheep with an analogous excretory defect. *Hepatology* 15, 1154–1159.

Koohmaraie, M., Shackelford, S.D., Wheeler, T.L., Lonergan, S.M. and Doumit, M.E. (1995) A muscle hypertrophy condition in lamb (callipyge) – characterization of effects on muscle growth and meat quality traits. *Journal of Animal Science* 73, 3596–3607.

Koohmaraie, M., Shackelford, S.D. and Wheeler, T.L. (1996) Effects of a beta-adrenergic agonist (l-644,969) and male sex condition on muscle growth and meat quality of callipyge lambs. *Journal of Animal Science* 74, 70–79.

Ladds, P.W., Briggs, G.D. and Foster, R.A. (1990) Epididymal aplasia in 2 rams. *Australian Veterinary Journal* 67, 457–458.

Lancaster, M.J. and Medwell, W.D. (1991) Neck wattles in lambs. *Australian Veterinary Journal* 68, 75–76.

Landauer, W. (1950) The Massachusetts Ancon sheep. *Journal of Heredity* 41, 88–95.

Landauer, W. and Chang, T.K. (1949) The Ancon or otter sheep. *Journal of Heredity* 40, 105–112.

Lapiere, C.M. and Nusgens, B.V. (1993) Ehlers–Danlos syndrome type VII-C, or human dermatosparaxis: the offspring of a union between basic and clinical research. *Archives of Dermatology* 129, 1316–1319.

Laws, L. and Saal, J.R. (1968) Lipidoses of the hepatic reticuloendothelial cells in a sheep. *Australian Veterinary Journal* 44, 416–417.

Lazovskii, A.A. (1983) [The genetic control of alopecia in sheep]. *Genetika* 19, 498–503.

Leipold, H.W. and Dennis, S.M. (1973) Atresia jejuni in a lamb. *Veterinary Record* 93, 644–645.

Leipold, H.W., Dennis, S.M., Schoneweis, D. and Guffy, M.M. (1972) Adactylia in Southdown lambs. *Journal of the American Veterinary Medical Association* 160, 1002–1003.

Linklater, K.A. (1994) Dandy–Walker malformation in lambs. *Veterinary Record* 135, 191.

Littlejohn, A.I. (1954) Entropion in newborn lambs. *Veterinary Record* 66, 211–214.

Luft, B. (1972a) Austosomal deletion in two ewes and two rams with brachygnathia superior. *Deutsche Tierarztliche Wochenschrift* 79, 327–330.

Luft, B. (1972b) Deletion of one arm of an acrocentric autosome of two ewes and two rams with brachygnathia superior. In: *Proceedings of the Seventh International Congress on Animal Reproduction and Artificial Insemination, Munich*, Vol. 2. Deutschen Gesellschaft für Zuchtungskunde, Bonn, pp. 1123–1127.

McFarland, L.Z. and Deniz, E. (1964) Extreme agnathia in a ewe lamb. *Cornell Veterinarian* 54, 541–544.

Mackie, J.T. and McIntyre, B. (1992) Congenital hypotrichosis in Poll Dorset sheep. *Australian Veterinary Journal* 69, 146–147.

McKusick, V.A. (1997) *Mendelian Inheritance in Man*, 12th edn. Johns Hopkins University Press, Baltimore.

McManus, T.J. and Lambe, C. (1972) Species differences in nucleoside metabolism of red cells. In: Gerlach, E., Moser, K., Deutsch, E. and Wilmanns, W. (eds) *International Symposium on Erythrocytes, Thrombocytes and Leucocytes.* Thieme, Stuttgart, pp. 135–139.

McTaggart, H.S., Ritchie, J.S.D. and Copeland, A.N. (1974) Redfoot disease of lambs. *Veterinary Record* 94, 153–159.

Manktelow, B.W. and Hartley, W.J. (1975) Generalised glycogen storage disease in sheep. *Journal of Comparative Pathology* 85, 139–145.

Markson, L.M., Terlecki, S., Shand, K.C., Sellers, K.C. and Woods, A.J. (1959)

Hypomyelinogenesis congenita in sheep. *Veterinary Record* 71, 269–271.

Matejka, M., Cribiu, E.P., Ricordeau, G., Eychenne, F., Durand, V., Lajous, D. and Darre, R. (1990) Evidence for rare heritable fragile site in Romanov sheep chromosome-8. *Reproduction in Domestic Animals* 25, 220–226.

Mauch, C., van der Mark, K., Helle, O., Mollenhauer, J., Pfaffle, M. and Krieg, T. (1988) A defective cell surface collagen-binding protein in dermatosparactic sheep fibroblasts. *Journal of Cell Biology* 106, 205–211.

Mayo, G.M.E. and Mulhearn, C.J. (1969) Inheritance of congenital goitre due to a thyroid defect in Merino sheep. *Australian Journal of Agricultural Research* 20, 533–547.

Miller, A.W. (1993) Hydrocephalus in a flock of Suffolk sheep. *Veterinary Record* 132, 228.

Minchev, P.V. (1976) Genetically conditioned congenital hypoplasia of ewe's mammary gland. *Doklady Bolgarskoi Akademii Nauk* 29, 1835–1837.

Montgomery, G.W., Crawford, A.M., Penty, J.M., Dodds, K.G., Ede, A.J., Henry, H.M., Pierson, C.A., Lord, E.A., Galloway, S.M., Schmack, A.E., Sise, J.A., Swarbrick, P.A., Hanrahan, V., Buchanan, F.C. and Hill, D.F. (1993) The ovine Booroola fecundity gene (FecB) is linked to markers from a region of human chromosome-4q. *Nature Genetics* 4, 410–414.

Montgomery, G.W., Lord, E.A., Penty, J.M., Dodds, K.G., Broad, T.E., Cambridge, L., Sunden, S.L.F., Stone, R.T. and Crawford, A.M. (1994) The Booroola fecundity (FecB) gene maps to sheep chromosome 6. *Genomics* 22, 148–153.

Morley, F.H.W. (1954) A new lethal factor in Australian Merino sheep. *Australian Veterinary Journal* 30, 237–240.

Munday, B.L. (1970) Epitheliogenesis imperfecta in lambs and kittens. *British Veterinary Journal* 126, xlvii.

Nakada, H., Jacobsen, P.F., Johnsen, R.D. and Kakulas, B.A. (1990) Positive immunostaining of ovine congenital progressive muscular dystrophy with antibody against N-terminal and C-terminal dystrophin. *Australian Veterinary Journal* 67, 271–272.

Nakano, T., Walker, B. and Young, B.A. (1994) Analysis of tissues from normal lambs and those with spider syndrome. *Canadian Journal of Animal Science* 74, 583–585.

Nasseri, A.A. and Prasad, M.C. (1990) Pathology of female reproductive organs in hypothyroidism in sheep. *Indian Veterinary Journal* 67, 111.

Nel, J.A. (1964) Hypotrichosis congenita in Karakul sheep. *South African Journal of Agricultural Science* 7, 875–877.

Nel, J.A. and Louw, D.J. (1953) The lethal factor in grey Karakul sheep. *Farming in South Africa* 28, 169–172.

Neuenschwander, S. and Pliska, V. (1990) Hemophilia in sheep and the use of sheep in blood coagulation research. *Journal of Animal Breeding and Genetics – Zeitschrift für Tierzuchtung und Zuchtungsbiologie* S5, 155–164.

Neuenschwander, S. and Pliska, V. (1994) Factor VIII in blood plasma of haemophilic sheep: analysis of clotting time-plasma dilution curves. *Haemostasis* 24, 27–35.

Nguyen, T.C. (1985) Evaluation of linkage between the R-O-i and C blood group systems in sheep. *Animal Blood Groups and Biochemical Genetics* 16, 13–17.

Nguyen, T.C. and Bunch, T.D. (1980) Blood groups and evolutionary relationships among domestic sheep (*Ovis aries*), domestic goat (*Capra hircus*), aoudad (*Ammotragus lervia*) and European mouflon (*Ovis musimon*). *Annales de Génétique et de Sélection Animale* 12, 169–180.

Nguyen, T.C., Morera, L., Llanes, D. and Leger, P. (1992) Sheep blood polymorphism and genetic divergence between french Rambouillet and Spanish Merino – role of genetic drift. *Animal Genetics* 23, 325–332.

Nicholas, F.W. (1996) *Introduction to Veterinary Genetics*. Oxford University Press, Oxford.

O'Toole, D., Jeffrey, M., Jones, T., Morgan, G. and Green, R. (1993) Pathology of renal dysplasia and bladder aplasia–hypoplasia in a flock of sheep. *Journal of Veterinary Diagnostic Investigation* 5, 591–602.

Owen, J.B., Whitaker, C.J., Axford, R.F.E. and Dewi, I.A. (1990) Expected consequences of the segregation of a major gene in a sheep population in relation to observations on the ovulation rate of a flock of Cambridge sheep. *Animal Production* 51, 277–282.

Parry, H.B. (1960) Scrapie: a transmissible hereditary disease of sheep. *Nature* 185, 441–443.

Parsonson, I.M., Della-Porta, A.J. and Snowdon, W.A. (1977) Congenital abnormalities in new born lambs after infection of pregnant sheep with akabane virus. *Infection and Immunity* 15, 254–262.

Patterson, D.S.P. and Sweasey, D. (1970) Lipid hexose : phosphorus ratio as an aid to the diagnosis of congenital myelin defects in lambs and piglets. *Acta Neuropathologia* 15, 318–326.

Phillips, P.H., Bunn, C.M. and Anderson, C.E. (1993) Ovine hereditary chondrodysplasia (spider syndrome) in Suffolk lambs. *Australian Veterinary Journal* 70, 73–74.

Pritchard, D.H., Napthine, D.V. and Sinclair, A.J. (1980) Globoid cell leucodystrophy in Polled Dorset sheep. *Veterinary Pathology* 17, 399–405.

Pritchard, G.C., Jeffrey, M., Welchman, D.D., Windsor, R.S. and Morgan, G. (1994) Multiple cases of Dandy–Walker malformation in three sheep flocks. *Veterinary Record* 135, 163–164.

Radomska, M.J. and Kozanecki, A. (1964) [Studies on the inheritance of the 'thread' defect in wool, and its effect on the production and some biological factors of wool in Polish Merino sheep]. *Roczniki Nauk Rolniczych. Seria B: Zootechniczna* 84, 533–544.

Ramadan, R.O. (1993) A case report – hemimelia and ectrodactyly in a najdi sheep. *Agri-Practice* 14, 30–32.

Rasmusen, B.A. (1958) Blood groups in sheep. I. The X–Z system. *Genetics* 43, 814–821.

Rasmusen, B.A. (1960) Blood groups in sheep. II. The B system. *Genetics* 45, 1405–1417.

Rasmusen, B.A. (1966) Linkage between the C and I blood group loci in sheep. *Genetics* 54, 356.

Rasmusen, B.A., Stormont, C. and Suzuki, Y. (1960) Blood groups in sheep. III. The A, C, D, and M systems. *Genetics* 45, 1595–1603.

Rendel, J. (1956) Heritability of multiple birth in sheep. *Journal of Animal Science* 15, 193–201.

Reuser, A.J.J. (1993) Molecular biology, therapeutic trials and animal models of lysosomal storage diseases – type-II glycogenosis as an example. *Annales de Biologie Clinique* 51, 218–219.

Roberts, J.A.F. (1926) Colour inheritance in sheep. II. The piebald pattern of the piebald breed. *Journal of Genetics* 17, 77–83.

Roberts, J.A.F. (1929) The inheritance of a lethal muscle contracture in sheep. *Journal of Genetics* 21, 57–69.

Rowett, M.A. and Fleet, M.R. (1993) Albinism in a Suffolk sheep. *Journal of Heredity* 84, 67–69.

Sagdullaev, U. (1964) [The inheritance of length of the light-coloured tips of fibres of Sur Karakuls]. *Referativnyi Zurnal Zivotnovodstvo Veterinarija* 8, abstract 8. 58. 134.

Sakul, H., Snowder, G.D. and Hemenway, K.J. (1996) Evaluation of techniques for correction of entropion in lambs. *Small Ruminant Research* 20, 187–191.

Savage, N.G. (1984) Infertility in a ram associated with a knobbed acrosome abnormality of the spermatozoa. *Canadian Veterinary Journal* 25, 126–127.

Schinckel, P.G. (1951) Birth coats of Australian Merino lambs. *Nature* 168, 698.

Schinckel, P.G. (1955) Inheritance of birthcoats in a strain of Merino sheep. *Australian Journal of Agricultural Research* 6, 595–607.

Schwartz, K.V. and Vogel, J.G. (1994) Unraveling the yarn of the ancon sheep. *BioScience* 44, 764–768.

Shelton, M. (1968) A recurrence of the Ancon dwarf in Merino sheep. *Journal of Heredity* 59, 267–268.

Short, B.F. (1958) A dominant felting lustre mutant fleece type in the Australian Merino sheep. *Nature* 181, 1414–1415.

Sides, R.H. and Banks, R.G. (1987) The genetics of carpet wool production: the Elliotdale breed. In: *Proceedings of the Sixth Conference, Australian Association of Animal Breeding and Genetics, University of Western Australia, Perth, 9–11 February 1987*. AAABG, Katanning, Western Australia, pp. 356–359.

Simpson, S.P. and Slee, J. (1988) The inheritance of non-response to noradrenalin in newborn Scottish Blackface lambs. *Genetical Research* 51, 65–69.

Singh, V.K., Mathur, P.B. and Chattopadhyay, S.K. (1970) Brachygnathia in sheep and its inheritance. *Current Science* 39, 187–188.

Skelly, B.J., Jeffrey, M., Franklin, R.J.M. and Winchester, B.G. (1995) A new form of ovine G(M1)-gangliosidosis. *Acta Neuropathologica* 89, 374–379.

Stegelmeier, B.L., James, L.F., Panter, K.E. and Molyneux, R.J. (1995) Serum swainsonine concentration and alpha-mannosidase activity in cattle and sheep ingesting *Oxytropis sericea* and *Astragalus lentiginosus* (locoweeds). *American Journal of Veterinary Research* 56, 149–154.

Swartz, H.A., Vogt, D.W., Michalski, K.A., Warren, J.E. and Kiesling, D. (1985) Chromosomal evaluation of a ewe lamb with atresia ani vaginalis. *American Journal of Veterinary Research* 46, 2145–2146.

Swatland, H.J. (1974) Developmental disorders of skeletal muscle in cattle, pigs and sheep. *Veterinary Bulletin* 44, 179–200.

Tanev, D.N. (1964) [The inheritance of doggy wool in fine-wooled crossbred sheep]. *Nauchni Trudove, Vissh Selskostopanski Institut 'Georgi Dimitrov', Zootekhnicheski Fakultet* 14, 95–99.

Tebbutt, S.J., Wardle, C.J.C., Hill, D.F. and Harris, A. (1995) Molecular analysis of the ovine cystic fibrosis transmembrane conductance regulator gene. *Proceedings of the National Academy of Sciences of the USA* 92, 2293–2297.

Teodoreanu, N. (1958) Hereditary ophthalmia in Tsigai lambs. In: *Proceedings of the Tenth International Congress of Genetics, Montreal (Abstracts)*, Vol. 2. pp. 292–293.

Teodoreanu, N., Gureu, L. and Mihsilesco, M. (1964) [Histological investigations on the sex glands and the thymus of the White Tzagai lambs affected with semi-lethal hereditary ophthalmia]. In: *Proceedings of the Fifth International Congress on Animal Reproduction and Artificial Insemination, Trento*, Vol. III. pp. 282–283.

Terlecki, S., Richardson, C., Bradley, R., Buntain, D., Young, G.B. and Pampiglione, G. (1978) A congenital disease of lambs clinically similar to inherited cerebellar cortical atrophy (daft lamb disease). *British Veterinary Journal* 134, 299–307.

Thomson, D., Davison, A.M. and Angus, K.W. (1973) Spontaneous glomerulonephritis in Finnish Landrace lambs. *Journal of Medical Microbiology* 6, xvii.

Thornton, R.N. and MacColl, D.J. (1985) A suspected inherited neuromuscular disease in Dorset Down sheep. *New Zealand Veterinary Journal* 33, 172–173.

Tibau i Font, J., Lauvergne, J.J., Boitard, M., Bougler, J. and Chabert, Y. (1982) *Inheritance of Horns in Sheep: A Computer-Generated Microform Bibliography.* Bulletin Technique du Département de Génétique Animale, INRA, Jouy-en-Josas. 83 pp.

Tucker, E.M. and Kilgour, L. (1970) An inherited glutathione deficiency and a concomitant reduction in potassium concentration in sheep red cells. *Experientia* 26, 203–204.

Tucker, E.M. and Kilgour, L. (1972) A glutathione deficiency in the red cells of certain Merino sheep. *Journal of Agricultural Science* 79, 515–516.

Tucker, E.M., Wright, P.C. and Young, J.D. (1977) Influence of arginase deficiency on amino acid concentrations in sheep erythrocytes with normal and with a defective system for aminoacids. *Journal of Physiology* 271, 47–48.

Turner, H.N., Hayman, R.H., Triffitt, L.K. and Prunster, R.W. (1962) Response to selection for multiple births in the Australian Merino: a progress report. *Animal Production* 4, 165–176.

Vaiman, D., Pailhoux, E., Payen, E., Saidimehtar, N. and Cotinot, C. (1995) Evolutionary conservation of a microsatellite in the Wilms Tumour (WT) gene: mapping in sheep and cattle. *Cytogenetics and Cell Genetics* 70, 112–115.

Vaiman, D., Koutita, O., Oustry, A., Elsen, J.M., Manfredi, E., Fellous, M. and Cribiu, E.P. (1996) Genetic mapping of the autosomal region involved in XX sex-reversal and horn development in goats. *Mammalian Genome* 7, 133–137.

Vasiliev, R.G. and Aripov, U.Kh. (1983) [Morphological traits of Karakul sheep of different colours]. *Zhivotnovodstvo* 5, 39–40.

Walde, I., Schuh, M. and Glawischnig, E. (1991) On the occurrence of ovine and bovine squamous cell carcinoma (cancer eye) in Austria. *Wiener Tierarztliche Monatsschrift* 78, 209–216.

Warwick, B.L., Davis, S.P., Berry, R.O. and Patterson, R.E. (1960) 'Silky': a dominant mutation in sheep. *Journal of Heredity* 51, 39–42.

Watson, W.A., Broadhead, G.D. and Kilpatrick, R. (1962) Congenital goitre in Dorset-Horn lambs. *Veterinary Record* 74, 506–509.

Watson, W.A., Barnett, K.C. and Terlecki, S. (1972) Progressive retinal degeneration (Bright blindness) in sheep: a review. *Veterinary Record* 91, 665–670.

Wendt, K., Pohl, I. and Mrosk, H. (1960) [Cryptorchidism in sheep and its economic importance]. *Archiv für Tierzucht* 3, 440–458.

West, D.M., Burbidge, H.M., Vermunt, J.J. and Arthur, D.G. (1995) Hereditary chondrodysplasia ('spider syndrome') in a New Zealand Suffolk lamb of American origin. *New Zealand Veterinary Journal* 43, 118–122.

Westlake, V.J., Jolly, R.D., Jones, B.R., Mellor, D.J., Machon, R., Zanjani, E.D. and Krivit, W. (1995) Hematopoietic cell transplantation in fetal lambs with ceroid-lipofuscinosis. *American Journal of Medical Genetics* 57, 365–368.

Wickham, G.A. (1978) Development of breeds for carpet wool production in New Zealand. *World Review of Animal Production* 14, 33–40.

Wiener, G. (1966) Genetic and other factors in the occurrence of swayback in sheep. *Journal of Comparative Pathology* 76, 435–447.

Wiener, G. (1987) The genetics of copper metabolism in animals and man. In: Howell, J.M. and Gawthorne, J.M. (eds) *Copper in Animals and Man*, Vol. 1. CRC Press, Boca Raton, Florida, pp. 45–61.

Wray, C., Mathieson, A.O. and Copland, A.N. (1971) An achondroplasia syndrome in South Country Cheviot sheep. *Veterinary Record* 88, 521–522.

Wriedt, C. (1925) Das Anconschaf. *Zeitschrift für Induktire Abstammungs Verebungslehre* 39, 281–286.
Wright, P.C., Young, J.D., Mangan, J.L. and Tucker, E.M. (1977) An inherited arginase deficiency in sheep erythrocytes. *Journal of Agricultural Science* 88, 765–767.
Wright, P.J., Galloway, D.B. and Clarke, I.J. (1996) Gonadotrophin secretion in ewes with bilateral gonadal hypoplasia. *Australian Veterinary Journal* 73, 157–158.
Xu, Z.C., Dunham, P.B., Dyer, B. and Blostein, R. (1993) Differentiation of Na+–K+ pumps of low-K+ sheep red blood cells is promoted by L(p) membrane antigens. *American Journal of Physiology* 265, C99–C105.
Young, J.D. (1983) Erythrocyte amino acid and nucleoside transport. In: Agar, N.S. and Board, P.G. (eds) *Red Blood Cells of Domestic Animals*. Elsevier, Amsterdam, pp. 271–290.
Zwiep, I.N. (1958) Congenital blindness in lambs. *Tijdschrift voor Diergeneeskunde* 83, 1220–1222.

Reviews (arranged chronologically)

Roberts, J.A.F. and Crew, F.A.E. (1925) The genetics of the sheep. *Bibliography of Genetics* 2, 263–286.
Warwick, B.L. (1931) *Breeding experiments with sheep and swine*. Ohio Agricultural Experimental Station Bulletin. Ohio Agricultural Experiment Station, Wooster, Ohio. 18–37.
Hutt, F.B. (1934) Inherited lethal characters in domestic animals. *Cornell Veterinarian* 24, 1–25.
Rae, A.L. (1956) The genetics of the sheep. *Advances in Genetics* 8, 190–265.
Schumann, H. (1957) Inherited malformations of the limbs in sheep and goats. *Wiener Tierarztliche Monatsschrift* 44, 93–106.
Hamori, D. (1966) Studies on genetical defects in herds of cattle, pigs and sheep. *Hungarian Agriculture Review* 15, 29–30.
Hutt, F.B. (1968) Genetic defects of bones and joints in domestic animals. *Cornell Veterinarian* 58 (Suppl.), 104–113.
Dennis, S.M. (1974) A survey of congenital defects of sheep. *Veterinary Record* 95, 488–490.
Saperstein, G., Leipold, H.W. and Dennis, S.M. (1975) Congenital defects of sheep. *Journal of the American Veterinary Medical Association* 167, 314–322.
Dennis, S.M. and Leipold, H.W. (1976) Diagnosing congenital defects in sheep. *Zuchthygiene* 11, 105–112.
Andresen, E. and Christensen, K. (1977) Genetics of disease in relation to preventive veterinary medicine. In: *Husdyrsykdommer i relasjon tilary. NK Vet-Symposium, Voksenasen, Oslo*. Nordisk Komite for Veterinaervitenskapelig Samarbeid, Oslo, pp. 15–53.
Ballarini, G. (1977) Hereditary diseases in veterinary practice. *Rivista di Zootecnia e Veterinaria* 2, 177–186.
Jolly, R.D. and Hartley, W.J. (1977) Storage diseases of domestic animals. *Australian Veterinary Journal* 53, 1–8.
Behrens, H. (1979) Hereditary diseases of sheep. In: Parey, P. (ed.) *Lehrbuch der Schafkrankheiten*, 2nd edn. Springer Verlag, Berlin, pp. 255–260.
Dennis, S.M. and Leipold, H.W. (1979) Ovine congenital defects. *Veterinary Bulletin* 49, 233–239.

Dennis, S.M. (1980) Congenital and inherited defects in sheep. In: Morrow, D.A. (ed.) *Current Therapy in Theriogenology*, W.B. Saunders, Philadelphia, pp. 924–927.

Stanik, J. (1980) Heredity of health and hereditary diseases in sheep husbandry. *Veterinarstvi* 30, 393–395.

Desnick, R.J., Patterson, D.F. and Scarpelli, D.G. (1982) *Animal Models of Inherited Metabolic Diseases*, 1st edn. Alan R. Liss, New York.

Fries, R. and Ruddle, F.H. (1986) Gene mapping in domestic animals. In: Augustine, P.C., Danforth, H.D. and Bakst, M.R. *Biotechnology for Solving Agricultural Problems*. Martinus Nijhoff, Dordrecht, pp. 19–37.

Basrur, P.K. and Yadav, B.R. (1990) Genetic diseases of sheep and goats. *Veterinary Clinics of North America – Food Animal Practice* 6, 779–802.

Adalsteinsson, S. (1991) Gene identification in sheep. In: Maijala K. (ed.) *Genetic Resources of Pig, Sheep and Goat*. Elsevier Science, Amsterdam, The Netherlands, pp. 365–378.

Elias, E. and Bennett, R. (1992) Congenital defects in Awassi fat-tailed lambs. *Small Ruminant Research* 8, 141–150.

Dennis, S.M. (ed.) (1993) Congenital abnormalities. *Veterinary Clinics of North America – Food Animal Practice* 9.

Dennis, S.M. (1993) Congenital defects of sheep. *Veterinary Clinics of North America – Food Animal Practice* 9, 203–217.

Hiraga, T. and Dennis, S.M. (1993) Congenital duplication. *Veterinary Clinics of North America – Food Animal Practice* 9, 145–161.

Huston, K. (1993) Heritability and diagnosis of congenital abnormalities in food animals. *Veterinary Clinics of North America – Food Animal Practice* 9, 1–9.

Jolly, R.D. (1993) Lysosomal storage diseases in livestock. *Veterinary Clinics of North America – Food Animal Practice* 9, 41–53.

Ladds, P.W. (1993) Congenital abnormalities of the genitalia of cattle, sheep, goats, and pigs. *Veterinary Clinics of North America – Food Animal Practice* 9, 127–144.

Saperstein, G. (1993) Congenital abnormalities of internal organs and body cavities. *Veterinary Clinics of North America – Food Animal Practice* 9, 115–125.

Steffen, D.J. (1993) Congenital skin abnormalities. *Veterinary Clinics of North America – Food Animal Practice* 9, 105–114.

Biochemical Genetics 5

L. DI STASIO
Department of Animal Science, Via Genova 6, 10126 Torino, Italy

Introduction 133
Red-cell Antigens 134
Serological analysis 134
Genetics 135
Blood-group systems 136
Blood and Milk Proteins 137
Electrophoretic analysis 138
Genetics 139
Polymorphic systems 140
Plasma systems 140
Erythrocyte systems 141
Milk systems 143
Practical Applications 143
References 144

Introduction

Genetic variability in a species is very common and the visible traits described in the previous chapters are the most evident examples. However, in the 1950s the introduction of some methods of analysis opened up a completely new area of investigation, making it possible to study the polymorphism of a wide range of proteins on the basis of their specific biochemical characteristics.

The different glycoproteins on the surface of the red cells can be identified by serological analysis for their antigenic properties, while the variants of many proteins in the blood, milk and body tissues can be detected by electrophoretic methods on the basis of their electric charge. With the rapid application of these techniques, it became evident that the amount of genetic variability was much greater than previously believed.

The genetic studies showed that the investigated proteins are under the control of alleles following a simple Mendelian inheritance, so that the identification of the protein polymorphism is an indirect identification of the polymorphism at the deoxyribonucleic acid (DNA) level.

Red-cell Antigens

Since the beginning of this century, it has been observed in different species that the erythrocytes from an individual can be agglutinated when mixed with the serum of some other individuals. However, it was in the 1940s that the research on blood groups in humans (Landsteiner and Wiener, 1940) and in cattle (Ferguson, 1941) gave rise to that branch of biology designated 'immunogenetics'. These studies revealed that the glycoproteins on the membrane of the red cells, or blood-group factors, exist in several molecular forms, characterized by antigenic properties and therefore identifiable by antigen–antibody reactions. From that time the research in the field of immunogenetics rapidly advanced in cattle and extended to other farm animals. The development in sheep was closely related to progress in cattle so that in the late 1950s and 1960s intensive studies were devoted to the detection of ovine antigenic factors and to the elucidation of their mechanism of inheritance. The interest in immunogenetics was further stimulated when it was demonstrated that some antigens are involved in basic physiological processes. Since then, little has been added to the knowledge on blood groups in sheep, apart from the discovery of new antigens.

Serological analysis

At first, naturally occurring antibodies were used, such as those identifying the antigens of the R system, but soon after, with the impetus of what was developing in cattle, many other antigens were detected by means of antibodies induced by isoimmunization (with animals of the same species) or heteroimmunization (with animals of different species). The classical procedure consists of a series of injections of the blood of one animal, the donor, into another animal, the recipient, lacking one or more of the antigens carried by the donor; the recipient will react with an immune response by producing antibodies against the foreign antigen(s). When donor and recipient differ for more than one antigen, the antiserum will contain more than one population of antibodies. In this case, an exhaustive absorption with erythrocytes of other animals is necessary to obtain a monospecific antiserum, which at this point is called the reagent.

The identification of the ovine blood-group factors was at first based on the agglutination reaction. As it was later demonstrated that most antibodies do not induce agglutination of the red cells while fixing the complement, the haemolytic test was introduced. Briefly, an aliquot of a diluted suspension of the erythrocytes to be tested is mixed with an aliquot of the reagent at the appropriate dilution; after 30 min of incubation, rabbit complement absorbed with sheep red cells (to remove natural heterohaemolysins for sheep) is added;

the mixture is incubated again, usually for 3 h, and then the results are recorded; the presence of the haemolytic reaction indicates that the red cells carry the antigen corresponding to that particular reagent. The sum of all the antigens carried by an individual is its blood group.

In earlier investigations, bovine isoimmune or natural antisera, already available for cattle blood typing, were also used to detect antigenic substances in sheep (Stormont, 1951; Rasmusen, 1957, 1960; Rasmusen *et al.*, 1960; Millot, 1966). In order to define the existing relationships between blood groups in sheep and cattle, Nguyen (1972) carried out a comprehensive study of the cross-reactions of 77 cattle reagents and 17 sheep reagents with bovine and ovine erythrocytes. The results revealed a surprisingly high degree of homology between the two species: 12 cattle reagents had antibodies enabling detection of ten ovine antigens and three ovine reagents could detect bovine antigens.

Details on the procedures used to produce the reagents and to perform the serological analysis can be found in the fundamental papers by Rasmusen (1957, 1960), Rasmusen *et al.* (1960) and Nguyen (1972).

Genetics

Concerning the genetic determinism of the immunological polymorphism, the segregation analysis allowed the assignment of the identified blood-group factors to seven different systems controlled by seven different loci, generally with multiple alleles. It is interesting to note that the largest allelic series have been detected for some loci controlling the blood-group systems: the *EAB* locus in cattle has more than 600 alleles (Bell, 1983) and the *EAB* locus in sheep about 100 (Nguyen, 1988). The alleles at a given locus can control the presence of a single antigen or of a group of antigens which are transmitted to the progeny as a single unit (phenogroup), or the absence of known antigens. All the alleles are codominant, except for the EAR^{O} and 'null' alleles, which behave as recessive.

The antigens and the reagents are conventionally named by letters. At first capital letters were used, by analogy with the terminology adopted in cattle. Subsequently, with the increase of the number of laboratories involved in the research on ovine blood groups, a comparison of the reagents produced by the different working teams and the standardization of the nomenclature turned out to be a major need. Therefore, a comparison test was organized in 1973, with 11 participating laboratories. As a result, 17 antigenic factors were officially recognized and rules for nomenclature were laid down for the first time (Nguyen, 1973). It was agreed to designate the seven blood-group systems with the letters A, B, C, D, M, R and X, and within each of the systems A, B, C, D and M to designate the factors by a, b, c and so on; for the R system the factor designation was not changed, while no decisions were made for the X system. After this initial step, the international recognition of the reagents and the standardization of the nomenclature have been achieved by comparison tests periodically organized by the International Society for Animal Genetics (ISAG).

Blood-group systems

At present 22 ovine blood-group factors in seven systems have been internationally recognized (Table 5.1). Many other specificities have been detected by different laboratories with experimental reagents (108 at the 1993 ISAG comparison test) and are awaiting confirmation; they are provisionally designated by the laboratory code followed by a number.

A system

The A system was described by Rasmusen *et al.* (1960). Two internationally recognized factors and two additional factors, provisionally named F/J16 and F/J19 (Nguyen and Ruffet, 1975), have been reported. At least five alleles (EAA^0 recessive, the others codominant) are known.

B system

The B system was described by Stormont *et al.* (1957) and Rasmusen (1960). It is the most complex system, with nine internationally accepted specificities and at least six additional antigens (F/J3, F/J4, F/J8, F/J26, F/J37, F/J38) detected by experimental reagents (Nguyen, 1988). About 100 alleles have been described, EAB^0 being recessive, the others codominant. The system is homologous to the complex B system in cattle in its serological and genetic features (Rasmusen, 1960; Nguyen and Ruffet, 1975).

C system

The C system was described by Rasmusen *et al.* (1960). There are two internationally recognized factors and two provisional factors (F/J5, F/J6), with about 20 alleles (EAC^0 recessive, the others codominant). It is homologous to the C system of cattle. The *EAC* locus is linked to the *Erythrocyte Antigen Secretor I* locus (Rasmusen, 1966; Nguyen, 1985) and the *Amino Acid Transport* locus (Tucker *et al.*, 1980).

Table 5.1. Loci for blood group systems.

System	Locus symbol	Antigens*	No. of alleles
Erythrocyte antigen A	*EAA*	a (A), b (K)	At least 5
Erythrocyte antigen B	*EAB*	a (P), b (B'), c (Y), d (N'), e (E'), f (E), g (O'), h (S), i (Ix)	About 100
Erythrocyte antigen C	*EAC*	a (C), b (Cx)	About 20
Erythrocyte antigen D	*EAD*	a (D), b	At least 3
Erythrocyte antigen M	*EAM*	a (M), b (L), c (M')	At least 3
Erythrocyte antigen R	*EAR*	R, O	2
Erythrocyte antigen X	*EAX*	X, Z	2

*The original names are in parentheses.

D system
The D system was described by Rasmusen *et al.* (1960). Unlike other sheep blood-group factors, the antigens of the D system are agglutinogens. Two antigens have been recognized (Anonymous, 1980).

M system
The M system was described by Rasmusen *et al.* (1960) and Tucker and Ellory (1970). Three antigens and at least three codominant alleles have been described. The system is homologous to the S system of cattle. It is associated with the polymorphism of potassium content in the red cells: the 'low-potassium' (LK) cells have the EAMb antigen, while the 'high-potassium' (HK) cells have the EAMa antigen. Therefore, it was postulated that the EAMb antigen inhibits active potassium transport in the cell, resulting in a low internal potassium ion (K^+) concentration. In fact, when LK cells are sensitized with the blood-group reagent anti-EAMb, the inhibitory action of the EAMb antigen is removed or reduced (Rasmusen and Hall, 1966a, b; Tucker, 1976). For more details see Ellory and Tucker (1983).

R system
The R system was the first polymorphic system detected in sheep by means of natural antibodies in animals lacking the R antigen and designated as group O (Ycas, 1949). The group O sheep were routinely identified by the absence of reaction with anti-R, but in some cattle sera Stormont (1951) discovered a natural antibody selectively reacting with the blood of all the sheep of group O and put forward the hypothesis that the corresponding antigen was determined by an allele recessive to the one that determines the presence of R antigen. It was also demonstrated that the R and O substances appear on the red cells of lambs only at the age of 1–2 and 3–4 weeks, respectively (Tucker, 1964). Secretion of the antigens depends on the genotype at the *EAR* locus and on the independent secretor locus *EASI* (Rendel, 1957): the recessive $EASI^i$ allele inhibits the secretion of the soluble R and O antigens and the sheep are classified as group i (i.e. with red cells negative for R and O). The EAR system is associated with the alkaline phosphatase system.

X system
The X system was described by Rasmusen (1957). Two antigens and two codominant alleles have been identified.

Blood and Milk Proteins

At the same time as the development of the blood-typing research, another powerful biochemical method, electrophoresis, enabled the detection of variability of proteins other than erythrocyte antigens. The term 'electrophoresis' generally defines the migration in a medium of a particle with an electric charge by the action of an electric field. In the case of proteins, the overall charge originates from the carboxyl (-COOH) and amino ($-NH_2$) groups, which can assume an electric charge according to the pH of the medium.

Actually, evidence of the electrophoretic migration of proteins dates back to the end of the last century. Since then, boundary electrophoresis in free liquid medium and subsequently zone electrophoresis in different stabilizing media, such as filter-paper or agar, has made possible great progress in protein characterization. However, completely new perspectives in the knowledge of protein variability were revealed with the introduction of starch gel as support medium (Smithies, 1955). In fact, extension of electrophoretic investigation at population level showed that many proteins exhibit polymorphism, with the existence of various molecular forms, called variants, generally differing in one or a few amino acid substitutions or sometimes a deletion. Therefore, the high resolving power and the relative simplicity of the method, together with the ease of the sample collection, contributed to the rapid, simultaneous application of electrophoretic analysis to all domestic species, sheep included.

When more sophisticated electrophoretic techniques were set up (such as polyacrylamide gel electrophoresis, electrofocusing, two-dimensional electrophoresis, etc.), further protein variability was identified.

Electrophoretic analysis

To perform electrophoretic analysis, a small amount of the sample is inserted into the support medium, the gel, with an appropriate pH depending on the protein to be tested. Then the ends of the gel are connected to a suitable electrolyte solution and the electric field is applied for the time necessary to obtain the best resolution. After the electrophoresis, the gel is stained with a general protein dye or with a specific enzymatic-staining solution. With this procedure, the different variants of a protein, if characterized by a different overall electric charge, can be identified on the basis of their different electrophoretic mobility.

The electrophoretic pattern of sheep β-lactoglobulin (LGB) represents a simple and clear example. As shown in Fig. 5.1, two variants can be detected by starch-gel electrophoresis, called A and B in order of decreasing mobility. The comparison of their primary structure explains the reason for their different electrophoretic behaviour. The B variant differs from the A variant in the replacement of the histidine in position 20 by tyrosine (Gaye *et al.*, 1986) and, as a consequence, the A variant has a more negative charge, which induces the more anodal migration.

However, it must be emphasized that the situation is not always as simple as it appears from these statements. In fact, electrophoretic migration also depends on a number of parameters other than the electric charge of the molecule: size, shape, concentration, degree of hydration and dissociation of the molecule; viscosity, pH, temperature and ionic strength of the medium; electric-field strength; and migration time. Moreover, the electrophoretic pattern is not always clearly interpretable, especially where enzymes are concerned. As such

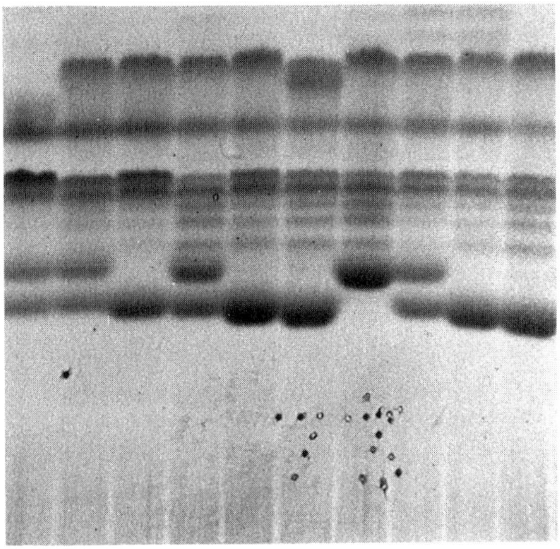

Fig. 5.1. β-Lactoglobulin (LGB) phenotypes.

a complex subject would require a separate discussion, readers interested in more details can refer to the excellent paper by Baker and Manwell (1983), where many aspects concerning the interpretation and significance of the electrophoretic variation of erythrocyte enzymes are considered.

Genetics

The segregation analysis showed that the protein variants are controlled by alleles at single loci, mostly codominant. In the case of codominance, the identification of an electrophoretic variant corresponds in practice to the identification of the responsible allele; that is, the observed phenotype corresponds to the genotype. Therefore, the polymorphic proteins are very suitable as genetic markers for several practical applications, which will be briefly discussed later.

Obviously, a locus can be useful only if its polymorphism is not too rare. But what is the meaning of 'too rare'? To focus attention on the loci that exhibit an appreciable variability, the geneticists chose an arbitrary cut-off point: a locus can be defined as polymorphic when the most common allele has a frequency of less than 0.99 or 0.95, so that the rarest allele cannot be maintained only by recurrent mutation.

Concerning the nomenclature, the protein variants and the responsible alleles are conventionally designated by letters, except for the variants of some systems (amino acid transport (TR), nucleoside phosphorylase (NP), X protein (XP)), which are designated by names (i.e. high, low, positive, negative). For the polymorphic proteins detected by electrophoresis, international recognition of the genetic variants is achieved by periodical comparison tests organized by ISAG.

Polymorphic systems

Over the past decades throughout the world a very large number of proteins in the blood and milk of many breeds of sheep have been analysed for polymorphism, so that it would be an enormous task to cover the whole subject. On the other hand, several reviews with detailed information and data can be found in the literature (McDermid *et al.*, 1975; Tucker, 1975; Tucker and Baker, 1989; Larsen *et al.*, 1992). For this reason and also because many of the observed phenotypic variations require further genetic investigation, only the systems for which genetic control is well established are considered in this chapter (Table 5.2). Readers interested in a more complete list of loci and alleles can refer to Tucker and Di Stasio (1996).

Plasma systems

Albumin

The albumin system was described by Efremov and Braend (1964). Of the three internationally recognized alleles, ALB^S is the most common in all the examined breeds. The locus is linked to the *vitamin D-binding protein* locus (Erhardt and Simianer, 1993).

Table 5.2. Loci for biochemical polymorphism.

System	Locus symbol	Alleles*
In the plasma:		
Albumin	*ALB*	*F, S, W, (D), (T), (V)*
Arylesterase	*ES*	*A, o*
Transferrin	*TF*	*A, B, C, D, E, G, P, U, V, H, K, (M), (N), (L), (X)*
In the erythrocytes:		
Amino acid transport	*TR*	*H, h*
Carbonic anhydrase	*CA2*	*F, S, (M)*
Haemoglobin-α_1	*HBA1*	*D, L, (A)*
Haemoglobin-β	*HBB*	*A, B, (E), (G), (H), (I)*
Haemoglobin-βC	*HBBC*	*C*
Malic enzyme 1	*ME1*	*F, S*
NADH diaphorase	*DIA*	*F, S*
Nucleoside phosphorylase	*NP*	*H, I*
Potassium transport	*KE*	*L, h*
X protein	*XP*	*P, n*
In the milk:		
Lactoglobulin β	*LGB*	*A, B, (C)*

*The alleles awaiting confirmation are in parentheses.
NADH, nicotinamide adenine dinucleotide.

Arylesterase

The arylesterase system was described by Lee (1964) and Tucker *et al.* (1967). Using starch-gel electrophoresis followed by a specific staining procedure (with α-naphthyl acetate and fast blue BB), two phenotypes with different staining intensity, designated 'positive' and 'negative', could be identified. The arylesterase activity appears during the first few weeks after birth. The genetic analysis showed that the negative phenotype is controlled by a recessive allele, generally predominant in the investigated breeds. Using other substrates, each of the two phenotypes could be further subdivided into three phenotypes, but their genetic determinism is only postulated (Gahne and Göransson, 1970).

Transferrin

The transferrin system shows the highest variability, with 15 variants noted: 11 have been internationally recognized, while the others are awaiting confirmation. Ashton (1958) first described the five more common variants (A, B, C, D, E). Subsequently other variants, rare or limited to one breed, were observed. For example, TF H and TF K have been found in certain Czechoslovakian breeds (Stratil, 1973) and TF L in Scottish Blackface × Welsh Mountain crosses (Archibald and Webster, 1986). The polymorphism is controlled by codominant alleles.

Others

Polymorphism has been also reported for alkaline phosphatase, α-2-macroglobulin, haemopexin, immunoglobulin, leucine aminopeptidase, lipoprotein-β, post-transferrin, prealbumin, protease inhibitor-α_1 and vitamin D-binding protein (see Tucker and Di Stasio, 1996). Further investigations are needed to confirm their genetic control.

Erythrocyte systems

Amino acid transport

Tucker and Kilgour (1970) described in Finnish Landrace sheep an inherited glutathione deficiency, a consequence of defective amino acid transport across the red-cell membrane. The rare allele TR^h, responsible for the reduced amino acid transport, behaves as recessive. The locus is closely linked to the *EAC* locus: erythrocytes with normal amino acid transport properties can be either Cb-positive or Cb-negative, while erythrocytes with defective properties are never Cb-negative (Tucker *et al.*, 1980). For more details see Young (1983).

Carbonic anhydrase

The carbonic anhydrase system was described by Tucker *et al.* (1967). The polymorphism is controlled by codominant alleles, with $CA2^S$ widespread. The rare allele $CA2^M$ has been reported only in Churra breed (Ordas and San Primitivo, 1983).

Haemoglobin-α_1

Rare polymorphism has been observed, with two internationally accepted and one provisional allele (Vascov and Efremov, 1967; Vestri *et al.*, 1982).

Haemoglobin-β

Harris and Warren (1955) first described two variants, A and B, determined by codominant alleles. Later, four additional provisional haemoglobin-β (HBB) variants were reported (John and John, 1977; Kilgour *et al.*, 1990; Manca *et al.*, 1993). The A and B variants differ for many functional properties: for example, in comparison with HBB B, HBB A has greater oxygen affinity and higher haemoglobin concentration and packed-cell volume.

Haemoglobin-β C   Van Vliet and Huisman (1964) and Braend *et al.* (1964) described a variant, HBBC C, expression of the βC chain and produced in the case of anaemia, instead of the normal HBB chain, only by sheep with HBB^A or HBB^G. The amount of HBBC C and the rate of production are related to the severity of the anaemia (see Garrick, 1983).

For more details on the biochemical, functional and genetic aspects concerning the haemoglobin polymorphism see Blunt and Huisman (1975) and Garrick and Garrick (1983).

Malic enzyme 1

The malic enzyme 1 system was described by Baker and Manwell (1977). Two variants have been recognized, controlled by codominant alleles.

Nicotinamide adenine dinucleotide diaphorase

The nicotinamide adenine dinucleotide (NADH) diaphorase system was described by Brewer *et al.* (1967). Two variants have been observed, controlled by codominant alleles.

Nucleoside phosphorylase

Tucker and Young (1976) described genetic variation in NP activity, with two phenotypes, designated 'high' and 'low'. The allele responsible for low enzymatic activity behaves as recessive.

Potassium transport

Kerr (1937) first reported the existence of quantitative differences in the erythrocyte potassium content, detected by biochemical assay. It was demonstrated that the polymorphism is genetically determined, with the allele for HK being recessive. The polymorphism is associated with the EAM blood-group system and develops completely a few weeks after birth. For more details, see Ellory and Tucker (1983).

X Protein
The X protein is an unidentified protein described by Tucker *et al.* (1967), who found two phenotypes, designated 'positive' and 'negative'. The negative phenotype is determined by a recessive allele, X^n. The electrophoretic bands appear gradually during the first few weeks after birth.

Others
Polymorphism has been also reported for arginase, catalase, γ-glutamyl cysteine synthetase, glucose phosphate isomerase, haemoglobin-α, lactate dehydrogenase A, nucleoside transport and reduced glutathione (see Tucker and Di Stasio, 1996). Further investigations are needed to confirm their genetic control.

Milk systems

Lactoglobulin β
Bell and McKenzie (1967) first described two variants, LGB A and LGB B, distributed in all the investigated breeds. In addition, a third rare variant, LBG C, has been identified (Erhardt, 1989).

Others
Polymorphism has been also reported for αs1-casein (King, 1966; Ferranti *et al.*, 1995).

Practical Applications

From the general description of the immunological and electrophoretic polymorphisms, it appears that they have in common three characteristics which are of basic importance in view of practical applications, both in animal breeding and in fundamental research: the phenotype of an individual with respect to these traits does not change during life (apart from a few exceptions); it can be identified by objective methods of analysis; and it is controlled by simple Mendelian inheritance.

For these reasons, the biochemical polymorphisms are extensively used throughout the world for parentage control (François *et al.*, 1992). The test is based on the possibility of detecting inconsistency between the genotype of an individual and those of the putative parents. The probability of excluding an individual as possible parent (usually the ram, even if cases of disputed maternity can exist) is dependent on the number of markers used as well as on the number and frequency of the alleles at each locus (Jamieson, 1965). To reveal pedigree errors is a fundamental step in animal breeding and selection: such errors preclude herd-book registration, and above all they reduce the accuracy of the estimates of heritability and breeding values based on progeny testing.

The marker loci are also very useful tools in population studies, because by means of the allele frequencies it is possible to characterize the genetic structure of the breeds and to compare them in order to get information on their origin, evolution and genetic relationships. The allele frequencies can also be used to calculate the degree of heterozygosity of the populations, which represents an indirect estimate of their overall genetic variability. These applications are especially valuable where rare breeds are concerned (Clarke *et al.*, 1989; Zanotti Casati *et al.*, 1990; Di Stasio *et al.*, 1992; Montgelard *et al.*, 1994).

Since the discovery of biochemical polymorphisms, many studies have been devoted to investigating the possible existence of genetic correlations with economically important traits, such as production and fertility (Fésüs, 1994), but, as a rule, no clear associations have been found in sheep. In this respect, DNA markers are expected to play a major role. For example, none of the 18 classical protein markers considered were genetically linked to the *Booroola* gene (Tate *et al.*, 1992), while linkage has been reported with microsatellite markers (Montgomery *et al.*, 1993).

Finally, one of the most fascinating aspects of the investigation of protein polymorphism is its important contribution to the knowledge of biochemical and physiological processes. Many examples can be found in sheep. The existence of different KE phenotypes has been useful for investigating cation transport in the cell, and further advances in the elucidation of this mechanism were made possible when it was shown that the LK type is associated with the EAMb antigen. The discovery of polymorphism of the amino acid and nucleoside transport systems opened the way to investigation of other metabolic processes. Haemoglobin switching provides a model for research on the regulatory mechanisms of gene expression.

References

Anonymous (1980) Report of the Sheep Comparison Test and Workshop (Wageningen, 1980). *Animal Blood Groups and Biochemical Genetics* 12, 226–227.

Archibald, A.L. and Webster, J. (1986) A new transferrin allele in sheep. *Animal Genetics* 17, 191–194.

Ashton, G.C. (1958) Polymorphism in the β-globulins of sheep. *Nature* 181, 849–850.

Baker, C.M.A. and Manwell, C. (1977) Heterozygosity of the sheep: polymorphism of 'malic enzyme', isocitrate dehydrogenase (NADP$^+$), catalase and esterase. *Australian Journal of Biological Science* 30, 127–140.

Baker, C.M.A. and Manwell, C. (1983) Electrophoretic variation of erythrocyte enzymes of domesticated mammals. In: Agar, N.S. and Board, P.G. (eds) *Red Blood Cells of Domestic Mammals.* Elsevier, Amsterdam, pp. 367–412.

Bell, K. (1983) The blood groups of domestic animals. In: Agar, N.S and Board, P.G. (eds) *Red Blood Cells of Domestic Mammals.* Elsevier, Amsterdam, New York, pp. 133–164.

Bell, K. and McKenzie, H.A. (1967) The whey proteins of ovine milk: β-lactoglobulins A and B. *Biochemica Biophysica Acta* 147, 123–134.

Blunt, M.H. and Huisman, T.H.J. (1975) The haemoglobins of sheep. In: Blunt, M.H. (ed.) *The Blood of Sheep. Composition and Function.* Springer Verlag, Berlin, pp. 155–183.

Braend, M., Efremov, G. and Helle, O. (1964) Abnormal haemoglobin in sheep. *Nature* 204, 700.

Brewer, G.J., Eaton, J.W., Knutsen, C.S. and Beck, C.C. (1967) A starch gel electrophoretic method for the study of diaphorase isozymes, and preliminary results with sheep and human erythrocytes. *Biochemical and Biophysical Research Communications* 29, 198–204.

Clarke, S.W., Tucker, E.M., Hall, S.J.G. (1989) Genetic polymorphisms and their relationships with inbreeding and breed structure in rare British sheep: the Portland, Manx Laghtan and Hebridean. *Conservation Biology* 3, 381–388.

Di Stasio, L., Rasero, R., Giaccone, P. and Fiandra, P. (1992) Valle del Belice sheep: genetic structure and relationships with other sheep populations reared in Sicily. *Agricoltura Mediterranea* 122, 66–69.

Efremov, G. and Braend, M. (1964) Haemoglobin, transferrin and albumins of sheep and goat. In: Matousek, J. (ed.) *Proceedings IXth European Conference on Animal Blood Groups and Biochemical Polymorphisms, Prague*, W. Junk Publishing, The Hague, The Netherlands, pp. 313–320.

Ellory, J.C. and Tucker, E.M. (1983) Cation transport in red blood cells. In: Agar, N.S. and Board, P.G. (eds) *Red Blood Cells of Domestic Mammals*. Elsevier, Amsterdam, pp. 291–314.

Erhardt, G. (1989) Evidence for a third allele at the β-lactoglobulin (β-Lg) locus of sheep milk and its occurrence in different breeds. *Animal Genetics* 20, 197–204.

Erhardt, G. and Simianer, H. (1993) Linkage between loci for serum albumin and vitamin D binding protein (GC) in sheep. *Animal Genetics* 24, 301–303.

Ferguson, L.C. (1941) Heritable antigens in the erythrocytes of cattle. *Journal of Immunology* 40, 213–242.

Ferranti, P., Malorni, A., Nitti, G., Laezza, P., Pizzano, R., Chianese, L. and Addeo, F. (1995) Primary structure of ovine αs1-caseins: localization of phosphorylation sites and characterization of genetics variants A, C and D. *Journal of Dairy Research* 62, 281–296.

Fésüs, L. (1994) The influence of haemoglobin types on reproduction and production in Hungarian Merino sheep. *Animal Genetics* 25 (Suppl. I), 95–97.

François, D., Bosher, M., Mériaux, J.C. and Nguyen, T.C. (1992) Les controles de filiation dans les populations. *INRA Productions Animales* (hors série), 273–276.

Gahne, B. and Göransson, M. (1970) The genetic control of arylesterase activity in the serum of Gotland sheep. *Animal Blood Groups and Biochemical Genetics* 1, 127–134.

Garrick, M.D. (1983) Haemoglobin switching. In: Agar, N.S. and Board, P.G. (eds) *Red Blood Cells of Domestic Mammals*. Elsevier, Amsterdam, pp. 209–226.

Garrick, M.D. and Garrick, L.M. (1983) Haemoglobins and globin genes. In: Agar, N.S. and Board, P.G. (eds) *Red Blood Cells of Domestic Mammals*. Elsevier, Amsterdam, New York, Oxford, pp. 165–207.

Gaye, P., Hue-Delahaie, D., Mercier, J.C., Soulier, S., Vilotte, J.L. and Furet, J.P. (1986) Ovine β-lactoglobulin messenger RNA: nucleotide sequence and mRNA levels during functional differentiation of the mammary gland. *Biochimie* 68, 1097–1107.

Harris, H. and Warren, F.L. (1955) Occurrence of electrophoretically distinct haemoglobins in ruminants. *Biochemical Journal* 60, 24.

Jamieson, A. (1965) The genetic of transferrins in cattle. *Heredity* 20, 419–441.

John, M.E. and John, M. (1977) A new haemoglobin β-chain variant in sheep. *Animal Blood Groups and Biochemical Genetics* 8, 183–190.

Kerr, S.E. (1937) Studies on the inorganic composition of blood. IV. The relationship of potassium in the acid soluble phosphorus fractions. *Journal of Biological Chemistry* 117, 227–235.

Kilgour, L., Dixon, S.C. and Tucker, E.M. (1990). Two new sheep haemoglobin, one of which is replaced by haemoglobin C in anaemia. *Animal Genetics* 21, 115–121.

King, J.W.B. (1966) The casein of sheep's milk. In: *Proceedings Xth European Conference on Animal Blood Groups and Biochemical Polymorphisms, Paris*, INRA, Institut Pasteur, Paris, pp. 427–431.

Landsteiner, K. and Wiener, A.S. (1940) An agglutinable factor in human blood recognized by immune sera for rhesus blood. *Proceedings of the Society for Experimental Biology and Medicine* 43, 223.

Larsen, B., Di Stasio, L. and Tucker, E.M. (1992) List of alleles for blood and milk polymorphisms in cattle, sheep and goats. *Animal Genetics* 23, 188–192.

Lee, R.M. (1964) Di-(2-chloroethyl)arylphosphatases: a study of their reaction with β-esterase, and of the genetic control of their hydrolysis in sheep. *Biochemical Pharmacology* 13, 1551–1568.

McDermid, E.M., Agar, N.S. and Chai, C.K. (1975) Electrophoretic variation of red cell enzyme systems in farm animals. *Animal Blood Groups and Biochemical Genetics* 6, 127–174.

Manca, L., Di Luccia, A., Pieragostini, E., Naitana, S. and Masala, B. (1993) Haemoglobin I: a new β-globin chain variant found in sheep of Italian breeds. *Animal Genetics* 24, 203–204.

Millot, P. (1966) Les groupes sanguins des bovins et des ovins: aperçu sur les groupes sériques bovins. *Revue de Pathologie comparée* 3, 209–219.

Montgelard, C., Nguyen, T.C. and Dubray, D (1994) Genetic variability in French populations of the Corsican mouflon (*Ovis ammon musimon*): analysis of 2 blood proteins and red-cell blood groups. *Génétique, Sélection, Evolution* 26, 303–315.

Montgomery, G.W., Crawford, A.M., Penty, J.M., Dodds, K.G., Ede, A.J., Henry, H.M., Pierson, C.A., Lord, E.A., Galloway, S.M., Schmack, A.E., Sise, J.A., Swarbrik, P.A., Hanrahan, V., Buchanan, F.C. and Hill, D.F. (1993) The ovine Booroola fecundity gene (Fec B) is linked to markers from a region of human chromosome 4q. *Nature Genetics* 4, 410–414.

Nguyen, T.C. (1972) Les groupes sanguins des ovins. I. Rélations entre les groupes sanguins des ovins et des bovins. *Annales de Génétique et de Sélection animale* 4, 363–374.

Nguyen, T.C. (1973) Report on the Sheep Blood Groups Workshop. *Animal Blood Groups and Biochemical Genetics* 4, 241–243.

Nguyen, T.C. (1985) Evaluation of linkage between the R-O-i and C blood group systems in sheep. *Animal Blood Groups and Biochemical Genetics* 16, 13–17.

Nguyen, T.C. (1988) Loci for blood polymorphism in sheep and goats: red cell blood groups and lymphocyte antigens. In: Lauvergne, J.J. (ed.) *Standardized Genetic Nomenclature for Sheep and Goats 1986*. Lavoisier, Paris, pp. 53–56.

Nguyen, T.C. and Ruffet, G. (1975) Les groupes sanguins des ovins. II. Facteurs antigéniques supplémentaires dans les systèmes A, B, C et M: estimation des fréquences 'alleliques' aux systèmes A, B, C, D, M et R dans les races françaises: Berrichon-du-Cher, Île de France et Texel. *Annales de Génétique et de Sélection animale* 7, 145–157.

Ordas, J.G. and San Primitivo, F. (1983) A new allele at the red cells carbonic anhydrase locus in sheep. *Animal Blood Groups and Biochemical Genetics* 14, 233–234.

Rasmusen, B. (1957) Blood groups in sheep. I. The X-Z system. *Genetics* 43, 814–821.

Rasmusen, B. (1960) Blood groups in sheep. II. The B system. *Genetics* 45, 1405–1417.

Rasmusen, B. (1966) Linkage between the C and I blood-group loci in sheep. *Genetics* 54, 356.
Rasmusen, B. and Hall, J.G. (1966a) Association between potassium concentration and serological type of sheep red blood cells. *Science* 151, 1551–1552.
Rasmusen, B. and Hall, J.G. (1966b) An investigation into the association between potassium levels and blood types in sheep and goats. In: *Proceedings Xth European Conference on Animal Blood Groups and Biochemical Polymorphisms.* INRA, Institut Pasteur, Paris, pp. 453–457.
Rasmusen, B., Stormont, C. and Suzuki, Y. (1960) Blood groups in sheep. III. The A, C, D and M systems. *Genetics* 45, 1595–1603.
Rendel, J. (1957) Further studies on some antigenic characters of sheep blood determined by epistatic action of genes. *Acta Agriculturae Scandinavica* 7, 224–259.
Smithies, O. (1955) Zone electrophoresis in starch gels: group variations in the serum proteins of normal human adults. *Biochemical Journal* 61, 629–641.
Stormont, C. (1951) An example of a recessive blood group in sheep. *Genetics* 36, 577–578.
Stormont, C., Suzuki, Y. and Rasmusen, B.A. (1957) Cross-reactions of ovine and bovine isoimmune antibodies in cattle and sheep blood-typing studies. *Journal of Animal Science* 16, 1102.
Stratil, A. (1973). Two transferrin variants and the effect of neuraminidase. *Animal Blood Groups and Biochemical Genetics* 4, 153–159.
Tate, M.L., Manly, H.C., Dodds, K.G. and Montgomery, G.W. (1992) Genetic linkage analysis between protein polymorphisms and the Fec^B major gene in sheep. *Animal Genetics* 23, 417–424.
Tucker, E.M. (1964) The appearance of R and O substances on the red cells of lambs. *Immunogenetics Letter* 3, 133–135.
Tucker, E.M. (1975) Genetic markers in the plasma and red blood cell. In: Blunt, M.H. (ed.) *The Blood of Sheep. Composition and Function.* Springer Verlag, Berlin, pp. 123–153.
Tucker, E.M. (1976) Some physiological aspects of genetic variation in the blood of sheep. *Animal Blood Groups and Biochemical Genetics* 7, 207–215.
Tucker, E.M. and Baker, C.M.A. (1989) Sheep allelic series 1987: blood and milk polymorhisms. In: Lauvergne, J.J. (ed.) *Standardized Genetic Nomenclature for Sheep and Goats 1987: Loci for Visible Traits Other than Colour and Blood and Milk Polymorphisms.* Bureau des Ressources Génétiques, Paris, pp. 79–152.
Tucker, E.M. and Di Stasio, L. (1996) Loci controlling blood and milk polymorphisms. In: Lauvergne, J.J., Dolling C.H.S. and Renieri, C. (eds) *Mendelian Inheritance in Sheep.* Università degli Studi di Camerino, Camerino, pp. 115–150.
Tucker, E.M. and Ellory, J.C. (1970) The M-L blood group system and its influence on red cell potassium levels in sheep. *Animal Blood Groups and Biochemical Genetics* 1, 101–112.
Tucker, E.M. and Kilgour, L. (1970) An inherited glutathione deficiency and a concomitant reduction in potassium concentration in sheep red cells. *Experientia* 26, 203–204.
Tucker, E.M. and Young, J.D. (1976) Genetic variation in the purine nucleoside phosphorylase activity of sheep red cells. *Animal Blood Groups and Biochemical Genetics* 7, 109–117.
Tucker, E.M., Suzuki, Y. and Stormont, C. (1967) Three new phenotypic systems in the blood of sheep. *Vox Sanguinis* 13, 246–262.
Tucker, E.M., Evans, N.R.S. and Kilgour, L. (1980) Close linkage between the C blood

group locus and the locus controlling amino acid transport in sheep erythrocytes. *Animal Blood Groups and Biochemical Genetics* 11, 119–125.

Van Vliet, G. and Huisman, T.H.Y. (1964) Changes in haemoglobin types of sheep as a response to anaemia. *Biochemical Journal* 93, 401–409.

Vascov, B. and Efremov, G. (1967) Fourth haemoglobin type in sheep. *Nature* 216, 593–594.

Vestri, R., Giordano, P.C. and Bernini, L.F. (1982) Duplication of the haemoglobin α-chain gene in sheep: characterization of a new -chain variant present in animal possessing the α^{Leu} and the $^{II}\alpha^{His}$ chains. *Biochemical Genetics* 21, 25–35.

Ycas, M.K.V. (1949) Studies of the development of a normal antibody and of cellular antigens in the blood of sheep. *Journal of Immunology* 61, 327–347.

Young, J.D. (1983) Erythrocyte amino acid and nucleoside transport. In: Agar, N.S. and Board, P.G. (eds) *Red Blood Cells of Domestic Mammals*. Elsevier, Amsterdam, New York, Oxford, pp. 271–290.

Zanotti Casati, M., Gandini, G.C. and Leone, P. (1990) Genetic variation and distances of five Italian native sheep breeds. *Animal Genetics* 21, 87–92.

Molecular Genetics of Sheep 6

B.C. POWELL
Department of Animal Science, University of Adelaide, Waite Campus, Glen Osmond, South Australia 5064, Australia

Introduction 149
Genome Structure 150
  Properties of repetitive sequences 151
Wool-keratin Genes 153
  Keratin intermediate-filament genes 153
  Keratin-associated protein genes 156
  Regulation of wool-keratin gene expression 160
  Hair and wool mutations 161
Metallothionein Genes 162
Lysozyme Genes 163
Globin Genes 165
Conclusions 171
Acknowledgements 171
References 172

Introduction

The molecular genetics of sheep is in its infancy. Fewer than 700 markers have been mapped in the sheep genome (see Chapters 10 and 11) and there is molecular information for only about 200 sheep genes, frequently just short pieces. In this review I shall discuss the sheep-wool keratin, globin, metallothionein and lysozyme genes, incorporating information from more advanced studies in other species as appropriate. The study of these gene families highlights features of genes, chromosome structure and genome evolution that have wide relevance. In discussing the molecular genetics of sheep, I shall first review chromosome structure and consider gene function against that perspective.

Genome Structure

The sheep karyotype contains many acrocentric chromosomes, in contrast to the human karyotype, where metacentrics are in abundance. This gross morphological difference indicates the magnitude of the large scale rearrangements that have occurred in mammalian evolution. It belies, however, the segmental nature of chromosome shuffling, which has resulted in the conservation of extensive regions between species (for reviews, see Eppig and Nadeau, 1995; Womack and Kata, 1995). Current genetic maps reveal, for example, 70 segments of homology between the human and cattle genomes. Interestingly, keratin genes on human chromosome 17 are part of a conserved segment of 47 centimorgans (cM) that includes at least 60 genes shared with mouse chromosome 11. From livestock gene-mapping studies, it seems that this homology segment may also be conserved in cattle and sheep (see later). Although our knowledge of mammalian molecular genetics derives mainly from work on the human and mouse genomes, it is applicable in broad outline to sheep.

Estimates of the number of mammalian genes range from 50,000 to 100,000 genes, dispersed among 3 billion base pairs (bp) of deoxyribonucleic acid (DNA). At an average size of 15,000 bp, genes occupy about half the genome (Lewin, 1994). However, most genes have exons and introns, and the actual coding capacity of all the genes requires about 5% of the genome. The other 95% includes unique and repetitive sequences in similar abundance. Repetitive DNA includes various types of satellite DNA and dispersed repeat families, known as short and long interspersed repeat families (SINEs and LINEs) (Fig. 6.1). In contrast to the essential conservatism of gene-coding regions, repetitive elements have undergone a relatively rapid evolution (Jobse *et al.*, 1995). This probably results from a combination of factors; repetitive elements do not encode essential gene functions and are therefore not under selection pressure, and they often have a simple repetitive sequence, whose alteration is promoted by recombination effects, such as mispairing and unequal crossing-over.

Repetitive sequences occur throughout the genome, frequently in introns. However, centromeres contain mostly α-satellite DNA and telomeres are composed of simple hexanucleotide repeats. Long tandem arrays of repeated DNA, like α-satellite DNA and telomere DNA, form a type of DNA complex called heterochromatin, which generally has a negative influence on gene expression (for a review, see Lohe and Hilliker, 1995). Surprisingly, heterochromatin seems to exert a long-range effect over gene expression, and chromosomal rearrangements or random integration of DNA, such as transgenes, can bring genes under its silencing influence, affecting their function in an erratic manner, which is known as position-effect variegation (Reuter and Spierer, 1992; Shaffer *et al.*, 1993). Deoxyribonucleic acid is also found in a form called euchromatin, which is more open and permissive for gene expression. Euchromatin harbours mostly unique DNA and dispersed repeats. These types of chromatin have different effects on gene regulation, exemplified by the globin loci (Craddock *et al.*, 1995). Interestingly, there are correlations between the

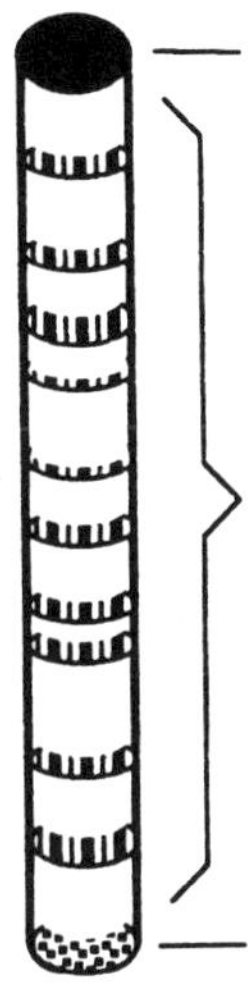

Fig. 6.1. A cartoon of a G-banded chromosome showing the location of repetitive sequences. Alpha satellite DNA is located at the centromere, simple hexanucleotide repeats at the telomere and other types of repeats are dispersed between.

two types of chromatin and the classical Giemsa (G) and reverse (R) banding techniques for karyotyping (for a review, see Gardiner, 1995). Giemsa-band DNA typically contains few genes, is involved in few chromosomal breakpoints or rearrangements and is SINE-rich but LINE-poor. Reverse-band DNA generally exhibits opposite properties.

Properties of repetitive sequences

Satellite DNA is located at the centromeres of chromosomes and consists of tandem arrays of repeats of variable sequence, often hundreds of kilobases (kb) in length, accounting for 10–20% of genome sequence (for a review, see Sunkel and Coelho, 1995). Short interspersed repeats and microsatellite repeats have also been found in satellite DNA (Skowronski *et al.*, 1984). Centromeric satellite DNA has undergone a rapid evolution such that repeats differ between species and even from one chromosome to another. However, consensus sequences for the binding of centromeric proteins appear to be conserved in the arrays and are probably one of the reasons that satellite DNA is essential for chromosome segregation during meiosis and mitosis.

Microsatellite and minisatellite DNA consists of repeats from 2 to 200 bp in length in tandem arrays of up to 10–20 kb. In their simplest form, satellites may be simple cytosine–adenosine (CA) repeats of no more than 30 bp, known as microsatellites. The inherent length polymorphisms of these types of satellite DNA, combined with the polymerase chain reaction (PCR), have revolutionized genome mapping.

Short interspersed repeat families are typically of the order of 100–500 bp in length, with up to 100,000 copies of each distributed throughout the genome (for a review, see Rogers, 1985). At least three types of SINEs have been characterized to date in the sheep genome. Two SINEs, originally known as artiodactyl (art) A and C, are approximately 115 bp and 75 bp, respectively, in size and are frequently found in multimeric combinations. These contain sequences that are homologous to the DNA polymerase-III-binding sites of transfer ribonucleic acid (tRNA) genes. The C repeat appears to be derived from a $tRNA^{gly}$ pseudogene and is usually found in combination with the A repeat. A longer SINE (art 2) of approximately 500 bp has 75 bp at one end that is similar to one end of the A repeat (Duncan, 1987). These SINEs are conserved in other artiodactyl genomes, and in a revised nomenclature (Lenstra *et al.*, 1993) are called Bov-A2 (A dimer), Bov-tA (AC combination) and Bov-B (art 2). They occur throughout the sheep keratin and sheep globin loci (Rogers, 1984; Spence *et al.*, 1985; Wilson *et al.*, 1988; Li *et al.*, 1991; Powell *et al.*, 1992; Hardison and Miller, 1993; Saban and King, 1994).

Long interspersed repeat families are usually a few kilobases in size and are found at a frequency of 10,000–100,000 copies in the genome. The L1 LINE occurs in the genome of many mammals. The L1 LINEs contain open reading frames homologous to retroviral reverse transcriptases and they have internal RNA polymerase II promoters and poly-A tracts, a typical example of the 'selfish' gene described by Dawkins (1976).

Short interspersed repeat families appear to be more frequent in regions of higher gene density and LINEs less frequent. As both SINEs and LINEs contain either DNA or RNA polymerase sites, they have the potential to move around the genome and randomly influence gene function. How widespread this might be is not certain (Britten, 1996), but there seem to be only a few repeats in each family that can actively move around (Deininger *et al.*, 1992). A SINE insertion affecting gene expression causes neurofibromatosis, a human genetic disease (Wallace *et al.*, 1991). Recombination between two SINEs at the β-globin locus may cause the gene deletions underlying hereditary persistence of fetal haemoglobin (Jagadeeswaran *et al.*, 1982). These types of events undoubtedly influence gene expression in the sheep genome, but as the medical imperatives are different they will probably remain undetected.

Tandem arrays of microsatellite DNA are prone to amplification and can increase in size from a few bp to hundreds of bp. This can have deleterious effects on gene function. The recent discoveries of trinucleotide repeat expansion as the cause of some human genetic diseases (Sutherland and Richards, 1995; Campuzano *et al.*, 1996) may foreshadow a more widespread effect of simple repeat expansion on gene dysfunction. As yet, however, there are no reports of similar events involving the ubiquitous CA dinucleotide repeat.

Wool-keratin Genes

Wool is produced from a follicle, a tiny intricate structure that is one of the most active organs in the body. Follicles are established in the later stages of embryonic development and undergo cycles of activity and dormancy throughout life. As a result of wool-industry and government investment, there have been significant advances in our understanding of wool proteins and genes.

Keratin proteins are the major components of hair and wool and bestow their structural and mechanical properties. Over the last decade, our knowledge of the complexity of these keratins has expanded, primarily through work on sheep-wool genes (for a review, see Powell and Rogers, 1994). Between 50 and 100 keratin genes are expressed during wool and hair growth and 13 families have been characterized so far (Crewther, 1976; Swart *et al.*, 1976; Powell *et al.*, 1983, 1992, 1995; Bertolino *et al.*, 1988, 1990; Wilson *et al.*, 1988; Frenkel *et al.*, 1989; McNab *et al.*, 1989; MacKinnon *et al.*, 1990; Kaytes *et al.*, 1991; Tobiasch *et al.*, 1992a, b; Fratini *et al.*, 1993, 1994; Yu *et al.*, 1993; Huh *et al.*, 1994; Powell and Beltrame, 1994; Winter *et al.*, 1994). We know the sequences of some 40 hair and wool proteins and of many of their genes, the majority of them derived from sheep.

The keratin proteins can be divided into two large groups, the keratin intermediate-filament (IF) proteins and the keratin-associated proteins (KAPs). In the wool fibre, these proteins are associated in a highly organized fashion, the keratin IF proteins forming 8–10 nm diameter filaments embedded in a matrix of the KAPs. There are two families of keratin IF, type I and type II, and the fundamental unit of the filament is formed by the pairing of the two types. This dictates that the genes are coordinately regulated. The KAP group is much more complex. Eleven families have been discovered, the first over 30 years ago, but how they interact is still a mystery.

The general picture of wool-keratin gene organization that has emerged in the last decade as more genes have been cloned reveals that the genes are clustered in families and are usually only a short distance apart (5–10 kb). Separate clusters of the two keratin IF gene families and five of the KAP gene families have been discovered and some of their chromosomal locations mapped. From comparative information, it is apparent that keratin gene organization and structure have been conserved during evolution. The close genetic linkage that frequently occurs within gene families suggests that 'global' regulatory domains might govern their expression.

Keratin intermediate-filament genes

The ancestry of the keratin IF genes can be traced back over a billion years to a probable lamin-like IF protein that became involved with nuclear compartmentalization (Blumenberg, 1988). Separate loci for the type I and type II families have been mapped in several genomes and there is evidence that they are associated with ancient chromosomal arrangements that date from the evolution of the chordates over 400 million years ago. The legacy of this can be seen in the syntenic linkages of *Wnt*, *Hox* and *Rar* genes at both the keratin type I and the type II IF loci in mouse and human chromosomes (Fig. 6.2), supporting the occurrence of a large-scale genomic duplication involving these homologous gene sets at an early point in evolution (Hart *et al.*, 1992; Nadeau *et al.*, 1992). Mapping of some of these loci in sheep and cattle suggests that the linkages will be conserved. In sheep, the keratin IF type I loci are located at 11q25–q29 and the type II loci at 3q14–q22 (Hediger *et al.*, 1991).

In structural organization, the wool-keratin IF genes are similar to the

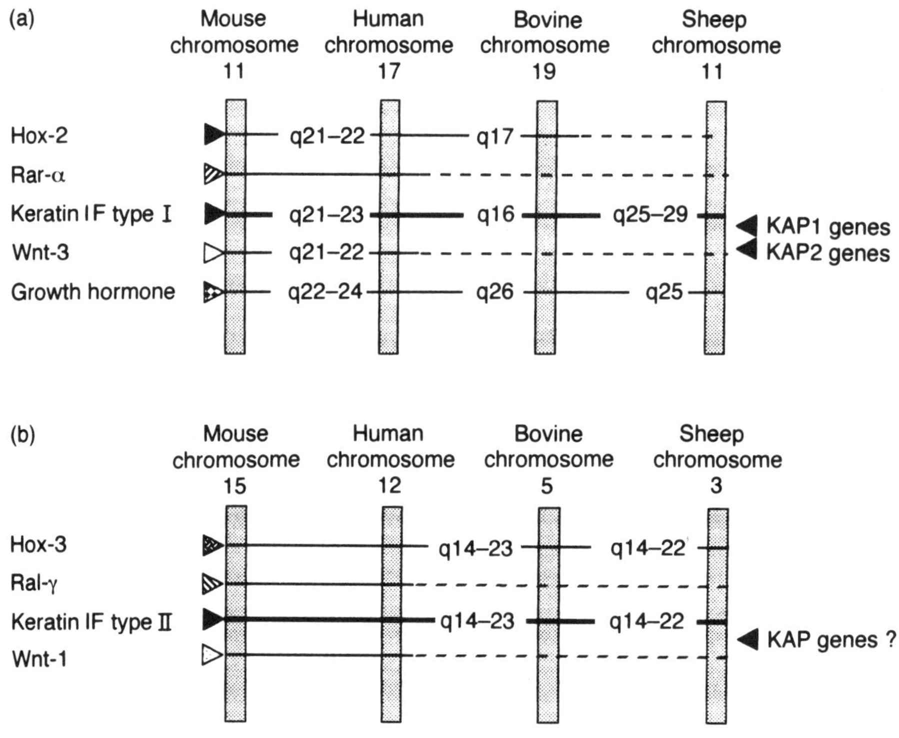

Fig. 6.2. Conservation of chromosome segments in mammalian evolution. Linkage maps for keratin, *Wnt*, *Hox*, *Rar* and growth hormone are shown. (a) Comparative maps for mouse chromosome 11, human chromosome 17, bovine chromosome 19 and sheep chromosome 11. (b) Comparative maps for mouse chromosome 15, human chromosome 12, bovine chromosome 5 and sheep chromosome 3. Known chromosomal assignments are connected by solid lines and inferred locations are shown by dashed lines. Map locations are given where known. The KAP1 and KAP2 assignments in sheep are based on genetic-linkage and restriction-mapping data; the presence of another KAP locus (KAP?) near the keratin type II locus is hypothetical (see text under 'Keratin-associated protein genes' for information). (Data are from Nadeau *et al.*, 1989, 1992; Elliott *et al.*, 1990; Hediger *et al.*, 1990, 1991; Fries *et al.*, 1991; Gunawardana and Fries, 1992; Hart *et al.*, 1992; Parsons *et al.*, 1994b; Eppig and Nadeau, 1995.) Hox, homoeobox; Wnt, wingless integration site (gene); Rar, retinoic acid receptor; KAP, keratin-associated protein.

human epidermal-keratin IF genes. Type I genes are 4–5 kb in size and contain six introns and type II genes are 7–9 kb in size with eight introns (Powell and Rogers, 1986; Wilson *et al.*, 1988; Kaytes *et al.*, 1991; Powell *et al.*, 1992; Powell and Beltrame, 1994). The type I and type II epidermal genes map to different chromosomes in all mammalian species (Lessin *et al.*, 1988; Romano *et al.*, 1988; Rosenberg *et al.*, 1988; Nadeau *et al.*, 1989; Popescu *et al.*, 1989; Fries *et al.*, 1991; Hediger *et al.*, 1991; Hart *et al.*, 1992), with the exception of a single human type I gene that maps to the type II locus (Waseem *et al.*, 1990; Yoon *et al.*, 1994). In several mapping studies, some of the keratin probes hybridized to chromosomal locations other than the two major keratin clusters (Romano *et al.*, 1988; Rosenberg *et al.*, 1988, 1991; Hediger *et al.*, 1991). These 'ectopic' sites may represent keratin pseudogenes derived from reverse transcription of messenger RNA (mRNA) and integrated at random into the genome or may represent bona fide orphan keratin genes.

Linkage between the hair- and epidermal-keratin genes has been established in mice by genetic studies (Compton *et al.*, 1991) and is expected to hold for other mammalian genomes. Recent data obtained from yeast artificial chromosomes (YACs) (Yoon *et al.*, 1994) and overlapping cosmid clones (Powell and Beltrame, 1994) are revealing the long-range organization of the keratin type II locus in more detail. Predictions from these data suggest that the region is between 500 and 600 kb in size and contains 20–30 genes (Powell and Beltrame, 1994; Yoon *et al.*, 1994), possibly including all the type II IF hair and epidermal genes (Fig. 6.3).

A 100 kb segment of DNA from the sheep-keratin type II locus has been cloned and extensively characterized (Powell *et al.*, 1992, 1995). Six type II genes, three wool and three wool-related, have been mapped in this region and much of the DNA sequenced. Genes encoding three of the four sheep-wool type II proteins are clustered within 40 kb and flanked by recently recognized wool-related genes, prompting speculation that the clustered wool genes are located in a domain that contains all the regulatory controls required for wool-follicle expression.

Eight epidermal-keratin type II IF genes, K2.1–K2.8,* have been isolated on a 630 kb YAC clone that maps to human 12q13 (Yoon *et al.*, 1994). The equivalent sheep type II region is 3q14–q22. The mapping data also reveal the probable existence of several related genes, possibly hair genes. Future mapping of this YAC should reveal the organizational structure of the type II locus in fine detail.

---

*In a proposed revision of the nomenclature for keratin proteins (Powell and Rogers, 1994), the keratin IF type I and II families are numbered according to the formula K1.n and K2.n, where K indicates keratin, 1 and 2 identify the keratin as type I or type II IF and n refers to the specific keratin component. Database gene symbols would be KRT1.n and KRT2.n. Existing catalogue numbers (Moll *et al.*, 1982) are retained as the n values, hence K1.14 specifies the type I IF keratin component currently referred to as K14, and K2.1 specifies the type II IF keratin currently referred to as K1. This system is simply extended to the hair-matrix or keratin-associated proteins as KAPm.n (database gene symbol, KRTAP) such that KAP1 identifies the high-sulphur B2 family (Powell and Rogers, 1994).

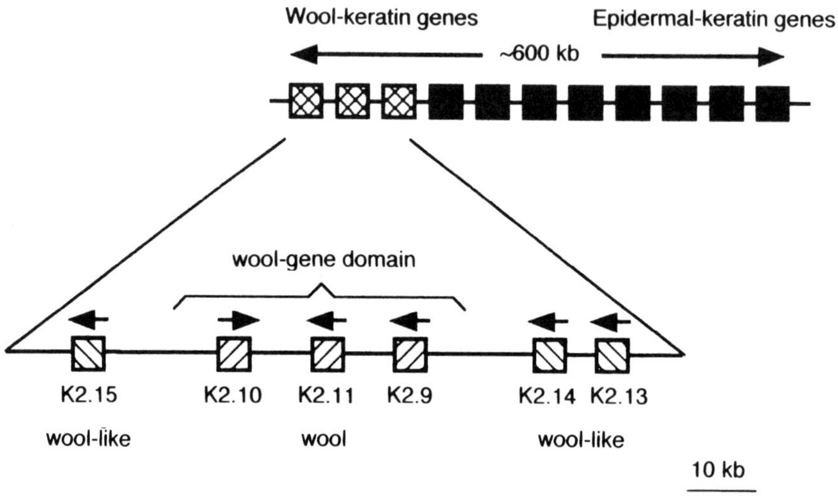

Fig. 6.3. Schematic diagram of a putative keratin IF type II locus at sheep chromosome 3q14–q22. The arrangement of wool-keratin and epidermal-keratin IF genes at the locus in the upper part of the diagram is arbitrary and is drawn to reflect approximate proportions of the two 'classes', rather than actual numbers (estimated to be 20–30) or disposition. The organization of the six sheep-wool and wool-like genes are from Powell and Beltrame (1994). The direction of transcription of the wool and wool-like genes is indicated by arrows. The locus diagram is based on mouse genetic linkage, human YAC and sheep cosmid data (Compton *et al.*, 1991; Powell and Beltrame, 1994; Yoon *et al.*, 1994).

Like the type II genes, the hair and epidermal type I genes are linked in mice (Compton *et al.*, 1991), but detailed information on their organization is not as advanced. Linkage of two to three wool and wool-related genes within 40 kb segments has been described (Powell *et al.*, 1986; Kaytes *et al.*, 1991). The human epidermal-keratin genes, K1.14 and K1.16, are linked within 10 kb of each other (Rosenberg *et al.*, 1988; Savtchenko *et al.*, 1990) and K1.13, K1.15 and K1.19 are located within a 55 kb segment of DNA (Filon *et al.*, 1994), but their relative order is unknown.

Keratin-associated protein genes

The KAP genes encode the proteins of the matrix that forms between the IFs during the differentiation of hair cortical cells. To date, 11 families of KAP genes have been identified, mainly in sheep. Each gene is quite small, between 0.6 and 1.5 kb in size, and the striking feature that primarily contributes to their small size is the absence of introns. This is a rare occurrence, and the KAP genes of hair and wool constitute one of the largest groups of intronless genes known.

The first KAP genes were characterized in 1983 (Powell *et al.*, 1983) and encoded three members of the sheep-wool high-sulphur B2 family. In our revised nomenclature, they are the KAP1 family (Powell and Rogers, 1994).

Their structure and organization have been typical of all other KAP genes characterized since then. The sequences of their coding regions are highly conserved and show indications of homogenization by gene conversion. Two genes (KAP1.1 and KAP1.4) are located approximately 2 kb apart and a third gene (KAP1.3), identified in a separate clone, is closely linked to a member of the KAP2 family (Fig. 6.4). Pulse-field gel electrophoresis using KAP1 family and KAP2 family probes has suggested that the two families may be linked as a supercluster of genes (B.C. Powell, A. Fratini and G.E. Rogers, unpublished data), in line with the initial discovery made in the genomic clones.

A recent interesting finding by Parsons *et al.* (1994b) suggests that a KAP1 gene is linked to the locus for keratin type I IF genes (KRT1@) on sheep chromosome 11 (see Fig. 6.2). From restriction fragment length polymorphism (RFLP) studies, they discovered linkage of at least one KAP1 gene to growth hormone and knowing the linkage between growth hormone and the KRT1 locus, they deduced the linkage of the KAP1 and KRT1 loci. The exact arrangement of the two loci is unknown, and KAP1 could be linked to KRT1 beyond growth hormone or they could be adjacent. With the possible existence of a cluster containing the KAP1 and KAP2 gene families, the KAP1 and KRT1@ loci linkage raises the exciting possibility of a supercluster of the KAP1, KAP2 and KRT1 gene families. Given the ancient conservation of the chromosome segments around the two keratin IF loci (see Fig. 6.2) and the possibility of common evolutionary origins for some of the KAP families, there is the interesting prospect that KAP genes may also be linked to the KRT2@ locus.

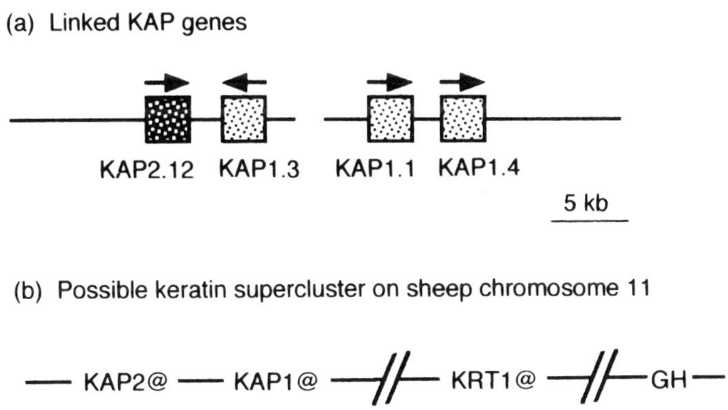

Fig. 6.4. Schematic diagram of the possible arrangement of the sheep KAP1 and KAP2 gene families. (a) Representation of two λ genomic clones carrying genes for the KAP1 and KAP2 families (B.C. Powell, unpublished; Powell *et al.*, 1983). The direction of transcription is indicated by arrows. The genes are not drawn to scale. (b) Hypothetical linkage of KAP1, KAP2 and keratin IF type I gene families and the growth hormone gene on sheep chromosome 11. This arrangement is based on cloning, pulse-field mapping and linkage data (B.C. Powell, unpublished; Powell *et al.*, 1983; Parsons *et al.*, 1994b). KRT1@, keratin IF type I locus; GH, growth hormone.

The KAP1 family contains at least four genes and a number of distinctive alleles (Powell *et al.*, 1983; Rogers *et al.*, 1994). These keratin-associated proteins range in size from 151 to 181 amino acids, differing mainly in the number of tandem copies (from two to five) of a ten-amino-acid repeat, SIQTSCCQPT, located in the central part of the proteins (Fig. 6.5). Six KAP1.3 (B2C) alleles have been identified, showing simple nucleotide and amino acid substitutions, but three of the four KAP1.1 alleles show more extensive differences, with three, four and five tandem copies of the ten-amino-acid repeat (Rogers *et al.*, 1994). Gene-conversion events appear to be maintaining a high level of sequence identity in the family. As the nucleotide sequences of the repeats are highly conserved, variation in tandem copy number probably occurs

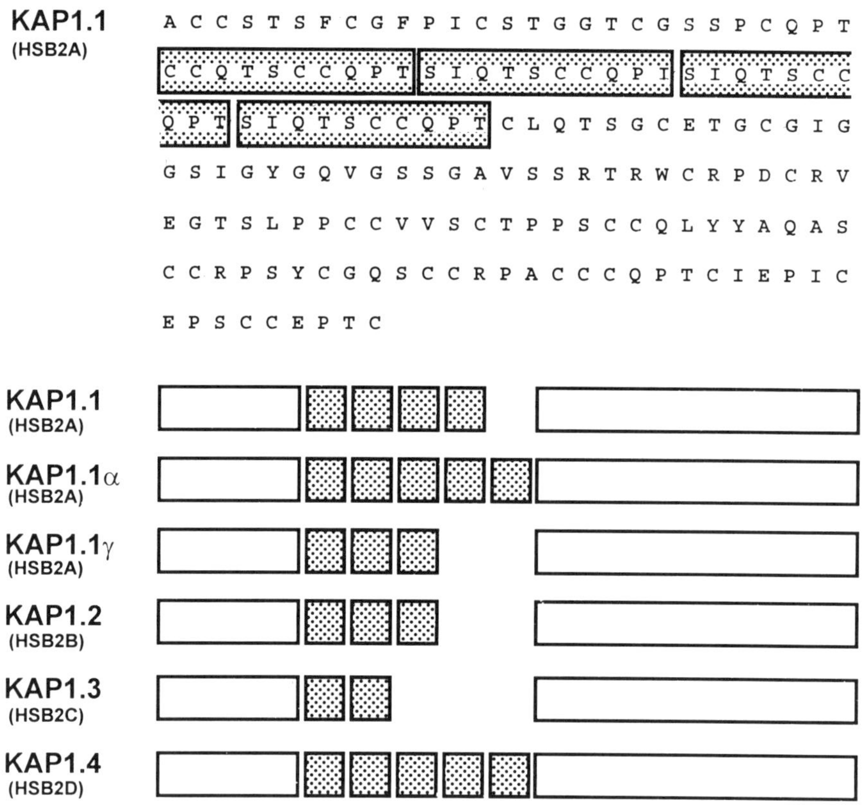

Fig. 6.5. Comparative sequences of the sheep-wool KAP1 cysteine-rich proteins. The KAP1.1 sequence (formerly high-sulphur (HS)B2A) is given in full, with the tandem decapeptide repeats boxed. The KAP1 family proteins are compared in modular format, highlighting the variable numbers of decapeptide repeats. The open boxes either side of the repeats symbolize the N-terminal and C-terminal segments of the protein, which differ in only a few amino acid positions between proteins. Four genes (KAP1.1–1.4) and two KAP1.1 alleles (KAP1.1α and 1.1γ) have been characterized. (Data are from Crewther, 1976; Powell *et al.*, 1983; Rogers *et al.*, 1994.)

through slipped mispairing and unequal crossing-over events during chromosome replication and recombination.

Two wool KAP gene families (KAP5 and KAP10) are specifically expressed late in follicle differentiation in the fibre cuticle (MacKinnon *et al.*, 1990; Jenkins and Powell, 1994; B.C. Powell, A. Nesci and R. Voyle, unpublished data). The genes encode strikingly different proteins, which are probably constituents of the exocuticle layer (for review, see Powell and Rogers, 1997). The KAP5 proteins range from 168 to at least 197 amino acids and have a distinctive amino acid composition (averaging 33 mol% cysteine and 27 mol% glycine), with a strikingly ordered structure of glycine-rich and cysteine-rich repeats of approximately ten and 20 amino acids, respectively. Differences in the number and modular arrangement of these repeats suggest that there have been segmented rearrangements during the evolution of this family (Jenkins and Powell, 1994). There appear to be several human and sheep genes but perhaps as many as fifteen mouse genes in this family (Powell *et al.*, 1991).

In both cuticle-keratin gene families some genes are known to be closely linked but their relationship to one another is not known. In the human genome, the KAP5 genes have been mapped to two sites on chromosome 11, at 11q13 and 11p15 (MacKinnon *et al.*, 1991), but their sheep homologues have not yet been identified. The loci were designated the KRN1 and KRN1-like loci, respectively, named before the KAP/KRTAP nomenclature was proposed. A gene-specific probe for the KAP5.1 gene, known to be linked to KAP5.3p, a pseudogene (Jenkins and Powell, 1992), mapped it to 11q13. Other members of the family, numbering up to ten genes in the human genome (Powell *et al.*, 1991), may reside at the 11p15 region. A number of genes have been mapped to the human 11p15 band, foremost among them the β-globin locus (for a review, see Junien and McBride, 1989). This region has been conserved in several other mammals (for references, see Zneimer *et al.*, 1990), although there appears to have been a chromosomal rearrangement in the cattle genome, such that some genes (including the β-globin locus) are located on chromosome 15 (Fries *et al.*, 1988) and others on the unassigned linkage group U7 (Womack and Moll, 1986). The discovery of their relative locations in the sheep genome will indicate whether the cattle arrangement predates the cattle/sheep divergence.

There are no mapping data for any of the other KAP gene families that encode cysteine-rich wool proteins (KAP3, 4, 9 and 11), other than a reported linkage of three genes of the KAP9 family within an 18 kb piece of mouse DNA (McNab *et al.*, 1989).

The KAP6 gene family that encodes glycine/tyrosine-rich proteins numbers at least nine genes, which, in the sheep genome, are located within a 1 Mb segment of DNA, three of them within 40 kb (Fratini *et al.*, 1993). Restriction fragment length polymorphism studies indicate that the KAP8 gene, which encodes a different glycine/tyrosine-rich protein, is linked to this group (Parsons *et al.*, 1994b) and located on sheep chromosome 1 (Wood *et al.*, 1992). The location of the KAP7 gene, which encodes a similar protein, is at present unknown, but it is expected to map to the same region. Significantly, recent RFLP studies have indicated linkage between the KAP6 locus and wool-fibre diameter (Parsons *et al.*, 1994a). As keratin-associated protein 6 genes encode structural proteins, it is anticipated that another gene linked to the KAP locus is responsible for the effect.

Close genetic linkage has thus far been a typical finding for genes in hair-keratin families and will probably apply to the other unmapped KAP gene families as well. In a broader perspective, if the results of the cloning and mapping strategies that have linked KAP1 and KAP2 loci extend to the other KAP gene families, there is the possibility of a supercluster containing several KAP families and the KRT1 family, probably totalling more than 60 genes.

Regulation of wool-keratin gene expression

Compared with many other systems, little is known about how wool-keratin genes are regulated. A 400 bp promoter from the wool-keratin type II gene, K2.10, targets LacZ expression to the hair-follicle cortex of transgenic mice (Keough *et al.*, 1995), and a 600 bp promoter from the mouse-hair KAP9.1 gene is also effective in targeting expression (McNab *et al.*, 1989). Essential regulatory elements are therefore located in these short keratin-gene promoters.

To identify sequences that may be important in regulating gene expression, comparisons of promoter sequences are invaluable, particularly when they can be made between evolutionarily distant species, revealing what has been termed a phylogenetic footprint (Gumucio *et al.*, 1992). In a comparative survey of 15 hair-keratin gene promoters, Powell *et al.* (1991) noted several highly conserved DNA sequences grouped with particular gene families. One sequence, CTTTGAAG (originally designated hair keratin 1 (HK1)), is strictly conserved in the promoters of all hair-keratin IF genes, with at least one copy located within 180–240 bp upstream of the transcription-start site of each gene (Powell *et al.*, 1991, 1992). It is also present in the hair KAP gene promoters, although not perfectly conserved, and corresponds to the consensus binding site $\mathrm{CTTTG}\frac{\mathrm{AA}}{\mathrm{TT}}$ for lymphoid enhancer factor 1 (LEF-1) (Zhou *et al.*, 1995). Binding of recombinant human LEF-1 to the sheep K2.9 and K2.10 promoters has been shown *in vitro* (Zhou *et al.*, 1995). Lymphoid enhancer factor 1 appears to create a bend in the DNA helix, facilitating interactions of proteins bound at different sites in a promoter (Giese *et al.*, 1992, 1995). Zhou *et al.* (1995) suggest that LEF-1 is involved in creating favourable transcriptional domains in the hair-keratin gene promoters required for maintenance of hair-keratin gene expression. Mutagenesis of the LEF-1 site in the wool K2.10 promoter severely reduces the level of expression in the follicles of transgenic mice but does not abolish it (R.A. Keough, B.C. Powell and G.E. Rogers, manuscript in preparation). Other transcription factors must bind elsewhere in the 400 bp K2.10 promoter and dictate follicle specificity. Consistent with this, a methodical screen of the K2.10 promoter identified the binding sites for several proteins (Keough, 1995).

Many of the hair-keratin gene promoters, and, indeed, many of the epidermal-keratin gene promoters, contain potential binding sites for two common transcription factor complexes, AP1 and AP2 (Powell *et al.*, 1991, 1992; Fratini *et al.*, 1993). The involvement of AP1 and AP2 in regulating hair-keratin gene expression is reinforced by the detection of FOS, one of the proteins of the AP1 complex, in hair-follicle keratinocytes (Fisher *et al.*, 1991). Direct binding of purified AP2 to a wool-keratin promoter has recently been shown (S.M. Dunn and B.C. Powell, unpublished data). There are several examples, including a number of epidermal-keratin genes (Byrne and Fuchs, 1993; Magnaldo *et al.*, 1993; Casatorres *et al.*, 1994; Lu *et al.*, 1994), of AP1 and AP2 sites associated with binding sites for other regulatory factors, implicating their involvement in complex control mechanisms. In this regard, it is notable that the LEF-1 site is situated next to AP1 and AP2 consensus binding sequences in the wool K2.9 and K2.10 genes (Powell *et al.*, 1992).

Hair and wool mutations

The keratin IF loci are part of chromosomal segments that have been conserved during mammalian evolution (see Fig. 6.2) (Hart *et al.*, 1992; Nadeau *et al.*, 1992). Several hair mutations map to these regions in mice; two (*bareskin* and *rex*) map to the KRT1 locus and five (*shaven*, *caracul*, *naked*, *velvet coat* and *hairy ears*) map to the KRT2 locus. Whether these mutations directly affect hair-keratin genes or regulatory genes that map to the same locations, such as the homoeobox 3 family genes, is not known.

Over 50 mouse-hair mutations have been documented (for a review, see Sundberg, 1994), but we know of only a few wool mutants. Two interesting sheep-wool mutations are the *lustre* mutation and the *N* gene mutation (for a review, see Ryder and Stephenson, 1968). *Lustre* is a single-gene, dominant mutation, which has been recorded a number of times (for references, see Blair, 1990); it confers a lustrous appearance to the fleece, reduces follicle density and results in a deficiency of glycine/tyrosine-rich wool proteins (Gillespie and Darskus, 1971). The *lustre* mutation may be a regulatory gene defect, because the expression of several genes encoding glycine/tyrosine-rich proteins are affected. The *naked* mutation in mice shows some interesting parallels; mice exhibit a marked reduction in the amount of the glycine/tyrosine-rich proteins and produce malformed hairs (Tenenhouse and Gold, 1976; Raphael *et al.*, 1982, 1984). *Naked* maps in the vicinity of a number of regulatory genes, near the KRT2 locus (Sundberg, 1994).

The sheep *N* gene also has pleiotropic effects, causing the formation of horns and affecting a number of wool properties, including fibre diameter and medullation (Dry, 1955a). Sheep carrying the *N* gene form the basis of the carpet-wool industry in Australia and New Zealand. The *N* gene was first recognized in New Zealand Romneys, and a number of alleles are now known (Robards *et al.*, 1993). A recessive gene, *nr*, not allelic to *N*, produces similar phenotypic effects (Dry, 1955b). Although horns and wool are visibly different, there are some developmental similarities between them and they both contain substantial amounts of keratin proteins (Marshall and Gillespie, 1977; Chapman, 1986). The identities of the *N* and *nr* genes are not known, but the similar effects they exert indicate that they act in the same pathway, reminiscent of growth factor/growth-factor receptor relationships.

Metallothionein Genes

Metallothioneins (MTs) are small, cysteine-rich proteins that bind heavy metals. They are believed to be involved in zinc and copper homoeostasis, heavy-metal detoxification (for review, see Hamer, 1986) and the prevention of oxidative damage (Tamai *et al.*, 1993). Zinc is important for the activity of zinc-finger transcription factors (Zeng *et al.*, 1991; Berg and Shi, 1996) and is required by over 300 enzymes. Incidentally, in a rather circular fashion, a zinc-finger transcription factor has been shown to regulate MT gene expression (Brugnera *et al.*, 1994). Both zinc and copper are vital elements and their deficiency, resulting from nutritional or genetic defects, can have dramatic consequences (for reviews, see Danks, 1991; Neldner, 1991). In the context of this review, it is notable that cross-linking of hair-keratin proteins in the last stages of fibre formation is dependent on copper and that, in general, fibre formation and wool properties are sensitive to both zinc and copper deficiencies (for a review, see Chapman, 1989). Additionally, a mutation in a copper transporter is responsible for the human genetic disease, Menkes syndrome, one of whose symptoms is steely hair (Danks, 1991; Mercer *et al.*, 1993).

Metallothionein genes contain three exons and are approximately 2 kb in size. The high level of sequence identity between genes and between species is indicative of a common evolutionary origin. However, the number of MT genes seems to vary enormously between mammals; mice have at least four genes (Quaife *et al.*, 1994), humans at least 16, six of them pseudogenes (West *et al.*, 1990; Quaife *et al.*, 1994; Stennard *et al.*, 1994) and sheep probably nine, including at least one pseudogene (Peterson *et al.*, 1988).

Large segments of the MT loci have now been cloned in humans, mice and sheep. A 67 kb segment of the sheep MT locus contains five genes encoding the major MT proteins (Fig. 6.6). Three functional MT-I genes are centrally located in the sheep cluster, flanked by a functional MT-II gene at one end and an MT-I pseudogene at the other end. The locations of the other putative sheep MT genes are unknown, but, based on the recent finding of the MT-IV gene in humans and

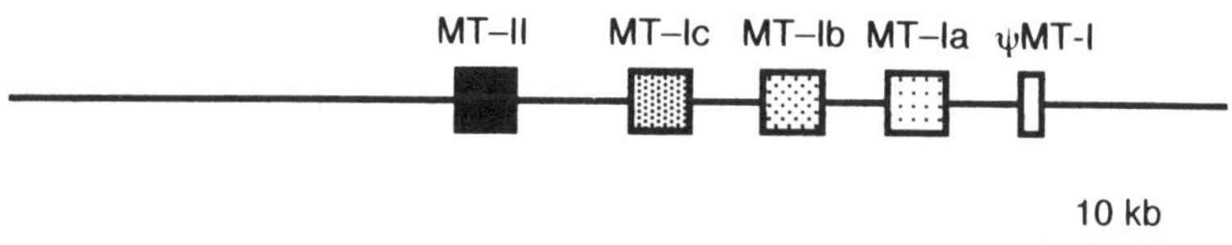

Fig. 6.6. Arrangement of the sheep metallothionein locus.

mice, 20 kb upstream of their MT-III genes (Quaife *et al.*, 1994), more sheep MT genes may lie in the vicinity of the existing cluster. The MT genes have not yet been mapped in sheep but their chromosomal assignments are known in several mammals, principally cattle (chromosome 18: Ryan and Womack, 1994), human (chromosome 16: West *et al.*, 1990) and mouse (chromosome 8: Fletcher *et al.*, 1991).

The sheep MT genes respond in a differential manner to heavy-metal inducers, and this can be explained in principle by the presence and degree of conservation of binding sites for various transcription factors and transcription-factor abundance. Metallothionein gene promoters contain a specific DNA motif, known as the metal-response element (MRE), usually present in the proximal promoter regions in multiple copies with slightly varying sequence and functionality (Karin and Holtgreve, 1984; Peterson and Mercer, 1986; Culotta and Hamer, 1989). For example, the mouse MT-I MREd element responds more efficiently than MREa, and when present in more than one copy produces a greater cooperative response (Culotta and Hamer, 1989). In sheep fibroblasts, the MT-II gene is expressed at a higher basal level than the MT-I genes, possibly the result of a more highly conserved MRE in its promoter (Peterson and Mercer, 1988). Expression of the MT genes, like the β-globin genes, appears to be regulated at a higher level by locus-control regions (LCRs) (see later) (Palmiter *et al.*, 1993).

In the development of sheep lung, only the MT-II gene is expressed, and then only transiently, whereas all four known sheep MT genes are expressed in complex patterns during liver development (Pitt *et al.*, 1992). The level of expression of MT-Ib reported by Pitt *et al.* (1992) differs from the early study conducted by Peterson *et al.* (1988) suggesting a polymorphic effect on MT gene expression between Australian and US sheep breeds.

It is clear that MT genes are regulated by metals and that MT proteins can bind metals but, despite being under investigation for 35 years, it is still not certain exactly what metallothioneins do. A double knockout of two of the four mouse MT genes (MT-I and MT-II) had no effect on mouse development apart from an increase in susceptibility to cadmium toxicity (Michalska and Choo, 1993; Masters *et al.*, 1994). Such a mild effect seems surprising. Other neutral knockouts have occurred with a variety of genes and they usually point to a functional redundancy in the pathway, but which proteins might cooperate with the MTs is not known.

Lysozyme Genes

The lysozyme genes of ruminants are a nice example of adaptive evolution, wherein new functions arise by gene duplication and mutation (for a review, see Irwin *et al.*, 1992). Most mammals have a single lysozyme gene, which functions in macrophages to destroy bacteria as part of immune-system activity. Ruminants have retained the original gene that works in the immune system but have evolved additional lysozyme genes to function in other tissues, notably stomach, kidney, tear and milk. There are a range of lysozyme genes in ruminants: two in camels and llamas and up to ten in advanced ruminants such as sheep, cow and goat. The chevrotain, a primitive ruminant, has an intermediate number.

Lysozyme is a bacteriolytic enzyme of 130 amino acids which hydrolyses the β-1,4 glycosidic bond of proteoglycans in bacterial cell walls. The adaptation of the primordial lysozyme gene to the ruminant stomach, where it facilitates foregut fermentation, is particularly interesting, because of the changes in gene-promoter function and protein stability required. In adapting to the ruminant stomach, lysozymes must work in an acidic environment, resisting acid hydrolysis and digestion by stomach proteases. The necessary amino acid changes seem to have occurred in two phases, a period of rapid mutation, presumably resulting from a strong positive selection pressure to function in the stomach, followed by a period of steady mutational gain (Irwin *et al.*, 1992). The initial period of rapid mutation is believed to have occurred before the radiation of the advanced ruminants, 30–40 million years ago.

Most of the lysozyme genes appear to reside at a single locus, a result of gene duplication along the chromosome. In the cow genome, they map to chromosome 5 in a 2–3 Mb segment of DNA, but a single lysozyme-like sequence also maps to chromosome 7 (Gallagher *et al.*, 1993). In the sheep genome, the major locus appears to be on chromosome 3 (Crawford *et al.*, 1995). Several cow lysozyme genes have been sequenced but, as yet, only complementary DNA (cDNA) data are available for sheep (Irwin and Wilson, 1990). The cow lysozyme genes have three introns, all occurring in the coding region of the genes.

At least three closely related lysozymes have been identified in sheep stomach (Irwin and Wilson, 1990) and a different lysozyme in sheep kidney (Ito *et al.*, 1993). There are a number of alleles of the sheep lysozymes (Irwin and Wilson, 1990; Davies and Maddox, 1994). The coding regions of the sheep-stomach lysozymes are very similar and seem to have evolved in a concerted fashion, most probably by gene conversion, whereas the 3′ non-coding regions probably evolved independently of each other (Irwin and Wilson, 1990). The sheep stomach lysozymes differ by at most seven amino acids and, as a group, exhibit about 40 different amino acid differences from the kidney enzyme. The stomach lysozymes have a low net charge (–2 to –3 at pH 7) compared with the kidney lysozyme (+2.5 at pH 7).

In the hope of identifying crucial changes in the evolution of lysozyme to function in the acidic environment of stomach, amino acid sequences from a range of ruminant and non-ruminant lysozymes have been compared. The sequence of the independently evolved langur-monkey stomach lysozyme has enabled comparisons to be made which are not biased by common ancestral heritage. A number of commonalities have been noted (Jolles *et al.*, 1990), but none are absolutely conserved or unique to all stomach lysozymes. A major change in the adaptation to function in the acidic environment of the stomach is anticipated to be a reduction in the number of acid-labile peptide bonds. The aspartyl–prolyl bond is the most labile of peptide bonds (Piszkiewicz *et al.*,

1970; Marcus, 1985), and one of them is present in all non-stomach lysozymes but not in the lysozymes of advanced ruminants. This change has been argued as an important evolutionary step in adaptation to stomach function. However, the aspartyl–prolyl peptide is present in the stomach lysozymes of the camel and langur, and its removal, therefore, does not seem to be essential for function in the stomach. A low net charge is considered a significant feature of enzyme adaptation to an acidic environment, but this criterion also does not rigorously apply. Low net charge is a property of ruminant lysozymes, but the langur enzyme has a higher charge than sheep kidney lysozyme.

However, there is a unique feature of stomach lysozymes, surprisingly not previously noted. One residue, a glutamyl residue at position 50, is completely conserved between the stomach enzymes and is not found in any non-stomach lysozyme. This charged residue might result in a conformational change, possibly important for lysozyme function in an acidic environment. *In vitro* mutagenesis of this residue and functional testing of the mutated lysozyme could indicate whether it has a significant effect on enzyme function and, therefore, its potential as a key evolutionary adaptation. In any event, the initial period of rapid mutation in the duplicated lysozyme gene(s), predicted from sequence comparisons, might reflect the need for several genetic mutations in the optimal adaptation of lysozyme to the stomach.

Globin Genes

The globin genes were among the first mammalian genes to be cloned and, as probably the best studied of mammalian gene families, they have advanced our understanding of gene organization, gene regulation and chromosome structure. In this, the study of sheep globins has made important contributions.

Two gene families produce the α- and β-globin polypeptides that combine to form tetrameric haemoglobin. Embryonic, fetal and adult forms of the globins are produced in a sequential manner to meet the oxygen requirements of the organism at each developmental period. Detailed studies have been made of the globin loci in the human, rabbit, mouse, sheep, cow and goat genomes, revealing that separate clusters for the α-globin and β-globin genes occur on different chromosomes (for reviews, see Higgs *et al.*, 1989; Townes and Behringer, 1990; Hardison and Miller, 1993). These studies have led to landmark findings in molecular biology, notably the concept of the LCR for eukaryotic gene regulation, and continue to improve our knowledge of how genes are regulated and interact in response to changing developmental programmes. Instrumental in this progress has been the study of globin-gene mutations, particularly the thalassaemias, which arise from deletions at the globin-gene loci.

The α- and β-globin genes had a common origin approximately 450 million years ago (Czelusniak *et al.*, 1982) and have retained a similar gene structure and organization, with ubiquitous and lineage-restricted transcription-factor binding sites responsible for their coordinate regulation. The genes are typically 1–2 kb in size, with three exons. Their chromosomal locations are known for a number of mammals but not yet for sheep. In the closely related cattle genome, the β-globin locus is located at 15q22–q27 and the α-globin locus is in linkage group U8, currently unassigned. Comparative analyses of the globin loci indicate that, since the mammalian radiation, globin genes have been prone to gene duplication, either as simple tandem gene duplications or as more extensive block duplications of several genes (Garner and Lingrel, 1988, 1989; Hardison and Miller, 1993; Ristaldi *et al.*, 1995). The sheep globin loci exemplify these events.

The structure of the sheep α-globin locus appears to be relatively simple compared with the β-globin locus. The ancestral mammalian globin family contained four genes, ζ–5′α–3′α–θ. In the sheep, only the α-globin genes have been identified thus far. However, like other mammalian α-globin loci, including the locus in the closely related goat genome (Hardison and Miller, 1993), the sheep locus probably also contains an embryonically expressed gene (ζ) located at the extreme 5′ end of the cluster. The existence of a θ gene at the 3′ end of the sheep cluster is more problematic, since it has not been found at the mouse, goat or horse α-globin loci.

The major sheep α-globin haplotype contains two tandem α-globin genes, α^{I} and α^{II} (Fig. 6.7). Duplications of the sheep α^{II} gene, but not α^{I}, occur reasonably frequently (Rando *et al.*, 1986). In a survey of 60 sheep, the α^{I}/α^{II} haplotype was found in 88% of animals, the $\alpha^{I}/\alpha^{II}/\alpha^{II}$ haplotype in 10% and the $\alpha^{I}/\alpha^{II}/\alpha^{II}/\alpha^{II}$ haplotype in 2% of the sample population (Rando *et al.*,

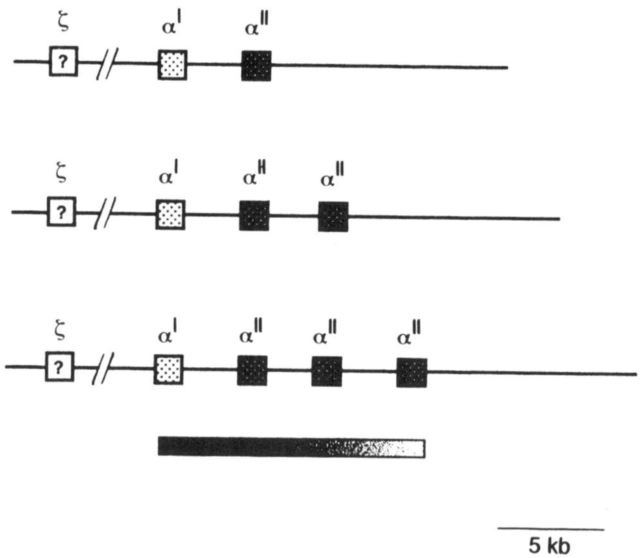

Fig. 6.7. Arrangement of the sheep α-globin genes in the different haplotypes. The graded bar signifies a decrease in expression levels from left to right, 5′ to 3′. The existence of the ζ-globin gene is postulated from the closely related goat α-globin locus.

1986). Both the α^{I} and α^{II} genes function in fetal and adult life, but they are expressed at different levels (Vestri *et al.*, 1991). The α^{I} mRNA and protein are approximately twice as abundant as the products from the 3′ α^{II} gene, a difference exaggerated for transcripts from the more 3′ α^{II} genes in the triplicated and quadruplicated haplotypes (Vestri *et al.*, 1994). This polarity effect suggests that the 3′ α^{II} genes are at a transcriptional disadvantage, perhaps signifying that the original duplication which generated the second α^{II} gene interrupted an enhancer or, alternatively, that competition for an enhancer located beyond the 5′ end of the cluster is sensitive to distance effects. Although an enhancer is classically defined as being able to act in a position-independent manner, their operative ranges have not been rigorously tested. In the human and mouse α-globin loci, an enhancer is known to lie 40 kb upstream of the embryonic gene (Higgs *et al.*, 1990; Jarman *et al.*, 1991; Kielman *et al.*, 1993), an arrangement that presumably also applies to the sheep α-globin cluster. Conceivably, the enhancer may be operating near the limit of its range, such that the promoters of α-globin genes duplicated to a more 3′ location in different sheep haplotypes cannot compete as effectively for its activity as closer genes. The polarity effect also occurs at the human α-globin locus, where the order of the α genes (5′ α^{II}, α^{I} 3′) is reversed with respect to the sheep locus. In this case the level of the α^{II} protein is approximately twice as abundant as the α^{I} protein (Liebhaber *et al.*, 1986). Furthermore, the expression-gradient phenomenon also appears to occur in some duplicated human α-globin and γ-globin haplotypes (for references, see Vestri *et al.*, 1994).

The β-globin locus has been studied more thoroughly than the α-globin locus. It is part of a syntenic group, including the *Ha-ras*, parathyroid hormone and lactate dehydrogenase genes, whose conservation has been documented during the evolution of the human, cat, mouse, rabbit and cattle genomes. Sequences related to the hair-cuticle KAP5 keratins also map in the vicinity of the human β-globin locus at 11p15 (MacKinnon *et al.*, 1991).

Extensive regions of the β-globin locus have been sequenced in several genomes: over 70 kb for the human, 55 kb for the mouse and 45 kb for the rabbit locus (for references, see Hardison and Miller, 1993) but only about 12 kb for the sheep locus (Kretschmer *et al.*, 1981; Garner and Lingrel, 1989; Saban and King, 1994). The ancestral mammalian β-globin cluster contained five genes, 5′ ε–γ–η–δ–β 3′. In the ancestor of sheep, goats and cattle, the γ-globin gene was lost, the δ-globin gene was inactivated and the remaining four-gene set, 5′ ε-ε(η)-ψβ(δ)-β 3′, was duplicated. This set underwent another duplication in the sheep/goat ancestor, resulting in loci that contain 12 β-globin genes, known as the A haplotype in sheep (Fig. 6.8). Subsequently, a deletion of one of the four gene sets seems to have occurred, such that some sheep have eight β-globin genes: the B haplotype (Garner and Lingrel, 1988, 1989). Haplotype frequencies in the population appear to differ, depending on geographical location. In some European sheep the two haplotypes are equally prevalent, whereas in Italian sheep the B haplotype predominates, being present in approximately 85% of the population (Hadjisterkotis *et al.*, 1995).

Of all the β-globin genes in the sheep haplotypes, only the two most 5′

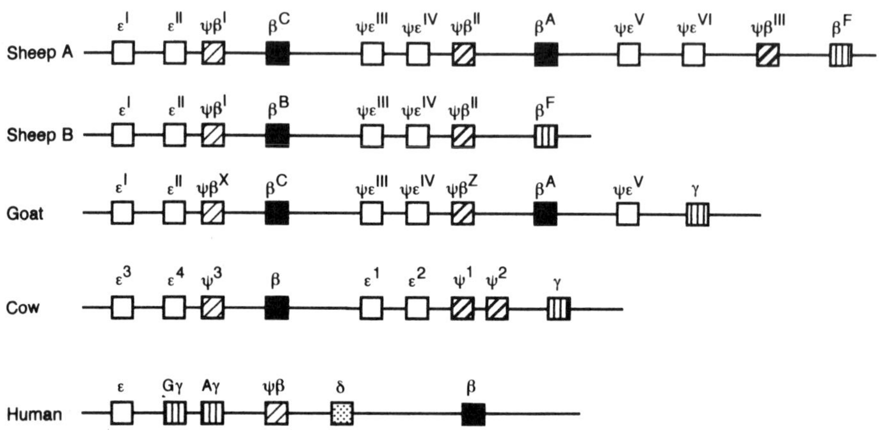

Fig. 6.8. Comparison of β-globin loci. Genes with similar expression characteristics in different species are coded alike. In the sheep A and B haplotypes, only the ε^{I}, ε^{II} and β genes are functional. (Compiled from Schimenti and Duncan, 1985; Garner and Lingrel, 1988, 1989; Hardison and Miller, 1993.)

embryonic genes (ε^{I} and ε^{II}) and the β genes remain functional. After the sheep β genes were duplicated, they evolved new developmental functions and came under different regulatory controls (Huisman *et al.*, 1969). The β^{F}-globin, which is common to both haplotypes, is expressed in the fetus. In the A haplotype, β^{C}-globin is expressed from birth to approximately 6 months of age and β^{A}-globin is expressed in mature adults. However, in goats, and probably in sheep, severe anaemia can reverse the switch from β^{C} to β^{A}. In the B haplotype, the β gene has mutated to a distinctive form, β^{B}, now distinguishable from the β^{C} and β^{A} genes of the A haplotype in sequence and regulation. The β^{B}-globin appears at birth, as the β^{F} gene is switched off, and performs the functions carried out by both β^{C} and β^{A} in the A haplotype.

The structure and regulation of the human β-globin locus is known in fine detail, and comparative studies suggest that regulation of the locus is fundamentally similar in all mammals (Hardison and Miller, 1993). Studies of the β-globin locus, particularly the interest in the nature of some thalassaemia mutations, in which far-upstream regions are deleted, led to the pioneering concept of the LCR as a major regulatory feature governing gene control (Forrester *et al.*, 1987; Grosveld *et al.*, 1987). The β-globin LCR is essential for high-level expression of the β-globin genes and is deleted in some thalassaemias. In the human locus, there are four prominent deoxyribonuclease (DNase) I hypersensitive (HS) sites located outside the cluster, 5–20 kb 5′ to the embryonic globin gene (Fig. 6.9). Deoxyribonuclease I HS sites represent exposed regions in the chromatin where transcription factors are bound, regulating gene activity. Locus-control region activity has been mapped to short core regions in the HS2, HS3 and HS4 sites (Philipsen *et al.*, 1990; Pruzina *et al.*, 1991; Talbot and

Grosveld, 1991). Multiple combinations of binding sites for lineage-restricted and ubiquitously expressed transcription factors have been discovered at all these sites, but the HS3 site appears to be of paramount importance for globin-gene regulation. Recent studies (Ellis *et al.*, 1996) have now shown that the HS3 site has a novel activity that opens and remodels chromatin structure, enabling transcription factors to bind and regulate globin-gene activity throughout the locus (Fig. 6.9).

Although there is no specific information for the postulated sheep globin LCR, extensive evidence has been presented describing the LCR at the closely

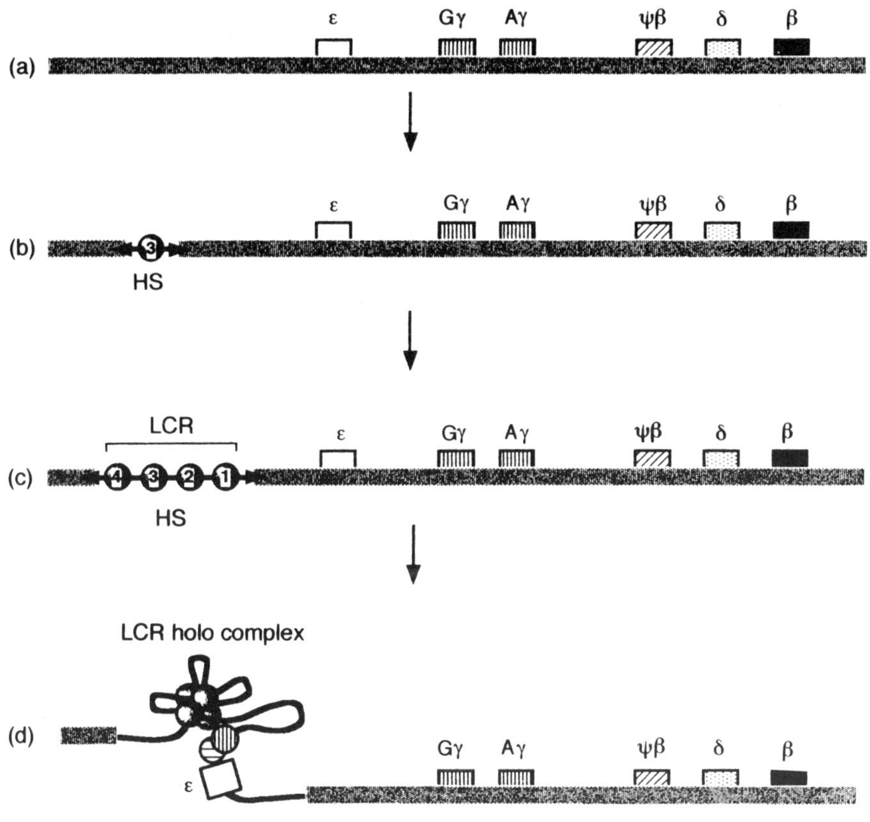

Fig. 6.9. Possible mechanism of β-globin LCR function. (a) The human β-globin locus is shown in a transcriptionally repressed state (symbolized by the wide line obscuring the genes). (b) Establishment of an active transcriptional domain at the locus is initiated by proteins binding to the HS3 site, with acquisition of concomitant DNase I nuclease sensitivity (symbolized by a thin line). Chromatin opening spreads from HS3 (arrows). (c) The entire LCR, comprising of four HS sites, is opened. (d) Proteins bound at the HS sites form a holocomplex, which interacts via DNA looping, with proteins bound at the promoters of the globin genes, depending upon the developmental stage. Interaction with the ε-globin gene promoter is shown. Not drawn to scale. (Adapted from Ellis *et al.*, 1996, by permission of Oxford University Press.)

related goat β-globin locus (Li *et al.*, 1990, 1991). Three HS sites are located 5–15 kb 5′ to the embryonic goat globin gene, corresponding to human HS1, HS2 and HS3, and they have been highly conserved during evolution (Hardison and Miller, 1993). Regulation of β-globin gene activity at the triplicated goat locus and the duplicated and triplicated sheep loci probably parallels that described for the simpler human β-globin locus. However, the 5′ and 3′ ends of the four-gene set that has been duplicated in the sheep and goat genomes have not been defined, and the possibility exists that some or all of the HS sites may also have been duplicated. This would represent a significant difference between the β-globin loci of artiodactyls and other mammals and would have interesting implications for their regulation. Another distinction between the sheep and human loci is the relative position of the fetal globin gene. In sheep, there is a single fetal gene (β^F) at the extreme 3′ end of the cluster, whereas in the human locus there are two fetal genes (Gγ and Aγ), which are located after the embryonic gene and before the adult gene, in the developmental order in which the genes are expressed. Thus, the coincidence of gene polarity and the developmental progression of gene switching at the human locus does not occur at the sheep locus.

The β-globin genes are controlled by the 5′ LCR, but the α-globin genes appear to be regulated in a somewhat different manner. There is a prominent DNase I HS site (HS-40) approximately 40 kb upstream of the embryonic α-globin gene, which is important for regulating α-globin gene expression (Higgs *et al.*, 1990; Jarman *et al.*, 1991) and which was originally believed to act like the β-globin LCR. However, recent evidence indicates that they are not functionally equivalent (Craddock *et al.*, 1995). The reasons for this may lie in the nature of the chromatin domain in which the genes reside, which has implications for chromosome structure and gene regulation in general (Hardison and Miller, 1993; Craddock *et al.*, 1995). The α-globin locus is located in a guanidine–cytidine (GC)-rich isochore, a segment of the chromosome some hundreds of kb in length that is early-replicating in all cell types, whereas the β-globin locus is located in a GC-poor isochore that is only early-replicating in erythroid cells. The α-globin locus is surrounded by widely expressed genes (Vyas *et al.*, 1992; Kielman *et al.*, 1993; Vickers *et al.*, 1993) and appears to be part of a region of relatively open chromatin structure in all cell types, but the β-globin locus is only found in an open chromatin configuration in erythroid cells, a direct result of LCR action. It has been suggested that the presence of widely expressed or housekeeping genes in the vicinity of the α-globin locus maintains the chromatin in an open configuration and that the role of the HS-40 site is to function as a strong enhancer for the α-globin genes (Craddock *et al.*, 1995). In contrast, the β-globin LCR actively opens the chromatin to promote β-globin gene transcription (Ellis *et al.*, 1996). These two mechanisms of gene control may occur regularly throughout genomes, according to the chromatin environment in which the genes are located, and may be identifiable by the G-banding pattern of chromosomes (Craddock *et al.*, 1995).

Conclusions

Evolution is slow and conservative, the products and processes of genome function being similar in different species. Although sheep molecular genetics is in its infancy, its progress will benefit enormously from advances in several other genome-mapping and sequencing projects, most notably the human genome project. In the next decade, the complete sequence of the human genome should be known and we can expect to see the rapid transfer of much of that information to related genomes, such as those of cattle and sheep. This will accelerate the acquisition of sheep molecular information, at present limited to many short fragments of genes and a handful of complete genes.

The economic rewards of wool have driven wool research for decades, producing significant advances in our knowledge of the wool proteins and genes but few insights into the developmental processes that establish and regulate wool-follicle activity. Because wool and hair are very similar, information from the analyses of the many documented mouse-hair mutations and the growing number of mouse-gene knockouts that affect hair development should augment the pace of this research in sheep.

There are some interesting parallels between the keratin IF and globin gene systems. In both cases, the functional protein product requires the obligatory pairing of two types of proteins, namely type I IF with type II IF and α-globin with β-globin. Several genes exist for each protein component, clustered in the genome and located on different chromosomes from the genes for the partner. Whether the regulatory mechanisms controlling the keratin IF genes recapitulate the mechanisms exhibited by either the α- or the β-globin loci is an intriguing question.

Globins are one of the foremost gene systems on which molecular research embarked, and interest in evolutionary processes has been one of the reasons why the globin loci of so many species have been studied. The pioneering research on the regulation of the α- and β-globin loci is revealing fundamental information on different chromosomal environments and how they influence gene expression; the LCR is a prime example of this. The markedly different gene organization of the sheep β-globin locus, where segmental duplication and triplication of sets of genes has occurred in the A and B haplotypes, promises to enhance our understanding of competitive gene control and the mechanism of genome evolution.

Acknowledgements

I would like to acknowledge the financial support provided by the International Wool Secretariat and also the support of my colleagues in the wool-keratin laboratory, particularly Simon Bawden, who shares my interest in higher-order gene control.

References

Berg, J.M. and Shi, Y. (1996) The galvanization of biology: a growing appreciation for the roles of zinc. *Science* 271, 1081–1085.

Bertolino, A.P., Checkla, D.M., Notterman, R., Sklaver, I., Schiff, T.A., Freedberg, I. and DiDona, G.J. (1988) Cloning and characterization of a mouse type I hair keratin cDNA. *Journal of Investigative Dermatology* 91, 541–546.

Bertolino, A.P., Checkla, D.M., Heitner, S., Freedberg, I.M. and Yu, D. (1990) Differential expression of type I hair keratins. *Journal of Investigative Dermatology* 94, 297–303.

Blair, H.T. (1990) Inheritance of a major gene for excessively lustrous wool in sheep. *Journal of Heredity* 81, 220–222.

Blumenberg, M. (1988) Concerted gene duplications in the two keratin gene families. *Journal of Molecular Evolution* 27, 203–211.

Britten, R.J. (1996) Cases of ancient mobile element DNA insertions that now affect gene regulation. *Molecular Phylogenetics and Evolution* 5, 13–17.

Brugnera, E., Georgiev, O., Radtke, F., Heuchel, R., Baker, E., Sutherland, G.R. and Schaffner, W. (1994) Cloning, chromosomal mapping and characterization of the human metal-regulatory transcription factor MTF-1. *Nucleic Acids Research* 22, 3167–3173.

Byrne, C. and Fuchs, E. (1993) Probing keratinocyte and differentiation specificity of the human K5 promoter *in vitro* and in transgenic mice. *Molecular and Cellular Biology* 13, 3176–3190.

Campuzano, V., Montermini, L., Molto, M.D., Pianese, L., Cossee, M., Cavalcanti, F., Monros, E., Rodius, F., Duclos, F., Monticelli, A., Zara, F., Canizares, J., Koutnikova, H., Bidichandani, S.I., Gellera, C., Brice, A., Trouillas, P., Demichele, G., Filla, A. and Defrutos, R. (1996) Friedreichs ataxia – autosomal recessive disease caused by an intronic GAA triplet repeat expansion. *Science* 271, 1423–1427.

Casatorres, J., Navarro, J.M., Blessing, M. and Jorcano, J.L. (1994) Analysis of the control of expression and tissue specificity of the keratin 5 gene characteristic of basal keratinocytes. *Journal of Biological Chemistry* 269, 20489–20496.

Chapman, R.E. (1986) Hair, wool, quill, nail, claw, hoof and horn. In: Bereiter-Hahn, J., Matoltsy, A.G. and Richards, K.S. (eds) *Biology of the Integument*, Vol. 2. Springer-Verlag, Berlin, pp. 293–317.

Chapman, R.E. (1989) Follicular malfunctions and resultant effects on wool fibres. In: Rogers, G.E., Reis, P.J., Ward, K.A. and Marshall, R.C. (eds) *The Biology of Wool and Hair.* Chapman and Hall, London, pp. 243–256.

Compton, J.G., Ferrera, D.M., Yu, D.-W., Recca, V., Freedberg, I.M. and Bertolino, A.P. (1991) Chromosomal localization of mouse hair keratin genes. *Annals of the New York Academy of Sciences* 642, 32–43.

Craddock, C.F., Vyas, P., Sharpe, J.A., Ayyub, H., Wood, W.G. and Higgs, D.R. (1995) Contrasting effects of alpha and beta globin regulatory elements on chromatin structure may be related to their different chromosomal environments. *European Molecular Biology Organization Journal* 14, 1718–1726.

Crawford, A.M., Dodds, K.G., Ede, A.J., Pierson, C.A., Montgomery, G.W., Garmonsway, H.G., Beattie, A.E., Davies, K., Maddox, J.F., Kappes, S.W., Stone, R.T., Nguyen, T.C., Penty, J.M., Lord, E.A., Broom, J.E., Buitkamp, J., Schwaiger, W., Epplen, J.T., Matthew, P. and Matthews, M.E. (1995) An autosomal genetic linkage map of the sheep genome. *Genetics* 140, 703–724.

Crewther, W.G. (1976) Primary structure and chemical properties of wool. In: Ziegler,

K. (ed.) *Proceedings of the 5th International Wool Textile Research Conference* Vol. I. German Wool Research Institute, Aachen, pp. 1–101.

Culotta, V.C. and Hamer, D.H. (1989) Fine mapping of a mouse metallothionein gene metal response element. *Molecular and Cellular Biology* 9, 1376–1380.

Czelusniak, J., Goodman, M., Hewett-Emmett, D., Weiss, M.L., Venta, P.J. and Tashian, R.E. (1982) Phylogenetic origins and adaptive evolution of avian and mammalian haemoglobin genes. *Nature* 298, 297–300.

Danks, D.A. (1991) Copper deficiency and the skin. In: Goldsmith, L.A. (ed.) *Physiology, Biochemistry, and Molecular Biology of the Skin*, Vol. 2. Oxford University Press, Oxford, pp. 1351–1361.

Davies, K.P. and Maddox, J.F. (1994) Ovine dinucleotide repeat polymorphism at the lysozyme (LYZ) locus. *Animal Genetics* 25, 204.

Dawkins, R. (1976) *The Selfish Gene*. Oxford University Press, London.

Deininger, P.L., Batzer, M.A., Hutchinson III, C.A. and Edgell, M.H. (1992) Master genes in mammalian repetitive DNA amplification. *Trends in Genetics* 8, 307–311.

Dry, F.W. (1955a) The dominant *N* gene in New Zealand Romney sheep. *Australian Journal of Agricultural Research* 6, 725–769.

Dry, F.W. (1955b) The recessive *N* gene in New Zealand Romney sheep. *Australian Journal of Agricultural Research* 6, 833–862.

Duncan, C.H. (1987) Novel Alu-type repeat in artiodactyls. *Nucleic Acids Research* 15, 1340.

Elliott, R.W., Lee, B.K. and Eicher, E.M. (1990) Localization of the growth hormone gene to the distal half of mouse chromosome 11. *Genomics* 8, 591–594.

Ellis, J., Tan-Un, K.C., Harper, A., Michalovich, D., Yannoutos, N., Philipsen, S. and Grosveld, F. (1996) A dominant chromatin-opening activity in 5′ hypersensitive site 3 of the human β-globin locus control region. *European Molecular Biology Organization Journal* 15, 562–568.

Eppig, J.T. and Nadeau, J.H. (1995) Comparative maps: the mammalian jigsaw puzzle. *Current Opinion in Genetics and Development* 5, 709–716.

Filon, M., Sarafian, V., Lussier, M., Belanger, C., Lapointe, L. and Royal, A. (1994) Arrangement of a cluster of three mouse type I keratin genes expressed sequentially during esophageal-type epithelial cell differentiation. *Genomics* 24, 303–310.

Fisher, C., Byers, M.R., Iadrola, M.J. and Powers, E.A. (1991) Patterns of epithelial expression of Fos protein suggest an important role in the transition from viable to cornified cell during keratinization. *Development* 111, 253–258.

Fletcher, C., Norman, D.J., Germond, E. and Heintz, N. (1991) A multilocus linkage map of mouse chromosome 8. *Genomics* 9, 737–741.

Forrester, W.C., Takegawa, S., Papayannopoulou, T., Stamatoyannopoulos, G. and Groudine, M. (1987) Evidence for a locus activation region: the formation of developmentally stable hypersensitive sites in globin-expressing hybrids. *Nucleic Acids Research* 15, 10159–10177.

Fratini, A., Powell, B.C. and Rogers, G.E. (1993) Sequence, expression and evolutionary conservation of a gene encoding a glycine/tyrosine-rich keratin-associated protein of hair. *Journal of Biological Chemistry* 268, 4511–4518.

Fratini, A., Powell, B.C., Hynd, P.I., Keough, R.A. and Rogers, G.E. (1994) Dietary cysteine regulates the levels of mRNAs encoding a family of cysteine-rich proteins of hair. *Journal of Investigative Dermatology* 102, 178–185.

Frenkel, M.J., Sleigh, M.J., Ward, K.A., Powell, B.C. and Rogers, G.E. (1989) The keratin BIIIB family: isolation of cDNA clones and structure of a gene and related pseudogene. *Genomics* 4, 182–191.

Fries, R., Hediger, R. and Stranzinger, G. (1988) The loci for parathyroid hormone and β-globin are closely linked and map to chromosome 15 in cattle. *Genomics* 3, 302–307.

Fries, R., Threadgill, D.W., Hediger, R., Gunawardana, A., Blessing, M., Jorcano, J.L., Stranzinger, G. and Womack, J.E. (1991) Mapping of bovine cytokeratin sequences to four different sites on three chromosomes. *Cytogenetics and Cell Genetics* 57, 135–141.

Gallagher, D.S., Jr, Threadgill, D.W., Ryan, A.M., Womack, J.E. and Irwin, D.M. (1993) Physical mapping of the lysozyme gene family in cattle. *Mammalian Genome* 4, 368–373.

Gardiner, K. (1995) Human genome organization. *Current Opinion in Genetics and Development* 5, 315–322.

Garner, K.J. and Lingrel, J.B. (1988) Structural organization of the β-globin locus of B-haplotype sheep. *Molecular Biology and Evolution* 5, 134–140.

Garner, K.J. and Lingrel, J.B. (1989) A comparison of the β^A- and β^B-globin gene clusters in sheep. *Journal of Molecular Evolution* 28, 175–184.

Giese, K., Cox, J. and Grosschedl, R. (1992) The HMG domain of lymphoid enhancer factor 1 bends DNA and facilitates assembly of functional nucleoprotein complexes. *Cell* 69, 185–195.

Giese, K., Kingsley, C., Kirshner, J.R. and Grosschedl, R. (1995) Assembly and function of a TCRα enhancer complex is dependent on LEF-1-induced DNA bending and multiple protein–protein interactions. *Genes and Development* 9, 995–1008.

Gillespie, J.M. and Darskus, R.L. (1971) Relation between the tyrosine content of various wools and their content of a class of proteins rich in tyrosine and glycine. *Australian Journal of Biological Science* 24, 1189–1197.

Grosveld, F., Assendelft, G.B.V., Greaves, D.R. and Kollias, G. (1987) Position-independent, high-level expression of the human β-globin gene in transgenic mice. *Cell* 51, 975–985.

Gumucio, D.L., Heilstedt-Williamson, H., Gray, T.A., Tarle, S.A., Shelton, D.A., Tagle, D.A., Slightom, J.L., Goodman, M. and Collins, F.S. (1992) Phylogenetic footprinting reveals a nuclear protein which binds to silencer sequences in the human γ and ε globin genes. *Molecular and Cellular Biology* 12, 4919–4929.

Gunawardana, A. and Fries, R. (1992) Assignment of the HOX2 and HOX3 gene clusters to the bovine chromosome regions 19q17-qter and 5q14-23. *Animal Genetics* 23, 161–165.

Hadjisterkotis, E., Manca, L., Naitana, S. and Masala, B. (1995) A comparative study on the haemoglobin polymorphism of domestic sheep of the islands of Chios, Cyprus and Sardinia. *Comparative Biochemistry and Physiology: Physiology* 112, 547–552.

Hamer, D.H. (1986) Metallothionein. *Annual Review of Biochemistry* 55, 913–951.

Hardison, R. and Miller, W. (1993) Use of long sequence alignments to study the evolution and regulation of mammalian globin gene clusters. *Molecular Biology and Evolution* 10, 73–102.

Hart, C.P., Compton, J.G., Langley, S.H., Hunihan, L., LeClair, K.P., Zelent, A., Roderick, T.H. and Ruddle, F.H. (1992) Genetic linkage analysis of the murine developmental mutant velvet coat (Ve) and the distal chromosome 15 developmental genes Hox-3.1, Rar-γ, Wnt-1 and Krt-2. *Journal of Experimental Zoology* 263, 83–95.

Hediger, R., Johnson, S.E., Barendse, W., Drinkwater, R.D., Moore, S.S. and Hetzel, J. (1990) Assignment of the growth hormone gene locus to 19q26-qter in cattle and to 11q25-qter in sheep by *in situ* hybridization. *Genomics* 8, 171–174.

Hediger, R., Ansari, H.A. and Stranzinger, G.F. (1991) Chromosome banding and gene localizations support extensive conservation of chromosome structure between cattle and sheep. *Cytogenetics and Cell Genetics* 57, 127–134.

Higgs, D.R., Vickers, M.A., Wilkie, A.O., Pretorius, I.M., Jarman, A.P. and Weatherall, D.J. (1989) A review of the molecular genetics of the human alpha-globin gene cluster. *Blood* 73, 1081–1104.

Higgs, D.R., Wood, W.G., Jarman, A.P., Sharp, J., Lida, J., Pretorius, I.M. and Ayuub, H. (1990) A major positive regulatory region located far upstream of the human α-globin gene locus. *Genes and Development* 4, 1588–1601.

Huh, N.-h., Kashiwagi, M., Konishi, C., Hashimoto, Y., Kohno, Y., Nomura, S. and Kuroki, T. (1994) Isolation and characterization of a novel hair follicle-specific gene, Hacl-1. *Journal of Investigative Dermatology* 102, 716–720.

Huisman, T.H.J., Lewis, J.P., Blunt, M.H., Adams, H.R., Miller, A., Dozy, A.M. and Boyd, E.M. (1969) Hemoglobin C in newborn sheep and goats: a possible explanation for its function and biosynthesis. *Pediatric Research* 3, 189–198.

Irwin, D.M. and Wilson, A.C. (1990) Concerted evolution of ruminant stomach lysozymes: characterization of lysozyme cDNA clones from sheep and deer. *Journal of Biological Chemistry* 265, 4944–4952.

Irwin, D.M., Prager, E.M. and Wilson, A.C. (1992) Evolutionary genetics of ruminant lysozymes. *Animal Genetics* 23, 193–202.

Ito, Y., Yamada, H., Nakamura, M., Yoshikawa, A., Ueda, T. and Imoto, T. (1993) The primary structures and properties of non-stomach lysozymes of sheep and cow, and implication for functional divergence of lysozyme. *European Journal of Biochemistry* 213, 649–658.

Jagadeeswaran, P., Tuan, D., Forget, B.G. and Weissman, S.M. (1982) A gene deletion ending at the midpoint of a repetitive DNA sequence in one form of hereditary persistence of fetal hemoglobin. *Nature* 296, 469–470.

Jarman, A.P., Wood, W.G., Sharpe, J.A., Gourdon, G., Ayyub, H. and Higgs, D.R. (1991) Characterisation of the major regulatory element upstream of the human α-globin gene cluster. *Molecular and Cellular Biology* 11, 4679–4689.

Jenkins, B.J. and Powell, B.C. (1992) Isolation and characterization of a sheep cysteine-rich cuticle keratin pseudogene. *DNA Sequence* 3, 181–184.

Jenkins, B.J. and Powell, B.C. (1994) Differential expression of genes encoding a cysteine-rich keratin family in the hair cuticle. *Journal of Investigative Dermatology* 103, 310–317.

Jobse, C., Buntjer, J.B., Haagsma, N., Breukelman, H.J., Beintema, J.J. and Lenstra, J.A. (1995) Evolution and recombination of bovine DNA repeats. *Journal of Molecular Evolution* 41, 277–283.

Jolles, J., Prager, E.M., Alnemri, E.S., Jolles, P., Ibrahimi, I.M. and Wilson, A.C. (1990) Amino acid sequences of stomach and non-stomach lysozymes of ruminants. *Journal of Molecular Evolution* 30, 370–382.

Junien, C. and McBride, O.W. (1989) Report of the committee on the genetic constitution of chromosome 11. *Cytogenetics and Cell Genetics* 51, 226–258.

Karin, M. and Holtgreve, H. (1984) Nucleotide sequence requirements for transient expression of human metallothionein-IIA-thymidine kinase fusion genes. *DNA* 3, 319–325.

Kaytes, P., McNab, A.R., Rea, T.J., Groppi, V., Kawabe, T.T., Buhl, A.E., Berolino, A.P., Hatzenbuhler, N.T. and Vogeli, G. (1991) Hair-specific keratins: characterization and expression of a mouse type I keratin gene. *Journal of Investigative Dermatology* 97, 835–842.

Keough, R.A. (1995) An investigation of hair follicle cell immortalisation and hair keratin gene regulation. PhD thesis, Adelaide.

Keough, R.A., Powell, B.C. and Rogers, G.E. (1995) Targeted expression of SV-40 T antigen in the hair follicle of transgenic mice produces an aberrant phenotype. *Journal of Cell Science* 108, 957–966.

Kielman, M.F., Smits, R., Devi, T.S., Fodde, R. and Bernini, L.F. (1993) Homology of a 130-kb region enclosing the alpha-globin gene cluster, the alpha-locus controlling region, and two non-globin genes in human and mouse. *Mammalian Genome* 4, 314–323.

Kretschmer, P.J., Coon, H.C., Davis, A., Harrison, M. and Nienhuis, A.W. (1981) Hemoglobin switching in sheep: isolation of the fetal γ-globin gene and demonstration that the fetal γ- and adult β^A-globin genes lie within eight kilobase segments of DNA. *Journal of Biological Chemistry* 256, 1975–1982.

Lenstra, J.A., van Boxtel, J.A., Zwaagstra, K.A. and Schwerin, M. (1993) Short interspersed nuclear element (SINE) sequences of the Bovidae. *Animal Genetics* 24, 33–39.

Lessin, S., Huebner, R.K., Isobe, M., Croce, C.M. and Steinert, P.M. (1988) Chromosomal mapping of human keratin genes: evidence of non-linkage. *Journal of Investigative Dermatology* 91, 572–578.

Lewin, B. (1994) *Genes V.* Oxford University Press, New York.

Li, Q., Zhou, B., Powers, P., Enver, T. and Stamatoyannopoulos, G. (1990) β-Globin locus activation regions: conservation of organization, structure and function. *Proceedings of the National Academy of Sciences of the USA* 87, 8207–8211.

Li, Q., Zhou, B., Powers, P., Enver, T. and Stamatoyannopoulos, G. (1991) Primary structure of the goat β-globin locus control region. *Genomics* 9, 488–499.

Liebhaber, S.A., Cash, F.E. and Ballas, S.K. (1986) Human α globin gene expression: the dominant role of the α-2 locus in mRNA and protein synthesis. *Journal of Biological Chemistry* 261, 15327–15333.

Lohe, A.R. and Hilliker, A.J. (1995) Return of the H-word (heterochromatin). *Current Opinion in Genetics and Development* 5, 746–755.

Lu, B., Rothnagel, J.A., Longley, M.A., Tsai, S.Y. and Roop, D.R. (1994) Differentiation-specific expression of human keratin 1 is mediated by a composite AP-1/steroid hormone element. *Journal of Biological Chemistry* 269, 7443–7449.

MacKinnon, P.J., Powell, B.C. and Rogers, G.E. (1990) Structure and expression of genes for a class of cysteine-rich proteins of the cuticle layers of differentiating wool and hair follicles. *Journal of Cell Biology* 111, 2587–2600.

MacKinnon, P.J., Powell, B.C., Rogers, G.E., Baker, E.G., MacKinnon, R.N., Hyland, V.J., Callen, D.F. and Sutherland, G.R. (1991) An ultrahigh-sulphur keratin gene of the human hair cuticle is located at 11q13 and cross-hybridizes with sequences at 11p15. *Mammalian Genome* 1, 53–56.

McNab, A.R., Wood, L., Theriault, N., Gierman, T. and Vogeli, G. (1989) An ultra-high sulphur keratin gene is expressed specifically during hair growth. *Journal of Investigative Dermatology* 92, 263–266.

Magnaldo, T., Bernerd, F., Freedberg, I.W., Ohtsuki, M. and Blumenberg, M. (1993) Transcriptional regulators of expression of K#16, the disease-associated keratin. *DNA and Cell Biology* 12, 911–923.

Marcus, F. (1985) Preferential cleavage at aspartyl–prolyl peptide bonds in dilute acid. *International Journal of Peptide and Protein Research* 25, 542–546.

Marshall, R.C. and Gillespie, J.M. (1977) The keratin proteins of wool, horn and hoof from sheep. *Australian Journal of Biological Science* 30, 389–400.

Masters, B.A., Kelly, E.J., Quaife, C.J., Brinster, R.L. and Palmiter, R.D. (1994) Targeted disruption of metallothionein I and II genes increases sensitivity to cadmium. *Proceedings of the National Academy of Sciences of the USA* 91, 584–588.

Mercer, J.F.B., Livingston, J., Hall, B., Paynter, J.A., Begy, C., Chandrasekharappa, S., Lockhart, P., Grimes, A., Bhave, M., Siemeiniak, D. and Glover, T.W. (1993) Isolation of a partial candidate gene for Menkes disease by positional cloning. *Nature Genetics* 3, 20–25.

Michalska, A.E. and Choo, K.H. (1993) Targeting and germ-line transmission of a null mutation at the metallothionein I and II loci in mouse. *Proceedings of the National Academy of Sciences of the USA* 90, 8088–8092.

Moll, R., Franke, W.W., Schiller, D.L., Geiger, B. and Krepler, R. (1982) The catalog of human cytokeratin polypeptides: patterns of expression of specific cytokeratins in normal epithelia, tumours and cultured cells. *Cell* 31, 11–24.

Nadeau, J.H., Berger, F.G., Cox, D.R., Crosby, J.L., Davisson, M.T., Ferrara, D., Fuchs, E., Hart, C., Hunihan, L., Lalley, P.A., Langley, S.H., Martin, G.R., Nichols, L., Philips, S.J., Roderick, T.H., Roop, D.R., Ruddle, F.H., Skow, L.C. and Compton, J.G. (1989) A family of type I keratin genes and the homeobox-2 gene complex are closely linked to the rex locus on mouse chromosome 11. *Genomics* 5, 454–462.

Nadeau, J.H., Compton, J.G., Giguere, V., Rossant, J. and Varmuza, S. (1992) Close linkage of retinoic acid receptor genes with homeobox- and keratin-encoding genes on paralogous segments of mouse chromosomes 11 and 15. *Mammalian Genome* 3, 202–208.

Neldner, K.H. (1991) The biochemistry and physiology of zinc metabolism. In: Goldsmith, L.A. (ed.) *Physiology, Biochemistry, and Molecular Biology of the Skin*, Vol. 2. Oxford University Press, Oxford, pp. 1329–1350.

Palmiter, R.D., Sandgren, E.P., Koeller, D.M. and Brinster, R.L. (1993) Distal regulatory elements from the mouse metallothionein locus stimulate gene expression in transgenic mice. *Molecular and Cellular Biology* 13, 5266–5275.

Parsons, Y.M., Cooper, D.W. and Piper, L.R. (1994a) Evidence of linkage between high-glycine–tyrosine keratin gene loci and wool fibre diameter in a Merino half-sib family. *Animal Genetics* 25, 105–108.

Parsons, Y.M., Piper, L.R. and Cooper, D.W. (1994b) Linkage relationships between keratin-associated protein (KRTAP) genes and growth hormone in sheep. *Genomics* 20, 500–502.

Peterson, M.G. and Mercer, J.F. (1986) Structure and regulation of the sheep metallothionein-Ia gene. *European Journal of Biochemistry* 160, 579–585.

Peterson, M.G. and Mercer, J.F. (1988) Differential expression of four linked sheep metallothionein genes. *European Journal of Biochemistry* 174, 425–429.

Peterson, M.G., Hannan, F. and Mercer, J.F. (1988) The sheep metallothionein gene family: structure, sequence and evolutionary relationship of five linked genes. *European Journal of Biochemistry* 174, 417–424.

Philipsen, S., Talbot, D., Fraser, P. and Grosveld, F. (1990) The β-globin dominant control region: hypersensitive site 2. *European Molecular Biology Organization Journal* 9, 2159–2167.

Piszkiewicz, D., Landon, M. and Smith, E.L. (1970) Anomalous cleavage of aspartyl–proline peptide bonds during amino acid sequence determinations. *Biochemical and Biophysical Research Communications* 40, 1173–1178.

Pitt, B.R., Brookens, M.A., Steve, A.R., Atlas, A.B., Davies, P., Kuo, S.M. and Lazo, J.S. (1992) Expression of pulmonary metallothionein genes in late gestational lambs. *Pediatric Research* 32, 424–430.

Popescu, N.C., Bowden, P.E. and DiPaolo, J.A. (1989) Two type II keratin genes are localized on human chromosome 12. *Human Genetics* 8, 109–112.

Powell, B.C. and Beltrame, J.S. (1994) Characterisation of a hair (wool) keratin intermediate filament gene domain. *Journal of Investigative Dermatology* 102, 171–177.

Powell, B.C. and Rogers, G.E. (1986) Hair keratin: composition, structure and biogenesis. In: Bereiter-Hahn, J., Matoltsy, A.G. and Richards, K.S. (eds) *Biology of the Integument*, Vol. 2. Springer-Verlag, Berlin, pp. 696–721.

Powell, B.C. and Rogers, G.E. (1994) Differentiation in hard keratin tissues: hair and related structures. In: Leigh, I., Watt, F. and Lane, E.B. (eds) *The Keratinocyte Handbook*. Cambridge University Press, Cambridge, pp. 401–436.

Powell, B.C. and Rogers, G.E. (1997) The role of keratin proteins and their genes in the growth, structure and properties of hair. In: Jolles, P. and Zahn, H. (eds) *Formation and Structure of Human Hair.* Birkhäuser, Verlag, Switzerland, pp. 59–148.

Powell, B.C., Sleigh, M.J., Ward, K.A. and Rogers, G.E. (1983) Mammalian keratin gene families: organization of genes coding for the B2 high-sulphur proteins of sheep wool. *Nucleic Acids Research* 11, 5327–5346.

Powell, B.C., Cam, G.R., Fietz, M.J. and Rogers, G.E. (1986) Clustered arrangement of keratin intermediate filament genes. *Proceedings of the National Academy of Sciences of the USA* 83, 5048–5052.

Powell, B.C., Nesci, A. and Rogers, G.E. (1991) Regulation of keratin gene expression in hair follicle differentiation. *Annals of the New York Academy of Sciences* 642, 1–20.

Powell, B.C., Crocker, L.A. and Rogers, G.E. (1992) Hair follicle differentiation: expression, structure and evolutionary conservation of the hair type II keratin intermediate filament gene family. *Development* 114, 417–434.

Powell, B.C., Arthur, J. and Nesci, A. (1995) Characterization of a gene encoding a cysteine-rich keratin associated protein synthesized late in rabbit hair follicle differentiation. *Differentiation* 58, 227–232.

Pruzina, S., Hanscombe, O., Whyatt, D. and Grosveld, F. (1991) Hypersensitive site 4 of the human β globin locus control region. *Nucleic Acids Research* 19, 1413–1419.

Quaife, C.J., Findley, S.D., Erickson, J.C., Froelick, G.J., Kelly, E.J., Zambrowicz, B.P. and Palmiter, R.D. (1994) Induction of a new metallothionein isoform (MT-IV) occurs during differentiation of stratified squamous epithelia. *Biochemistry* 33, 7250–7259.

Rando, A., Ramunno, L. and Masina, P. (1986) Variation in α-globin loci in sheep. *Molecular Biology and Evolution* 3, 168–176.

Raphael, K.A., Chapman, R.E., Frith, P.E. and Pennycuick, P.R. (1982) The structure of hair and follicles of mice carrying the naked (*N*) gene. *Genetical Research Cambridge* 39, 139–148.

Raphael, K.A., Marshall, R.C. and Pennycuick, P.R. (1984) Protein and amino acid composition of hair from mice carrying the naked (*N*) gene. *Genetical Research Cambridge* 44, 29–38.

Reuter, G. and Spierer, P. (1992) Position effect variegation and chromatin proteins. *BioEssays* 14, 605–612.

Ristaldi, M.S., Casula, S., Rando, A. and Vestri, R. (1995) Sheep alpha-globin gene sequences – implications for their concerted evolution and for the down-regulation of the 3′ genes. *Journal of Molecular Evolution* 40, 349–353.

Robards, G., Champion, S., Kurtz, N. and Kurtz, L. (1993) A brief history of the development in Australasia of speciality carpet wool sheep. *Wool Technology and Sheep Breeding* 41, 379–387.

Rogers, G.E. (1984) Studies on keratin multigene families. In: Campbell, P.N. and Phelps, C.F. (eds) *Biochemical Society Symposium*. Symposium 49, Biochemical Society, London, pp. 85–108.

Rogers, G.R., Hickford, J.G.H. and Bickerstaff, R. (1994) Polymorphism in two genes for B2 high sulphur proteins of wool. *Animal Genetics* 25, 407–415.

Rogers, J.H. (1985) The origin and evolution of retrosposons. *International Review of Cytology* 93, 187–279.

Romano, V., Bosco, P., Rocchi, M., Costa, G., Leube, R.E., Franke, W.W. and Romeo, G. (1988) Chromosomal assignments of human type I and type II cytokeratin genes to different chromosomes. *Cytogenetics and Cell Genetics* 48, 148–151.

Rosenberg, M., RayChaudhury, A., Shows, T.B., LeBeau, M.M. and Fuchs, E.V. (1988) A group of type I keratin genes on human chromosome 17: characterization and expression. *Molecular and Cellular Biology* 8, 722–736.

Rosenberg, M., Fuchs, E., Lebeau, M.M., Eddy, R.L. and Shows, T.B. (1991) Three epidermal and one simple epithelial type II keratin genes map to human chromosome 12. *Cytogenetics and Cell Genetics* 57, 33–38.

Ryan, A.M. and Womack, J.E. (1994) Assignment of the metallothionein 2 gene (MT2A) to bovine chromosome 18 by somatic cell analysis. *Animal Genetics* 25, 196.

Ryder, M.L. and Stephenson, S.K. (1968) *Wool Growth*. Academic Press, London.

Saban, J. and King, D. (1994) Sequence of the sheep fetal beta globin gene and flanking region. *Biochimica et Biophysica Acta* 1218, 87–90.

Savtchenko, E.S., Tomic, M., Ivker, R. and Blumenberg, M. (1990) Three parallel linkage groups of human acidic keratin genes. *Genomics* 7, 394–407.

Schimenti, J.C. and Duncan, C.H. (1985) Structure and organization of the bovine beta-globin genes. *Molecular Biology and Evolution* 2, 514–525.

Shaffer, C.D., Wallrath, L.L. and Elgin, S.C.R. (1993) Regulating genes by packaging domains: bits of heterochromatin in euchromatin? *Trends in Genetics* 9, 35–37.

Skowronski, J., Plucienniczak, A., Bednarek, A. and Jaworski, J. (1984) Bovine 1.709 satellite: recombination hotspots and dispersed repeated sequences. *Journal of Molecular Biology* 177, 399–416.

Spence, S.E., Young, R.M., Garner, K.J. and Lingrel, J.B. (1985) Localization and characterization of members of a family of repetitive sequences in the goat β globin locus. *Nucleic Acids Research* 13, 2171–2186.

Stennard, F.A., Holloway, A.F., Hamilton, J. and West, A.K. (1994) Characterisation of six additional human metallothionein genes. *Biochimica et Biophysica Acta* 1218, 357–365.

Sundberg, J.P. (ed.) (1994) *Handbook of Mouse Mutations with Skin and Hair Abnormalities*. CRC Press, Boca Raton, Florida.

Sunkel, C.E. and Coelho, P.A. (1995) The elusive centromere: sequence divergence and functional conservation. *Current Opinion in Genetics and Development* 5, 756–767.

Sutherland, G.R. and Richards, R.I. (1995) Simple tandem DNA repeats and human genetic disease. *Proceedings of the National Academy of Sciences of the USA* 92, 3636–3641.

Swart, L.S., Joubert, F.J. and Parris, D. (1976) Homology in the amino acid sequences of the high sulphur proteins from keratins. In: Ziegler, K. (ed.) *Proceedings of the 5th International Wool Textile Research Conference*, Vol. II. German Wool Research Institute, Aachen, pp. 254–263.

Talbot, D. and Grosveld, F. (1991) The 5′ HS2 of the β globin locus control region enhances transcription through the interaction of a multimeric complex binding at two functionally distinct NF-E2 sites. *European Molecular Biology Organization Journal* 10, 1391–1398.

Tamai, K.T., Gralla, E.B., Ellerby, L.M., Valentine, J.S. and Thiele, D.J. (1993) Yeast and mammalian metallothioneins functionally substitute for yeast copper–zinc superoxide dismutase. *Proceedings of the National Academy of Sciences of the USA* 90, 8013–8017.

Tenenhouse, H.S. and Gold, R.J.M. (1976) Loss of a homologous group of proteins in a dominantly inherited ectodermal malformation. *Biochemical Journal* 159, 149–160.

Tobiasch, E., Schweizer, J. and Winter, H. (1992a) Structure and site of expression of a murine type II hair keratin. *Molecular Biology Reports* 16, 39–47.

Tobiasch, E., Winter, H. and Schweizer, J. (1992b) Structural features and sites of expression of a new 65kD and 48kD hair-related keratin pair associated with a special type of parakeratotic epithelial differentiation. *Differentiation* 50, 163–178.

Townes, T.M. and Behringer, R.R. (1990) Human globin locus activation region (LAR): role in temporal control. *Trends in Genetics* 6, 219–223.

Vestri, R., Pieragostini, E., Yang, F., Gregorio, P.D., Rando, A. and Masina, P. (1991) Expression of triplicated and quadruplicated α-globin genes in sheep. *British Journal of Haematology* 77, 110–116.

Vestri, R., Pieragostini, E. and Ristaldi, M.S. (1994) Expression gradient in sheep alpha–alpha and alpha–alpha–alpha globin gene haplotypes – messenger RNA levels. *Blood* 83, 2317–2322.

Vickers, M.A., Vyas, P., Harris, P.C., Simmons, D.L. and Higgs, D.R. (1993) Structure of the human 3-methyladenine DNA glycosylase gene and localization close to the 16p telomere. *Proceedings of the National Academy of Sciences of the USA* 90, 3437–3441.

Vyas, P., Vickers, M.A., Simmons, D.L., Ayyub, H., Craddock, C.F. and Higgs, D.R. (1992) *Cis*-acting sequences regulating expression of the human alpha-globin cluster lie within constitutively open chromatin. *Cell* 69, 781–793.

Wallace, M.R., Anderson, L.B., Saulino, A.M., Gregory, P.E., Glover, T.W. and Collins, F.S. (1991) A *de novo* Alu insertion results in neurofibromatosis type 1. *Nature* 353, 864–866.

Waseem, A., Gough, A.C., Spurr, N.K. and Lane, E.B. (1990) Localization of the gene for human simple epithelial keratin 18 to chromosome 12 using polymerase chain reaction. *Genomics* 7, 188–194.

West, A.K., Stallings, R., Hildebrand, C.E., Chiu, R., Karin, M. and Richards, R.I. (1990) Human metallothionein genes: structure of the functional locus at 16q13. *Genomics* 8, 513–518.

Wilson, B.W., Edwards, K.J., Sleigh, M.J., Byrne, C.R. and Ward, K.A. (1988) Complete sequence of a type I-microfibrillar keratin gene. *Gene* 73, 21–31.

Winter, H., Siry, P., Tobiasch, E. and Schweizer, J. (1994) Sequence and expression of murine type I hair keratins mHa2 and mHa3. *Experimental Cell Research* 212, 190–200.

Womack, J.E. and Kata, S.R. (1995) Bovine genome mapping: evolutionary influence and the power of comparative genomics. *Current Opinion in Genetics and Development* 5, 725–733.

Womack, J.E. and Moll, Y.D. (1986) Gene map of the cow: conservation of linkage with mouse and man. *Journal of Heredity* 77, 2–7.

Wood, N.J., Phua, S.H. and Crawford, A.M. (1992) A dinucleotide repeat polymorphism at the glycine- and tyrosine-rich keratin locus in sheep. *Animal Genetics* 23, 391.

Yoon, S.-J., LeBlanc-Straceski, J., Ward, D., Krauter, K. and Kucherlapati, R. (1994) Organization of the human keratin type II gene cluster at 12q13. *Genomics* 24, 502–508.

Yu, J., Yu, D.-W., Checkla, D.M., Freedberg, I.M. and Bertolino, A.P. (1993) Human hair keratins. *Journal of Investigative Dermatology* 101, 56S–59S.

Zeng, J., Vallee, B.L. and Kagi, J.H.R. (1991) Zinc transfer from transcription factor IIIA fingers to thionein clusters. *Proceedings of the National Academy of Sciences of the USA* 88, 9984–9988.

Zhou, P., Byrne, C., Jacobs, J. and Fuchs, E. (1995) Lymphoid enhancer factor 1 directs hair follicle patterning and epithelial cell fate. *Genes and Development* 9, 570–583.

Zneimer, S.M., Crawford, D., Schneider, N.R. and Beutler, B. (1990) Mapping of the human ribonuclease inhibitor gene (RNH) to chromosome 11p15 by *in situ* hybridization. *Genomics* 8, 175–178.

Molecular Genetics of Immune Molecules

7

W.R. HEIN
Basle Institute for Immunology, Grenzacherstrasse 487, Postfach CH-4005, Basle, Switzerland

Introduction 183
Immunoglobulins 184
T-cell Receptors 186
Major Histocompatibility Complex Antigens 188
  Sheep major histocompatibility complex class I genes 188
  Sheep major histocompatibility complex class II genes 189
Cytokines and Their Receptors 190
Other Immune Molecules 191
Conclusions 192
Acknowledgements 193
References 193

Introduction

The recognition and effector functions of the immune system are mediated by specialized molecules, which may be anchored in the plasma membrane, dissolved in the cytoplasm or secreted by its constituent cells. A large number of different molecules expressed in these compartments are now known to play a role in immunity, and, in the past few years, significant advances have been made in characterizing some of these in sheep. These advances will be reviewed briefly in this chapter. The majority of the data now available have been derived from the analysis of complementary deoxyribonucleic acid (cDNA) clones. This has been a useful way of identifying transcribed genes, deducing the primary structure of immunologically important proteins and giving early indications of the likely number and complexity of some of the gene families involved. While in most respects the structure of individual molecules resembles closely those in other mammals, a few interesting differences, which might have functional significance, have also emerged. So far, there is less information available about the genomic structure and organization of genes encoding immune molecules in sheep.

Immunoglobulins

The immunoglobulins (Igs) are important as both recognition and effector molecules. When expressed on the surface of B lymphocytes, Ig molecules comprise heavy and light chains, each of which contains constant and variable regions. The unique combination of variable regions in the Ig molecules expressed by an individual B cell forms the structural basis for the specificity of antigen recognition. Although this specificity of recognition is preserved after secretion, fluid-phase Igs also promote a number of effector functions that are mediated by interactions between the Ig constant regions and appropriate receptors on other cells. These effector functions are not influenced by the antigen specificity of Igs but do differ between Igs of different classes, i.e. Ig molecules which have different constant regions.

The Igs are encoded at three distinct loci for the heavy chain (*IGH*), κ light chain (*IGK*) and λ light chain (*IGL*). Each locus contains tandem arrays of genes encoding variable (*V*), joining (*J*) and constant (*C*) regions, although the number and arrangement of individual genes is highly variable. The *IGH* locus contains additional diversity (*D*)-region genes and multiple *C*-region gene segments. B cells are able to change expression from IgM to one of the downstream *C*-region genes by a process termed isotype switching. Sheep B lymphocytes express four major classes of heavy chains (IgM, IgG, IgA and IgE) and two classes of light chains (IgK and IgL). So far, there is no evidence in sheep for a fifth heavy-chain class, IgD. This gene is expressed in humans and rodents but is absent from chickens and rabbits, and probably from cattle, so it may well be the case that sheep also lack IgD. Three sheep heavy chains, namely IgM (Patri and Nau, 1992; Hein and Dudler, 1993a), IgG (Foley and Beh, 1989; Clarkson *et al.*, 1993) and IgE (Engewerda *et al.*, 1992), and both IgK (Foley and Beh, 1992) and IgL (Foley and Beh, 1989) light chains have been cloned (Table 7.1). The *IGKC* gene contains a *Taq*I polymorphism (Parsons *et al.*, 1993) and occurs on sheep chromosome 3p22–p17 (Broad *et al.*, 1995).

The overall structures of the constant and variable regions of sheep Igs are similar to those of other mammals and will not be discussed in detail here, since they are well described in many standard textbooks. One interesting exception concerns the structure of the constant region of sheep IgG. There are usually several distinct subtypes of IgG in mammals, and so far two have been cloned in sheep, *IGGC1* and *IGGC2*. The structure of *IGGC1* is comparable to other mammalian IgG molecules, but the hinge region of *IGGC2* in sheep, and also in other ruminants, is shortened. Modelling studies indicate that the close positioning of the Fab and Fc faces produces a compact molecule with little scope for angular rotational movements of Fab relative to Fc (Clarkson *et al.*, 1993). The resulting steric hindrance may limit the accessibility of the molecule to Fc receptors expressed on other cells and could explain some of the functional differences observed between sheep IgG isotypes. For example, sheep IgG1, but not IgG2, is selectively transported into colostrum, milk and other exocrine secretions, and sheep alveolar macrophages bind monomeric

Table 7.1. List of published sheep immunoglobulin and T-cell receptor sequences.

Sequence	Reference
Immunoglobulins	
Constant-region genes	
IGMC	Patri and Nau, 1992; Hein and Dudler, 1993a
IGGCI	Foley and Beh, 1989; Clarkson *et al.*, 1993
IGGC2	Clarkson *et al.*, 1993
IGEC	Engewerda *et al.*, 1992
IGLC	Foley and Beh, 1989
IGKC	Foley and Beh, 1992
Variable-region genes	
IGHV	Engewerda *et al.*, 1992; Patri and Nau, 1992; Dufour *et al.*, 1996
IGLV	Foley and Beh, 1989; Reynaud *et al.*, 1991, 1995
IGKV	Foley and Beh, 1992
T-cell receptors	
Constant-region genes	
TCRAC	Hein *et al.*, 1991
TCRBC	Grossberger *et al.*, 1993
TCRGC	Hein *et al.*, 1990; Hein and Dudler, 1993b
TCRDC	Hein *et al.*, 1990
Variable-region genes	
TCRAV	Hein *et al.*, 1991
TCRBV	Grossberger *et al.*, 1993
TCRGV	Hein and Dudler, 1993b
TCRDV	Hein and Dudler, 1993b
Invariant-component genes	
CD3G	Hein and Tunnacliffe, 1990
CD3D	Hein and Tunnacliffe, 1990
CD3E	Hein and Tunnacliffe, 1993
TCRZ	Hein and Tunnacliffe, 1993

IgG1 but not IgG2 (see discussion in Clarkson *et al.*, 1993).

Provisional estimates of the numbers of gene families and individual gene segments encoding constant and variable regions of sheep Igs can be made. Sheep appear to contain single *IGMC*, *IGEC* and *IGKC* genes. Two allelic variants of *IGMC* have been identified, but nothing is known about their relative frequencies (Hein and Dudler, 1993a). As mentioned above, there are at least two *IGGC* genes, and Southern blot analysis indicates the presence of a multiple, but not completely resolved, number of *IGLC* genes (Foley and Beh, 1989). Present indications suggest the presence of about 90–100 *IGLV* genes, 60–70 of which are functional with the remainder being pseudogenes (Reynaud *et al.*, 1995). There are probably no more than about ten functional *IGKV* genes (Foley and Beh, 1992; W.R. Hein and L. Dudler, unpublished data). The ratio of functional *IGLV* to *IGKV* genes in the germ line is therefore of the order of

6 : 1, slightly lower than the observed phenotypic ratio expressed on peripheral sheep B cells, which is about 9 : 1. Factors that might in part account for this discrepancy, such as whether or not there is a preferential order of rearrangement at one or other of the loci during B-cell ontogeny or whether there is a biased selection of B cells, have not yet been examined experimentally.

Nine germ-line heavy-chain *V*-region segments, six of which are potentially functional and three of which are pseudogenes, have recently been isolated from sheep DNA (Dufour *et al.*, 1996). The *IGHV* segments belong to a homogeneous gene family (> 80% sequence identity) that is related to the human VH4 family and to the murine VH1 subgroup. Southern-blot experiments indicate that the nine genes isolated constitute the major part of the *IGHV* locus. All expressed sheep *IGHV* segments examined are closely homologous to the germ-line sequences, so that the sheep *IGHV* repertoire seems to be derived from a relatively small germ-line gene family (Dufour *et al.*, 1996).

Perhaps the most interesting feature that so far distinguishes sheep Igs from those of many other mammals is the genetic mechanism used to diversify the primary repertoire. The ileal Peyer's patch (IPP) is the major B-cell-generating organ in sheep and is the source of most peripheral B cells (Griebel and Hein, 1996). During B-cell development in the IPP, sheep *IGLV* genes accumulate somatic mutations in the DNA encoding the complementarity-determining regions (CDRs), the site of contact between antibody molecules and antigens (Reynaud *et al.*, 1991). Oddly, the mutation process is not affected by exposure to either foreign antigens or T cells, as is the case with hypermutation occurring in germinal centres. Moreover, the patterns of nucleotide replacements suggest that the mutations are targeted mechanistically to the CDR codons independently of antigen selection (Reynaud *et al.*, 1995). In view of the reasonably large number of germ-line Ig light-chain genes, sheep should be capable of producing a large primary repertoire by conventional gene rearrangement, and it seems paradoxical that they make extensive use of a specialized hypermutation mechanism. Therefore, the reasons why sheep use this specialized system of somatic diversification and the advantages that it confers remain somewhat obscure at present.

T-cell Receptors

T-cell receptors (Tcrs) are the functional analogue of surface Igs and confer specificity of antigen recognition to T cells. There are three distinct loci of Tcr genes, α–δ (*TCRA–D*), β (*TCRB*) and γ (*TCRG*), each of which contains tandem arrays of individual gene segments, as for the Ig loci. The *TCRD* locus is nested within the *TCRA* locus. Like Igs, Tcrs are assembled during ontogeny by the ordered rearrangement of discrete *V*, *D* and *J* gene segments, which then become spliced to a *C*-region gene. In their mature form, Tcr proteins are expressed on the cell surface in two distinct heterodimeric forms, the αβ Tcr and γδ Tcr.

Sheep possess a single *TCRAC* and a single *TCRDC* gene (Hein *et al.*,

1990, 1991), probably two *TCRBC* genes (Grossberger *et al.*, 1993) and at least five *TCRGC* genes (Hein *et al.*, 1990; Hein and Dudler, 1993b). The sheep *TCRB* locus has been mapped to chromosome 4q32–qter (Pearce *et al.*, 1995). All five of the known *TCRGC* genes are expressed at the transcript level, but their usage is developmentally regulated. The *TCRGC2* and *TCRGC4* genes have only been detected during fetal development, *TCRGC5* is used perinatally and *TCRGC1* is only used postnatally. The most frequently used *TCRGC* gene is *TCRGC3* and transcripts are abundant from early stages of fetal development until well into adult life (Hein and Dudler, 1993b). The regulated expression of an unusually large number of *TCRGC* genes is a feature so far unique to sheep, although cattle are likely to be similar (Ishiguro and Hein, 1994). The sheep Tcr Cγ proteins are also distinguished by their heterodox structure, in that they are the longest Tcr chains so far identified. They contain additional sequence elements in the connecting peptide region of the constant domains and have either one or, in some cases, two additional cysteine residues (Hein *et al.*, 1990; Hein and Dudler, 1993b). It is not known whether these unusual structural features confer special functional properties to the sheep γδ Tcr.

The sheep *TCRAV* and *TCRBV* gene repertoires have not been extensively characterized so far, but the early indications suggest that they are likely to resemble their human and rodent counterparts, in that there is a large numer of different *V*-gene segments at these loci (Hein *et al.*, 1991; Grossberger *et al.*, 1993). More information is available about the number of *TCRGV* and *TCRDV* genes (Hein and Dudler, 1993b). In comparison with other animals, sheep contain a large number of these genes. So far, 12 different *TCRGV* gene segments have been identified and, based on their sequence homologies, these comprise seven families. Four distinct families of *TCRDV* genes are known and 28 different *V*-gene segments have been isolated as cDNA. Twenty five of these are members of the *TCRDV1* family, and they seem to have arisen by extensive gene duplication. Cattle contain a comparable number of closely homologous *V* genes at these loci, so that the duplications probably accumulated in mammalian lineages preceding the speciation of ruminants. The total number of *V* genes at these loci is not known with any certainty. However, based on their relative representation in the cDNA clones sequenced so far and on the analysis of genomic DNA by Southern hybridization, a provisional estimate of about 20 *TCRGV* genes and perhaps 40–60 *TCRDV* genes seems justified. So far, six *TCRGJ* and three *TCRDJ* gene segments have been identified (Hein and Dudler, 1993b; W.R. Hein and L. Dudler, unpublished data).

There is nothing unusual about the primary structure of individual *TCRGV* and *TCRDV* genes and they conform to the normal mammalian prototypes. However, the pool of sheep *TCRDV* genes seems to have been selected in some way on an evolutionary time-scale. By comparing the sequences, two short, distinctly hypervariable regions have been localized and these are reminiscent of the CDR regions found more typically in *IGV* genes. A similar, albeit less distinctly hypervariable, region occurs in sheep *TCRGV* genes (Hein and Dudler, 1993b). The presence of these regions suggests a possible functional role in the recognition of antigens although there is so far no evidence to either confirm or refute this notion.

In addition to the heterodimers that mediate antigen recognition, the Tcr contains several invariant proteins, the CD3γ, CD3δ, CD3ε and Tcrζ chains, which play important roles in the assembly of the receptor and in the transmission of signals to the cytoplasm. The Tcrz chain is frequently referred to as CD3z but it is unrelated to the other CD3 subunits. All of the sheep molecules have been cloned as cDNA (Hein and Tunnacliffe, 1990, 1993). There are minor differences between the structures of CD3 proteins in sheep and those in other species and these have been reviewed recently (Ishiguro and Hein, 1994). All of the invariant components are encoded by single genes. As in humans and mice, the sheep *CD3E*, *CD3G* and *CD3D* genes are closely linked on the chromosome; they have many structural features in common and probably developed by gene duplication. The sheep *TCRZ* gene has distinct structural features and is located in the genome remote from the *CD3* genes (Hein and Tunnacliffe, 1990, 1993). It has recently been mapped to chromosome 1p14–p11 (Ansari *et al.*, 1994).

Major Histocompatibility Complex Antigens

The major histocompatibility complex (MHC) is an assembly of genes encoding both cell-surface and other molecules, many of which play crucial roles in the immune system. The MHC is usually subdivided into three regions, termed MHC class I, II and III. The MHC class I and II regions are further subdivided into a and b subclasses. MHC class Ia and IIa genes are highly polymorphic and encode molecules that bind and present peptides to CD8$^+$ and CD4$^+$ T cells, respectively, and these are frequently referred to as the 'classical MHC' molecules. The MHC class Ib, IIb and III regions are less polymorphic and encode relatively invariant molecules, which also play a role in the immune system.

Collectively, the use of serological techniques, heterologous or homologous DNA probes, cDNA and genomic cloning and production of monoclonal antibodies indicate that the sheep MHC system (*Ovar-Mhc*) contains class I, class II and class III genes (Chardon *et al.*, 1985). The *Ovar-Mhc* region has been localized to chromosome 20q15–q23 (Mahdy *et al.*, 1989). However, in comparison with humans and rodents and, indeed, several other domestic animals, the overall structure and organization of the *Ovar-Mhc* region remains poorly characterized.

Sheep major histocompatibility complex class I genes

The products of sheep class I genes were identified first using serology to define segregating antigenic specificities. Three independent studies commencing in the 1970s each identified 11–13 lymphocyte surface antigens controlled by a single genetic system. The distribution of antigens in the sheep populations studied indicated that the system was controlled by autosomal codominant genes of at least two linked loci (Millot, 1979; Stear and Spooner, 1981; Cullen *et al.*, 1982). The lymphocyte antigens were later confirmed as *Ovar-MhcI* determinants by showing that the segregation of restriction fragment length polymorphism (RFLP) banding patterns detected with human MHC class I cDNA probes paralleled the haplotype segregation established by serological typing (Chardon *et al.*, 1985).

Three different sheep class I MHC proteins were identified by monoclonal antibodies (Puri *et al.*, 1987a). The three species of protein would require no more than two loci since the differences between them could be allele- rather than locus-specific. However, the subsequent isolation of five distinct classes of *Ovar-MhcI* cDNAs from the thymus of a single animal (Grossberger *et al.*, 1990) requires the expression of at least three loci. A dinucleotide repeat polymorphism has been defined within *Ovar-MhcI* (Groth and Wetherall, 1994). The predicted structure of *Ovar-MhcI* molecules is similar to that of the class Ia proteins of other species. No information is currently available about sheep class Ib proteins.

Sheep major histocompatibility complex class II genes

The MHC class II molecules are polymorphic glycoproteins, consisting of non-covalently linked α (~35 kDa) and β (~28 kDa) subunits. Most of the polymorphism occurs in the second exons of the genes, encoding either the $\alpha 1$ or the $\beta 1$ domains, the regions of the molecule involved in peptide binding. Class II genes are classified into different families using a nomenclature adapted from humans and are denoted *DQ*, *DR*, *DP*, etc.

Early studies of the *Ovar-MhcII* region by genomic Southern analysis using human leucocyte antigen gene (*HLA*) probes identified a complex pattern of cross-hybridizing bands, which suggested that sheep contained homologues of *DQ* and *DR* genes but probably not *DP* (Chardon *et al.*, 1985; Scott *et al.*, 1987). Other experiments furnished immunochemical evidence for the existence of several subsets of sheep class II molecules (Puri *et al.*, 1987a, b; Dutia *et al.*, 1990). Overall, this early work indicated that the *Ovar-MhcII* region was complex and that the use of cross-species DNA probes had limited usefulness in terms of identifying individual genes, because they seemingly hybridized indiscriminately to many closely related genes.

Subsequently, there has been a more concerted effort to characterize *Ovar-MhcII* genes by first screening either genomic or cDNA libraries, subcloning and sequencing genes of interest and then expressing the genes in appropriate cells, so that the identity of different genes and the fine specificities of monoclonal antibodies could be verified. By mapping cosmid clones derived from two unrelated heterozygous sheep using human and mouse probes, Deverson *et al.* (1991) identified seven distinct α genes and 24 distinct β or β-related sequences. Consistent with earlier predictions, *DQ* and *DR* homologues were detected, but not *DP*.

Not surprisingly, the *Ovar-DQ* and *Ovar-DR* regions have been subjected to the most scrutiny. Scott *et al.* (1991a) determined the sequences of two genes, *Ovar-DQA1* and *Ovar-DQA2*. Both genes were transcribed and both were polymorphic, although the number of alleles detected varied with the haplotype examined. The same group reported the sequences of one *Ovar-DQB* gene and an *Ovar-DRB* pseudogene which lacked a defined exon 2 (Scott *et al.*, 1991b). A dinucleotide-repeat polymorphism is present in the *DRB* pseudogene (Blattman and Beh, 1992). Four new alleles of *DRA*, *DQA1* and *DQA2* genes were identified and sequenced by Fabb *et al.* (1993). The *Ovar-DRA* region appears to contain limited polymorphism, since the sequence described by these authors differed at only two amino acids from another *DRA* sequence reported independently (Ballingall *et al.*, 1992) and only a limited number of alleles were detected by RFLP analysis (Escayg *et al.*, 1993; Fabb *et al.*, 1993). A far greater level of polymorphism exists at the *Ovar-DQB* subregion and 12 sequences are known. A comparison of sheep sequences with homologues from other animals indicates that the *DQB* gene and allele lineages predate the speciation of ruminants (van Oorschot *et al.*, 1994).

The level of polymorphism is therefore highest among *Ovar-DQA* and *Ovar-DQB* genes. These subregions have been mapped from overlapping cosmid clones and exist as two linked loci on a linear tract of 130 kb DNA (Wright and Ballingall, 1994). The loci are 22 kb apart and each consists of a *DQA* and a *DQB* gene in a tail-to-tail orientation. The *DQA1* and *DQB1* genes are separated by 11 kb while the *DQA2* and *B2* genes are 25 kb apart. Several of the cloned *Ovar-MhcII* genes have been expressed following transfection into mouse L cells and then used to analyse the fine specificities of monoclonal antibodies identifying sheep MHC class II proteins. Collectively, this work has identified cross-locus specificities and defined the fine specificities of antibodies (Ballingall *et al.*, 1992, 1995).

A number of other *Ovar-MhcII* genes of the class IIb type have also been identified. The *DY* subregion of the sheep MHC contains an *A/B* gene pair (Wright *et al.*, 1994), an *Ovar-DNA* gene has been isolated and sequenced (Wright *et al.*, 1995a) and a sheep orthologue of the *HLA-DOB* gene identified (Wright *et al.*, 1995b). The functional roles of these genes remain more obscure and, in some cases, they are expressed at either very low levels or not at all.

Cytokines and Their Receptors

Cytokines function as soluble or membrane-attached messenger molecules in the immune system and exert important regulatory roles. A number of sheep cytokines have been cloned in the past few years as genomic DNA or cDNA transcripts. The overall strategic aim of this work was to secure a source of recombinant sheep cytokines, which could then be used as vaccine adjuvants or immunomodulators. There has been less emphasis on gaining insights into cytokine genes. Although some cytokines are composed of more than one protein subunit, each protein is encoded by a single gene. Limited degrees of polymorphism have been detected in the genes for interferon-γ (Penty and Hill, 1991) and interleukin 3 (IL-3) (Hawken *et al.*, 1994). The cytokine genes are therefore perhaps unlikely to be as useful to geneticists as are other multigenic or highly polymorphic markers. The roles of sheep cytokines, their molecular characterization and their potential usefulness have been reviewed recently (McInnes, 1993). Sheep cytokine sequences published so far are listed in Table 7.2.

Of the cytokine receptors, only the IL-2 receptor α (IL-2Rα) chain (CD25) has been cloned in sheep (Verhagen *et al.*, 1992; Bujdoso *et al.*, 1992). The *IL2RA* gene was recently mapped by *in situ* hybridization to sheep chromosome 13q12–q15 (Ansari *et al.*, 1995) and shown to contain a polymorphic dinucleotide-repeat microsatellite (Matthew and Maddox, 1994).

Other Immune Molecules

Two additional sets of molecules expressed on cells of the immune system have been cloned in sheep. The CD1 molecules are members of a family that has structural and sequence homology to MHC class I. They are cell-surface heterodimers, consisting of an α chain (43–49 kDa) associated with a β_2-microglobulin light chain. In humans, five distinct CD1 loci have been

Table 7.2. List of published sheep cytokine and cytokine receptor sequences.

Sequence	Reference
Cytokine genes	
OTPI	Imakawa *et al.*, 1987; Klemann *et al.*, 1990
EPO	Fu *et al.*, 1993
IFNA	Whaley *et al.*, 1991
IFNG	McInnes *et al.*, 1990; Radford *et al.*, 1991
GMCSF	McInnes and Haig, 1991; O'Brien *et al.*, 1991
GCSF	O'Brien *et al.*, 1994
TNFA	Young *et al.*, 1990; Green and Sargan, 1991; Nash *et al.*, 1991
TGFB1	Woodall *et al.*, 1994
IL1A	Andrews *et al.*, 1991
IL1B	Fiskerstrand and Sargan, 1990; Andrews *et al.*, 1991
IL2	Goodall *et al.*, 1990; Seow *et al.*, 1990; Bujdoso *et al.*, 1995
IL3	McInnes *et al.*, 1993
IL4	Seow *et al.*, 1993
IL7	Barcham *et al.*, 1995
IL8	Legastelois *et al.*, 1994; Seow *et al.*, 1994
IL10	Dutia *et al.*, 1994; Martin *et al.*, 1995
Cytokine-receptor genes	
IL2RA	Bujdoso *et al.*, 1992; Verhagen *et al.*, 1992

identified, *CD1A*, *B*, *C*, *D* and *E*, which encode serologically defined CD1a, b, c, d and e molecules. In recent experiments, a human *CD1C* probe and a sheep CD1 clone isolated early in the work were used to screen a number of sheep cDNA libraries. Four different cDNA transcripts were identified by sequencing, but they were all homologues of human *CD1B*. The analysis of genomic sheep DNA by Southern hybridization, using a homologous probe, showed that the ovine genome contains at least seven *CD1* genes (Ferguson *et al.*, 1996). The CD1 molecules function in antigen presentation but in a way that does not involve the binding of peptide fragments, as in the classical MHC I ligands. Instead, CD1 molecules seem to play novel roles in the presentation of lipid or glycolipid material to non-MHC-restricted T cells, perhaps including the γδ T cells. Since sheep have an unusually large number of γδ T cells and an extensive repertoire of *TCRG* and *TCRD* genes, the seemingly preferential duplication of genes at one of the *CD1* loci could be of functional importance.

Sheep contain another extensive family of genes that appear to be expressed exclusively in γδ T cells. The proteins, identified first in sheep, were initially called T19 but are now also referred to as WC1 (workshop cluster 1). The analysis of sheep genomic DNA by Southern hybridization and sequencing indicates the presence of perhaps as many as 50 *T19* genes (O'Keeffe *et al.*, 1994; Walker *et al.*, 1994). Due to sequence similarities, the T19 proteins are clearly members of the scavenger-receptor superfamily, but so far their precise role(s) in sheep γδ T cells remain cryptic.

Conclusions

From the foregoing, one or two general conclusions can be drawn. There has been a concerted effort, over the last 5–6 years in particular, to characterize immune molecules of sheep at the molecular level. The bulk of the work has been done by relatively few research groups and all involved should feel justifiably gratified that this degree of progress was accomplished with relatively little overlap of effort, although a certain amount of duplication always remains unavoidable. Sheep geneticists and other researchers now have access to a useful array of molecular probes for several groups of immune molecules and this should facilitate future studies.

Obviously, however, much more remains to be done. It can be anticipated that, in the years ahead, additional molecules will be cloned and the panel of available probes will expand. It is also necessary to begin the arduous task of characterizing the germ-line structure of gene families and groups that are now defined as isolated cDNA or genomic sequences, so that an overall 'view' of the *Ovar-Mhc*, *Ovar-Ig* and *Ovar-Tcr* loci, to mention just three examples, can eventually be achieved. Ideally, efforts to define the physical interrelationships of sheep genes encoding immune molecules should be coordinated with broader genome-mapping approaches.

Acknowledgements

I thank Drs Alan Young and Louis DuPasquier for critically reviewing the manuscript. The Basle Institute for Immunology was founded and is supported by F. Hoffmann–La Roche Ltd, Basle, Switzerland.

References

Andrews, A.E., Barcham, G.J., Brandon, M.R. and Nash, A.D. (1991) Molecular cloning and characterisation of ovine IL-1a and IL-1b. *Immunology* 74, 453–460.

Ansari, H.A., Pearce, P.D., Maher, D.W. and Broad, T.E. (1994) Regional assignment of conserved reference loci anchors unassigned linkage and syntenic groups to ovine chromosomes. *Genomics* 24, 451–455.

Ansari, H.A., Pearce, P.D., Maher, D.W. and Broad, T.E. (1995) Human chromosome 10 loci map to three different sheep chromosomes. *Mammalian Genome* 6, 46–48.

Ballingall, K.T., Wright, H., Redmond, J., Dutia, B.M., Hopkins, J., Lang, J., Deverson, E.V., Howard, J.C., Puri, N. and Haig, D. (1992) Expression and characterization of ovine major histocompatibility complex class II (OLA-DR) genes. *Animal Genetics* 23, 347–359.

Ballingall, K.T., Dutia, B.M., Hopkins, J. and Wright, H. (1995) Analysis of the fine specificities of sheep major histocompatibility complex class II-specific monoclonal antibodies using mouse L-cell transfectants. *Animal Genetics* 26, 79–84.

Barcham, G.J., Andrews, A.E. and Nash, A.D. (1995) Cloning and expression of a cDNA encoding ovine interleukin 7. *Gene* 154, 265–269.

Blattman, A.N. and Beh, K.J. (1992) Dinucleotide repeat polymorphism within the ovine major histocompatibility complex. *Animal Genetics* 23, 392.

Broad, T.E., Burkin, D.J., Cambridge, L.M., Jones, C., Lewis, P.E., Morse, H.G., Geyer, D., Pearce, P.D., Ansari, H.A. and Maher, D.W. (1995) Six loci mapped onto human chromosome 2p are assigned to sheep chromosome 3p. *Animal Genetics* 26, 85–90.

Bujdoso, R., Sargan, D., Williamson, M. and McConnell, I. (1992) Cloning of a cDNA encoding the ovine interleukin-2 receptor 55-kDa protein, CD25. *Gene* 113, 283–284.

Bujdoso, R., Williamson, M., Roy, D., Hunt, P., Blacklaws, B., Sargan, D. and McConnell, I. (1995) Molecular cloning and expression of DNA encoding ovine interleukin 2. *Cytokine* 7, 223–231.

Chardon, P., Kirszenbaum, M., Cullen, P.R., Geffrotin, C., Auffray, C., Strominger, J.L., Cohen, D. and Vaiman, M. (1985) Analysis of the sheep MHC using HLA Class I, II, and C4 cDNA probes. *Immunogenetics* 22, 349–358.

Clarkson, C.A., Beale, D., Coadwell, J.W. and Symons, D.B.A. (1993) Sequence of ovine Ig $\gamma 2$ constant region heavy chain cDNA and molecular modelling of ruminant IgG isotypes. *Molecular Immunology.* 30, 1195–1204.

Cullen, P.R., Bunch, C., Brownlie, J. and Morris, P.J. (1982) Sheep lymphocyte antigens: a preliminary study. *Animal Blood Groups and Biochemical Genetics* 13, 149–159.

Deverson, E.V., Wright, H., Watson, S., Ballingall, K., Huskisson, N., Diamond, A.G. and Howard, J.C. (1991) Class II major histocompatibility complex genes of the sheep. *Animal Genetics* 22, 211–225.

Dufour, V., Malinge, S. and Nau, F. (1996) The sheep Ig variable region repertoire consists of a single VH family. *Journal of Immunology* 156, 2163–2170.

Dutia, B.M., Hopkins, J., Allington, M.P., Bujdoso, R. and McConnell, I. (1990) Characterization of monoclonal antibodies specific for alpha- and beta-chains of sheep MHC class II. *Immunology* 70, 27–32.

Dutia, B.M., Hunt, P., Sargan, D.R., Dalziel, R.G. and Hopkins, J. (1994) Sequence of the sheep interleukin-10-encoding cDNA. *Gene* 149, 393–394.

Engewerda, C.R., Sandeman, R.A., Stuart, S.J. and Sandeman, R.M. (1992) Isolation and sequence of sheep immunoglobulin E heavy-chain complementary DNA. *Veterinary Immunology and Immunopathology* 34, 115–126.

Escayg, A.P., Montgomery, G.W., Hickford, J.G. and Bullock, D.W. (1993) A *Bgl*II RFLP and the ovine MHC class II DRA locus. *Animal Genetics* 24, 217.

Fabb, S.A., Maddox, J.F., Gogolin-Ewens, K.J., Baker, L., Wu, M.-J. and Brandon, M.R. (1993) Isolation, characterization and evolution of ovine major histocompatibility complex class II *DRA* and *DQA* genes. *Animal Genetics* 24, 249–255.

Ferguson, E.D., Dutia, B.M., Hein, W.R. and Hopkins, J. (1996) The ovine CD1 gene family contains at least four *CD1B* homologues. *Immunogenetics* 44, 86–96.

Fiskerstrand, C. and Sargan, D. (1990) Nucleotide sequence of ovine interleukin-1 beta. *Nucleic Acids Research* 18, 7165.

Foley, R.C. and Beh, K.J. (1989) Isolation and sequence of sheep IgH and L chain cDNA. *Journal of Immunology* 142, 708–711.

Foley, R.C. and Beh, K.J. (1992) Analysis of immunoglobulin light chain loci in sheep. *Animal Genetics* 23, 31–42.

Fu, P., Evans, B., Lim, G.B., Moritz, K. and Wintour, E.M. (1993) The sheep erythropoietin gene: molecular cloning and effect of hemorrhage on plasma erythropoietin and renal/liver messenger RNA in adult sheep. *Molecular and Cellular Endocrinology* 93, 107–116.

Goodall, J.C., Emery, D.C., Perry, A.C.F., English, L.S. and Hall, L. (1990) Complementary DNA cloning of interleukin 2 by PCR. *Nucleic Acids Research* 18, 5883.

Green, I.R. and Sargan D.R. (1991) Sequence of the cDNA encoding tumor necrosis factor-α: problems with cloning by inverse PCR. *Gene* 109, 203–210.

Griebel, P.J. and Hein, W.R. (1996) Expanding the role of Peyer's patches in B-cell ontogeny. *Immunology Today* 17, 30–39.

Grossberger, D., Hein, W. and Marcuz, A. (1990) Class I major histocompatibility complex cDNA clones from sheep thymus: alternative splicing could make a long cytoplasmic tail. *Immunogenetics* 32, 77–87.

Grossberger, D., Marcuz, A., Fichtel, A., Dudler, L. and Hein, W.R. (1993) Sequence analysis of sheep T-cell receptor β chains. *Immunogenetics* 37, 222–226.

Groth, D.M. and Wetherall, J.D. (1994) Dinucleotide repeat polymorphism within the ovine major histocompatibility complex class I region. Animal Genetics 25, 61.

Hawken, R.J., Davies, K.P. and Maddox, J.F. (1994) A dinucleotide repeat in the ovine interleukin-3 gene of limited polymorphism. *Animal Genetics* 25, 286.

Hein, W.R. and Dudler, L. (1993a) Nucleotide sequence of the membrane form of sheep IgM and identification of two Cμ allotypes. *Molecular Immunology* 30, 783–784.

Hein, W.R. and Dudler, L. (1993b) Divergent evolution of T cell repertoires: extensive diversity and developmentally regulated expression of the sheep γδ T cell receptor. *EMBO Journal* 12, 715–724.

Hein, W.R. and Tunnacliffe, A. (1990) Characterization of the CD3 γ and δ invariant subunits of the sheep T cell antigen receptor. *European Journal of Immunology* 20, 1505–1511.

Hein, W.R. and Tunnacliffe, A. (1993) Invariant components of the sheep T-cell antigen receptor: cloning of the CD3 ε and Tcr ζ chains. *Immunogenetics* 37, 279–284.

Hein, W.R., Dudler, L., Marcuz, A. and Grossberger, D. (1990) Molecular cloning of sheep T cell receptor γ and δ chain constant regions: unusual primary structure of γ chain hinge segments. *European Journal of Immunology* 20, 1795–1804.

Hein, W.R., Marcuz, A., Fichtel, A., Dudler, L. and Grossberger, D. (1991) Primary structure of the sheep T-cell receptor α chain. *Immunogenetics* 34, 39–41.

Imakawa, K., Anthony, R.V., Kazemi, M., Marotti, K.R., Polites, H.G. and Roberts, R.M. (1987) Interferon-like sequence of ovine trophoblast protein secreted by embryonic trophectoderm. *Nature* 330, 377–379.

Ishiguro, N. and Hein, W.R. (1994) The T cell receptor. In: Goddeeris, B.M.L. and Morrison, W.I. (eds) *Cell-Mediated Immunity in Ruminants*. CRC Press, Boca Raton, Florida, pp. 59–73.

Klemann, S.W., Imakawa, K. and Roberts, R.M. (1990) Sequence variability among ovine trophoblast interferon complementary DNA. *Nucleic Acids Research* 18, 6724.

Legastelois, I., Greenland, T., Arnaud, P., Mornex, J.F. and Cordier, G. (1994) Sequencing of the ovine interleukin-8-encoding cDNA using the polymerase chain reaction. *Gene* 150, 367–369.

McInnes, C.J. (1993) Current research on ovine cytokines. *British Veterinary Journal* 149, 371–386.

McInnes, C.J. and Haig, D.M. (1991) Cloning and expression of a cDNA encoding ovine granulocyte–macrophage colony-stimulating factor. *Gene* 105, 275–279.

McInnes, C.J., Logan, M., Redmond, J., Entrican, G. and Baird, G.D. (1990) Molecular cloning of the ovine gamma-interferon cDNA using the polymerase chain reaction. *Nucleic Acids Research* 18, 4012.

McInnes, C., Haig, D. and Logan, M. (1993) The cloning and expression of the gene for ovine interleukin-3 (multi-CSF) and a comparison of the *in vitro* hematopoietic activity of ovine IL-3 with ovine GM-CSF and human M-CSF. *Experimental Hematology* 21, 1528–1534.

Mahdy, E.A., Makinen, A., Chowdhary, B.P., Andersson, L. and Gustavsson, I. (1989) Chromosomal localization of the ovine major histocompatibility complex (OLA) by *in situ* hybridization. *Hereditas* 111, 87–90.

Martin, H.M., Nash, A.D. and Andrews, A.E. (1995) Cloning and characterisation of an ovine interleukin-10-encoding cDNA. *Gene* 159, 187–191.

Matthew, P. and Maddox, J.F. (1994) A polymorphic dinucleotide repeat microsatellite detected in the ovine interleukin 2 receptor alpha (IL2RA, CD25) gene. *Animal Genetics* 25, 200.

Millot, P. (1979) Genetic control of lymphocyte antigens in sheep: the *OLA* complex and two minor loci. *Immunogenetics* 9, 509–534.

Nash, A.D., Barcham, G.J., Brandon, M.R. and Andrews, A.E. (1991) Molecular cloning, expression and characterisation of ovine TNF-α. *Immunology and Cellular Biology* 69, 273–283.

O'Brien, P.M., Rothel, J.S., Seow, H.F. and Wood, P.R. (1991) Cloning and sequencing of the complementary DNA for ovine granulocyte–macrophage colony-stimulating factor. *Immunology and Cellular Biology* 69, 51–55.

O'Brien, P.M., Seow, H.F., Rothel, J.S. and Wood, P.R. (1994) Cloning and sequencing of an ovine granulocyte colony-stimulating factor cDNA. *DNA Sequence* 4, 339–342.

O'Keeffe, M.A., Metcalfe, S.A., Glew, M.D., Bowden, T., McInnes, S., Kimpton, W.G., Cahill, R.N.P., Hein, W.R. and Walker, I.D. (1994) Lymph node homing cells biologically enriched for γδ T cells express multiple genes from the T19 repertoire. *International Immunology* 6, 1687–1697.

Parsons, Y.M., Cooper, D.W., Piper, L.R. and Beh, K.J. (1993) A *Taq*I polymorphism at the immunoglobulin kappa light chain locus in Merino sheep. *Animal Genetics* 25, 54.

Patri, S. and Nau, F. (1992) Isolation and sequence of a cDNA coding for the immunoglobulin μ chain of the sheep. *Molecular Immunology* 29, 829–836.

Pearce, P.D., Ansari, H.A., Maher, D.W. and Broad, T.E. (1995) Five regional localizations to the sheep genome: first assignments to chromosomes 5 and 12. *Animal Genetics* 26, 171–176.

Penty, J.M. and Hill, D.F. (1991) An *Rsa* I restriction fragment length polymorphism at the ovine locus for gamma-interferon. *Animal Genetics* 22, 439.

Puri, N.K., Gogolin-Ewens, K.J. and Brandon, M.R. (1987a) Monoclonal antibodies to sheep MHC class I and class II molecules: biochemical characterization of three class I gene products and four distinct populations of class II molecules. *Veterinary Immunology and Immunopathology* 15, 59–86.

Puri, N.K., Walker, I.D. and Brandon, M.R. (1987b) N-terminal amino acid sequence analysis of the A and B polypeptides from four distinct subsets of sheep major histocompatibility complex class II molecules. *Journal Immunology* 236, 2996–3002.

Radford, A.J., Hodgson, A.L.M., Rothel, J.S. and Wood, P.R. (1991) Cloning and sequencing of the ovine gamma-interferon gene. *Australian Veterinary Journal* 68, 82–84.

Reynaud, C.-A., Mackay, C.R., Müller, R.G. and Weill, J.C. (1991) Somatic generation of diversity in a mammalian primary lymphoid organ: the sheep ileal Peyer's patches. *Cell* 64, 995–1005.

Reynaud, C.-A., Garcia, C., Hein, W.R. and Weill, J.C. (1995) Hypermutation generating the sheep immunoglobulin repertoire is an antigen-independent process. *Cell* 80, 115–125.

Scott, P.C., Choi, C.-L. and Brandon, M.R. (1987) Genetic organization of the MHC class II region. *Immunogenetics* 25, 116–122.

Scott, P.C., Gogolin-Ewens, K.J., Adams, T.E. and Brandon, M.R. (1991a) Nucleotide sequence, polymorphism, and evolution of ovine MHC class II *DQA* genes. *Immunogenetics* 34, 69–79.

Scott, P.C., Maddox, J.F., Gogolin-Ewens, K.J. and Brandon, M.R. (1991b) The nucleotide sequence and evolution of ovine MHC class II *B* genes: *DQB* and *DRB*. *Immunogenetics* 34, 80–87.

Seow, H.F., Rothel, J.S., Radford, A.J. and Wood, P.R. (1990) The molecular cloning of ovine interleukin 2 gene by the polymerase chain reaction. *Nucleic Acids Research* 18, 7175.

Seow, H.F., Rothel, J.S. and Wood, P.R. (1993) Cloning and sequencing an ovine interleukin-4-encoding cDNA. *Gene* 124, 291–293.

Seow, H.F., Yoshimura, T., Wood, P.R. and Colditz, I.G. (1994) Cloning, sequencing, expression and inflammatory activity in skin of ovine interleukin-8. *Immunology and Cellular Biology* 72, 398–405.

Stear, M.J. and Spooner, R.L. (1981) Lymphocyte antigens in sheep. *Animal Blood Groups and Biochemical Genetics* 12, 265–276.

van Oorschot, R.A.H., Maddox, J.F., Adams, L.J. and Fabb, S.A. (1994) Characterization and evolution of ovine MHC class II *DQB* sequence polymorphism. *Animal Genetics* 25, 417–424.

Verhagen, A.M., Andrews, A.E., Brandon, M.R. and Nash, A.D. (1992) Molecular cloning, expression and characterization of the ovine IL-2Rα chain. *Immunology* 76, 1–9.

Walker, I.D., Glew, M.D., O'Keeffe, M.A., Metcalfe, S.A., Clevers, H.C., Wijngaard, P.L.J., Adams, T.E. and Hein, W.R. (1994) A novel multi-gene family of sheep $\gamma\delta$ T cells. *Immunology* 83, 517–523.

Whaley, A.E., Carroll, R.S. and Imakawa, K. (1991) Cloning and analysis of a gene encoding ovine interferon alpha-II. *Gene* 106, 281–282.

Woodall, C.J., McLaren, L.J. and Watt, N.J. (1994) Sequence and chromosomal localisation of the gene encoding ovine latent transforming growth factor-beta 1. *Gene* 150, 371–373.

Wright, H. and Ballingall, K.T. (1994) Mapping and characterization of the *DQ* subregion of the ovine MHC. *Animal Genetics* 25, 243–249.

Wright, H., Ballingall, K.T. and Redmond, J. (1994) The *DY* sub-region of the sheep MHC contains an *A/B* gene pair. *Immunogenetics* 40, 230–234.

Wright, H., Redmond, J., Wright, F. and Ballingall, K.T. (1995a) The nucleotide sequence of the sheep MHC class II *DNA* gene. *Immunogenetics* 41, 131–133.

Wright, H., Redmond, J. and Ballingall, K.T. (1995b) The sheep orthologue of the *HLA-DOB* gene. *Immunogenetics* 43, 76–79.

Young, A.J., Hay, J.B. and Chan, J.Y.C. (1990) Primary structure of ovine tumor necrosis factor alpha complementary DNA. *Nucleic Acids Research* 18, 6723.

8 Genetics of Disease Resistance and Vaccine Response

H.W. RAADSMA,[1] G.D. GRAY[2] and R.R. WOOLASTON[3]

[1]Centre for Sheep Research and Extension, Department of Animal Health, The University of Sydney, Camden, NSW 2570, Australia; [2]The Department of Animal Science, The University of New England, Armidale, NSW 2351, Australia; [3]CSIRO, Division of Animal Production, Pastoral Research Laboratory, Armidale, NSW 2350, Australia

Introduction 200
Genetic Improvement of Resistance 200
- Resistance 201
- Resilience 202
- Reduced treatment costs 202

Genetic Resistance to Internal Parasites 203
- Use of faecal worm egg count as a measurement of resistance 203
- Differences between breeds 205
- Interrelationships among resistance to different genera of parasites 205

Genetic Resistance to Other Diseases 206
Important Diseases for Which We Need More Information 208
Genetic Relationships among Disease Traits 209
Resistance and Indirect Selection 209
- Resistance to internal parasites and indirect selection 210
- Resistance to fleece-rot/fly strike and indirect selection 210
- Genetic markers 211

Increasing Resistance and Production 213
- Breeding programmes that do not include disease resistance 214
- Inclusion of disease resistance in breeding programmes 214
- Choice of breed, strain and blood-line 215

Genetic Control of Response to Vaccination 215
Conclusions 217
References 217

Introduction

Domesticated sheep worldwide are faced with potential risk of disease. In different environments, some diseases have become more important and others less important depending on the climate, the standard of husbandry and the ability of humans to avoid, control and treat the problems created. It is unlikely, however, that there are any sheep production systems in which losses from disease are not a significant consideration. In a production system, the relative importance of infectious diseases depends on their impact on welfare and production. Where conventional control measures, such as vaccination and chemotherapy, have been either ineffective, unsustainable or uneconomic, genetic approaches have been considered. For parasitic gastroenteritis and, to a lesser extent, fly-strike, the problem is serious, as chemotherapy is becoming less effective with multiple resistance of parasites to chemicals. Furthermore, prospects for vaccination are not encouraging. Thus there has been a research emphasis, reflected in this chapter, on genetic approaches to control these diseases. Some of this research has provided basic information on the genetics of disease resistance, but it has always been motivated by the need to exploit this variation for practical purposes: to breed 'disease-resistant' flocks or to elucidate mechanisms of resistance for the development of other control methods.

This overview is devoted to describing genetic variation in resistance to sheep diseases, genes associated with resistance and procedures for selecting resistant animals for breeding purposes. The emphasis is on internal parasites, for it is one of the most ubiquitous constraints on sheep production systems, because resistance to anthelmintics is widespread and because our state of knowledge for practical breeding programmes has advanced rapidly in recent years. Only the major diseases which affect production are detailed here. For more detailed information on the extensive list of other diseases relevant to sheep production, the reader is referred to a comprehensive review by Gray *et al.* (1995).

Genetic Improvement of Resistance

Animal breeders interested in shifting the host–pathogen balance to their advantage are confronted with the problem of defining the most relevant measure of resistance. Woolaston and Baker (1996) described three approaches taken by sheep breeders and researchers to improving ovine resistance to nematode parasites. The principles are applicable to all diseases, but the most appropriate approach depends on the disease of interest.

Resistance

The fundamental definition of 'disease resistance' includes freedom from clinical signs of disease after challenge. Resistant animals remain free from clinical signs of disease, whereas susceptible animals become affected after being subject to the same disease challenge. The expression of disease resistance as a binomial trait holds for most diseases in sheep where clinical signs are well defined. Furthermore, the relative scale of susceptibility is usually extended by incorporating some form of linear score reflecting increasing severity of disease. In the analysis of binomial traits, such as disease resistance, it is also common practice to adopt a threshold model, as described by Dempster and Lerner (1950) and Falconer (1965).

A notable departure from the above definition applies to internal parasites. Hosts carrying reduced parasite burdens are said to be resistant. In the case of roundworm parasites in sheep, this means resistant animals carry fewer worms. Increased resistance to internal parasites is achieved by the host in a number of ways: reduced establishment of incoming larvae, arrested or delayed development, accelerated expulsion of adult worms and reduced fecundity of female worms. The main indicator of resistance is usually taken as an indirect measure of parasite numbers, in this case the faecal worm egg count (FEC), described in more detail below.

Taking breeding for resistance to internal parasites as an example, the advantages are as follows.

- There will be fewer diseased animals in the flock which suffer adverse effects of parasitism.
- Resistant animals will be transmitting fewer pathogenic agents, which means reduced pasture contamination. This advantage may not be so great with contagious diseases, such as foot-rot.
- The trait indicating resistance (FEC) is simple to measure and testing protocols can be designed to minimize both the period of infection and the adverse effects on the host.
- It is extremely variable and moderately heritable and thus amenable to reasonably rapid genetic progress.

One of the main difficulties with this approach is that it is very difficult to assign an economic value to reducing the proportion of diseased animals in the flock or the concentration of pathogens, relative to the value of genetic change in other traits, such as fleece weight, body weight, reproductive rate or average fibre diameter. It has also been argued, in the context of internal parasites, that increasing the proportion of resistant animals in a flock will increase the selective pressure on the parasite to adapt genetically, but evidence for this is equivocal (Windon, 1991; Woolaston and Eady, 1995).

Resilience

Hosts suffering few or no adverse effects on productivity during infection are defined as resilient. As a breeding goal, resilience has the advantage that no direct measures of the relative pathogen loads of animals are necessary. A significant disadvantage, however, is that true resilience is impractical to measure, because costly and time-consuming measurements of production are necessary. Unless production levels are known, with and without a disease challenge, it is not possible to readily determine how an animal with superior production during infection would perform in the absence of infection. Thus the breeder has no clear idea of the extent to which a selection programme is improving the frequency of desirable production genes *per se* or genes for tolerance to disease. In an unchanging environment, this may be of little consequence. In reality, however, where the degree of challenge is not constant, this can have marked effects on the relevance of a breeding programme. Resilient hosts, unlike their resistant counterparts, may have no favourable effect on the pathogen's concentration, such as the level of pasture contamination. Traits defined specifically to describe resilience to internal parasites have been found to have low heritability (Albers *et al.*, 1987) and usually require a prolonged period of infection, which is generally unacceptable to breeders of élite animals.

The advantages of selection for 'resilience' are perhaps more relevant in the case of foot-rot, where Egerton and Raadsma (1991) argued that, under selective breeding for increased resistance to foot-rot, sheep exposed to virulent strains of *Dichelobacter nodosus* would cope in a similar fashion to unselected sheep exposed to benign isolates of the pathogen. In this scenario, the impact of the presence of the disease on production is minimal and sheep would not require any managerial inputs to control the disease.

Reduced treatment costs

Some New Zealand workers investigated the utility of traits which were defined to measure the needs of individual animals for anthelmintic treatment over a certain period (Bisset *et al.*, 1994). Although these workers termed the trait 'resilience', it differs from the above definition, because it is determined not by a quantitative measure of production with and without challenge, but a subjective assessment of the need for anthelmintic treatment. Hence Woolaston and Baker (1996) described the approach as breeding for reduced treatment costs. With relatively straightforward assumptions about the cost of anthelmintic treatments, both current and future, it is simple to derive an economic value for this trait. Unfortunately, the trait appears to have low heritability and be uncorrelated with resistance, and this limits its usefulness (Bisset *et al.*, 1994). In a subsequent study (Bisset *et al.*, 1997), these workers modified the assessment procedure in an attempt to include weight gain under challenge as a quantitative assessment of the need for treatment. This modification, however, did not improve the heritability and, further, the testing protocol reduced the ability to detect genetic variation in other traits. Clearly, though, for some diseases, such as blowfly strike, resistance (reduced prevalence) and a reduced requirement for treatment are virtually indistinguishable traits.

The three approaches described above could be used for any disease that affects production, incurs treatment costs and can be measured on a linear scale. However, all of these methods suffer the obvious disadvantage that the disease must be present before an assessment can be made of the host's ability to deal with the threat.

Genetic Resistance to Internal Parasites

Evidence acquired in the last 40 years has shown clearly that a significant proportion of variation in resistance to internal parasites is genetically determined. For nematode infections of sheep, the proportion of variation that can be attributed to genetic effects (heritability) varies between 0.11 and 0.55 (Table 8.1) (reviewed by Morris *et al.*, 1995b; Woolaston and Eady, 1995). At such levels, there are prospects for including resistance in practical breeding programmes (Gray, 1991) and selected sheep have been shown to exhibit resistance under conditions of field challenge (Gray *et al.*, 1992; Morris *et al.*, 1995b). Heritability estimates for resilience are within the range of 0.03–0.14 (Albers *et al.*, 1987; Bisset *et al.*, 1994) and are generally lower than for resistance. Similarly, heritability estimates for 'reduced number of treatments' are also low (Morris *et al.*, 1995b).

The major features of the potential to exploit this genetic variation in resistance to internal parasites are demonstrated in single trait selection flocks. Such long-term flocks have been established for *Haemonchus contortus* (Woolaston and Piper, 1996), *Trichostrongylus colubriformis* (Windon, 1991) and *Ostertagia* spp. (Cummins *et al.*, 1991). In all cases, a change in FEC has been achieved in the desired direction, closely associated with reduced parasite burdens as a consequence of reduced establishment and parasite fecundity. No information is available on genetic change in resistance and its effect on expulsion rate of parasites.

Use of faecal worm egg count as a measurement of resistance

The benefits of selection for resistance accrue from the phenotypic effects of having fewer worms. This is predicted to lead to reduced impact on production, lower requirement for chemical control and reduced contamination of pasture by infective larvae. Other consequences will depend on the sign and magnitude of the genetic correlation between faecal worm egg count and other traits in the breeding objective. Sheep breeders will not select solely on the basis of FEC but will make use of production information combined in a formal or informal index. However, FEC remains the principal trait that can reliably estimate

Table 8.1. Summary of published estimates of heritability (± s.e.) of resistance to internal parasite infections in a range of sheep breeds, measured as faecal egg counts (FEC) (adapted in part from Eady *et al.*, 1996).

Breed	Age (months)	Counts FEC	Heritability	Source
Natural mixed infection				
Romney	5–8	1	0.34 ± 0.19	Watson *et al.*, 1986
Romney	5–8	2	0.53 ± 0.15	Baker *et al.*, 1991
Romney	5–8	1	0.27 ± 0.07	Bisset *et al.*, 1992
Romney	5	1	0.13 ± 0.07	McEwan *et al.*, 1992
Romney	4	1	0.39 ± 0.13	Morris *et al.*, 1993
Romney	6	1	0.46 ± 0.14	Morris *et al.*, 1993
Red Maasai, Dorper	10	2	0.20 ± 0.08	Baker *et al.*, 1994
Polish Longwool	6, 8	2	0.28 ± 0.16	Gruner and Lantier, 1995
Merino	6, 12	2	0.42 ± 0.14	Cumins *et al.*, 1991
Merino	10, 22	1	0.16 ± 0.12	Eady *et al.*, 1996
Artificial Haemonchus contortus				
Romanov	6–10	6	0.55	Gruner and Lantier, 1995
Merino	18	1	0.23 ± 0.13	Piper, 1987
Merino	4–5	2	0.30 ± 0.10	Albers *et al.*, 1987
Merino	4–5	1	0.29 ± 0.03	Woolaston and Piper, 1996
Merino	6–8	1	0.30 ± 0.13	Eady *et al.*, 1996
Hungarian Merino	6–7	4	0.49 ± 0.17	Sreter *et al.*, 1994
Artificial Trichostrongylus colubriformis				
Merino	4–5	5	0.41 ± 0.04	Woolaston *et al.*, 1991
Merino	5–13	1	0.20 ± 0.11	Eady *et al.*, 1996
Natural Ostertagia circumcincta				
Scottish Blackface	1–2	2	0.00 ± –	Bishop *et al.*, 1997
Scotish Blackface	3–6	4	0.33 ± 0.15	Bishop *et al.*, 1997

resistance. Many others have been investigated, but FEC is used in commercial breeding programmes, both in Australia (Nemesis: Anonymous, 1994) and New Zealand (WormFEC: McEwan *et al.*, 1995).

Not all questions about the practical use of FEC in breeding programmes have been addressed. Nutrition, stress, climate and physiological state can affect the interaction between worms and sheep (Bundy and Golden, 1987) and repeatability of FEC in a number of studies has been shown to be highly variable – from zero (Karlsson *et al.*, 1991) to 0.86 (Baker *et al.*, 1991) under field conditions – and sensitive to changes in management (Kelly and Gray, 1995). This is of importance to the industry, as the repeatability has a profound impact on the response to selection and the number of measurements required for optimal genetic progress. R.R. Woolaston and R.G. Windon (unpublished data) showed that, under carefully controlled conditions, the repeatability of FEC during challenge with *T. colubriformis* measured 2 weeks apart is high (about 0.8), but it falls progressively to about 0.6 as the period between measures increases to 8 weeks. The genetic correlations between measures at different times appear to be essentially unity (for details, see Morris *et al.*, 1995b; Woolaston and Eady, 1995; Bishop *et al.*, 1997) suggesting that, if breeders wish to obtain more accurate predictions of the genes for resistance, repeated measures taken further apart in time will be more useful than measures recorded close together. This is particularly relevant where measurements are made across consecutive challenges, where repeatability estimates of resistance are low to moderate (0.1–0.3) (Eady, 1995; Morris *et al.*, 1995b; Bishop *et al.*, 1996), but genetic correlations are almost unity, suggesting that the same genes are involved in resistance under repeat challenge.

The use of alternative indicators for resistance following challenge with nematode parasites in sheep has been reviewed by Eady (1995), including circulating eosinophil count, circulating antibody levels directed against parasite-specific antigens, faecal antigen test and packed cell volume in *H. contortus* infections. As heritability estimates are no higher than those for FEC and genetic correlations with these indicator traits are substantially less than unity, they are less efficient selection criteria than FEC in direct selection programmes. Their role will only be important if it can be shown that they reflect additional components of genetic variation in resistance not covered by FEC.

Differences between breeds

There are good examples among small ruminants, where knowledge of the breed has predictive value for resistance to parasitic infection. Such resistant sheep breeds include the Red Maasai (Baker, 1995), Florida Native and St Croix (Zajac, 1995). There have been many studies of the differences between breeds. These studies vary greatly in their size, the procedures by which breeds were sampled and the nature of the environment in which the animals were kept before and during testing. Nevertheless, for these three breeds at least, there is substantial evidence for their relative resistance. Resistance of very few breeds has been thoroughly assessed and there may be others of potential economic value with high levels of resistance that are worth investigating.

Interrelationships among resistance to different genera of parasites

Of considerable importance to sheep breeders interested in genetic improvement of parasite resistance is the issue of 'broad-spectrum' host resistance. Are sheep selected for resistance to one particular internal parasite equally resistant to other internal parasites? Although the genetic correlations between resistance to individual species of nematodes are not known, indirect evidence for favourable correlations between resistance to different species is provided by comparisons of selection lines selected for resistance to a single parasite species (Woolaston and Eady, 1995). Similar findings were indicated by Morris *et al.* (1995b), who reported on 'broad-based' differences between sire lines to the majority of important abomasal and small intestinal species in a temperate climate.

Most of the information on genetic resistance to internal parasites has focused on nematode infections and no information is available on within-flock genetic differences to trematode and cestode infections. Bouix *et al.* (1992) reported a heritability of 0.1–0.2 for coccidiosis, a parasitic condition that causes wastage and chronic scouring in young sheep when managed under intensive conditions with high stocking rates.

Dags, the build-up of faeces around the breech of sheep, particularly in wooled breeds, is a problem in its own right, causing soiling of saleable wool and a predisposition to fly strike, and is an indicator of intestinal infection. In Romneys in New Zealand, the presence and intensity of dags have been shown to be heritable in the range 0.06–0.50 (Morris *et al.*, 1995b). What is less clear is the relationship between dags and resistance to parasites, which are sometimes the causative agents of the diarrhoea that results in dag formation (Bisset *et al.*, 1994, 1996; Larsen *et al.*, 1995). In the absence of knowledge on the genetic correlation with resistance, the use of the dag score is unsuitable as an indicator of resistance, but it may be suitable as a criterion to reduce the prevalence of dags.

Genetic Resistance to Other Diseases

For bacterial, viral, ectoparasite and metabolic diseases, there is also evidence of genetic variation in resistance, but as a general rule this has been much less intensively investigated than resistance to internal parasites. The major diseases that affect production and for which we have sufficient preliminary information to consider selection programmes include: fly strike, fleece-rot, dermatophilosis, foot-rot and facial eczema (Table 8.2). Notable diseases which may cause serious disruption to production systems and for which conventional health-management strategies are suboptimal include: pasteurellosis, mastitis, salmonellosis, Johne's disease, blue tongue, sheep pox, tick-borne diseases and, to a lesser extent, trypanosomiasis. For this last group of diseases, insufficient information is available on the relevant genetic parameters or breed differences to consider inclusion in genetic improvement programmes.

These diseases have a number of features which contrast with internal parasites and affect how resistance is exploited.

Firstly, the close genetic relationship between resistance to fleece-rot and fly strike in Merino sheep allows breeders to make progress in resistance to both

Table 8.2. Heritability estimates of resistance to disease, excluding internal parasites.

Disease	Breed	Trait	Heritability	Reference
Fly strike	Merino	Incidence	0.10–0.58	Raadsma, 1991
Fleece-rot	Merino	Severity, incidence	0.05–0.80	Raadsma, 1995
Dermatophilosis	Merino	Severity, incidence	0.05–0.44	Lewer *et al.*, 1987; Raadsma *et al.*, 1992; Woolaston *et al.*, 1995
Foot-rot	Merino	Severity, incidence	0.15–0.29	Raadsma, 1995
	Romney	Severity Liability	0.14 0.17–0.28	Skerman *et al.*, 1988
	Mixed	Incidence Liability	0.03–0.34 0.07–0.55	Baker *et al.*, 1986
Facial eczema	Romney	γ-Glutamyl-transferase	0.45 ± 0.03	Morris *et al.*, 1995b
Ryegrass staggers	Romney × Coopworth	Incidence Liability	0.07 ± 0.02 0.27 ± 0.08	Morris *et al.*, 1994; Morris *et al.*, 1995a

by selection against fleece-rot alone. Advantages of using fleece-rot as a selection criterion include higher levels of expression to optimize selection differentials and less costly selection procedures. Both the reduced production penalty associated with fleece-rot and the ease of measurement make fleece-rot the preferred selection trait for resistance to both diseases. It is likely that a similar genetic relationship exists between resistance to dermatophilosis and fly strike and that selection against dermatophilosis will also reduce fly strike.

Secondly, in contrast to internal parasites, considerable variation has been found between strains and blood lines within the Merino breed in resistance to fleece-rot, fly strike and, to a lesser extent, dermatophilosis. Preliminary data also suggest differences between Merino blood lines within the fine-wool spectrum in resistance to foot-rot (H.W. Raadsma, A.A. Swan and I.P. Purvis, unpublished). This provides breeders with the option of strain and blood-line replacement to capture relatively large and rapid changes in resistance. As with breed substitution, the implications for all other production traits obviously need to be considered before making such changes.

Thirdly, for some bacterial diseases, including foot-rot, dermatophilosis, pasteurellosis, salmonellosis and some clostridial diseases, vaccines are available, but their efficacy is suboptimal, protection is short-lived and often a proportion of vaccinated sheep show inadequate responses. Exploitation of genetic mechanisms that affect vaccine response may be an alternative option to enhancing the resistance of sheep to these diseases. This aspect is discussed further on.

As with internal parasites, it is difficult to measure resistance and other production aspects simultaneously, because invariably disease has to be present for the host to express its inherent resistance. For many of the bacterial diseases, both reduced prevalence of disease and reduced severity are suitable end-points. This is not the case for fly strike, where resistance is only expressed as a binomial trait.

The advantages of selection for enhanced resistance to fly strike can be demonstrated in long-term selection flocks selected for and against resistance to fleece-rot. In one such flock (Raadsma, 1991), the resistant line requires preventive jetting less frequently than the susceptible line and, if left unjetted, would require fewer animals to be treated on an individual basis. Ponzoni (1984) devised several treatment scenarios to evaluate relative economic values for inclusion of fly-strike resistance in breeding programmes, ranging from no treatment of struck sheep to full preventive treatment, with decreasing economic incentive for selection as the intensity of alternative treatments increased. As discussed by Raadsma and Rogan (1987), the importance of disease resistance is more realistically seen in the long term, when resistant flocks would require no or little disease management. Clearly, any increase in labour costs or a desire to decrease reliance on chemicals for treatment would favour the development of such easy care features.

The advantages of selection for increased resistance to foot-rot are even less clear. In the case of this contagious disease, eradication of the essential agent, *D. nodosus*, from flocks of sheep has been shown to be feasible. However, the chance of successful eradication greatly decreases as the number of flocks and flock size increase. There are few benefits from unsuccessful eradication campaigns, and the risk of reintroduction of the essential agent on successful eradication is real. There are thus no cumulative benefits from conventional control and eradication programmes without the need for continued inputs to control this disease.

Important Diseases for Which We Need More Information

Although viral diseases are a major constraint to sheep production in many parts of the world, little is known about the genetic control of responsiveness to viral infection or vaccination (Kitching, 1991; Eaton and Gray, 1995). Jeggo *et al.* (1986) showed that Soay sheep infected with blue-tongue virus died before showing any clinical signs, whereas Dorset Horn and Merinos developed severe clinical signs. Similarly, Nimbkar and Nimbkar (1995) highlighted potential breed differences during severe outbreaks of sheep pox, where morbidity and mortality rates may exceed 70% and 20%, respectively, under smallholder conditions.

The importance of tick-borne diseases in sheep under tropical conditions may be underrated, and no information is available on innate resistance of indigenous breeds (Baker, 1995).

Bacterial diseases of potential economic significance which cannot be effectively controlled through vaccination include Johne's disease (*Mycobacterium paratuberculosis*), mastitis (mixed bacterial infections), salmonellosis, enterotoxaemia (*Clostridium perfringens*, type D), pasteurellosis (predominantly *Pasteurella haemolytica*) and foot abscess (*Fusobacterium necrophorum*). Although anecdotal reports indicate differences between breeds in resistance, systematic breed comparisons are lacking and within-breed studies are scarce.

Lantier and colleagues (Lantier *et al.*, 1990; Gruner and Lantier, 1995) described investigations of the equivalent to the murine *Ity/LPs/Bcg* gene complex and resistance to *Salmonella abortusovis* in sheep. These studies highlight how information on genetic mechanisms of resistance to disease in other species can be applied to sheep.

Genetic Relationships among Disease Traits

The major diseases which have attracted research attention are all favoured by conditions of high rainfall. It is thus highly likely that sheep grazed under such conditions would be exposed to internal-parasite challenge, fleece-rot and fly strike, and foot-rot in infected flocks. Information on genetic correlations between resistance to these diseases is of interest, not only to predict correlated responses when selecting for resistance to one disease, but in the eventuality that breeding for 'broad-based' disease resistance may be possible. At present, the literature is devoid of detailed and reliable estimates. Where preliminary information is available, multiple diseases were monitored on the same animals and results may therefore have been biased by the influence of one disease on the expression of the other. Results to date indicate that, at best, the genetic correlations between internal-parasite resistance and resistance to many of the other diseases described here are neutral or slightly favourable (Eady *et al.*, 1994; Raadsma *et al.*, 1995; Woolaston *et al.*, 1995). Similarly, genetic relationships between resistance to fleece-rot, dermatophilosis and foot-rot (essentially all diseases of the skin) appear to be essentially neutral. It is unlikely that strong favourable or unfavourable genetic relationships exist between resistance to the important diseases in sheep, but more reliable information is required to refine selection decisions which aim to include resistance to more than one disease.

Resistance and Indirect Selection

Understanding the mechanisms responsible for disease resistance offers us several advantages in disease control. Firstly, it allows us to identify processes that could be used as indirect selection criteria. Secondly, it may allow us to identify genes that contribute to resistance and hence could be used as genetic markers. Thirdly, it highlights features of the parasite–pathogen relationship that could be targeted by non-genetic means. Finally, it offers us an insight into mechanisms that could be utilized in other host–pathogen systems.

Although it is generally accepted that the major diseases which adversely affect production are under polygenic control, this does not exclude the possibility that genes of relatively large effect could operate. In a polygenic model with moderate heritabilities, many factors (most of which are non-genetic) contribute to resistance. The problem is to separate the genetic from the non-genetic factors. This requires specialized experimental resources. Most of our understanding of genetic mechanisms of disease resistance in sheep comes from comparisons of animals in lines selected for and against a particular disease. Even then, it is not always possible to distinguish gene effects from other effects and single-gene effects from other polygenic effects. Most experiments designed for detailed investigation of disease resistance lack sufficient power to detect the effects of interest. Eady (1995) and O'Meara and Raadsma (1995) stressed the genetic and practical requirements that need to be met before indicator traits can be considered useful indirect selection criteria.

Resistance to internal parasites and indirect selection

Immune responses are certainly implicated in the genetic resistance of sheep to worm infections (Gray and Gill, 1993), and lines of sheep successfully selected on the basis of response to vaccination against *T. colubriformis* (Windon, 1991) provide further evidence that there is genetic variation in the hosts' ability to respond to vaccination, as well as to infection. In lines of selected sheep, immune or inflammatory defence mechanism have been shown to be associated with resistance to the parasite (Gill, 1991; Windon, 1991; O'Meara *et al.*, 1992; Colditz *et al.*, 1996). A characteristic of the immune response to nematode infection is an accumulation of mucosal mast cells (MMC) and eosinophils in the gastrointestinal tract (Rothwell, 1989). The MMC accumulation is antigen-dependent and driven by the T lymphocytes, and these MMCs undergo changes to globule leucocytes during infection (Huntley *et al.*, 1984). Genetically resistant sheep have high numbers of globule leucocytes relative to non-resistant counterparts (Presson *et al.*, 1988). As the immune response to helminths becomes better understood, the possibility of using these traits as indicators of resistance will increase. At present, there are no indirect selection criteria that will reliably predict genetic merit of resistance to internal parasites without the need for direct challenge.

Resistance to fleece-rot/fly strike and indirect selection

This complex is the only other disease for which a number of indirect selection criteria have been evaluated and identified, and these have recently been reviewed by O'Meara and Raadsma (1995). Potentially useful fleece traits and skin immunological responses include greasy-wool colour, fibre-diameter variability and skin inflammatory response to *Lucilia cuprina* antigens. However, when direct and indirect selection strategies were considered in breeding programmes that included other production traits, indirect selection proved no more useful than direct selection alone (Atkins, 1987). The reasons for this were twofold. Firstly, the moderate prevalence of fleece-rot is sufficient to make informed (direct) selection decisions. Secondly, unfavourable genetic relationships between indirect selection traits and other production objectives reduce the perceived value of indirect selection traits, which have no economic value in their own right.

For other diseases covered in this chapter, no indirect selection traits of value have been identified.

Genetic markers

Searches for associations between genetic markers and disease resistance have mainly focused on resistance to nematodes. Such studies also require specialized genetic resources to follow segregation between marker and resistance.

Haemoglobin type

Sheep have two major alleles (A and B) for haemoglobin and several studies have suggested that animals with haemoglobin type AA (HbAA) are more resistant than HbAB, which are more resistant than HbBB, to infection with *H. contortus* (Allonby and Urquhart, 1976; Altaif and Dargie, 1978; Preston and Allonby, 1979). However, other workers have been unable to confirm this association (Radhakrishnan *et al.*, 1972; Riffkin and Yong, 1984; Albers and Gray, 1986; Kassai *et al.*, 1990). Therefore the general usefulness of haemoglobin type as a predictive marker for resistance, at least to *H. contortus*, is limited.

Major histocompatibility complex associations

Hein (Chapter 7) provided an account of the structure and function of genes located within the major histocompatibility complex (MHC). As MHC genes are important in immune function, they have been targeted as candidate genes for disease resistance. Polymorphism of MHC class I (associated with CD8-cell differentiation and cytotoxic T lymphocytes) and class II molecules (associated with antigen-presenting and CD4-cell differentiation) has been investigated for its association with nematode parasitism. Studies of the MHC of sheep have been reviewed by Wetherall and Groth (1992) and Hohenhaus and Outteridge (1995). Outteridge *et al.* (1985, 1986, 1988) reported an association between serologically defined class I antigens of the ovine MHC and response to challenge with *T. colubriformis*. Similar observations were made by Stear *et al.* (1995) who reported an association between class I MHC and resistance to *Ostertagia circumcincta*. Cooper *et al.* (1989), however, found no evidence for an association between class I antigens of the ovine MHC and susceptibility to *H. contortus* infection in the progeny of six rams. R.R. Woolaston, G.D. Gray, H.W. Raadsma and P.M. Outteridge (unpublished data) also failed to confirm an association between resistance to internal parasites in 400 progeny of 40 rams, following natural infection at pasture. Furthermore Douch and Outteridge (1989) and Outteridge (1991) showed that the putative resistance ovine leucocyte antigen (OLA) type does not confer increased resistance in all cases. Although the hypothesis that OLA type affects resistance cannot be dismissed, it remains unresolved whether this association is parasite-specific or, indeed, if it is a consequence of linkage disequilibrium with other genes in this complex. Until these issues are resolved, selective breeding for a particular MHC class I antigen is not recommended.

Evidence for class II gene association with parasite resistance in sheep is scarce. Hulme *et al.* (1991) reported that restriction fragment length polymorphism (RFLP) analyses, using cross-hybridizing human MHC DQA, DQB and DRB complementary deoxyribonucleic acid (cDNA) probes with restriction enzymes *Taq*I, *Pvu*II and *Hin*dIII showed that genes in, or closely linked to, the MHC have a significant effect on resistance to *T. colubriformis* infection. Blattman *et al.* (1993) showed no association between MHC class II polymorphism, detected by similar human probes, with resistance to *H. contortus* in a single family of parasite-resistant Merino sheep. Recently Stear and colleagues (Schwaiger *et al.*, 1995; Buitkamp *et al.*, 1996) have shown an association between low FEC after natural *Ostertagia* infection and MHC class II polymorphism in Scottish Blackface sheep. It is postulated that at least one 'resistance' locus in linkage disequilibrium with *DRB1* and *DY* is responsible and could serve as a marker for selective breeding (Stear *et al.*, 1995). However, the role of this marker in other breeds and in relation to other parasite species is unknown. Once again, until these issues are resolved, selective breeding for a particular MHC class II haplotype is not recommended.

Anonymous DNA markers

As a consequence of the enormous developments in molecular genetics of the sheep, the number of potential markers which could be used to scan the entire genome for chromosomal regions of interest is almost limitless (Nicholas and Blattman, 1995; see also Chapter 11). The application of knowledge from mapped microsatellite markers offers a powerful tool for designing linkage experiments that will point to the chromosomal location of genes with a major additive effect on disease resistance. Linkage studies can be complex, because they require crosses to be set up in which the segregation of both the putative marker and the trait of interest can be detected. This usually involves intercrossing two inbred lines, but in ruminants such experiments have been established by crossing breeds and lines that differ widely in resistance to disease. Resistant breeds have been identified for many disease traits, and genetically divergent populations (such as long-term selection lines or families with extreme performance) are also being exploited. For example, the resistant Red Maasai and Dorper breeds used in detection of resistance to *H. contortus* (Baker, 1995) and the lines selected for divergent response to challenge with *T. colubriformis* (D.J. Hulme and K.J. Beh, personal communication) have formed the basis of so-called quantitative trait loci (QTL) studies. Another breed suitable for future investigation is the Indonesian thin-tail sheep, which appears to have extreme resistance to infections with *Fasciola gigantica*. In such studies, information on gene(s) and/or their action will be useful for understanding not only disease resistance in sheep but also resistance to similar diseases in other ruminants.

At the time of writing, no QTL experiments that cover the entire genome have been completed in sheep for disease resistance. It should be noted that continued work on evaluation of candidate genes in parallel to anonymous marker work is warranted, as design limitations may exclude the detection of relevant genes, due to excessive polymorphism (i.e. class II MHC genes). Candidate-gene approaches, as used by Blattman and Beh (1994), also benefit from the transfer of knowledge generated in other species, for which the information available is often greater than it is for sheep.

Nicholas and Blattman (1995) highlighted the many pitfalls and constraints in experiments designed to detect genetic markers. These apply equally to some of the experiments reported here. A constraint on most of the past and current marker-detection studies remains that the significance of associations between markers and disease traits relies on appropriate statistical models or methods which are themselves still a major focus of research.

Exploitation of genetic markers

If a genetic marker for a particular trait is identified, under what circumstances should that marker be included as an additional selection criterion in a selection programme for that trait? This question has been considered (Nicholas and Blattman, 1995) and the answer is that a marker should be used in practice when the financial benefit from the extra response obtained from using the marker is greater than the additional cost associated with the marker. Genetic markers are likely to increase in number and in their usefulness for identifying resistant breeding animals and there are good reasons to believe they may be deployed successfully in breeding programmes.

Increasing Resistance and Production

No breeding programme should have disease resistance as its sole objective, whether the approach is to focus on selection within a breed or replacement with more suitable breeds. Realistic objectives will include disease resistance with other important production traits, and be expressed in units of production or in value of production. Therefore, genetic relationships of resistance with other traits in the objective, such as live-weight gain, fleece weight, fleece quality and resistance to other diseases, all need to be considered. At present, estimates of such relationships are scarce, and we only have indications of the likely genetic relationships between disease resistance and important production traits (Table 8.3). It should be noted that most genetic correlations available to date between disease resistance and production traits have been obtained from measurements on the same animals. These estimates may thus be affected by the environmental effect of disease on production.

Table 8.3. Summary of best available information on genetic correlations between production traits and disease resistance in young sheep.

Trait	Nematode resistance	Fleece-rot/fly strike	Foot-rot
Body weight	Neutral to slightly favourable	Neutral to slightly unfavourable	Unknown
Clean-fleece weight	Neutral to slightly unfavourable	Neutral to slightly unfavourable	Neutral to slightly unfavourable
Average fibre diameter	Neutral	Neutral to slightly favourable	Neutral

Breeding programmes that do not include disease resistance

For Merino improvement programmes which do not include disease resistance as a breeding objective, preliminary predictions can be made on the likely consequences for resistance to the major diseases. Programmes with a strong emphasis on clean-fleece weight may lead to a small increase in susceptibility to disease. Programmes with a strong emphasis on reducing fibre diameter are likely to see no changes in resistance to gastrointestinal nematodes, a slight improvement in resistance to fleece-rot and fly strike and no major change in resistance to foot-rot. Selection for increased body weight is likely to lead to a favourable effect on nematode resistance, a small decrease in resistance to fleece-rot and fly strike and an unknown effect on resistance to foot-rot. It should be noted that in all cases the likely changes in resistance/susceptibility are small.

Inclusion of disease resistance in breeding programmes

Despite the difficulty in assigning relative economic values for resistance to all major diseases in sheep and the uncertainty about genetic correlations between production traits and disease resistance, some attempts have been to made to include resistance to individual diseases. The relevant issues have been addressed by Ponzoni (1984) with respect to fleece-rot and fly strike, while the logistics of including resistance to gastrointestinal nematodes in breeding programmes have been discussed by a number of workers (Albers *et al.*, 1987; Piper and Barger, 1988; Bisset *et al.*, 1994, 1996; Woolaston, 1994; McEwan *et al.*, 1995; Morris *et al.*, 1995a; Pocock *et al.*, 1995; Woolaston and Eady, 1995; Woolaston and Baker, 1996). Strategies dealing with other diseases are few. Using details presented by Woolaston (1994) as an example for resistance to internal parasites, the following points emerge.

- Inclusion of resistance in a breeding programme need not greatly compromise gains in other traits.
- Irrespective of relative economic values assigned to internal-parasite resistance, optimum progress in parasite resistance will be approximately

between 50 and 70% of maximum gain, when considering other economically important traits.

- When multistage selection is used, it is preferable to cull for production traits at the first stage and resistance at the second stage, rather than vice versa.
- Repeated measurement of resistance offers considerable advantages over a single measurement, provided it can be done cost-effectively, e.g. on a reduced number of samples.

As pointed out by Woolaston (1994), though, in drawing such conclusions, assumptions about the economic value of FEC are important but not necessarily critical for breeders already committed to reducing FEC. Of overriding importance are assumptions about genetic parameters, as they affect all conclusions.

Choice of breed, strain and blood-line

The optimal choice of a resistant breed will depend not only on its relative resistance but also on the other commercial requirements of the producer. If resistant sheep breeds are better producers of meat or wool than breeds in current use, a change to the resistant breed may be advantageous. Comparison of the resistant Red Maasai breed with the susceptible Dorper breed has shown that, in Kenya, Red Maasai sheep are as productive as Dorpers (Baker *et al.*, 1993). It is important when comparing resistant and susceptible breeds in a parasitized environment that all aspects of productivity are considered, including reproductive, survival and production characteristics, measured under realistic commercial conditions.

Breed substitution is unlikely to be an option for apparel fine-wool production. However, within-breed exploitation of strain and blood-line differences in disease resistance within the Merino breed is practised by breeders for resistance to fleece-rot and fly strike. This option is currently of limited value in the case of resistance to internal parasites, given the limited variation between flocks of Merinos. Detailed comparisons of available blood-lines for disease resistance under commercial conditions are still a practical limitation, although such comparisons have been established for other production traits (Hygate and Atkins, 1988).

Genetic Control of Response to Vaccination

In sheep, a number of bacterial, viral and parasitic diseases can be partly or fully controlled by vaccination. If vaccination controls a disease for at least 12 months or more, it may offer a cheap and easy form of control. For example, whole-cell preparations or the principal toxoids of clostridia, anthrax and *Corynebacterium pseudotuberculosis* are effective immunogens and can be easily incorporated in multivalent (multistrain) bacterial vaccines. Similarly, vaccination with the L3-stage larvae of the lungworm (Lightowlers *et al.*, 1994) offers lifelong protection in vaccinated animals. However, for a large number of diseases, vaccines are only partly effective. Despite aggressive adjuvants (immune stimulants), some sheep fail to show adequate responses and remain unprotected. There have been relatively few investigations in sheep detailing environmental and genetic factors governing immune response following vaccination.

In two small-scale investigations using sheep, Nguyen (1983) and Berggren-Thomas *et al.* (1987) reported a moderate to high heritability for antibody response to chicken red blood cells and ovalbumin, respectively. Although such antigens are used traditionally as so-called 'complex-model antigens' (Biozzi *et al.*, 1982; Van der Zijpp, 1983; Warner *et al.*, 1987), their relationship with protective antibody responses to well-defined immunogens is doubtful. Raadsma *et al.* (1996) reported genetic aspects of protective antibody responses to multivalent foot-rot and clostridial vaccines, which are commercial vaccines commonly used in sheep. Heritability estimates for antibody response to the defined immunogens of *D. nodosus* in the foot-rot vaccine ranged from 0.24 to 0.58 for the nine major serogroups. For two clostridial organisms, namely *Clostridium tetani* (tetanus) and *Clostridium chauvoei* (blackleg), heritability estimates for vaccine response were lower (0.12 and 0.24, respectively). It is not possible to compare directly across vaccines, given the differences in vaccine composition and in age at vaccination. These results, nevertheless, show that significant additive genetic variation exists within Merino sheep for vaccine response.

Genetic correlations between antibody response to specific antigens within and between vaccines are also of interest to establish the importance of 'specific' and 'general' immune-response genes which are implicated in humoral (antibody) immune response. Results available to date show that, at least for bacterial vaccines, the degree of common genetic control of antibody response is dependent on the antigen examined (Raadsma *et al.*, 1996) and can range from 0 to over 85%. This has implications in programmes attempting to exploit 'general' immune response genes for increased humoral response. As vaccine antigens become better defined for both health and production traits, it may become evident that individual sheep vary in their ability to respond optimally to the full range of antigens.

Genetic correlations between vaccine response and disease resistance are also of interest as they provide information on the potential for indirect selection. Once again, there is little information on correlations between vaccine response and innate resistance after challenge without vaccination. For foot-rot Raadsma *et al.* (1995) report low genetic correlations between vaccine response and innate resistance. Similar estimates have yet to be obtained for other diseases.

Conclusions

From our current knowledge on disease resistance in sheep, it is evident that genetic variation exists for almost every case where it has been investigated. Resistance should only be included in breeding programmes for diseases which have the potential to reduce production and for which alternative methods of control are, or may be, economically unsustainable. To a large extent, this has been reflected in the global research effort over the last 20 years, which has focused strongly on resistance of sheep to internal parasites.

Although the long-term benefits of genetic change in disease resistance under commercial conditions have not been well documented, the potential changes that can be made are clearly shown in single-trait selection lines. Once again, this has mainly been done for resistance to gastrointestinal parasites, fly strike, foot-rot and facial eczema in sheep, and methodologies are now available for breeders to improve resistance to these diseases. In the case of resistance to internal parasites, selection for reduced FEC remains the recommended selection strategy, and it can be shown that this need not adversely affect genetic change in other important production traits.

Continued focus on alternative selection criteria, including DNA markers, is warranted, and it may have additional spin-offs in our understanding of host–pathogen relationships and thus present us with new avenues for disease control.

Finally, the enormous diversity of breeds, which are kept under highly varied climatic conditions and husbandry systems worldwide, has not been investigated rigorously enough. Not only do breed differences offer a rapid means of capturing potentially large genetic changes in disease resistance, but they also provide us with extreme experimental resources to increase our understanding of disease resistance in sheep.

References

Albers, G.A.A. and Gray, G.D. (1986) Breeding for worm resistance: a perspective. *International Journal of Parasitology* 17, 559–566.

Albers, G.A.A., Gray, G.D., Piper, L.R., Barker, J.S.F., Le Jambre, L.F. and Barger, I.A. (1987) The genetics of resistance and resilience to *Haemonchus contortus* in young Merino sheep. *International Journal of Parasitology* 17, 1355–1363.

Allonby, E.W. and Urquhart, G.M. (1976) A possible relationship between haemonchosis and haemoglobin polymorphism in Merino sheep in Kenya. *Research in Veterinary Science* 20, 212–214.

Altaif, K.I. and Dargie, J.D. (1978) Genetic resistance to helminths: the influence of breed and haemoglobin type on the response of sheep to primary infections with *Haemonchus contortus*. *Parasitology* 77, 161–175.

Anonymous (1994) *Nemesis Breeders Worm Control Network – Recommendations for Breeding Sheep for Worm Resistance* CSIRO, Armidale.

Atkins, K.D. (1987) Resistance to fleece rot and body strike: its role in a breeding objective for Merino sheep. In: *Proceedings of the Sheep and Blowfly and Flystrike*

Management Workshop. Department of Agriculture, Trangie, New South Wales, pp. 3.1–3.7.

Baker, R.L. (1995) Genetics of disease resistance in small ruminants in Africa. In: Gray, G.D, Woolaston, R.R. and Eaton, B.T. (eds) *Breeding for Resistance to Infectious Diseases in Small Ruminants.* ACIAR Monograph No. 34, ACIAR, Canberra, Australia, pp. 119–138.

Baker, R.L., Skerman,T.M. and Dobbie, J.L. (1986) Inheritance of footrot and footscald in sheep. *Journal of Animal Science* 63 (Suppl. S1), 196.

Baker, R.L., Watson, T.G., Bisset, S.A., Vlassoff, A. and Douch, P.G.C. (1991) Breeding sheep in New Zealand for resistance to internal parasites: research results and commercial application. In: Gray, G.D. and Woolaston, R.R. (eds) *Breeding for Disease Resistance in Sheep.* Australian Wool Corporation, Melbourne, pp. 19–32.

Baker, R.L., Reynolds, L., Mwamachi, D.M., Audho, J.O., Magadi, M. and Miller, J.E. (1993) Genetic resistance to gastrointestinal parasites in Dorper and Red Maasai × Dorper lambs in coastal Kenya. In: *Proceedings of the 11th SR–CRSP Scientific Workshop*, 3–4 March, Nairobi, Kenya.

Baker, R.L., Mwamachi, D.M., Audho, J.O. and Thorpe, W. (1994) Genetic resistance to gastrointestinal nematode parasites in Red Maasai sheep in Kenya. *Proceedings of the 5th World Congress on Genetics Applied to Livestock Production* 20, 277–280.

Berggren-Thomas, P.L., Kaattari, S., Hohenboken, W.D., Shrestha, J.N.B. and Heany, D.P. (1987) Inheritance of active and acquired immunity in sheep. *Journal of Animal Science* 64, 1302–1312.

Biozzi, G. (1982) Correlation between genetic regulation of immune responsiveness and host defence against infection and tumours. *European Journal of Clinical Investigation* 12, 373–376.

Bishop, S.C., Bairden, K., McKellar, Q.A., Park M. and Stear, M.J. (1997) Genetic parameters for faecal egg count following mixed, natural, predominantly *Ostertagia circumcincta* infection and relationships with liveweight in young lambs. *Animal Production* (in press).

Bisset, S.A., Vlassoff, A., Morris, C.A., Southey, B.R., Baker, R.L. and Parker, A.G.H. (1992) Heritability of and genetic correlations among faecal egg counts and productivity traits in Romney sheep. *New Zealand Journal of Agricultural Research* 35, 51–58.

Bisset, S.A., Morris, C.A., Squire, D.R., Hickey, D.M. and Wheeler, M. (1994) Genetics of resilience to nematode parasites in Romney sheep. *New Zealand Journal of Agricultural Research* 37, 521–534.

Bisset, S.A., Morris, C.A., Squire, D.R. and Hickey, D.M. (1997) Genetics of resilience to nematode parasites in young Romney sheep – use of weight gain under challenge to assess individual anthelmintic requirements. *New Zealand Journal of Agricultural Research* (in press).

Blattman, A.N. and Beh, K.J. (1994) T-cell receptor and immunoglobulin gene polymorphisms and resistance to *Haemonchus contortus* in sheep. *Journal of Animal Breeding and Genetics* 111, 65–74.

Blattman, A.N., Hulme, D.J., Kinghorn, B.P., Woolaston, R.R., Gray G.D. and Beh, K.J. (1993) A search for associations between major histocompatability complex restriction fragment length polymorphism bands and resistance to *Haemonchus contortus* in sheep. *Animal Genetics* 24, 277–282.

Bouix, J., Gruner, L., Vu Tien Khang, J., Luffau,G. and Yvore, P. (1992) Recherches sur la résistance des ovins au parasitisme interne: mise en evidence d'une composante génétique. *Bulletin des Groupements Techniques Vétérinaires* 5, 91–99.

Buitkamp, J., Filmether, P., Ludt, C., Fries, R. and Stear, M. (1996) Major histocompatibility complex alleles are associated with faecal egg counts following natural, predominantly *Ostertagia circumcincta* infection. In: *Proceedings 25th International Conference on Animal Genetics*, Tours, France, B043 (abstract).

Bundy, D.A.P. and Golden, M.H.N. (1987) Interaction between nutrition and parasitism in sheep. *Parasitology* 95, 623.

Colditz, I.G., Watson, D.L., Gray G.D. and Eady, S.J. (1996) Some relationships between age, immune responsiveness and resistance to parasites. *International Journal of Parasitology* 26, 869–878.

Cooper, D.W., Van Oorschot R.A.H., Piper, L.R. and Le Jambre, L.F. (1989) No association between the ovine leucocyte antigen (OLA) system in the Australian Merino and susceptibility to *Haemonchus contortus* infection. *International Journal of Parasitology* 19, 695–697.

Cummins, L.J., Thompson, R.L., Yong, W.K., Riffkin, G.G., Goddard, M.E., Gallinan, A.P.L. and Saunders, M.J. (1991) Genetics of *Ostertagia* selection lines. In: Gray, G.D. and Woolaston, R.R. (eds) *Breeding for Disease Resistance in Sheep*. Australian Wool Corporation, Melbourne, pp. 11–18.

Dempster, E.R. and Lerner, I.M. (1950) Heritability of threshold characters. (With an Appendix by Alan Robertson). *Genetics* 35, 212–236.

Douch, P.G.C. and Outteridge, P.M. (1989) The relationship between ovine lymphocyte antigens and parasitological and production parameters in Romney sheep. *International Journal for Parasitology* 19, 35–41.

Eady, S.J. (1995) Phenotypic traits associated with resistance to internal parasites. In: Gray, G.D, Woolaston, R.R. and Eaton, B.T. (eds) *Breeding for Resistance to Infectious Diseases in Small Ruminants*. ACIAR Monograph No. 34, ACIAR, Canberra, Australia, pp. 219–236.

Eady, S.J., Woolaston, R.R. and Mortimer, S.I. (1994) Internal parasite resistance of Merino flocks selected for production. *Proceedings 5th World Congress on Genetics Applied to Livestock Production* 20, 289–292.

Eady, S.J., Woolaston, R.R., Mortimer, S.I., Lewer, R.P., Raadsma, H.W., Swan, A.A. and Ponzoni, R.W. (1996) Resistance to nematode parasites in Merino sheep: sources of genetic variation. *Australian Journal of Agricultural Research* 47, 895–915.

Eaton B.T. and Gray G.D (1995) Genetic variation in resistance to viruses. In: Gray, G.D, Woolaston, R.R. and Eaton, B.T. (eds) *Breeding for Resistance to Infectious Diseases in Small Ruminants*. ACIAR Monograph No. 34, ACIAR, Canberra, Australia, pp. 1–14.

Egerton, J.R. and Raadsma, H.W. (1991) Breeding sheep for resistance to footrot. In: Axford, R.F.E. and Owen, J.B. (eds) *Breeding for Disease Resistance in Farm Animals*. CAB International, Wallingford, pp. 347–370.

Falconer, D.S. (1965) The inheritance of liability to certain diseases estimated from the evidence among relatives. *Annals of Human Genetics* 29, 51–76.

Gill, H.S. (1991) Genetic control of acquired resistance to haemonchosis in Merino lambs. *Parasite Immunology* 13, 617–628.

Gray, G.D. (1991) Breeding for resistance to trichostrongyle nematodes in sheep. In: Axford, R.F.E. and Owen, J.B. (eds) *Breeding for Disease Resistance in Farm Animals*. CAB International, Wallingford, pp. 100–125.

Gray, G.D. and Gill, H.S. (1993) Host genes, parasites and parasitic infections. *International Journal of Parasitology* 23, 485–494.

Gray, G.D., Barger, I.A., LeJambre, L.F., and Douch, P.G.C. (1992) Parasitological and immunological responses of genetically resistant Merino sheep on pastures contaminated with parasitic nematodes. *International Journal for Parasitology* 22, 417–425.

Gray, G.D., Woolaston, R.R. and Eaton, B.T. (eds) (1995) *Breeding for Resistance to Infectious Diseases in Small Ruminants.* ACIAR Monograph No. 34, ACIAR, Canberra, Australia, 322 pp.

Gruner, L. and Lantier, F. (1995) Breeding for resistance to infectious diseases in small ruminants in Europe. In: Gray, G.D, Woolaston, R.R. and Eaton, B.T. (eds) *Breeding for Resistance to Infectious Diseases in Small Ruminants.* ACIAR Monograph No. 34, ACIAR, Canberra, Australia, pp. 99–117.

Hohenhaus, M.A. and Outteridge, P.M. (1995) The immunogenetics of resistance to *Trichostrongylus colubriformis* and *Haemonchus contortus* parasites in sheep. *British Veterinary Journal* 151, 119–139.

Hulme, D.J., Windon, R.G., Nicholas, F.W. and Beh, K.J. (1991) Association between MHC Class II RFLP and *Trichostrongylus* resistance. In: Gray, G.D. and Woolaston, R.R. (eds) *Breeding for Disease Resistance in Sheep.* Australian Wool Corporation, Melbourne, pp. 115–120.

Huntley, J.F., Newlands, G. and Miller, H.R.P. (1984) The isolation and characterisation of globule leukocytes: their derivation from mucosal mast cells in parasitised sheep. *Parasite Immunology* 6, 371–390.

Hygate, L. and Atkins, K.D. (1988) Utilisation of information across wether trials. *Proceedings of the Australian Association of Animal Breeding and Genetics* 7, 414–417.

Jeggo M.J., Corteyn, A.H., Taylor W.P., Davidson, W.L. and Gorman, B.M. (1986) Virulence of bluetongue virus for British sheep. *Research in Veterinary Science* 42, 24–28.

Karlsson, L.J.E., Macleod, I.M., Leelawardeena, D.H., Sissoev, K. and Simmons, J. (1991) Selection for nematode resistance in sheep in Mediterranean environments resistance. In: Gray, G.D. and Woolaston, R.R. (eds) *Breeding for Disease Resistance in Sheep.* Australian Wool Corporation, Melbourne, pp. 131–138.

Kassai, T., Fesus, L., Hendrikx, W.M.L., Akats, C., Fok, E., Redl, P., Takacs, E., Nilsson, R., Van Leeuwen, M.A.W., Jansen, J., Bernadina, W.E. and Frankena, K. (1990) Is there a relationship between haemoglobin genotype and the innate resistance to experimental *Haemonchus contortus* infection in Merino lambs? *Veterinary Parasitology* 37, 61–77.

Kelly, M.J. and Gray, G.D (1995) Environmental effects on the repeatability of resistance to worms in sheep. *Proceedings of the Australian Association of Animal Breeding and Genetics* 11, 84–88.

Kitching, R.P. (1991) Genetic control of acute viral diseases. In: Axford, R.F.E. and Owen, J.B. (eds) *Breeding for Disease Resistance in Farm Animals.* CAB International, Wallingford, pp. 315–328.

Lantier, F., Tien, J.V., Oswald, I.P., Bioux, J., Guerrin, G., Nguyen, T.C., Gros, P., Shurr, E., Skamene, E., Boivin, R., Pepin, M. and Pardon, P. (1990) Resistance to abortusovis infection: is there an equivalent to the *Ity* gene in sheep? *Annales de Recherches Vétérinaires* 21, 305–306.

Larsen, J.W.A., Vizard, A.L., Ware, J.K.W. and Anderson, N. (1995) Diarrhoea due to trichostrongylid larvae in Merino sheep: repeatability and differences between bloodlines. *Australian Veterinary Journal* 72, 196–197.

Lewer, R.P, Gherardi, S.G. and Sutherland, S.S. (1987) Realised heritability estimates for resistance to dermatophilosis in Merino sheep. In: McGuirk, B.J. (ed.) *Merino Improvement Programs in Australia.* Australian Wool Corporation, Melbourne, pp. 347–350.

Lightowlers, M.W., Lloyd, S. and Soulsby, L. (1994) Vaccination against animal parasites. *Veterinary Parasitology* 54, 1–3, 177–204.

McEwan, J.C., Mason, P., Baker, R.L., Clarke, J.N., Hickey, S.M. and Turner, K. (1992) Effect of selection for productive traits on internal parasite resistance in sheep. *Proceedings of the New Zealand Society of Animal Production* 52, 53–56.

McEwan, J.C., Dodds, K.G., Watson, T.G., Greer, G.J., Hosking, B.C. and Douch, P.G.C. (1995) Selection for resistance to roundworms by the new Zealand sheep breeding industry; the WormFEC service. *Proceedings of the Australian Association of Animal Breeding and Genetics* 11, 70–73.

Morris, C.A., Watson, T.G., Baker, R.L, Hurford, A.P. and Hosking, B.C. (1993) Repeatability estimates and selection flock effects for faecal nematode egg counts in Romney breeding ewes. *Proceedings of the New Zealand Society of Animal Production* 53, 227–229.

Morris, C.A., Towers, N.R., Wheeler, M. and Wesselink, C. (1994) Selection for or against facial excema in Romney sheep, as monotored by serum cncentrations of a liver enzyme. *New Zealand Journal of Agricultural Research* 38, 211–219.

Morris, C.A., Towers, N.R., Wheeler, M. and Amyes, N.C. (1995a) A note on the genetics of resistance or susceptibility to ryegrass staggers in sheep. *New Zealand Journal of Agricultural Research* 38, 367–371.

Morris, C.A., Watson, T.G., Bisset, S.A, Vlassoff, A. and Douch, P.G.C. (1995b) Breeding sheep in New Zealand for resistance or resilience to nematode parasites. In: Gray, G.D, Woolaston, R.R. and Eaton, B.T. (eds) *Breeding for Resistance to Infectious Diseases in Small Ruminants.* ACIAR Monograph No. 34, ACIAR, Canberra, Australia, pp. 77–98.

Nguyen, T.C. (1983) The immune response in sheep: analyses of age, sex, and genetic effects on the quantitative antibody response to chicken red blood cells. *Veterinary Immunology and Immunopathology* 5, 237–245.

Nicholas, F.W. and Blattman, A.N. (1995) Application of genetic markers for parasites. In: Gray, G.D, Woolaston, R.R. and Eaton, B.T. (eds) *Breeding for Resistance to Infectious Diseases in Small Ruminants.* ACIAR Monograph No. 34, ACIAR, Canberra, Australia, pp. 259–274.

Nimbkar, C. and Nimbkar, B.V. (1995) Genetic variation in resistance to infectious disease in small ruminants in India. In: Gray, G.D, Woolaston, R.R. and Eaton, B.T. (eds) *Breeding for Resistance to Infectious Diseases in Small Ruminants.* ACIAR Monograph No. 34, ACIAR, Canberra, Australia, pp. 139–152.

O'Meara, T.J., and Raadsma, H.W. (1995) Phenotypic and genetic indicators of resistance to ectopathogens. In: Gray, G.D, Woolaston, R.R. and Eaton, B.T. (eds) *Breeding for Resistance to Infectious Diseases in Small Ruminants.* ACIAR Monograph No. 34, ACIAR, Canberra, Australia, pp. 187–218.

O'Meara, T.J., Nesa, M., Raadsma, H.W., Saville, D.G. and Sandeman, R.M. (1992) Variation in skin inflammatory responses between sheep bred for resistance or susceptibility to fleece rot and blowfly strike. *Research in Veterinary Science* 52, 205–210.

Outerridge, P.M. (1991) Ovine lymphocyte antigens and resistance to parasites. In: Gray, G.D. and Woolaston, R.R. (eds) *Breeding for Disease Resistance in Sheep*. Australian Wool Corporation, Melbourne, pp. 30–38.

Outteridge, P.M., Windon, R.G. and Dineen, J.K. (1985) An association between a lymphocyte antigen in sheep and the response to vaccination against the parasite *Trichostrongylus colubriformis*. *International Journal of Parasitology* 15, 121–127.

Outteridge, P.M., Windon, R.G., Dineen, J.K. and Smith, E.F. (1986) The relationship between ovine lymphocyte antigens and faecal egg count of sheep selected for responsiveness to vaccination against *Trichostrongylus colubriformis. International Journal of Parasitology* 16, 369–374.

Outteridge, P.M., Windon, R.G. and Dineen, J.K. (1988) An ovine lymphocyte antigen marker for acquired resistance to *Trichostrongylus colubriformis. International Journal of Parasitology* 18, 853–858.

Piper, L.R. (1987) Genetic variation in resistance to internal parasites. In: McGuirk, B.J. (ed.) *Merino Improvement Programs in Australia.* Australian Wool Corporation, Melbourne, pp. 351–363.

Piper, L.R and Barger, I.A. (1988) Resistance to gastro-intestinal strongyles: feasibility of a breeding program. In: *Proceedings of the Third World Congress on Sheep and Beef Cattle Breeding*, Paris, Vol. 2. INRA, Paris, France, pp. 593–611.

Pocock, M.J., Eady, S.J and Abbott, K.A. (1995) Nemesis in action – breeding for worm resistance. *Proceedings of the Australian Association of Animal Breeding and Genetics* 11, 74–78.

Ponzoni, R.W. (1984) The importance of resistance to fleece rot and body strike in the breeding objective of Australian Merino sheep. *Wool Technology and Sheep Breeding* 32, 33–40.

Presson, B.L., Gray, G.D. and Burgess, S.K. (1988) The effect of immunosuppression with dexamethasone on *Haemonchus contortus* infections in genetically resistant Merino sheep. *Parasite Immunology* 10, 675–680.

Preston, J.M. and Allonby, E.W. (1979) The influence of haemoglobin phenotype on the susceptibility of sheep to *Haemonchus contortus* infection in Kenya. *Research in Veterinary Science* 26, 140–144.

Raadsma, H.W. (1991) Resistance to fleece rot and flystrike. In: Axford, R.F.E. and Owen, J.B. (eds) *Breeding for Disease Resistance in Farm Animals.* CAB International, Wallingford, pp. 263–290.

Raadsma, H.W. (1995) Genetic variation in resistance to bacteria. In: Gray, G.D, Woolaston, R.R. and Eaton, B.T. (eds) *Breeding for Resistance to Infectious Diseases in Small Ruminants.* ACIAR Monograph No. 34, ACIAR, Canberra, Australia, pp. 15–42.

Raadsma, H.W. and Rogan, I.M. (1987) Genetic variation in resistance to blowfly strike. In: McGuirk, B.J. (ed.) *Merino Improvement Programs in Australia.* Australian Wool Corporation, Melbourne, pp. 321–340.

Raadsma, H.W., Sandeman, R.M., Sasiak, A.B., Engwerda, C.R. and O'Meara, T.J. (1992) Genetic inprovement in resistance to body strike in Merino sheep: where are we at with indirect selection? *Proceedings of the Australian Association of Animal Breeding and Genetics* 10, 143–146.

Raadsma, H.W., Attard, G.A., Nicholas, F.W. and Egerton, J.R. (1995) Disease resistance in Merino sheep 4. Genetic variation in immunological responsiveness to fimbrial *Dichelobacter nodosus* antigens and its relationship with resistance to footrot. *Journal of Animal Breeding Genetics* 112, 349–372.

Raadsma, H.W., Attard, G.A., Nicholas, F.W. and Egerton, J.R. (1996) Disease resistance in Merino sheep 5. Genetic heterogeneity in response to vaccination with *Dichelobacter nodosus* and clostridial antigens. *Journal of Animal Breeding and Genetics* 113, 181–199.

Radhakrishnan, C.V., Bradley, R.E. and Loggins, P.E. (1972) Host responses of worm-free Florida Native and Rambouillet lambs experimentally infected with *Haemonchus contortus. American Journal of Veterinary Research* 33, 817–823.

Riffkin, G.G. and Yong, W.K. (1984) Recognition of sheep which have innate resistance to trichostrongylid nematode parasites. In: Dineen, J.K. and Outteridge, P.M. (eds) *Immunologic Approaches to the Control of Endoparasites.* CSIRO, Melbourne, pp. 30–40.

Rothwell, T.L.W. (1989) Immune exclusion of parasitic nematodes from the alimentary tract. *International Journal of Parasitology* 19, 139–168.

Schwaiger, F.-W., Gostomski, D., Stear, M.J., Duncan, J.L., McKellar Q.A., Epplen, J.T. and Buitkamp, J. (1995) An ovine major histocompatability complex DRB1 allele is associated with low faecal egg counts following natural, predominantly OC infection. *International Journal of Parasitology* 25, 815–822.

Skerman, T.M., Johnson, D.L., Kane, D.W. and Clarke, J.N. (1988) Clinical footsclad and footrot in a New Zealand flock: phenotypic and genetic parameters. *Australian Journal of Agricultural Research* 39, 907–916.

Sreter, T., Kassai, T. and Takacs, E. (1994) The heritability and specificity of responsiveness to infection with *Haemonchus contortus* in sheep. *International Journal for Parasitology* 24, 871–876.

Stear, M.J., Bairden, K., Bishop, S., Buitkamp, J., Epplen, J.T., Gostomski, D., McKellar, Q.A., Schwaiger, F.-W. and Wallace, D.S. (1995) An ovine lymphocyte antigen is associated with reduce faecal egg counts in four month-old lambs following natural, predominantly *Ostertagia circumcincta* infection. *International Journal for Parasitology* 26, 423–428.

Van der Zijpp A.J. (1983) Breeding for immune responsiveness and disease resistance. *World Poultry Science Journal* 39, 118–131.

Warner, C.M., Meeker, D.L. and Rothschild, M.F. (1987). Genetic control of immune responsiveness: a review of its use as a tool for selection for disease resistance. *Journal of Animal Science* 64, 394–406.

Watson, T.G., Baker, R.L. and Harvey, T.G. (1986) Genetic variation in resistance to tolerance to internal nematode parasites in strains of sheep at Rotomahana. *Proceedings of the New Zealand Society of Animal Production* 46, 23–26.

Wetherall, J.D. and Groth, D.M. (1992) The major histocompatibility complex and parasite immunity. In: Yong, W.K. (ed.) *Animal Parasite Control Utilising Biotechnology.* CRC Press, Boca Raton, Florida, pp. 353–386.

Windon, R.G. (1991) Genetic control of host responses involved in resistance to the gastrointestinal nematodes of sheep. In: Axford, R.F.E. and Owen, J.B. (eds) *Breeding for Disease Resistance in Farm Animals.* CAB International, Wallingford, pp. 162–186.

Woolaston, R.R. (1994) Preliminary evaluation of strategies to breed Merinos for resistance to roundworms. *Proceedings of the Fifth World Congress on Genetics Applied to Livestock Production* 20, 281–284.

Woolaston, R.R. and Baker, R.L. (1996) Prospects of breeding small ruminants for resistance to internal parasites. *International Journal of Parasitology* 26, 845–856.

Woolaston, R.R. and Eady, S.J. (1995) Australian research on genetic resistance to nematode parasites bacteria. In: Gray, G.D, Woolaston, R.R. and Eaton, B.T. (eds) *Breeding for Resistance to Infectious Diseases in Small Ruminants.* ACIAR Monograph No. 34, ACIAR, Canberra, Australia, pp. 53–75.

Woolaston, R.R., Eady, S.J., Ponzoni, R.W., Lewer, R.P., Gifford, D.R. and Ancell, P.M.C. (1995) Genetic variation in resistance to dermatophilosis in sheep. *Proceedings of the Australian Association of Animal Breeding and Genetics* 11, 126–129.

Woolaston, R.R. and Piper, L.R. (1996). Selection of Merino sheep for resistance to *Haemonchus contortus*: genetic variation. *Animal Science* 62, 451–460.

Woolaston, R.R., Windon, R.G. and Gray, G.D. (1991) Genetic variation in resistance to internal parasites in Armidale experimental flocks. In: Gray, G.D, Woolaston, R.R. and Eaton, B.T. (eds) *Breeding for Resistance to Infectious Diseases in Small Ruminants.* ACIAR Monograph No. 34, ACIAR, Canberra, Australia, pp. 1–9.

Zajac, A.M. (1995) Genetic resistance to infectious disease in small ruminants: North America and the Caribbean. In: Gray, G.D, Woolaston, R.R. and Eaton, B.T. (eds) *Breeding for Resistance to Infectious Diseases in Small Ruminants.* ACIAR Monograph No. 34, ACIAR, Canberra, Australia, pp. 153–166.

Molecular Biology and Genetics of Scrapie in Sheep

9

N. HUNTER
Institute for Animal Health, BBSRC/MRC Neuropathogenesis Unit, West Mains Road, Edinburgh EH9 3JF, UK

Introduction	225
Clinical Signs and Pathology	226
The Importance of the Prion Protein	226
The Use of the Scrapie-modified Form of the Prion Protein in Scrapie Diagnosis	227
The Genetics of Scrapie Susceptibility in Sheep	228
Experimental scrapie	228
Natural scrapie	230
Involvement of Other Genes in Scrapie Control	233
Modes of Transmission of Scrapie	234
Conclusions	237
References	237

Introduction

Scrapie is a fatal transmissible degenerative disease affecting the central nervous system of sheep and goats. Experimental evidence suggests that there are different strains of scrapie (Bruce and Dickinson, 1987), although no infectious agent has been characterized. There are many hypotheses on the aetiology of the disease, partly because of its unusual physicochemical resistances (for example, Brown *et al.*, 1990), which have led to suggestions that the infectious agent may be made up mostly or completely of protein. A disease which is both the result of and transmitted by such a rogue protein has been variously described as a prion disease (Prusiner *et al.*, 1982), an infectious amyloidosis (Brown *et al.*, 1991) or simply a genetic mutation (Parry, 1960; Brown *et al.*, 1991; Collinge and Palmer, 1994). Others, citing evidence that scrapie carries its own genetic information, believe it to be a virus (Diringer, 1995; Manuelidis *et al.*, 1995) or a virino (mostly protein but with a small nucleic acid component) (Somerville, 1991). Many years of effort have gone into the search for the deoxyribonucleic acid (DNA) or ribonucleic acid (RNA) believed to form part of such structures, so far without success. Despite what many textbooks say, the nature of the agent is not finally resolved and, indeed, a full explanation of the characteristics of scrapie and other similar diseases of mammals (transmissible spongiform encephalopathies (TSEs)) presents problems, whichever theory is being proposed.

This chapter will concentrate on what is known about scrapie in sheep, rather than in the laboratory rodent models, and will present the conclusions from recent molecular-genetic studies in the light of earlier ideas.

Clinical Signs and Pathology

Typical clinical signs of scrapie in sheep (Dickinson, 1976) start insidiously with mildly impaired social behaviour, followed by locomotor incoordination or ataxia with trembling. Pruritis can result from the animal attempting to relieve what seems to be an intense itching by scratching against fence posts or by biting the affected area, and these clinical signs can last from 2 weeks to 6 months. In a description of scrapie-affected sheep from Shetland between 1985 and 1991 (Clark and Moar, 1992), most animals in one particular group showed signs of pruritis and emaciation, others had pruritis, emaciation and hyperaesthesia and others showed all these signs plus ataxia. These authors have also described sheep apparently with scrapie and such a short clinical course that they are simply 'found dead'. In a report of clinical signs in some Japanese sheep (Onodera and Hayashi, 1994), some animals (Suffolks and Corriedales) showed signs of pruritis but others (Corriedales) had died for no obvious reason – scrapie was diagnosed after histopathological examination. These differences in both symptoms and clinical phase may simply be due to breed characteristics or may indicate the presence in the field of different strains of scrapie.

Common lesions in the brain include neuronal degeneration, with the formation of vacuoles and proliferation of astroglial cells but no demyelination or other overt inflammatory responses. These features develop in the later stages of the incubation period and it is very difficult to detect (by histopathology) those animals with scrapie that are not visibly affected by the disease.

The Importance of the Prion Protein

The only consistent marker of scrapie, and one which is used as a hallmark of all TSEs, is an abnormal form of a neuronal surface glycoprotein protein called prion protein (PrP). Unlike normal PrP (PrP^{C}), the function of which is unknown, the aggregated form is partially resistant to proteases and is a disease-specific marker, usually represented as PrP with the superscript giving the disease, e.g. PrP^{Sc} for scrapie in sheep and PrP^{BSE} for bovine spongiform encephalopathy (BSE) in cattle. Also, PrP^{Sc} is closely associated with scrapie itself, as preparations enriched for infectivity contain high concentrations of aggregated PrP. Although some reports have suggested a dissociation between PrP fibrils and infectivity (Braig and Diringer, 1985; Manuelidis *et al.*, 1987), the usual finding is a partial copurification of PrP^{Sc} and infectivity. Hope and Manson (1991) reported a ratio of 100,000 PrP molecules to one infectious particle but pointed out that minor components which also copurify with infectivity need to be eliminated before concluding that PrP is the sole component of the infectious agent.

In one of the major scrapie aetiology hypotheses, PrP^{Sc} is itself the infecting agent or 'prion'. Arising from an infection or from a mutant PrP gene, PrP^{Sc} would act as a catalyst in the conversion of endogenous PrP^{C} into yet more PrP^{Sc}, thus destroying the normal function of the protein and resulting in degenerative disease (Prusiner *et al.*, 1990). Natural scrapie in sheep tends to be familial in appearance and has been said to result from a recessive gene, the protein product of which causes disease (Parry, 1984), and this 'genetic disease' hypothesis fits quite well with the 'prion disease' hypothesis. However, the disease, once produced, is often transmissible experimentally, which is not the case with proved genetic diseases (for example, thalassaemia (Rund *et al.*, 1991)), so the protein-only or genetic aetiology of scrapie remains unprecedented and other ideas must also be considered.

In an alternative hypothesis, the normal protein (PrP) would act as a receptor for the scrapie agent (Hunter, 1991). Mutant forms of the PrP would have a stronger affinity for the agent than the wild-type protein, resulting in disease only in the mutant genotype. Such a close association between PrP sequence and successful infection would fit the genetic data just as well as the 'genetic disease' hypothesis. Indeed, it is almost impossible to tell the difference between a simple genetic disease and genetic control over susceptibility to a widespread infection.

Proponents of yet another hypothesis believe that an independent scrapie genome, probably a small nucleic acid, would confer scrapie strain characteristics while depending on a 'coat' of PrP^{Sc} for protection from host defence mechanisms (Bruce and Dickinson, 1987). This agent is called a virus, unconventional virus or virino.

The Use of the Scrapie-modified Form of the Prion Protein in Scrapie Diagnosis

The abnormal disease-associated form of PrP is used both to diagnose a TSE at post-mortem and to search for infected tissues within the body of an animal. There are at least two methods for detection of PrP^{Sc}. Firstly, the PrP fibrils themselves can be visualized under the electron microscope and are referred to as scrapie-associated fibrils (SAF) (Dawson *et al.*, 1987; Clark *et al.*, 1994) or prion rods. Secondly, the protein can be detected either on blots made from electrophoresis gels or in sections of tissue by means of PrP-specific antibodies

(Western blots or immunocytochemistry) (Mohri *et al.*, 1992; Vankeulen *et al.*, 1995).

The alternative method of detecting infection, by mouse bioassay, is both expensive and time-consuming, because of the long incubation periods involved. However, it can be a more sensitive assay. The initial finding of PrP^{Sc} in placental tissue of a single scrapie-affected sheep (Ikegami *et al.*, 1991) was treated with some caution. However, recently, in another case in which PrP^{Sc} was detected in brain but not in spleen or placenta of a scrapie-affected pregnant sheep, infectivity was transmitted from the brain and placental homogenates to laboratory mice (Onodera *et al.*, 1993). Bioassay in mice can therefore be more sensitive than protein determination.

Also, PrP^{Sc} has been found in tissues of healthy sheep which have been exposed to scrapie. For example, Onodera *et al.* (1994) detected PrP^{Sc} in the spleen and lymph nodes of four apparently healthy sheep in Japan. Such studies are aimed at a preclinical diagnostic test and hope eventually to find PrP^{Sc} accumulating in a tissue easily available for biopsy.

The Genetics of Scrapie Susceptibility in Sheep

Experimental scrapie

Studies on the genetic control in sheep of the incidence of scrapie have been greatly facilitated by the work of Alan Dickinson, who selected two lines, positive and negative, of Cheviot sheep (Neuropathogenesis Unit (NPU) Cheviots) for differing incubation periods following injection with a source of scrapie called SSBP/1 (Dickinson, 1976). The gene that controls the experimental scrapie incubation period in these sheep is called *Sip* (for scrapie incubation period) and has two alleles, sA and pA (Dickinson and Outram, 1988). Negative-line Cheviots are Sip^{pApA} and survive subcutaneous (s.c.) injection of SSBP/1. Positive-line sheep are Sip^{sAsA} or Sip^{sApA} (Sip^{sA} is partially dominant (Foster and Hunter, 1991)) and these develop SSBP/1 scrapie from s.c. inoculation in 150–400 days. Different sources of scrapie-infected inocula have different transmission characteristics; for example, although SSBP/1 (termed an A-group isolate) has shorter incubation periods in Sip^{sA} carriers (positive-line) than in Sip^{pApA} (negative-line) sheep, CH1641 (a C-group scrapie isolate) and BSE have shorter incubation periods in a proportion of negative-line sheep (explained by PrP genetics (see later)) than in positive-line animals (Foster and Dickinson, 1988a; Foster *et al.*, 1993).

The *Sip* gene is also believed to control the incidence of natural scrapie, but in this case the 'susceptible' allele (Sip^{sA}) has been said to be recessive (Foster and Dickinson, 1988b). Others have found that natural scrapie affects only those animals homozygous for a recessive allele of a single gene (Gordon, 1966; Parry, 1984) and, because of such studies, Foster and Dickinson (1988b) said that *Sip* 'must have an unusual mechanism of action because the allele which is dominant in response to subcutaneously injected SSBP/1 scrapie is recessive in the natural disease'. This idea has caused problems for some molecular biologists trying to understand scrapie genetics; however, it could simply be that the sA allele is recessive in control of shortest survival times with both natural and experimental scrapie. More detailed elucidation of the action of the scrapie susceptibility/resistance gene may come from the study of the genetics of the PrP gene (see later).

Recent evidence has strongly implied that the *Sip* gene, as defined by classical genetics, is either closely linked to or is identical to the PrP gene, which has been extensively characterized by molecular genetics. Polymorphic variants of the PrP gene have been associated with the incidence of both experimental and natural TSEs in sheep (Goldmann *et al.*, 1991; Hunter *et al.*, 1994), goats (Goldmann *et al.*, 1995), laboratory mouse models (Hunter *et al.*, 1992) and humans (Collinge *et al.*, 1991). In NPU Cheviot sheep, a single PrP allele (encoding valine at codon 136 (V_{136})) is linked to Sip^{sA}; however, there are so many Sip^{pA}-related PrP alleles (all encoding alanine at codon 136 (A_{136}) but differing at other codons) that it has become necessary to use PrP genotype rather than *Sip* terminology (Tables 9.1 and 9.2).

In NPU Cheviot sheep, therefore, V_{136} on both PrP alleles (VV_{136}) is linked

Table 9.1. Sheep prion protein (PrP) variants found in Neuropathogenesis Unit Cheviot sheep.

	PrP codon			
Sip allele	136	141	154	171
sA	V*	L	R	Q
pA	A	L	R	Q
pA	A	L	H	Q
pA	A	L	R	R
pA	A	F	R	Q

*Single-letter animo acid code: A, alanine; F, phenylalanine; H, histidine; L, leucine; Q, glutamine; R, arginine; V, valine.

Table 9.2. Additional sheep prion protein (PrP) codon variants known at October 1995.

PrP codon			
112	137	171	211
M/T*	M/T	H	R/Q

*Single-letter animo acid code: H, histidine; M, methionine; Q, glutamine; R, arginine; T, threonine.

to short incubation periods (167 ± 5 days) following s.c. inoculation with SSBP/1, whereas those with longer incubation periods (322 ± 16 days) are heterozygotes with one V_{136} allele and one A_{136} allele (VA_{136}) (Goldmann *et al.*, 1994). Alanine encoded on both alleles (AA_{136}) is linked with resistance to s.c. challenge with SSBP/1. Incidence of scrapie in NPU Cheviots following challenge with CH1641 or BSE, however, does not associate primarily with codon 136 variation. Instead, animals encoding glutamine at codon 171 (Q_{171}) on both PrP alleles (QQ_{171}) succumb to intracerebral (i.c.) inoculation, whereas those having one allele with arginine (R_{171}) have a much longer incubation period (Goldmann *et al.*, 1994).

Although SSBP/1 targets animals carrying V_{136}, a minor influence of codon 171 is seen since VA_{136} animals which are also RQ_{171} have a longer incubation period (364 ± 17 days) than those which are QQ_{171} (260 ± 15 days). Similarly with CH1641 and BSE, although the major effect on incubation period depends on the codon 171 genotype (QQ_{171} being the most susceptible genotype), animals encoding V_{136} (VA_{136}) have longer incubation periods than AA_{136} sheep (Goldmann *et al.*, 1994). Taken as a whole, the multiple PrP alleles, linked with incubation period data (summarized in Table 9.3), now fully explain the rather confusing data based simply on linking the experimental incubation period with the two alleles of *Sip*.

Natural scrapie

To date, every PrP polymorphism shown to be associated with the incidence of various sources of experimental scrapie in sheep has also been shown to be

Table 9.3. Prion protein (PrP) genotypes of Neuropathogenesis Unit Cheviots and incidence of experimental scrapie (data simplified and taken from Goldmann *et al.*, 1994, and W. Goldmann, personal communication, 1995).

Infection source	Route	Sip genotype	PrP genotype	Incubation time (days)
SSBP/1	s.c.	sAsA	$VV_{136}QQ_{171}$	170
		sApA	$VA_{136}QQ_{171}$	260
		sApA	$VA_{136}RQ_{171}$	364
		pApA	$AA_{136}QQ_{171}$	Survive
		pApA	$AA_{136}RQ_{171}$	Survive
		pApA	$AA_{136}RR_{171}$	Survive
BSE	i.c.	sAsA	$VV_{136}QQ_{171}$	Not done
		sApA	$VA_{136}QQ_{171}$	802
		sApA	$VA_{136}RQ_{171}$	1874
		pApA	$AA_{136}QQ_{171}$	464
		pApA	$AA_{136}RQ_{171}$	1905
		pApA	$AA_{136}RR_{171}$	Not done

s.c., subcutaneous; i.c., intracerebral.

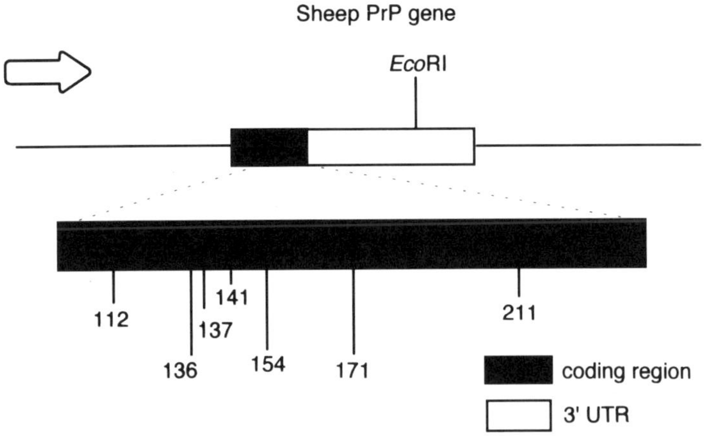

Fig. 9.1. Map of the sheep PrP gene. Within the rectangle, the hatched area represents the protein-coding region and the numbers indicate the polymorphic codons detailed in Tables 9.1 and 9.2. The clear area represents the downstream (3') untranslated region (UTR), which contains a polymorphic *Eco*RI restriction site. The arrow indicates the direction of transcription.

associated either with the incidence or with the age of onset of natural scrapie. With reference to the sheep PrP gene map in Fig. 9.1, this includes the non-coding region *Eco*RI restriction site (Hunter *et al.*, 1989, 1991) and the coding region amino acid polymorphisms at codon 136 (Goldmann *et al.*, 1991; Maciulis *et al.*, 1992; Hunter *et al.*, 1993, 1994; Laplanche *et al.*, 1993; Belt *et al.*, 1995; Clouscard *et al.*, 1995; Ikeda *et al.*, 1995) and codons 154 and 171 (Laplanche *et al.*, 1993; Goldmann *et al.*, 1994; Westaway *et al.*, 1994; Belt *et al.*, 1995; Clouscard *et al.*, 1995; Ikeda *et al.*, 1995; Hunter *et al.*, 1996). Sources of experimental scrapie appear to target sheep primarily according to genotype at either codon 136 or 171 and it is becoming clear that natural scrapie may do so as well. Although this may not surprise those who believe that scrapie arises primarily from the genome, 'infectious disease' supporters would find this remarkable, despite the fact that the dose of infectivity and the route and the time of infection would be unknown and uncontrolled in the natural disease.

The NPU Cheviot sheep flock is undergoing an outbreak of natural scrapie. Between 1986 and May 1995, 45 cases occurred in the positive line but none in the negative line. Lifespans of scrapie-affected animals range from 497 to 2250 days and 35 of these sheep have been PrP-genotyped. Genotypes of the entire flock are known and the primary factor associated with incidence of disease in this natural scrapie outbreak is the PrP genotype at codon 136 (Table 9.4). The genotype $V_{136}R_{154}Q_{171}$ is an absolute requirement for scrapie to develop and all homozygotes $VV_{136}RR_{154}QQ_{171}$ sheep are at high risk (Hunter *et al.*, 1996). No scrapie cases have occurred in *Sip*^{pApA} (or PrP AA_{136}) sheep. The $VA_{136}RR_{154}QQ_{171}$ animals also succumb to natural scrapie, with longer survival times than the homozygotes, but not all VA_{136} animals are susceptible – the

Table 9.4. Prion protein (PrP) genotypes of Neuropathogenesis Unit Cheviots and occurrence of natural scrapie.

PrP genotype	*n*	Mean age (days) at death (scrapie) (range)	Mean lifespan (days) (no scrapie) (range)
$VV_{136}RR_{154}QQ_{171}$	22*	804 (497–971)	–
$VA_{136}RR_{154}QQ_{171}$	4*	1208 (1107–1287)	–
$VA_{136}RR_{154}RQ_{171}$	34	–	1961 (1125–3435)
$VA_{136}HR_{154}QQ_{171}$	10	–	1647 (1180–2610)
AA_{136} This genotype never gets natural scrapie in this flock			

*Cases occurring since 1990.

$VA_{136}RQ_{171}$ subgroup is apparently resistant – so the $V_{136}R_{154}Q_{171}$ allele does not always link with disease and the $A_{136}R_{154}R_{171}$ allele gives survival advantage. There is also some survival advantage in the histidine and arginine at codon 154 (HR_{154}) rather than the RR_{154} genotype, an effect also observed in experimental SSBP/1 challenge (W. Goldmann, unpublished). The association of R_{171} or H_{154} with increased resistance has also been reported in Île-de-France sheep, another breed in which susceptibility is strongly associated with V_{136} (Laplanche *et al.*, 1993).

The NPU Cheviot outbreak of natural scrapie is thus unlike that described in France in Lacaune sheep (Clouscard *et al.*, 1995), in which codon 171 variation seemed to be of primary importance in determining scrapie incidence. Of 58 scrapie-affected Lacaunes, 76% were $AA_{136}QQ_{171}$, 14% were $VA_{136}QQ_{171}$ and 10% were $AA_{136}QH_{171}$. These genotypes were found in only 25% of healthy Lacaunes, with the remaining 75% encoding at least one R_{171} allele.

In Suffolk sheep, like Lacaunes, codon 136 has limited polymorphism and the PrP V_{136} allele, although it can be found (N. Hunter, unpublished), is extremely rare. As a result, the PrP genetics appears to be much simpler than in breeds that do encode V_{136} (Hunter *et al.*, 1994; Westaway *et al.*, 1994). In a study of US Suffolk sheep affected by scrapie (Westaway *et al.*, 1994), 31 scrapie sheep from 23 flocks of origin were all (100%) QQ_{171}, significantly different from the healthy sample of 69 animals from three scrapie-free flocks, which were 51% QQ_{171}, 43% RQ_{171} and 6% RR_{171}. Because, in the healthy sample, 39 animals of QQ_{171} genotype were more than 4 years of age, this was said formally to exclude the idea that the QQ_{171} genotype alone causes disease. Therefore, QQ_{171} seemed simply to control susceptibility in this case, with R_{171} linked to resistance. However, in other studies of Suffolk sheep (Ikeda *et al.*, 1995; N. Hunter, unpublished), scrapie cases have been found in some RQ_{171} Suffolks. It is not yet clear what distinguishes these animals from healthy RQ_{171} Suffolks.

These associations of scrapie with PrP genotype at either codon 136 or codon 171 may not be breed-dependent. Scrapie has been described in two different groups of Romanov sheep. In one of these, scrapie was associated with

V_{136} alleles (Laplanche *et al.*, 1993); however, in a different group of Romanovs, the association of disease was with QQ_{171} genotypes (Clouscard *et al.*, 1995). The difference between the two groups of Romanovs is intriguing. The scrapie outbreak described in Clouscard *et al.* (1995) and associated with QQ_{171} may have resulted not from a 'natural' infection but from some form of coincident 'contamination' associated with a challenge of the animals with nematode larvae. The strains of scrapie, the infecting dose or the route of infection may have been different in each Romanov group and may have influenced precisely which PrP genotypes became affected by disease. This experiment also raises the possibility that nematode larvae play a role in spreading scrapie between sheep and between flocks.

In Texel sheep in the Netherlands (Belt *et al.*, 1995), in a study involving 34 sheep from 18 different flocks, many different genotypes were affected by scrapie: $VV_{136}RR_{154}QQ_{171}$ (9%), $VA_{136}RR_{154}QQ_{171}$ (35%), $VA_{136}RR_{154}QH_{171}$ (44%), $VA_{136}RR_{154}RQ_{171}$ (3%) and $AA_{136}RR_{154}QQ_{171}$ (9%). Single-flock studies tend to give simpler disease association with codon 136 or codon 171 or both together, either because of the genetic make-up of the flock or because only a single scrapie source is acting (N. Hunter, unpublished). Sheep from many flocks may be affected by as many strains of scrapie having different PrP codon 'tropisms'. Multiflock studies are of great interest but may not give a clear picture.

Most studies so far have agreed on the protective effect associated with R_{171}, especially when homozygous (Laplanche *et al.*, 1993; Westaway *et al.*, 1994; Belt *et al.*, 1995; Clouscard *et al.*, 1995; Hunter *et al.*, 1996); however, there is a recent report from Japan of a single scrapie-affected Suffolk sheep of RR_{171} genotype (Ikeda *et al.*, 1995). It may be that Japanese scrapie targets PrP genotypes differently from European and US scrapie, but the sheep may also be different genetically. The same study (Ikeda *et al.*, 1995) reported a link of R_{171} with a particular series of *Eco*RI and *Hin*dIII restriction fragment length polymorphisms (RFLP) not found so far in British Suffolks (N. Hunter, unpublished observations). The implications of this finding are not clear at the moment; nor is the disease linkage (if any) of alleles encoding H_{171}, which have been found in both scrapie-affected and healthy sheep at low frequencies.

Involvement of Other Genes in Scrapie Control

Parry (1984) believed that natural scrapie was caused by a recessive gene. In another study, showing the importance of the *Sip* gene in control of natural scrapie (Foster and Dickinson, 1988b), the results were interpreted to suggest that, although the sA allele of *Sip* was dominant in controlling experimental scrapie, it was recessive in the case of the natural disease. Most sheep PrP geneticists would probably now say that Parry's recessive gene is likely to be PrP, which, in the Suffolk breed which he studied, is linked to scrapie incidence in a recessive manner – PrP QQ_{171} animals are at greatest risk of scrapie. The apparent conflict in the idea proposed by Alan Dickinson (Foster and Dickinson, 1988b) between 'dominant' susceptibility in experimental disease and 'recessive' susceptiblity in natural disease can be at least partially resolved by addressing the unusual characteristics of SSBP/1, the scrapie source used to define the *Sip* locus. This source targets sheep according to PrP codon 136 genotype, with V_{136} acting in a dominant manner, and with such a minor effect from the codon 171 genotype that it is not obvious when only *Sip* genotype is considered (Goldmann *et al.*, 1994). In outbreaks in which natural scrapie also targets codon 136, it does so within the QQ_{171} genotype group (recessive). Within the QQ_{171} group, the allele $V_{136}R_{154}Q_{171}$ is dominant, as there are, in many outbreaks, twice as many $VA_{136}RR_{154}QQ_{171}$ scrapie cases as there are $VV_{136}RR_{154}QQ_{171}$ (Hunter *et al.*, 1994). This effect is not reliant on the numbers of each allele within the affected flock.

Despite this, there is still room to consider the effects of other genes. It is not really clear, for example, why the $AA_{136}RR_{154}QQ_{171}$ genotype, which is apparently resistant to natural scrapie in Cheviots (Hunter *et al.*, 1996), is highly susceptible in Suffolks (Westaway *et al.*, 1994). This does not mean that VV_{136} Suffolks are scrapie-resistant – Suffolks which carry the V_{136} PrP allele have been found at low frequency among scrapie-affected sheep. One study found three $VA_{136}RR_{154}QQ_{171}$ and one $VV_{136}RR_{154}QQ_{171}$ sheep, all of which died of scrapie at the usual age of 3-4 years (N. Hunter, unpublished). It does suggest, however, that there may be additional factors involved in control of scrapie incidence, whether part of the PrP gene-expression control regions or not.

Genes known to be linked to incidence of other diseases, such as the major histocompatibility complex (MHC), may also have some influence on scrapie. Natural scrapie incidence in Île-de-France sheep has been said to be controlled by a recessive gene called *Scr* (Millot *et al.*, 1988), which is linked to the OLA (ovine lymphocyte antigen) complex (sheep equivalent of MHC) in Île-de-France sheep. However, an examination of sheep class I lymphocyte antigens in NPU Cheviot sheep found no linkage of scrapie with any of the markers tested (Hunter *et al.*, 1996). This area warrants further investigation.

Modes of Transmission of Scrapie

Despite the fact that scrapie is transmissible, there is still support for the 'genetic disease' hypothesis. If this aetiology is correct, it would have to encompass the many different genotypes that have been found in scrapie-affected sheep of different breeds and from different flocks (see 'The Genetics of Scrapie Susceptibility in Sheep' above). This is not unprecedented. In a study of a population of Jews in Kurdistan (Rund *et al.*, 1991), genetically isolated groups had 13 different mutations, within the same gene, which caused the same disease: thalassaemia. However, in one study of NPU Cheviots, three sheep of the most susceptible genotype ($VV_{136}RR_{154}QQ_{171}$) which would normally be expected to die from natural scrapie at under 1000 days of age (Hunter *et al.*, 1996) are alive at the time of writing at over 1800 old (Foster *et al.*, 1996). These animals were part of an embryo transfer (ET) experiment, but the only difference between them and the sheep in the main flock was the extremely stringent husbandry conditions used at the time of their birth to keep the environment clean. This suggests an infectious aetiology. It is undoubtedly possible that these animals will eventually succumb to scrapie; however, the age of onset of disease cannot be said to be determined simply by the PrP gene sequence in this case.

In considering infectious aetiologies, it is important also to consider the mode of transmission. Natural scrapie is so dependent on PrP genotype of the sheep for its expression as a disease that much of the older (pre-PrP) literature is now difficult to interpret, as the genotypes of the animals were not known with any certainty. Looking for infectivity in the body tissues of an animal completely resistant to scrapie but born to scrapie-affected parents is both possible and meaningless.

It is part of the dogma of scrapie that the disease in sheep is maternally transmitted – in other words, that a sheep acquires its infection from its mother around the time of birth. This literature has been reviewed recently (Ridley and Baker, 1995) in a timely article calling maternal transmission a 'myth'. However, in talking of Caxton's *The Golden Legend*, a book on Christian saints, Robertson Davies (Davies, 1994) said:

> Respect for the *Legend* did not mean a rejection of the Renaissance or subsequent adventures of the mind. Rather it meant a sympathetic understanding of a past age, which might still, under examination have much to tell us that would enlarge our lives.

He said it so much better than I could! Although only now can we begin to carry out sensible experiments on animals of known genotype and scrapie susceptibility, the older data should not be ignored, as they suggest areas that deserve further investigation.

If maternal transmission of scrapie does occur, the placenta is the prime suspect as the vehicle bringing scrapie infectivity to the developing lamb. There is evidence that scrapie can be transmitted from placental tissue samples (Pattison *et al.*, 1972; Onodera *et al.*, 1993), but there have also been transmission failures. Telling the difference between genuine maternal transmission and lateral transmission by close contact during the perinatal period is difficult. It is not always possible to know when an animal has acquired its infection, but, with the natural scrapie in the NPU Cheviot flock, it seems likely that most animals are infected at about the same age. All animals within this flock are genotyped within a few months of birth and the ages at death from scrapie are constant with apparently 100% of $VV_{136}RR_{154}QQ_{171}$ animals dying at 700–900 days of age and over 90% of heterozygotes $VA_{136}RR_{154}QQ_{171}$ dying at 1100–1200 days of age (Table 9.4). No other genotypes within the flock are affected. The difference in age at death from natural scrapie between homozygotes and heterozygotes, although expected with experimental challenge, is not always seen in other flocks (Hunter *et al.*, 1994). However, three NPU Cheviot $VV_{136}RR_{154}QQ_{171}$ animals have survived (so far) penned with other animals from the same experiment in areas apart from the main flock for over 1800 days. Stringently clean procedures were used around the time of their birth and, because these procedures are not used in the main flock and there seems to be no other difference between the two groups of animals, this could suggest that the perinatal period is a dangerous time for a susceptible lamb. Ten other $VV_{136}RR_{154}QQ_{171}$ sheep, subject to the same conditions as the survivors and penned with them, did succumb to natural scrapie, so either the 'over-the-top' procedures may not have been optimal for prevention of scrapie or there is another source of infection. What this source might be, however, is not at all obvious, although the recent suggestions that nematodes (Clouscard *et al.*, 1995) or mites (Wisniewski *et al.*, 1996) are involved in transmission of scrapie warrant further investigation. It is interesting that the three surviving $VV_{136}RR_{154}QQ_{171}$ sheep do not seem to have contracted scrapie by horizontal infection from the ten scrapie-affected animals of the same genotype kept in close proximity in the same pens. One of the simplest explanations for the survival of 23% (three out of 13) of the highly susceptible sheep against a background of 100% susceptibility in the main flock is that, by use of very clean conditions, the load of infection in the perinatal period has been reduced to a level which is just about a threshold required for subsequent disease. Ten of the sheep were above the threshold and got scrapie, three sheep were below the threshold and survived. The perinatal period is now being further investigated by comparing Caesarean-derived and bottle fed lambs with naturally birthed animals.

Embryo transfer (ET) procedures have also been investigated in the hope of bypassing any infection cycle. Although work with the NPU Cheviot flock was compromised by natural scrapie in the flock (Foster *et al.*, 1992, 1996), it is clear that ET procedures, coupled with stringent cleanliness at birth, do not prevent scrapie from occurring in all sheep. These experiments generated the three surviving $VV_{136}RR_{154}QQ_{171}$ animals discussed above and the ten $VV_{136}RR_{154}QQ_{171}$ sheep that did develop scrapie (nine at a mean age of 826 ± 24 days and one at 1267 days). This group of ten animals was not protected by ET procedures, even following International Embryo Transfer Society (IETS) protocols (Stringfellow and Seidel, 1990). A similar experiment set up in the USA (Foote *et al.*, 1993) is difficult to interpret because of lack of PrP-genotype information. However, one of the groups involved Cheviot sheep inoculated with SSBP/1 scrapie, and it has already been shown that these animals give similar incubation periods to those found in NPU Cheviots injected with SSBP/1 (Maciulis *et al.*, 1992). The incubation periods in scrapie-inoculated donors and recipients ranged from 6.1 months (183 days) to 14.9 months (447 days), with some survivors to 25 months (750 days) and this suggests that a mixture of PrP genotypes was present in the sheep (see 'Experimental scrapie' above and Table 9.3). Of the positive-control lambs (gestated and born naturally), two out of nine developed scrapie at 31 months (930 days) and 49 months (1470 days) of age and could be of PrP genotype VV_{136} and VA_{136}, respectively. If correct, this represents a V_{136} frequency of at least 17%. None of the 22 ET-derived Cheviot lambs succumbed to scrapie, with ages at death ranging from 74.5 months (2235 days) to 96 months (2880 days). It is, of course, possible that there were no susceptible lambs among the ET group, but there is at least a suggestion that ET proceedures have protected lambs from scrapie in this case. Further experiments are being carried out to clarify the situation.

Conclusions

Research into scrapie in sheep has reached a very exciting point. Firstly, the new PrP genetic information is being used to allow informed breeding for resistance to scrapie in commercial sheep flocks (Dawson, 1994; Hosie and Dawson, 1996). Secondly, the confusion of previous generations of scrapie experiments about modes of transmission is at last about to be clarified, using sheep of predefined susceptibility.

References

Belt, P.B.G.M., Muileman, I.H., Schreuder, B.E.C., Bos-de Ruijter, J., Gielkens, A.L.J. and Smits, M.A. (1995) Identification of five allelic variants of the sheep PrP gene and their association with natural scrapie. *Journal of General Virology* 76, 509–517.

Braig, H.R. and Diringer, H. (1985) Scrapie: concept of a virus-induced amyloidosis of the brain. *EMBO Journal* 4, 2309–2312.

Brown, P., Liberski, P., Wolff, A. and Gajdusek, D. (1990) Resistance of scrapie infectivity to steam autoclaving after formaldehyde fixation and limited survival after ashing at 360°C: practical and theoretical implications. *Journal of Infectious Diseases* 161, 467–472.

Brown, P., Goldfarb, L. and Gajdusek, D. (1991) The new biology of spongiform encephalopathy: infectious amyloidoses with a genetic twist. *Lancet* 337, 1019–1022.

Bruce, M.E. and Dickinson, A.G. (1987) Biological evidence that scrapie agent has an independent genome. *Journal of General Virology* 68, 79-89.

Clark, A.M. and Moar, J.A. (1992) Scrapie: a clinical assessment. *Veterinary Record* 130, 377–378.

Clark, A.M., Dawson, M. and Scott, A.C. (1994) Scrapie-associated fibrils in found dead sheep. *Veterinary Record* 134, 650–651.

Clouscard, C., Beaudry, P., Elsen, J.M., Milan, D., Dussaucy, M., Bounneau, C., Schelcher, F., Chatelain, J., Launay, J.M. and Laplanche, J.L. (1995) Different allelic effects of the codons 136 and 171 of the prion protein gene in sheep with natural scrapie. *Journal of General Virology* 76, 2097–2101.

Collinge, J. and Palmer, M.S. (1994) Human prion diseases. *Baillière's Clinical Neurology* 3, 241–255.

Collinge, J., Palmer, M.S. and Dryden, A.J. (1991) Genetic predisposition to iatrogenic Creutzfeldt–Jakob disease. *Lancet* 337, 1441–1442.

Davies, R. (1994) *The Cunning Man.* McClelland and Stewart, Toronto.

Dawson, M.M. (1994) Genotyping service for Swaledale sheep. *Veterinary Record* 134, 583–584.

Dawson, M.M., Mansley, L.M., Hunter, A.R., Stack, M.J. and Scott, A.C. (1987) Comparison of scrapie associated fibril detection and histology in the diagnosis of natural sheep scrapie. *Veterinary Record* 121, 591.

Dickinson, A.G. (1976) Scrapie in sheep and goats. In: Kimberlin, R. (ed.) *Slow Virus Diseases of Animals and Man.* North-Holland, Amsterdam, pp. 209–241.

Dickinson, A.G. and Outram, G.W. (1988). Genetic aspects of unconventional virus infections: the basis of the virino hypothesis. In: Bock, G. and Marsh, J. (eds) *Novel Infectious Agents and the Central Nervous System.* Wiley Interscience, Chichester, pp. 63–83.

Diringer, H. (1995) Proposed link between transmissible spongiform encephalopathies of man and animals. *Lancet* 346, 1208–1210.

Foote, W.C., Clark, W., Maciulis, A., Call, J.W., Hourrigan, J., Evans, R.C., Marshall, M.R. and Decamp, M. (1993) Prevention of scrapie transmission in sheep, using embryo-transfer. *American Journal of Veterinary Research* 54, 1863–1868.

Foster, J.D. and Dickinson, A.G. (1988a) The unusual properties of CH1641 a sheep passaged isolate of scrapie. *The Veterinary Record* 136, 5–8.

Foster, J.D. and Dickinson, A.G. (1988b) Genetic control of scrapie in Cheviot and Suffolk sheep. *Veterinary Record* 123, 159.

Foster, J.D., Hope, J. and Fraser, H. (1993) Transmission of bovine spongiform encephalopathy to sheep and goats. *The Veterinary Record* 133, 339–341.

Foster, J.D. and Hunter, N. (1991) Partial dominance of the sA allele of the Sip gene for controlling experimental scrapie. *Veterinary Record* 128, 548–549.

Foster, J.D., Hunter, N., Williams, A., Mylne, M., McKelvey, W., Hope, J., Fraser, H. and Bostock, C. (1996) Further observations on the transmission of scrapie in experiments using embryo transfer. *Veterinary Record* 138, 559–562.

Foster, J.D., McKelvey, W.A., Mylne, M.J., Williams, A., Hunter, N., Hope, J. and Fraser, H. (1992) Studies on maternal transmission of scrapie in sheep by embryo transfer. *Veterinary Record* 130, 341–343.

Goldmann, W., Hunter, N., Benson, G., Foster, J.D. and Hope, J. (1991) Different scrapie-associated fibril proteins (PrP) are encoded by lines of sheep selected for different alleles of the Sip gene. *Journal of General Virology* 72, 2411–2417.

Goldmann, W., Hunter, N., Smith, G., Foster, J. and Hope, J. (1994) PrP genotype and agent effects in scrapie – change in allelic interaction with different isolates of agent in sheep, a natural host of scrapie. *Journal of General Virology* 75, 989–995.

Goldmann, W., Martin, T., Hunter, N., Hughes, S., Smith, G., Hughes, K., Foster, J., Dawson, M. and Hope, J. (1995) *Novel Polymorphisms in Caprine PrP Protein and Their Association with Scrapie Incubation Period.* Prion Diseases, Göttingen, Germany.

Gordon, W. (1966) Variation in susceptibility of sheep to scrapie and genetic implications. In: *Report of Scrapie Seminar, 1964.* ARS 91–53, US Department of Agriculture, Washington DC, pp. 53–67.

Hope, J. and Manson, J. (1991) The scrapie fibril protein and its cellular isoform. In: Chesebro, B. (ed.) *Transmissible Spongiform Encephalopathies: Scrapie, BSE and Related Disorders.* Current Topics in Microbiology and Immunity 172. Springer-Verlag, Berlin, pp. 57–74.

Hosie, B. and Dawson, M. (1996) Scrapie genotyping for Suffolk sheep. *Veterinary Record* 138, 215-216.

Hunter, N. (1991) Scrapie and GSS – the importance of protein. *Trends in Neurosciences* 14, 389–390.

Hunter, N., Foster, J.D., Dickinson, A.G. and Hope, J. (1989) Linkage of the gene for the scrapie-associated fibril protein (PrP) to the Sip gene in Cheviot sheep. *Veterinary Record* 124, 364–366.

Hunter, N., Foster, J.D., Benson, G. and Hope, J. (1991) Restriction fragment length polymorphisms of the scrapie-associated fibril protein (PrP) gene and their association with susceptibility to natural scrapie in British sheep. *Journal of General Virology* 72, 1287–1292.

Hunter, N., Dann, J.C., Bennett, A.D., Somerville, R.A., McConnell, I. and Hope, J. (1992) Are Sinc and the PrP gene congruent? Evidence from PrP gene analysis in Sinc congenic mice. *Journal of General Virology* 73, 2751–2755.

Hunter, N., Goldmann, W., Benson, G., Foster, J.D. and Hope, J. (1993) Swaledale sheep affected by natural scrapie differ significantly in PrP genotype frequencies from healthy sheep and those selected for reduced incidence of scrapie. *Journal of General Virology* 74, 1025–1031.

Hunter, N., Goldmann, W., Smith, G. and Hope, J. (1994) The association of a codon 136 PrP gene variant with the occurrence of natural scrapie. *Archives of Virology* 137, 171–177.

Hunter, N., Foster, J., Goldmann, W., Stear, M., Hope, J. and Bostock, C. (1996) Natural scrapie in a closed flock of Cheviot sheep occurs only in specific PrP genotypes. *Archives of Virology* 141, 809–824.

Ikeda, T., Horiuchi, M., Ishiguro, N., Muramatsu, Y., Kai-Uwe, G. and Shinagawa, M. (1995) Amino acid polymorphisms of PrP with reference to onset of scrapie in Suffolk and Corriedale sheep in Japan. *Journal of General Virology* 76, 2577–2581.

Ikegami, Y., Ito, M., Isomura, H., Momotani, E., Sasaki, K., Muramatsu, Y., Ishiguro, N. and Shinagawa, M. (1991) Preclinical and clinical-diagnosis of scrapie by detection of PrP protein in tissues of sheep. *Veterinary Record* 128, 271–275.

Laplanche, J.L., Chatelain, J., Westaway, D., Thomas, S., Dussaucy, M., Brugerepicoux, J. and Launay, J.M. (1993) PrP polymorphisms associated with natural scrapie discovered by denaturing gradient gel-electrophoresis. *Genomics* 15, 30–37.

Maciulis, A., Hunter, N., Wang, S., Goldmann, W., Hope, J. and Foote, W.C. (1992) Polymorphisms of a scrapie-associated fibril protein (PrP) gene and their association with susceptibility to experimentally induced scrapie in Cheviot sheep in the United States. *American Journal of Veterinary Research* 53, 1957–1960.

Manuelidis, L., Sklaviadis, T. and Manuelidis, E.E. (1987) Evidence suggesting that PrP is not the infectious agent in Creutzfeldt-Jakob disease. *EMBO Journal* 6, 341–347.

Manuelidis, L., Sklaviadis, T., Akowitz, A. and Fritch, W. (1995) Viral particles are required for infection in neurodegenerative Creutzfeldt–Jakob-disease. *Proceedings of the National Academy of Sciences USA* 92, 5124–5128.

Millot, P., Chatelain, J., Dautheville, C., Salmon, D. and Cathala, F. (1988) Sheep major histocompatibility (OLA) complex: linkage between a scrapie susceptibility/resistance locus and the OLA complex in Île-de-France sheep progenies. *Immunogenetics* 27, 1–11.

Mohri, S., Farquhar, C.F., Somerville, R.A., Jeffrey, M., Foster, J. and Hope, J. (1992) Immunodetection of a disease specific PrP fraction in scrapie-affected sheep and BSE-affected cattle. *Veterinary Record* 131, 537–539.

Onodera, T. and Hayashi, T. (1994) Diversity of clinical signs in natural scrapie cases occurring in Japan. *Japan Agricultural Research Quarterly* 28, 59–61.

Onodera, A., Ikeda, T., Horiuchi, M., Ishiguro, N., Onuma, M., Hirano, N., Mikami, T., Honda, E., Hirai, K., Kai, K., Yugi, H., Muramatsu, Y. and Shinagawa, M. (1994) Survey of natural scrapie in Japan – analysis of RFLP types of the PrP gene and detection of PrP^{sc} mainly in Suffolk sheep. *Journal of Veterinary Medical Science* 56, 627–632.

Onodera, T., Ikeda, T., Muramatsu, Y. and Shinagawa, M. (1993) Isolation of scrapie agent from the placenta of sheep with natural scrapie in Japan. *Microbiology and Immunology* 37, 311–316.

Parry, H. (1960) Scrapie: a transmissible hereditary disease of sheep. *Nature, London* 185, 441–443.

Parry, H. (1984) *Scrapie.* Academic Press, London

Pattison, I.H., Hoare, M.N., Jebbett, J.N. and Watson, W.A. (1972) Spread of scrapie to sheep and goats by oral dosing with foetal membranes from scrapie-affected sheep. *Veterinary Record* 90, 465–468.

Prusiner, S.B., Groth, D.F., McKinley, M.P., Cochran, S.P., Bowman, K.A. and Masiarz, F.R. (1982) Scrapie agent is a novel proteinaceous infectious particle. *Federation Proceedings* 41, 695.

Prusiner, S.B., Scott, M., Foster, D., Pan, K.-M., Groth, D., Mirenda, C., Torchia, M., Yang, S.-L., Serban, D., Carlson, G.A., Hoppe, P.C., Westaway, D. and DeArmond, S.J. (1990) Transgenetic studies implicate interactions between homologous PrP isoforms in scrapie prion replication. *Cell* 63, 673–686.

Ridley, R.M. and Baker, H.F. (1995) The myth of maternal transmission of spongiform encephalopathy. *British Medical Journal* 311, 1071–1075.

Rund, D., Cohen, T., Filon, D., Dowling, C.E., Warren, T., Barak, I., Rachmilewitz, E., Kazazian, E. and Oppenheim, A. (1991) Evolution of a genetic disease in an ethnic isolate: β-thalassemia in the Jews of Kurdistan. *Proceedings of the National Academy of Sciences USA* 88, 310–314.

Somerville, R.A. (1991) The transmissible agent causing scrapie must contain more than protein. *Reviews in Medical Virology* 1, 131–139.

Stringfellow, D.A. and Seidel, S.M. (eds) (1990) *Manual of the International Embryo Transfer Society.* 2nd edn. Champaign, Illinois, 41 pp.

Vankeulen, L.J.M., Schreuder, B.E.C., Meloen, R. H., Poelenvandenberg, M., Mooijharkes, G., Vromans, M.E.W. and Langeveld, J.P.M. (1995) Immunohistochemical detection and localization of prion protein in brain-tissue of sheep with natural scrapie. *Veterinary Pathology* 32, 299–308.

Westaway, D., Zuliani, V., Cooper, C.M., Dacosta, M., Neuman, S., Jenny, A.L., Detwiler, L. and Prusiner, S.B. (1994) Homozygosity for prion protein alleles encoding glutamine-171 renders sheep susceptible to natural scrapie. *Genes and Development* 8, 959–969.

Wisniewski, H.M., Sugurdarson, S., Rubenstein, R., Kascsak, R.J. and Carp, R.I. (1996) Mites as vectors for scrapie. *Lancet* 347, 1114.

Cytogenetics: Physical Chromosome Maps

10

T.E. BROAD,[1] H. HAYES,[2] and S.E. LONG[3]

[1]*AgResearch Invermay Agricultural Centre, Private Bag 50034, Mosgiel, New Zealand;* [2]*Laboratoire de Génétique biochimique et de Cytogénétique, INRA-CRJ, F-78352 Jouy-en-Josas, France;* [3]*Cytogenetic and Reproduction Unit, Department of Clinical Veterinary Science, University of Bristol, Langford House, Langford, Bristol BS18 7DU, United Kingdom*

Introduction 242

Chromosome Preparation and Banding 242

- Lymphocyte cultures using mitogens 242
- Fibroblast cultures from fetal or adult tissues 243
- Advantages of lymphocyte and fibroblast cultures 243
- Synchronization of cell cultures for preparing elongated chromosomes 244
- Preparation of chromosome spreads 244
- Introduction to banding methods 245
- 'Structural' banding methods 245
- 'Dynamic' banding methods 257
- Other specialized banding techniques 260

Standard Karyotype and Chromosome Aberrations 261

- Development of a standardized nomenclature for sheep chromosomes 261
- Chromosome aberrations 262

In Situ Hybridization Techniques 267

- Introduction 267
- *In situ* hybridization procedures 267
- Chromosome painting 270

Somatic-cell Hybrids 271

- Production of cell hybrids 271
- Use of cell hybrids for gene mapping 272

Physical Map of the Sheep Genome 273

- Current physical map of the sheep genome 273
- Comparison with the cattle-gene map 277
- Comparison with the human- and mouse-gene maps 278

Emerging Techniques and Future Perspectives 279
Acknowledgements 280
References 280

Introduction

The historical development, techniques and findings of classical cytogenetics of the sheep have been extensively and recently reviewed by Halnan (1989), Popescu (1989) and Long (1990). Since then, the rapidly advancing process of gene mapping has placed increasingly urgent demands on the techniques of classical cytogenetics for higher resolution, as well as more accurate and repeatable procedures for the identification of chromosomes and definition of their structure. Human cytogenetics continues to be the 'engine' for these developments, as it was for the classical cytogenetic procedures used in the study of domestic animals. This review extends and updates the earlier reviews on sheep cytogenetics and those on the developing gene map of the sheep (Echard, 1990; Broad and Hill, 1994; Echard *et al.*, 1994). While it is concerned more with results than with the process, it recognizes that these are intimately interconnected. The procedural components are, however, leavened with summary details of their mode of action so that readers might be better equipped to understand and weigh the results.

As gene mapping, particularly genetic-linkage mapping, discovers markers on different regions of the sheep genome linked to traits of economic and biological importance, the focus will shift from the chromosomes to the level of chromosomal bands and below. High-resolution physical mapping will become all-important for the fine mapping of these regions and for the isolation and identification of the genes involved. Classical cytogenetics will develop the molecular tools and approaches to provide these insights. In sheep cytogenetics, this transition from the classical to the molecular is now under way. Therefore, in the final section of this review possible future developments in molecular cytogenetics will be briefly outlined for elucidating the structure and organization of the sheep genome.

Chromosome Preparation and Banding

The successful study of chromosomes depends on the preparation of good-quality slides with a large number of well-banded chromosome spreads.

Lymphocyte cultures using mitogens

B and T lymphocytes in blood are stimulated to divide by the lectin mitogens concanavalin A (ConA), pokeweed mitogen (PWM) and phytohaemagglutinin (PHA), the latter being used less widely now. Cultures may be established from whole blood (e.g. Moorhead *et al.*, 1960), from lymphocyte preparations separated by gravity sedimentation or from lymphocytes enriched by sedimentation equilibrium (Boyum, 1968) in a density gradient of Ficoll-Paque (Pharmacia) or Lymphoprep (Nycomed).

Lymphocytes are grown in various types of culture medium, e.g. Ham's F12, TC199, RPMI 1640 or MEM-α, supplemented with antibiotics (e.g. penicillin 100 IU ml^{-1}, streptomycin 100 μg ml^{-1}), an antimycotic (e.g. fungizone 0.25 μg ml^{-1}), fetal calf serum (FCS) (10%) and a mitogen (ConA or PWM 10 μg ml^{-1}). Cultures are typically incubated at 37°C for 3 days with ConA and 4–5 days with PWM.

Fibroblast cultures from fetal or adult tissues

In primary cultures, connective-tissue or fibroblast-like cells usually overgrow the epithelial cells; thus the resulting cultures are commonly referred to as fibroblast cultures. For cytogenetic applications, primary cell cultures from fetal tissues are preferred, as they have a higher mitotic index and a larger potential number of generations than those from adult tissues, about 50 versus 20.

Fetuses may be obtained at the slaughterhouse at various stages of development, but are preferred at one-third of gestation. Fetal lung appears to be a very suitable tissue to sample. For adult tissues, skin biopsies are conveniently taken from live animals, but other tissues may be removed during the first few days post-mortem.

Cultures may be established from tissue fragments, as described for mouse embryos (Adams, 1981). Cells are typically cultured in MEM–FCS 10% (Eagle's minimum essential medium with Earle's salts), supplemented with sodium bicarbonate (2.2 g l^{-1}) and antibiotics. Fetal primary cultures can be maintained for 15–20 passages (six to eight for adult cells) but, after seven to ten passages (three to four for adult cells), their mitotic index decreases and they become less suitable for cytogenetic applications. However, cells from low-passage primary cultures may be frozen at –80°C or preferably in liquid nitrogen for later use.

Advantages of lymphocyte and fibroblast cultures

Most laboratories develop strong preferences for particular culturing and chromosome-preparation procedures. Some find it difficult to obtain reproducible spreads with a high mitotic index from lymphocyte cultures, whereas others prefer such cultures for Q banding of sheep chromosomes. However, as blood is easy to collect and culture, it is preferred for the routine analysis of a large number of animals, e.g. to search for chromosomal abnormalities. When fresh blood is not available and for *in situ* hybridization, which requires slides with many well-banded chromosome spreads, fibroblast cultures often give more reproducible results than blood cultures.

Synchronization of cell cultures for preparing elongated chromosomes

Chromosomes prepared from normal cultures are usually in the middle to late stage of metaphase, because colchicine or colcemid are routinely used to arrest cells at this stage of the cell cycle. Elongated chromosomes, usually at prometaphase and early metaphase, are prepared from cell cultures arrested with drugs that reversibly inhibit deoxyribonucleic acid (DNA) replication. High-resolution banding patterns can be produced on such elongated chromosomes. However, the mitotic index of synchronized cultures is generally lower than that of unsynchronized cultures. Several protocols have been reported for preparing such chromosomes (e.g. Adams, 1981), but the most efficient use either thymidine or methotrexate.

Thymidine at concentrations over 1 mM almost completely inhibits DNA replication (Xeros, 1962). This inhibition can be reversed by changing the culture medium to a drug-free medium. Thymidine has low toxicity and can be used for two rounds of synchronization, producing an increased number of synchronized cells. Efficient protocols for synchronizing lymphocyte and fibroblast cultures with a single or a double thymidine block are reported, respectively, by Viegas-Pequignot and Dutrillaux (1978) and Hayes *et al.* (1991).

Methotrexate (amethopterin) also inhibits DNA replication but is more toxic than thymidine and can, therefore, be used for only one cycle of synchronization. The most commonly used protocol is that of Yunis (1976). Synchronization with methotrexate produces a more homogeneous population of mostly early-metaphase cells than the thymidine block, but the number of cells with very elongated chromosomes is much smaller.

Preparation of chromosome spreads

Prior to harvest, cells are typically incubated with 0.04 $\mu g\ ml^{-1}$ colchicine or colcemid for 1–2 h in unsynchronized cultures to increase the yield of chromosomes and for 20–30 min in synchronized cultures to improve their spreading. After cell harvest, hypotonic treatment is performed with either potassium chloride (KCl) (0.075 M) or in FCS diluted 1 : 6 with ethylenediamine tetra-acetic acid (EDTA) (0.4 $\mu g\ ml^{-1}$) filtered through a 0.45 μm membrane. Potassium chloride produces better chromosomes for G-banding by trypsin using Giemsa (GTG banding), whereas FCS–EDTA is preferred for reverse (R)-banding chromosomes. After fixation in methanol–acetic acid (1 : 3), the cell suspension is dropped on to clean microscope slides. Each laboratory has its own practice about whether to use wet or dry slides and the height for dropping the cell suspension. Slides may be stored at –20°C or –80°C for several years.

Introduction to banding methods

Since the 1970s, when it was discovered that certain stains and treatments could produce a characteristic and reproducible pattern of bands on individual chromosomes, karyotyping has developed and refined its precision. Banding makes it possible to distinguish all the chromosomes of mammalian species, to describe chromosomal rearrangements in detail, to localize markers on chromosomes and to compare karyotypes among closely or distantly related species. The mechanism underlying the banding phenomenon is not yet well understood but has been the object of many reports (e.g. Comings, 1978; Holmquist *et al.*, 1982; Babu and Verma, 1987; Korenberg and Rykowsky, 1988; Bickmore and Sumner, 1989; Holmquist, 1992; Saitoh and Laemmli, 1994). However, while chromosome identification has been the major objective of banding, increasing insights into the structure and organization of chromosomes are being gained from banding studies.

Banding techniques can be classified into two groups depending on the treatment used. The 'structural' group is based on the intrinsic nature of the chromosomes and is applied to fixed chromosomal preparations, while the 'dynamic' group of techniques is based on a modification of the chromosomal DNA induced by the incorporation of a thymidine analogue during cell culture.

'Structural' banding methods

Q-banding by fluorescence using quinacrine

Q-banding by fluorescence using quinacrine (QFQ banding) was the first reported banding technique, described by Caspersson *et al.* (1968). Bright and dull regions of fluorescence induced by blue ultraviolet (UV) light are revealed along the length of chromosomes by quinacrine mustard. Preferential binding and bright fluorescence are found in DNA regions rich in adenine–thymine (AT) base pairs (Weisblum and de Haseth, 1972). As the intensity of fluorescence progressively fades with exposure to UV light, this exposure should be kept to a minimum and immediate photography is necessary for identification of chromosomes. The advantage of the technique is that it does not alter chromosome structure, thus permitting sequential treatments such as G-banding, nucleolar organizer region (NOR) staining and *in situ* hybridization.

Q-banding is a robust and reproducible technique for sheep chromosomes, producing a fluorescent banding pattern similar to G-banding. Some laboratories find that landmark Q-bands are easier to distinguish than G-bands in sheep, and the variation in staining intensity of sheep chromosomes, particularly chromosomes 11 and 24, which are very faint, assists in their discrimination in the karyotype (e.g. Hediger *et al.*, 1991a; Ansari *et al.*, 1993a).

G-banding by trypsin using Giemsa

G-banding by trypsin using Giemsa is taken as the staining technique for producing reference karyotypes among mammalian species. The mechanism of G-banding is still not well understood, but some authors suggest that it depends on differences in protein composition along the chromosome (Dutrillaux *et al.*, 1972; Bostock and Sumner, 1978), while others suggest that trypsin treatment reorganizes chromatin, unmasking the meiotic chromomere pattern (Comings, 1978). The GTG- and QFQ-banding patterns are very similar, if not identical, although the mechanisms of their staining are quite different, but the GTG-banding patterns are stable. The age and quality of the slides greatly influence the quality of G-banding. It is preferable to age the slides for a week before banding and to adjust the duration of trypsin treatment for each batch of slides. It is generally found that improved results are obtained with slides prepared using hypotonic treatment with KCl.

The GTG-banded karyotypes of sheep are reported in Ford *et al.* (1980), Long (1985b), Matjeka *et al.* (1987), Menscher *et al.* (1989), Ansari *et al.* (1993a, 1996), Kaftanovskaya and Serov (1994) and Iannuzzi *et al.* (1995).

Nucleolar organizer region staining

Nucleolar organizer regions contain ribosomal ribonucleic acid (RNA) genes and are involved in the formation of nucleoli. They can be revealed by treating the chromosomes under alkaline conditions (assumed to extract DNA, RNA and histones), followed by treatment with silver nitrate, which stains the proteins associated with NORs. A procedure that works well in sheep is described in Di Berardino *et al.* (1979), based on the method of Bloom and Goodpasture (1976).

Clusters of silver grains are observed in the regions carrying transcriptionally active NORs on chromosomes identified either by QFQ banding before NOR staining or by R-banding of 5-bromo-2′-deoxyuridine (BrdU)-containing chromosomes using acridine orange (RBA) banding after NOR staining. In sheep, the telomeric regions of chromosomes, 1p, 2q, 3q, 4q and 25q, bear NORs (Henderson and Bruere, 1979; Mayr and Czaker, 1981), which are now given the official symbols RNR1, RNR2, RNR3, RNR4 and RNR5, respectively (Table 10.1).

C-banding

This selectively stains constitutive heterochromatin at all stages of the cell cycle. The method of Sumner (1972), using barium hydroxide treatment, is preferred as it is more easily controlled than that using sodium hydroxide (NaOH). It consists of successive treatments under acid and then alkaline and saline conditions at high temperatures, which are expected to preferentially extract DNA from chromosomal regions other than C bands, leaving the constitutive heterochromatin, which is then stained intensively by Giemsa (Comings *et al.*, 1973).

In sheep, the centromeric regions of the telocentrics stain more intensely than those of the metacentric chromosomes, which may have very little staining

Table 10.1. List of loci physically mapped on to the sheep genome (December 1995).

Symbol*	Name	Chr	Mode†	First author of reference
A2M	*α_2-Macroglobulin*	3q26–q35	I	Graphodatsky *et al.* (1993)
A2MR	*α-Macroglobulin receptor (low-density lipoprotein receptor-related protein)*	3q	SR	Broad *et al.* (1993b)
ACO1	*Aconitase 1, soluble*	U2	SZ	Imam-Ghali *et al.* (1987); Saidi-Mehtar *et al.* (1989, 1991)
		2p	SY	This review
ADA	*Adenosine deaminase*	U15	SZ	Saidi-Mehtar *et al.* (1981)
		13	SY	This review
ADH2	*Alcohol dehydrogenase 2 (class 1), β-polypeptide*	6cen	P‡	Lumsden *et al.* (1995)
AK1	*Adenylate kinase 1*	3p	SZ	Burkin *et al.* (1993)
ALB	*Albumin*	6q	SR	Broad *et al.* (1994b)
ALDOB	*Aldolase B, fructose-bisphosphatase*	2(U11)	SP	Payen *et al.* (1995)
AMELX	*Amelogenin, X-linked*	X	I	Shaw *et al.* (1993)
AMH	*Anti-mullerian hormone*	U19	SR	Imam-Ghali *et al.* (1991b)
		U19	SP	Payen *et al.* (1995)
		5	SY	This review
ANXL	*Annexin-like*	3p	SR	Broad *et al.* (1995e)
APOB	*Apoliprotein B*	3p	SR	Broad *et al.* (1995e)
ARAF1	*Murine sarcoma 3611 viral (v-raf) oncogene homologue 1*	X	SR	Saidi-Mehtar *et al.* (1989, 1991)
BLVR	*Bovine leukaemia virus receptor*	5q15	I	Popescu *et al.* (1995)
C1R	*Complement component 1R*	3q	SR	Broad *et al.* (1993b)
CA2	*Carbonic anhydrase II*	9q	SR	Broad *et al.* (1995f)
CA2L	*Carbonic anhydrase II-like*	1q	SR	Broad *et al.* (1995f)
CALB1	*Calbindin 1*	9q	SR	Broad *et al.* (1995f)
CD3Z	*CD3 antigen, ζ-polypeptide*	1p14–p11	I	Ansari *et al.* (1994c)
		1p	SR	Brad *et al.* (1995b)
CFTR	*Cystic-fibrosis transmembrane-conductance regulator*	4	F	Tebbutt *et al.* (1995)
CHRNB1	*Cholinergic receptor, nicotinic, β_1 polypeptide (muscle)*	11q14–q22	I	Pearce *et al.* (1993)
CHRND	*Cholinergic receptor, nicotinic, δ-polypeptide*	2q42–qter	I	Ansari *et al.* (1994b)
		2	SR	Broad *et al.* (1995d)

continued overleaf

Table 10.1. Continued

Symbol*	Name	Chr	Mode†	First author of reference
CHRNG	*Cholinergic receptor, nicotinic, γ-polypeptide*	U11	SR	Tabet-Aoul *et al.* (1992)
		2	SY	This review
COL2A1	*Collagen, type II, α_1 (primary osteoarthritis)*	U4	SR	Imam-Ghali *et al.* (1991a, 1992)
COL3A1	*Collagen, type III, α_1 (Ehlers–Danlos syndrome type IV, autosomal dominant)*	2q12–q14	I	Ansari *et al.* (1994b)
		2	SR	Broad *et al.* (1995d)
COL6A1	*Collagen, type VI, α_1*	1q	SR	Broad *et al.* (1995a)
CP	*Ceruloplasmin*	1q	SR	Broad *et al.* (1995a)
CRH	*Corticotrophin-releasing hormone*	9q23–q28	I	Ansari *et al.* (1993b); Broad *et al.* (1995f)
		9	SR	Broad *et al.* (1995f)
CSF2	*Colony-stimulating factor 2 (granulocyte–macrophage)*	5q13–q15	F	Hawken *et al.* (1995)
CSN1S1	*Casein, α-S1*	6	SR	Broad *et al.* (1994b)
CSN1S2	*Casein, α-S2*	6q32	I	Hayes *et al.* (1993a,b)
		6	SR	Broad *et al.* (1994b)
CSN2	*Casein, β*	U18	SR	Imam-Ghali *et al.* (1991b)
		6q22–q31	I	Ansari *et al.* (1992a,)
		4q32	I	Hayes *et al.* (1992a,b)
		6	SR	Broad *et al.* (1994b)
CSN3	*Casein, κ*	U18	SR	Imam-Ghali *et al.* (1991b)
		6	SY	This review
CYP11B	*Cytochrome P450, subfamily X1B*	9	SR	Broad *et al.* (1995f)
CYP17	*Cytochrome P450, subfamily XVII*	22q21–q23	I	Ansari *et al.* (1995)
CYP19	*Cytochrome P450, subfamily XIX (aromatization of androgens)*	7 (U3)	SP	Payen *et al.* (1995)
D1S1	*DNA segment: CSRD1613, BTA D1S1*	1q36–qter	I	Drinkwater *et al.* (1994)
D6S1	*DNA segment: OarJMP8*	6q34–q35	I	Pearce *et al.* (1994b)
D6S2	*DNA segment: BM1329, BTA D6S14*	6cen	F	Lumsden *et al.* (1995)
D6S3	*DNA segment: 260B5*	6	F	van Stijn *et al.* (1995)
D7S1	*DNA segment: GMBT19, BTA D10S1*	7q15–q23	I	Georges *et al.* (1991); Gunawardana (1991)
D8S1	*DNA segment: CSSM25, BTA D9S16*	8q17–q21	F	Johnson *et al.* (1995)

Table 10.1. Continued

Symbol*	Name	Chr	Mode†	First author of reference
D9S1	*DNA segment: GMBT6, BTA D14S1*	9q13–q17	I	Gunawardana (1991)
D10S1	*DNA segment: RP11*	10q24–q32	I	Ansari *et al.* (1993b); Broad *et al.* (1995g)
		10	SC	Broad *et al.* (1995g)
D10S2	*DNA segment: AGLA226, BTA D12S20*	10	SP	Broad *et al.* (1995g)
D10S3	*DNA segment: OarDB3*	10	SP	Broad *et al.* (1995g)
D10S4	*DNA segment: OarHH41*	10	SP	Broad *et al.* (1995g)
D10S5	*DNA segment: OarVH58*	10	SP	Broad *et al.* (1995g)
D10S6	*DNA segment: TGLA441*	10	SP	Broad *et asl.* (1995g)
D11S1	*DNA segment: GMBT22, BTA D19S1*	11q17–qter	I	Georges *et al.* (1991); Gunawardana (1991)
D11S2	*DNA segment: cos1*	11qter	I	Monteagudo *et al.* (1992a)
D15S1	*DNA segment: INRA046, BTA D15S5*	15	I	Vaiman *et al.* (1995)
D17S1	*DNA segment: GMBT11, BTA D26S1*	17q23–qter	I	Gunawardana (1991)
D18S1	*DNA segment: GMBT16, BTA D21S1*	18q21–qter	I	Georges *et al.* (1991); Gunawardana (1991)
D22S1	*DNA segment: GMBT11, BTA D26S1*	22q11–q21	I	Gunawardana (1991)
DYS1	*DNA segment: OY1.1*	Y	I	Jorgensen *et al.* (1992)
DYS2	*DNA segment: OY1.11*	Y	I	Jorgensen *et al.* (1992)
DYZ1	*DNA segment: BRY.1, BTA DYZ7*	Y	I	Matthews and Reed (1991)
EEF2	*Eukaryotic translation elongation factor 2*	5q15	F	Hayes *et al.* (1995)
EGF	*Epidermal growth factor*	6	SR	Montgomery *et al.* (1994)
ELN	*Elastin*	24q16–qter	I	Ansari *et al.* (1992a,b)
ENO1	*Enolase 1*	U1	SZ	Saidi-Mehtar *et al.* (1981); Imam-Ghali *et al.* (1987)
		12	SY	This review
ESD	*Esterase D*	3p34–p22	I	Graphodatsky *et al.* (1993)
		10	SZ, CS	Broad *et al.* (1995g)
ESR	*Oestrogen receptor*	8q25–q27	I	Ansari *et al.* (1994c)
EVI1	*Ectopic viral integration site 1*	1q	SR	Broad *et al.* (1995a)
F10	*Coagulation factor X*	10q33–qter	I	Pearce *et al.* (1995)
F8C	*Coagulation factor VIIIc, procoagulant component (haemophilia A)*	Xq24–q33	I	Backfisch (1991); Backfisch *et al.* (1994)

continued overleaf

Table 10.1. Continued

Symbol*	Name	Chr	Mode†	First author of reference
FGF2	*Fibroblast growth factor 2 (FGFB)*	17q23–q25	I	Pearce *et al.* (1995)
FGG	*Fibrinogen, γ-polypeptide*	17q12–q13	I	Johnson *et al.* (1993)
FN1	*Fibronectin 1*	U11	SR	Tabet-Aoul *et al.* (1992)
		2q41–q44	I	Ansari *et al.* (1994b)
		2	SR	Pearce *et al.* (1994b); Broad *et al.* (1995d)
FPGS	*Folylpolyglutamate synthase*	3p	SC	Burkin *et al.* (1993)
FRA9A	*Fragile site 9A*	9q11	CY	Matejka *et al.* (1990)
FSHB	*Follicle-stimulating hormone, β*	15q24–qter	I	Hediger *et al.* (1990b, 1991b)
		15	SP	Vaiman *et al.* (1995)
FUT	*Fucosyltransferase*	15q15	F	Oulmouden *et al.* (1995)
G6PD	*Glucose-6-phosphate dehydrogenase*	X	SZ	Saidi-Mehtar *et al.* (1981)
GAPD	*Glyceraldehyde-3-phosphate dehydrogenase*	3, U4	SZ	Saidi-Mehtar *et al.* (1989, 1991)
GART	*Phosphoribosyl amidoimidazole synthetase, Phosphoribosylglycinamide formyltransferase, Phosphoribosylglycinamide synthase*	1q, U10	SC	Broad *et al.* (1984); Burkin *et al.* (1993)
GGTA1	*Glycoprotein, α-galactosyltransferase 1*	3p26	F	Hayes *et al.* (1995)
GGTB2	*Glycoprotein-4-β-galactosyltransferase 2*	2	SR	Pearce *et al.* (1994b); Broad *et al.* (1995a)
GH1	*Growth hormone 1*	11q25–qter	I	Hediger *et al.* (1990a)
GJA3	*Gap-junction protein, α_3, 46 kD (connexin 46)*	10	SR	Broad *et al.* (1995g)
GJB2	*Gap-junction protein, β_2, 26 kD (connexin 26)*	10	SR	Broad *et al.* (1995g)
GLA	*Galactosidase, α*	X	SZ	Saidi-Mehtar *et al.* (1981)
GLUT2	*Glucose transporter 2*	1q	SR	Broad *et al.* (1995a)
GPI	*Glucose phosphate isomerase*	U12	SZ	Saidi-Mehtar *et al.* (1979, 1981)
		14q22–q14	I	Chowdhary *et al.* (1991a,b)
GRO1	*Gro 1 oncogene*	6	SR	Broad *et al.* (1994a)
GSN	*Gelsolin*	2pter–p24	I	Ansari *et al.* (1993b); Broad *et al.* (1995d)

Table 10.1. Continued

Symbol*	Name	Chr	Mode†	First author of reference
GSR	*Glutathione reductase*	U16	SZ	Imam-Ghali *et al.* (1987)
		24	SZ	Burkin *et al.* (1993)
		26	SY	This review
HBG	*Haemoglobin, γ*	U17	SZ	Saidi-Mehtar *et al.* (1987)
		15	SY	This review
HEXA	*Hexosaminidase A (α-polypeptide)*	7q13–q22	I	Ansari *et al.* (1992a); Broad *et al.* (1995c)
HMGCR	*Hydroxymethylglutaryl-CoA reductase*	7	SR	Broad *et al.* (1995c)
HPRT	*Hypoxanthine phosphoribosyltransferase*	X	SC	Saidi-Mehtar *et al.* (1981); Tucker *et al.* (1981)
IAPP	*Islet amyloid polypeptide*	3q	SR	Broad *et al.* (1993b)
IDH1	*Isocitrate dehydrogenase 1, soluble*	U11	SZ	Saidi-Mehtar *et al.* (1981)
		2	SZ	Tabet-Aoul *et al.* (1992); Broad *et al.* (1995d)
IF	*Complement factor I*	6	SR	Broad *et al.* (1994b)
IFNAR	*Interferon receptor, α*	1q	SR	Broad *et al.* (1995a)
IFNT	*Interferon, trophoblast*	2p15	F	Iannuzzi *et al.* (1993)
IFNW	*Interferon, ω*[1]	2p15	F	Iannuzzi *et al.* (1993)
IGF1	*Insulin-like growth factor 1 (somatomedin C)*	3, U4	SR	Imam-Ghali *et al.* (1991b)
IGF2	*Insulin-like growth factor 2 (somatomedin A)*	21q21–qter	I	Ansari *et al.* (1994c)
IGHML	*Immunoglobulin heavy-chain M-like*	3p23	F	Hayes and Petit (1993)
IGKC	*Immunoglobulin, κ chain*	3p	SR	Broad *et al.* (1995e)
IL2RA	*Interleukin 2 receptor, α*	13q12–q15	I	Ansari *et al.* (1995)
IL3	*Interleukin 3 (colony-stimulating factor, multiple)*	5q13–q15	F	Hawken *et al.* (1995)
IL4	*Interleukin 4*	5q13–q15	F	Hawken *et al.* (1995)
IL5	*Interleukin 5 (colony-stimulating factor, eosinophil)*	5q13–q15	F	Hawken *et al.* (1995)
IL7	*Interleukin 7*	9	SR	Broad *et al.* (1995f)
INHA	*Inhibin, α*	2q41–q43	F	Brunner *et al.* (1995b)
INHBA	*Inhibin, βA*	4q27–q31	I	Ansari *et al.* (1994c)
		4q27	F	Brunner *et al.* (1995a)

continued overleaf

Table 10.1. Continued

Symbol*	Name	Chr	Mode†	First author of reference
INHBB	*Inhibin, βB*	2q31–q33	F	Goldammer *et al.* (1995)
ITPA	*Inosine triphosphatase-A (nucleoside triphosphate pyrophosphatase)*	U15	SZ	Saidi-Mehtar *et al.* (1989; 1991)
		13	SY	This review
KIT	*Hardy Zuckerman 4 feline sarcoma viral (v-kit) oncogene homologue*	6	SR	Broad *et al.* (1994b)
KRT1@	*Keratin 1 cluster (KRTA)*	11q25–q29	I	Hediger *et al.* (1991a)
KRTAP8	*Keratin-associated protein 8 (high gly-tyr type I, F)*	1q	SP	Wood *et al.* (1992)
KRT2@	*Keratin 2 cluster (KRTB)*	3q14–q22	I	Hediger *et al.* (1991a)
LALBA	*Lactalbumin, α*	3, U4	SR	Imam-Ghali *et al.* (1991b)
		3q21	I	Hayes *et al.* (1993c)
LAMC1	*Laminin, γ_1 (LAMB2)*	12q22–q24	I	Pearce *et al.* (1995)
LCK	*Lymphocyte-specific protein tyrosine kinase*	2	SR	Broad *et al.* (1995d); Pearce *et al.* (1994b)
LDHA	*Lactate dehydrogenase A*	U13	SZ	Saidi-Mehtar *et al.* (1981)
		21	SY	This review
LDHB	*Lactate dehydrogenase B*	3q	SZ	Saidi-Mehtar *et al.* (1981); Jones *et al.* (1985); Burkin *et al.* (1993)
LGB	*Lactoglobulin, β*	3p28	F	Hayes and Petit (1993)
LHCGR	*Luteinizing hormone/chorionic gonadotrophin receptor*	3p	SR	Broad *et al.* (1995e)
LPL	*Lipoprotein lipase*	2	SR	Broad *et al.* (1995d)
LPO	*Lactoperoxidase*	U20	SR	Imam-Ghali *et al.* (1991a)
		11q13	F	Hayes *et al.* (1993c)
MBP	*Myelin basic protein*	23q11–q12.3	I	Ansari *et al.* (1994c)
MDH2	*Malate dehydrogenase 2, NAD (mitochondrial)*	U14	SZ	Saidi-Mehtar *et al.* (1981)
		24	SY	This review
ME1	*Malic enzyme 1, soluble*	U8	SZ	Saidi-Mehtar *et al.* (1981)
		8	SY	This review
MOS	*Moloney murine sarcoma viral (v-mos) oncogene homologue*	9	SR	Burkin *et al.* (1993)

Table 10.1. Continued

Symbol*	Name	Chr	Mode†	First author of reference
MPI	*Mannose phosphate isomerase*	U9	SZ	Saidi-Mehtar and Hors-Cayla (1981)
		18	SY	This review
MX1	*Myxovirus (influenza) resistance 1, homologue of murine (interferon-inducible protein p78)*	1q	SR	Broad *et al.* (1995a)
MYC	*Avian myelocytomatosis viral (v-myc) oncogene homologue*	9	SR	Burkin *et al.* (1993)
MYCN	*Avian myelocytomatosis viral (v-myc) related oncogene, neuroblastoma-derived*	3p	SR	Broad *et al.* (1995e)
MYF5	*Myogenic factor 5*	3q	SR	Broad *et al.* (1993b)
MYHL	*Myosin 1-like heavy chain*	3q	SR	Broad *et al.* (1993b)
NEB	*Nebulin*	2	SR	Broad *et al.* (1995d)
NEFL	*Neurofilament, light polypeptide*	2	SR	Broad *et al.* (1995a)
NGFB	*Nerve growth factor, β-polypeptide*	1p24–p21	I	Ansari *et al.* (1994c)
NP	*Nucleoside phosphorylase*	U3	SZ	Saidi-Mehtar and Hors-Cayla (1981)
		7	SZ	Burkin *et al.* (1993)
NRAMP	*Natural resistance-associated macrophage protein (might include leishmaniasis)*	2	W§	Pitel *et al.* (1994a,b)
		2q36–q41	I	Pitel *et al.* (1995)
NTS	*Neurotensin*	3q12–q14	I	Wood *et al.* (1993)
		3q	SR	Wood *et al.* (1993)
ODC1	*Ornithine decarboxylase*	3p	SR	Broad *et al.* (1995e)
OLA	*Major histocompatibility complex*	U6	SZ	Millot *et al.* (1981)
		20q15–q23	I	Mahdy *et al.* (1989); Hediger *et al.* (1991a)
OTC	*Ornithine carbamoyltransferase*	X	SR	Saidi-Mehtar *et al.* (1989, 1991)
PAH	*Phenylalanine hydroxylase*	3q	SR	Burkin *et al.* (1993)
PDEA	*Phosphodiesterase, cyclic GMP (rod receptor), α-polypeptide*	5q31–q33	I	Pearce *et al.* (1995)
PDEB	*Phosphodiesterase, cyclic GMP (rod receptor), β-polypeptide*	6	SR	Broad *et al.* (1994b)
		6q33–qter	I	Lord *et al.* (1994)

continued overleaf

Table 10.1. Continued.

Symbol*	Name	Chr	Mode†	First author of reference
PDGFRA	*Platelet-derived growth factor receptor, α*	6	SR	Montgomery *et al.* (1994)
PEPB	*Peptidase B*	U4	SZ	Saidi-Mehtar *et al.* (1981)
		3	SZ	Jones *et al.* (1985)
PEPC	*Peptidase C*	U1	SZ	Saidi-Mehtar *et al.* (1981, 1987)
		12	SY	This review
PGD	*Phosphogluconate dehydrogenase*	U1	SZ	Saidi-Mehtar *et al.* (1981, 1987)
		12	SY	This review
PGK1	*Phosphoglycerate kinase 1*	X	SZ	Saidi-Mehtar *et al.* (1981)
PGM1	*Phosphoglucomutase 1*	1p	SZ	Broad *et al.* (1995b)
PGM2	*Phosphoglucomutase 2*	6	SZ	Burkin *et al.* (1993)
PGM3	*Phosphoglucomutase 3*	U8	SZ	Saidi-Mehtar *et al.* (1981)
		8	SY	This review
PI7	*Protease inhibitor 7 (protease nexin I)*	2	SR	Carter *et al.* (1995)
PKM2	*Pyruvate kinase, muscle*	7, U3	SZ	Saidi-Mehtar and Hors-Cayla (1981)
		7	SR	Broad *et al.* (1995c)
PLP	*Proteolipid protein*	X	SR	Saidi-Mehtar *et al.* (1989, 1991)
POMC	*Proopiomelanocortin (adrenocorticotrophin/ β-lipotrophin)*	3p	SR	Broad *et al.* (1995e)
PPAT	*Phosphoribosylpyrophosphate amidotransferase*	6	SC	Burkin *et al.* (1991; 1993)
PRACFT	*Phosphoribosylamino-imidazolecarboxamide formyltransferase (AICAR transformylase)/IMP cyclohydrolase (inosinicase)*	2	SC	Broad *et al.* (1995d)
PRKCB1	*Protein kinase C, β_1-polypeptide*	24q13–q18	I	Ansari *et al.* (1993b)
PRLR	*Prolactin receptor*	16q17	F	Hayes *et al.* (1995)
RB1	*Retinoblastoma 1 (including osteosarcoma)*	10q13	F	Hayes *et al.* (1993c)
		10	SR	Broad *et al.* (1995g)
RCP	*Red cone pigment (colour-blindness, protan)*	X	SR	Saidi-Mehtar *et al.* (1989, 1991)
RHO	*Rhodopsin*	19q23–qter	I	Ansari *et al.* (1994c)

Table 10.1. Continued.

Symbol*	Name	Chr	Mode†	First author of reference
RNR1	*Ribosomal DNA*	1pter	CY	Ansari *et al.* (1991b); Di Meo *et al.* (1993); Henderson (1979); Henderson and Bruere (1977, 1979, 1980); Mayr and Czaker (1981); Mayr *et al.* (1985); Moreno-Millan and Rodero-Franganillo (1990)
		1pter	I	Phua *et al.* (1991); Ansari *et al.* (1992b)
RNR2	*Ribosomal DNA*	2qter	CY	*RNR1* plus Ansari *et al.* (1991a)
		2qter	I	*RNR1*
RNR3	*Ribosomal DNA*	3qter	CY	*RNR1* plus Broad *et al.* (1993b)
		3qter	I	*RNR1*
RNR4	*Ribosomal DNA*	4qter	CY	*RNR1* plus Ansari *et al.* (1992b)
		4qter	I	*RNR1*
RNR5	*Ribosomal DNA*	25qter	CY	*RNR2*
		25qter	I	*RNR1*
RPS4X	*Ribosomal protein S4, X-linked*	X	I	Shaw *et al.* (1993)
S100B	*S100 protein, β-polypeptide (neural)*	1q	SR	Broad *et al.* (1995a)
SFTP2	*Surfactant-associated protein 2*	2	SR	Broad *et al.* (1995d)
SHMT	*Serine hydroxymethyl-transferase*	3q	SZ	Jones *et al.* (1985); Burkin *et al.* (1993)
SI	*Sucrase-isomaltase*	1q	SR	Broad *et al.* (1995a)
SOD1	*Superoxide dismutase 1, soluble*	U10	SZ	Broad *et al.* (1984)
		1q	SR	Burkin *et al.* (1993)
SOX2	*SRY-related HMG-box gene 2*	1	SP	Payen *et al.* (1995)
		1q33	F	Hayes *et al.* (1995)
SPP1	*Secreted polypeptide 1*	6	SR	Montgomery *et al.* (1994)
SRY	*Sex determining region Y*	Y	SP	Payen *et al.* (1995)
STS	*Steroid sulphatase*	X	I	Shaw *et al.* (1993)
SYN1	*Synapsin I*	X	SR	Saidi-Mehtar *et al.* (1989, 1991)

continued overleaf

Table 10.1. Continued

Symbol*	Name	Chr	Mode†	First author of reference
TAC2R	*Tachykinin 2 receptor*	25q14–q23	I	Ansari *et al.* (1995)
TCRB	*T-cell receptor, β*	4q32–qter	I	Pearce *et al.* (1995)
TCRD	*T-cell receptor, δ*	7	SR	Broad *et al.* (1995c)
TF	*Transferrin*	1q42–q45	I	Broad *et al.* (1995a)
		1q	SR	Broad *et al.* (1995a); Burkin *et al.* (1993)
TGFB1	*Transforming growth factor, β_1*	8 or 16	F	Woodall *et al.* (1994)
TNP1	*Transition protein 1 (during histone-to-protamine replacement)*	2	SR	Pitel *et al.* (1994b)
TPI	*Triose phosphate isomerase 1*	3q	SR	Broad *et al.* (1993b)
TPI1L	*Triose phosphate isomerase 1-like*	6q	SR	Broad *et al.* (1993b)
TSHB	*Thyroid-stimulating hormone, β*	1p	SR	Broad *et al.* (1995b)
UBE1	*Ubiquitin activating enzyme 1*	X	I	Shaw *et al.* (1993)
UMPS	*Uridine monophosphate synthase*	1q	SC	Burkin *et al.* (1993)
UOX	*Urate oxidase*	1p	SR	Broad *et al.* (1995b)
VIL1	*Villin 1*	U11	SR	Tabet-Aoul *et al.* (1992)
		2	SR	Broad *et al.* (1995d)
WT1	*Wilms' tumour 1*	15	SP	Payen *et al.* (1995); Vaiman *et al.* (1995)
ZFX	*Zinc-finger protein, X-linked*	X	P	Aasen and Medrano (1990)
ZFY	*Zinc-finger protein, Y-linked*	Y	P	Aasen and Medrano (1990)
ZNF146	*Zinc-finger protein 146*	14q24	F	Hayes *et al.* (1995)
ZNF164	*Zinc-finger protein 164*	17q24	F	Hayes *et al.* (1995)

*Underlined loci have also been mapped in cattle (Eggen and Fries, 1995). Loci in bold font are those designated as anchored reference loci by O'Brien *et al.* (1993) for comparative genome mapping.

†The abbreviations used for the mode of mapping are as follows: CY, cytological (banding or staining); I/F, isotopic/fluorescence *in situ* hybridization; SC, somatic-cell hybrid, complementation (nutritional selection); SP, somatic-cell hybrid, polymerase chain reaction; SR, somatic-cell hybrid, restriction enzyme and Southern hybridization analysis; SY, synteny with other markers, or conserved synteny with other species; SZ, somatic-cell hybrid, isozyme analysis.

‡ P, PCR from mapped clones or DNA fragments; § W, flow sorted chromosomes, PCR or Southern hybridization analysis.

in some preparations because their constitutive heterochromatin is in very small blocks (Evans *et al.*, 1973; Schnedl and Czaker, 1974; Glahn-Luft and Wassmuth, 1977).

Molecular studies have suggested a possible basis for the weak C-banding of the metacentric chromosomes of sheep. Centromeres of mammalian chromosomes are known to contain highly repetitive DNA sequences, which are termed 'satellite DNAs' because of their initial discovery by density gradient centrifugation of DNA and their abundance in the genome. Sheep satellites I (1.714 g ml^{-1}) and II (1.724 g ml^{-1}) have been isolated (Kurnit *et al.*, 1978; Buckland, 1983; Reisner and Bucholtz, 1983; Bottema *et al.*, 1994; Novak, 1984), and localized to the centromeres by isotopic (Kurnit *et al.*, 1978; Novak, 1984) and fluorescence (Bottema *et al.*, 1994) *in situ* hybridization (FISH). Whereas satellite I hybridized predominantly to the centromeres of sheep teloocentric chromosomes, satellite II was localized at the centromeres of both metacentric and telocentric chromosomes of sheep (Kurnit *et al.*, 1978). Little hybridization of either class of satellite was seen to the sex chromosomes. This observation has recently been confirmed and extended by Burkin *et al.* (1996), who suggested that the (reduced) amounts of satellite I in the sheep metacentric chromosomes correlates with the order in which these chromosomes were formed by Robertsonian fusion during the evolution of sheep from goats and cattle. Differences in the number, organization and composition of subfamilies of satellites I and II in the sheep metacentric and sex chromosomes were also found by Burkin *et al.* (1996).

'Dynamic' banding methods

General remarks on staining 'dynamic' bands

'Dynamic' banding techniques depend on the incorporation of 5-bromo-2′-deoxyuridine (BrdU) (a structural analogue of thymidine) into DNA during a defined interval of the DNA replication (S)-phase, i.e. into either early-replicating DNA for G-banding or late-replicating DNA for R-banding (Dutrillaux *et al.*, 1973; Latt, 1973; Dutrillaux and Viegas-Pequignot, 1981; Drouin *et al.*, 1990). After staining, BrdU-containing chromosomal segments appear as unstained regions, whereas those that do not contain BrdU are intensely stained or brightly fluorescent, depending on the staining method used. In synchronized cultures, most cells will be at a similar stage of the cell cycle and therefore have a similar pattern and extent of BrdU incorporation and hence banding pattern. The quality of R- and G-bands obtained using these techniques is usually very good. The BrdU incorporation produces elongated chromosomes with more bands and a better-defined pattern than the 'structural' banding methods. However, cultures must be incubated in the dark to prevent photodegradation of BrdU.

The fluorochrome–photolysis–Giemsa method for RBG or GBG banding

R- or G-banding of BrdU-containing chromosomes using Giemsa (RBG or

GBG) is achieved by the fluorochrome–photolysis–Giemsa (FPG) method of Perry and Wolff (1974). Chromosomes prepared after BrdU incorporation are treated with Hoechst 33258 and exposed to black light, which induces photolysis of the BrdU-containing DNA, resulting in their reduced affinity for Giemsa stain. Only the regions free of BrdU are, therefore, stained.

A sheep karyotype, numbered according to the 1995 (Texas) standard, prepared by FPG staining is shown in Fig. 10.1. Other FPG sheep karyotypes have been published by Hayes *et al.* (1991) and Iannuzzi *et al.* (1995).

Propidium iodide staining for RBP or GBP banding R- or G-banding of BrdU-containing chromosomes using propidium iodide (RBP or GBP) is especially useful for the simultaneous observation of RBP or GBP banding and fluorescent hybridization signals (Lemieux *et al.*, 1992). Propidium iodide is a fluorochrome that stains chromosomal DNA red, but its staining mechanism has not been elucidated. Slides are stained with propidium iodide and mounted with *p*-phenylenediamine (PPD) at pH 11 (rather than pH 7) to prevent fluorescent fading and exposed to UV light. Under these alkaline conditions and exposure to UV light, BrdU-containing regions fluoresce red while regions without BrdU fade quickly and appear dark.

An RBP-banded sheep karyotype numbered according to the 1995 (Texas) standard is shown in colour in Plate 1.

Acridine orange staining for RBA or GBA banding R- or G-banding of BrdU-containing chromosomes using acridine orange (RBA or GBA) (Couturier *et al.*, 1973) produces a fluorescent banding pattern. With single-stranded DNA and DNA containing BrdU, acridine polymerizes and fluoresces dark red, while with double-stranded DNA and BrdU-free DNA it remains monomeric and fluoresces green.

Before the 1995 (Texas) standardization, RBA-banded sheep karyotypes were published by Matjeka *et al.* (1988), Di Berardino *et al.* (1989) and Iannuzzi *et al.* (1995). Subsequent to the 1995 standardization, an RBA karyotype was published by Ansari *et al.* (1996).

Preparation of 5-bromo-2′-deoxyuridine-containing chromosomes for R- and G-banding

R-banding Cell cultures synchronized with either thymidine or methotrexate are incubated after the release of this block with BrdU (10 μg ml^{-1}) and 5-fluoro-2′-deoxyuridine (FDU) (0.5 μg ml^{-1}) (Dutrillaux *et al.*, 1973; Viegas-Pequignot and Dutrillaux, 1978). As the cell cycle is blocked in the middle of the S phase, incorporation of BrdU into the DNA begins when the cell cycle resumes and continues during the second half of the S phase. The FdU enhances the incorporation of BrdU into DNA by inhibiting thymidylate synthase.

Karyotypes of RBG-banded early metaphase and prometaphase chromosomes of the sheep were published by Hayes *et al.* (1991).

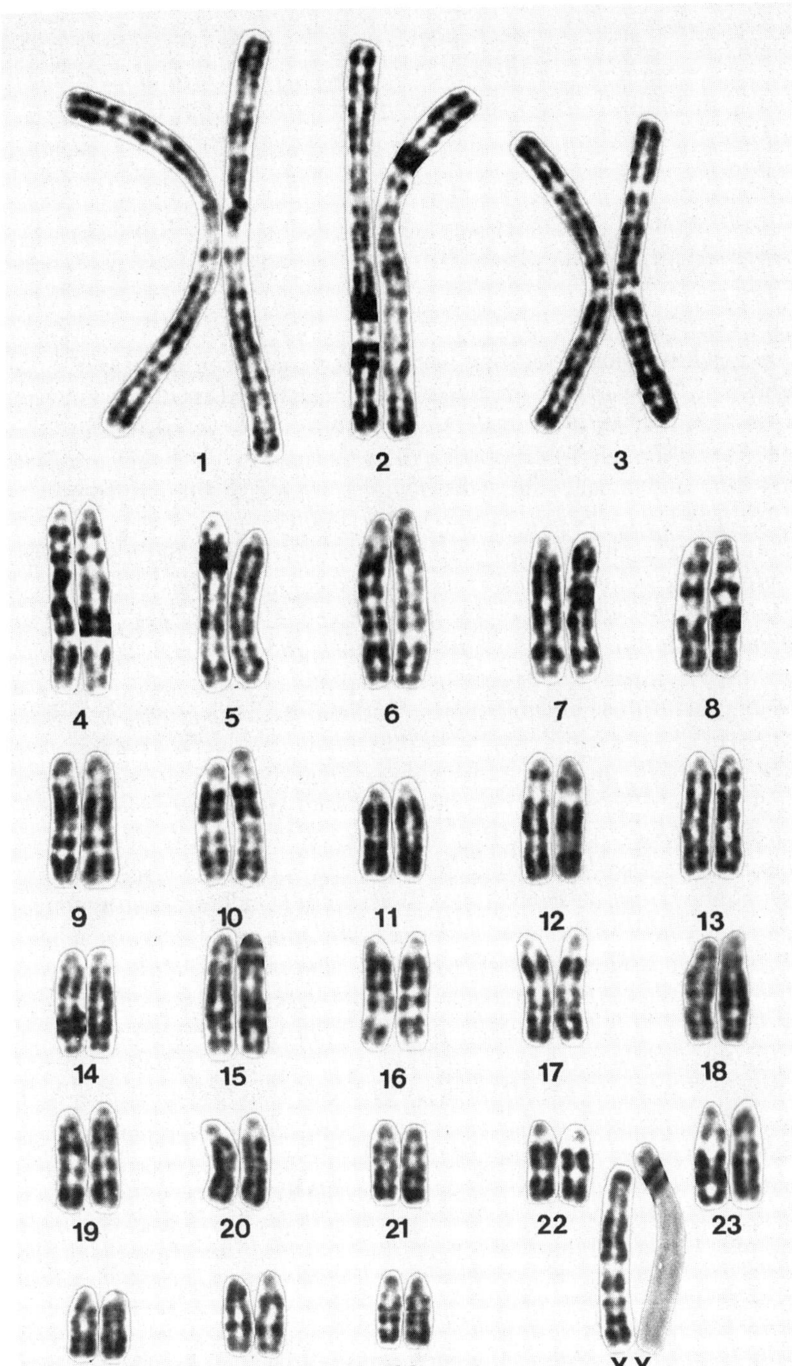

Fig. 10.1. An RBG-banded sheep karyotype (prepared by Dr Helene Hayes).

G-banding Cells are blocked with methotrexate in the presence of 10 μg ml^{-1} BrdU. After release of the block, cells are incubated in medium without BrdU, which is, therefore, incorporated only into early-replicating DNA.

Other specialized banding techniques

Fragile sites

These are chromosomal loci that have an increased incidence of gaps or breaks (Sutherland, 1979). In humans, fragile sites are classified as common or rare. The difference is that all individuals are probably homozygous for the common fragile sites, although the proportion of metaphases in which they are expressed varies from individual to individual, while rare fragile sites are seen in only a small proportion of the population (Sutherland, 1991).

Most fragile sites are not expressed spontaneously but require specific tissue-culture conditions before they can be recognized. Thus, depending on the conditions under which they are expressed, rare fragile sites can be further subdivided into folate-sensitive, distamycin-A-inducible and BrdU-requiring. Common fragile sites can be divided into aphidicolin-inducible, 5-azacytidine-inducible and BrdU-inducible (Sutherland and Richards, 1995). For details of the mechanisms of their induction, see Sutherland and Hecht (1985).

In the human, the most important rare fragile site is that at Xq27.3, which is associated with a familial form of mental retardation known as the fragile-X syndrome (Davies, 1989). Other fragile sites have no proved clinical significance. They have been implicated as having a role in oncogenesis, but this is still contentious (Le Beau, 1986). Alternatively, it has been suggested that fragile sites could be the points at which chromosome breaks occur to form constitutional rearrangements (Hecht and Hecht, 1984a, b). It is the possibility of this latter role that has stimulated interest in fragile sites in domestic animals, since a number of folate-senstive sites are at known breakpoints of reciprocal translocations in the pig (Ronne, 1995).

Few investigations have been made of fragile sites in the sheep. A heritable fragile site in chromosome 8 has been reported in Romanov sheep (Matejka *et al.*, 1990), located at 8q1.1. (under the 1995 (Texas) nomenclature, this is now chromosome 9q1.1). The same group of workers found presumptive fragile sites on chromosomes 8, 12 and 20 in Booroola Australian Merinos × Merino d'Arles crossbred sheep (Cribiu *et al.*, 1991) but did not identify band locations. Chromosome 8 was also identified as carrying a gap in both chromatids in a ram of the Rasa Aragonesa breed in Spain (Pascual and Arruga, 1995). This gap was present in all 30 cells examined after growth in conventional lymphocyte culture and was tentatively located at 8q12–q13, i.e. near the fragile site reported by Matjeka *et al.* (1990). The International Conference on the Standardization of Domestic Animal Karyotypes (ISCNDA) 1989 (Di Berardino *et al.*, 1990) chromosome numbering was used in these reports: this now corresponds to sheep chromosome 9 in the new 1995 (Texas) standard. Presumptive fragile sites have been induced in metacentric and acrocentric chromosomes of sheep using 5, 10 and 15 μg ml^{-1} of FdU (Ali and Long, 1995), with an incidence ranging from 0 to 150 per 100 cells. Identification of fragile sites using data from a single individual is difficult, but a statistical model for this has been discussed by Bohm *et al.* (1995).

G11-alkaline Giemsa staining
Used for distinguishing between sheep and hamster chromosomes in somatic-cell hybrids, this is covered below, under 'Somatic-cell Hybrids'.

Standard Karyotype and Chromosome Aberrations

The diploid number of chromosomes in the domestic sheep (*Ovis aries*) is 54. The autosomes consist of three pairs of large metacentric chromosomes and 23 pairs of telocentric chromosomes. The X is the largest acrocentric chromosome and the Y is a very small metacentric chromosome, which usually looks like a small, square dot. The X chromosome can be recognized in unbanded preparations because of the presence of small short arms, but the telocentric autosomes cannot be distinguished without differential staining techniques.

Development of a standardized nomenclature for sheep chromosomes

The first international meeting to discuss the standardization of nomenclature of chromosomes of domestic animals was held in Reading, England, in 1976. At that time, the art of G-banding chromosomes was very new, but already a number of polymorphisms and abnormalities had been detected. Standard karyotypes were needed to describe these correctly and were agreed for cattle, sheep, goats, pigs, horses, cats and rabbits. The conclusions from this meeting were published in 1980 (Ford *et al.*, 1980). These standards have remained the basis for future, more detailed, standards.

At the Fourth European Colloquium on Cytogenetics of Domestic Animals held in Uppsala, Sweden, in 1980, coordinators were appointed to facilitate the development of the standards for each of the species. Subsequently, a more detailed G-band karyotype with the recommended landmarks and regions for each sheep chromosome was produced (Long, 1985b).

With the development of other banding techniques, particularly R-bands, new standards were required. This was discussed at the Second International Conference on the Standardization of Domestic Animal Karyotypes, held in Jouy-en-Josas, France, in 1989 (Di Berardino *et al.*, 1990). At this meeting the sheep R-band standard karyotype was agreed and the cattle GTG and RBA standards were correlated by means of sequential Q to G and Q to R staining. Since the RBA and RBG patterns in the sheep correlated well with those in cattle, the GTG and R-bands for sheep were correlated by inference from the cattle correlations.

Mistakes in this correlation were clarifed in 1994 (Ansari *et al.*, 1994c), following sequential Q to G (Ansari *et al.*, 1993a) and Q to R (Ansari *et al.*, 1996) band preparations of sheep chromosomes. A new, 1995 (Texas) standard G-band karyotype was agreed at the Ninth North American Colloquium on Domestic Animal Cytogenetics and Gene Mapping, held at College Station, Texas, and this was correlated with the cattle standard on the basis of similar band patterns and gene loci (Table 10.2 gives the new standard and correlations).

Chromosome aberrations

The types of chromosome abnormalities and the terminology used for both structural and numerical anomalies have been described by Long (1995). Compared with cattle and pigs, relatively few chromosome abnormalities have been recognized in sheep (Table 10.3). Many of these have been reviewed by Long (1990). Vijh *et al.* (1994) have recently reviewed chromosome abnormalities associated with congenital malformations and reproductive disorders in sheep.

As well as being of great value to genetic, developmental and reproductive studies, chromosome abnormalities have been, and will continue to be, of particular significance for the identification of ovine chromosomes and for refining the sheep karyotype.

Centric-fusion (Robertsonian) translocations

These are the most important and most extensively investigated chromosome abnormality of sheep. The first was reported by Bruere in 1969 and during the succeeding 25 years a total of five different centric-fusion translocations have been discovered (Table 10.3). All were found in Romney-derived sheep in New Zealand. Only the t_1 (*rob* 6;24) has been found outside New Zealand and this was in a British Romney Marsh sheep (Bruere *et al.*, 1978), a breed from which several New Zealand breeds were developed (Bruere *et al.*, 1976).

During the 1970s, extensive breeding programmes were carried out on animals with the t_1, t_2 and t_3 translocations, either singly or in combination, to assess their effect on fertility (Bruere, 1974, 1975; Bruere and Chapman, 1974; Long, 1978a). (All three translocations were renumbered to be consistent with the Reading standard by Ansari *et al.* (1993a). This numbering is consistent also with the 1995 (Texas) standardization. The t_2 translocation is *rob* 9;10 and t_3 is *rob* 7;25.) These studies included effects on male meiosis (Bruere *et al.*, 1972; Bruere 1974, 1975; Bruere and Chapman, 1974; Chapman and Bruere, 1975, 1977; Long, 1978b) and limited studies on preimplantation embryos (Long, 1977). The meiotic studies showed individual ram variation in the degree of non-disjunction, but it was higher in translocation-carrying animals. The work was reviewed by Bruere and Ellis (1979) and the conclusion was that, although

Table 10.2. Comparison of the 1995 (Texas) G-band standard with previous sheep standards and cattle G-band standard chromosomes.

1995 (Texas)*	Reading picture†	Reading text†	*Hereditas* (1985)‡	ISCNDA 1989 (R-band)§	Cattle 1995 (Texas)¶
1p	1p	1p	1p	1p	3
1q	1q	1q	1q	1q	1
2p	2p	2p	2p	2p	8
2q	2q	2q	2q	2q	2
3p	3p	3q	3p	3p	11
3q	3q	3p	3q	3q	5
4	4	4	4	6	4
5	5	5	5	5	7
6	6	6	6	4	6
7	7	7	7	7	10
8	9	8	8	8	9
9	8	9	9	9	14
10	10	10	10	10	12
11	11	11	11	11	19
12	12	12	12	12	16
13	13	13	1	13	13
14	14	14	14	14	18
15	15	15	15	15	15
16	16	16	16	16	20
17	17	17	17	17	17
18	18	18	18	18	21
19	19	20	20	20	22
20	20	19	19	19	23
21	21	21	24	21	29
22	22	22	22	22	26
23	23	23	21	23	24
24	24	24	25	26	25
25	25	25	23	25	28
26	26	26	26	24	27
X	X	X	X	X	X
Y	Y	Y	Y	Y	Y

*Standard G-, Q- and R-banded idiograms of the domestic sheep (*Ovis aries*): homology with cattle (*Bos taurus*). Report of the 1995 Committee for the Standardization of the Sheep Karyotype.
†Reading conference (Ford *et al.*, 1980).
‡Long (1985b).
§ISCNDA 1989 (Di Berardino *et al.*, 1990).
¶9th North American Colloquium on Domestic Animal Cytogenetics and Gene Mapping (Popeseu *et al.*, 1996).

there was some disjunction during meiosis in rams heterozygous for centric-fusion translocations, this was not reflected in reduced fertility. It was suggested that there was prezygotic selection and that the unbalanced spermatids failed to mature and take part in fertilization.

Meiotic studies of rams heterozygous for the t_1, t_2 and t_3 translocations using the electron microscope have confirmed that trivalents are formed during meiosis and that levels of non-disjunction are marginally raised compared with normal animals (Dai *et al.*, 1994a, b, c). They showed that, during meiosis, the trivalent results in delayed synaptonemal pairing, which may then interfere with other bivalents. However, there is generally synaptic adjustment, so that by late prophase these complex configurations are resolved. The later in prophase the resolution occurs, the more likely it is that fertility will be impaired. Compared with normal sheep, rams heterozygous for the t_1 translocation had significantly higher levels of synaptonemal-complex abnormalities (Dai *et al.*, 1994b) and it was suggested that the high levels of aneuploid cells in t_3 heterozygous rams

Table 10.3. Chromosomal abnormalities in domestic sheep (*Ovis aries*).

Abnormality	Chromosomes	First report
Centric-fusion translocations	t_1 (6;24)*	Bruere (1969)
	t_2 (9;10)*	Bruere and Mills (1971)
	t_3 (7;25)*	Bruere *et al.* (1972)
	t_4 (5;8)	Pearce *et al.* (1994a)
	t_5 (8;22)	Pearce *et al.* (1994a)
Reciprocal translocations	$1p^-;19q^+$†	Glahn-Luft and Wassmuth (1978)
	$13q^-;19q^+$‡	Anamthawat-Jonsson *et al.* (1992)
	$2q^+;3q^-$	Slota *et al.* (1986)
Autosomal aneuploidy	53,XY/54,XY	Dunn and Roberts (1972)
Sex-chromosome aneuploidy	53,XO	Zartman *et al.* (1981)
	53,XO/54,XX	Baylis *et al.* (1984)
	55,XXY	Bruere *et al.* (1969a)
	54,XY/55,XYY	Moraes *et al.* (1980)
Deletions	Numerous	See Long (1990)
Autosomal duplication	20§	Matejka and Cribiu (1989)
Intersexes	54,XX	Dain (1972)
Freemartins	54,XX/54,XY	Gerneke (1967)
Testicular feminization	54,XY	Bruere *et al.* (1969b)
True hermaphrodites	54,XX	Fayrer-Hosken *et al.* (1992)

Chromosomes are numbered according to the 1995 (Texas) standard. Original designation for: *t_1 = (5;26) and t_2 = (8;11) in Bruere *et al.* (1974). The numbers cited in the table are from Ansari *et al.* (1993a) (which are consistent with those of the 1995 standardization).
†$1p^-;19q^+ = 1p^-;19q^+$ in Glahn-Luft and Wassmuth (1978).
‡$13q^-;19q^+ = 13q^-;20q^+$ in Anamthawat-Jonsson *et al.* (1992).
§20 = 19 in Matejka and Cribiu (1989).

may have been due to the influence of the NORs, which can lead to delayed pairing (Dai *et al.*, 1994c).

During the investigation of centric-fusion translocations and their effect on fertility, a few individuals were found with phenotypic abnormalities. These included hourglass testes (Bruere and Mills, 1971), segmental aplasia (Bruere *et al.*, 1972) and testicular hypoplasia (Long, 1978a). However, in a study of 1995 offspring, less than 1% showed obvious phenotypic abnormalities (Bruere and Ellis, 1979) and it is not thought that the reproductive-tract anomalies are directly due to the action of the translocation chromosomes.

Compared with the other centric-fusion translocations, the t_4 (*rob* 5;8) and t_5 (*rob* 8;22) translocations, reported by Ansari *et al.* (1993a) and Pearce *et al.* (1994a), have been little studied.

The reason for the occurrence of centric-fusion translocations in sheep in New Zealand is not known. One suggestion has been that they arose due to close inbreeding during the development of the new breeds. This hypothesis was investigated by Long (1985a), who examined 258 closely inbred sheep belonging to rare breeds in Britain. None of these showed centric-fusion translocations.

What may be a case of a heterozygous Robertsonian translocation involving chromosomes 6 and 26 has been reported in an Indian Deccani ram that was a cryptorchid (Murthy *et al.*, 1995).

Reciprocal translocations

Only three have been reported. The first (described as $1p^-;20q^+$ but designated $1p^-;19q^+$ in the 1995 (Texas) nomenclature) was found in a ram in Germany. Heterozygous animals produced fewer twins and triplets than normal and it was suggested that this was due to the early death of embryos with unbalanced chromosome complements (Glahn-Luft and Wassmuth, 1980). Flow-cytometric analysis of sperm from animals heterozygous for the 1;19 translocation produced five peaks. These corresponded to a normal, balanced and unbalanced chromosome status (Lewalski *et al.*, 1993) and examination of embryos sired by rams heterozygous for the 1;19 translocation showed that some were partially monosomic and some trisomic for the translocated segment of chromosome 1 and that there was indeed some retardation of embryo growth at the 12–15-day stage (Lewalski *et al.*, 1994).

A second reported reciprocal translocation ($13q^-;19q^+$), also involving chromosome 19 (but reported as 20 under the *Hereditas* nomenclature in Long (1985b)), was found in Icelandic sheep (Anamthawat-Jonsson *et al.*, 1992). This translocation was also presumed to cause a reduction in fertility due to early embryonic death and had in fact been detected because of reduced fertility in a particular line of animals (Adalsteinsson and Hallgrimsson, 1977).

No information is available on the fertility of animals found in Poland with the $2q^+;3p^-$ translocation (Slota *et al.*, 1986).

Autosomal aneuploidy

There have been two reports of mixoploidy with a presumptive autosomal monosomic cell line. One was in an acephalic–acardiac monster (Dunn and Roberts, 1972) and the other was in a congenitally malformed lamb with multiple abnormalities of the lumbar spinal cord (Zhang *et al.*, 1992). Unfortunately, in neither case was the missing autosome positively identified.

Sex chromosome aneuploidy

X-chromosome monosomy has been described only twice (Zartman *et al.*, 1981; Baylis *et al.*, 1984), and on both occasions the animals presented as barren ewes. The XXY, Klinefelter-like, complement has been reported in seven rams in New Zealand (Bruere *et al.*, 1969a; Bruere and Kilgour, 1974). All showed normal libido (Kilgour and Bruere, 1970; Bruere and Kilgour, 1974) but had small hypoplastic scrotal testes. Aneuploidy of the Y chromosome is even rarer than that of the X. However, there has been one report of 54,XY/55,XYY mixoploidy in a Polworth ram in Brazil that showed no abnormality of the external genitalia (Moraes *et al.*, 1980).

Deletions

There have been a number of reports of chromatid and sometimes chromosome deletions, which may or may not be associated with developmental anomalies (for review, see Long, 1990). None of these early reports contained definite identification of the chromosome anomaly by means of differential staining so that the significance of these findings remains to be verified.

Autosomal duplications

There has been one report of an inherited anomaly in chromosome number 20 (Matejka and Cribiu, 1989) in Île-de-France sheep. This consisted of a larger-than-normal short arm, which stained positively on C-banding, indicating heterochromatin.

Intersexes

The most common form of intersexuality in sheep is freemartinism, which occurs in 1–6.85% of mixed-sex multiple births (Cribiu *et al.*, 1990; Long, 1990). The condition is similar to that in cattle (for review, see Marcum, 1974) and the animals are genetic females. Major placental anastomosis was first discovered by Alexander and Williams (1964) and cytogenetic evidence of bone marrow sex-chromosome chimerism was first described by Gerneke (1967). The animals are thus 54,XX/54,XY blood chimeras. The percentage of XX and XY cells varies greatly (Gill and Davies, 1991; Szatkowska *et al.*, 1993; Szatkowska, 1995). The ovine freemartin has inhibited development of the Müllerian-duct system and enhanced development of the Wolffian-duct system, as in cattle, but the extent of masculinization often seems to be greater than in cattle (Gill and Davies, 1991). This cannot be attributed to the fact that masculine freemartins are easier to detect (Smith, 1996). Descriptions of anatomical findings have been given by a number of authors (Bruere and

Macnab, 1968; Dain and Tucker, 1970; Long, 1980; Matejka *et al.*, 1985; Cribiu *et al.*, 1990; Smith, 1996).

The fertility of rams born twin to freemartins has been little investigated. One report (Rynkiewicz-Szatkowska, 1992) described the semen of two rams that were blood chimeras (54,XX/54,XY). One animal had normal semen but the other had a low sperm count. Whether this was due to chimerism is not known.

Other types of intersexes are rare. Bruere *et al.* (1969b) described a male pseudohermaphrodite, which was almost certainly a case of testicular feminization (Lyon and Hawkes, 1970), and Dain (1972) mentioned another male pseudohermaphrodite, which was a genetic female.

There have been two reports of true hermaphrodites in domestic sheep. That reported by Fayrer-Hosken *et al.* (1992) was a genetic female and, when the ovotestis was removed, it cycled normally, became pregnant and produced triplets! A second case (Smith *et al.*, 1996) was found post-mortem at an abattoir. This animal had a female tract, an abdominal testis on the left and an ovary, with a corpus luteum and Graafian follicle, on the right. However, the genotype was not determined.

Another hermaphrodite has been described in Argali-mouflon × Mexican desert bighorn (Bunch *et al.*, 1991), which was an XX/XXY mosaic.

In Situ Hybridization Techniques

Introduction

In situ hybridization is the most direct and precise technique for localizing genes or DNA segments on chromosomes and is of central importance in genomic mapping. Radioactive *in situ* hybridization was first described in 1969 by Pardue and Gall and by John *et al.* Since then, much progress has been made with the advent of FISH procedures and improved chromosome identification techniques. These have been the subject of many reviews (e.g. Henderson, 1982; Pardue, 1985; Lichter *et al.*, 1991; Trask, 1991; Leitch *et al.*, 1994). Loci mapped on sheep chromosomes by these techniques are listed in Table 10.3. The localization of DNA sequences in relation to the centromeric regions of sheep chromosomes has been briefly reviewed above, under 'C-banding'.

In situ *hybridization procedures*

These, based on the use of radioactive and non-radioactive probes, follow a stepwise procedure, which is now briefly described.

Preparation of chromosomes

Any of the methods described earlier, under 'Chromosome Preparation and Banding', are suitable for radioactive *in situ* hybridization, but, for FISH, BrdU-containing chromosome spreads are preferred in various laboratories.

Probe labelling

Several different methods may be used, e.g. nick translation (Rigby *et al.*, 1977), random priming (Feinberg and Vogelstein, 1983) and the polymerase chain reaction (PCR) (Lo *et al.*, 1988; Richard *et al.*, 1994) using tritium-, biotin- or digoxygenin-coupled nucleotides. Tritium was the first widely used label but is now being replaced by biotin and digoxygenin because of their obvious advantages, which include a more rapid and simpler process, enhanced precision of localization of their signals and the simultaneous detection of several probes on the same chromosome preparation.

Nick translation permits the labelling of larger quantities of probe than random priming, e.g. 500 ng to 1 μg compared with 50 ng. Usually, the type of labelling selected depends on the size of the probe used. For example, tritium labelling by nick translation or random priming is used for sequences ranging in size from 0.5 to 10 kb, particularly for small complementary DNA (cDNA) probes.

In general, biotin labelling by nick translation requires probes at least 2–3 kb long. Usually, digoxygenin labelling by nick translation uses probes at least 5 kb long. Nick-translation labelling kits are now available from several manufacturers, with or without labelled nucleotides, and should be used according to the supplied instructions for optimal results. Plate 2a shows an example of a FISH to sheep chromosome 3q21 using the nick-translated biotin-labelled probe, which is a goat-genomic DNA fragment of 11.5 kb of the *α-lactalbumin* gene (*LALBA*) (Hayes *et al.*, 1993c).

For short probes, ranging from 0.25 to 1.5 kb, labelled with biotin or digoxygenin-coupled nucleotides, PCR-based procedures produce a high specific activity of labelling, which is essential for their detection. For PCR, primers specific for the amplified sequence are necessary or, if the sequence is inserted into a plasmid, universal or cloning-cassette primers may be used. Plate 2b is an example of the *in situ* hybridization of a PCR–biotin-labelled probe of 0.94 kb, which is a bovine cDNA fragment of the *glycoprotein α-galactosyltransferase* gene (*GGTA1*), to sheep chromosome 3p26 by Hayes *et al.* (1995a).

Pretreatment of chromosome preparations with ribonuclease A

This eliminates RNA from the preparations, which may be a source of background hybridization.

Denaturation of chromosomal deoxyribonucleic acid

Standard treatments, which include alkaline pH or high temperature, alter chromosome morphology, making their identification and the regional localization of the probe difficult or impossible. Therefore, most protocols use formamide, which lowers the temperature of denaturation to 70°C.

Preparation and denaturation of the probes

After labelling, non-incorporated nucleotides are removed by gel filtration, using Sephadex columns. The DNA probe is precipitated with ethanol and dried to remove all traces of the solvent. If the probe is overdried, it is difficult to resuspend subsequently in the hybridization mixture.

The probe concentration in the hybridization mixture is adjusted according to its nature and the type of labelling used. Thus, short non-radioactive PCR-labelled probes are resuspended at a concentration of 0.05–1 ng μl^{-1} and single-copy sequences in plasmids or cosmids at a concentration of 6–12 ng μl^{-1}, while those in yeast artificial chromosomes (YACs) require a much higher concentration of 50–200 ng μl^{-1}. For radioactive probes, the optimal concentration depends on their specific activity, but it usually equates to about 1 million counts per minute per slide. In all cases, a major aim is to limit the background as much as possible.

Hybridization
This is performed at 37°C in a moist chamber for 1–3 days.

Posthybridization washing
Chromosome preparations hybridized with radioactive probes usually have more background problems than those with non-radioactive probes. Therefore, it is necessary to increase the number and duration of washing steps in radioactive *in situ* hybridizations. Examples of washing procedures employed are reported in Henderson (1982), Pardue (1985), Harper and Chan (1986) and Leitch *et al.* (1994). However, in practice, the stringency of the washing steps has to be determined empirically in different laboratories and often in different experiments to reach an optimal ratio between specific signal and background 'noise'. If the background is unusually high, insufficient washing may not be the only problem and the quality of probes, slides, glassware and reagents should be checked. When background seems to be due to the probe, it may be reduced by including increasing concentrations of competitor, repetitive DNA or by decreasing the probe concentrations.

Detection of the hybridization signals and chromosome identification
Using radioactive probes These are detected by autoradiography (see Pardue, 1985, for experimental details).

When chromosomes are banded before *in situ* hybridization by QFQ staining, up to 100 good-quality spreads will already have been photographed and their positions on the slides will have been recorded. The silver grains over the chromosomes on these spreads can be manually drawn on to photographic prints made of these metaphases or the metaphases can be photographed a second time.

When posthybridization banding is performed by staining of BrdU-containing chromosomes (in which the incubation at 87°C is shortened or omitted), up to 100–150 metaphase spreads with silver grains over chromosomes are first photographed on Agfa Ortho 25 film at 25 ASA. After banding, these metaphases are rephotographed.

In both pre- and posthybridization banding procedures, the positions of silver grains over each chromosome from all photographed metaphases are then recorded on a histogram. The statistical significance of the major peak can be compared with the background hybridization over the remainder of the genome by χ^2 analysis (e.g. Ewens *et al.*, 1992). The position of the major peak localizes the probe to a particular region of a chromosome.

Using non-radioactive probes Biotin-labelled probes can be detected with antibiotin antibody (Viegas-Pequignot *et al.*, 1989) or with avidin, which is a glycoprotein with a high affinity for biotin (Cherif *et al.*, 1989). Digoxygenin-labelled probes are detected by immunocytochemistry.

Simultaneous observation of the banding pattern and the hybridization sites is possible if BrdU-containing chromosomes are used and stained with propidium iodide, as described above, under 'Preparation of 5-bromo-2′-deoxyuridine-containing chromosomes for R- and G-banding' and as seen in Plates 2a and 2b. Under fluorescence microscopy, using a 100-watt mercury-vapour lamp and a 450–490 nm transmission-block filter, both the fluorescein and the propidium iodide are excited. The hybridized DNA probe, which fluoresces yellow-green, and the RBP or GBP bands on the chromosomes, which fluoresce red, can be seen simultaneously and photographed with Ektachrome 400 ASA film (Kodak). To improve the quality of the prints, slides made on this film can be enlarged with a slide duplicator, using either a high contrast film Ektachrome 100 Plus Professional (EPP 5005) or Ektachrome SE duplicating film SO366 for colour slides and prints or Ilford Pan F 50 ASA film for black-and-white prints.

The number of metaphase spreads scored for data analysis depends on the probe size and the background level. With small probes (0.5–3 kb) and low background, recurrent single- and double-spot signals are recorded and 50–100 spreads with signal are observed. With small probes (0.5–3 kb) but high background, analysis is tedious and time-consuming, as it may be necessary to observe 200–300 spreads with signal. Sometimes, in these cases, the signal is not statistically significantly different from the background and no result is obtained. Thus, when high levels of background are observed, it may frequently not be worth continuing the experiment.

In the case of large probes (10–40 kb) and low background, observation of 20 spreads with single- or double-spot signals is sufficient. With such probes and high background, it is necessary to record recurrent double-spot signals on at least 20–30 spreads.

Chromosome painting

This is essentially FISH, using whole-chromosome-specific DNA libraries as the probe. It was first used in the human to detect chromosome anomalies and identify marker chromosomes in tumour cells (Cremer *et al.*, 1988) and in cells of patients with genetic diseases (Lichter *et al.*, 1988a, b; Pinkel *et al.*, 1988).

Whole-chromosome-specific DNA libraries are obtained by separating chromosomes by flow cytometry and are commercially available only for human chromosomes at present. However, human paints have been used to hybridize to metaphase spreads from other species (heterologous chromosome painting) and identify chromosomal segments with conserved synteny among species of different mammalian orders (Jauch *et al.*, 1992; Wienberg *et al.*, 1992; Scherthan *et al.*, 1994; Hayes, 1995; Rettenberger *et al.*, 1995; Solinas-Toldo *et al.*, 1995). The procedure used is largely similar to that described above, under '*In situ* hybridization procedures', with a few modifications, as reported in Hayes (1995). No human–sheep or the painting of sheep libraries on to the chromosomes of other species has yet been reported.

Two other forms of chromosome painting warrant a brief mention. The first, involving the use of total genomic DNA as a paint, has been used to identify sheep chromosomes in sheep × hamster-cell hybrids (see below, under 'Production of cell hybrids'). The second, based on the use of the *Alu* family of repetitive sequences as a paint, produces R-band-like patterns on the chromosomes of human and mouse (e.g. Korenberg and Rykowski, 1988). In sheep and cattle, this technique was not found to produce discrete R-like bands (Rajcan-Separovic and Sabour, 1993).

Somatic-cell Hybrids

Production of cell hybrids

Human × rodent and cattle × rodent somatic-cell hybrids have been powerful tools in the gene mapping of the human and cattle and have spurred the development of sheep- × rodent-cell hybrids for mapping the sheep genome. To produce such cell hybrids, sheep lymphocytes or fibroblasts have been fused with permanent, mutant cell lines from mice or hamsters. Cell fusions have been performed using Sendai virus or polyethylene glycol. The permanent rodent cell lines also carry mutations producing enzyme deficiencies. When these enzyme deficiencies are complemented by genes on chromosomes retained from the sheep cells, the interspecies cell hybrids are able to proliferate in selective culture medium. The rodent cell lines confer on the cell hybrids the capacity for permanent proliferation.

Most sheep chromosomes are eliminated from the hybrid cells, but some are retained. By using established cell cloning techniques, cell-hybrid clones containing different sheep chromosomes can be propagated. In some cell hybrids, the sheep chromosomes are stable, being retained for a long period over many cell divisions. However, in others, the retained sheep chromosomes may be highly fragmented, with some fragments being translocated into, or fused with, the rodent chromosomes. The alkaline Giemsa, or G11, cytogenetic staining technique (Bobrow *et al.*, 1972), which stains sheep chromosomes light blue and hamster chromosomes magenta, is used to distinguish sheep from rodent chromosomes and to distinguish between intact retained sheep chromosomes and fragments. Fluorescence-painting of the chromosomes in a cell hybrid with labelled sheep-genomic DNA can also be used for this purpose. Conventional banding is then used to identify the sheep chromosomes, but, usually, it is not possible to identify the chromosomal origin of fragmented chromosomes. Recently, inter-*Alu* PCR and PCR analysis using chromosome-specific microsatellite markers have been used by Burkin *et al.* (1995) to confirm the identity of specific sheep chromosomes in cell hybrids.

Sheep-cell hybrids were produced initially by Cianfriglia (1979) and by the Paris–Algerian group of Saidi-Mehtar and coworkers (e.g. Saidi-Mehtar *et al.*, 1979, 1981; Saidi-Mehtar and Hors-Cayla, 1981; Imam-Ghali *et al.*, 1991b). Then others were generated by the Cambridge group of Tucker and colleagues (1981). In the period from 1982 to 1995, the Denver, USA–New Zealand groups of Jones, Burkin and Broad produced a large number of sheep- × hamster-cell hybrids (Jones *et al.*, 1985; Burkin *et al.*, 1993; Broad *et al.*, 1994). Sheep-cell hybrids have also been generated by groups in Zaragosa, Spain (Dr M.V. Arruga, personal communication) and Novosibirsk, Russia (Dr O. Serov, personal communication) (Brascamp *et al.*, 1991).

The cell hybrids originating in Paris and Denver have been the foundation for most of the sheep gene-mapping assignments to date. The Denver hybrids have been particularly useful, as the sheep chromosomes have been cytogenetically identified. Recently, a panel of sheep-cell hybrids that covers the entire sheep genome and contains 17 of the sheep chromosomes in a monochromosomal state has been created, using auxotrophic hamster cells from the Eleanor Roosevelt Institute, Denver, USA (Burkin *et al.*, 1995).

Use of cell hybrids for gene mapping

Sheep genes on the retained chromosomes, or their fragments, are detected by DNA hybridization with probes from known genes; by screening the cell hybrids for sheep-gene proteins, enzymes and surface antigens; and by PCR amplification of genetic markers in the cell-hybrid DNA. When identifiable sheep chromosomes are present, mapping depends on finding a high level of positive concordance between the presence of a particular marker and a specific chromosome, as well as negative concordance for the lack of the same markers in cell hybrids that do not contain that sheep chromosome. In the case of cell hybrids retaining fragments of sheep chromosomes, mapping is usually unable to establish the chromosomal location of markers, but can identify those which cosegregate and are therefore likely to be on the same chromosome. Such markers are said to be syntenic. This depends on using a much larger panel of cell hybrids than is required in the former case and on statistical analysis using criteria of association (Chevalet and Corpet, 1986).

Genomic mapping by cell-hybrid analysis can, therefore, establish the chromosomal location of genes, or their synteny, but is not usually able to establish the relative order of genes on a chromosome or their orientation relative to each other or to such chromosomal landmarks as the centromere or telomere. (For this, a panel of 'primary' cell hybrids, established from individuals with cytogenetically defined deletions or translocations affecting a specific chromosome, or a panel of 'secondary' cell hybrids, derived by fragmenting a cell hybrid containing a single foreign chromosome, is required. Neither type of panel exists for sheep.) As the location of markers on the sheep chromosomes is established, regions of chromosomes with homologous genes in other species are being defined simply by comparison of species gene maps. These comparisons help to accelerate the process of mapping as they identify other genes that are present adjacent to these markers in the conserved chromosomal regions, which may also be present in similar regions of the sheep genome.

Physical Map of the Sheep Genome

Current physical map of the sheep genome

Summarized in Fig. 10.2, the physical map updates the 1993 map published by Echard *et al.* (1994). One or more loci have now been reported to be physically mapped on all of the chromosomes of the sheep. The numbering of chromosomes, as well as their bands and regions in the sheep idiogram shown in Fig. 10.2, is in accordance with the new 1995 (Texas) standard (Dr H.A. Ansari, AgResearch Grasslands, Palmerston North, New Zealand, personal communication). Table 10.1 (see pp. 247–256) lists the mapped loci and the method of their mapping. Because of recent revisions to the sheep idiogram, some previously reported regional localizations of loci (cited in Table 10.1) differ from those shown in Fig. 10.2.

The genetic nomenclature of all mapped loci follows the human and Committee on Genetic Nomenclature of Sheep and Goats (COGNOSAG) nomenclature guidelines, as previously explained (Echard *et al.*, 1994). The name and symbol of the newly mapped *phosphoribosylaminoimidazolecarboxamide formyltransferase* locus (*PRACFT*) (see Table 10.1) was assigned after consultation with the human-genome nomenclature committee.

The total number of physically mapped loci on the sheep genome stands at 201 as of 1 December 1995. In comparison, 107 loci (of which 97 were physically mapped) were reported on the sheep genome in June 1993 and just 41 in August 1989 (Echard, 1990). The use of somatic cell hybrids was responsible for most (126, or about two-thirds) of the assignments, and 89 of the loci were regionally assigned to sheep chromosomes by *in situ* hybridization. (These numbers exceed the total number of mapped loci because several loci have been physically mapped by more than one technique.) Since 1993, 24 of the *in situ* assignments have been by FISH, reflecting the growing use of this technique in animals.

As shown in Fig. 10.2, chromosomes 1, 2 and 3 are the best mapped, with 21, 24 and 31 loci, respectively. Sheep chromosome 6 is also relatively densely

Fig. 10.2. (over) Physically mapped loci in sheep (December 1995). The idiogram, constructed by Dr H.A. Ansari, AgResearch Grasslands, Palmerston North, New Zealand, is in accordance with the 1995 (Texas) standard sheep karyotype. Loci whose chromosomal location is ambiguous (that is, are reported on two chromosomes) are marked with an asterisk.

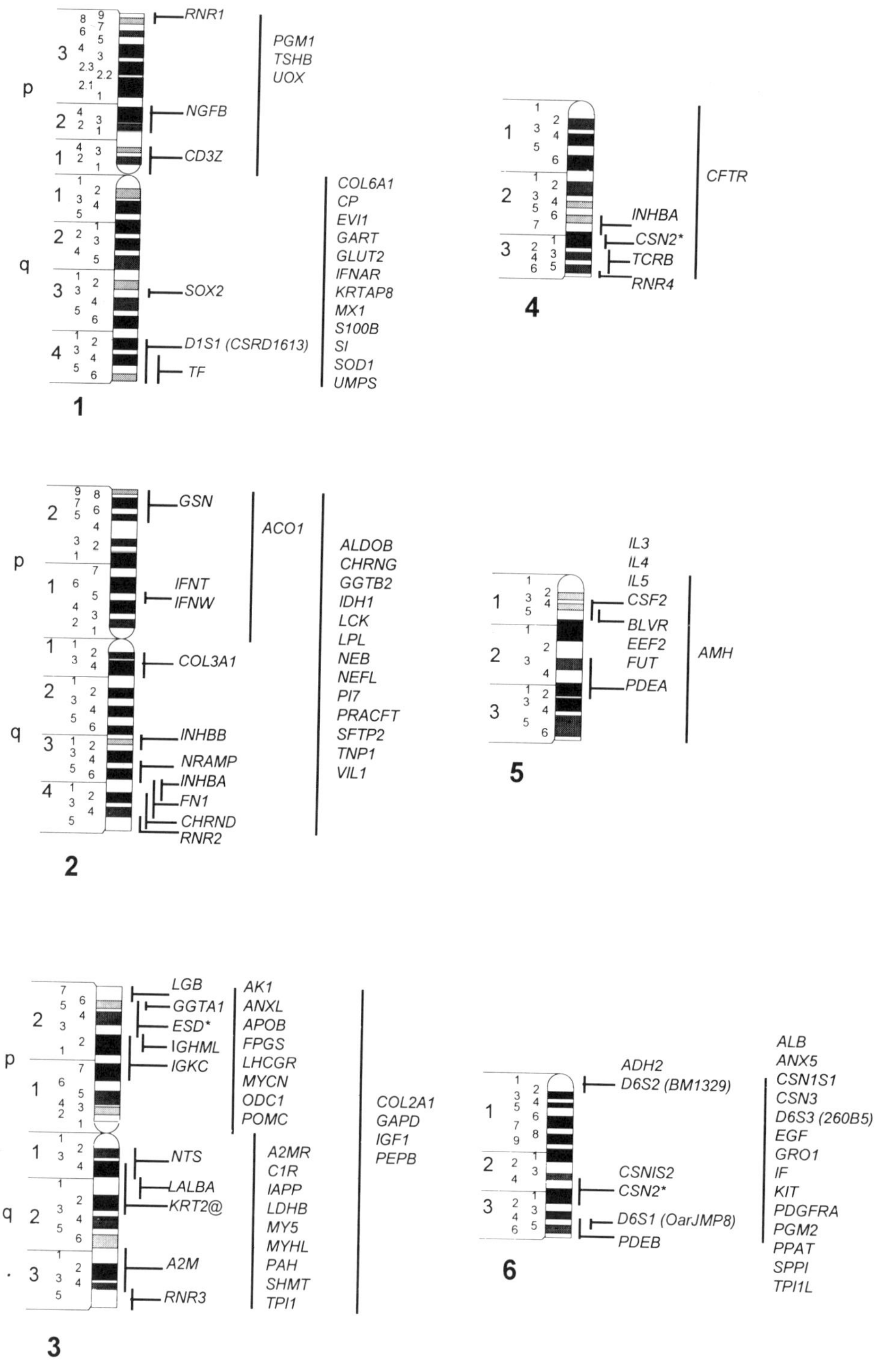
RNR1
NGFB
CD3Z
PGM1
TSHB
UOX
COL6A1
CP
EVI1
GART
GLUT2
IFNAR
KRTAP8
MX1
S100B
SI
SOD1
UMPS
SOX2
D1S1 (CSRD1613)
TF
1
CFTR
INHBA
CSN2*
TCRB
RNR4
4
GSN
ACO1
IFNT
IFNW
COL3A1
INHBB
NRAMP
INHBA
FN1
CHRND
RNR2
ALDOB
CHRNG
GGTB2
IDH1
LCK
LPL
NEB
NEFL
PI7
PRACFT
SFTP2
TNP1
VIL1
2
IL3
IL4
IL5
CSF2
BLVR
EEF2
FUT
PDEA
AMH
5
LGB
GGTA1
ESD*
IGHML
IGKC
AK1
ANXL
APOB
FPGS
LHCGR
MYCN
ODC1
POMC
NTS
LALBA
KRT2@
A2M
RNR3
A2MR
C1R
IAPP
LDHB
MY5
MYHL
PAH
SHMT
TPI1
COL2A1
GAPD
IGF1
PEPB
3
ADH2
D6S2 (BM1329)
CSNIS2
CSN2*
D6S1 (OarJMP8)
PDEB
ALB
ANX5
CSN1S1
CSN3
D6S3 (260B5)
EGF
GRO1
IF
KIT
PDGFRA
PGM2
PPAT
SPPI
TPI1L
6

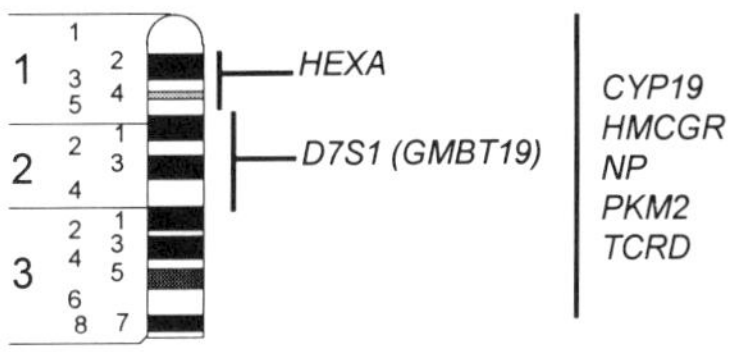

7

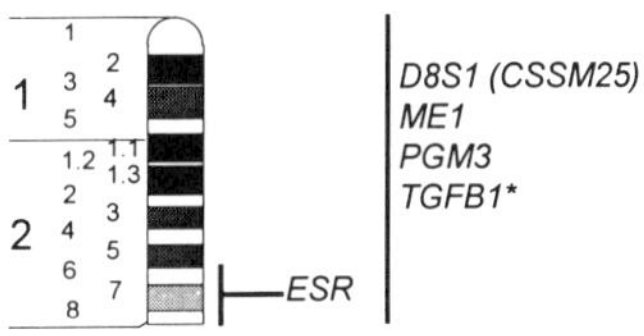

8

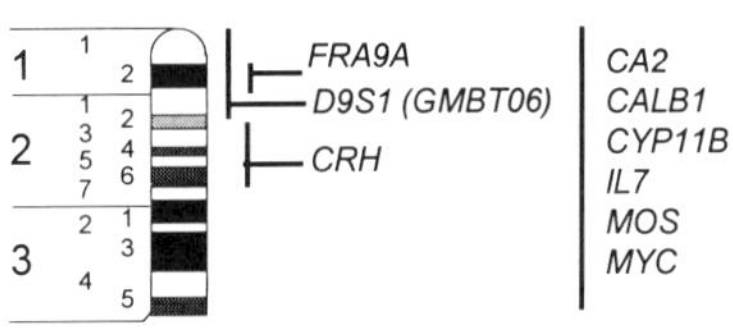

9

10

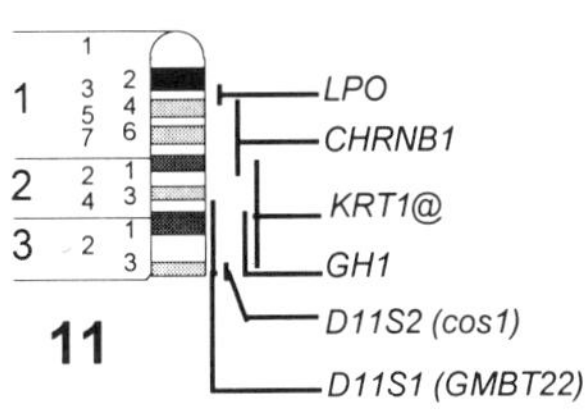

11

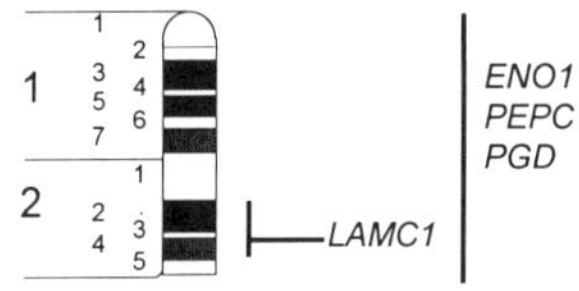

12

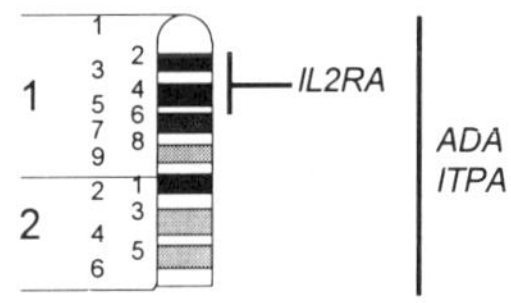

13

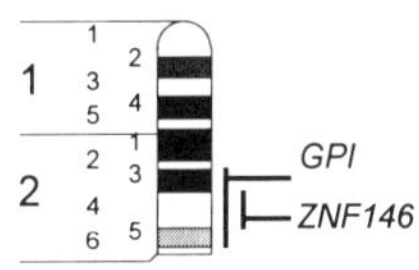

14

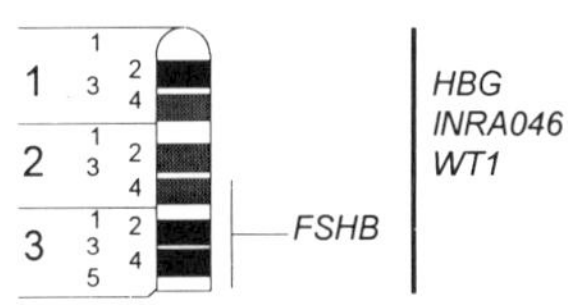

15

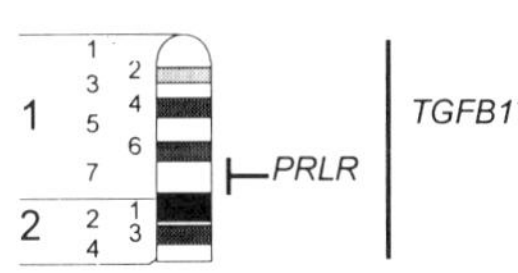

16

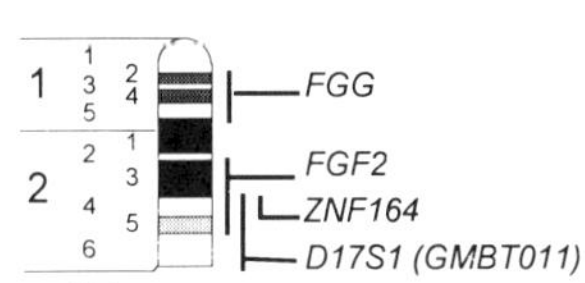

17

continued overleaf

Fig. 10.2. continued

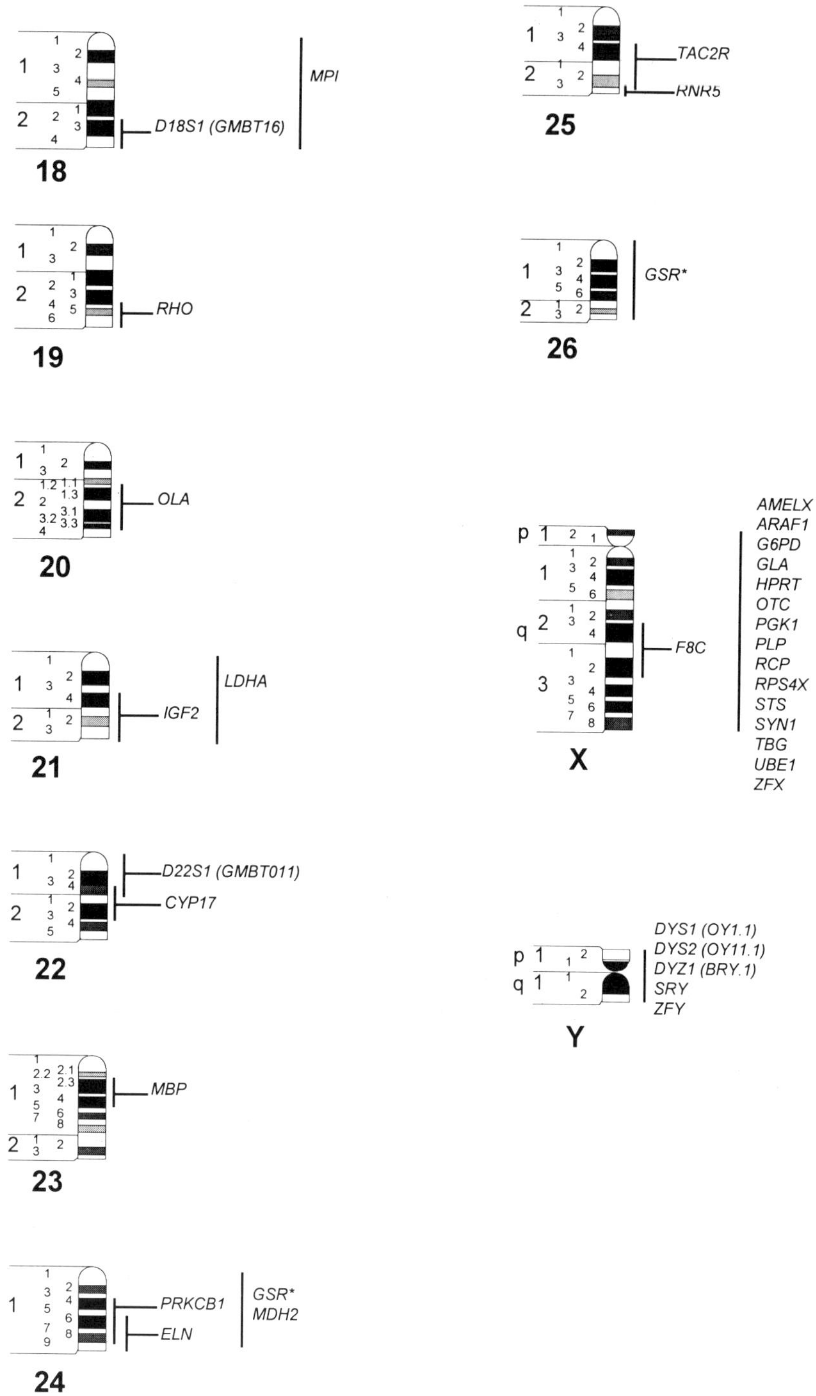

mapped, with 20 loci, because it carries the Booroola fecundity gene. A total of 13 loci, which have been previously assigned by cell-hybrid analysis to sheep syntenic groups whose chromosomal location was unknown (Echard *et al.*, 1994), have been tentatively assigned here to sheep chromosomes (see Table 10.1). These assignments have been made by comparison with their mapped location in the cattle genome, except for the *peptidase C* gene (*PEPC*). This gene is syntenic in sheep and humans with the *enolase 1* gene (*ENO1*) and the *phospogluconate dehydrogenase* gene (*PGD*). It is shown to be on sheep chromosome 12, whereas in cattle it is reported to be on chromosome 2 (Eggen and Fries, 1995).

There are four instances where genes are shown on two different chromosomes: *esterase D* (*ESD*) on sheep chromosome 3p34–p22 and chromosome 10; *casein-β* (*CSN2*) on sheep chromosome 4q22–q31 and 6q22–q31; *glutathione reductase* (*GSR*) on sheep chromosomes 24 and 26; and *transforming growth factor-β_1* (*TGFB1*) on sheep chromosomes 8 and 16 (see Table 10.1 and Fig. 10.2). On Fig. 10.2 these loci are marked with an asterisk. The assignment of *ESD* to two chromosomes by different mapping techniques remains an unresolved ambiguity, although its location on sheep chromosome 10 appears to be consistent with its conserved synteny on human chromosome 13 (see Broad *et al.*, 1995g). In the case of *CSN2*, the different assignments are due entirely to nomenclature differences between earlier standard karyotypes, and it has been agreed that it is on sheep chromosome 6 (see above, under 'Standard Karyotype and Chromosome Aberrations'). Further cytogenetic characterization of sheep- × hamster-cell hybrids has revealed that the reported location of *GSR* on sheep chromosome 24 is an error and that it is on sheep 26 (T.E. Broad, unpublished). Finally, *TGFB1* was reported to be on either sheep chromosome 8 or 16 by Woodall *et al.* (1994).

Most loci physically assigned to the sheep genome have been mapped by using DNA probes from other species with conserved coding sequences with known functions, i.e. type 1 loci. Sheep chromosome 10 currently has more microsatellite (type 2) loci assigned to it than other sheep chromosomes: these have also been mapped to this chromosome by linkage analysis (see Chapter 11).

All reported physical and linkage assignments of loci on the sheep genome are now publicly accessible in the sheep-genome database SheepBase maintained at AgResearch Invermay Agricultural Centre, Mosgiel, New Zealand (URL; http://dirk.invermay.cri.nz) and at other sites in Europe and the USA (Sise *et al.*, 1996; see also Chapter 11).

Comparison with the cattle-gene map

The cattle map, with 452 loci physically assigned markers (estimated from Eggen and Fries, 1995), has nearly 2.5 times more such markers than the sheep map. As in sheep, the main mapping technique has been cell-hybrid analysis, with 74% of assignments, but the proportion of assignments by *in situ* hybridization is 26% in cattle (Eggen and Fries, 1995), compared with 42% in sheep. However, of the 201 loci on the sheep physical map, 112 have been mapped in cattle (underlined in Table 10.1). With two exceptions, concerning the localizations of the NORs (*RNR1–5*) and the syntenic group, comprising *ENO1*, *PGD* and *PEPC*, these confirm the cytogenetic evidence (see above, under 'Standard Karyotype and Chromosome Aberrations') for a very high level of genetic similarity between sheep and cattle. As explained in Echard *et al.* (1994), *RNR3*, which is on sheep chromosome 3qter, is not on its equivalent cattle chromosome, number 5, but is instead on cattle chromosome 11qter; and *PEPC*, which has a syntenic relationship with *ENO1* and *PGD* in sheep and human, is in a different syntenic group in cattle.

As well as these regions of chromosomal reshuffling, it appears that two small regions of the cattle genome may also be duplicated or rearranged in sheep. Thus, the anonymous bovine minisatellite GMBT16, which has been mapped by *in situ* hybridization to cattle chromosome 17, hybridizes to two sheep chromosomes, numbers 17 and 22 (see *D17S1* and *D22S1* in Table 10.1). The other example is derived from linkage-mapping evidence (see Chapter 11). Crawford *et al.* (1995) have shown that bovine microsatellites assigned to sheep chromosome 9 map not only to its equivalent cattle chromosome, number 14, but are also derived from cattle chromosome 9, which is the equivalent of sheep chromosome 8. Sheep chromosome 9 has an additional subcentromeric Giemsa-positive band compared with cattle chromosome 14 (see above, under 'Standard Karyotype and Chromosome Aberrations'). It is tempting to speculate that this band may have been derived from cattle chromosome 9 during evolution.

Comparison with the human- and mouse-gene maps

Compared with the cattle-gene map and particularly the very densely mapped human and mouse genomes, the sheep genome is sparsely mapped. However, of the 327 conserved type 1 loci designated by O'Brien *et al.* (1993) as being suitable for establishing cross-reference points between the genomes of mammals, 87 have now been mapped on the sheep genome (in bold type in Table 10.1). They extend the evidence that large regions of the cattle (O'Brien *et al.*, 1993) and sheep genomes retain conserved synteny with extensive regions of the human genome. However, it should be remembered that such data do not permit conclusions to be drawn about the relative order of genes in these species, so do not rule out the possibility of rearrangements resulting in different orders of loci within these evolutionarily conserved chromosomal segments in the various mammalian species. Evidence for such rearrangements has recently been reported by Lord *et al.* (1996), who found differences in the orders of loci on sheep chromosome 6 compared with those in conserved chromosomal segments of cattle, pig, mouse and human.

Emerging Techniques and Future Perspectives

A YAC library containing large fragments of sheep DNA, some over a megabase, has been constructed (Broom and Hill, 1994), and is being used increasingly as a source of probes for *in situ* hybridization (e.g. Hawken *et al.*, 1995; Lumsden *et al.*, 1995; Tebbutt *et al.*, 1995) and microsatellites for linkage mapping (see Chapter 11). It is clear that this use will increase. Sheep YACs will almost certainly be used for developing a contig map, that is, a map constructed of overlapping, non-chimeric, YAC clones, for particular regions of special interest in the sheep genome. For the human, YAC contig maps already exist for several entire chromosomes.

The sheep YAC library was developed from total-genomic DNA. However, chromosome-specific libraries are being constructed from sheep- × hamster-cell hybrids containing single sheep chromosomes (Dr D.F. Hill, AgResearch Molecular Biology Unit, Department of Biochemistry, University of Otago, Dunedin, New Zealand, personal communication). Other chromosome-specific libraries are being produced from individual sheep chromosomes isolated by flow cytometry (Dr D.J. Burkin, Department of Pathology, University of Cambridge, UK, personal communication). Like the YAC libraries, these sheep libraries will be used for the isolation and mapping of genes and DNA segments on to sheep chromosomes. They will also be a source of probes for chromosome painting.

Fluorescence *in situ* hybridization procedures (such as those described above, under '*In Situ* Hybridization Techniques') will be the basis of the fine mapping of regions of the sheep genome. In the mapping of the human genome, the simultaneous localization of several probes, each labelled with a different fluorochrome, is routinely performed in several laboratories. The cost of image enhancement and processing technologies will soon not be the limiting factor for such multi-FISH assignments to be made in sheep. In the human, too, high-resolution FISH mapping is performed on interphase nuclei and on metaphase chromosomes that have been delicately treated so that their constituent DNA strands are able to unwind and extend. It will be only a matter of time before these powerful techiques are applied to sheep chromosomes.

Just as the availability of sheep YAC clones has extended the reach and distance of the mapping of regions of the sheep genome, so too could the development of PCR-based FISH enable the mapping of shorter fragments. This procedure uses short oligonucleotide primers to anneal to their complementary sites in the DNA and generates a labelled (fluorescent) signal in the region spanning these by PCR directly on the slide. Already being used in pig and other species, its development is under way in the sheep (Dr C. Bottema, Waite Institute, University of Adelaide, Australia, personal communication).

Microdissection has been used to generate region-specific DNA libraries of particular chromosomes of the human and mouse. This technique is also being developed for studying particular sheep chromosomes (Dr A.M. Crawford, AgResearch Molecular Unit, University of Otago, New Zealand, personal communication).

As the mapping of the sheep genome progresses, evolutionary comparisons with the genomes of other related artiodactyls and of more distantly related mammals will become increasingly more detailed and interesting. These comparisons will hasten the application of new and emerging molecular-cytogenetic techniques to the sheep and contribute to a greater understanding of the intimate interrelationships between its biological appearance and physiology with the organization and molecular architecture of its genome.

Acknowledgements

T.E. Broad was supported by a grant from the New Zealand Foundation for Research, Science and Technology. T.E. Broad gratefully acknowledges the expert computing assistance of Dr Neville Jopson, AgResearch Invermay, New Zealand, in producing Fig. 10.2 (in which the idiogram was initially drawn by Dr Helal Ansari), and other members of AgResearch, particularly Dr Diana Hill, Dr Peter Fennessy, Dr Helal Ansari and Dr Phillip Pearce, for their critical reading of the draft manuscript and their comments. H. Hayes thanks Drs B. Dutrillaux and P. Popescu for their constant interest and helpful discussions. We are also indebted to Dr Genevieve Echard for her guidance and encouragement throughout.

References

Aasen, E. and Medrano, J.F. (1990) Amplification of the ZFY and ZFX genes for sex identification in humans, cattle, sheep and goats. *Biotechnology* 8, 1279–1281.

Adalsteinsson, S. and Hallgrimsson, S. (1977) Inherited fertility depression in Icelandic sheep. *Journal of Agricultural Research in Iceland* 9, 77–82.

Adams, R.L.P. (1981) *Cell Culture for Biochemists*, 1st edn. Elsevier, Amsterdam, New York, Oxford, 292 pp.

Alexander, G. and Williams, D. (1964) Ovine freemartins. *Nature* 201, 1296–1298.

Ali, A. and Long, S.E. (1995) Preliminary report on the induction of fragile sites in sheep chromosomes using fluorodeoxyuridine (FUdR). In: *Proceedings of the 9th North American Colloquium on Domestic Animal Cytogenetics and Gene Mapping*, College Station, Texas. pp. 54–55.

Anamthawat-Jonsson, K., Long, S.E., Basrur, P.K. and Adalsteinsson, S. (1992) Reciprocal translocation (13;20)(q12;q22) in an Icelandic sheep. *Research in Veterinary Science* 52, 367–370.

Ansari, H., Broad, T., Crellen, L., Davidson, B., Lewis, P., McNabb, L., Maher, D., Malcolm, A., Pearce, P., Phua, S. and Wood, N. (1991a) Regional assignment of three genes to sheep chromosome 2. In: *Proceedings of the Australasian Gene Mapping and Molecular Genetics Symposium, Dunedin*, p. 39.

Ansari,-H.A., Phua, S.H., Wood, N.J., Pearce, P.D., Crellen, L.E., Davidson, B.P., Maher, D.W. and Broad, T.E. (1991b) Physical mapping of rDNA genes in sheep by *in situ* hybridization. In: *Proceedings of the 7th North American Colloquium on Domestic Animal Cytogenetics and Gene Mapping*. Philadelphia, Pennsylvania, p. 128.

Ansari, H.A., Broad, T.E., Burkin, D.J., Cambridge, L., Carpenter, M.A., Davidson, B.P.,

Jones, C.A., Lewis, P.E., Maher, D.W., Malcolm, A.A., McNabb, L.R., Pearce, P.D., Phua, S.H. and Wood, N.J. (1992a) Regional assignment of five genes onto sheep chromosomes. In: *Abstracts of the 10th European Colloquium on the Cytogenetics of Domestic Animals, Utrecht*, p. 27.

Ansari, H.A., Pearce, P.D., Maher, D.W., Malcolm, A.A., Wood, N.J., Phua, S.H. and Broad, T.E. (1992b) Resolving ambiguities in the sheep karyotype: a molecular cytogenetic approach. In: *Proceedings of the 10th European Colloquium on the Cytogenetics of Domestic Animals*. Utrecht, Netherlands, pp. 21–25.

Ansari, H.A., Pearce, P.D., Maher, D.W., Malcolm, A.A. and Broad, T.E. (1993a) Resolving ambiguities in the karyotype of the domestic sheep (*Ovis aries*). *Chromosoma* 102, 340–347.

Ansari, H.A., Pearce, P.D. and Maher, D.W. (1993b) Regional assignment of anchored reference loci to sheep chromosomes. In: *Proceedings of the 8th North American Colloquium on Domestic Animal Cytogenetics and Gene Mapping*. Guelph, Canada, pp. 59–61.

Ansari, H.A., Bosma, A.A., Bunch, T.D., Long, S.E. and Popescu, C.P. (1994a) Clarification of chromosome nomenclature in the sheep (*Ovis aries*): report of the committee for the standardization of the sheep karyotype. *Cytogenetics and Cell Genetics* 67, 114–115.

Ansari, H.A., Pearce, P.D., Maher, D.W., Wood, N.J., Phua, S.H. and Broad, T.E. (1994b) Regional mapping of loci from human chromosome 2q to sheep chromosome 2q. *Genomics* 20, 122–124.

Ansari, H.A., Pearce, P.D., Maher, D.W. and Broad, T.E. (1994c) Regional assignment of loci mapped in human, cattle and mouse anchors unassigned linkage and syntenic groups to ovine chromosomes. *Genomics* 24, 451–455.

Ansari, H.A., Pearce, P.D., Maher, D.W and Broad, T.E. (1995) Human chromosome 10 loci map to three different sheep chromosomes. *Mammalian Genome* 6, 46–48.

Ansari, H.A., Maher, D.W., Pearce, P.D. and Broad, T.E. (1996) Resolving ambiguities in the karyotype of domestic sheep (*Ovis aries*). II. G-, Q-, and R-banded idiograms and chromosome-specific molecular markers. *Chromosoma* 10, 62–67.

Babu, A. and Verma, R.S. (1987) Chromosome structure: euchromatin and heterochromatin. *International Review of Cytology* 108, 1–60.

Backfisch, W. (1991) Kartierung und Charakterisierung des Faktor VIII Gens. Sowie die Tragerinnendiagnose der Hamophile a biem Schaf. PhD dissertation, ETH Number 9579, Swiss Institute for Technology, Zurich, 88 pp.

Backfisch, W., Neuenschwander, S., Giger, U., Stranzinger, G. and Pliska,V. (1994) Carrier detection of ovine hemophilia A using an RFLP marker, and mapping the factor VIII gene on the ovine X-chromosome. *Journal of Heredity* 85, 474–478.

Baylis, M.S., Wayte, D.M. and Owen, J.B. (1984) An XO/XX mosaic sheep with associated gonadal dysgenesis. *Research in Veterinary Science* 36, 125–126.

Bickmore, W.A. and Sumner, A.T. (1989) Mammalian chromosome banding – an expression of genome organization. *Trends in Genetics* 5, 144–148.

Bloom, S.E. and Goodpasture, C. (1976) An improved technique for selective silver staining of nucleolar organizer regions in human chromosomes. *Human Genetics* 34, 199–206.

Bobrow, M., Madan, K. and Pearson, P.L. (1972) Staining of some specific regions of human chromosomes, particularly the secondary constriction of no. 9. *Nature New Biology* 238, 122–124.

Bohm, U., Dahm, P.F., McAllister, B.F. and Greenbaum, I.F. (1995) Identifying chromosomal fragile sites from individuals: a multinomial statistical model. *Human Genetics* 95, 249–256.

Bostock, C.J. and Sumner, A.T. (1978) Substructure of chromosomes and the banding phenomena. In: *The Eukaryotic Chromosome*. North-Holland, Amsterdam, pp. 375–405.

Bottema, C.D.K., Cerin, H., Rizza, S., Cook, J., Webb, G.C. and Fowler, C.J.S. (1994) Isolation of centromeric DNA sequences from cattle and sheep. In: *Proceedings of the 5th World Congress on Genetics Applied to Livestock Production*, University of Guelph, Guelph, Canada. Vol. 21, pp. 63–65.

Boyum, A. (1968) Separation of leucocytes from blood and bone marrow. *Scandinavian Journal of Clinical Laboratory Investigation* 21 (Suppl. 97), 31.

Brascamp, E.W., Haley, C.S., Fries, R., Grosclaude, F., Hanset, R. and Womack, J.E. (1991) Genome mapping in farm animals, a survey of current research: report of a working group of the European Association for Animal Production. In: *42nd Annual Meeting of the European Association for Animal Production, 8–12 September, Berlin*. pp. 1–22.

Broad, T.E. and Hill, D.F. (1994) Mapping the sheep genome: practice, progress and promise. *British Veterinary Journal* 150, 237–252.

Broad, T.E., Jones, C. and Patterson, D. (1984) Gene mapping in the sheep: synteny of SOD1, PRGS and PAIS. *Cytogenetics and Cell Genetics* 37, 427.

Broad, T.E., Burkin, D.J., Jones, C., Lewis, P.E., Ansari, H.A. and Pearce, P.D. (1993b) Mapping of MYF5, CIR, MYHL, TPI1, IAPP, A2MR and RNR onto sheep chromosome 3q. *Animal Genetics* 24, 415–419.

Broad, T.E., Ansari, H.A., Burkin, D.J., Cambridge, L.M., Carpenter, M.A., Gleeson, A.J., Jones, C., Lewis, P.E., Maher, D.W. and Pearce, P.D. (1994a) Physical mapping of the sheep genome: six new chromosomal assignments by *in situ* hybridization, and sixteen more loci mapped using an expanded panel of sheep × hamster cell hybrids. *Animal Genetics* 25 (Suppl. 2), 45.

Broad, T.E., Burkin, D.J., Cambridge, L.M., Maher, D.W., Lewis, P.E., Ansari, H.A., Pearce, P.D. and Jones, C. (1994b) Seven loci from human chromosome 4 map onto sheep chromosome 6: a proposal to restore the original nomenclature of this chromosome. *Mammalian Genome* 5, 429–433.

Broad, T.E., Burkin, D.J., Cambridge, L.M., Maher, D.W., Lewis, P.E., Ansari, H.A., Pearce, P.D. and Jones, C. (1995a) Four human chromosome 3q and four human chromosome 21 loci map onto sheep chromosome 1q. *Mammalian Genome* 6, 202–205.

Broad, T.E., Burkin, D.J., Cambridge, L.M., Carpenter, M.A. and Jones, C. (1995b) Four loci from human chromosome 1 are physically assigned to sheep chromosome 1p. *Mammalian Genome* 6, 551–553.

Broad, T.E., Burkin, D.J., Cambridge, L.M., Pearce, P.D., Maher, D.W., Ansari, H.A. and Jones, C. (1995c) Physical assignment of loci to sheep chromosome 7 confirms its homology to cattle chromosome 10. *Mammalian Genome* 6, 749–750.

Broad, T.E., Lewis, P.E., Burkin, D.J., Gleeson, A.J., Carpenter, M.A., Jones, C., Pearce, P.D., Maher, D.W. and Ansari, H.A. (1995d) Thirteen loci physically assigned to sheep chromosome 2 by cell hybrid analysis and *in situ* hybridization. *Mammalian Genome* 6, 862–866.

Broad, T.E., Burkin, D.J., Cambridge, L.M., Jones, C., Lewis, P.E., Morse, H.G., Geyer, D., Pearce, P.D., Ansari, H.A. and Maher, D.W. (1995e) Six loci mapped onto human chromosome 2p are assigned to sheep chromosome 3p. *Animal Genetics* 26, 85–90.

Broad, T.E., Burkin, D.J., Cambridge, L.M., Maher, D.W., Lewis, P.E., Ansari, H.A.,

Pearce, P.D. and Jones, C. (1995f) Assignment of five loci from human chromosome 8q onto sheep chromosome 9. *Cytogenetics and Cell Genetics* 68, 102–106.

Broad, T., Lambeth, M., Burkin, D., Jones, C., Pearce, P., Maher, D. and Ansari H. (1995g) Sheep chromosome 10 has extensive conserved synteny with cattle chromosome 12 and human chromosome 13. In: *Proceedings of the 6th Australasian Gene Mapping Workshop, Dunedin*. Abstract 61.

Broom, M.F. and Hill, D.F. (1994) Construction of a large-insert yeast artificial chromosome library from sheep DNA. *Mammalian Genome* 5, 817–819.

Bruere, A.N. (1969) Male sterility and an autosomal translocation in Romney sheep. *Cytogenetics* 8, 209–218.

Bruere, A.N. (1974) The segregation patterns and fertility of sheep heterozygous and homozygous for three different Robertsonian translocations. *Journal of Reproduction and Fertility* 41, 453–464.

Bruere, A.N. (1975) Further evidence of normal fertility and the formation of balanced gametes in sheep with one or more different Robertsonian translocations. *Journal of Reproduction and Fertility* 45, 323–331.

Bruere, A.N. and Chapman, H.M. (1974) Double translocation heterozygosity and normal fertility in domestic sheep. *Cytogenetics and Cell Genetics* 13, 342–351.

Bruere, A.N. and Ellis, P.M. (1979) Cytogenetics and reproduction of sheep with multiple centric fusions (Robertsonian translocations). *Journal of Reproduction and Fertility* 57, 363–375.

Bruere, A.N. and Kilgour, R. (1974) Normal behaviour patterns and libido in chromatin-positive Klinefelter sheep. *Veterinary Record* 95, 436–439.

Bruere, A.N. and Mcnab, J. (1968) A cytogenetical investigation of six intersex sheep, shown to be freemartins. *Research in Veterinary Science* 9, 170–180.

Bruere, A.N. and Mills, R.A. (1971) Observations on the incidence of Robertsonian translocations and associated testicular changes in a flock of New Zealand Romney sheep. *Cytogenetics* 10, 260–272.

Bruere, A.N, Marshall, R.B. and Ward, D.P.J. (1969a) Testicular hypoplasia and XXY sex chromosome complement in two rams: the ovine counterpart of Klinefelter's syndrome in man. *Journal of Reproduction and Fertility* 19, 103–108.

Bruere, A.N., McDonald, M.F. and Marshall, R.B. (1969b) Cytological analysis of an ovine male pseudohermaphrodite and the possible role of the Y chromosome in cryptorchidism of sheep. *Cytogenetics* 8, 148–157.

Bruere, A.N., Chapman, H.M. and Wyllie, D.R. (1972) Chromosome polymorphism and its possible implications in the select Drysdale breed of sheep. *Cytogenetics* 11, 233–246.

Bruere, A.N., Zartman, D.L. and Chapman, H.M. (1974) The significance of the G-bands and C-bands of three different Robertsonian translocations of domestic sheep. *Cytogenetics and Cell Genetics* 13, 479–488.

Bruere, A.N., Chapman, H.M., Jaine, P.M. and Morris, R.M. (1976) Origin and significance of centric fusions in domestic sheep. *Journal of Heredity* 67, 149–154.

Bruere, A.N., Evans, E.P., Burtenshaw, M.D. and Brown, B.B. (1978) Centric fusion polymorphism in Romney Marsh sheep of England. *Journal of Heredity* 69, 8–10.

Brunner, R.M., Goldhammer, T., Hiendleder, S., Jager, C. and Schwerin, M. (1995a) Mapping the gene encoding for inhibin βA (INHBA) to chromosome 4q26 in sheep. *Mammalian Genome* 6, 308–309.

Brunner, R.M., Goldhammer, T., Hiendleder, S., Jager, C. and Schwerin, M. (1995b) Comparative mapping of the gene coding for inhibin-α (INHA) to chromosome 2 in sheep and cattle. *Mammalian Genome* 6, 309.

Buckland, R.A. (1983) Comparative structure and evolution of goat and sheep satellite I DNAs. *Nucleic Acids Research* 11, 1349–1360.

Bunch, T.D., Callan, R.J., Maciulis, A., Dalton, J.C., Figueroa, M.R., Kunzler, R. and Olson, R.E. (1991) Hermaphroditism in a wild sheep: a case report. In: *Proceedings of the 7th North American Colloquium on Domestic Animal Cytogenetics and Gene Mapping*, Philadelphia, Pennsylvania, p. 57.

Burkin, D.J., Broad, T.E. and Jones, C. (1991) Isolation and characterization of sheep × hamster somatic cell hybrids selectively retaining a sheep translocation chromosome. *Cytogenetics and Cell Genetics* 58, 2124.

Burkin, D.J., Morse, H.G., Broad, T.E., Pearce, P.D., Ansari, H.A., Lewis, P.E. and Jones, C. (1993) Mapping the sheep genome: production of characterized sheep × hamster cell hybrids. *Genomics* 16, 466–472.

Burkin, D.J., Broad, T.E., Crawford, A.M., and Jones, C. (1995) Mapping the sheep genome: towards the completion of a well characterized sheep–hamster cell hybrid mapping panel. In: *9th North American Colloquium on Domestic Animal Cytogenetics and Gene Mapping*, College Station, Texas, p. 52.

Burkin, D.J., Broad, T.E. and Jones, C. (1996) The chromosomal distribution and organization of sheep satellite I and II centromeric DNA using characterized sheep–hamster somatic cell hybrids. *Chromosome Research* 4, 49–55.

Carter, R.E., Cerosaletti, K.M., Burkin, D.J., Fournier, R.E.K., Jones, C., Greenberg, B.D., Citron, B.A. and Festoff, B.W. (1995) The gene for the serpin thrombin inhibitor (PI7), protease nexin I, is located on human chromosome 2q33–q35 and on syntenic regions in the mouse and sheep genomes. *Genomics* 27, 196–199.

Caspersson, T., Farber, S., Foley, G.E., Kudynowski, J., Modest, E.J., Simonsson, E., Wagh, V. and Zech, L. (1968) Chemical differentiation among metaphase chromosomes. *Experimental and Cell Research* 49, 219–222.

Chapman, H.M. and Bruere, A.N. (1975) The frequency of aneuploidy in the secondary spermatocytes of normal and Robertsonian translocation-carrying rams. *Journal of Reproduction* 45, 333–342.

Chapman, H.M. and Bruere, A.N. (1977) Chromosome morphology during meiosis of normal and Robertsonian translocation-carrying rams (*Ovis aries*). *Canadian Journal of Genetics and Cytology* 19, 93–102.

Cherif, D., Bernard, O. and Berger, R. (1989) Detection of single-copy genes by nonisotopic *in situ* hybridization on human chromosomes. *Human Genetics* 81, 358–362.

Chevalet, C. and Corpet, F. (1986) Statistical decision rules concerning synteny or independence between markers. *Cytogenetics and Cell Genetics* 43, 132–139.

Chowdhary, B.P., Harbitz, I., Davies, W. and Gustavsson, I. (1991a) Chromosomal localization of the glucose phosphate isomerase (GPI) gene in cattle, sheep and goat by *in situ* hybridization – chromosomal banding homology versus molecular conservation in Bovidae. *Hereditas* 114, 161–170.

Chowdhary, B.P., Harbitz, I., Davies, W. and Gustavsson, I. (1991b) Tentative chromosomal assignment of the glucose phosphate isomerase gene in cattle, sheep, and goat by *in situ* hybridization. *Génétique, Sélection et Evolution* 23 (Suppl. 1), 91–95.

Cianfriglia, M. (1979) Cytogenetic characterization of Chinese hamster–sheep somatic cell hybrids. *Bollettino della Societa Italiana di Biologia Sperimentale (Napoli)* 55, 2201–2207.

Comings, D.E. (1978) Mechanisms of chromosome banding and implications for chromosome structure. *Annual Review of Genetics* 12, 25–46.

Comings, D.E., Avelino, E., Okada, T.A. and Wyandt, H.E. (1973) The mechanism of

C- and G-banding of chromosomes. *Experimental and Cell Research* 77, 469–493.

Couturier, J., Dutrillaux, B. and Lejeune, J. (1973) Etude des fluorescences spécifiques des bandes R et des bandes Q des chromosomes humains. *Compte Rendu de l'Académie des Sciences (Paris)* 276, 339–342.

Crawford, A.M., Dodds, K.G., Ede, A.J., Pierson, C.A., Montgomery, G.W., Garmonsway, H.G., Beattie, A.E., Davies, K., Maddox, J.F., Kappes, S.W., Store, R.T., Nguyen, T.C., Penty, J.M., Lord, E.A., Broom, J.E., Buitkamp, J., Schwaiger, W., Epplen, J.T., Matthew, P., Matthews, M.E., Hulme, D.J., Beh, K.J., McGraw, R.A. and Beattie, C.W. (1985) An autosomal genetic linkage map of the sheep genome. *Genetics* 140, 703–724.

Cremer, T., Lichter, P., Borden, J., Ward, D.C. and Manuelidis, L. (1988) Detection of chromosome aberrations in metaphase and interphase tumour cells by *in situ* hybridization using chromosome-specific library probes. *Human Genetics* 80, 235–246.

Cribiu, E.P., Durand, V. and Chaffaux, St. (1990) Fréquence du freemartinisme chez des agnelles booroola Mérinos d'Arles. *Recueil de Médecine Vétérinaire* 166, 919–922.

Cribiu, E.P., Durand, V., Saget, O. and Chaffaux, S. (1991) Cytogenetic investigations in Booroola Merinos d'Arles sheep (preliminary results). In: *Major Genes for Reproduction in Sheep, 2nd International Workshop for the Booroola Gene, Toulouse, 16–18 July 1990*. pp. 273–279.

Dai, K., Dollin, A.E. and Gillies, C.B. (1994a) Synaptonemal complex analysis of domestic sheep (*Ovis aries*) with Robertsonian translocations. I. Pachytene karyotype substaging of normal sheep. *Genome* 37, 672–678.

Dai, K., Gillies, C.B. and Dollin, A.E. (1994b) Synaptonemal complex analysis of domestic sheep (*Ovis aries*) with Robertsonian translocations. II. Trivalent and pairing abnormalities in Massey I and Massey II heterozygotes. *Genome* 37, 679–689.

Dai, K., Gillies, C.B. and Dollin, A.E. (1994c) Synaptonemal complex analysis of domestic sheep (*Ovis aries*) with Robertsonian translocations. III. Deficient pairing and NOR role in Massey II heterozygotes. *Genome* 37, 802–808.

Dain, A.R. (1972) Differences in chromosome lengths between male and female sheep. *Nature* 237, 455–457.

Dain, A.R. and Tucker, E. M. (1970) Cytogenetic, anatomical and blood group studies of sheep twin chimaeras. *Proceedings of the Royal Society of London B* 175, 183–200.

Davies, K.A. (ed.) (1989) *The Fragile X Syndrome*. Oxford University Press, Oxford.

Di Berardino, D., Arrighi, F.E. and Kieffer, N.M. (1979) Nucleolus organizer regions in two species of Bovidae. *Journal of Heredity* 70, 47–50.

Di Berardino, D., Hayes, H., Fries, R. and Long, S.E. (eds) (1990) The Second International Conference on Standardization of Domestic Animal Karyotypes. *Cytogenetics and Cell Genetics* 53, 65–78.

Di Berardino, D., Lioi, M.B., Miranda, C., Di Milia, A., D'Agostino, M.G. and Matassino, D. (1989) R-banding pattern of the prometaphase chromosomes of the domestic sheep *Ovis aries*. *Génétique, Sélection et Evolution* 21, 1–10.

Di Meo, G.P., Iannuzzi, L., Peracatti, A. and Ferrara, L. (1993) Identification of nucleolus organizer regions in sheep by sequential GBG/AgNOR and RBG/AgNOR techniques. In: *Proceedings of the 8th North American Colloquium on Domestic Animal Cytogenetics and Gene Mapping*. Guelph, Canada, p. 21.

Drinkwater, R.D., Johnson, S.E., Barendse, W., Harrison, B., Womack, J.E. and Hetzel, D.J. (1994) Isolation and mapping of a single-locus minisatellite sequence marker to

1q36-qter and synteny group U10 in cattle and to 1q36-qter in sheep. *Genomics* 19, 149–151.
Drouin, R., Lemieux, N. and Richer, C.-L. (1990) Analysis of DNA replication during S-phase by means of dynamic chromosome banding at high resolution. *Chromosoma* 99, 273–280.
Dunn, H.O. and Roberts, S.J. (1972) Chromosome studies of an ovine acephalic–acardiac monster. *Cornell Veterinarian* 62, 425–431.
Dutrillaux, B. and Viegas-Pequignot, E. (1981) High resolution R and G banding on the same preparation. *Human Genetics* 57, 93–95.
Dutrillaux, B., Finaz, C., de Grouchy, J. and Lejeune, J. (1972) Comparison of banding patterns of human chromosomes obtained with heating, fluorescence, and proteolytic digestion. *Cytogenetics* 11, 113–116.
Dutrillaux, B., Laurent, C., Couterier, J. and Lejeune, J. (1973) Coloration par l'acridine orange de chromosomes préalablement traités par le 5-bromo-deoxyuridine (BUDR). *Compte Rendu de l'Académie des Sciences (Paris)* 276, 3179–3181.
Echard, G. (1990) The gene map of the sheep (*Ovis aries*) $2N = 54$. In: O'Brien, S.J. (ed.) *Genetic Maps*. Cold Spring Harbor Laboratory Press, Cold Spring Harbor, New York State, pp. 4.118–4.120.
Echard, G., Broad, T.E., Hill, D. and Pearce, P. (1994) Present status of the ovine gene map (*Ovis aries*): comparison with the bovine map (*Bos taurus*). *Mammalian Genome* 5, 324–332.
Eggen, A. and Fries, R. (1995) An integrated cytogenetic and meiotic map of the bovine genome. *Animal Genetics* 26, 215–236.
Evans, H.J., Buckland, R.A. and Sumner, A.T. (1973) Chromosome homology and heterochromatin in goat, sheep and ox studied by banding techniques. *Chromosoma* 42, 383–402.
Ewens, W.J., Griffiths, R.C., Ethier, S.N., Wilcox, S.A. and Graves, J.A.M. (1992) Statistical analysis of *in situ* hybridization data: derivation and use of the z_{max} test. *Genomics* 12, 675–682.
Fayrer-Hosken, R.A., Huber, T.L., Miller, D.M., Thompson, F.N. and Caudal, A.B. (1992) Infertility in a ewe as the result of ovo-testis. *Journal of American Veterinary Medicine Association* 200, 1528–1529.
Feinberg, A.P. and Vogelstein, B. (1983) A technique for radiolabeling DNA restriction endonuclease fragments to a high specific activity. *Analytical Biochemistry* 132, 6–13.
Ford, C.E., Pollock, D.L. and Gustavsson, I. (eds) (1980) Proceedings of the First International Conference for the standardization of banded karyotypes of domestic animals, Reading, England (1976). *Hereditas* 92, 145–162.
Georges, M., Gunawardana, A., Threadgill, D.W., Lathrop, M., Olsaker, I., Mishra, A., Sargeant, L.L., Schoeberlein, A., Steele, M.R., Terry, C., Threadgill, D.S., Zhao, X., Holm, T., Fries, R. and Womack, J.E. (1991) Characterization of a set of variable number tandem repeat markers conserved in Bovidae. *Genomics* 11, 24–32.
Gerneke, W.H. (1967) Cytogenetic investigations on normal and malformed animals with special reference to intersexes. *Onderstepoort Journal of Veterinary Research* 34, 219–300.
Gill, J.J.B. and Davies, D.A.R. (1991) XX/XXY chromosomal chimerism in infertile sheep of the Cambridge breed. *Génétique, Sélection et Evolution* 23 (Suppl. 1), 126s–130s.
Glahn-Luft, B. and Wassmuth, R. (1977) C banding in sheep with balanced homologous inter chromosomal translocations. *Annales de Génétique et Sélection animale* 9, 540.

Glahn-Luft, B. and Wassmuth, R. (1978) G-banding in sheep with heterozygous and homozygous translocations. *Annales de Génétique et Sélection animale* 10, 594.

Glahn-Luft, B. and Wassmuth, R. (1980) The influence of the 1/20 translocation in sheep on the efficiency of reproduction. In: *Proceedings of the 31st Annual Meeting of the European Association for Animal Production, 1–4 Sept.* Abstract.

Goldammer, T., Brunner, R.M., Hiendleder, S. and Schwerin, M. (1995) Comparative mapping of sheep inhibin subunit β_b to chromosome 2 in sheep and cattle by fluorescence *in situ* hybridization. *Animal Genetics* 26, 199–200.

Graphodatsky, A.S., Biltueva, L.S., Filippov, V.A., Lushnikova, T.P., Shumny, T.V., Ermolaev, V.I. (1993) Localization of ESD and A2M genes to sheep chromosome 3 by *in situ* hybridization. *Cytogenetics and Cell Genetics* 62, 156–158.

Gunawardana, A. (1991) Physical mapping polymorphic DNA markers in the Bovidae family. (A comparative study in cattle, sheep and goat.) PhD dissertation, ETH No. 9556, Swiss Federal Institute of Technology, Zurich, 85 pp.

Halnan, C.R.E. (1989) *Cytogenetics of Animals.* CAB International, Wallingford, UK, 521 pp.

Harper, M.M. and Chan, L. (1986) Chromosomal fine mapping of apolipoprotein genes by *in situ* hybridization to metaphase chromosomes. *Methods in Enzymology* 128, 863–876.

Hawken, R., Broom, M., van Stijn, T., Lumsden, J., Maddox, J. and Broad, T. (1995) Clustering of the IL3, IL4, IL5, and CSF2 genes on ovine chromosome 5. In: *Proceedings of the 6th Australasian Gene Mapping Workshop and New Zealand Genetical Society, Dunedin.* Abstract 70.

Hayes, H. (1995) Chromosome painting with human chromosome-specific DNA libraries reveals the extent and distribution of conserved segments in bovine chromosomes. *Cytogenetics and Cell Genetics* 71, 168–174.

Hayes, H. and Petit, E. (1993) Mapping of the β-lactoglobulin gene and of an immunoglobulin M heavy chain-like sequence to homoeologous cattle, sheep and goat chromosomes. *Mammalian Genome* 4, 207–210.

Hayes, H., Petit, E. and Dutrillaux, B. (1991) Comparison of RBG-banded karyotypes of cattle, sheep and goats. *Cytogenetics and Cell Genetics* 57, 51–55.

Hayes, H., Petit, E., Lemieux, N. and Dutrillaux, B. (1992a) Chromosomal localization of ovine beta-casein gene by non isotopic *in situ* hybridization and R-banding. In: *Proceedings of the 10th European Colloquium on the Cytogenetics of Domestic Animals.* Utrecht, Netherlands, pp. 89–93.

Hayes, H., Petit, E., Lemieux, N. and Dutrillaux, B. (1992b) Chromosomal localization of ovine beta-casein gene by non isotopic *in situ* hybridization and R-banding. *Cytogenetics and Cell Genetics* 61, 286–288.

Hayes, H., Petit, E., Bouniol, C. and Popescu, P. (1993a) Localization of the alpha-S2-casein gene (CASAS2) to the homoeologous cattle, sheep and goat chromosomes 4 by *in situ* hybridization. *Cytogenetics and Cell Genetics* 64, 281–285.

Hayes, H., Petit, E., Bouniol, C. and Popescu, P. (1993b) Conformation of the localization of the alpha-S2-casein gene (CSN1S2) to chromosome 4 in cattle, sheep, goat and a cow with a t(4;8) translocation by *in situ* hybridization. In: *Proceedings of the 8th North American Colloquium on Domestic Animal Cytogenetics and Gene Mapping.* Guelph, Canada, p. 167.

Hayes, H.C.; Popescu, P. and Dutrillaux, B. (1993c) Comparative gene mapping of lactoperoxidase, retinoblastoma, and α-lactalbumin genes in cattle, sheep and goats. *Mammalian Genome* 4, 593–597.

Hayes, H., Le Chalony, C., Goubin, G., Mercier, D., Payen, E., Bignon, C. and Kohno,

K. (1995) Localization of ZNF164, ZNF146, GGTA1, SOX2, PRLR and EEF2 on homoeologous cattle, sheep and goat chromosomes by fluorescent *in situ* hybridization and comparison with the human gene map. *Cytogenetics and Cell Genetics* 72, 342–346.

Hecht, F. and Hecht, B.K. (1984a) Fragile sites and chromosome breakpoints in constitutional rearrangements. I. Amniocentesis. *Clinical Genetics* 26, 169–173.

Hecht, F. and Hecht, B.K. (1984b) Fragile sites and chromosome breakpoints in constitutional rearrangements. II. Spontaneous abortions, stillbirths, and newborns. *Clinical Genetics* 26, 174–177.

Hediger, R., Johnson, S., Barendse, W., Drinkwater, R., Moore, S. and Hetzel, J. (1990a) Assignment of the growth hormone gene locus to 19q26-qter in cattle and 11q25-qter in sheep by *in situ* hybridization. *Genomics* 8, 171–174.

Hediger, R., Johnson, S. and Hetzel, D.J.S. (1990b) Localization of the beta-subunit of follicle stimulating hormone in cattle and sheep by *in situ* hybridization. *Animal Genetics* 22 (Suppl. 1), 81.

Hediger, R., Ansari, H.A. and Stranzinger, G.F. (1991a) Chromosome banding and gene localizations support extensive conservation of chromosome structure between sheep and cattle. *Cytogenetics and Cell Genetics* 57, 127–134.

Hediger, R., Johnson, S. and Hetzel, D.J.S. (1991b) Localization of the beta-subunit of follicle stimulating hormone in cattle and sheep by *in situ* hybridization. *Animal Genetics* 22, 237–244.

Henderson, A.S. (1982) Cytological hybridization to mammalian chromosomes. *International Review of Cytology* 76, 1–46.

Henderson, L.M. (1979) A silver-staining study of the nucleolus organizer regions of the chromosomes of domestic sheep (*Ovis aries*). PhD thesis, Massey University, Palmerston North, New Zealand, 193 pp.

Henderson, L.M. and Bruere, A.N. (1977) Association of nucleolus organizer chromosomes in domestic sheep (*Ovis aries*) shown by silver staining. *Cytogenetics and Cell Genetics* 19, 326–334.

Henderson, L.M. and Bruere, A.N. (1979) Conservation of nucleolus organizer regions during evolution in sheep, goat, cattle and aoudad. *Canadian Journal of Genetics and Cytology* 21, 1–8.

Henderson, L.M. and Bruere, A.N. (1980) Inheritance of Ag-stainability of the nucleolus organizer regions in domestic sheep *Ovis aries*. *Cytogenetics and Cell Genetics* 26, 1–6.

Holmquist, G.P. (1992) Review article: chromosome bands, their chromatin flavors, and their functional features. *American Journal of Human Genetics* 51, 17–37.

Holmquist, G.P., Gray, M., Porter, T. and Jordan, J. (1982) Characterization of Giemsa dark- and light-band DNA. *Cell* 31, 121–129.

Iannuzzi, L., Gallagher, D.S., Di Meo, G.P., Ryan, A.M., Ferrara, L. and Womack, J.E. (1993) Chromosomal localization of omega and trophoblast interferon genes in goat and sheep by fluorescent *in situ* hybridization. *Journal of Heredity* 84, 301–304.

Iannuzzi, L., Di Meo, G.P., Perucatti, A. and Ferrara, L. (1995) G- and R-banding comparison of sheeep (*Ovis aries* L) chromosomes. *Cytogenetics and Cell Genetics* 68, 85–90.

Imam-Ghali, M., Cals, M., Saidi-Mehtar, N., Guerin, G. and Martin, P. (1991a) Sheep gene mapping by somatic cell hybridization: two additional markers (COL2A1 and LPO).*Cytogenetics and Cell Genetics* 58, 2127.

Imam-Ghali, M., Saidi-Mehtar, N. and Guerin, G. (1991b) Sheep gene mapping:

additional DNA markers included (CASB, CASK, LALBA, IGF-1 and AMH). *Animal Genetics* 22, 165–172.

Imam-Ghali, M., Saidi-Mehtar, N. and Guerin, G. (1992) Sheep gene mapping: synteny between COL2A1 and the LDHB–PEPB–TPI–GAPD–LALBA–IGF1 group. *Animal Genetics* 23, 529–532.

Imam-Ghali, M., Saidi-Mehtar, N. and Hors-Cayla, M.-C. (1987) Sheep gene mapping by somatic cell hybridization: synteny of PEPC, ENO1 and PGD. *Cytogenetics and Cell Genetics* 46, 632–633.

Jauch, A., Wienberg, J., Stanyon, R., Arnold, N., Tofanelli, S., Ishida, I. and Cremer, T. (1992) Reconstruction of genomic rearrangements in great apes and gibbons by chromosome painting. *Proceedings of the National Academy of Sciences of the USA* 89, 8611–8615.

John, H.A., Birnstiel, M.L. and Jones, K.W. (1969) RNA-DNA hybrids at the cytological level. *Nature* 223, 234–238.

Johnson, S.E., Barendse, W. and Hetzel D.J.S. (1993) The γ-fibrinogen gene maps to chromosome 17 in both cattle and sheep. *Cytogenetics and Cell Genetics* 62, 176–180.

Johnson, S.E., Moore, S.S., Mackinnon, R., Hetzel, D.J.S. and Barendse, W. (1995) The cosmid CSSM25 assigns syntenic group U2 to bovine chromosome-9 and is localized to ovine chromosome-8. *Mammalian Genome* 6, 529–531.

Jones, C., Morse, H.G., Geyer, D. and Broad, T.E. (1985) Gene mapping in the sheep: assignment of LDHB, SHMT and PEPB to chromosome M3. *Cytogenetics and Cell Genetics* 40, 662.

Jorgensen, C.B., Gallagher, D.S., Schwerin, M. and Thomsen, P.D. (1992) Chromosomal distribution and use of some bovine male-specific repeats. *Animal Genetics* 23 (Suppl. 1), 79.

Kaftanoskaya, H.M. and Senov, O.L. (1994) High resolution GTG-banded chromosomes of cattle, sheep and goat: a comparative study. *Journal of Heredity* 85, 395–400.

Kilgour, R. and Bruere, A.N. (1970) Behaviour patterns in chromatin-positive Klinefelter's syndrome in sheep. *Nature* 225, 71–72.

Korenberg, J.E. and Rykowski, J.R. (1988) Human genome organization: Alu, LINEs and the molecular structure of metaphase chromosome bands. *Cell* 53, 391–400.

Kurnit, D.M., Brown, F.L. and Maio, J.J. (1978) Mammalian repetitive DNA sequences in a stable Robertsonian system. *Cytogenetics and Cell Genetics* 21, 145–167.

Latt, S.A. (1973) Microfluorometric detection of deoxyribonucleic acid replication in human metaphase chromosomes. *Proceedings of the National Academy of Sciences of the USA* 70, 3395–3399.

Le Beau, M.M. (1986) Chromosomal fragile sites and cancer specific rearrangements. *Blood* 67, 849–858.

Leitch, A.R., Schwarzacher, T., Jackson, D. and Leitch, I.J. (1994) *In Situ Hybridization: a Practical Guide*, 1st edn. BIOS Scientific Publishers, Oxford, 118pp.

Lemieux, N., Dutrillaux, B. and Viegas-Pequignot, E. (1992) A simple method for simultaneous R- or G-banding and fluorescence *in situ* hybridization of small single copy genes. *Cytogenetics and Cell Genetics* 59, 311–312.

Lewalski, H., Otto, F.J., Kranert, T. and Wassmuth, R. (1993) Flow cytometric detection of unbalanced ram spermatozoa from heterozygous 1;20 translocation carriers. *Cytogenetics and Cell Genetics* 64, 286–291.

Lewalski, H., Meinecke-Tillmann, S. and Meinecke, B. (1994) Preimplantation development of ovine embryos sired by heterozygous carriers of the 1;20 translocation. In:

Christensen, K. (ed.) *Proceedings of the 11th European Colloquium of Cytogenetics of Domestic Animals*. Copenhagen, pp. 87–91.

Lichter, P., Cremer, T., Borden, J., Manuelidis, L. and Ward, D.C. (1988a) Delineation of individual human chromosomes in metaphase and interphase cells by *in situ* hybridization using recombinant DNA libraries. *Human Genetics* 80, 224–234.

Lichter, P., Cremer, T., Tang, C.-J.C., Watkins, P.C., Maneulidis, L. and Ward, D.C. (1988b) Rapid detection of human chromosome 21 aberrations by *in situ* hybridization. *Proceedings of the National Academy of Sciences of the USA* 85, 9664–9668.

Lichter, P., Boyle, A.L., Cremer, T. and Ward, D.C. (1991) Analysis of genes and chromosomes by nonisotopic *in situ* hybridization. *Genetic Analysis and Technical Applications* 8, 24–35.

Lo, Y.-M.D., Mehal, W.Z. and Fleming, K.A. (1988) Rapid production of vector free biotinylated probes using the polymerase chain reaction. *Nucleic Acids Research* 16, 8719.

Long, S.E. (1977) Cytogenetic examination of pre-implantation blastocysts of ewes mated to rams heterozygous for the Massey I (t_1) translocation. *Cytogenetics and Cell Genetics* 18, 82–89.

Long, S.E. (1978a) Reproductive performance of ewes mated to rams heterozygous for the Massey I (t_1) centric fusion (Robertsonian) translocation. *Veterinary Record* 102, 399–401.

Long, S.E. (1978b) Chiasma counts and non-disjunction frequencies in a normal ram and in rams carrying the Massey I (t_1) Robertsonian translocation. *Journal of Reproduction and Fertility* 53, 353–356.

Long, S.E. (1980) Some pathological conditions of the reproductive tract of the ewe. *Veterinary Record* 106, 175–176.

Long, S.E. (1985a) Cytogenetic examination of some rare breeds of sheep in Great Britain. *Journal of Dairy Science* 68 (Suppl. 1), 250.

Long, S.E. (1985b) Standard nomenclature for the G-band karyotype of the domestic sheep (*Ovis aries*). *Hereditas* 103, 165–170.

Long, S.E. (1990) Chromosomes of sheep and goats. In: McFeely, R.A. (ed.) *Domestic Animal Cytogenetics*. Advances in Veterinary Science and Comparative Medicine 34, Academic Press, New York, pp. 109–129.

Long, S.E. (1995) Cytogenetics. In: Meredith, M.J. (ed.) *Animal Breeding and Infertility*. Blackwell Science, Oxford, pp. 39–61.

Lord, E.A., Montgomery, G.W., Penty, J.M., Dodds, K.G., Ansari, H.A., Henry, H.M., Pierson, C.A. and Crawford, A.M. (1994) A genetic linkage map for sheep chromosome 6. In: *Proceedings of the 5th Australasian Gene Mapping Workshop*. Armidale.

Lord, E.A., Lumsden, J.M., Dodds, K.G., Henry, H.M., Crawford, A.M., Ansari, H.A., Pearce, P.D., Maher, D.W., Stone, R.T., Kappes, S.M., Beattie, C.W. and Montgomery, G.W. (1996) The linkage map of sheep chromosome 6 compared with orthologous regions in other species. *Mammalian Genome* 7, 373–376.

Lumsden, J., van Stijn, T., Lord, E., Broom, M. and Montgomery, G. (1995) A YAC contig of the ADH2 region on sheep chromosome 6. In: *Proceedings of the 6th Australasian Gene Mapping Workshop and New Zealand Genetical Society*. Dunedin, New Zealand, abstract 73.

Lyon, M.F. and Hawkes, S.G. (1970) X-linked gene for testicular feminization in the mouse. *Nature* 227, 1217–1219.

Mahdy, E.A., Makinen, A., Chowdhary, B.P., Andersson, L. and Gustavsson, I. (1989) Chromosomal localization of the ovine major histocompatibility complex (OLA) by *in situ* hybridization. *Hereditas* 111, 87–90.

Marcum, J.B. (1974) The freemartin syndrome. *Animal Breeding Abstracts* 42, 227–242.

Matjeka, M. and Cribiu, E.P. (1987) Idiogramme et representation schematique des bandes G des chromosomes du mouton domestique (*Ovis aries* L). *Génétique, Sélection et Evolution* 19, 113–126.

Matjeka, M.and Cribiu, E.P. (1989) Observation d'un nouveau variant chromosomique chez le mouton domestique (*Ovis aries* L.). *Génétique, Sélection et Evolution* 21, 11–15.

Matjeka, M., Cribiu, E.P., Chaffaux, S., Schmitt, O. and Ricordeau, G. (1985) Deux cas de freemartinisme chez des moutons Romanov (*Ovis aries*). *Génétique, Sélection et Evolution* 17, 419–434.

Matjeka, M., Cribiu, E.P. and Popescu, C.P. (1988) Diagrammatic representation of RBA-banded karyotype of domestic sheep (*Ovis aries*). *Journal of Heredity* 79, 71–74.

Matjeka, M., Cribiu, E.P., Ricordeau, G., Eychenne, F., Durand, V., Lajous, D. and Darre, R. (1990) Evidence for a heritable fragile site in Romanov sheep chromosome 8. *Reproduction in Domestic Animals* 25, 220–226.

Matthews, M.E. and Reed, K.C. (1991) A DNA sequence that is present in both sexes of artiodactyla is repeated on the Y chromosome of sheep, cattle and goats. *Cytogenetics and Cell Genetics* 56, 40–44.

Mayr, B. and Czaker, R. (1981) Variable positions of nucleolus organizer regions in Bovidae. *Experientia* 37, 564–565.

Mayr, B., Schweizer, D., Mendelak, M., Krutzler, J., Schleger, W., Kalat, M. and Auer, H. (1985) Levels of conservation and variation of heterochromatin and nucleolus organizers in Bovidae. *Canadian Journal of Genetics and Cytology* 27, 665–682.

Menscher, S.H., Bunch, T.D. and Maciulis, A. (1989) High resolution G-banded karyotype and idiogram of the goat: a sheep-goat G-banded comparison. *Journal of Heredity* 80, 150–155.

Millot, P., Saidi-Mehtar, N. and Hors-Cayla, M.-C. (1981) Complexe d'histocompatibilité ovin (OLA) et marques enzymatiques dans des hybrides hamster × mouton. I. Asynténie entre PGM3-ME1 et OLA. *Annales de Génétique* 24, 82–88.

Monteagudo, L.V., Pitel, F., Tejedor, M.T., Yerle, M., Arruga, M.V. and Gellin, J. (1992a) *In situ* hybridization of cosmids in sheep. In: *Proceedings of the 10th European Colloquium on the Cytogenetics of Domestic Animals.* Utrecht, Netherlands, pp. 151–155.

Montgomery, G.W., Lord, E.A., Penty, J.M., Dodds, K.G., Broad, T.E., Cambridge, L., Sunden, S.L.F., Stone, R.T. and Crawford, A.M. (1994) The Booroola fecundity (FecB) gene maps to sheep chromosome 6. *Genomics* 22, 148–153.

Moorhead, P.S., Nowell, P.C., Mellman, W.J. and Battips, D.M. (1960) Chromosome preparations of leucocytes cultured from human peripheral blood. *Experimental and Cell Research* 20, 613–616.

Moraes, J.C.F, Mattevi, M.S. and Ferreira, J.M.M. (1980) Chromosome studies in Brazilian rams. *Veterinary Record* 107, 489–490.

Moreno-Millan, F. and Rodero-Franganillo, A. (1990) Nucleolar organizer regions, types of association and identification of carrier chromosome in domestic sheep. *Génétique, Sélection et Evolution* 22, 273–277.

Murthy, S.R.S., Jayashankar, M.R., Nagaraja, C.S., Govindaiah, M.G. and Sridhara-Murthy, S.R. (1995) Association of cryptorchidism with autosomal translocation in Deccani sheep. *Indian Journal of Small Ruminants* 1, 47–48.

Novak, U. (1984) Structure and properties of a highly repetitive DNA sequence in sheep. *Nucleic Acids Research* 12, 2343–2350.

O'Brien, S.J., Womack, J.E., Lyons, L.A., Moore, K.J., Jenkins, N.A. and Copeland, N.G. (1993) Anchored reference loci for comparative genome mapping in mammals. *Nature Genetics* 3, 103–112.

Oulmouden, A., Vaiman, D., Oustry, A., Cribiu, E.P. and Julien, R. (1995) Localization of fucosyl-transferase gene to bovine and goat chromosome 7 and sheep chromosome 5. *Mammalian Genome* 6, 760–761.

Pardue, M.-L. (1985) *In situ* hybridization. In: Hames, B.D. and Higgins, S.J. (eds) *Nucleic Acid Hybridization – a Practical Approach*. IRL Press, Oxford, pp. 179–202.

Pardue, M.-L. and Gall, J.C. (1969) Molecular hybridization of radiocative DNA to the DNA of cytological preparations. *Proceedings of the National Academy of Sciences of the USA* 64, 600–604.

Pascual, I. and Arruga, M.V. (1995) Identification of a gap in sheep. In: *Proceedings of the 9th North American Colloquium on Domestic Animal Cytogenetics and Gene Mapping*. College Station, Texas, p. 49.

Payen, E., Saidi-Mehtar, N., Pailhoux E. and Cotinot, C. (1995) Sheep gene mapping: assignment of ALDOB, CYP19, WT and SOX2 by somatic cell hybrid analysis. *Animal Genetics* 26, 331–339.

Pearce, P.D., Ansari, H.A., Maher, D.W., Malcolm, A.A., Wood, N.J., Phua, S.H. and Broad, T.E. (1993) Assignment of the β subunit of the muscle nicotinic acetylcholine receptor to sheep chromosome 11. *Cytogenetics and Cell Genetics* 64, 59–61.

Pearce, P.D., Ansari, H.A., Maher, D.W., Malcolm, A.A., Stewart-Scott, I.A. and Broad, T.E. (1994a) New Robertsonian translocation chromosomes in domestic sheep (*Ovis aries*), *Cytogenetics and Cell Genetics* 67, 137–140.

Pearce, P.D., Ansari, H.A., Maher, D.W., Broad, T.E., Cambridge, L.M., Lewis, P.E., Burkin, D.J. and Jones, C. (1994b) The assignment of fourteen new loci to ovine chromosomes. In: *Proceedings of the 5th Australasian Gene Mapping Workshop*. Armidale.

Pearce, P.D., Ansari, H.A., Maher, D.W. and Broad, T.E. (1995) Five regional localizations to the sheep genome: first assignments to chromosomes 5 and 12. *Animal Genetics* 26, 171–176.

Perry, P. and Wolff, S. (1974) New Giemsa method for differential staining of sister chromatids. *Nature* 261, 156–158.

Phua, S., Wood, N., Ansari, H., Broad, T., Crellen, L., Davidson, B., Lewis, P., McNabb, L., Maher, D., Malcolm, A. and Pearce, P. (1991) Physical mapping of sheep ribosomal RNA loci by *in situ* chromosomal hybridization. In: *Proceedings of the Australasian Gene Mapping and Molecular Genetics Symposium*. Dunedin, New Zealand, pp. 55.

Pinkel, D., Landegent, J., Collins, C., Fuscoe, J., Segraves, R., Lucas, J. and Gray, J. (1988) Fluorescence *in situ* hybridization with human chromosome-specific libraries: detection of trisomy 21 and translocations of chromosome 4. *Proceedings of the National Academy of Sciences of the USA* 85, 9138–9142.

Pitel, F., Lantier, I., Tabet-Aoul, K., Saidi-Mehtar, N., Elsen, J.-M., Lantier, F. and Gellin, J. (1994a) TNP1 gene: a sheep polymorphic marker for a conserved region among mammals. *Mammalian Genome* 5, 390–392.

Pitel, F., Lantier, I., Riquet, J., Lanneluc, I., Tabet-Aoul, K., Saidi-Mehtar, N., Lantier, F. and Gellin, J. (1994b) Cloning, sequencing, and localization of an ovine fragment of the *NRAMP* gene, a candidate for the ITY/LSH/BCG gene. *Mammalian Genome* 5, 834–835.

Pitel, F., Cribiu, E.P., Yerle, M., Lahbib-Mansais. Y., Lanneluc, I., Lantier, F. and Gellin, J. (1995) Regional localization of the ovine NRAMP gene to chromosome 2q41–q42

by *in situ* hybridization. *Cytogenetics and Cell Genetics* 70, 116–118.
Popescu, C.P. (1989) *Cytogénétique des Mammifères d'élévage*. INRA, Paris, 119 pp.
Popescu, C.P., Boscher, J., Hayes, H.C., Ban, J. and Kettmann, R. (1995) Chromosomal localization of the BLV receptor candidate gene in cattle, sheep, and goats. *Cytogenetics and Cell Genetics* 69, 50–52.
Popescu, C.P., Long, S., Riggs, P., Womack, J., Schmutz, S., Fries, R. and D.S. Gallagher (1996) Standardization of cattle karyotype nomenclature: report of the committee for the standardization of the cattle karyotype. *Cytogenetics and Cell Genetics* 74, 259–261.
Rajcan-Separovic, E. and Sabour, M.P. (1993) Fluorescence *in situ* hybridization of bovine *Alu*-like sequences to bovine and ovine chromosomes. *Genome* 36, 984–986.
Reisner, A.H. and Bucholtz, C.A (1983) Apparent relatedness of the main component of ovine 1.714 satellite DNA to bovine 1.715 satellite DNA. *EMBO Journal* 2, 1145–1149.
Rettenberger, G., Lett, C., Zechner, U., Kunz, J., Vogel, W. and Hameister, H. (1995) Visualization of the conservation of synteny between humans and pigs by heterologous chromosomal painting. *Genomics* 26, 372–378.
Richard, F., Vogt, M., Muleris, M., Malfoy, B. and Dutrillaux, B. (1994) Increased FISH efficiency using APC probes generated by direct incorporation of labeled nucleotides by PCR. *Cytogenetics and Cell Genetics* 65, 169–171.
Rigby, P.W.J., Diechman, M., Rhodes, C. and Berg, P. (1977) Labeling deoxyribonucleic acid to high specific activity *in vitro* by nick translation with DNA polymerase I. *Journal of Molecular Biology* 113, 237–251.
Ronne, M. (1995) Localization of fragile sites in the karyotype of *Sus scrofa domestica*: present status. *Hereditas* 122, 153–162.
Rynkiewicz-Szatkowska, I. (1992) Reproductive performance of XX/XY chimeric rams. In: *Proceedings of the 10th European Colloquium on Cytogenetics of Domestic Animals*. Utrecht, Netherlands, pp. 213–217.
Saidi-Mehtar, N. and Hors-Cayla, M.-C. (1981) Sheep gene mapping by somatic cell hybridization. III. Synteny between pyruvate kinase M2 (PKM2) and nucleoside phosphorylase (NP) in domestic sheep. *Annales de Génétique* 24, 148–151.
Saidi-Mehtar, N., Hors-Cayla, M.C., Van Cong, N. and Benne, F. (1979) Sheep gene mapping by somatic cell hybridization. *Cytogenetics and Cell Genetics* 25, 200.
Saidi-Mehtar, N., Hors-Cayla, M.-C. and Van Cong, N. (1981) Sheep gene mapping by somatic cell hybridization: four syntenic groups: ENO1–PGD, ME1–PGM3, LDHB–PEPB–TPI and G6PD–PGK–GALA. *Cytogenetics and Cell Genetics* 30, 193–204.
Saidi-Mehtar, N., Goossens, M., Hors-Cayla, M.-C. and Rosa, J. (1987) Sheep gene mapping by molecular hybridization of ovine gamma-globulin and enzyme markers. *Cytogenetics and Cell Genetics* 46, 686.
Saidi-Mehtar, N., Imam-Ghali, M., Heuertz, S. and Hors-Cayla, M.-C. (1989) Sheep gene mapping by somatic cell hybridization. *Cytogenetics and Cell Genetics* 51, 1072.
Saidi-Mehtar, N., Imam-Ghali, M., Heuertz, S. and Hors-Cayla, M.-C. (1991) Sheep gene mapping by somatic cell hybridization. *Génétique, Sélection et Evolution* 23 (Suppl. 1), 196s–200s.
Saitoh, Y. and Laemmli, U.K. (1994) Metaphase chromosome structure: bands arise from a differential folding path of the highly AT-rich scaffold. *Cell* 76, 609–622.
Scherthan, H., Cremer, T., Arnason, U., Weier, H.-U., Lima de Faria, A. and Fronicke,

L. (1994) Comparative chromosome painting discloses homologous segments in distantly related mammals. *Nature Genetics* 6, 342–347.

Schnedl, W. and Czaker, R. (1974) Centromeric heterochromatin and comparison of G-banding in cattle, goat and sheep chromosomes (Bovidae). *Cytogenetics and Cell Genetics* 13, 246–255.

Shaw, J., Watson, J.M., Wilcox, S. and Graves, J.A.M. (1993) Location of genes of sheep and cattle sex chromosomes. In: *Proceedings of the 4th Australasian Gene Mapping Workshop*. Adelaide, Australia.

Sise, J.A., Hillyard, A.L. and Montgomery, G.W. (1996) The sheep gene map database (SheepBase) is now available on the World Wide Web. *Mammalian Genome* 7, 1.

Slota, F., Danielak, B. and Kozubska, A. (1986) Structural rearrangement of metacentric chromosomes in the ram of the Polish Wrzosowka breed. In: *Abstracts of the 7th European Colloquium on Cytogenetics of Domestic Animals, Warsaw, Poland,* p. 37.

Smith, K.C. (1996) Reproductive abnormalities in ewes. *PhD thesis, University of Bristol,* 221 pp.

Smith, K.C., Long, S.E. and Parkinson, T.J. (1996) A case of true hermaphroditism in the sheep. *Veterinary Record* 138, 497–498.

Solinas-Toldo, S., Lengauer, C. and Fries, R. (1995) Comparative genome map of human and cattle. *Genomics* 27, 489–496.

Sumner, A.T. (1972) A simple technique for demonstrating centromeric heterochromatin. *Experimental and Cell Research* 75, 304–306.

Sutherland, G.R. (1979) Heritable fragile sites on human chromosomes. I. Factors affecting expression in lymphocyte culture. *American Journal of Human Genetics* 31, 125–135.

Sutherland, G.R. (1991) Chromosomal fragile sites. *Genetic Analysis: Techniques and Application* (*GATA*) 8, 161–166.

Sutherland, G.R. and Hecht, F. (1985) *Fragile Sites on Human Chromosomes*. Oxford University Press, New York, 271 pp.

Sutherland, G.R. and Richards, R.I. (1995) The molecular basis of fragile sites in human chromosomes. *Current Opinion in Genetics and Development* 5, 323–327.

Szatkowska, I. (1995) Frequency of lymphocytic XX/XY chimerism in Leine sheep coming from heterosexual twin and multiple births. *Journal of Applied Genetics* 36, 373–378.

Szatkowska, I., Udala, J. and Chomiczewska-Mazaraki, A. (1993) Leukocyte chimaerism in an infertile Polish Merino ewe. *Medycyna Waterynaryjna* 49, 565–567.

Tabet-Aoul, K., Pitel, F., Lantier, I., Gros, P., Saidi-Mehtar, N. and Lantier, F. (1992) Mapping of the ovine genome by somatic cell hybrid cell lines: the IDH1, FN1, VIL1 and CHRNG syntenic group is conserved in sheep. *Animal Genetics* 23 (Suppl. 1), 96.

Tebbutt, S., Broom, M., van Stijn, T., Montgomery, G. and Hill, D. (1995) The ovine cystic fibrosis gene locus. In: *Proceedings of the 6th Australasian Gene Mapping Workshop and New Zealand Genetical Society, Dunedin,* abstract 28.

Trask, B.J. (1991) Fluorescence *in situ* hybridization: applications in cytogenetics and gene mapping. *Trends in Genetics* 7, 149–154.

Tucker, E.M., Dain, A.R., Wright, L.J. and Clarke, S.W. (1981) Culture of sheep × mouse hybridoma cells *in vitro*. *Hybridoma* 1, 77–86.

Vaiman, D., Pailhoux, E., Payen, E., Saidi-Mehtar, N. and Cotinot, C. (1995) Evolutionary conservation of a microsatellite in the Wilms Tumour (WT) gene: mapping in sheep and cattle. *Cytogenetics and Cell Genetics* 70, 112–115.

Van Styn, T., Lumsden, J., Montgomery, G. and Broom, M. (1995) Fluorescence *in situ* hybridization with YAC clones by physical mapping in sheep. In: *Proceedings of the 6th Australasian Gene Mapping Workshop and New Zealand Genetical Society Conference*. University of Otago, Dunedin, abstract 82.

Viegas-Pequignot, E. and Dutrillaux, B. (1978) Une méthode simple pour obtenir des prophases et des prométaphases. *Annales de Génétique* 21, 122–125.

Viegas-Pequignot, E., Dutrillaux, B, Magdelenat, H. and Coppey-Molsan, M. (1989) Mapping of single-copy DNA sequences on human chromosomes by *in situ* hybridization with biotinylated probes: enhancement of detection sensitivity by intensified-fluorescence digital–imaging microscopy. *Proceedings of the National Academy of Sciences of the USA* 86, 582–586.

Vijh, R.K., Sinha, Y.K.P., Tantia, M.S. and Sahai, R. (1994). Cytogenetics of sheep with special reference to chromosomal abnormalities associated with congenital malformations and reproductive disorders – a review. *Agricultural Reviews, Karnal* 15, 43–56.

Weisblum, B. and de Haseth, P.L. (1972) Quinacrine, a chromosome stain specific for deoxyadenylate-deoxythymidylate rich regions in DNA. *Proceedings of the National Academy of Sciences of the USA* 69, 629–632.

Wienberg, J., Stanyon, R., Jauch, A. and Cremer, T. (1992) Homologies in human and *Macaca fuscata* chromosomes revealed by *in situ* suppression hybridization with human chromosome-specific DNA libraries. *Chromosoma* 101, 265–270.

Wood, N.J., Phua, S.H. and Crawford, A.M. (1992) A dinucleotide repeat polymorphism at the glycine-and tyrosine-rich keratin locus in sheep. *Animal Genetics* 23, 391.

Wood, N.J., Ansari, H.A., Broad, T.E., Burkin, D.J., Jones, C.A., Lewis, P.E., Maher, D.W., Pearce, P.D. and Phua, S.H. (1993) Regional assignment of the neurotensin locus in sheep. *Mammalian Genome* 4, 541–543.

Woodall, C.J., McLaren, L.J. and Watt, N.J. (1994) Sequence and chromosomal localisation of the gene encoding ovine latent transforming growth factor-β1. *Gene* 150, 371–373.

Xeros, N. (1962). Deoxyriboside control and synchronization of mitosis. *Nature* 194, 682.

Yunis, J.J. (1976) High resolution of human chromosomes. *Science* 191, 1268–1270.

Zartman, D.L., Hinesley, L.L. and Gnatkowski, M.W. (1981) A 53, X female sheep (*Ovis aries*). *Cytogenetics and Cell Genetics* 30, 54–58.

Zhang, T.Q, Buoen, L.C., Rossow, K.D., Weber, A.F. and Ruth, G.R. (1992) Mosaic autosomal monosomy in a lamb with congenital defects. In: *Proceedings of the 10th European Colloquium on Cytogenetics of Domestic Animals*, Utrecht, Netherlands, pp. 285–287.

The Sheep Linkage Map

11

G.W. MONTGOMERY and A.M. CRAWFORD
AgResearch Molecular Biology Unit, Department of Biochemistry and Centre for Gene Research, University of Otago, PO Box 56, Dunedin, New Zealand

Introduction 297
Gene Maps and Markers 298
The Linkage Map 299
SheepBase 328
Applications of the Linkage Map 337
Single-gene searches 337
Polygenic traits 338
Studies in natural populations 340
Comparative Aspects and Genome Evolution 340
Future Prospects 341
References 342

Introduction

Most characters of economic importance to sheep breeders show continuous variation. Consequently, breeders searching for improved production have to embark on expensive and time consuming breeding programmes with selection decisions supported by quantitative genetic analysis. The advances in quantitative genetics that have improved rates of genetic gain in breeding programmes are reviewed in other chapters.

In the last decade, powerful new methods for studying genes and their inheritance have been applied in sheep. This exciting research on genome organization and function will radically change our understanding of sheep genetics. This in turn will have an impact on breeding programmes and sheep husbandry practices.

Linkage mapping is one area of rapid progress, and the publication of the first autosomal linkage map for the sheep in 1995 (Crawford *et al.*, 1995) represents an important milestone. The objectives of this chapter are to discuss the development of the sheep linkage map, list sets of single locus markers available, provide the current map and review applications of the map.

Gene Maps and Markers

Genetic linkage was first observed early this century by Beateson and Punnett as exceptions to Mendel's law of independent assortment (see Peters, 1959). When genes or genetic markers are near each other on the same chromosome, they are not inherited independently in families but cosegregate at meiosis. This phenomenon is called genetic linkage and, with a high density of markers spread over each chromosome, the markers can be assembled into a linkage map of the genome (Botstein *et al.*, 1980). Cosegregating markers are placed into linkage groups, with the proportion of recombinants (crossover events) detected between linked markers being used as a measure of the distance between them. The requirements for linkage mapping are, therefore, large numbers of informative genetic markers and large families in which to study the inheritance of each marker. The development of deoxyribonucleic acid (DNA)-based methods provided markers with sufficient density to assemble a linkage map for sheep.

The first method for the rapid detection of DNA polymorphisms – restriction fragment length polymorphism (RFLPs) – was published in 1980 (Botstein *et al.*, 1980). Since that time, a number of other methods have been developed and used widely in sheep (Table 11.1). These methods include DNA fingerprinting (Crawford *et al.*, 1993), microsatellites (Crawford *et al.*, 1990), random amplified polymorphic DNA (RAPD) (Cushwa *et al.*, 1996) and single-stranded conformational polymorphisms (SSCPs) (Penty *et al.*, 1995). DNA markers used for the construction of genetic maps generally fall into two classes (O'Brien *et al.*, 1993). The first class (type I) are markers associated with gene sequences that are conserved across mammalian species. The second class (type II) are highly polymorphic but usually anonymous DNA segments. The most common markers used in map construction in mammals are the microsatellites. These markers use the polymerase chain reaction to detect variation in simple dinucleotide repeats and have a number of advantages over other classes of marker (Weber and May, 1989). Large numbers of markers can be cloned, they are highly variable (have many alleles), they require only small amounts of DNA for analysis and methods can be easily automated. The disadvantage of microsatellites is that most are anonymous DNA fragments (type II markers) and work only in closely related species.

Methods for the construction of linkage maps must take account of mating systems and the different features of DNA marker systems available. Interspecies hybrids provide efficient mapping pedigrees that make use of fixed differences in DNA sequence between species. It is possible to construct high density maps of both conserved type I loci and the highly polymorphic type II loci in these crosses. Interspecies hybrids are being used successfully in the construction of maps for mice (Copeland *et al.*, 1994) and deer (Tate *et al.*, 1995).

The more common approach uses conventional pedigrees and many highly polymorphic DNA markers. We have used this approach for construction of the linkage map for sheep, but have combined this with mapping conserved genes. The large number of alleles available for microsatellite markers means sufficient individuals are informative in the pedigrees to construct high resolution maps. Conserved genes can also be accurately located on the map by linkage to the microsatellite markers.

There were two principal sources for the microsatellite markers used in the framework map. Markers were cloned from sheep genomic libraries or screened from the set of microsatellites used for construction of the cattle linkage map. The set of microsatellite markers used for the construction of the sheep linkage map, together with the primer sequences and source, are listed in Table 11.1. Most of the microsatellite markers used for the framework map were type II markers.

Restriction frament length polymorphisms detected using complementary DNA (cDNA) probes for conserved genes (Lord *et al.*, 1996) provide one source of type I markers to anchor maps of divergent species (O'Brien *et al.*, 1993). Approximately 40% of cDNA probes from cattle and humans identified RFLP markers in sheep under a standard set of conditions (Montgomery *et al.*, 1992). Restriction frament length polymorphism (RFLP) markers currently available for sheep, together with their chromosomal location (if known) and the sources, are listed in Table 11.2. Microsatellites associated with gene sequences (Moore *et al.*, 1992) and SSCPs (Penty *et al.*, 1995) also provide a valuable source of type I markers for map construction and these are listed in Table 11.3. Table 11.4 lists the type II markers used for this map.

The Linkage Map

Prior to 1994, the genetic linkage map for sheep was very rudimentary, with only 17 markers assigned to seven linkage groups (Rasmusen, 1996; Tucker *et al.*, 1980; Leveziel *et al.*, 1991; Penty and Montgomery, 1991a; Crawford *et al.*, 1992). Following a search for the Booroola fecundity (*FecB*) gene (Montgomery

Table 11.1. Characteristics of DNA marker types currently used in the analysis of the sheep genome.

Marker name	Marker type	DNA requirement	Easily automated	Genetic variability
RFLP	I	High	No	Low
Minisatellite	II	High	No	High
Microsatellite	II (mostly)	Low	Yes	High
RAPD	II	Low	Yes	Low
SSCP	I	Low	Yes	Low
AFLP	II	Low	Yes	High

Table 11.2. RFLP markers published for sheep.

Locus symbol	Locus name	Probe	Enzyme	Reference	Sheep location
ACO2	Aconitase 2		*Pst*I	Ryan and Womack (1994b)	–
ADRA1C	Adrenergic, α_{1C}receptor	alpha1c	*Pvu*II	Montgomery *et al.* (1995a)	–
AHSG	α_2-HS-glycoprotein	B4	*Bgl*II	Montgomery *et al.* (1995a)	–
ALB	Albumin	SSA1	*Eco*RI	Montgomery *et al.* (1993)	6
ALDOA	Aldolase A		*Eco*RV	Broom *et al.* (1996)	24
ANX5	Annexin V	PAPI–1.6	*Bgl*II	Montgomery *et al.* (1995a)	6
APOB	Apolipoprotein B	λapoB10	*Taq*I	Phua *et al.* (1994b)	3p
ATP1B1	ATPase, Na^+, K^+-transporting β_1-polypeptide	pbNKA1	*Taq*I	Montgomery *et al.* (1995a)	–
ATP7A	ATPase, Cu^{++}-transporting α-polypeptide (Menkes syndrome)	B8.1(Mc1)	*Eco*RV	Potts *et al.* (1994b)	X
BMP3	Bone morphogenetic protein 3	BMP3-315	*Asp7*	Montgomery *et al.* (1995a)	6
BCHE	Butyrylcholinesterase	BCHE	*Hind*	Montgomery *et al.* (1995a)	–
C1R	Complement subcomponent C1r	pHC1r/253	*Pst*I	Phua and Wood (1995c)	3q
C4	Complement component 4		*Taq*I	Groth *et al.* (1990)	–
CA2	Carbonic anhydrase II	pHCAII38-3	*Taq*I	Phua *et al.* (1994d)	9
CAL1L	Calpactin 1, light polypeptide	p11	*Taq*I	Montgomery *et al.* (1995a)	–
CD5	CD5 antigen	pBCD5	*Hind*	Montgomery *et al.* (1995a)	–
CCNA	Cyclin A	CycA	*Taq*I	Lord *et al.* (1996)	6
CGA	Glycoprotein hormones, α-polypeptide	pALPHA-1	*Bst*EII	Penty *et al.* (1991c)	–
COL1A1	Collagen, type 1, α_1	DPP-1	*Pst*I	Montgomery *et al.* (1995a)	–
COL3A1	Collagen, type III, α_1	pIII-55	*Eco*RI	Phua *et al.* (1992b)	2q12–q21
–	Cone transducin, subunit	pML1	*Pvu*II	Montgomery *et al.* (1995a)	–

CNP	2',3'-cyclic nucleotide 3'-phosphodiesterase	–	*Msp*1	Penty *et al.* (1991a)	–
CRH	Corticotrophin-releasing hormone	pcRF31	*Pvu*II	Sise and Montgomery (1992)	9q23–q28
CSN1S1	Casein, α_{S1}		*Bgl*I	Leveziel *et al.* (1991)	6
CSN1S1	Casein, α_{S1}		*Taq*I	Di Gregorio *et al.* (1991)	
CSN1S2	Casein, α_{S2}		*Pvu*II	Leveziel *et al.* (1991)	6
CSN1S2	Casein, α_{S2}		*Eco*RV	Di Gregorio *et al.* (1991)	
CSN1S2	Casein, α_{S2}	218	*Eco*RI	Phua *et al.* (1992a)	
CSN2	Casein, β		*Bgl*I	Leveziel *et al.* (1991)	6q22–q31
CSN2	Casein, β		*Taq*I	Di Gregorio *et al.* (1991)	
CSN3	Casein, κ		*Pvu*II	Leveziel *et al.* (1991)	6
CSN3	Casein, κ		*Hind*III	Di Gregorio *et al.* (1991)	
CYP17	Cytochrome P450, subfamily XVII (steroid 17α-hydroxylase)	p45017a	*Pst*I	Montgomery *et al.* (1995a)	22q22–q31
CYP21	Cytochrome P450, subfamily XXI (steroid 21-hydroxylase)	p450C21	*Pst*I	Montgomery *et al.* (1995a)	–
DMD	Duchenne muscular dystrophy	Ca1A	*Pvu*II	Galloway *et al.* (1995)	X
DQA	Ovine major histocompatibility complex class II region, DQa chain		Various	Hulme *et al.* (1993) Litchfield *et al.* (1993)	20q15–q23
DQB	Ovine major histocompatibility complex class II region, DQb chain		Various	Scott *et al.* (1991) Blattman *et al.* (1993) Grain *et al.* (1993) Hulme *et al.* (1993)	20q15–q23
DRA	Ovine major histocompatibility complex class II region, DRa chain		*Bgl*II	Escayg *et al.* (1993)	20q15–q23
DRB	Ovine major histocompatibility complex class II region, DRb chain		Various	Scott *et al.* (1991) Blattman *et al.* (1993) Grain *et al.* (1993) Litchfield *et al.* (1993)	20q15–q23

continued overleaf

Table 11.2. Continued.

Locus symbol	Locus name	Probe	Enzyme	Reference	Sheep location
EGF	Epidermal growth factor	phEGF121	*Sca*I	Montgomery *et al.* (1993)	6
ELN	Elastin	pcBE85	*Pst*I	Phua *et al.* (1993c)	24q16–qter
ENO1	Enolase 1 (α)		*Eco*RI	Schmahl *et al.* (1993)	U1
EPO	Erythropoietin		*Eco*RV	Broom *et al.* (1996)	24
ESD	Esterase D	pESD-14.1.1	*Eco*RI	Phua *et al.* (1994a)	3p24–p26
FDXR	Ferredoxin reductase	pBAR16	*Taq*I	Montgomery *et al.* (1995a)	–
FGF2	Fibroblast growth factor 2 (basic)	pJJ11-1	*Xba*I	Penty and Montgomery (1993)	17q23–q25
FGF5	Fibroblast growth factor 5		*Pvu*II	Lord *et al.* (1996)	6
FN1	Fibronectin 1	pFH1	*Hind*	Montgomery *et al.* (1995a)	2q41–q44
FSA	Follistatin	FS1.1	*Pst*I	Tisdall and Penty (1991)	–
FSHB	Follicle-stimulating hormone, β-polypeptide	–	*Msp*I	Montgomery *et al.* (1990)	15q24–qter
FSHR	Follicle-stimulating hormone receptor	pKH28	*Pst*I	Montgomery *et al.* (1995d)	3
GCG	Glucagon	bg13A7	*Msp*I	Penty and Montgomery (1991c)	–
GGTB2	Glycoprotein 4-β-galactosyl-transferase	7A	*Msp*I	Montgomery *et al.* (1995a)	2
GH1	Growth hormone 1		*Bam*HI	Valinsky *et al.* (1990)	11q25–qter
GH1	Growth hormone 1		*Taq*I	Parsons *et al.* (1992a)	
GH1	Growth hormone 1		*Pvu*II	Gootwine *et al.* (1993)	
GNRHR	Gonadotrophin-releasing hormone receptor		*Taq*I	Montgomery *et al.* (1995b)	6
GPI	Glucose phosphate isomerase	pGPI8R	*Taq*I	Penty *et al.* (1995)	14q22–q24
GRO1	GRO1 oncogene (melanoma growth-stimulating activity, a)	pBC-KC	*Eco*RV	Montgomery *et al.* (1995a)	6
GSN	Gelsolin	M1D	*Eco*RI	Phua *et al.* (1993b)	2pter–p24
GSR	Glutathione reductase	pGRH-B	*Pst*I	Phua and Wood (1995a)	24

HBA	Haemoglobin α_2		*Bcl*I	Broom *et al.* (1996)	24
HBB	Haemoglobin β		*Bam*HI	Di Gregorio *et al.* (1987)	15
IF	I factor (complement)	HI1971	*Sca*I	Montgomery *et al.* (1993)	6
IFNG	Interferon-γ	–	*Rsa*I	Penty and Hill (1991)	–
IGFI	Insulin-like growth factor 1	pOAIGF1per	*Bgl*II	Demmer and Galloway (1993)	3
IGHE	Immunoglobulin, ε		*Taq*I	Clarke and Beh (1992)	–
IGJ	Immunoglobulin J polypeptide	#0857	*Taq*I	Montgomery *et al.* (1995a)	–
IGKC	Immunoglobulin, κ constant region		*Taq*I	Parsons *et al.* (1994c)	–
IGKV	Immunoglobulin, κ variable region		*Taq*I	Foley and Beh (1992)	–
IGLV	Immunoglobulin, λ variable region		*Taq*I	Foley and Beh (1992)	–
IL4R	Interleukin-4 receptor		*Taq*I	Broom *et al.* (1996)	24
IL2RB	Interleukin-2 receptor, β	pUCb 30SB	*Dra*I	Montgomery *et al.* (1995a)	–
INHA	Inhibin, α		*Taq*I	Sise *et al.* (1991)	2
INHBA	Inhibin, βA		*Taq*I	Hiendleder *et al.* (1992)	4q27–q31
JUNB	jun B proto-oncogene	465.20	*Bst*XI	Montgomery *et al.* (1995a)	–
KIT	Hardy–Zuckerman 4 feline sarcoma viral (*v-kit*) oncogene homologue	phckit-171	*Eco*RI	Montgomery *et al.* (1993)	6
KRT1.1	Keratin 1.1 (wool microfibril component 8C1 type 1)		*Msp*I	Rogers *et al.* (1993b)	–
KRTAP1.3	Keratin-associated protein 1.3 (B2C high-sulphur keratin)		*Bsr*I	Rogers *et al.* (1993a)	11q
KRTAP5	Keratin-associated protein 5.1 (ultra-high-sulphur keratin)		*Bam*HI	Parsons *et al.* (1992b)	–
LGB	Lactoglobulin, β		*Hin*dIII	Ali *et al.* (1990)	3p28
LHB	Luteinizing hormone β-polypeptide	LH-7	*Dra*I	Penty *et al.* (1995)	14
LHCGR	Luteinizing-hormone receptor	pKH26	*Eco*RV	Montgomery *et al.* (1995d)	3
MAG	Myelin-associated glycoprotein	–	*Apa*I	Penty *et al.* (1995)	14

continued overleaf

Table 11.2. Continued.

Locus symbol	Locus name	Probe	Enzyme	Reference	Sheep location
MBP	Myelin basic protein	pHBP-1	*Pst*I	Phua *et al.* (1994e)	–
MET	Met proto-oncogene	pmet5	*Taq*I	Montgomery and Sise (1990)	–
MET	Met proto-oncogene	pmetH	*Bgl*II	Montgomery *et al.* (1995a)	–
MET	Met proto-oncogene	pmetD	*Taq*I	Montgomery *et al.* (1995a)	–
MYCN	Avian myelocytomatosis viral (*v-myc*)-related oncogene neuroblastoma-derived	pNB-1	*Bgl*II	Montgomery *et al.* (1995a)	3p
MMP1	Matrix metalloproteinase 1 (interstitial collagenase)		*Taq*I	Shay and Muggli-Cockett (1993)	–
MYF5	Myogenic factor 5	Myf-s/18-7	*Eco*RI	Phua and Wood (1993)	3q
MYHL	Myosin 1 heavy-chain-like protein	cHNKp1	*Pst*I	Phua *et al.* (1993d)	3q
NTS	Neurotensin	bhNT12	*Msp*I	Penty *et al.* (1991b)	3q12–q14
PAH	Phenylalanine hydroxylase	PAH247	*Taq*I	Phua and Wood (1995b)	3q
PDEB	Phosphodiesterase, cyclic GMP (rod receptor), β-polypeptide	pGEM2 + cDNAbPDE	*Sca*I	Montgomery *et al.* (1995a)	6q33–q36
PDGFRA	Platelet-derived growth-factor receptor, α-polypeptide	α17B	*Pvu*II	Montgomery *et al.* (1993)	6
PDHA1	Pyruvate dehydrogenase E1 α subunit	PDHE1α1c	*Hin*dIII	Galloway *et al.* (1996)	X
PHKA1	Phosphorylase kinase, α_1	pαPHK-D1	*Bgl*II	Potts *et al.* (1994a)	X
PHKG2	Phosphorylase kinase, γ_2 (testis)		*Eco*RV	Broom *et al.* (1996)	24
PLAT	Plasminogen activator, tissue	pdeltaRIPA0	*Taq*I	Phua *et al.* (1994c)	–
	Polyubiquitin	pCL208	*Pst*I	Penty and Montgomery (1991b)	–

POMC	Pro-opiomelanocortin		*Eco*RI	Ryan and Womack (1994a)	3p
PRKCB1	Protein kinase C, β_1-polypeptide	–	*Msp*I	Broom and Penty (1993)	24q13–qter
PRM1	Protamine 1	pBP1	*Eco*RI	Phua *et al.* (1993a)	–
PRM2	Protamine 2		*Nco*I	Broom *et al.* (1996)	24
PRNP	Prion protein		*Eco*RI	Hunter *et al.* (1989)	10q24–q32
PLP	Proteolipid protein	PLP1	*Msp*I	Sise *et al.* (1992)	X
RHO	Rhodopsin	bd20	*Pst*I	Phua *et al.* (1993e)	19q23–qter
SCF	Stem-cell factor	–	*Bgl*II	Tisdall *et al.* (1996)	3q
SPN	Sialophorin		*Pst*I *Pvu*II	Broom *et al.* (1996)	24
SPP1	Secreted phosphoprotein 1	OP-12	*Taq*I	Montgomery *et al.* (1993)	6
TBG	Thyroxine-binding globulin	λcTBG8	*Msp*I	Hanrahan *et al.* (1992)	X
TIMP1	Tissue inhibitor of metalloproteinase 1	pBGR19	*Apa*I	Hanrahan *et al.* (1994)	X
UCP	Uncoupling protein	–	*Pvu*II	Tou *et al.* (1992)	–
UOX	Urate oxidase	pUO-p	*Pst*I	Montgomery *et al.* (1995a)	1p
XIST	Inactive X-specific transcript	pXISTcow1	*Bam*H1	van Stijn and Galloway (1995)	X

ATPase, adenosine triphosphatase; GMP, guanosine monophosphate.

Table 11.3. A description of the type I microsatellite and SSCP markers used for the sheep genetic linkage map.

Marker abbreviation	Marker name	Primer sequences (5'–3')	Chromosome assignment	Informative meioses	Annealing temperature (°C)	Reference
ADCYAP1	Adenylate-cyclase-activating polypeptide	CCAGACGCCGACTTCGCCGAGG GCCTGAAGTCCACTGAGAAGAAAGGAG	23	153	60	Wood and Phua (1993)
ADCYC	Calmodulin-independent adenylcyclase	AAAGTGACACAACAGCTTCTCCAG AACGCGTGTCCTAGTTTGGCTGTG	15	167	55	Moore and Byrne (1993b)
CD5	CD5	CAACTAGTCATTCTTGAGGGATTTTC ATACGGAAAAAATACCAGATTAGACACT	21	191	TD‡	Matthew and Maddox (1994b)
ELN	Elastin	CGGGAACAGGGCCAGCAGC CTGGGTCTGACTGGACTGGAC	24	154	65	Moore *et al.* (1992)
FSH	Follicle-stimulating hormone	TGGGATATAGACTTAGTGGC CAGTTTCTAAGGCTACATGGT	15	18	63	Moore *et al.* (1992)
HBB	β-Haemoglobin	GATATAAAAAGAAGACCCAGTAG TACCTGAGTCATATGTAATATTCC	15	78	TD	Wood *et al.* (1993)
HMH1R	Histamine H1 receptor	GGCTTCAACTCACTGTAACACATT TTCTTCAAGTATCACCTCTGTGGCC	12	193	60	Moore and Byrne (1993a)
IL2RA	Interleukin-2 receptor α	AGCAGAGGTACAGGTGGTAAGCA GATATGCCTTGGAGAAGGTAGCGTAT	13	158	58	Matthew and Maddox (1994a)
IL3	Interleukin 3	AGCCTCTCTATCTTGCATCTGCTC TCATCTGAGTTCAAGGAGCCCT	5	53	57	Hawken and Maddox (1994)
KAP8	Glycine- and tyrosine-rich keratin	ACACTTGGAATTCAATAAATATGTGTGG CCTGGCAGGCCACAGTTCAC	1	188	63	Ede and Crawford (1995)
LHB	Luteinizing hormone β-polypeptide	TATCACTTTCACCACCAGCATCTGCG GCGGCAGGATGACAGGCAGCACC	14	SSCP§	63	Penty *et al.* (1995)

LYZ	Lysozyme	ATGGATGTTCACCAGTTATTGTGGT AGTGAGCAAGGCATTACAGCATG	3	114	55	Davies and Maddox (1994)
MT2	Metallothionein	TGAGGAGGATTCGTTACCGTCTGAG CCGAGGGGCAGCGGATTAAG	14	187	57	Hawken and Maddox (1994)
OCAM	Opioid-binding cell-adhesion moleculae	CCTGACTATAATGTACAGATCCCTC GCAGAATGACTAGGAAGGATGGCA	21	165	55	Moore *et al.* (1992)
OLADRB	Ovine DRB gene	TCTCTGCAGCACATTTCCTGG CGTACCCAGAGAG/TTGAGTGAAGTATC	20	179	59	Schwaiger *et al.* (1993)
OLADRBps	Ovine DRB pseudogene	CTGCCAATGCAGAGACACAAGA GTCTGTCTCCTGTCTTGTCATC	20	173	62	Blattman and Beh (1992)
OMHC1	Ovine MHC class 1	ATCTGGTGGGCTACAGTCCATG GCAATGCTTTCTAAATTCTGAGGAA	20	179	60	Groth and Wetherall (1994)
RBP3	Interphotoreceptor retinoid-binding protein	CTATGATCACCTTCTATGCTTCC CCCTAAATACTACCATCTTAGAAG	25	122	50	Barendse *et al.* (1994)
RHO	Rhodopsin	CTCCTCACACGGCTGCTGGG GTTAACTACAGTCTGTCCAGCTC	19	104	60	Moore *et al.* (1992)
TCRB	T-cell receptor B	CATTCTCCTCTTCTGGGTGG GCTGGCACAGAGATATCAYGGC	4	205	59	Buitkamp *et al.* (1993)

‡TD, touchdown annealing procedure 60°C × 3; 57°C × 3; 54°C, × 3; 51°C, × 3; 48°C, × 20.
§SSCP not analysed in the IMF. All other markers listed are microsatellites.

Table 11.4. A description of the type II microsatellite markers used for the sheep genetic linkage map.

Marker name*	Primer sequences (5'–3')†	Chromosome assignment	Informative meioses‡	Annealing temperature§	Reference
AGLA226	AACAAGCCATGCTGAATGGTCTTATGTG GAAATGCAGTGTTGTCAGCCAAGC	10	189	TD	Georges and Massey (1992)
AGLA285	CTTGACCTGCAACAGGCATGCAG GCCAATCCCATTAAAGTTGAAGGCAG	13	65	TD	Georges and Massey (1992)
AGLA293	GTCTGAAATTGGAGGCAATGAGGC CCCAAGACAACTCAAGTCAAAGGACC	3	113	TD	Georges and Massey (1992)
ARO28	GATTTCTCTAGTGAGTAACA TCTTGCCCAGATGTTCTTAG	2	91	50	Avraham *et al.* (1993)
BL25	AACAGTGGCAATGGAAGTGG AGTCAGGATCTAGTGGGTGAGTG	25	105	52	Bishop *et al.* (1994)
BL41	CCTCGCCATCTTTATTCCT AAGATCAACTTATTCCTCACAGTGG	1	135	57	Bishop *et al.* (1994)
BM1225	TTTCTCAACAGAGGTGTCCAC ACCCCTATCACCATGCTCTG	16	180	61	Bishop *et al.* (1994)
BM1227	CACCAGTGATATTGGCTTATGG GGAAGAAACACTTCCAAACCC	8	166	TD	Bishop *et al.* (1994)
BM1237	TCATCTTGGGCATAAGACAGG ATTGTTCCCAGCATCTTAGAGG	7	142	TD	Bishop *et al.* (1994)
BM1258	GTATGTATTTTTCCCACCCTGC GAGTCAGACATGACTGAGCCTG	20	141	63	Bishop *et al.* (1994)
BM1303	CTTGGGAAAAATTGCCAGC CTCTGCCGCTTGCCTCTC	19	160	51	Bishop *et al.* (1994)
BM1314	TTCCTCCTCTTCTCTCCAAAC ATCTCAAACGCCAGTGTGG	22	199	TD	Bishop *et al.* (1994)

BM1329	TTGTTTAGGCAAGTCCAAAGTC AACACCGCAGCTTCATCC	6	142	63	Bishop *et al.* (1994)
BM143	ACCTGGGAAGCCTCCATATC CTGCAGGCAGATTCTTTATCG	6	132	63	Bishop *et al.* (1994)
BM1558	TGAGGAAAGCCTTGGCAG ACTGGGCCTAGCTCCTTCTC	19	203	TD	Bishop *et al.* (1994)
BM1815	AGAGGATGATGGCCTCCTG CAAGGAGACAAGTCAAGTTCCC	20	153	60	Crawford *et al.* (1995)
BM1818	AGCTGGGAATATAACCAAAGG AGTGCTTTCAAGGTCCATGC	20	176	53	Bishop *et al.* (1994)
BM1824	GAGCAAGGTGTTTTTCCAATC CATTCTCCAACTGCTTCCTTG	1	171	63	Bishop *et al.* (1994)
BM1853	AGCCTTTTGTAGGTGTTCATTG ATGGGGTTGCAAAGAGTCAG	5	90	61	Bishop *et al.* (1994)
BM1861	TGTTGGCCTCGTCTATGACC CTTCTTGATGACCTTCTCCACC	3	96	55	Crawford *et al.* (1995)
BM1905	GTCCATGGGTTCACAAAGAG ACGCCTGCTGATGCTGTAG	20	35	TD	Bishop *et al.* (1994)
BM203	GGGTGTGACATTTTGTTCCC CTGCTCGCCACTAGTCCTTC	26	149	53	Bishop *et al.* (1994)
BM2113	GCTGCCTTCTACCAAATACCC CTTCCTGAGAGAAGCAACACC	2	151	55	Bishop *et al.* (1994)
BM226	ATTGCCTTGTCCGTGTATCC CCGGCTGAATTGCTATAAGC	23	156	55	Bishop *et al.* (1994)
BM2504	CAGCTTTCCATCCCCTTTC CTCCCATCCCAAACACAGAC	8	138	55	Bishop *et al.* (1994)
BM2818	TTCTGTGGTTGAAGAGTGTTCC CAATGGCTAAGAGGTCCAGTG	3	177	53	Bishop *et al.* (1994)

continued overleaf

Table 11.4. Continued.

Marker name*	Primer sequences (5'–3')†	Chromosome assignment	Informative meioses‡	Annealing temperature§	Reference
BM302	GAATTCCCATCACTCTCTCAGC GTTCTCCATTGAACCAACTTCA	9	99	53	Bishop *et al.* (1994)
BM304	CTGGTGTTCCTTTCATATCAACC GGCACGTACTAACCTGTAAAACC	3	119	53	Bishop *et al.* (1994)
BM3413	TCCCTGGTAACCAATGAATTC CAATGGATTTGACCCTCCC	18	159	55	Bishop *et al.* (1994)
BM3628	CTGAGATGGACTCAGGGAGG GTTGGATTGGAAAGGTTAGGC	19	124	55	Bishop *et al.* (1994)
BM4005	AGTCCATGGGACCACAAAAG ACTGTACAGCAAAGTGATTCAGAT	24	156	55	Bishop *et al.* (1994)
BM4025	TCGAATGAACTTTTTTGGCC CACTGACTATGTGACTTTGGGC	12	172	55	Bishop *et al.* (1994)
BM4107	AGCCCCTGCTATTGTGTGAG ATAGGCTTTGCATTGTTCAGG	16	177	54	Bishop *et al.* (1994)
BM415	GCTACAGCCCTTCTGGTTTG GAGCTAATCACCAACAGCAAG	6	137	52	Bishop *et al.* (1994)
BM4208	TCAGTACACTGGCCACCATG CACTGCATGCTTTTCCAAAC	8	198	55	Bishop *et al.* (1994)
BM4311	TCCACTTCTTCCCTCATCTCC GAAGTATATGTGTGCCTGGCC	6	196	63	Bishop *et al.* (1994)
BM4505	TTATCTTGGCTTCTGGGTGC ATCTTCACTTGGGATGCAGG	22	185	TD	Bishop *et al.* (1994)
BM4513	GCGCAAGTTTCCTCATGC TCAGCAATTCAGTACATCACCC	9	184	57	Bishop *et al.* (1994)

BM4621	CAAATTGACTTATCCTTGGCTG TGTAACATATGGGCTGCATC	6	138	58	Bishop *et al.* (1994)
BM4630	TGGTTTGCTGGTGAGATTAGG TTCATAATGGGGTGAGTGAGTG	9	197	53	Bishop *et al.* (1994)
BM6041	GGCTGCTGCATGTCAGTG GACTTGAGCTCCTCCAGGG	22	173	55	Bishop *et al.* (1994)
BM6108	TTCTAATGTAGAGCAAAGTGATTGA TGTAGGAGGGACAGATTGGG	10	72	TD	Bishop *et al.* (1994)
BM6438	TTGAGCACAGACACAGACTGG ACTGAATGCCTCCTTTGTGC	1	175	60	Bishop *et al.* (1994)
BM6444	CTCTGGGTACAACACTGAGTCC TAGAGAGTTTCCCTGTCCATCC	2	197	TD	Bishop *et al.* (1994)
BM6506	GCACGTGGTAAAGAGATGGC AGCAACTTGAGCATGGCAC	1	155	TD	Bishop *et al.* (1994)
BM6526	CATGCCAAACAATATCCAGC TGAAGGTAGAGAGCAAGCAGC	26	132	60	Bishop *et al.* (1994)
BM7109	CAGGTAAAAGAGCGGCTTTG CAGCTTCATGCCCTAGAAGG	14	185	54	Bishop *et al.* (1994)
BM719	TTCTGCAAATGGGCTAGAGG CACACCCTAGTTTGTAAGCAGC	12	183	61	Bishop *et al.* (1994)
BM741	GCCCCTGAAGGAATGGTG CCAAAAGGTCCTATCTCCAAA	5	92	TD	Bishop *et al.* (1994)
BM757	TGGAAACAATGTAAACCTGGG TTGAGCCACCAAGGAACC	9	98	55	Bishop *et al.* (1994)
BM8125	CTCTATCTGTGGAAAAGGTGGG GGGGGTTAGACTTCAACATACG	17	156	55	Bishop *et al.* (1994)
BM827	GGGCTGGTCGTATGCTGAG GTTGGACTTGCTGAAGTGACC	3	172	55	Bishop *et al.* (1994)

continued overleaf

Table 11.4. Continued.

Marker name*	Primer sequences (5'–3')†	Chromosome assignment	Informative meioses‡	Annealing temperature§	Reference
BM848	TGGTTGGAAGGAAAACTTGG CCTCTGCTCCTCAAGACAC	15	143	50	Bishop *et al.* (1994)
BM861	TTGAGCCACCTGGAAAGC CAAGCGGTTGGTTCAGATG	X	171	55	Bishop *et al.* (1994)
BM864	TGGTAGAGCAATATGAAGGCC GGAAATCCAAGAAAGAGGGG	1	172	55	Bishop *et al.* (1994)
BMC1009	GCACCAGCAGAGAGGACATT ACCGGCTATTGTCCATCTTG	3	86	55	Bishop *et al.* (1994)
BMC1206	GGGTGGCTATGACTCCAGTG GGTCCAGCCTTCCACCAC	21	163	55	Bishop *et al.* (1994)
BMC3224	CCATCACTGCTATTCTACCTCC CACAGCCAATTTCTGATTTCA	21	142	56	Bishop *et al.* (1994)
BMC4203	GCAAATGTAAGCTGAAGGCC CCATAGGGTCGCAAAGAGTC	6	130	60	Lord *et al.* (1996)
BMS360	ACAAAACCACTTTCTTAGCAAACA CTGGGTCTTCATGGTAGGGA	6	273	60	Stone *et al.* (1995)
BMS483	GGTATGAGACCAGGTGTGGG CAGGGCCACATTTCCAAG	6	90	60	Stone *et al.* (1995)
BMS690	CATAGGGATATGTTGTGCATCC TCAAAGAAACTTCAAGCCAGC	6	83	60	Stone *et al.* (1995)
BR3510	GCTGGTGGGTTGTTTACCAC ACCCCGTGGACTGTAGTCTG	15	191	TD	Bishop *et al.* (1994)
BR6504	AATACCATGGACAGAGGAGCC AGGTTCAGGTCACCACAATACC	12	124	TD	Bishop *et al.* (1994)

CSSM13	TCATACCTTAAGAGAGTCGAG CTATCAGACAGACAGTCAGACAGTCAG	21	183	55	Moore *et al.* (1994)
CSSM18	TGTGCATAATTTGTGTCCGTCCGGA AGGAATTCCCTCTAGAAAAGCAGGC	18	160	TD	Moore *et al.* (1994)
CSSM25	GTAGTTATCAAAATAAGAATGCTT GGTCGAAGAGAGTCAGACATGACT	8	107	TD	Moore *et al.* (1994)
CSSM31	CCAAGTTTAGTACTTGTAAGTAGA GACTCTCTAGCACTTTATCTGTGT	23	168	55	Moore *et al.* (1994)
CSSM32	TTATTTTCAGTGTTTCTAGAAAAC TATAATATTGCTATCTGGAAATCC	1	20	55	Moore *et al.* (1994)
CSSM43	AAAACTCTGGGAACTTGAAAACTA GTTACAAATTTAAGAGACAGAGTT	26	145	55	Moore *et al.* (1994)
CSSM47	TCTCTGTCTCTATCACTATATGGC CTGGGCACCTGAAACTATCATCAT	2	111	55	Moore *et al.* (1994)
ETH225	GATCACCTTGCCACTATTTCCT ACATGACAGCCAGCTGCTACT	9	191	50	Steffen *et al.* (1993)
ETH3	GAACCTGCCTCTCCTGCATTGG ACTCTGCCTGTGGCCAAGTAGG	11	79	TD	Steffen *et al.* (1993)
HEL10	ATCTGCCTGAAGCCAGTCAC GGTTTCCTGCACCTGCATGA	11	167	TD	Kaukinen and Varvio (1993)
HUJ223	CATCCCTGATGGAAGCAGGG CTCTCCCGGTCTGCTTTGCA	17	128	TD	Shalom *et al.* (1994)
HUJ614	CGCCAGGCATGGTGAAGTCG CCCAGCACAGTACAGGCTGC	19	101	55	Barendse *et al.* (1994)
HUJ616	TTCAAACTACACATTGACAGGG GGACCTTTGGCAATGGAAGG	13	152	TD	Barendse *et al.* (1994)
HUJ625	AGCAGCATGAAGAGAGTCCC GAGGTCACATACCCATCAAGC	12	143	58	Barendse *et al.* (1994)

continued overleaf

Table 11.4. Continued.

Marker name*	Primer sequences (5'–3')†	Chromosome assignment	Informative meioses‡	Annealing temperature§	Reference
ILSTS002	TCTATACACATGTGCTGTGC CTTAGGGGTGAAGTGACACG	14	113	TD	Kemp *et al.* (1992)
ILSTS004	CTTAAAATCTGTCTTTCTTCC TAGTGTGTATTAGGTTTCTCC	1	187	50	Bishop *et al.* (1994)
ILSTS005	GGAAGCAATGAAATCTATAGCC TGTTCTGTGAGTTTGTAAGC	7	118	50	Brezinsky *et al.* (1993a)
ILSTS008	GAATCATGGATTTTCTGGGG TAGCAGTGAGTGAGGTTGGC	9	140	50	Kemp *et al.* (1993)
ILSTS011	GCTTGCTACATGGAAAGTGC CTAAAATGCAGAGCCCTACC	9	154	50	Brezinsky *et al.* (1993b)
ILSTS013	CTTGATCCTTATAGAACTGG ACACAAAATCAGATCAGTGG	8	70	50	Brezinsky *et al.* (1993b)
ILSTS017	GTCCCTAAAATCGAAATGCC GCATCTCTATAACCTGTTCC	X	693	63	Kemp *et al.* (1993)
ILSTS019	AAGGAACCTCATGTAGAAGC ACTTTTGGACCCTGTAGTGC	21	201	TD	Kemp *et al.* (1993)
ILSTS020	AACAAGTGCTTTGTTGGTGC GATAAAAGCACATCTGCTGC	7	104	TD	Kemp *et al.* (1993)
INRA11	CGAGTTTCTTTCCTCGTGGTAGGC GCTCGGCACATCTTCCTTAGCAAC	1	189	55	Bishop *et al.* (1994)
INRA13	GCACAGTGACCTCTCAATAAATGC CCACTATTCTTGCCTGAAGAATCC	12	18	50	Vaiman *et al.* (1992)
INRA30	ATGCAAATCTGCTAAATCACCTAT CCCAACTCTCACATCCAGAT	X	186	TD	Ciampolini *et al.* (1993)

JP15	GGAAATACCTTATCTTTCATTCTTGACTGTGG CCTTCTTTCTCATTGCTAACTTATATTAAATATCC	21	112	63	Crawford *et al.* (1995)
MAF109	GGAAGATTAGAACTTTCATATATCTTTAAACTC TAATTGAATTTGAAGTGTATATGCCTAAATGC	1	184	58	Swarbrick and Crawford (1992)
MAF18	TTTTCCTTCAGAGCCAAGAGGACAAG GTATACTGCCTCTCTGCGTGATGGG	13	118	63	Crawford *et al.* (1990)
MAF209	GATCACAAAAAGTTGGATACAACCGTGG TCATGCACTTAAGTATGTAGGATGCTG	17	144	63	Buchanan and Crawford (1992a)
MAF214	GGGTGATCTTAGGGAGGTTTTGGAGG AATGCAGGAGATCTGAGGCAGGGACG	16	151	58	Buchanan and Crawford (1992d)
MAF23	GTGGAGGAATCTTGACTTGTGATAG GGCTATAGTCCATGGAGTCGCAG	3	143	63	Swarbrick *et al.* (1991a)
MAF33	GATCTTTGTTTCAATCTATTCCAATTTC GATCATCTGAGTGTGAGTATATACAG	9	178	58	Buchanan and Crawford (1992c)
MAF35	TCAAGAATTTTGGAGCACAATTCTGG AGTTACAAATGCAAGCATCATACCTG	23	131	63	Swarbrick *et al.* (1991a)
MAF36	TTGCGAAAAGTTGGACACAATTGAGC CATATACCTGGGAGGAATGCATTACG	22	168	63	Swarbrick *et al.* (1991c)
MAF4	AAAGGTCTTGACTCTGTGTCGG TCCACGGGGTCTCAAAGAGTCG	1	167	65	Buchanan *et al.* (1991a)
MAF45	ATTGACACTTCAGTAAGTTAACAATGG CAGACACAACTGAGCAACTAGCGC	X	778	60	Swarbrick *et al.* (1992)
MAF48	GGAAACCAAAGCCACTTTTCAGATGC AGACGTGACTGAGCAACTAAGTACG	Unknown	124	63	Buchanan *et al.* (1991b)
MAF50	GTAGACTACTCATGAAAATCAGGTCTTAGG GGGACATGCAGCTATACACTTGAG	4	103	63	Swarbrick *et al.* (1992b)
MAF64	AATAGACCATTCAGAGAAACGTTGAC CTCATGGAATCAGACAAAAGGTAGG	1	198	64	Swarbrick *et al.* (1991b)

continued overleaf

Table 11.4. Continued.

Marker name*	Primer sequences (5'–3')†	Chromosome assignment	Informative meioses‡	Annealing temperature§	Reference
MAF65	AAAGGCCAGAGTATGCAATTAGGAG CCACTCCTCCTGAGAATATAACATG	15	174	60	Buchanan *et al.* (1992)
MAF70	GCAGGACTCTACGGGGCCTTTGC CACGGAGTCACAAAGAGTCAGACC	4	205	65	Buchanan and Crawford (1992c)
MAF92	TAGAATGTCATGTTCTCAGCATTCCC AACCCATGAATCATCTCTAACTACCTC	22	165	65	Crawford *et al.* (1991)
McM150	CCACTTGGAGTGAAAATGAGACA AGGAAAATCTTCCGGAGCTAAAC	16	144	TD3	Crawford *et al.* (1995)
McM152	GGAACTCTCATAGTTTCCCACTCC CCTAGAAGCCTGGCTAAAATGTG	13	189	TD2	Hulme *et al.* (1994)
McM218	CACTAAAAGCTTATGAAAGTTCCAGC GATCCTAGCATCAGTCTCCAGATG	4	190	TD3	Crawford *et al.* (1995)
McM42	CATCTTTCAAAAGAACTCCGAAAGTG CTTGGAATCCTTCCTAACTTTCGG	9	135	55	Hulme *et al.* (1994)
McM527	GTCCATTGCCTCAAATCAATTC AAACCACTTGACTACTCCCCAA	5	180	50	Hulme *et al.* (1994)
McM64	TACAGTCCATGGGGTCACAAGAG TCTGAATCTACTCCCTCCTCAGAGC	2	195	50	Hulme *et al.* (1994)
McM73	CTCTTCATTCTGCAAAAGTTTGTCAC GCTTGTGAGATGAACAATAAGTCATAGG	4	161	55	Hulme *et al.* (1994)
OarAE101	TAAGAAATATATTTGAAAAAACTGTATCTCCC TTCTTATAGATGCACTCAAGCTAGG	6	122	63	Montgomery *et al.* (1993)
OarAE119	CTCAGCAAATGGTTCCTGGGCACC TTTTATAGTGAGGTGACCACTTGATG	19	151	60	Penty *et al.* (1993)

OarAE129	AATCCAGTGTGTGAAAGACTAATCCAG GTAGATCAAGATATAGAATATTTTTCAACACC	5	100	60	Penty *et al.* (1993)
OarAE133	AGCCAGTAGGCCCTCACCAGG CCAACCATTGGCAGCGGGAGTGTGG	X	860	65	Hanrahan *et al.* (1994)
OarAE16	CTTTTTAATGGCTCGGTAATATTCCTC CATCAGAGGAATGGGTGAAGACGTGG	13	140	58	Penty *et al.* (1993)
OarAE25	AATCTAAGTGTTCATCAATAGATGAATGG GTGGCTAACACAGCAATGGG	X	764	65	Hanrahan *et al.* (1994)
OarAE54	TACTAAAGAAACATGCTCCCAC AGAAACATTTATTCTTATCCTCAGTG	25	185	64	Penty *et al.* (1993)
OarAE57	CAGACTCTTATCTACTGCCCTGCC GGGAGAGAATGAAGGAAGGGCTGC	1	193	63	Penty *et al.* (1993)
OarAE64	TGCAAGAAGGGCAGACCTTGGAG CAGACCACTCTCTTCCCTCCACG	7	186	63	Ede *et al.* (1994)
OarCP101	TCATACCTTAAGAGAGTCGAG CTATCAGACAGACAGTCAGACAGTCAG	20	114	55	Crawford *et al.* (1995)
OarCP125	GCAAATAGCCTCTTGTATGATCCTTGG ACCAAAACAAGACCTTTATTTTTCATGG	6	284	63	Lord *et al.* (1996)
OarCP131	ATGTGCCACAACTAAGACCAGCAC CCAGGGAAGTCCCATACAATTAACC	X	209	63	Galloway *et al.* (1996)
OarCP134	TAAGTGGGATTTCATCTTTCCGGGATC TGTCAAGGAAACAGGTGGTGTAAG	18	209	50	Pierson *et al.* (1994)
OarCP16	CTGCAATAACCCTTAACCTCTGCTTAC GTGTGAAGAATAGAGGGCTGGTAGC	17	40	63	Ede *et al.* (1995a)
OarCP20	GATCCCCTGGAGGAGGAAACGG GGCATTTCATGGCTTTAGCAGG	21	186	63	Ede *et al.* (1995a)
OarCP21	CATGCTGTTTATTCTTGTTTGTATGTGC CTGCTACTGTTAACCTCCCTCC	8	76	63	Ede *et al.* (1995a)

continued overleaf

Table 11.4. Continued.

Marker name*	Primer sequences (5'–3')†	Chromosome assignment	Informative meioses‡	Annealing temperature§	Reference
OarCP23	GGGGTCACAAAGAGTCAGACATGC CTATGCTTAACATGTTCTAGAACACG	16	45	63	Ede *et al.* (1995a)
OarCP26	GGCCTAACAGAATTCAGATGATGTTGC GTCACCATACTGACGGCTGGTTCC	4	197	63	Ede *et al.* (1995a)
OarCP34	GCTGAACAATGTGATATGTTCAGG GGGACAATACTGTCTTAGATGCTGC	3	186	63	Ede *et al.* (1995b)
OarCP38	CAACTTTGGTGCATATTCAAGGTTGC GCAGTCGCAGCAGGCTGAAGAGG	10	99	55	Ede *et al.* (1995b)
OarCP43	TATGGCTGAGGAATATTTTCTGTGTTG GAATGGATAAAGAAGATGTGAGACTGAG	3	142	63	Ede *et al.* (1995b)
OarCP49	CAGACACGGCTTAGCAACTAAACGC GTGGGGATGAATATTCCTTCATAAGG	17	139	63	Ede *et al.* (1995b)
OarCP73	AAAACTGAGAAATATTCAGATGCAAC TAAACGTCCATCAACAGAGGAAGGG	20	78	63	Ede *et al.* (1995b)
OarCP79	AAAACCATCTCAAAAGATCACCGTCCC GGCCAGCACTACTCTTATAGCCAG	2	169	65	Ede *et al.* (1995b)
OarCP88	GGTCTTTTCTTGTCAGGAATGATGGATGC TCTGACAAACCAGTCATTTCC	19	121	50	Pierson *et al.* (1994)
OarCP9	CCTGATTTCTTATGGATAAGCAAGAAC GAGGAGCCTGTGGCTAAATGAGG	9	167	63	Ede *et al.* (1995a)
OarCP93	AAAATTTCGCTCCCTCCCTCTCTTTCC AGTCGGATATATGTATGTGTGTATGG	Unknown	194	55	Pierson *et al.* (1994)
OarCP99	GTTTACCACTAAGCCACCAGGGAAGC CCTTACACTCTGATAGAGGATATTTAC	16	114	60	Ede *et al.* (1995b)

OarDB3	CTATACAGTTTCTGCTGTATAGNAAAGTG CAGGTGCCAAGTATTACATGC	10	139	63	Bancroft (1993)
OarDB6	GACATGACTAAAGCAATTTAGCATGC TGGACTACAGTCCATAGCGTCTC	1	165	63	Bancroft (1993)
OarEL01	CTGCTGCCACCTTACTCTGCTC TTTTCTAACATGAAGGCTCCCCAC	24	67	63	Lord *et al.* (1996)
OarEL03	GCTTCCTTGGAAAGGAAAGCAGTC AGACAAGTTCTGCAGATACCCTGC	6	238	63	Lord *et al.* (1996)
OarEL13	CTGTAATAACTTATATGGGAAAAGTC ACCTGTTACTTTCAGGTGTACAG	6	207	63	Lord *et al.* (1996)
OarFCB11	GGCCTGAACTCACAAGTTGATATATCTATCAC GCAAGCAGGTTCTTTACCACTAGCACC	2	140	63	Buchanan and Crawford (1993)
OarFCB128	CAGCTGAGCAACTAAGACATACATGCG ATTAAAGCATCTTCTCTTTATTTCCTCGC	2	149	63	Buchanan and Crawford (1993)
OarFCB129	GCGACTTAGCAGCAGCAGCATCC CATCAAGAGATGAATGAGTAAAGAAGATG	3	182	63	Buchanan *et al.* (1994)
OarFCB193	TTCATCTCAGACTGGGATTCAGAAAGGC GCTTGGAAATAACCCTCCTGCATCCC	11	55	65	Buchanan and Crawford (1993)
OarFCB20	AAATGTGTTTAAGATTCCATACA GTG GGAAAACCCCCATATATACCTATAC	2	161	63	Buchanan *et al.* (1994)
OarFCB226	CTATATCYYGCCTTTCCCTTTTTCCTGC GTGAGTCCCATAGAGCATAAGCTC	2	178	63	Buchanan *et al.* (1994)
OarFCB266	GGCTTTTCCACTAGCTTTACATAGGAGTG CACCACATACCAAACACACAGCCTGC	25	180	63	Buchanan and Crawford (1993)
OarFCB304	CCCTAGGAGCTTTCAATAAAGAATCGG CGCTGCTGTCAACTGGGTCAGGG	19	131	63	Buchanan and Crawford (1993)
OarFCB48	GAGTTAGTACAAGGATGACAAGAGGCAC GACTCTAGAGGATCGCAAAGAACCAG	17	113	55	Buchanan *et al.* (1994)

continued overleaf

Table 11.4. Continued.

Marker name*	Primer sequences (5'–3')†	Chromosome assignment	Informative meioses‡	Annealing temperature§	Reference
OarFCB5	GACCTGACCCTTACTCTCTTCACTC AAGTTAATTTTCTGGCTGGAAAACCC	3	158	63	Buchanan *et al.* (1994)
OarHH22	CAACAGGACCTTGAAAACCACACC CTCAGTTTAATTCCATAGACCAACAGG	21	117	63	Ede *et al.* (1994)
OarHH29	GAGTCTCCCGGGAGCCCCATTAGC TTCTTTCTCTGTCTGTATCTCTCTCTCCC	9	113	55	Crawford *et al.* (1995)
OarHH30	CTCAGTCTCAACTTTGTTCCTCTATAGC GAAAGCTAAGGCTGAACATTGTGCCC	2	194	50	Pierson *et al.* (1994)
OarHH35	AATTGCATTCAGTATCTTTAACATCTGGC ATGAAAATATAAAGAGAATGAACCACACGG	4	177	63	Henry *et al.* (1993)
OarHH41	TCCACAGGCTTAAATCTATATAGCAACC CCAGCTAAAGATAAAAGATGATGTGGGAG	10	185	63	Henry *et al.* (1993)
OarHH44	GGTATCAGGACCGCTCAAGTCCCTCC GGTGGTCTTCCTCTTGGTGAGAC	Unknown	27	63	Henry *et al.* (1993)
OarHH47	TTTATTGACAAACTCTCTTCCTAACTCCACC GTAGTTATTTAAAAAAATATCATACCTCTTAAGG	18	176	63	Henry *et al.* (1993)
OarHH55	GTTATTCCATATTCTTTCCTCCATCATAAGC CCACACAG AGCAACTAAAACCCAGC	6	153	63	Montgomery *et al.* (1993)
OarHH56	GCAACCCACTCATCTCTCCGTGTC GAAAACTTAAGTTCCAGCTATTAA AATAGC	20	156	60	Ede *et al.* (1994)
OarHH62	TAATGAGTCAAACACTACTGAGAGAC AATATATAAAGAGAAAAGCTGGGGTGCC	16	125	63	Ede *et al.* (1994)
OarHH64	CGTTCCCTCACTATGGAAAGTTATATATGC CACTCTATTGTAAGAATTTGAATGAGAGC	4	130	55	Henry *et al.* (1993)

OarJMP1	CAAAGAGTCAGACATGACTGAACGAG CAAAAGGACAAAGCTTGTCTGAGGCAC	6	278	63	Lumsden *et al.* (1996)
OarJMP11	GGTCTGCAAAGAGCAGGACGTGAC CTGATCTGGACATGGTGTGGGCTC	26	139	TD	Lumsden *et al.* (1996)
OarJMP12	CAAAGAATTGGCCACAACTGAGTGAC GCCTGCTTTCTTTGCGTACAACTG	6	102	TD	Lumsden *et al.* (1996)
OarJMP23	GTATCTTGGGAGCCTGTGGTTTATC GTCCCAGATGGGAATTGTCTCCAC	26	174	TD	Lumsden *et al.* (1996)
OarJMP29	GTATACACGTGGACACCGCTTTGTAC GAAGTGGCAAGATTCAGAGGGGAAG	24	175	TD	Lumsden *et al.* (1996)
OarJMP36	CCCACTTTCTGGAAGGCAGAAATG CTTATTGTGTTTTCTGCCAGGGAG	6	167	63	Lumsden *et al.* (1996)
OarJMP4	GTAACGAGGCATGGTGCCTACAAC CCAACATCTTGCTATGTGCATGTATG	6	291	63	Lumsden *et al.* (1996)
OarJMP51	CATGCATATAGTGGCAGGTAGGAC CTCCGCTCATGTCTCATCCCTATTC	26	139	TD	Lumsden *et al.* (1996)
OarJMP58	GAAGTCATTGAGGGGTCGCTAACC CTTCATGTTCACAGGACTTTCTCTG	26	182	TD	Lumsden *et al.* (1996)
OarJMP8	CGGGATGATCTTCTGTCCAAATATGC CATTTGCTTTGGCTTCAGAACCAGAG	6	190	63	Lumsden *et al.* (1996)
OarMP4	GTCTTGTCCATCCATACAAGGG GGCAGATTCTTTACCACTGAGCC	X	806	63	Galloway *et al.* (1996)
OarVH110	CTCTAGAGGATCACAGAGAGTCGG GCAGAAACATTTTTTTCCTTCAATATAGTTTCCC	21	170	58	Hanrahan *et al.* (1993)
OarVH116	AGTGTGACTAGAGAACTAAATTTTGAAGGTC TATTTTTCCATCAAAAAGAACTCTATAGGGC	17	157	63	Hanrahan *et al.* (1993)
OarVH34	TCGTAAGAGTGGACACAACTGAGCG CGCAGTATTTAGTCCTTTTAATAATGGC	3	147	63	Pierson *et al.* (1993)

continued overleaf

Table 11.4. Continued.

Marker name*	Primer sequences (5'–3')†	Chromosome assignment	Informative meioses‡	Annealing temperature§	Reference
OarVH4	GTTAATTTACAGTGTTGTGTTAGTTTCAGC CTCTAGAGGATCTTGTAATAATCTATAAAGG	17	108	58	Ede *et al.* (1994)
OarVH41	CCCCAGTTTGTAATCTGGGTGACAG CGTAGGGTGTAAACGGTTTAGTGTCAG	Unknown	88	63	Pierson *et al.* (1993)
OarVH54	ATGGTTACTGAATGGCTGCCTAACCC CCTTAGGAACTAATGTGCACTTGTATGTG	18	138	50	Pierson *et al.* (1994)
OarVH58	CAGATCTCCTTGCAGCCAAGGGAC ATTAAAGAAAAAGAAAGTTAAAATCAGAACC	10	150	63	Pierson *et al.* (1993)
OarVH61	GATGATAAAGTTCACATGTGAAAAAGTCC AAGGTACTCAGATTTTGCTGAATCTGC	Unknown	49	63	Pierson *et al.* (1993)
OarVH72	CTCTAGAGGATCTGGAATGCAAAGCTC GGCCTCTCAAGGGGCAAGAGCAGG	25	102	62	Pierson *et al.* (1993)
RM106	TGCAGTTCCGACCATGGTGGA CAAATGCTATCATTTCTTGGACAA	16	178	50	Kossarek *et al.* (1993d)
RM32	GTTTGCAAAGGAATGTTATAATTGAA AAAACATTAACCTCAGAAACAGAA	2	18	55	Bishop *et al.* (1994)
RM4	CAGCAAAATATCAGCAAACCT CCACCTGGGAAGGCCTTTA	15	167	55	Kossarek *et al.* (1993a)
RM6	CTACAATATCTGGTCACTGGA GATCACCATATTTATGAGATGG	5	92	50	Kossarek *et al.* (1993b)
RM65	CTTTGTAACTAACTCTACTGCAG TCTGACTCCCCCTCCCTGCAAG	1	123	55	Bishop *et al.* (1994)
RM67	TGAGTAATGCAATAGATACAGTATT GCTTTGGCCATATGAAGAGCTTT	4	109	50	Kossarek *et al.* (1993c)

RM96	TCGCAAAAAGTTGGACAAGACT TTAGCAGGGTGCCTGACACTT	3	95	55	Bishop *et al.* (1994)
TGLA10	CTAAATTTATCCCACTGTGGCTCTT CAATCTGCAGTAGCATACATCCTTG	2	183	55	Georges and Massey (1992)
TGLA110	GTATTTTCAGACAGCCCGCTGGTGTC CAGCATTTACTTATACACACTCACCTG C	2	140	TD	Georges and Massey (1992)
TGLA116	TGGAGAAGATTTGGCTGTGTACCCA GCACAGTAATAAGAGTGATGGCAGA	4	49	55	Georges and Massey (1992)
TGLA122	AATCACATGGCAAATAAGTACATAC CCCTCCTCCAGGTAAATCAGC	18	145	TD	Georges and Massey (1992)
TGLA13	CATTTAATATACATATATGACTATGTGCC GAGCCTCTTTTTTACTTGTGCTCC	2	164	TD	Georges and Massey (1992)
TGLA137	GTTGACTTGTTAATCACTGACAGCC CCTTAGACACACGTGAAGTCCAC	5	180	55	Georges and Massey (1992)
TGLA15	GGTGACTGTACAACACTGTGAATG TAGGCTATGGTCCAAGGGTTCAC	X	770	60	Georges and Massey (1992)
TGLA176	GCTTTAAGAACTGCTCTCAAGGCAC CTGGGTGAGAAATGGGTAGTAGG	5	209	TD	Georges and Massey (1992)
TGLA23	GCCCCAGTTGATTTATTAAAAACAAATGC CCATTCATGCAAACTGAGACACAAGC	13	122	TD	Georges and Massey (1992)
TGLA231	GCTGCAAAGAGTTGGACAGAACTGAGC CTCCATTTCCCTTTGGTTTGTAAAGAC	17	207	TD	Georges and Massey (1992)
TGLA272	GATCAAAGCATGGAGGCCAGAGAGAGAG GGCGGTTAGAGGCTTGCACGC	7	170	TD	Georges and Massey (1992)
TGLA303	CTTGTGTGCCAGACCCAGGAATCC CATAAGTCAAAGTAACAGTTTAGATGTCC	5	113	TD	Georges and Massey (1992)
TGLA322	CATGCCACCTCTTGTCTGAAA CTTTAACATGGTTTAAATGACTATT	17	163	50	Georges and Massey (1992)

continued overleaf

Table 11.4. Continued.

Marker name*	Primer sequences (5'–3')†	Chromosome assignment	Informative meioses‡	Annealing temperature§	Reference
TGLA328	AAAGAAAGGCTGATATTGTAGCATC CTAACTGTGAGATTGCTTTGTAGTA	X	175	50	Georges and Massey (1992)
TGLA334	CAAACCCGAGGCTATGGCATTCGC CAACTCAGGTAGTCCAATTCCAATGC	12	188	TD	Georges and Massey (1992)
TGLA340	CCAGGCTGGCCACGCACCTCCATC C GAAGGAGCCCAGAAGAGATCCCTG	3	74	TD	Georges and Massey (1992)
TGLA357	GCAGAGTCTGAGTTTAAACTTCTCTAACACC GAGGGCAAAAAG GTTTGGGGTGTATGG	14	194	TD	Georges and Massey (1992)
TGLA377	GACTGTCATTATCTTCCAGCGGAG GATCTCTGGTTGAAATGGCCAGCAG	2	138	TD	Georges and Massey (1992)
TGLA387	CAAAGTCTTAGAATAAACTGGATGG GTCCCTTTGTTTACTTTGATAAAAC	20	142	53	Georges and Massey (1992)
TGLA415	GGTCACAAAGAGTCAGACACAACTGAGC CCTCCTCTGTTATCAATAGAGATATGC	1	173	TD	Georges and Massey (1992)
TGLA436	CTACTGACAGATGATTAGATAAAGAAGATGG ATTCTTTTGTATGGCTGAATGATATTCC	3	182	TD	Georges and Massey (1992)
TGLA441	CACAACTGGTAAAATGGCAGTGGCAG GGCCAAGTCATCTGAATTTAATATACAG	10	164	TD	Georges and Massey (1992)
TGLA444	CTCACAGCACAAAGCAGATGTTT TAGTGTTAGAGCAAAATTAATCTTC	10	72	50	Georges and Massey (1992)
TGLA48	AAATGTTTTATCTTGACTACTAAGC ACATGACTCTGCCATAGAGCAT	5	66	55	Georges and Massey (1992)
TGLA49	AGAAAAGGAATAATGAGACAGATTA GGCAGGACTTCACTCTTTTTCA	1	192	TD	Georges and Massey (1992)

TGLA51	CGCAGAGTCAGACGTGACTGAAGC CTTTTAAGGTTATCTGCCACCTAAGAC	11	102	TD	Georges and Massey (1992)
TGLA53	CAGCAGACAGCTGCAAGAGTTAGC CTTTCAGAAATAGTTTGCATTCATGCAG	12	148	TD	Georges and Massey (1992)
TGLA54	CTCAATATTTTGCAATAACATATAAGG CTCAATATTTTGCAATAACATATAAGG	X	556	63	Georges and Massey (1992)
TGLA58	CAGCAGCAGTGTTGGCTCCAAGAGC GTGCATGGACAATAATTCTACTCTCCAC	3	175	TD	Georges and Massey (1992)
TGLA67	GTGTCAAGTAAGACTGTTCAAA TATGGAATTGCAAAGAATTGGAT	3	95	55	Georges and Massey (1992)
TGLA68	ATCTTACTTACCTTCTCAGAGCT GGGACAAAATTTTACATATACACTT	X	556	55	Georges and Massey (1992)
TGLA72	ATGGGTGGGTCTGAATGAGGTGG GGGCATGAGTAGGTATTTCACAGG	X	581	60	Georges and Massey (1992)
TGLA73	GCTTCTTTCTCTTTAAATTCTATATGG GAGAGGAGAATCACCTAGAGAGGC	8	171	TD	Georges and Massey (1992)
TGLA75	GATAGTTTAACTTGAGGAGATTGCC GACAAAACCAGGTAAATCTTTCTAC	15	74	TD	Georges and Massey (1992)
TGLA77	GATGAGCTGCTATTTCTCTTCTAATGC CAGGGATGGTTCCCATAGCAACTGC	3	91	TD	Georges and Massey (1992)
TGLA89	GTGTTATAACAGAAAGTTCTCATTA TATTGCAGGCAGACTCTTTACC	X	551	60	Georges and Massey (1992)
TGLA94	CATCAAAACAGTGAAGGATGATTGCCAG CGAATCTCTTCTAGGGATTGAGACTGTG	11	92	TD	Georges and Massey (1992)
URB002	(unpublished)	17	168	55	Crawford *et al.* (1995)
URB037	(unpublished)	2	209	55	Crawford *et al.* (1995)
URB038	(unpublished)	1	191	TD	Crawford *et al.* (1995)

continued overleaf

Table 11.4. Continued.

Marker name*	Primer sequences (5'–3')†	Chromosome assignment	Informative meioses‡	Annealing temperature§	Reference
UWCA4	CAGCTTGAAATGAGTATGTCACC CAACAAGAAAGCAGAGACTCC	18	173	58	Sun *et al.* (1993)
UWCA5	CTTCCCTGTGTTAGGCTATGT CAAGAGAAGACAAGACTACAGA	14	70	TD	Sun *et al.* (1994)
UWCA9	CCTTCTCTGAATTTTTGTTGAAAGC GGACAGAAGTGAGTGACTGAGA	8	163	TD	Sun *et al.* (1994)
VIAS-S5	CCCAAGAGGCAATTATCATATTTCTGC GGTTTTTAGTTTCAGTTTACATATATATTATACATG	2	132	55	Primmer and Matthews (1993)

*Markers derived from cattle are shown in italics.

†Many of the primer sequences for the TGLA and AGLA markers were designed from the sequence provided in the patent (Georges and Massey, 1992), so they may be different from the published bovine primer sequences for the same marker.

‡Informative meioses for the autosomal markers were derived from the AgResearch International Mapping Flock (Crawford *et al.*, 1995). Informative meioses for markers on chromosomes 6 and X include other flocks (see Galloway *et al.*, 1996; Lord *et al.*, 1996).

§TD, touchdown annealing procedure 60°C, × 3; 57°C, × 3; 54°C, × 3; 51°C, × 3; 48°C, × 20. TD2, touchdown annealing procedure 52°C, × 1; 53°C, × 5l; 55°C, × 24. TD3, touchdown annealing procedure 49°C, × 1; 50°C, × 5; 52°C, × 24.

et al., 1993), microsatellites and RFLPs used in the study were analysed for segregation in the half-sib pedigrees. This search identified 19 linkage groups, containing a total of 52 markers (Crawford *et al.*, 1994). Of the 19 linkage groups, 13 could be assigned to sheep chromosomes.

A linkage map containing 232 markers was published a year later (Crawford *et al.*, 1995). This map was constructed from segregation analysis in the AgResearch International Mapping Flock (IMF). The IMF comprised nine three-generation pedigrees. Five breeds contributed to the pedigrees: Texel, Coopworth, Perendale, Romney and Merino. The different breed crosses were used to try to maximize the heterozygosity of the F1 generation. The total number of informative meioses possible in the IMF was 222. DNA from these animals has been distributed to 14 laboratories worldwide, so that all markers can be mapped in a common pedigree.

The current sex-averaged genetic linkage map of the sheep genome is presented in Fig. 11.1. The sheep genome was covered with 31 linkage groups, ranging in size from 2.0 to 195.8 centimorgans (cM). The map is presented in two parts: the main framework map, comprising 174 markers that could be ordered unequivocally at odds greater than 1000 : 1, and accessory markers, shown to the right of the main framework map. Where an accessory marker was analysed in the IMF and the odds were greater than 10 : 1 for placement within a particular interval on the framework map, its position is indicated by a thickened line. The linkage map was constructed with multipoint linkage analysis, using the BUILD option of CRI-MAP (Lander and Green, 1987). Genotypes resulting in double recombinants (CHROMPIC option) for the most likely order were rechecked. In addition, the genotypes of families were rechecked when their proportion of recombinants for a particular pair of markers was atypical (from a contingency table analysis). Genotypes were also rechecked for markers that considerably increased the map length. The rechecking involved rescoring the autoradiographs and, if there was any ambiguity, the whole pedigree was genotyped again. The resulting map positions, from the BUILD option of CRI-MAP, were then checked, using the FLIPS options, to ensure that the odds in favour of the order presented were at least 1000 : 1 over the alternative orders checked. CRI-MAP used in the generation of all maps of domestic animal species, uses the Kosambi mapping function.

Where linkage groups have been published for pedigrees other than the IMF, combined maps were calculated where the data files could be merged (Lord *et al.*, 1996), or equivalent phase known meioses were calculated from the published data to locate markers on the framework map. Locations calculated from other pedigrees are shown as a dashed line to the right of the main framework.

All of the cattle-derived markers used on the sheep linkage map have been assigned to cattle chromosomes (A. Eggen and R. Fries, 1995). If the group contained markers only assigned in cattle, the chromosomal banding homology between cattle and sheep chromosomes (Hediger *et al.*, 1991) was used to provisionally assign the group to a sheep chromosome.

The size of the sheep genome has been estimated from meiotic chromosomes of normal rams (Chapman and Bruere, 1977). From 34 cells in which every bivalent was clear, between 42 and 75 chiasmata per cell were counted. This gave a mean total number of 55.4 chiasmata per diploid cell at diakinesis–metaphase I. This corresponds to a minimum length of 2770 cM for the autosomal map. Total coverage of the current map was 2070 cM, with an average distance between framework markers of 14.4 cM, and corresponds to approximately 75% coverage of the sheep autosomes.

Reasonable coverage of most chromosomes was achieved, the exceptions being chromosomes 5, 7, 11 and 13, which have linkage groups spanning less than 50 cM. Fourteen anonymous microsatellites that were informative in the IMF remained unlinked to any other marker. Eight of these markers have been placed on the cattle linkage map and are shown below the corresponding sheep linkage groups on Figure 11.1. All unlinked markers showed heterozygous males, indicating that they were autosomal or pseudoautosomal in origin.

Significant differences in recombination rates between the sexes were observed in the framework map. The model that allowed for sex-specific recombination rates at each interval fitted the data significantly better ($P < 0.001$) than the model assuming that the recombination rate was the same for both sexes. The male recombination rate was higher in 63% of the 143 intervals tested which is significantly different from 50% ($P < 0.01$). The total male map length was 2413 cM, whereas the female map length was 1907 cM. In other mammals studied to date, the female shows the higher recombination rate. For example, in the human, the female recombination is 1.7 times that for males. The recombination rate for rams is approximately 25% higher than for ewes.

SheepBase

Sheep mapping information must be regularly updated to keep pace with the rapid progress extending both the physical and the linkage maps. The ideal way

Fig. 11.1. (see opposite) A genetic linkage map of the sheep genome. The consensus linkage map of the sheep genome comprises the main framework map, with markers ordered unequivocally at odds greater than 1000:1 (Crawford *et al.*, 1995), and accessory markers, shown to the right of the framework map. A vertical line indicates their most likely position. Where an accessory marker was analysed in the IMF and the odds were greater than 10:1 for placement within a particular interval on the framework map, its position is indicated by a thickened line. The map was constructed using multipoint-linkage analysis (Crawford *et al.*, 1995; Galloway *et al.*, 1996; Lord *et al.*, 1996). All distances were estimated using the Kosambi mapping function and distances for the autosomes are sex-averaged. The position of loci linked to markers on the map in other studies is shown by a broken line. To aid in the comparison of the sheep and cattle chromosomes the cytogenetically equivalent (Hediger *et al.*, 1991) cattle chromosome(s) is/are shown in parentheses below each of the sheep chromosomes. The position of the centromere is shown by an oval on the linkage map for those chromosomes where its position could be determined.

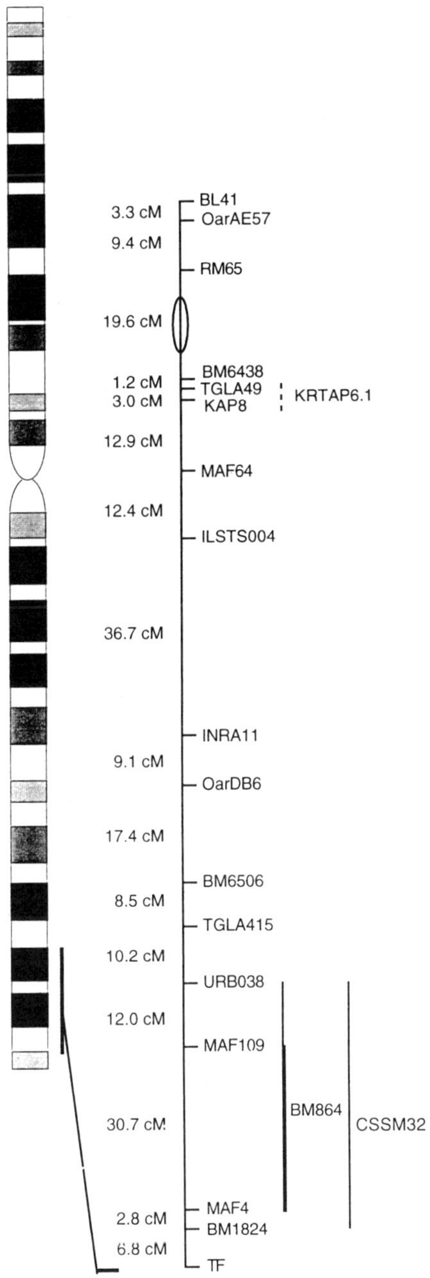

Sheep Chr 1 (195.8 cM)
(q arm = cattle chr 1
p arm = cattle chr 3)

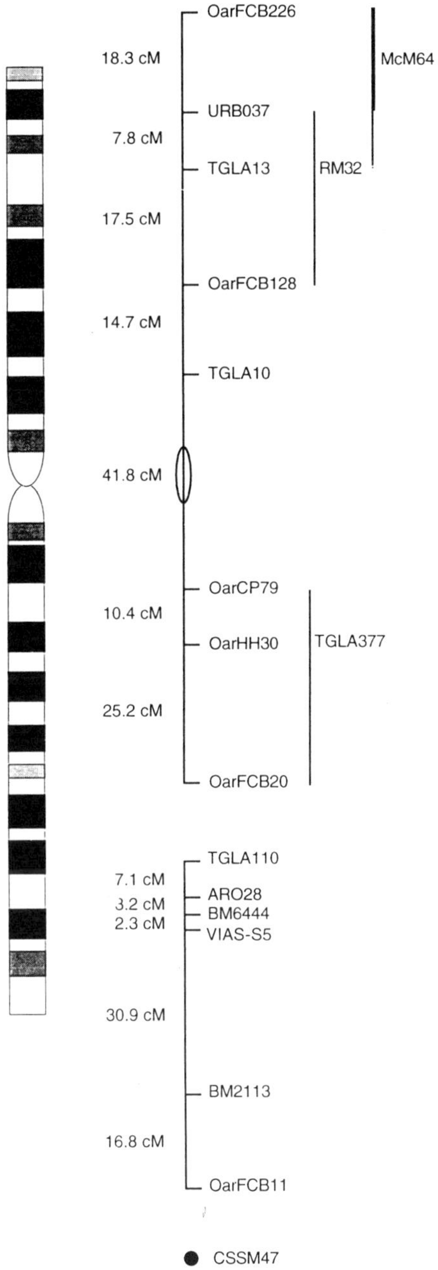

Sheep Chr 2 (60.3 + 141.7 cM)
(q arm = cattle chr 2
p arm = cattle chr 8)

Fig. 11.1. continued

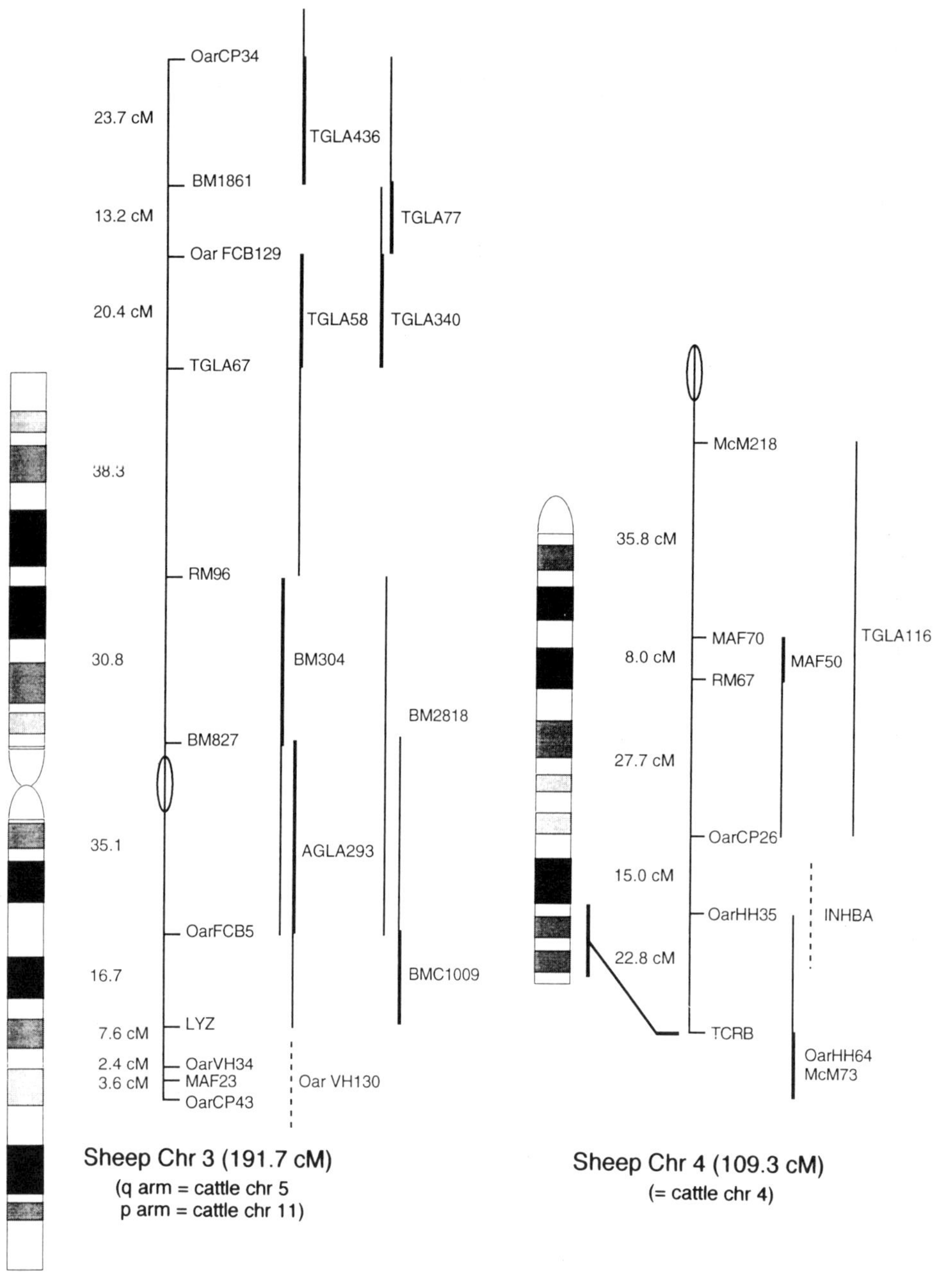

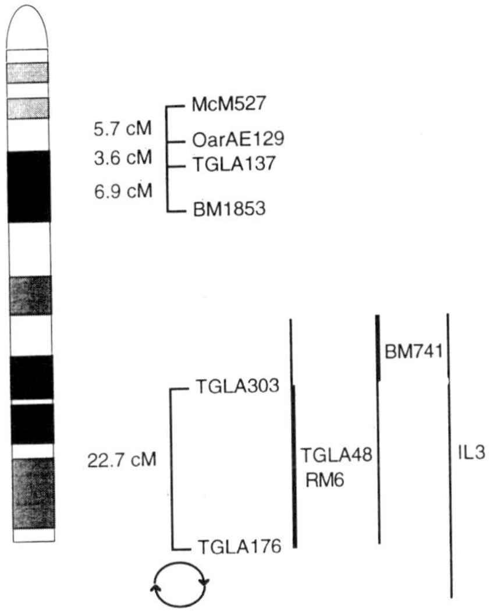

Sheep Chr 5 (16.2 cM + 22.7 cM)
(= cattle chr 7)

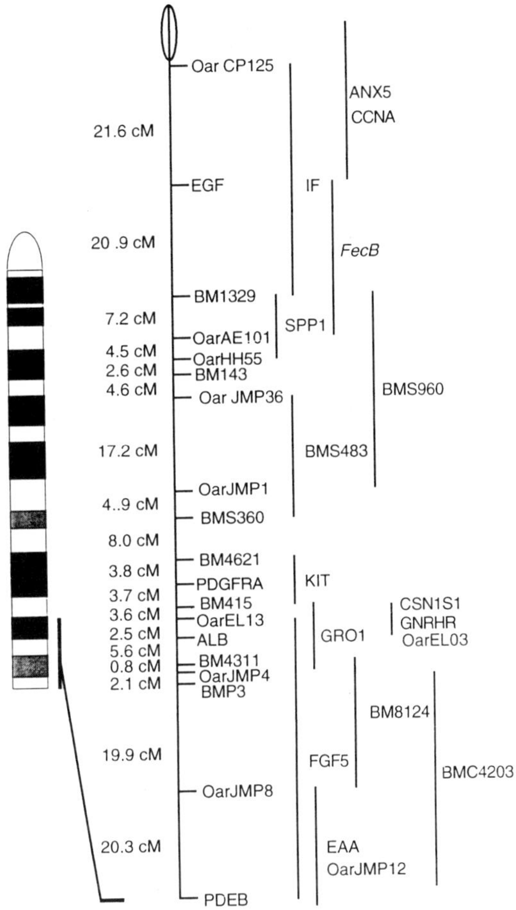

Sheep Chr 6 (153.9 cM)
(= cattle chr 6)

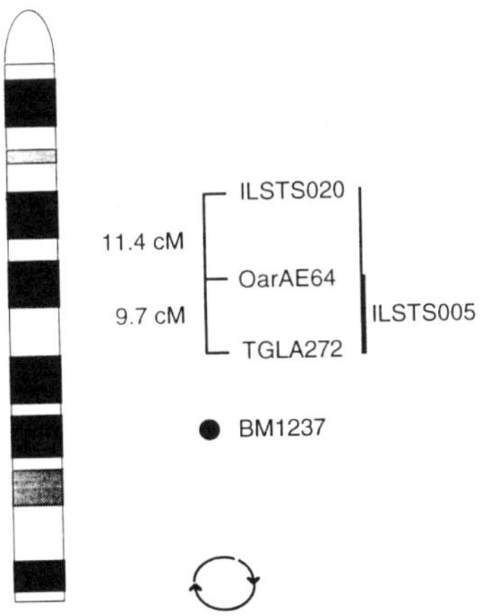

Sheep Chr 7 (21.1 cM)
(= cattle chr 10)

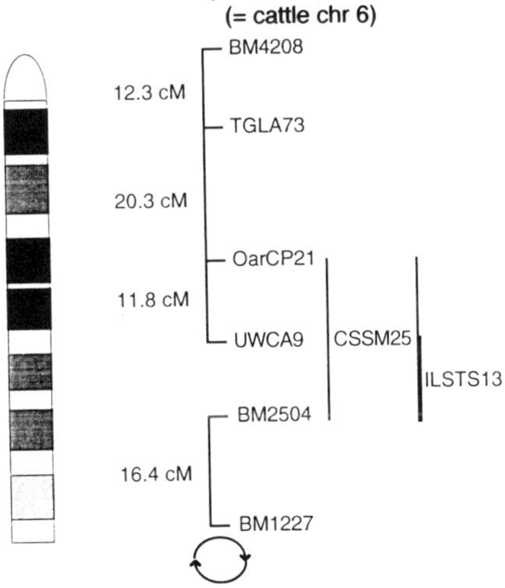

Sheep Chr 8 (44.4 + 16.4 cM)
(= most of cattle chr 9)

Fig. 11.1. continued

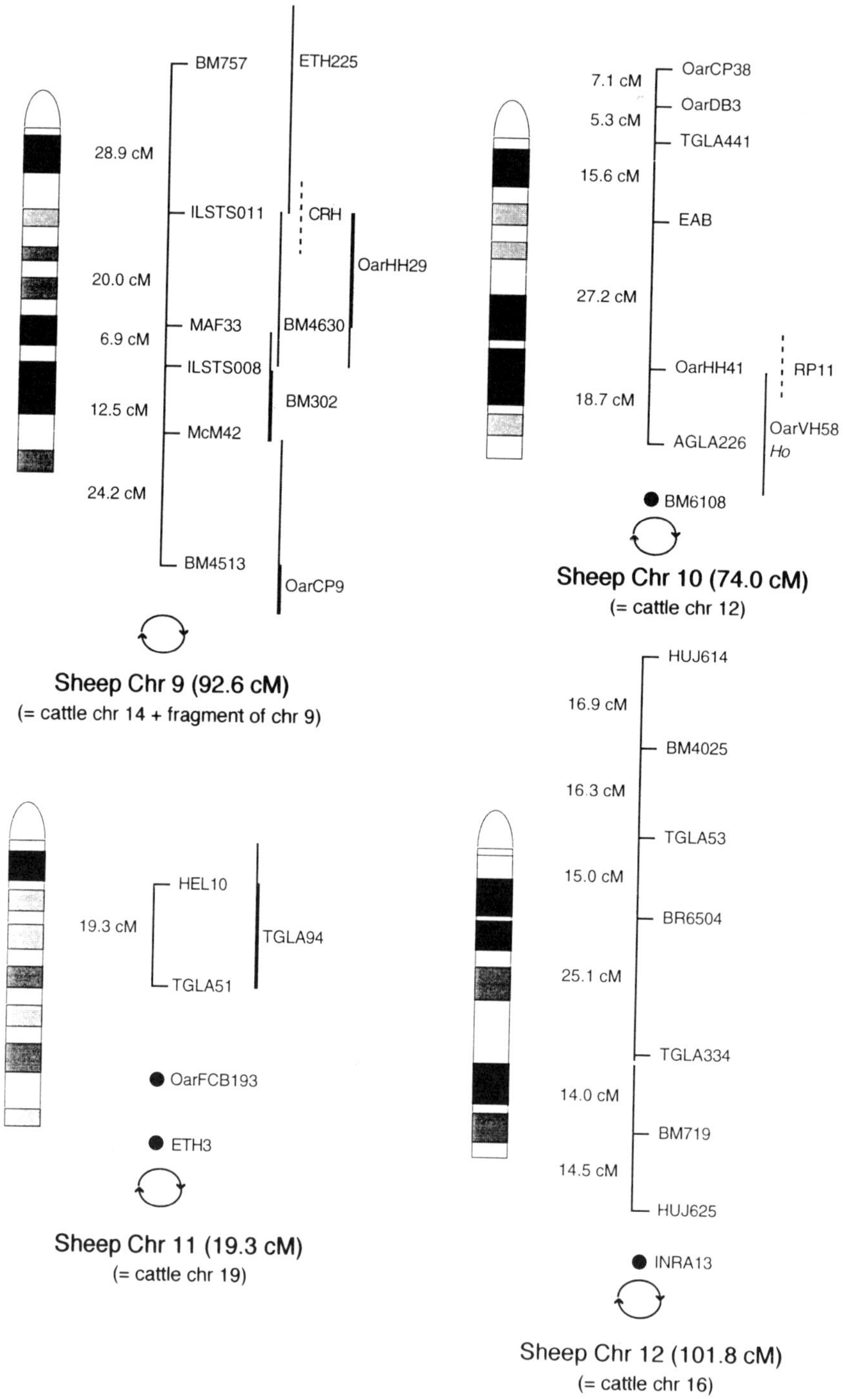

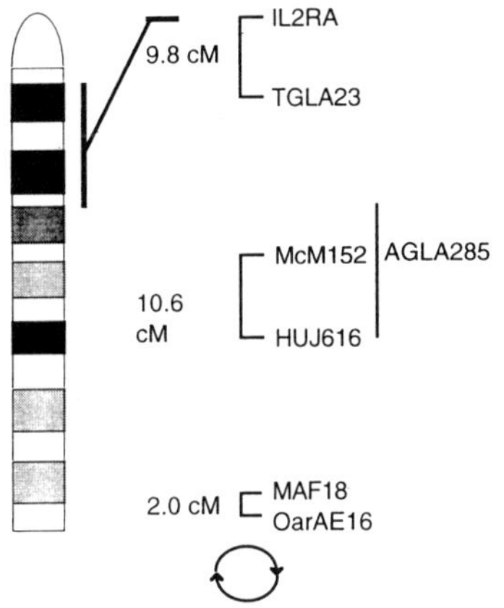

Sheep Chr 13 (2.0 + 10.6 + 9.8 cM)
(= cattle chr 13)

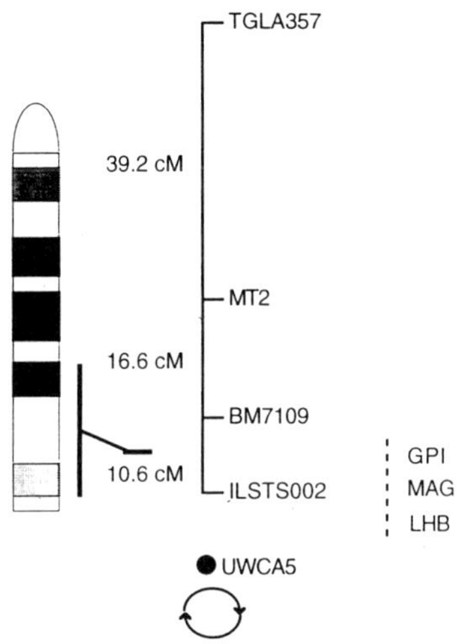

Sheep Chr 14 (66.3 cM)
(= cattle chr 18)

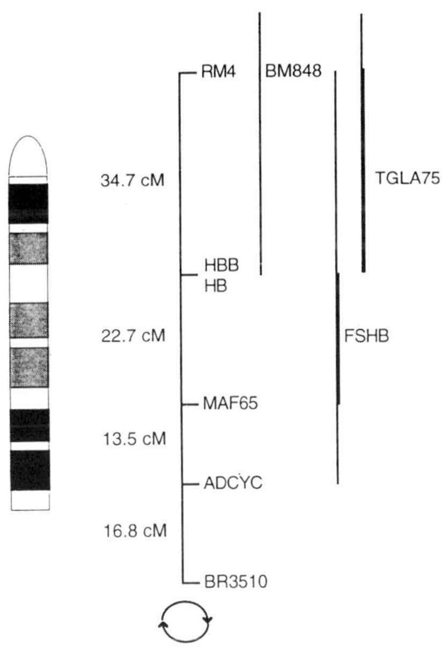

Sheep Chr 15 (87.6 cM)
(= cattle chr 15)

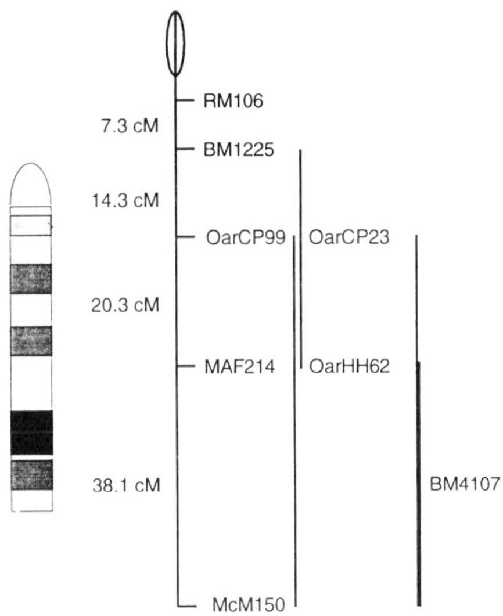

Sheep Chr 16 (80.0 cM)
(= cattle chr 20)

Fig. 11.1. continued

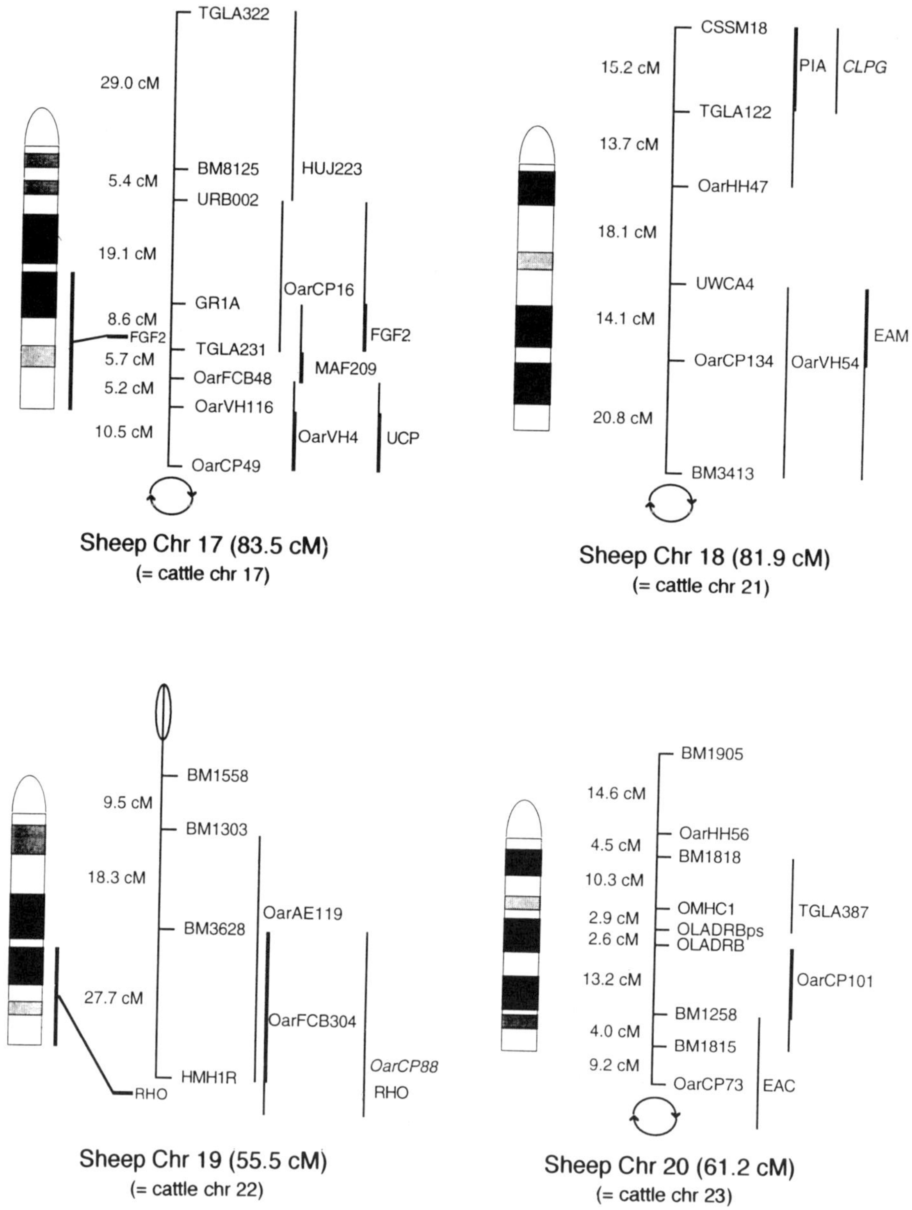

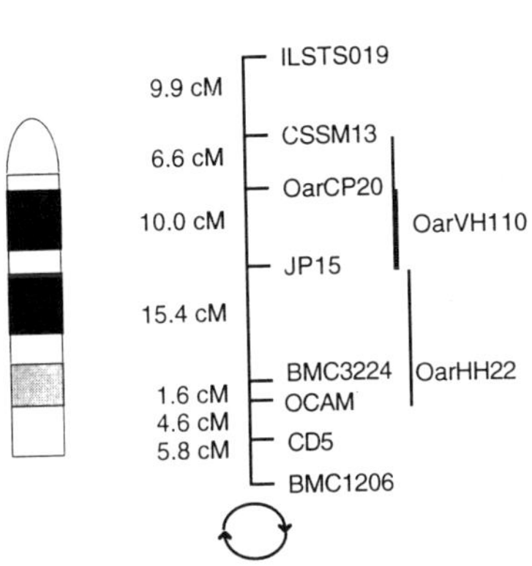

Sheep Chr 21 (53.9 cM)
(= cattle chr 25)

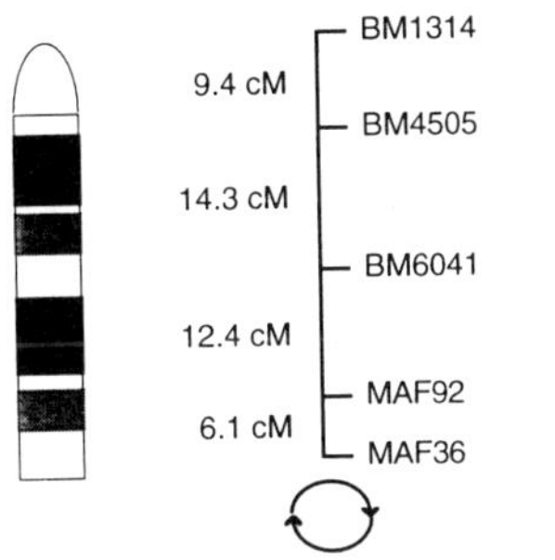

Sheep Chr 22 (42.2 cM)
(= cattle chr 26)

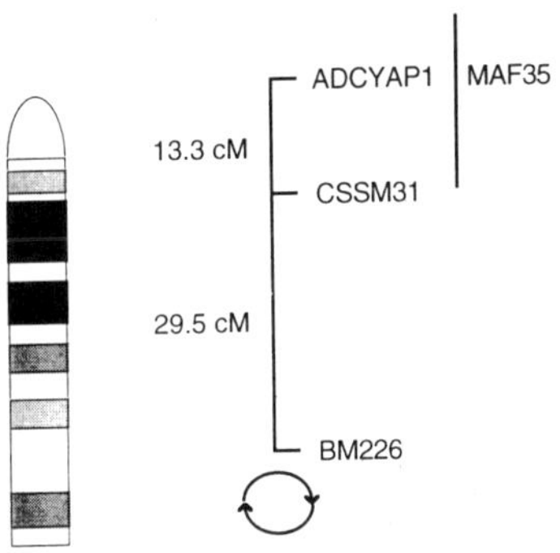

Sheep Chr 23 (42.8 cM)
(= cattle chr 24)

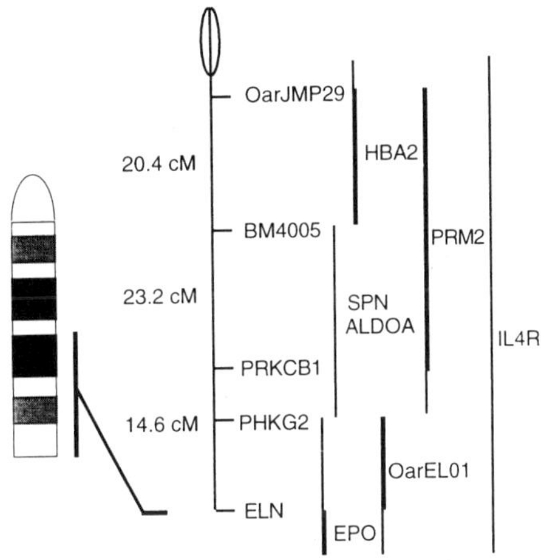

Sheep Chr 24 (57.6 cM)
(= cattle chr 29)

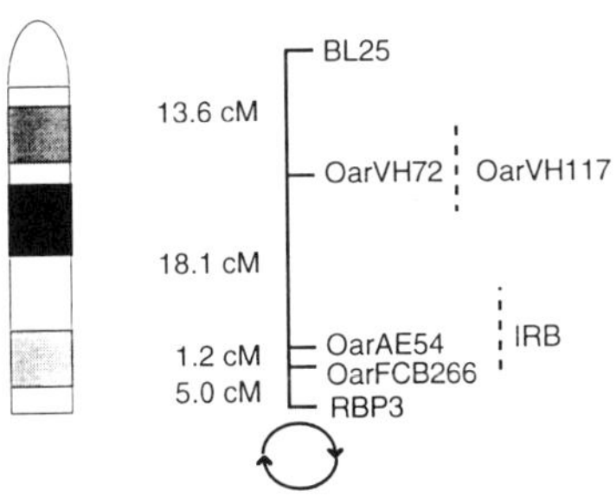

Sheep Chr 25 (37.9 cM)
(= cattle chr 28)

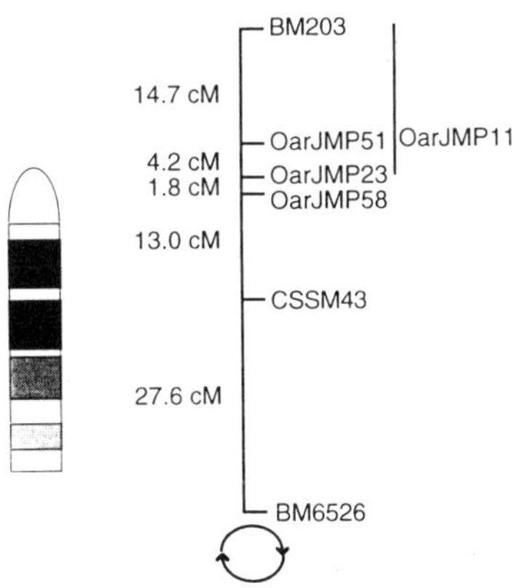

Sheep Chr 26 (61.3 cM)
(= cattle chr 27)

Fig. 11.1. continued

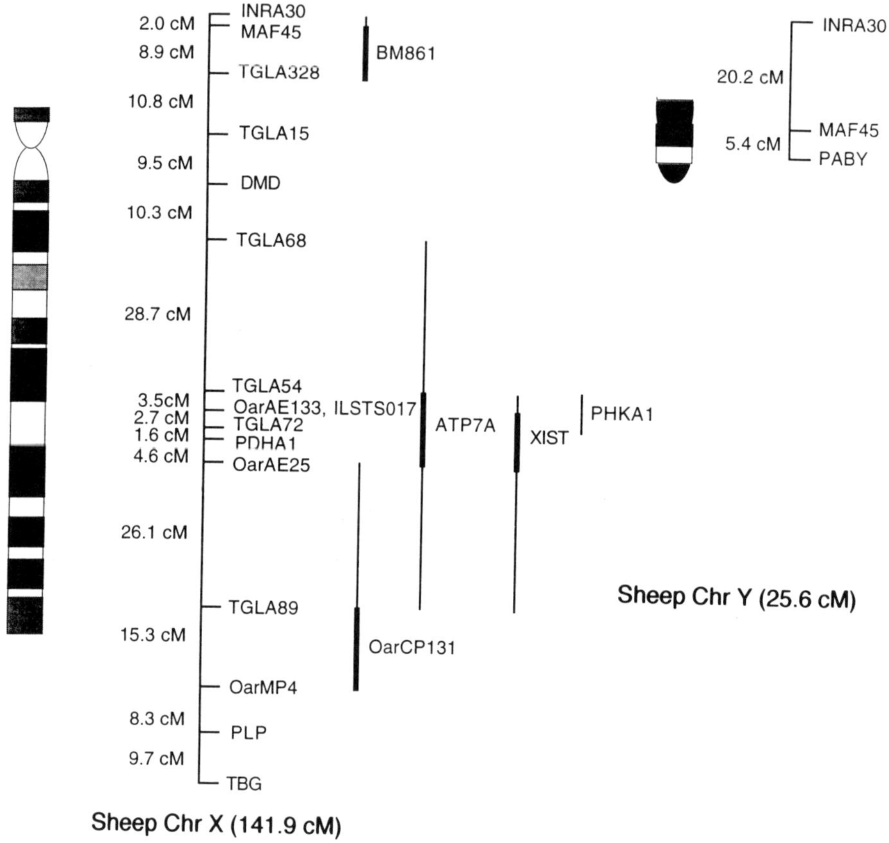

to present the most recent maps is through a computer database available on the World Wide Web (WWW). Additions and corrections made to the database are immediately accessible to the international community. We have developed SheepBase as the computer database for the sheep genome (Sise *et al.*, 1996). SheepBase provides an up-to-date compilation of published data, including both physical and linkage maps, together with information on individual loci and associated references. The software for the database and WWW interface were developed by Dr Alan Hillyard, Jackson Laboratory, and Dr Jian Hu and Mr Chris Mungall, Roslin Institute.

The database is currently held at AgResearch, Invermay Agricultural Centre, Mosgiel, New Zealand. SheepBase is accessible from AgResearch (New Zealand), the National Agriculture Library (NAL) (USA) and the Roslin Institute (UK).

AgResearch [http://dirk.invermay.cri.nz]

NAL [http://tetra.gig.usda.gov:8400]
Roslin Institute [http://www.ri.bbsrc.ac.uk/sheepmap/]

An international editorial committee for SheepBase has been established: Tom Broad, AgResearch, Invermay Agricultural Centre, New Zealand (broadte @agresearch.cri.nz), Noelle Cockett, Utah State University, USA (fanoelle@ cc.usu.edu), Grant Montgomery, AgResearch, University of Otago, New Zealand (grant.montgomery@stonebow.otago.ac.nz) and Frank Nicholas, University of Sydney, Australia (frankn@doolittle.vetsci.su.oz.au).

The current version of SheepBase contains 460 loci gathered from 105 publications. There are approximately 180 type I loci and 280 type II loci currently in the database. There are two linkage maps available: the framework map (SMAP) is based on the genetic linkage map published by Crawford *et al.* (1995). Genetic distance between loci is given in cM and the calculation is based on the Kosambi mapping function. Orientation of the linkage maps is not specified.

Cytogenetic anchoring of the linkage groups to a particular chromosome is based on one or more markers in the linkage group being assigned to the chromosome by one of three methods: (i) direct *in situ* hybridization, (ii) somatic-cell hybrid analysis; or (iii) assignment to a cytogenetically equivalent (Reading 1976 standard) bovine chromosome.

A second consensus map (SMC) is also presented,with additional published linkage data added to the framework data in SMAP where there are markers in common. The SMC is equivalent to the mapping information presented in this chapter.

Applications of the Linkage Map

Single-gene searches

One of the first applications of DNA markers has been the search for linkage between markers and single genes recognized by their phenotypic effects. The Booroola fecundity gene increases ovulation rate and litter size. In 1993 the *FecB* locus was linked to markers from a region of human chromosome 4q (Montgomery *et al.*, 1993). Subsequent mapping experiments assigned the locus to sheep chromosome 6 (Montgomery *et al.*, 1994).

Two other genes affecting phenotypic traits have been located on the sheep linkage map. The *Callipyge* (*CLPG*) locus increases muscle mass in the pelvic limb of affected individuals. This locus was linked to markers on sheep chromosome 18 (Cockett *et al.*, 1994). Recently, the *Horns* (*Ho*) locus, segregating in crosses between Merino and Romney sheep, was mapped to the distal end of chromosome 10 (Montgomery *et al.*, 1996).

The first two searches for *FecB* and *CLPG* used markers before they had been assembled into a map. Linkage to *FecB* was identified after screening 145 markers and took 3 years of marker development (Crawford *et al.*, 1991; Montgomery *et al.*, 1993). There are a number of single genes identified by their phenotypic effects (see Chapter 4) that could be located with suitable pedigrees. Several disease genes are being studied as models of human genetic disease. A sheep model for the human neurodegenerative disease neuronal ceroid lipofuscinosis (CLN) (Batten's disease) is being analysed (Broom *et al.*, 1996). The sheep gene has been excluded from sheep chromosome 24, the region syntenic to the common human form of CLN on HSA16.

Provided suitable family material is available, future searches for single gene traits will be far more efficient. Prior information on the likely location of the gene may come from studies in other species. An efficient search strategy can be designed and individual chromosomes searched with the existing markers.

Polygenic traits

Most traits of interest to sheep farmers show continuous variation. The observed variation may be due to a few loci with several alleles of large effect or many loci with alleles having small effects (polygenic variation). Progress has been made recently in segregation analysis to distinguish between these two possibilities (see Chapter 17). Methods are also being developed to use the genetic map to analyse complex traits (Tanksley, 1993).

Genetic linkage to markers can be used to detect and estimate the effects of individual polygenes contributing to a trait showing quantitative variation (quantitative trait loci (QTL)). The possibility of searching for individual polygenes was suggested in 1961 (Thoday, 1961) and the development of highly polymorphic markers with known order and distance on each chromosome has provided a powerful resource to search for these. Differences in the quantitative trait values are compared in two groups of offspring that inherit alternative marker alleles from a heterozygous parent. The analysis is carried out within half-sib or full-sib families for an outbred species such as the sheep. Large family sizes of more than 100 segregating progeny are necessary to detect QTL (Tanksley, 1993).

Quantitative traits for growth, fatness and the length of the small intestine were mapped to a region on chromosome 4 in pigs (Andersson *et al.*, 1994) in a cross of wild and domestic pigs. From the initial experiment, it was not possible to locate accurately the position of the genes responsible for the effects or to decide whether the effects were due to a single locus or several linked loci. Identifying the loci once the chromosomal region is identified will be difficult. Studies on the polygenes present on pig chromosome 4 are continuing, with the analysis of further markers and the results from specific matings of animals that are recombinant within the region.

Experiments to map polygenes associated with a number of traits in sheep are in progress. Merino and Romney sheep have been crossed in a double backcross design to map genes associated with wool traits (Wuliji *et al.*, 1995). Segregation of wool fibre diameter and fleece weight in the Merino backcrosses did not reveal any evidence for major genes affecting these traits. DNA samples have been collected and genetic markers are currently being typed in the flock.

Many economically important traits, as well as being multigenic, are expensive or difficult to measure. Consequently, phenotypic selection for these traits has often been limited for practical reasons. Resistance to sheep diseases, such as foot-rot or facial eczema, and pests, such as nematode parasites, lice and blowflies, are all difficult to measure, and expensive breeding stock must be exposed to the disease. Many carcass quality traits cannot be measured easily until the animal is slaughtered, unless expensive methods, such as computer-aided tomography (CAT) scanning, are employed. Analysis of the polygenic nature for these traits is under way in a number of laboratories, with the hope that useful genetic markers will be found to help current phenotypic selection methods.

The analysis of 'candidate' genes provides an alternative approach to the search using random markers. Significant associations have been reported between genetic markers for the high-glycine–tyrosine keratin (*KRTAP6*) and acidic fibroblast growth factor (*FGF1*) loci and fibre diameter (Parsons *et al.*, 1994b; Robinson *et al.*, 1995). The *FGF1* locus was also associated with fleece weight. These observations need to be confirmed in other pedigrees. If allelic variation at these loci is associated with fibre diameter in the sheep population, they would provide markers for selection. Enzymes involved in the oxidation of free radicals, such as the superoxide dismutases and glutathione peroxidase reductase systems are being examined in relation to resistance for facial exzema. The major histocompatibility complex (MHC) has been examined for associations with resistance to infections with nematode parasites (Wetherall and Groth, 1992).

One significant finding of the sheep linkage mapping to date has been the demonstration that sheep are genetically very heterogeneous. The average number of coinformative meioses using adjacent framework markers in the IMF was 118. This compares favourably with the cattle reference population of Bishop *et al.* (1994), which was approximately twice the size but contained an average of only 81 coinformative meioses per marker. The estimate from Bishop *et al.* (1994) includes all the markers and not just framework markers. If we include all the markers in the estimate for sheep, we expect an average of 88 coinformative meioses per marker, which is still more than cattle from a sheep pedigree half the size.

The IMF was generated using a variety of breeds in order to produce a high level of genetic heterogeneity. Subsequent studies (Buchanan *et al.*, 1994) have shown that the extent of genetic diversity within the New Zealand Romney population and the Australasian Merino populations was similar to that in a mixed population of six different breeds, including those used to generate the IMF. The situation in cattle is different, since *Bos taurus* purebreed populations have a significantly lower level of heterozygosity than crossbreeds or *B. taurus*/*Bos indicus* crosses (Bishop *et al.*, 1994). This finding has major implications for QTL searches.

At this stage, it would appear that QTL searches within sheep populations will require less than half the number of genotypes to generate the same linkage information as cattle. For those traits, such as leanness, growth rate and resistance to some diseases, where the genetic basis is likely to be similar in sheep and cattle, it will be considerably more efficient to use sheep populations for QTL searches.

Studies in natural populations

Microsatellite markers are also proving useful in ecological studies to determine parentage in wild sheep populations, such as Rocky Mountain Bighorns *Ovis canadiensis* (Forbes *et al.*, 1995) and Soays (Gulland *et al.*, 1993). It is hoped that the use of DNA markers can be extended beyond measuring reproductive success to look at other aspects important in the survival of wild populations such as parasite burdens and behaviour.

Comparative Aspects and Genome Evolution

The development of the sheep map is being made together with analysis of genomes for a wide range of species. The human genome is rapidly being mapped to the resolution of a single base pair. The entire genome has been cloned in manageable size pieces and arrayed end to end (Cohen *et al.*, 1993). Portions of many of the genes have been sequenced and are currently being located on the map. These advances have required many developments both in the technology and in the storage and analysis of the information. The mouse genome is also well mapped, with more reliable gene orders than the current human map. However, the mouse has fewer regions with the detailed maps now available for human chromosomes.

Despite progress in the sheep genetic map described in these last two chapters, the resolution of maps in the farmed animals is a long way behind that in the mouse and the human. However, all mammalian cells have similar amounts of DNA. The structures of the genomes are remarkably similar and the technology is the same. Consequently, the sheep genome project gains enormously from and also contributes to the developments in other species. The four major areas are development of techniques, comparative information on the locations of individual genes, mapping of genes affecting production traits and comparative genome organization.

The sheep linkage map was developed in collaboration with a number of laboratories across the world. In particular, there was close collaboration between the AgResearch Molecular Biology Unit and US Department of Agriculture (USDA), Clay Center. Sheep and cattle microsatellite markers were swapped between the two laboratories and tested for use in the reciprocal species. About one-third of the microsatellites identified in cattle will amplify polymorphic loci in sheep (Moore *et al.*, 1991). Consequently, half of the markers on the framework map for sheep also appear on the cattle map (Crawford *et al.*, 1995). This strategy reduced the time and cost of marker development for the sheep map, and also provided direct comparisons of the order and distance of markers on the two maps. In general, the order and distances of markers on the two maps are highly conserved. Comparison of the physical maps also demonstrates a high degree of conservation (see Chapter 8).

Comparisons of the linkage maps shows that, in addition to the Robertsonian fusions that have generated the sheep metacentric chromosomes OOV1–3, there is one other translocation (Crawford *et al.*, 1995). Sheep chromosome 9 contains markers derived from cattle chromosome 14 (BBO14) and one end of cattle chromosome 9 (BBO9). The BBO9 markers BM1227 and ETH225, which are adjacent on the cattle map (Bishop *et al.*, 1994), are both informative in sheep. Two-point linkage analysis between these two markers in sheep shows no evidence of linkage (LOD = –2.0 at 30 cM), indicating that the chromosomal breakpoint occurred somewhere between BM1227 and ETH225.

The sheep map is also remarkably similar to the deer genetic map being developed using an interspecific hybrid (Tate *et al.*, 1995). There are many fixed DNA polymorphisms between the deer species, and the pedigrees are particularly valuable for mapping gene loci (type I markers). The current deer map has over 200 known genes ordered into linkage groups. The linkage map for the region equivalent to sheep chromosome OOV24 in deer is identical to the sheep linkage map for this region (Broom *et al.*, 1996). Work is in progress to increase the number of common markers on the sheep and deer maps.

Gene orders are also preserved in more distantly related species. Thus maps of the human and the mouse are also important sources of comparative information. They can be used to check the accuracy of mapping information. They are particularly useful for evaluating potential 'candidate' genes in cloning projects. Genes can be excluded because they map to a different chromosome (Montgomery *et al.*, 1995c) or can be tested in pedigrees if they fall in the region of interest (Montgomery *et al.*, 1995b).

The effort of mapping complex traits can also be shared between species. In the human and the mouse, the emphasis is on mapping genes associated with disease. This includes mapping genes associated with diabetes and fat metabolism. Some of these genes are being tested to see whether they are implicated in energy metabolism and carcass composition in livestock. Study of these traits in livestock may also provide insights into genetic variation in the human. The high heterozygosity observed in sheep (Crawford *et al.*, 1995) and the strain variation available may mean that mapping genes for milk production in sheep would be more efficient than projects in cattle. In any case, progress in mapping genes for lactation in cattle could be applied to selection in sheep.

Future Prospects

The next framework linkage map will be published within a year. This is expected to have over 600 markers, with extensive coverage of all except the Y chromosome. From these markers will be drawn a set of evenly spaced markers that are highly informative and useful in most pedigrees for routine mapping projects. It will be possible to automate genotyping with these markers using the existing technology of DNA-sequencing machines. The combined developments will allow the rapid location of single genes.

The chromosomal regions containing many of the polygenes with moderate effects will also be located. This has yet to be demonstrated for sheep and will require major commitments to appropriate phenotypic measurements, in addition to the marker genotyping. At present, the cost of a single genotype averages about NZ$5. For most parentage tests and marker-assisted selection, several genotypes must be analysed. Consequently, the current cost of genotyping is too high for applications in sheep industries in most countries. The average cost of genotyping must come down for the effective use of markers and maps in the sheep industry.

These developments should be sought from within the livestock genome projects, but may come from genome projects in other species. The utility of the sheep map must be enhanced by increasing the number of genes on the map. The rapid developments, particularly in the human genome project, mean that partial DNA sequence will be known for most mammalian genes in the near future. The functional analysis of all these genes will take longer than the sequencing projects.

The challenge for sheep industries around the world is to capitalize on the incredible wealth of genetic information now emerging. There will be new and unforeseen opportunities for selection, which will improve existing farming practices and lead to the development of sustainable alternative production systems.

References

Ali, S., McClenaghan, M., Simons, J.P. and Clark, A.J. (1990) Characterisation of the alleles encoding ovine ß-lactoglobulins A and B. *Gene* 91, 201–207.

Andersson, L., Haley, C.S., Ellegren, H., Knott, S.A., Johansson, M., Andersson, K., Andersson-Eklund, L., Edfors-Lilja, I., Fredholm, M., Hansson, I., Hakansson, J. and Lundstrom, K. (1994) Genetic mapping of quantative trait loci for growth and fatness in pigs. *Science* 263, 1771–1774.

Avraham, A., Band, M., Yoffe, O., Shani, M. and Ron, M. (1993) Bovine dinucleotide repeat polymorphism at the ARO28 locus. *Animal Genetics* 24, 147.

Bancroft, D. (1993) Genetic variation and fitness in Soay sheep. PhD thesis, University of Cambridge.

Barendse, W., Armitage, S.M., Kossarek, L.M., Shalom, A., Kirkpatrick, B.W., Ryan, A.M., Clayton, D., Li, L., Neibergs, H.L., Zhang, N., Grosse, W.M., Weiss, J., Creighton, P., McCarthy, F., Ron, M., Teale, A.J., Fries, R., McGraw, R.A., Moore, S.S., Georges, M., Soller, M., Womack, J.E. and Hetzel, D.J.S. (1994) A genetic linkage map of the bovine genome. *Nature Genetics* 6, 227–238.

Bishop, M.D., Kappes, S.M., Keele, J.W., Stone, R.T., Sunden, S.L.F., Hawkins, G.A., Solinas Tolodo, S., Fries, R., Grosz, M.D., Yoo, J. and Beattie, C.W. (1994) A genetic linkage map for cattle. *Genetics* 136, 619–639.

Blattman, A.N. and Beh, K.J. (1992) Dinucleotide repeat polymorphism within the ovine major histocompatibility complex. *Animal Genetics* 23, 392.

Blattman, A.N., Hulme, D.J., Kinghorn, B.P., Woolaston, R.R., Gray, G.D. and Beh, K.J. (1993) A search for associations between major histocompatibility complex restriction fragment length polymorphism bands and resistance to *Haemonchus contortus* infection in sheep. *Animal Genetics* 24, 277–282.

Botstein, D., White, R.L., Skolnick, M. and Davis, R.W. (1980) Construction of a genetic linkage map in man using restriction fragment length polymorphisms. *American Journal of Human Genetics* 32, 314–331.

Brezinsky, L., Kemp, S.J. and Teale, A.J. (1993a) ILSTS005 – a polymorphic bovine microsatellite. *Animal Genetics* 24, 73.

Brezinsky, L., Kemp, S.J. and Teale, A.J. (1993b) Five polymorphic bovine microsatellites. *Animal Genetics* 24, 75–76.

Broom, J.E. and Penty, J.M. (1993) A Msp I RFLP at the ovine protein kinase C beta locus. *Animal Genetics* 24, 146.

Broom, J.E., Tate, M.L. and Dodds, K.G. (1996) Linkage mapping in sheep and deer identifies a conserved pecora ruminant linkage group orthologous to two regions of HSA16 and a portion of HSA7q. *Genomics* 33, 308–364.

Buchanan, F.C. and Crawford, A.M. (1992a) Ovine dinucleotide repeat polymorphism at the MAF209 locus. *Animal Genetics* 23, 183.

Buchanan, F.C. and Crawford, A.M. (1992b) Ovine dinucleotide repeat polymorphism at the MAF70 locus. *Animal Genetics* 23, 185.

Buchanan, F.C. and Crawford, A.M. (1992c) Ovine dinucleotide repeat polymorphism at the MAF33 locus. *Animal Genetics* 23, 186.

Buchanan, F.C. and Crawford, A.M. (1992d) Ovine dinucleotide repeat polymorphism at the MAF214 locus. *Animal Genetics* 23, 394.

Buchanan, F.C. and Crawford, A.M. (1993) Ovine microsatellites at the OarFCB11, OarFCB128, OarFCB193, OarFCB266 and OarFCB304 loci. *Animal Genetics* 24, 145.

Buchanan, F.C., Swarbrick, P.A. and Crawford, A.M. (1991a) Ovine dinucleotide repeat polymorphism at the MAF4 locus. *Animal Genetics* 22, 373–374.

Buchanan, F.C., Swarbrick, P.A. and Crawford, A.M. (1991b) Ovine dinucleotide repeat polymorphism at the MAF48 locus. *Animal Genetics* 22, 379–380.

Buchanan, F.C., Swarbrick, P.A. and Crawford, A.M. (1992) Ovine dinucleotide repeat polymorphism at the MAF65 locus. *Animal Genetics* 23, 85.

Buchanan, F.C., Adams, L.J., Littlejohn, R.P., Maddox, J.F. and Crawford, A.M. (1994) Determination of evolutionary relationships among sheep breeds using microsatellites. *Genomics* 22, 397–403.

Buitkamp, J., Schwaiger, F.-W. and Epplen, J.T. (1993) Vb6 T-cell receptor elements in artiodactyls: conservation and germline polymorphisms. *Mammalian Genome* 4, 504–510.

Chapman, H.M. and Bruere, A.N. (1977) Chromosome morphology during meiosis of normal and Robertsonian translocation–carrying rams (*Ovis aries*). *Canadian Journal of Genetics and Cytology* 19, 93–102.

Ciampolini, R.K., Goudarzi, D., Viaman, D. and Leveziel, H. (1993) A new bovine dinucleotide repeat microsatellite: Microsatellite INRA 30. *Animal Genetics* 24, 221.

Clarke, R.A. and Beh, K.J. (1992) Taq I reveals a polymorphism in sheep when probed with the ovine IgE gene. *Animal Genetics* 23, 482.

Cockett, N.E., Jackson, S.P., Shay, T.L., Nielsen, D., Moore, S.S., Steele, M.R., Barendse, W., Green, R.D. and Georges, M. (1994) Chromosomal localization of the callipyge gene in sheep (*Ovis aries*) using bovine DNA markers. *Proceedings of the National*

Academy of Sciences of the USA 91, 3019–3023.

Cohen, D., Chumakov, I. and Weissenbach, J. (1993) A first generation physical map of the human genome. *Comptes Rendus de l'Academie des Sciences Serie Iii Science de la Vie*, 316, 1484–1488.

Copeland, N.G., Jenkins, N.A., Gilbert, D.J., Eppig, J.T., Maltias, L.J., J.C., M., Dietrich, W.F., Weaver, A., Lincoln, S.E., Steen, R.G., Stein, L.E., Nadeau, J.H. and Lander, E.S. (1994) A genetic linkage map of the mouse: current applications and future prospects. *Science* 262, 57–66.

Crawford, A.M., Buchanan, F.C. and Swarbrick, P.A. (1990) Ovine dinucleotide repeat polymorphism at the MAF18 locus. *Animal Genetics* 21, 433–434.

Crawford, A.M., Buchanan, F.C. and Swarbrick, P.A. (1991) The use of dinucleotide repeat or microsatellites as genetic markers in domestic animals. *Proceedings of the New Zealand Society of Animal Production* 51, 79–83.

Crawford, A.M., Montgomery, G.W., Brown, T.M., Swarbrick, P.A., Penty, J.M. and Mathias, H.C. (1992) Linkage mapping of microsatellite loci on the ovine genome. *Animal Genetics* 23, 95 (abstract).

Crawford, A.M., Swarbrick, P.A., Buchanan, F.C. and Dodds, K.G. (1993) DNA fingerprinting analysis of Booroola pedigrees: a search for linkage to the Booroola gene. *Theoretical and Applied Genetics* 87, 271–277.

Crawford, A.M., Montgomery, G.W., Pierson, C.A., Brown, T., Dodds, K.G., Sunden, S.L.F., Henry, H.M., Ede, A.J., Swarbrick, P.A., Berryman, T., Penty, J.M. and Hill, D.F. (1994) Sheep linkage mapping: nineteen linkage groups derived from the analysis of paternal half-sib families. *Genetics* 137, 1–7.

Crawford, A.M., Dodds, K.G., Ede, A.J., Pierson, C.A., Montgomery, G.W., Garmonsway, H.G., Beattie, A.E., Davies, K., Maddox, J.F., Kappes, S.W., Stone, R.T., Nguyen, T.C., Penty, J.M., Lord, E.A., Broom, J.E., Buitkamp, J., Schweiger, W., Epplen, J.T., Mathew, P., Mathews, M.E., Hulme, D.J., Beh, K.J., McGraw, R.A. and Beattie, C.W. (1995) An autosomal genetic linkage map of the sheep genome. *Genetics* 703–724.

Cushwa, W.T., Dodds, K.G., Crawford, A.M. and Medrano, J.F. (1996) Identification and genetic mapping of RAPD markers to the sheep genome. *Mammalian Genome* 7, 580–585.

Davies, K.P. and Maddox, J.F. (1994) Ovine dinucleotide repeat polymorphism at the lysozyme (LYZ) locus. *Animal Genetics* 25, 204.

Demmer, J. and Galloway, S.M. (1993) A Bgl II RFLP at the ovine locus for insulin-like growth factor 1 (IGF1). *Animal Genetics* 24, 76.

Di Gregorio, P., Rando, A. and Masina, P. (1987) Differences in the DNA restriction patterns between sheep with HbA and HbB. *Animal Genetics* 18, 241–247.

Di Gregorio, P., Rando, A., Pieragostini, E. and Masina, P. (1991) DNA polymorphism at the casein loci in sheep. *Animal Genetics* 22, 21–30.

Ede, A.J. and Crawford, A.M. (1995) Mutations in the sequence flanking the microsatellite at the glycine- and tyrosine-rich keratin locus (KAP8) prevent the amplification of some alleles. *Animal Genetics* 26, 43–44.

Ede, A.J., Pierson, C.A., Henry, H.M. and Crawford, A.M. (1994) Ovine microsatellites at the OarAE64, OarHH22, OarHH56, OarHH62 and OarVH4 loci. *Animal Genetics* 25, 51–52.

Ede, A.J., Pierson, C.A. and Crawford, A.M. (1995a) Ovine microsatellites at the OarCP9, OarCP16, OarCP20, OarCP21, OarCP23 and OarCP26 loci. *Animal Genetics* 26, 128–129.

Ede, A.J., Pierson, C.A. and Crawford, A.M. (1995b) Ovine microsatellites at the

OarCP34, OarCP38, OarCP43, OarCP49, OarCP73, OarCP79 and OarCP99 loci. *Animal Genetics* 26, 130–131.

Eggen, A. and Fries, R. (1995) An integrated cytogenetic and meiotic map of the bovine genome. *Animal Genetics* 26, 215–236.

Escayg, A.P., Montgomery, G.W., Hickford, J.G. and Bullock, D.W. (1993) A *Bgl*II restriction fragment length polymorphism at the ovine MHC class II DRA locus. *Animal Genetics* 24, 217.

Foley, R.C. and Beh, K.J. (1992) Analysis of immunoglobulin light chain loci in sheep. *Animal Genetics* 23, 31–42.

Forbes, S.H., Hogg, J.T., Buchanan, F.C., Crawford, A.M. and Allendorf, F.W. (1995) Microsatellite evolution in congeneric mammals: domestic and bighorn sheep. *Molecular Biology and Evolution* 12, 1106–1113.

Galloway, S., Cambridge, L., Hanrahan, V., Dodds, K., Crawford, A. and Hill, D. (1995) A microsatellite and RFLP map of the sheep X chromosome. In: *Proceedings of the 6th Australasian Gene Mapping Workshop*. New Zealand Genetical Society, Dunedin, abstract 69.

Galloway, S.M., Hanrahan, V., Dodds, K.G., Potts, M.D., Crawford, A.M. and Hill, D.F. (1996) A microsatellite and RFLP map of the sheep X chromosome. *Genome Research* 6, 667–677.

Georges, M. and Massey, J. (1992) *Polymorphic DNA Markers in Bovidae*. WO Publication No. 92/13120. World Intellectual Property Organisation, Geneva.

Gootwine, E., Sise, J.A., Penty, J.M. and Montgomery, G.W. (1993) The duplicated gene copy of the ovine growth hormone gene contains a Pvu II polymorphism in the second intron. *Animal Genetics* 24, 319–321.

Grain, F., Nain, M.C., Labonne, M.P., Lantier, F., Lechopier, P., Gebuhrer, L., Asso, J., Maddox, J. and Beutel, H. (1993) Restriction fragment length polymorphism of DQB and DRB class-ii genes of the ovine major histocompatibility complex. *Animal Genetics* 25, 377–384.

Groth, D.M., Lintorn-Terry, E., Outteridge, P., Windon, R.G. and Wetherall, J.D. (1990) Taq I RFLP polymorphism of the ovine complement component C4 gene. *Nucleic Acids Research* 18, 3102.

Groth, D.M. and Weatherall, J.D. (1994) Dinucleotide repeat polymorphism within the ovine major histocompatibility complex class I region. *Animal Genetics* 25, 61.

Gulland, F.M., Albon, S.D., Pemberton, J.M., Moorcroft, P.R. and Clutton-Brock, T.H. (1993) Parasite associated polymorphism in a cyclic ungulate population. *Proceedings of the Royal Society London, B, Biological Science* 254, 7–13.

Hanrahan, V., Sise, J.A. and Hill, D.F. (1992) A Msp I polymorphism at the ovine thyroxine binding globulin locus (TBG). *Animal Genetics* 23, 293.

Hanrahan, V., Ede, A.J., Pierson, C.A. and Hill, D.F. (1993) Ovine microsatellites at the OarVH98, OarVH110, OarVH116, OarVH117 and OarVH130 loci. *Animal Genetics* 24, 223.

Hanrahan, V., Penty, J.M. and Montgomery, G.W. (1994) An *Apa* I polymorphism at the ovine tissue inhibitor of metalloproteinase locus (TIMP). *Animal Genetics* 25, 287.

Hawkin, R.J. and Maddox, J.F. (1994) A polymorphic tetra-nucleotide repeat in the ovine metallothionein II gene. *Animal Genetics* 25, 366.

Hawkin, R.J., Davies, K.P. and Maddox, J.F. (1994) A dinucleotide repeat in the Interleukin 3 gene of limited polymorphism. *Animal Genetics* 25, 286.

Hediger, R., Ansari, H.A. and Stranzinger, G.F. (1991) Chromosome banding and gene localizations support extensive conservation of chromosome structure between cattle and sheep. *Cytogenetics and Cell Genetics* 57, 127–134.

Henry, H.M., Penty, J.M., Pierson, C.A. and Crawford, A.M. (1993) Ovine microsatellites at the OarHH35, OarHH41, OarHH44, OarHH47 and OarHH64 loci. *Animal Genetics* 24, 222.

Hiendleder, S., Leyhe, B., Jaeger, C. and Wassmuth, R. (1992) A Taq I polymorphism at the ovine βA-inhibin (INHBA) locus. *Animal Genetics* 23, 291.

Hulme, D.J., Nicholas, F.W., Windon, R.G., Brown, S.C. and Beh, K.J. (1993) The MHC Class II region and resistance to an intestinal parasite in sheep. *Journal of Animal Breeding and Genetics* 110, 459–472.

Hulme, D.J., Silk, J.P., Redwin, J.M., Barendse, W. and Beh, K.J. (1994) Ten polymorphic ovine microsatellites. *Animal Genetics* 25, 434.

Hunter, N., Foster, J.D., Dickinson, A.G. and Hope, J. (1989) Linkage of the gene for the scrapie-associated fibril protein (PrP) to the Sip gene in Cheviot sheep. *Veterinary Record* 124, 384–386.

Kaukinen, J. and Varvio, S.-L. (1993) Eight polymorphic bovine microsatellites. *Animal Genetics* 24, 148.

Kemp, S.J., Brezinsky, L. and Teale, A.J. (1992) ILSTS002: a polymorphic bovine microsatellite. *Animal Genetics* 23, 184.

Kemp, S.J., Brezinsky, L. and Teale, A.J. (1993) A panel of bovine ovine and caprine polymorphic microsatellites. *Animal Genetics* 24, 363–366.

Kossarek, L.M., Grosse, W.M., Dietz, A.B., Womack, J.E., McGraw, R.A. (1993a) Bovine dinucleotide repeat polymorphism RM004. *Journal of Animal Science* 71, 3175.

Kossarek, L.M., Grosse, W.M., Finlay, O., McGraw, R.A. (1993b) Bovine dinucleotide repeat polymorphism RM006. *Journal of Animal Science* 71, 3176.

Kossarek, L.M., Su, X., Grosse, W.M., Finlay, O., Barendse, W., Hetzel, D.J.S., Dietz, A.B., Womack, J.E., McGraw, R.A. (1993c) Bovine dinucleotide repeat polymorphism RM067. *Journal of Animal Science* 71, 3178.

Kossarek, L.M., Su, X., Grosse, W.M., Finlay, O., Barendse, W., Hetzel, D.J.S., McGraw, R.A. (1993d) Bovine dinucleotide repeat polymorphism RM106. *Journal of Animal Science* 71, 3180.

Lander, E.S. and Green, P. (1987) Construction of multilocus genetic maps in humans. *Proccedings of the National Academy of Sciences USA* 84, 2263–2267.

Leveziel, H., Metenier, L., Guerin, G., Cullen, P., Provot, C., Bertaud, M. and Mercier, J.C. (1991) Restriction fragment length polymorphisms of ovine casein genes: close linkage between the α_{s1}, α_{s2}, β- and κ-casein loci. *Animal Genetics* 22, 1–10.

Litchfield, A.M., Raadsma, H.W., Hulme, D.J., Brown, S.C., Nicholas, F.W. and Egerton, J.R. (1993) Disease resistance in Merino sheep. 2. RFLPs in Class II MHC and their association with resistance to footrot. *Journal of Animal Breeding and Genetics* 110, 321–334.

Lord, E.A., Lumsden, J.M., Dodds, K.G., Henry, H.M., Crawford, A.M., Ansari, H.A., Pearce, P.D., Maher, D.W., Stone, R.T., Kappes, S.M., Beattie, C.W. and Montgomery, G.W. (1996) The linkage map of sheep chromosome 6 compared with orthologous regions in other species. *Mammalian Genome* 7, 373–376.

Lumsden, J.M., Lord, E.A. and Montgomery, G.W. (1996) Characterisation and linkage mapping of ten sheep microsatellites from a sheep × hamster cell hybrid. *Animal Genetics* 27, 203–206.

Matthew, P. and Maddox, J.F. (1994a) A polymorphic dinucleotide repeat microsatellite detected in the ovine interleukin 2 receptor alpha (IL2RA, CD25). *Animal Genetics* 25, 200.

Matthew, P. and Maddox, J.F. (1994b) A polymorphic microsatellite detected in the ovine CD5 gene. *Animal Genetics* 25, 371.

Montgomery, G.W. and Sise, J.A. (1990) Extraction of DNA from sheep white blood cells. *New Zealand Journal of Agricultural Research* 33, 437–441.

Montgomery, G.W., Tate, M.L., Sise, J.A. and Hill, D.F. (1990) The Booroola F gene is excluded from close genetic linkage with either haemoglobin β or follicle stimulating hormone β loci. *Proceedings of the New Zealand Society for Animal Production* 50, 411–415.

Montgomery, G.W., Sise, J.A., Penty, J.M., Tou, H.M. and Hill, D.F. (1992) Sheep linkage mapping: restriction fragment length polymorphism detection with heterologous cDNA probes. *Animal Genetics* 23, 411–416.

Montgomery, G.W., Crawford, A.M., Penty, J.M., Dodds, K.G., Ede, A.J., Henry, H.M., Pierson, C.A., Lord, E.A., Galloway, S.M., Schmack, A.E., Sise, J.A., Swarbrick, P.A., Hanrahan, V., Buchanan, F.C. and Hill, D.F. (1993) The ovine Booroola fecundity gene (*FecB*) is linked to markers from a region of human chromosome 4q. *Nature Genetics* 4, 410–414.

Montgomery, G.W., Lord, E.A., Penty, J.M., Dodds, K.G., Broad, T.E., Cambridge, L., Sunden, S.L.F., Stone, R.T. and Crawford, A.M. (1994) The Booroola fecundity (FecB) gene maps to sheep chromosome 6. *Genomics* 22, 149–153.

Montgomery, G.W., Penty, J.M., Henry, H.M., Sise, J.A., Lord, E.A., Dodds, K.G. and Hill, D.F. (1995a) Sheep linkage mapping: RFLP markers for comparative mapping studies. *Animal Genetics* 26, 249–259.

Montgomery, G.W., Penty, J.M., Lord, E.A., Brooks, J. and McNeilly, A.S. (1995b) The gonadotrophin-releasing hormone receptor maps to sheep chromosome 6 outside of the region of the *FecB* locus. *Mammalian Genome* 6, 436–438.

Montgomery, G.W., Penty, J.M., Lord, E.A. and Broom, M.F. (1995c) The search for the Booroola (*FecB*) gene. *Journal of Reproduction and Fertility* Suppl. 49, 113–121.

Montgomery, G.W., Tate, M.L., Henry, H.M., Penty, J.M. and Rohan, R.M. (1995d) The follicle-stimulating hormone receptor and luteinizing hormone receptor genes are closely linked in sheep and deer. *Journal of Molecular Endocrinology* 15, 259–265.

Montgomery, G.W., Henry, H.M., Dodds, K.G., Beattie, A.E., Wuliji, T. and Crawford, A.M. (1996) Mapping the *Horns* (*Ho*) locus in sheep: a further locus controlling horn development in domestic animals. *Journal of Heredity* 87, 358–363.

Moore, S.S. and Byrne, K. (1993a) Dinucleotide polymorphism at the bovine histamine H1 receptor locus. *Animal Genetics* 24, 72.

Moore, S.S. and Byrne, K. (1993b) Dinucleotide polymorphism at the bovine calmodulin independent adenylcyclase locus. *Animal Genetics* 24, 150.

Moore, S.S., Sargent, L.L., King, T.J., Mattick, J.S., Georges, M. and Hetzel, D.J.S. (1991) The conservation of dinucleotide microsatellites among mammalian genomes allows the use of heterologous PCR primer pairs in closely related species. *Genomics* 10, 654–660.

Moore, S.S., Barendse, W., Berger, K.T., Armitage, S.M. and Hetzel, D.J.S. (1992) Bovine and ovine DNA microsatellites from the EMBL and GENBANK databases. *Animal Genetics* 23, 463–467.

Moore, S.S., Byrne, K., Berger, K.T., Barendse, W., McCarthy, F., Womack, J.E. and Hetzel, D.J.S. (1994) Characterisation of 65 bovine microsatellites. *Mammalian Genome* 5, 84–90.

O'Brien, S.J., Womack, J.E., Lyons, L.A., Moore, K.J., Jenkins, N.A. and Copeland, N.G. (1993) Anchored reference loci for comparative genome mapping in mammals. *Nature Genetics* 3, 103–112.

Parsons, Y.M., Cooper, D.W., Piper, L.R. and Adams, T.E. (1992a) RFLPs at the ovine locus for growth hormone. *Animal Genetics* 23, 571.

Parsons, Y.M., Cooper, D.W., Piper, L.R. and Powell, B.C. (1992b) A *Bam* HI polymorphism in ovine wool cuticle keratin sequences. *Animal Genetics* 23, 569.

Parsons, Y.M., Piper, L.R. and Cooper, D.W. (1994b) Linkage relationships between keratin associated protein (KRTAP) genes and growth hormone in sheep. *Genomics* 20, 500-502.

Parsons, Y.M., Cooper, D.W. and Piper, L.R. (1994a) Evidence of linkage between high-glycine-tyrosine keratin gene loci and wool fibre diameter in a Merino half-sib family. *Animal Genetics* 25, 105–108.

Parsons, Y.M., Cooper, D.W., Piper, L.R. and Beh, K.J. (1994c) A *Taq* I polymorphism at the immunoglobulin kappa light chain locus in Merino sheep. *Animal Genetics* 25, 54.

Penty, J.M. and Hill, D.F. (1991) A Rsa I restriction fragment length polymorphism at the ovine locus for gamma-interferon. *Animal Genetics* 22, 439.

Penty, J.M. and Montgomery, G.W. (1991a) Genetic linkage between α-inhibin and fibronectin 1 in sheep. In:*Australasian Gene Mapping and Molecular Genetics Symposium, Dunedin, New Zealand*, New Zealand Genetical Society,p. 65.

Penty, J.M. and Montgomery, G.W. (1991b) A *Pst* I restriction fragment length polymorphism at the ovine locus for poly-ubiquitin. *Animal Genetics* 22, 438.

Penty, J.M. and Montgomery, G.W. (1991c) A *Msp* I restriction fragment length polymorphism at the ovine locus for glucagon. *Animal Genetics* 22, 440.

Penty, J.M. and Montgomery, G.W. (1993) An Xba RFLP at the ovine basic fibroblast growth factor locus (FGFB). *Animal Genetics* 24, 329.

Penty, J.M., Tou, H.M. and Hill, D.F. (1991a) A *Msp* I restriction fragment length polymorphism at the ovine locus for 2',3'-cyclic nucleotide 3'-phosphohydrolase. *Animal Genetics* 22, 196.

Penty, J.M., Tou, H.M. and Hill, D.F. (1991b) A *Msp* I restriction fragment length polymorphism at the ovine locus for neurotensin. *Animal Genetics* 22, 197.

Penty, J.M., Sise, J.A. and Montgomery, G.W. (1991c) Restriction fragment length polymorphisms at the ovine locus for the α-subunit of the pituitary glycoprotein hormones. *Animal Genetics* 22, 198.

Penty, J.M., Henry, H.M., Ede, A.J. and Crawford, A.M. (1993) Ovine microsatellites at the OarAE16, OarAE54, OarAE57, OarAE119 and OarAE129 loci. *Animal Genetics* 24, 219.

Penty, J.M., Lord, E.A., Dodds, K.G., Galloway, S.M. and Montgomery, G.W. (1995) Linkage of LHB and MAG to GPI on sheep chromosome 14. *Mammalian Genome* 6, 299–300.

Peters, J.A. (1959) *Classic Papers in Genetics*. Prentice Hall, Inglewood Cliffs, New Jersey.

Phua, S.H. and Wood, N.J. (1993) The myogenic factor 5 (MYF5) locus in sheep carries a rare *Eco* RI RFLP. *Animal Genetics* 24, 220.

Phua, S.H. and Wood, N.J. (1995a) Haplotypes of restriction polymorphisms as informative genetic markers at the ovine glutathione reductase (GSR) locus. *Animal Genetics* 27, 122–123.

Phua, S.H. and Wood, N.J. (1995b) A high frequency *Taq* I polymorphism at the ovine phenylalanine hydroxylase (PAH) locus. *Animal Genetics* 27, 124.

Phua, S.H. and Wood, N.J. (1995c) A polymorphism detected in sheep using a complememt component C1r (C1R) probe. *Journal of Animal Science* 3, 629.

Phua, S.H., Collins, L., Lewis, P., Wood, N., McNabb, L., Broad, T. and Pearce, P. (1992a) An *Eco* RI restriction fragment length polymorphism at the ovine alpha-s2-casein (CASAS2) locus. *Animal Genetics* 23, 475.

Phua, S.H., Collins, L., Wood, N., McNabb, L., Davidson, B., Broad, T. and Lewis, P. (1992b) An *Eco* RI restriction fragment length polymorphism at the ovine locus for α1 type III collagen (COL3A1). *Animal Genetics* 23, 477.

Phua, S.H., Collins, L.J., Wood, N., Davidson, B.P. and McNabb, L.R. (1993a) A low frequency *Eco* RI RFLP at the ovine protamine 1 (PRM1) locus. *Animal Genetics* 24, 142.

Phua, S.H., Cambridge, L.M. and Wood, N.J. (1993b) *Eco* RI reveals a polymorphism at the ovine gelsolin (GSN) locus. *Animal Genetics* 24, 143.

Phua, S.H., Wood, N.J. and Davidson, B.P. (1993c) A *Pst*I RFLP at the ovine elastin (ELN) locus. *Animal Genetics* 24, 143.

Phua, S.H., Collins, L.J., Wood, N.J. and Davidson, B.P. (1993d) A three-allele *Pst* I RFLP at the ovine locus for myosin I heavy chain-like protein (M1HC). *Animal Genetics* 24, 150.

Phua, S.H., Cambridge, L.M. and Wood, N.J. (1993e) *Pst* I enzyme reveals a low frequency polymorphism at the rhodopson (RHO) locus in sheep. *Animal Genetics* 24, 220.

Phua, S.H., Wood, N.J. and Collins, L.J. (1994a) The ovine esterase D (ESD) locus carries a high-frequency Eco RI polymorphism. *Animal Genetics* 25, 288.

Phua, S.H., Cambridge, L.M. and Wood, N.J. (1994b) *Taq* I and *Msp* I restriction polymorphisms at the ovine apolipoprotein B (APOB) locus. *Animal Genetics* 25, 289.

Phua, S.H., Wood, N.J. and Collins, L.J. (1994c) A high-frequency biallelic *Taq* I RFLP at the PLAT (plasminogen activator, tissue) locus in sheep. *Animal Genetics* 25, 293.

Phua, S.H., Wood, N.J. and Collins, L.J. (1994d) A compound *Taq* I polymorphism in sheep revealed by a carbonic anhydrase II (CA2) probe. *Animal Genetics* 25, 372.

Phua, S.H., Wood, N.J. and Hagemann, L.J. (1994e) A *Pst*I restriction fragment length polymorphism in sheep detected with a human myelin basic protein (MBP) probe. *Journal of Animal Science* 72, 3266.

Pierson, C.A., Hanrahan, V., Ede, A.J. and Crawford, A.M. (1993) Ovine microsatellites at the OarVH34, OarVH41, OarVH58, OarVH61 and OarVH72 loci. *Animal Genetics* 24, 224.

Pierson, C., Ede, A.J. and Crawford, A.M. (1994) Ovine microsatellite at the OarHH30, OarHH51, OarVH54, OarCP88, OarCP93, and OarCP134 loci. *Animal Genetics* 25, 294–295.

Potts, M.D., Hanrahan, V. and Hill, D.F. (1994a) A *Bgl*II polymorphism at the ovine phosphorylase kinase alpha subunit locus (PHKA1). *Animal Genetics* 25, 288.

Potts, M.D., Hanrahan, V. and Hill, D.F. (1994b) An *Eco*RV polymorphism at the ovine Menkes disease locus (ATP7A). *Animal Genetics* 25, 293.

Primmer, C.R. and Matthews, M.E. (1993) Ovine dinucleotide repeat polymorphism at the VIAS-S5 locus. *Animal Genetics* 24, 146.

Rasmusen, B. (1966) Linkage between the C and I blood groups in sheep. *Genetics* 54, 356.

Robinson, N., Ward, W. and Matthews, M. (1995) Allelic variation linked to FGF1 is associated with reduced wool fibre diameter and greasy fleece weight in sheep. In: *Proceedings of the 6th Australasian Gene Mapping Workshop, Dunedin*, abstract 12.

Rogers, G.R., Hickford, J.G., Bickerstaffe, R. and Woods, J.L. (1993a) *Bsr*I RFLP in the gene for the ovine B2C high-sulphur wool protein. *Animal Genetics* 24, 69.

Rogers, G.R., Hickford, J.G.H. and Bickerstaffe, R. (1993b) *Msp*I RFLP in the gene for a Type I intermediate filament wool gene. *Animal Genetics* 24, 218.

Ryan, A.M. and Womack, J.E. (1994a) Assignment of the proopiomelanocortin (POMC) gene to Bta 11 and Rflps in cattle and sheep. *Animal Genetics* 25, 58.

Ryan, A.M. and Womack, J.E. (1994b) A *Pst*I restriction fragment length polymorphism in the ovine *Aconitase* 2 gene. *Journal of Animal Science* 72, 1076.

Schmahl, G., Leyhe, B., Hiendleder, S. and Wassmuth, R. (1993) An *Eco*RI polymorphism at the ovine enolase 1 (ENO1) locus. *Animal Genetics* 24, 70.

Schwaiger, F.W., Buitkamp, J., Weyer, E. and Epplen, J.T. (1993) Typing of artiodactyl MHC-DRB genes with the help of intronic simple repeated DNA sequences. *Molecular Ecology* 2, 55–59.

Scott, P.C., Gogolin-Ewens, K.J., Adams, T.E. and Brandon, M.R. (1991) Nucleotide sequence, polymorphism, and evolution of ovine MHC class II *DQA* genes. *Immunogenetics* 34, 69–79.

Shalom, A., Mosig, M.O., Barendse, W., Friedman, A. and Soller, M. (1994) Dinucleotide repeat polymorphism at the bovine HUJ246, HUJII77, HUJ223, HUJVI74, and HUJI75. *Animal Genetics* 25, 56.

Shay, T.L. and Muggli-Cockett, N.E. (1993) A *Taq*I restriction fragment length polymorphism in the ovine collagenase gene. *Journal of Animal Science* 71, 1355.

Sise, J.A. and Montgomery, G.W. (1992) A *Pvu*II polymorphism at the ovine corticotrophin releasing hormone (CRH) locus. *Animal Genetics* 23, 86.

Sise, J.A., Tou, H.M. and Montgomery, G.W. (1991) A *Taq*I polymorphism at the ovine α-Inhibin locus. *Animal Genetics* 22, 195.

Sise, J.A., Hanrahan, V. and Hill, D.F. (1992) A *Msp*I polymorphism at the ovine proteolipid protein locus (PLP). *Animal Genetics* 23, 292.

Sise, J.A., Hillyard, A.L. and Montgomery, G.W. (1996) The sheep gene map database (SheepBase) is now available on the World Wide Web. *Mammalian Genome* 7, 1.

Steffen, P., Eggen, A., Dietz, A.B., Womack, J.E., Stranzinger, G. and Fries, R. (1993) Isolation and mapping of polymophic microsatellites in cattle. *Animal Genetics* 24, 121–124.

Stone, R.T., Pulido, J.C., Duyk, G.M., Kappes, S.M., Keele, J.W. and Beattie, C.W. (1995) A small-insert bovine genomic library highly enriched for microsatellite repeat sequences. *Mammalian Genome* 6, 714–724.

Sun, H.S., Dentine, M.R. and Kirkpatrick, B.W. (1993) A polymorphic microsatellite (UWCA4) detected on bovine chromosome 21. *Animal Genetics* 24, 149.

Sun, H.S., Huff, B.M., Dentine, M.R. and Kirkpatrick, B.W. (1994) Polymorphic bovine microsatellites UWCA5, UWCA7 and UWCA9. *Animal Genetics* 25, 59.

Swarbrick, P.A. and Crawford, A.M. (1992) An ovine dinucleotide repeat polymorphism at the MAF109 locus. *Animal Genetics* 23, 84.

Swarbrick, P.A., Buchanan, F.C. and Crawford, A.M. (1990) Ovine dinucleotide repeat polymorphism at the MAF23 locus. *Animal Genetics* 21, 191.

Swarbrick, P.A., Buchanan, F.C. and Crawford, A.M. (1991a) Ovine dinucleotide repeat polymorphism at the MAF35 locus. *Animal Genetics* 22, 369–370.

Swarbrick, P.A., Buchanan, F.C. and Crawford, A.M. (1991b) Ovine dinucleotide repeat polymorphism at the MAF64 locus. *Animal Genetics* 22, 375–376.

Swarbrick, P.A., Buchanan, F.C. and Crawford, A.M. (1991c) Ovine dinucleotide repeat polymorphism at the MAF36 locus. *Animal Genetics* 22, 377–378.

Swarbrick, P.A., Howes, J. and Crawford, A.M. (1992a) Ovine dinucleotide repeat polymorphism at the MAF50 locus. *Animal Genetics* 23, 187.

Swarbrick, P.A., Howes, J. and Crawford, A.M. (1992b) MAF45, a highly polymorphic marker for the pseudoautosomal region of the sheep genome, is not linked to the *FecX*^{I} (Inverdale) gene. *Genomics* 13, 849–851.

Tanksley, S.D. (1993) Mapping polygenes. *Annual Reviews of Genetics* 27, 205–233.

Tate, M.L., Mathius, H.C., Fennessy, P.F., Dodds, K.G., Penty, J.M. and Hill, D.F. (1995) A new gene mapping resource: interspecies hybrids between Pere David's deer (*Elaphurus davidianus*) and red deer (*Cervus elaphus*). *Genetics* 139, 1383–1391.

Thoday, J.M. (1961) Location of polygenes. *Nature* 191, 368–370.

Tisdall, D.J. and Penty, J.M. (1991) A *Pst* I restriction fragment length polymorphism at the ovine locus for follistatin. *Animal Genetics* 22, 437.

Tisdall, D.J., Quirke, L.D. and Galloway, S.M. (1996) Ovine stem cell factor gene is located within a syntenic group on chromosome 3 conserved across mammalian species. *Mammalian Genome* 7, 472.

Tou, H.M., Penty, J.M. and Montgomery, G.W. (1992) A *Pvu*II restriction fragment length polymorphism at the ovine uncoupling protein locus. *Animal Genetics* 23, 83.

Tucker, E.M., Evans, N.R.S. and Kilgour, L. (1980) Close linkage between the C blood group locus and the locus controlling amino acid transport in sheep erythrocytes. *Animal Blood Groups and Biochemical Genetics* 11, 119–125.

Vaiman, D., Osta, R., Mercier, D., Grohs, C. and Leveziel, H. (1992) Characterisation of five new bovine dinucleotide repeats. *Animal Genetics* 24, 537–541.

Valinsky, A., Shani, M. and Gootwine, E. (1990) Restriction fragment length polymorphism in sheep at the growth hormone locus is the result of variation in gene number. *Animal Biotechnology* 1, 135–144.

van Stijn, T. and Galloway, S. (1995) A *Bam*H1 polymorphism at the ovine inactive X-specific transcript locus (XIST). *Animal Genetics* 26, 279.

Vos, P., Hogers, R., Bleeker, M., Reijans, M., van de Lee, T., Hornes, M., Frijters, A., Pot, J., Peleman, J., Kuiper, M. and Zabeau, M. (1995) AFLP: a new technique for DNA fingerprinting. *Nucleic Acids Research* 23, 4407–4414.

Weber, J.L. and May, P.E. (1989) Abundant class of human DNA polymorphisms which can be typed using the polymerase chain reaction. *American Journal of Human Genetics* 44, 388–396.

Wetherall, J.D. and Groth, D.M. (1992) The major histocompatibility complex and parasite immunity. In: (ed.) *Animal Parasite Control Utilising Biotechnology.* CRC Press Boca Rarton, Florida, pp. 353–386.

Wood, N.J. and Phua, S.H. (1993) A dinucleotide repeat polymorphism at the adenylate cyclase activating polypeptide locus in sheep. *Animal Genetics* 24, 329.

Wood, N.J., Phua, S.H. and Davidson, B.P. (1993) Polymorphism in an A- and T-rich element in theβ-haemoglobin locus in sheep.*Animal Genetics* 24, 71.

Wuliji, T., Montgomery, G.W., Dodds, K.G., Andrews, R.A., Beattie, A., Neill, P., Turner, P. and Rogers, J. (1995) Establishing a flock for gene mapping in wool traits. *Proceedings of the New Zealand Society of Animal Production* 55, 285–288.

Genetics of Behaviour 12

G.N. HINCH
Department of Animal Science, University of New England, Armidale, NSW 2351, Australia

Introduction 353
Feeding Behaviour 355
  Grazing behaviour 355
  Voluntary feed intake 356
Social Behaviour and Organization 357
  Associations 358
  Handling 359
  Mechanisms 361
Reproductive Behaviour 362
  Male behaviours 362
  Female behaviours 363
Maternal Behaviour 363
  Breed comparisons 364
  Selection criteria for improved maternal behaviour 366
Conclusion 368
Acknowledgements 369
References 369

Introduction

The study of behavioural genetics can begin from a genetic, physiological or a behavioural bias. The first two approaches use behaviour as a means for understanding the workings of genes or physiological systems (Plomin *et al.*, 1990). In the third case, behaviour is seen as being of intrinsic interest or relevance. The question that remains to be answered is: how do genes control or change behaviour? For those interested in animal production, the question of how genes control and change behaviour is of central concern and this chapter will focus on this problem by asking about the presence of genetic variation for behavioural traits of sheep and the form which this variation may take.

Evidence that behavioural traits are heritable has been found in many species, including insects (Roff and Mousseau, 1987) and rodents (DeFries *et al.*, 1974), the heritabilities tending to be 0.3 or less. Genetic variation in the behavioural traits for domestic livestock has been identified mostly in traits related to well-being and adaptability to intensification (Goddard, 1980) and most estimates of heritability range from 0.25 to 0.50 (Craig, 1994). For example, Willham *et al.* (1963) examined avoidance learning behaviour in piglets and was able to show large maternal and non-additive genetic variance. Their estimates of heritability from sire and dam components was 0.45 ± 0.12. Additive genetic variation has been identified in other behavioural traits such as broodiness in hens (Craig, 1994), fearfulness and distraction in dogs (see review by Mackenzie *et al.*, 1986), open-field behaviour in pigs (Beilharz and Cox, 1967) and social dominance and aggression in chickens (Guhl *et al.*, 1960; Craig *et al.*, 1965).

There is little information on the genetics of behaviour in ruminant farm animals. Faure (1994) indicates this could be for at least three reasons.

- Behaviour traits have low repeatability, possibly due to the fact that many behaviours are part of an interaction with other organisms/conspecifics.
- Behaviour is difficult to measure and therefore studies of multiple traits tend to have been conducted on small sample sizes and have large standard errors associated with genetic parameter estimations
- Behavioural traits are often not normally distributed and therefore accurate estimates of genetic parameters is difficult.

Given these limitations it may not be surprising that we do not have a large body of data on the genetics of behaviour of sheep. However, behavioural traits have been central to the use of sheep since they were first domesticated.

It seems likely that the initial phases of domestication were associated with the choice of animals with behavioural traits suited to people's requirements. Hale (1962) has listed some of the behaviours favouring domestication (Table 12.1) and Tennessen and Hudson (1981) have largely confirmed Hale's conclusions, identifying the behavioural traits differentiating between domesticated and non-domesticated ungulates as: non-selective feeders, occupying open habitats and being socially advanced, non-territorial species living in groups of 15–100 animals.

The behavioural patterns described by these authors are apparent in the

Table 12.1. Behavioural characteristics favouring domestication (adapted from Hale, 1962).

Group structure	Large hierarchically structured groups
Sexual behaviour	Promiscuous mating behaviour
Parent–young interactions	Praecocial young establishing a bond at an early age
Responses to humans	Short flight distances
Habitat requirements	Herbivorous and adaptable to a wide range of environmental conditions

Ovis species, but one could equally argue that selection for these behavioural traits has continued as sheep have been required to adapt to changes in management and handling. To what extent the selection for domestication on behavioural characteristics and subsequent selection have altered variation in behavioural traits is difficult to predict. Certainly, much of the evidence for variation in behavioural traits of sheep confounds genetic, maternal and environmental effects. In the review that follows, the primary behaviours of the sheep ethogram – feeding behaviours, social behaviours, reproductive and maternal behaviours – will be examined for evidence of genetic variation.

Feeding Behaviour

The grazing behaviour of modern sheep breeds appears to be similar to that of their wild counterparts which live in a wide range of environments. It seems that sheep are very adaptable in terms of the range of feeds which they will eat ranging from seaweed (Paterson and Colemen, 1982) to cacti (Lynch *et al.*, 1992) and most grasses and forbs. Because of this adaptability it seems unlikely that significant variability in feed choice will be identifiable as due to genotype.

Sheep do show preferences for particular areas within plant communities, possibly associated with changes in the quality and characteristics of the feed available or previous experience of the plants. This is most obvious in large areas of free range, and evidence of selectivity was shown in one study (Gluesing and Balph, 1980), where animals were willing to walk 50% further to locate a preferred plant species.

Grazing behaviour

Sharafeldin and Shafie (1965) examined the grazing behaviour of a variety of breeds and found that the indigenous Ossimi breed was better adapted to walking in the high solar radiation load of the hot arid environment of Egypt compared with Texels, Russian Caucasian Merinos and German Merinos. There were also differences in feed selectivity, Texels being the most selective. In contrast, in a more benign environment Berggren-Thomas and Hohenboken (1986) found no significant differences between the grazing time and distance travelled during grazing for crossbred ewes from four sire breeds (Suffolk, Clun Forest Dorset, Polpay and Border Leicester). These results would suggest that breed differences in grazing behaviours may well be a function of environmental conditions and reflect breed thermoregulatory characteristics rather than innate behavioural differences.

This was also the conclusion of Dudzinski and Arnold (1979), who recorded the grazing behaviour of Romney, Dorset Horn, Cheviot, Suffolk, Southdown and Border Leicester ewes. They found differences in the patterns of grazing in relation to daylight and also in total grazing times; Suffolk animals grazed for the longest periods (8.1 h) and Cheviots and Southdowns the least

(5.5 and 5.4 h respectively). The authors concluded that these differences were a reflection of differences in physiological response to the relatively hot environment of Western Australia. Border Leicesters were the most reactive to environmental change and Dorset Horns the least.

There are few data on the variation in grazing times within flocks and perhaps this is not a behavioural measure of importance, as there is considerable evidence that sheep can adapt to shortened grazing times by increases in grazing intensity and changes in bite size. There appear to be no estimates of the genetic contribution to variation in grazing intensity of sheep but heritability of this trait in dairy cattle has been reported to be not significantly different from zero (Macha and Olsarova, 1986).

The adaptability of grazing behaviour implies that there is considerable capacity for sheep to adapt their grazing behaviour to meet their voluntary intake drive. Variation in total grazing time and intake is further confounded by the selective capacities of sheep for different plant species and plant parts. Considerable variation in this selectivity between individuals of the same breed has been reported (Arnold, 1981; Lynch *et al.*, 1992). Those breeds with a greater 'intake drive' appear to be less selective, but there are no definitive experiments to show this clearly. It seems likely that selection for increases in voluntary feed intake may reduce selectivity, with some implications for sheep grazing heterogeneous rangeland environments.

Voluntary feed intake

In a study of the feed intake and grazing behaviour of Border Leicester × Merino, Merino, Corriedale and Dorset Horn ewes, Arnold (1975) found that breeds were similar in their behavioural responses to changes in physiological state (pregnancy and lactation) and to changes in pasture conditions although there were some differences in digestible organic-matter intake (DOMI) kg^{-1} live weight, favouring Border Leicester × Merino and Corriedale ewes. An earlier study by Langlands (1973) had also shown sire-breed effects on milk and herbage intake of lambs from Merino dams and sired by Merino, Border Leicester, Corriedale, Dorset Horn and Southdown rams, the larger breeds having higher intakes.

Lee *et al.* (1992) measured the DOMI from pasture by Merino ewes by using controlled-release devices containing a chromium marker. They found that, at the phenotypic level, intake was positively related to live weight and that this was consistent across years. Estimate of heritability for DOMI was 0.22 ± 0.08 which is slightly lower than estimates for stall-fed cattle. Further analysis of a larger data set (Lee *et al.*, 1995) showed that between-flock variation (15 Merino blood-lines) was relatively small after adjustment for live weight and that the heritability of DOMI, estimated using a sire-model restricted maximum likelihood (REML) analysis, was 0.12 ± 0.07 when data were pooled across seasons.

This heritability estimate for DOMI is considerably lower than that reported (0.70 ± 0.12) for intensively reared lambs (Cameron, 1988), but Lee *et

Table 12.2. Heritability estimates (bold) and genetic correlations (SE*) for digestible organic matter intake (DOMI) and live weight (adapted from Lee *et al.*, 1995).

Trait	DOMI	DOMI kg^{-1}	Live weight
DOMI	0.12 ± 0.07	–0.29 ± 0.22	0.73 ± 0.31
DOMI kg^{-1}		**0.10 ± 0.07**	–0.42 ± 0.27
Live weight			**0.67 ± 0.09**

*SE, standard error.

al. (1995) argued that this is not unexpected, as the opportunity for diet selection in grazing animals is greater than in housed animals and consequently the genetic variation in feed intake as a proportion of total variation is likely to be smaller.

Genetic correlations between DOMI and intake kg^{-1} live weight and live weight are shown in Table 12.2. A moderate to strong positive correlation of DOMI with live weight was the most clear relationship.

These data clearly indicate some genetic variation in feed intake but would also confirm the earlier suggestions that environmental factors and the innate adaptability of grazing behaviour in sheep make these traits inappropriate for selection. However, social organization will influence the utilization of pasture and, given that breed differences do exist in spatial distribution patterns, it seems likely that between- and within-breed variation in spatial characteristics may be a more appropriate selection criterion to be used in altering pasture utilization. These behaviours will be examined more closely in the following section.

Social Behaviour and Organization

Social behaviours of sheep are largely dominated by the flocking nature of the species, which allows it to be easily manipulated by humans and was possibly the characteristic that was predominant in initiating domestication. Kilgour (1976) describes the sheep as:

> a defenceless, vigilant, tight flocking, wool covered ruminant; evolved within a mountain grassland habitat, displaying a follower type dam–offspring relationship with imitation between young and old in establishing range, showing a seasonal breeding and separate adult male sub group structure.

This to a large degree defines the behavioural characteristics of sheep. It also indirectly highlights that the species, at least at first appearance, is relatively homogeneous in terms of behavioural traits. As argued by Lynch *et al.* (1992), there is a continuum of social organization differences but a surprising consistency in the basic social behavioural patterns.

The social behaviours exhibited by sheep include both individual agonistic and amicable behaviours, as well as organization of the flock. A review of the agonistic behaviours for various wild and domestic sheep has been presented by Lynch *et al.* (1992). It would seem that there is little breed variation in these behaviours, although there may be some seasonal differences in incidence associated with the seasonality of mating behaviour. It might be expected that domestication and the associated reduction in male competition may have altered the need for a full repertoire of aggressive behaviours, but these behaviours remain in domestic sheep breeds and are seen in the context of multiple joining. For example, Fowler and Jenkins (1976) have clearly shown that dominance is related to mating success by reducing flock fertility by 28% when the dominant ram was vasectomized.

Aggressiveness as a trait of the individual could be important in determining dominance and reproductive success in rams and, as a fitness trait, it could be argued that it is likely to be heritable. However, there do not appear to be studies to confirm this in sheep, and selection based on traits associated with androgen production (which is known to increase aggression), such as testis size, do not seem to have altered the incidence of aggressive behaviours.

The individual trait of leadership can also be seen in sheep flocks, but its repeatability is low. In a study by Lynch *et al.* (1989), it was shown that in young Merino and Merino × Border Leicester ewes there was low repeatability of leadership measures and that leadership was not related to fear measurements. Whether this is true in older animals is unclear but most reports suggest that if leadership does occur, it is dominated by a few older animals and may be a learnt behaviour. Some evidence that leadership traits may be heritable comes from reports of a 'leader' sheep breed in Iceland (Jonmundsson *et al.*, 1994). These animals are described as a particular strain whose instinct is to run or walk in front of the flock. They are described as 'calm' in normal husbandry operations, with morphological traits that differ from ordinary Icelandic sheep. Further information is needed to determine if the characteristics of this 'breed' are associated with leadership or an alternative trait, such as fearfulness.

Associations

There is increasing evidence for associations between pairs of animals (usually measured in terms of nearest neighbour distance) in a diversity of domestic breeds (Arnold, 1985), but how long these associations last is unclear and repeatability is unknown. Certainly there is much anecdotal evidence of tag numbers often being together during husbandry routines. Clear cut associations are most clearly seen in mixed age or breed flocks. Dorset and Merino animals appear to form strong associations and Shillito-Walser *et al.* (1981) have presented evidence that these associations are learnt, but their persistence seems to vary: for example, Lawrence (1990) reported that associations between Scottish Blackface ewes and their lambs were almost non-existent by 12 months, while Hinch *et al.* (1990) showed that associations between Merino ewes and lambs could last for at least 2.5 years.

Flocking, group size and home range

The mechanisms that contribute to the flocking nature of sheep are unclear but may be associated with the desire to maintain visual contact with their flock mates. Flock structures seem to vary with breed, ranging from the relatively uniform but wide dispersal of British hill breeds to the tight-flocking structure of the Merino. This is reflected in nearest-neighbour distances, ranging from as low as 1.5 m in the Merino to 7.5 m in Scottish Blackface (Arnold, 1985). Breeds also vary in the size of groups that are formed, and this may be an alternative measure of gregariousness, as Merinos have been reported to form larger groups (33) than Dorsets (10) under the same management conditions (Arnold and Pahl, 1967).

Home range – the area over which an animal normally travels – has also been used to describe differences in the organization of sheep breeds. Hunter and Davies (1963) were the first to report a home range for family groups of Scottish Blackface ewes, and subsequently a number of studies have identified family groupings, particularly in the less gregarious breeds. It is possible that grazing and organization behavioural patterns are learnt, as suggested by the cross-fostering study of Key and MacIver (1980), although this is far from certain.

Gregariousness may be an interesting trait to examine in the context of selection for behavioural traits that could improve adaptability to handling and intensive management. Anecdotal reports would suggest that higher levels of gregarious behaviour (shorter nearest-neighbour distances) may be negatively correlated with desirable maternal behaviours; however, we have few measures of within-flock variation in nearest-neighbour distances and, as it is markedly influenced by environmental conditions, other selection criteria associated with handling may be more useful.

Handling

The social behavioural patterns of sheep have a number of implications for the ease of handling of this species. The strong follower and flocking drives of sheep, their strong flight reaction and the dominant effect of vision on responses seem to be characteristic of all sheep breeds.

There are few studies that have evaluated the 'handling' behaviour of breeds in the same environment. Whateley *et al.* (1974) compared a number of breeds and breed crosses under husbandry conditions of mustering, shearing, yarding, time to restraint, etc. They identified clear breed differences and it would appear that the less gregarious breeds were those ranked as the most difficult to handle for most, but not all, responses measured (Table 12.3).

There do not appear to be any definitive studies of breed differences in response to different yarding situations, but the techniques used by Hutson (1984) could be adapted to allow a closer examination of the variation that exists in 'handling responses'.

Table 12.3. Ranking of breeds on the basis of ease of handling during various husbandry routines (from Whateley *et al.*, 1974).

Routine	Easiest to handle	Most difficult to handle
Reaction to dogs	Perendale and Merino × Romney	Dorset × Romney
Mustering	Perendale and Coopworth	Romney
Yard movement	Perendale and Coopworth	Romney
Catching from pens	Coopworth and Perendale	Romney and crosses
Drafting	Perendale	Romney
Drenching	Romney	Perendale and Coopworth
Crutching (holding)	Romney	Perendale and Coopworth

Response to confinement and fearfulness

We have little information to indicate the extent of variation that exists in sheep responses to confinement. It appears that the more individualistic, less gregarious sheep breeds are less able to cope with restricted space and, in the same context, the more gregarious animals are less able to cope with individual confinement.

Studies of 'fearfulness' in sheep have shown that there is some variability in this trait, both within flocks and between breeds. A few studies have used open-field tests to quantify behavioural differences. Torres-Hernandez and Hohenboken (1979) examined eight crossbred groups for 'emotional behavioural' responses to a dog. Romney crossbreds had the highest scores for vocalization, ambulation and investigative behaviours, while Colombia crossbreds were more prone to foot-stamping. In another study, Le Neindre *et al.* (1993) examined differences between Romanov, Merino and Romanov × Merino cross ewes in their behavioural responses to isolation, with or without feed or in the presence or absence of a human. They found that the two pure breeds differed in most behaviours measured. Romanov animals appeared more reactive, moving more often, bleating less, feeding less and showing a less olfactory approach to humans. The crossbred animals were similar to the Romanov ewes, but not for all behaviours. The authors reported that the tests were highly repeatable 'within animal' and showed that reactivity in an open-field test correlated well with fear of humans, which suggests that there is genetic variation, which can be exploited to advantage to alter adverse handling responses.

At the Institute of Cytology and Genetics in Siberia, a behavioural test based on a form of fearfulness assessment (defensiveness on human approach to supplementarily fed animals) has been developed to examine the relationship between the degree of domestication (as exhibited by lack of fearfulness) and production traits. These studies were based on the hypothesis that, during the process of domestication, selection would have been for reduced reactivity to stressful situations, type-1 animals showing minimal movement away from humans and type-3 animals a strong adversive response.

Flocks were divided into three 'type' groups based on a defensive score taken at between 10 and 12 months of age, and Belyaev and Martynova (1974) reported evidence of behavioural polymorphism in the flock of Altai sheep studied. The trait measured was repeatable in the sense that 92% of animals classified as type 1 at 4–5 months remained type 1 when classified again at 12 months of age (Martynova *et al.*, 1975). For behavioural type, there was a significant correlation of 0.34 between mothers and daughters.

Further studies (Stakan *et al.*, 1976) reported phenotypic relationships with production, type 1 animals having 11% more wool, 24% greater weaning rate and 14.5% greater milk production than type 3 ewes. Using the same behavioural classification system, Lankin *et al.* (1979, 1980) also showed a difference between the behaviour types in corticosteroid response to weaning, type 1 animals showing non-significant increases 30 min after separation, compared with large increases in type 3 animals.

More recently, Lankin (1987) and Lankin and Nedelchev (1989) have examined the distribution of the behavioural types in various breeds. They have found marked differences, with the Ostfriesian and Blackhead Persian having a much greater proportion of type 1 animals than the Awassi and Karakul breeds. However, it is difficult to be certain of the basis of these differences, as the management system of rearing was not standardized.

Mechanisms

To maintain social structures, sheep need mechanisms of recognition and, although it is clear that vision is the dominant sense used by sheep, we have no information on the relative contribution of genetic and environmental components to recognition. Kendrick and Baldwin (1987) have shown clearly that cerebral cortex responses differ according to different shapes, animals being more responsive to other animals with horns (size) and to dogs. Further studies have revealed that recognition of humans is strongest when a person is facing or moving away but relatively weak if a human is in a quadrupedal position. It seems that brain cells respond to an 'innate' concept of what a human shape is and also respond differently according to the direction of movement. To date, these data have been collected using domesticated breeds and it is interesting to speculate as to breed differences that may exist, particularly between 'wild' breeds, which may perceive humans in a different manner, and domesticated breeds.

Auditory signals are used by sheep in a number of settings, but predominantly between mother and young. Shillito-Walser (1978) has suggested that auditory cues are used primarily for orientation. However, there are clear differences in sonographic characteristics of ewes from different breeds (Clun Forest, Finnish, Jacob, Dalesbred and Soay) (Shillito-Walser and Hague, 1980) and lambs can identify breed according to auditory cues. Embryo-transfer studies would suggest that the recognition of sound is learnt, but further testing is needed to quantify this.

It seems that there is good evidence for the existence of genetic variation for some social characteristics of sheep, particularly for gregariousness and fearfulness, but we have few estimates of the genetic parameters for these traits or of correlations with production traits. Certainly, any correlations of these social characters with behaviours associated with reproductive success (e.g. maternal behaviour) are of importance, as selection for improvements in handling or confinement responses may have implications for other aspects of production. This issue will be examined further when selection criteria for maternal behaviour of ewes are examined.

Reproductive Behaviour

Male behaviours

A number of breed comparisons of reproductive behaviour have been made over the years, but a word of caution is required in interpretation, as it is clear from recent studies that experiences in early life can have a profound effect on the later sexual behaviour of rams. Having said this, there is some evidence that breed variation does exist in male reproductive behaviours. Breed comparisons of ram sexual behaviour have shown differences between breeds associated with season and changes in androgen levels (e.g. Dufour *et al*., 1984). Lindsay and Ellsmore (1968) reported some seasonal differences in the response of three different ram breeds (Merino, Dorset Horn and Border Leicester) to oestrous ewes in pen tests. Border Leicesters showed the most seasonal variation. Mickelsen *et al.* (1982) also found breed differences in libido score. Scores changed throughout the year, being lowest in spring. On average, libido was highest for Suffolk and Lincoln rams and lowest for Polypay rams; however, it is possible that these rankings may have been confounded with rearing conditions.

In a study of libido in German Mutton Merino and Ossimi rams, no differences between breeds or crossbreds were noted (Galal *et al*., 1978), nor were there differences between Romney and Cheviot rams tested by Kelly *et al.* (1975). In contrast, Hulet *et al.* (1962) found Rambouillet rams to be more active than Targhee or Columbia rams, and Land and Sales (1977) found that young Merinos had a consistently lower intensity of mating behaviour (measured as frequency of mounts and copulations) than Finnish Landrace or crossbred rams. Breed differences in incidence of other reproductive behaviours, such as nosing, nudging, pawing, flehmen (lip curl) and mounting, have also been reported (Bernon and Shrestha, 1984).

Within-breed studies have shown low heritability estimates for serving capacity of 20-month-old Merino rams. Heritability estimates for a variety of serving-capacity measures (number of serves, libido index) reported by Purvis (1985) were low and often negative (−0.095 to 0.002 ± 0.10). He reported a high level of sexual inactivity in these inexperienced Merino rams and found that 37.9% failed to serve during testing. Of the rams that were active, repeatability of serving-capacity scores was low (0.27). It would seem that, at least in young inexperienced rams, the amount of additive genetic variation in male reproductive behaviours is probably small.

Tulley and Burfening (1983) studied the variation in mating behaviour of Rambouillet rams from lines selected for increased and decreased prolificacy, as well as a randomly selected line. The number of mounts and services declined from the high-prolificacy line through to the low line, suggesting some additive variation in this trait in mature rams.

Female behaviours

Breed differences in seasonality of breeding are well known in ewes (Lynch *et al.*, 1992), and estimates of variability in this and associated traits have shown repeatabilities in lambing intervals of 0.2 and heritabilities of 0.2 as well (Mendel and Pirchner, 1994).

Oestrous activity of Romney, Coopworth and Perendale ewes was monitored in 1.5- and 4.5-year-old ewes by Kelly *et al.* (1976) who found breed differences in the onset of oestrus of about 1 week, while the length of the oestrous cycle was similar.

The data reviewed for both rams and ewes suggest that there is little additive genetic variation in reproductive behaviours, with the possible exception of that associated with seasonality. Possibly of greater practical significance to reproductive success is the variation that has been reported in maternal behaviour.

Maternal Behaviour

Maternal behaviour is a trait of sheep mainly associated with rearing ability and, as such, has been incorporated into many breed comparisons as a factor contributing to variation in lamb survival.

From a behavioural perspective, maternal behaviour has a number of components, including the initial maternal interest at parturition, the behaviour of the lamb in seeking the udder and the bonding of ewe and lamb and their recognition of one another. These components, plus a number of other traits, contribute to ‘rearing ability’ and are highly variable, being influenced by a number of climatic and managerial influences.

Maternal behaviour has been shown to differ between breeds in a variety of environments and has been claimed to be associated with rearing ability of breeds. The maternal component of rearing ability has been shown to be repeatable (0.1–0.15) and heritable (0.10) for both Merinos and other breeds and breed crosses (Fogarty *et al.*, 1982; Piper *et al.*, 1982) and selection programmes for rearing ability in Merinos have improved lamb survival (Atkins, 1980), but the contribution of maternal behaviour to rearing ability is unknown, although it would seem likely that its contribution will be greater in breeds bearing a large number of twins, where maternal behavioural problems have been identified (Alexander *et al.*, 1983).

Forgarty (1984) suggests that there is substantial heterosis expressed for reproductive traits and that maternal heterosis, expressed in crossbred dams, tends to be higher than individual heterosis for reproductive traits. He concluded that heterosis of the various components is cumulative for lambs weaned per ewe joined. Heterosis for rearing ability and lamb survival has been demonstrated (5–25%) and is generally higher in primiparous ewes with twin lambs (e.g. Iwan *et al.*, 1971). Whether heterosis is present for maternal behavioural traits is unknown; however, there are clear maternal behaviour advantages favouring crossbred ewes, which will be discussed later.

In one of the few studies to have quantified genetic parameters for maternal and lamb behaviours, Alexander and Stevens (1992) found, in a Merino flock, that lamb survival to 3 days was a heritable character of the ewe and not the lamb. The major ewe behavioural characteristics, genetically correlated with survival, included the length and difficulty of the birth process in single lambs and the likelihood of ewe–lamb separations for twins.

Breed comparisons

Breed comparisons of the maternal behaviour of mature ewes are common, with a wide variety of breeds being compared by Whateley *et al.* (1974). They found that, under the same management system, overall mothering ability (Table 12.4) was poorest in Romney and Merino ewes and was best in Romney × Border Leicester crosses. Romney ewes also ranked poorly in terms of shelter-seeking at lambing, while Merino ewes were more likely to desert their lambs. Cheviot ewes were quick to leave their lambs at tagging but were also quick to return after the shepherd had departed.

Alexander *et al.* (1983) compared the behaviour of twin-bearing Merino,

Table 12.4. Lambing behaviour characteristics of breeds lambing in New Zealand (% of ewes lambing) (from Whateley *et al.*, 1974).

Breed	Romney	Border × Romney	Dorset × Romney	Perendale	Cheviot	Merino
Mothering problems	7	3	4	3	2	1
At tagging						
Stood with lamb	52	78	82	59	41	11
Stood close to lamb	37	19	16	33	28	38
Deserted and returned	5	2		7	14	28
Deserted	5	1	1	2	12	23
Mothering score						
Good	57	89	88	80	82	39
Average	36	10	11	19	13	36
Poor	7	1	1	1	5	26

Dorset Horn and Border Leicester ewes in Australia and Romney ewes in New Zealand. At lamb-tagging Dorset Horn and Romney ewes were more likely to remain with or near their lambs, while Merinos were more likely to retreat. Merino ewes did not ensure that they were accompanied by their lambs and permanent separations were much more common in this breed; it would seem that Merino ewes are 'satisfied' with one lamb!

In a detailed examination of behaviour during the neonatal period of Merino and Border Leicester × Merino ewes, Edonu (1987) was able to show clear differences in grooming behaviour and lamb vigour in these breeds. Crossbred ewes were more vocal than Merinos, as were their lambs, and the ewes spent more time grooming their twin lambs. Differences were also apparent in the strength of bond between ewe and lamb(s) in terms of activity, bleating and searching, all of which were higher in the crossbred ewes during the first week post-partum.

Alexander *et al.* (1990) measured 28 aspects of behaviour that could be related to lamb survival in Merino, a selected line of Border Leicester (Glen Vale) and Border Leicester ewes. The Glen Vales appeared superior in maternal behaviour, particularly in their ability to remain with their lambs (Table 12.5). A 'twin care' test, involving measuring the time taken for a ewe to move to one of her lambs, which had been tethered 20 m away, was suggested as a good basis for a selection programme to improve maternal care in twinning ewes.

In another breed comparison O'Conner and Lawrence (1992) found differences between Mule (Bluefaced Leicester × Scottish Blackface) and Scottish Blackface lambs in 'udder-seeking' behaviours immediately post-partum, but not in ewe behaviours.

Generally, the ewe is seen as the primary player in the development of the bond between ewe and lamb, as the ewe normally learns to recognize her offspring within 1–2 h of parturition. However, studies in addition to that of O'Conner and Lawrence (1992) have suggested that the lamb also has a role in the maintenance of the bond. For example, Stevens *et al.* (1984) examined the contribution that the lamb makes to separation of twin lambs of Merino ewes. Lambs were pure Merino or Border Leicester × Merino genotypes and significantly more (58 vs. 35%) pure Merino lambs were separated from their dams than crossbreds. The difference could not be related to birth weight or time spent at the birth site but would appear to be associated with easier arousal

Table 12.5. Incidence (%) of separation of twin lambs from ewes where ewe behaviour was at fault (from Alexander *et al.*, 1990).

	Border Leicester	Glen Vale	Border × Merino	Merino
Ewes separated from their lambs for any period	53	20	28	48
Ewes permanently separated from their lambs	18	6	10	9
Ewes temporarily separated from their lambs	43	17	20	43

and coordination of the crossbred lambs, allowing them to follow their dams more easily. More recently, the studies of Nowak and Lindsay (1990) have also shown differences in the rate of learning between lambs of different breeds (crossbred vs. Merino), suggesting a possible mechanism for breed differences in responsiveness to their dam. Embryo transfer studies may help to clarify further the role of ewe and lamb breed in the maintenance of the ewe–lamb 'bond'.

The greater vigour and arousal of crossbred lambs suggest that they are better able to respond to and reinforce maternal care and bonding. Such positive effects are not necessarily related to behaviours such as teat seeking time or time to stand, which are sometimes poorly correlated with survival. Lindsay *et al.* (1990) have argued that crossbred lambs have a high frequency of vocalization, which acts to reinforce recognition by the ewe. Whether such a criterion could be used for selection is unknown. A series of studies by Shillito-Walser and her colleagues suggest that breeds may differ in their use of sound. In a comparison of Clun Forest, Finnish Landrace, Jacob and Soay lambs, Shillito and Alexander (1975) found that Clun Forest lambs were most able to identify their mothers by sound and Jacob lambs the least able.

Lamb behaviours commonly encompassed under the term vigour also appear to vary between breeds, as well as within breeds; however, within-breed comparisons can be confounded by factors such as birth injury and temperature, which suppress normal behaviours during the early neonatal period. Slee and Springbett (1986) reported differences among ten sheep breeds in the length of time lambs took to stand following parturition and in the time to reach the udder. Feral and hill breeds (Soay, Borerary Blackface, Scottish Blackface, Welsh Mountain and Cheviot) reached the udder earlier than the others while Finn and Merino lambs were particularly slow (Table 12.6). These data suggest that udder-seeking behaviour is associated with good resistance to hypothermia when evaluated across breeds. Resistance to hyperthermia is moderately heritable but there are no reports of genetic correlations between this trait and behaviour.

Selection criteria for improved maternal behaviour

Isolation and temperament

Isolation of the ewe from the flock is a characteristic indicator of parturition in many wild breeds (Rocky mountain bighorn, soay, mouflon) but does not appear consistently in domestic breeds. Isolation behaviour ensures that ewes can remain at the birth site for long periods after birth and wild sheep have been reported to remain at the birth site for a number of days; in contrast, most domestic breeds stay for only a few hours.

In a comparison of Merino strains, one selected for improved rearing ability and the other a random bred control, Szanter-Coddington (1994) reported that the strain selected for improved rearing ability stayed on the birth site slightly longer than the random bred ewes. In the same lines, Kilgour and Szanter-

Table 12.6. Time (min) required for lambs to stand for 3 min or to reach the udder (from Slee and Springbett, 1986).

Breed	Time to stand	Time to reach udder
Scottish Blackface	17.5	30.9
Welsh Mountain	19.2	24.3
Cheviot	24.0	35.8
Soay and Boreray	17.7	30.9
Oxford	27.2	31.2
Border Leicester	29.6	56.1
Southdown	20.1	53.8
Merino	19.3	72.6
Finnish Landrace	54.0	94.6

Coddington (1995) found no strain differences in reactivity assessed in an open-field test but did find differences in responses to the presence of a human in a form of 'open-field' test. The ewes selected for rearing ability moved about less, confirming the suggestion of Lindsay *et al.* (1990) that the suppression of locomotive activity around the time of parturition is likely to have a positive effect on rearing. The authors suggested that it is possible that such a test may be able to be used to select within a breed for improved maternal ability.

Putu (1988) evaluated the reactivity of three Merino genotypes, using an open-field temperament test, and correlated this with subsequent maternal behaviour. The within-flock correlation between temperament and time spent at the birth site was 0.60 and that between temperament and the incidence of ewe/lamb separation was 0.30. He found that the ewes that were least active in the test spent more time on the birth site and also vocalized more often.

Murphy *et al.* (1994) also found that the relationship between temperament, measured at weaning, and maternal behaviour was strong in Merinos. Repeatabilities of between 0.52 and 0.77 were reported for temperament measures, and the correlation coefficients between these measures and maternal behaviours are presented in Table 12.7. Further studies are continuing, using the movement response as a selection criterion to quantify the amount of additive genetic variation that exists for such traits and to monitor correlated responses in maternal behaviours.

Grooming

Breeds differ in the degree that they use visual and auditory cues to discriminate between own and alien lambs (Shillito-Walser *et al.* 1982); however, all breeds appear to be reliant on olfactory cues to allow initial bond formation (Levy and Poindron, 1987). Grooming behaviour may well play a role in the development of this bond as well as in the encouragement of suckling. Inactivity of the ewe next to a lamb that has not suckled will suppress lamb activity (Lynch *et al.*, 1992). Whether selection for improved grooming, behaviour within a breed is

Table 12.7. Correlation coefficients between measures of temperament and maternal behaviour (from Murphy *et al.*, 1994).

Temperament/maternal behaviour	Distance from ewe to lamb at tagging	Time spent by the ewe on the birth site	Time spent by the ewe licking her lamb
Time in zones in open field	–0.35	0.44	0.11
Zones crossed in open field	–0.04	–0.10	0.12
Number of high pitched bleats	0.10	–0.13	0.14
Subjective score (1–10)	–0.04	0.09	0.10
Movement response to enclosure in a box	0.30	–0.03	–0.39

possible is difficult to say. There is evidence of breed differences in the time of commencement of grooming particularly in primiparous ewes, and there also appears to be within-breed variation in grooming traits, such as those measured by Poindron *et al.* (1984) in Romanov and Île-de-France ewes.

Hill breeds appear to show less variability in maternal behaviour and are clustered around the 'good' end of the maternal behaviour spectrum. In contrast, more intensively reared breeds (e.g. Suffolk) show a wider range in maternal behaviour, often being slower to groom and more easily distracted. Some studies have been unable to show that variation in individual maternal traits is related to lamb survival, but a composite score used by O'Conner *et al.* (1985) shows that a good relationship exists, particularly for ewes with larger litters.

These reports would confirm the suggestions of Alexander *et al.* (1983) and Piper *et al.* (1980) that the variability in maternal behaviour, both between and within breeds, may be capable of being exploited by selection to improve lamb survival in a variety of environments. Theoretically, selection is possible; however, there are a number of confounding issues that need to be taken into account, including nutrition interactions and litter size (Lindsay *et al.*, 1990). Further studies are necessary to identify behavioural or endocrine measures (Keverne and Kendrick, 1990) that might be useful as selection criteria to exploit the genetic variation which does exist in this complex of behaviours.

Conclusion

This review highlights the fact that a large proportion of the evidence for genetic variation in behavioural traits of sheep is based on breed comparisons. For the few estimates of genetic parameters that are available, heritability estimates are low, as are the majority of phenotypic and genetic correlations. This confirms the observation made at the beginning of this chapter for other species and highlights the problems that exist in the recording of behaviour where measurements are dependent on interaction with other animals. However, the impact of new genetic techniques on our capacity to characterize behavioural traits may alter this situation. Recent reports of specific deoxyribonucleic acid (DNA) fingerprint bands associated with differences in isolation-induced aggressive behaviours in mice (Schenider-Stock and Epplen, 1996) suggest that there may be genetic markers for some behavioural traits, and studies in farm animals for more complex traits, such as maternal behaviour, may be fruitful.

For the majority of environments in which sheep are found, they appear well adapted, and further improvements in behaviours associated with farm production efficiency are most likely to be associated with selection for behavioural traits that will confer greater adaptability in particular management systems. In extensive-management contexts, this is most likely to be via selection for improved maternal behaviours, while, for intensive-rearing systems, selection for changes in social behaviours, such as gregariousness or fearfulness, are the most likely choices. Certainly, Hohenboken's conclusion in 1986 that we do not know enough about the heritability and phenotypic and genetic correlations of behavioural traits of farm animals appears to be still true for sheep.

Acknowledgements

My grateful thanks go to Dr J.J. Lynch for helpful comments on the manuscript and to Dr A.B. Lawrence for some perceptive comments. I would also like to thank Dr A. Ruvinsky for his help with the translation of papers from Russian.

References

Alexander, G. and Stevens, D. (1992) *Analysis of 5 Years' Data on Genetic Aspects of Ewe and Lamb Behavioural Traits Concerned with Lamb Survival.* Final report to the Australian Wool Corporation on project CPB074. Australian Wool Corporation, Melbourne.

Alexander, G., Stevens, D., Kilgour, R., de Langen, H., Mottershead, B.E. and Lynch, J.J. (1983) Separation of ewes from twin lambs: incidence in several sheep breeds. *Applied Animal Ethology* 10, 301–317.

Alexander, G., Stevens, D., Bradley, L.R. and Barwick, S.A. (1990) Maternal behaviour in Border Leicester, Glen Vale (Border Leicester derived) and Merino sheep. *Australian Journal of Experimental Agriculture* 30, 27-38

Arnold, G.W. (1975) Herbage intake and grazing behaviour in ewes of four breeds at different physiological states. *Australian Journal of Agricultural Research* 26, 1017–1024.

Arnold, G.W. (1981) Grazing behaviour. In: Morley, F.H.W. (ed.) *Grazing Animals.* World Animal Science, Elsevier, Amsterdam, pp. 79–104.

Arnold, G.W. (1985) Association and social behaviour. In: Fraser, A.F. (ed.) *Ethology of Farm Animals.* Elsevier, Amsterdam, pp. 233–246.

Arnold, G.W. and Pahl, P.J. (1967) Subgrouping in sheep flocks. *Proceedings of the Ecological Society of Australia* 2, 183–189.

Atkins, K.D. (1980) Selection for skin folds and fertility. *Proceedings of the Australian Society of Animal Production* 13, 174–176.

Beilharz, R.G. and Cox, D.F. (1967) Genetic analysis of open field behaviour in swine. *Journal of Animal Science* 26, 988–990.

Belyaev, D.K. and Martynova, V.N. (1974) Some characteristics of behaviour in fine-woolled sheep and their correlation with reproductive features. In: *1st World Congress on Genetics Applied to Livestock Production*, Vol. 1. Graficas Orbe, Madrid, pp. 1039–1042.

Berggren-Thomas, B. and Hohenboken, W.D. (1986) The effects of sire breed, forage availability and weather on the grazing behaviour of crossbred ewes. *Applied Animal Behaviour Science* 15, 217–228.

Bernon, D.E. and Shrestha, J.N.B. (1984) Sexual activity patterns in rams. *Canadian Journal of Comparative Medicine* 48, 42–46.

Cameron, N.D. (1988) Genetic relationships between growth and feed intake in performance tested ram lambs: an investigation of variance component estimation procedures. *Journal of Animal Breeding and Genetics* 105, 431–440.

Craig, J.V. (1994) Genetic influences on behaviour associated with well-being and productivity in livestock. In: *Proceedings of the 5th World Congress on Genetics Applied to Livestock Production*, Vol. 20. Guelph, Ontario, pp. 150–157.

Craig, J.V., Ortman, L.L. and Guhl, A.M. (1965) Genetic selection for social dominance ability in chickens. *Animal Behaviour* 13, 114–131.

DeFries, J.C., Hegmann, J.P. and Halcomb, R.A. (1974) Response to 20 generations of selection for open-field activity in mice. *Behavioural Biology* 11, 481–495.

Dudzinski, M.L. and Arnold, G.W. (1979) Factors influencing the grazing behaviour of sheep in a Mediterranean climate. *Applied Animal Ethology* 5, 125–144.

Dufour, J.J., Fahmy, M.H. and Minvielle, F. (1984) Seasonal changes in breeding activity, testicular size, testosterone concentration and seminal characteristics in rams with short and long breeding season. *Journal of Animal Science* 58, 416–422.

Edonu, K.E. (1987) Studies on the post-parturient behaviour of ewes and lambs. Masters of Rural Science thesis, University of New England, Armidale, Australia, 137 pp.

Faure, J.M. (1994) Behavioural genetics: an overview. In: Smith, C. *et al.* (eds) *Proceedings of 5th World Congress on Genetics Applied to Livestock Production*, Vol. 20. Guelph, Ontario, p. 129.

Fogarty, N.M. (1984) Breeding for reproductive performance. In: Lindsay, D.R. and Pearce, D.T. (eds) *Reproduction in Sheep*. Australian Academy of Science and Australian Wool Corporation, Canberra, pp. 226–233.

Fogarty, N.M., Dickerson, G.E. and Young, L.D. (1982) Genetic parameters for reproduction in sheep. *Proceedings of Australian Society of Animal Production* 14, 435–438.

Fowler, D.G. and Jenkins, L.D. (1976) The effects of dominance and infertility of rams on reproductive performance. *Applied Animal Ethology* 2, 327–337.

Galal, E.S., El-Gamal, A.A., Aboul-Naga, A. and El-Fouly, M.A. (1978) Male reproductive characteristics of Merino and Ossimi sheep and their crosses. *Animal Production*. 27, 261–267.

Gluesing, E.A. and Balph, D.F. (1980) An aspect of feeding behaviour and its importance in grazing systems. *Journal of Range Management* 33, 426–427.

Goddard, M.E. (1980) Behaviour genetics and animal production. In: Wodzicka-Tomaszewska, M., Edey, T.N. and Lynch, J.J. (eds) *Behaviour in Relation to*

Reproduction, Management and Welfare of Farm Animals. University of New England, Armidale, pp. 29–36.

Guhl, A.M., Craig, J.V. and Mueller, C.D. (1960) Selective breeding and aggressiveness in chickens. *Poultry Science* 39, 970–980.

Hale, E.P. (1962) Domestication and the evolution of behaviour. In: Hafez, E.S.E. (ed.) *Behaviour of Domestic Animals*. Baillière, Tindall & Cox, London, pp. 21–53.

Hinch, G.N., Lynch, J.J., Elwin, R.L. and Green, G.C. (1990) Long-term associations between Merino ewes and their offspring. *Applied Animal Behaviour Science* 27, 93–103.

Hohenboken, W.D. (1986) Inheritance of behavioural characteristics in livestock: A review. *Animal Breeding Abstracts* 54, 623–639.

Hulet, C.V. Ercanbrack, S.K., Price, D.A., Blackwell, R.L. and Wilson, L.O. (1962) Mating behaviour of the ram in the multi-sire pen. *Journal of Animal Science* 21, 870–874.

Hunter, R.F. and Davies, G.E. (1963) The effect of method of rearing on the social behaviour of Scottish Blackface hoggets. *Animal Production* 5, 183–194.

Hutson, G.D. (1984) Spacing behaviour of sheep in pens. *Applied Animal Behaviour Science* 12, 111–119.

Iwan, L.G., Jefferies, B.C. and Turner, H.N.(1971) Estimation of heterosis in Merino and Corriedale crosses with sheep. *Australian Journal of Agricultural Research* 22, 521–535.

Jonmundsson, J.V., Birgisson, L.G. and Adalsteinsson, S. (1994) Leader sheep in Iceland. In: Smith, C. *et al.* (eds) *Proceedings of the 5th World Congress on Genetics Applied to Livestock Production* Vol. 20. Guelph, Ontario, pp. 162–164.

Kelly, R.W., Allison, A.J. and Shackell, G.H. (1975) Libido testing and subsequent mating performance in rams. *Proceedings of the New Zealand Society of Animal Production* 35, 204–211.

Kelly, R.W., Allison, A.J. and Shackell, G.H. (1976) Seasonal variation in oestrous and ovarian activity of five breeds of ewes in Otago. *New Zealand Journal of Experimental Agriculture* 4, 209–214

Kendrick, K.M. and Baldwin, B.A.(1987) Cells in the temporal cortex of conscious sheep can respond preferentially to the sight of faces. *Science* 236, 448–450.

Keverne, E.B. and Kendrick, K.M. (1990) Neurochemical changes accompanying parturition and their significance for maternal behaviour. In: Krasnegor, N.A. and Bridges, R.S. (eds) *Mammalian Parenting*. Oxford University Press, Oxford, pp. 281–304.

Key, C. and MacIver, R.M. (1980) The effects of maternal influences on sheep: Breed differences in grazing, resting and courtship behaviour. *Applied Animal Ethology* 6, 33–48.

Kilgour, R. (1976) Sheep behaviour: its importance in farming systems, handling, transport and pre-slaughter treatment. In: *Proceedings of the Sheep Assembly and Transport Workshop*. West Australian Department Agriculture, Perth, pp. 64–84.

Kilgour, R.J. and Szanter-Coddington, M.R. (1995) Arena behaviour of ewes selected for superior mothering ability differs from that of unselected ewes. *Animal Reproduction Science* 37, 133–141.

Land, R.B. and Sales, D.I. (1977) Mating behaviour and testis growth of Finnish Landrace, Tasmanian Merino and crossbred rams. *Animal Production* 24, 83–90.

Langlands, J.P. (1973) Milk and herbage intakes by grazing lambs born to Merino ewes and sired by Merino, Border Leicester, Corriedale, Dorset Horn and Southdown rams. *Animal Production* 16, 285–291.

Lankin, V.S. (1987) Polymorphism of passive defence behaviour in milk and course wool sheep. In: Stakan, G.A. (ed.) *Genetics and Selection of Animals.* Nauka, Novosibirsk, pp. 104–113 (in Russian).

Lankin, V.S. and Nedelchev, D. (1989) Behavioural reactions and increase in productivity of milk sheep. *Agricultural Biology* 2, 83–88 (in Russian).

Lankin, V.S., Stakan, G.A. and Naumenko, E.V. (1979) Domestication behaviour in sheep. I. Relationships of functional status of hypothalamo-adrenal system with behaviour and age in fine-wool sheep. *Genetica 15*, 891–899 (in Russian).

Lankin, V.S., Stakan, G.A. and Naumenko, E.V.(1980) Domestication behaviour in sheep. III. Hereditary determination of the endocrine function of the pituitary–adrenal system in fine-wool sheep with different behaviour types. *Genetica* 16, 1089–1095 (in Russian).

Lawrence, A.B. (1990) Mother–daughter and peer relationships of Scottish hill sheep. *Animal Behaviour* 39, 481–486.

Lee, G.J., Atkins, K.D. and Mortimer, S.I. (1992) Pasture intake by Merino ewes and its relationship with liveweight and ultrasonic fat depth. *Proceedings of the Australian Association of Animal Breeding and Genetics* 10, 456–457.

Lee, G.J., Atkins, K.D. and Mortimer, S.I. (1995) Variation between Merino ewes in pasture intake 2. Within-flock genetic parameters for intake and some production traits. *Livestock Production Science* 41, 143–150.

Le Neindre, P., Poindroin, P., Trillat, G. and Orheur, P (1993) Influence of breed on reactivity of sheep to humans. *Genetics, Selection and Evolution* 25, 447–458.

Levy, F. and Poindron, P. (1987) The importance of amniotic fluids for the establishment of maternal behaviour in experienced and inexperienced ewes. *Animal Behaviour* 35, 1188–1192.

Lindsay, D.R. and Ellsmore, J. (1968) The effect of breed, season and competition on mating behaviour of rams. *Australian Journal of Experimental Agriculture and Animal Husbandry* 8, 649–652.

Lindsay, D.R., Nowak, R., Putu, I.G. and McNeill, D. (1990) Behavioural interactions between the ewe and her young at parturition: a vital step for the lamb. In: Oldham, C.M., Martin, G.B. and Purvis, I.W. (eds) *Reproductive Physiology of Merino Sheep: Concepts and Consequences.* School of Agriculture, University of Western Australia, Perth, pp. 191–205.

Lynch, J.J., Hinch, G.N., Bouissou, M.F., Elwin, R.L., Green, G.C. and Davies, H.I. (1989) Social organisation in young Merino and Merino × Border Leicester ewes. *Applied Animal Behaviour Science* 22, 49–63.

Lynch J.J., Hinch, G.N. and Adams, D.B. (1992) *The Behaviour of Sheep: Biological Principles and Implications for Production.* CAB International, Wallingford CSIRO Publications, Melbourne.

Macha, J. and Olsarova, J. (1986) Variability and heritability of grazing intensity in cattle. *Acta Universitatis Agriculturae Brno* 34, 313–320.

Mackenzie, S.A., Oltenacu, E.A.B. and Houpt, K.A. (1986) Canine behavioural genetics – a review. *Applied Animal Behaviour Science* 15, 365–393.

Martynova, V.N., Stakan, G.A., Soskin, A.A. and Chernov, L.L. (1975) Variation and inheritance of behaviour in fine-wool sheep. 1 Formation of individual behaviour of sheep in the course of ontogenesis. *Genetica* 11, 31–39 (in Russian).

Mendel, C. and Pirchner, F. (1994) Aseasonality in mountain sheep. *Journal of Animal Breeding and Genetics* 111, 451–462.

Mickelsen, W.D., Paisley, L.G. and Dahmen, J.J. (1982) Seasonal variations in scrotal circumference, sperm quality and sexual ability in rams. *Journal of the American*

Veterinary Medicine Association 181, 376–380.

Murphy, P., Purvis, I., Lindsay, D., Le Neindre, P., Orgeur, P. and Poindronm, P. (1994) Measures of temperament are highly repeatable in Merino sheep and some are related to maternal behaviour. *Proceedings of the Australian Society of Animal Production* 20, 247–250.

Nowak, R. and Lindsay, D.R. (1990) Effect of genotype and litter size on discrimination of mothers by their twelve-hour-old lambs. *Behaviour* 115, 1–13.

O'Conner, C.E. and Lawrence, A.B. (1992) Relationship between lamb vigour and ewe behaviour at parturition. *Animal Production* 54, 361–366.

O'Conner, C.E.,Jay, N.P., Nicol, A.M. and Beatson, P.R. (1985) Ewe maternal behaviour score and lamb survival. *Proceedings of the New Zealand Society of Animal Production* 45, 159–162.

Paterson, I.W. and Coleman, C.D. (1982) Activity patterns of seaweed-eating sheep on North Ronaldsay, Orkney. *Applied Animal Ethology* 8, 137–146.

Piper, L.R., Bindon, B.M., Atkins, K.D. and McGuirk, B.J. (1980) Genetic variation in ovulation rate in Merino ewes aged 18 months. *Proceedings to Australian Society of Animal Production* 13, 409–412.

Piper, L.P., Hanrahan, J.P., Evans, R. and Bindon, B.M. (1982) Genetic variation in individual and maternal components of lamb survival in Merinos. *Proceedings of the Australian Society of Animal Production* 14, 29–30.

Plomin, R., Defries, J.C. and McClearn, G.E. (1990). *Behavioural Genetics: a Primer*, 2nd edn. Freeman, New York.

Poindron, P., Raksanyi, I., Orgeur, P. and LeNeindre, P. (1984) Comparison of maternal behaviour at parturition in primiparous or multiparous ewes of Ile-de-France, Prealpus du Sud and Romanov breeds. *Genetics, Selection and Evolution.* 16, 503–522.

Purvis, I.W. (1985) *Genetic relationships between male and female reproductive traits.* PhD thesis, University of New England, Armidale, Australia.

Putu, I.G. (1988) Maternal behaviour in Merino ewes during the first two days after parturition and survival of lambs. PhD thesis, University of Western Australia.

Roff, D.A. and Mousseau, T.A. (1987) Quantitative genetics and fitness: lessons from *Drosophila. Heredity* 58, 103–118.

Schenider-Stock, R. and Epplen, J.T. (1996) Congenic AB mice: a novel means for studying the (molecular) genetics of aggression. *Behaviour Genetics* 25, 475–482.

Sharafeldin, M.A. and Shafie, M.M. (1965) Animal behaviour in the sub-tropics II. Grazing behaviour of sheep. *Netherlands Journal of Agricultural Science 13*, 239–247.

Shillito, E.E. and Alexander, G. (1975) Mutual recognition amongst ewes and lambs of four breeds of sheep (*Ovis aries*). *Applied Animal Ethology* 1, 151–165.

Shillito-Walser, E.E. (1978) A comparison of the role of vision and hearing in ewes finding their own lambs. *Applied Animal Ethology* 4, 71–79.

Shillito-Walser, E.E. and Hague, P. (1980) Variations in the structure of bleats from sheep of four different breeds. *Behaviour* 75, 22–35.

Shillito-Walser, E.E., Willadson, S. and Hague, P. (1981) Pair associations between lambs of different breeds born to Jacob and Dalesbred ewes after embryo transplantation. *Applied Animal Ethology* 7, 351–358.

Shillito-Walser, E.E., Walters, E. and Ellison, J. (1982) Observations on vocalization of ewes and lambs in the field. *Behaviour* 91, 190–203.

Slee, J. and Springbett, A. (1986) Early post-natal behaviour in lambs of ten breeds. *Applied Animal Behaviour Science* 15, 229–240.

Stakan, G.A., Martynova, V.N., Soskin, A.A. and Chernov, L.L. (1976) Variation and inheritance of behaviour in fine-wool sheep. II Study of the inheritance of domestic behaviour and correlations between the behaviour and productive characters in fine-wool sheep. *Genetica* 12, 35–43 (in Russian).

Stevens, D., Alexander,G., Mottershead, B.E. and Lynch, J.J. (1984) Role of the lamb in post-partum separation of ewes from twin lambs. *Proceedings of the Australian Society of Animal Production* 15, 751.

Szanter-Coddington, M.R. (1994) Genetic improvement of lamb survival in the Australian Merino: behavioural and physiological indicators of maternal ability. Masters of Rural Science thesis, University of New England, Armidale, Australia, 138 pp.

Tennessen, T. and Hudson, R.J. (1981) Traits relevant to the domestication of herbivores. *Applied Animal Ethology* 7, 87–102.

Torres-Hernandez, G. and Hohenboken, W. (1979) An attempt to assess traits of emotionality in crossbred ewes. *Applied Animal Ethology* 5, 71–83.

Tulley, D. and Burfening, P.J. (1983) Libido and scrotal circumference of rams as affected by season of the year and altered photoperiod. *Theriogenology* 20, 435–448.

Whateley, J., Kilgour, R. and Dalton, D.C. (1974) Behaviour of hill country sheep breeds during farming routines. *Proceedings of the New Zealand Society of Animal Production* 34, 28–36

Willham, R., Cox, D.F. and Karas, G.G. (1963) Genetic variation in a measure of avoidance learning in swine. *Journal of Comparative Physiology and Psychology* 56, 294–297.

Biology and Genetics of Reproduction 13

I.W. PURVIS and M. HILLARD
CSIRO Division of Animal Production, Pastoral Research Laboratory, Armidale, New South Wales, Australia

Introduction	375
Biology of Early Reproductive Development	376
Gonadal development	376
Phenotypic sexual determination	378
Development of the hypothalamopituitary gonadal axis	379
Components of Reproductive Rate in the Ewe	380
Introductory remarks	380
Genetic variation in components of reproduction in the female	381
Male Reproduction	386
Biology of male reproduction	386
Genetics of male reproduction	386
Male Measures of Female Reproductive Performance	387
Selection Experiments on Components of Reproduction Rate	388
Estimation of Genetic Parameters for Reproduction Rate	388
Conclusions	389
Acknowledgements	389
References	390

Introduction

Reproduction is the fundamental element of the evolutionary process. Charles Darwin (1859) argued that animals and, indeed, all living organisms would increase indefinitely, were not their numbers reduced by natural selection. Part of the variability between animals is found in their ability to reproduce and survive. The other essential feature that Darwin identified was that animals pass on their genes to the next generation and so the concept of heritability is an integral part of the evolution of organisms. This chapter is about this very process in the domestic sheep – genetic control of reproductive form and function.

The study of reproduction in the domestic sheep is one of the most fascinating aspects of sheep genetics. One of the primary reasons for this is that, for many of the measures of reproductive efficiency, there are large differences between breeds, and between ewes and between rams within breeds and within flocks. On first inspection, there is, therefore, a large potential for changing reproductive processes by genetic means. However, only part of the observed variation (phenotypic variation) is due to genetic differences and, as a result, achieving permanent change by genetic means has presented a real challenge to geneticists.

In this chapter, the focus is on reproductive processes and control mechanisms expressed primarily on a whole-animal basis. Aspects of molecular and cellular reproduction are covered in other chapters in this book. As sheep populations throughout the world are primarily utilized for the supply of meat, fibre or dairy products, a natural concept that underlies this chapter is one of reproductive efficiency. This concept is vitally important in the evaluation of all livestock systems and is an integral part of concerns, on a global scale, about the capacity of world food resources to cope with the anticipated population growth in the next century. Several studies aimed at evaluating the relative impact of different biological traits on the efficiency of sheep production have identified components of reproductive efficiency as critical factors, and more so than growth or weight traits (Large, 1970; Wang and Dickerson, 1991; Greef *et al.*, 1995).

Reproductive efficiency also plays an important role in the conservation of breeds and strains of livestock that have become threatened for reasons of fashion or because alternative technologies have made them redundant. Reproductive performance and efficiency in flocks that are maintained under adverse environmental conditions in a village or on a smallholding can be absolutely vital to the economic well-being of families and larger communal groups.

Thus, the objective of this chapter is to review the basic biology of reproduction in the sheep and to examine variability of reproduction rate, with the object of identifying genetic differences and thus highlighting opportunities for improving reproduction by manipulation of the genotype. In reviewing the existing knowledge, reference to the literature has been restricted to key papers and recent reviews.

Biology of Early Reproductive Development

Gonadal development

The sheep fetus is initiated with the appearance of the mesonephros at approximately 20 days after conception. By 23 days after mating, the bipotential genital ridge is evident and is being colonized by primordial germ cells. Gonadal sex determination, believed to be initiated by a putative transcriptional regulatory protein encoded by the sex-determining region Y (SRY) gene on the Y chromosome (see Chapter 15), must occur before day 30–35 in the sheep fetus. At this time, the first morphological distinction between the fetal testis and ovary is possible, with the appearance of the testicular cords, containing germ cells (prespermatogonia) associated with some somatic cells (fetal Sertoli cells), which are then compartmentalized within a basal lamina.

The molecular processes that determine the fate of the bipotential gonad, both upstream and downstream of *SRY* function, are largely unknown. Downstream of *SRY* gonadal sex determination, the fate of primordial germ cells is dependent on critical interactions with somatic cells. In the fetal testis, germ cells excluded from the testicular cord (tubular) compartment degenerate, while prespermatogonia enclosed in the tubular compartment divide mitotically (prespermatogenesis), followed by interphase arrest, with no further activity until the second wave of mitosis begins, immediately before the onset of the meiotic divisions in the pubertal phase of rapid testicular enlargement. Intertubular steroid (testosterone)-producing cells of the testis (Leydig cells) develop shortly after formation of testicular cords.

In the fetal ovary, mitosis of primordial germ cells (termed oogonia) continues until about day 120 of gestation. Initiation of germ-cell meiosis begins around day 55 and oocyte formation then continues until day 120. Primordial follicles (i.e. oocytes surrounded by a single layer of flattened granulosa cells) first appear around day 75, at which time the maximum number of germ cells is present in the ovary (9×10^6). From day 75 to day 90, a massive loss of germ cells occurs (approximately 80%), due largely to atresia during meiotic prophase. Follicular growth is evident by day 100 (i.e. the granulosa layer differentiates to cuboidal and begins to proliferate) and antral follicles are present by birth. Recruitment of primordial follicles into the growing pool continues throughout life (McNatty *et al.*, 1995). In the oocyte, meiosis is arrested in the dictyate stage of prophase I. Oocytes are maintained in prophase I (the germinal-vesicle stage) until the final signal for follicle maturation and ovulation (the luteinizing hormone (LH) surge) is received or until the follicle becomes atretic (Turnbull *et al.*, 1977; see Scaramuzzi *et al.*, 1993, for review) (Fig. 13.1).

At the birth of the male lamb, the seminiferous tubules occupy about 80% of testicular volume. Relatively little development then occurs until the pubertal stage, when rapid testicular growth occurs and spermatogenesis is established. Spermatogenesis involves the transformation of undifferentiated stem cells into highly differentiated spermatozoa. Each stage of this development involves a specific and constant association of germ-cell types and occurs in a coordinated manner, such that localized parts of seminiferous tubules can be observed in the same stage of the spermatogenic cycle. These stages follow each other in succession along the length of a tubule, which, on a whole-testis basis, results in a continuous and relatively constant production of spermatozoa by post-pubertal males (Courot *et al.*, 1970). Thus the postpubertal phase is an ongoing period of slow testicular development, during which the animal maintains an adult rate of spermatogenesis.

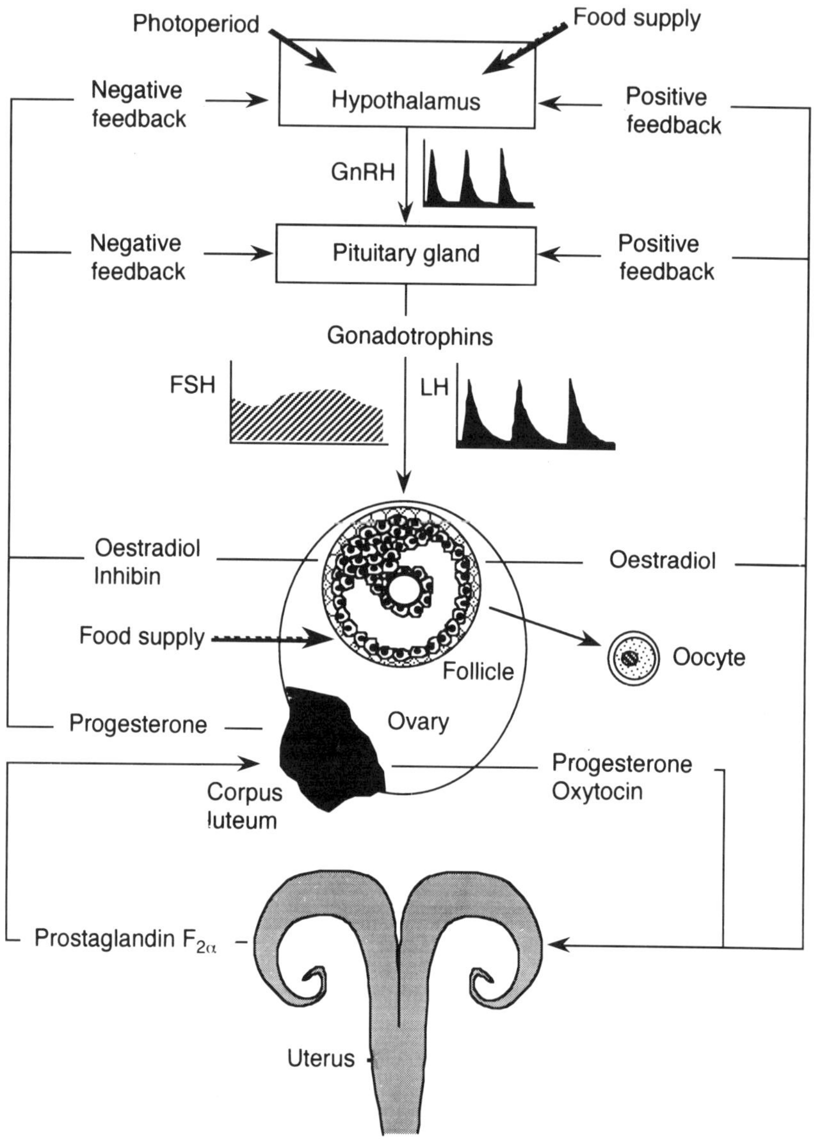

Fig. 13.1. Hormonal regulation of the oestrous cycle, from Scaramuzzi *et al.*, 1993. FSH, follicle-stimulating hormone; GnRH, gonadotrophin-releasing hormone; LH, luteinizing hormone.

Phenotypic sexual determination

Regardless of sex chromosome genotype, before the differentiation of gonadal sex, each embryo/fetus has the potential to develop both male (Wolffian ducts) and female (Müllerian ducts) accessory reproductive systems. The mechanisms required to ensure normal phenotypic sexual development are controlled by the presence or absence of hormones produced by the fetal gonads subsequent to gonadal sex differentiation. Immediately after formation of the testicular cords, fetal Sertoli cells produce anti-Müllerian hormone (AMH), which actively induces the regression of the Müllerian-duct system, thereby preventing the development of female reproductive organs. Subsequently, testosterone produced by the fetal testicular Leydig cells induces the Wolffian ducts to form the epididymides, vasa efferentia and seminal vessicles. Formation of the male external genitalia requires the bioconversion of testosterone to 5α-dihydrotestosterone in the target tissue. In the developing female fetus, the ovaries do not produce either AMH or testosterone, thereby creating a permissive environment for the differentiation of the Müllerian ducts (to form the fallopian tubes, the uterine horns and body and the upper cervix) and the passive regression of the Wolffian ducts, respectively. Steroid content (particularly oestradiol) of the fetal ovary remains low (< 10 mg $ovary^{-1}$) from day 55 until the end of gestation (Smith *et al.*, 1993; McNatty *et al.*, 1995, for review). Interestingly, *in vitro* cultures of sheep ovaries (but not testes) at days 31 and 47 of gestation release measurable concentrations of oestradiol into the media, suggesting that unknown cells of the fetal ovary may actively produce oestradiol at about the time sexual differentiation begins (Mauléon, 1978).

Development of the hypothalamopituitary gonadal axis

Cells and fibres that stain immunohistochemically for gonadotrophin-releasing hormone (GnRH) first appear in the fetal preoptic–hypothalamic area between days 40 and 43 of gestation. Immunoreactive fibres are evident in the median eminence by day 49 and morphological maturation of the GnRH neuron system is complete around mid-gestation (see Caldani *et al.*, 1995, for review).

Pituitary gonadotrophic cells have been identified as early as day 49–50 (Messaoud-Toumi *et al.*, 1993). Luteinizing-hormone-positive cell numbers increase between days 100 and 120 and remain high through to term, while follicle-stimulating hormone (FSH)-containing cells increase until about day 130 and then fall before birth. Immunoactive FSH and LH have been detected in pituitary extracts from day 55. From days 75 to 135 of gestation, the pituitary contents and concentrations of bioactive and immunoreactive FSH increase with increasing age in sheep fetuses, but are higher in females than in males (Phillips *et al.*, 1992).

Plasma LH and FSH concentrations are higher at midgestation (days 75 to 130) than during early gestation (from days 55 to 60) or immediately before birth, and levels are lower in male than in female fetuses. This indicates that the testis, but not the ovary, exerts negative feedback on the hypothalamus–pituitary by midgestation. Furthermore, the possibility that a sexually dimorphic GnRH pulse generator exists from early development is indicated by increases in LH pulse frequency and mean plasma FSH after castration of male but not female sheep fetuses from about 110 days of gestation (Matwijiw and

Faiman, 1989, 1991; Mesiano *et al.*, 1991). The late-gestational suppression of circulating gonadotrophic hormones has been ascribed to increased sensitivity to negative feedback of placentally derived steroids on the hypothalamus–pituitary, together with an increase in inhibitory factors within the central nervous system. This results in reduced secretion of GnRH from the fetal hypothalamus. The additional suppression observed in the male fetus in late gestation has been ascribed to the higher levels of circulating inhibin produced by the fetal testis, rather than circulating testosterone (see Brooks *et al.*, 1995; Jenkin *et al.*, 1995, for reviews).

Suppression of fetal male gonadotrophic secretion (chronic GnRH-agonist treatment) during the second half of gestation results in reduced neonatal testicular weight, due principally to fewer Sertoli cells in the testicular cords. Although compensatory growth occurs during the first 2 months postnatally, testosterone secretion appears to be compromised at puberty (Brooks *et al.*, 1995; Thomas *et al.*, 1995). Similar treatment of female fetuses appears to have no effect on ovarian mass or morphology. These results, together with earlier studies of the effects of fetal pituitary oblation in sheep (Liggins and Kennedy, 1968), suggest that, in contrast to the fetal testis, fetal ovarian development is not critically dependent on gonadotrophic stimulation during the latter half of gestation.

Once the lamb is born, the further development of the reproductive organs of both sexes comes under strong influence of environmental factors. These developments will be considered in the subsequent sections on the genetic variability in components of male and female reproductive characteristics.

Components of Reproductive Rate in the Ewe

Introductory remarks

Traditionally, consideration of ewe reproduction has involved those traits that could be assessed by simple inspection of animals at pasture. Thus, in early studies, fertility was described as the proportion of ewes that lambed of those joined to the ram(s), and estimation of prolificacy or litter size required observation at lambing. An appreciation of the potential significance of embryo mortality (EM) in the days following fertilization led to a changed definition of fertility. By the simple expedient of using marking crayons on rams and vasectomized rams and hence identifying those ewes in the flock that were mated and those that subsequently returned to be mated, a more sophisticated partition of ewe reproductive components evolved. Still further refinement was achieved through technological advances, such as laparoscopy (Thimonier and Mauléon, 1969) and ultrasonic scanning (Wilkins and Fowler, 1984). The capacity to identify whether ewes were ovulatory and to allocate a measure of ovulation rate by counting the number of corpora lutea on the surface of the ovaries was a vital advance in acquiring the capacity to prioritize objectively the contribution of the various stages in the reproductive process to overall prolificacy and mortality. Similarly, the use of ultrasonic scanning to identify the number of embryos carried during the different stages of gestation has been important in providing an increased capacity to identify sources of reproductive loss and hence to develop genetic and management strategies for reducing such losses.

Despite attempts to use sophisticated technologies to quantify the relative contribution of different factors at the end of gestation, observation of ewes during lambing is still the only practical method of allocating maternal pedigrees. Increased accuracy of maternal pedigrees will be provided in the future by the use of DNA techniques (Jeffreys *et al.*, 1985).

Genetic variation in components of reproduction in the female

Puberty

Puberty can best be defined as the time at which reproduction first becomes possible, as characterized by the release of germ cells (Dyrmundsson, 1973). The first visible signs of oestrus are not always accompanied by ovulation. Puberty should be distinguished from sexual maturity, which can be defined as when the ewe first expresses its full reproductive potential by achieving the rearing of maximal litter size.

Considerable variation exists between and within breeds in the age and live weight at which first oestrus is expressed. In many of the prolific breeds of sheep, ewes show first visible signs of oestrus at ages as early as 4–5 months after birth and are capable of lambing within a year (see reviews on prolific breeds in Fahmy, 1996). In contrast, spring-born Merino ewe lambs rarely reach puberty in their first year, although those of the Booroola Merino strain do begin cycling at a significantly earlier age (Bindon and Piper, 1982). Where nutrition is suboptimal and ewes do not achieve puberty in the first year, their first oestrus can be delayed until the beginning of the next spring.

Within-breed variation in age and weight at attainment of puberty has also been documented. Fogarty (1984) quotes estimates of heritability for age at puberty ranging from 0.1 to 0.26. However, there is likely to be a complex interrelationship between age and live weight and attainment of fully functional sexual function that would need to be elucidated in order to understand fully the consequences of longer-term selection for reduced age at which puberty is attained. Although early puberty can be valuable in intensive enterprises where accelerated lambing is an objective, for many extensive situations early puberty can lead to unplanned pregnancies in young ewes.

Oestrus expression, seasonality and fertility

The duration of the breeding season is also important to breeders who are interested in flexible mating and lambing schedules and in accelerated lamb-production systems. In an evolutionary sense, seasonal breeding may be interpreted as an adaptation that would achieve the goal of delivering lambs at a time when there is the highest probability of having favourable weather and nutritional conditions. There is increasing interest in optimizing reproductive performance at times other than normal for a climatic region or for a particular breed. This has been stimulated by the imperatives of international meat markets, which demand delivery of meat products in every month of the year. These developments have stimulated interest in the seasonality and length of the breeding season in sheep.

Breeds differ enormously in the length and timing of their breeding season. It is likely that this variation is a result of the wide range of environments and latitudes in which sheep breeds have evolved. Some highly seasonal breeds, such as the Suffolk, display a 'deep' anoestrus. The onset of this period occurs under the influence of inhibitory photoperiods, and the activity of the mechanism driving the pulses of LH is reduced. This pulse generator becomes more sensitive to negative feedback by oestradiol and, in the absence of the preovulatory surge of LH, the final stages of follicular maturation, and thus ovulation and oestrus, are effectively blocked (Matos and Thomas, 1990).

Prolonged sexual seasons are a feature of several of the more prolific breeds. The Romanov, Finnish Landrace and D'Man breeds are almost completely aseasonal (Bindon *et al.*, 1996). A feature of all Indonesian breeds, including the thin- and fat-tailed types, is that they too can breed all year round and have mean lambing intervals as low as 200 days (Bradford and Inounu, 1996).

Differences between ewes in the onset of the breeding season is a feature of most breeds and this has been shown to have a moderate genetic component (Hanrahan and Quirke, 1986; Fahmy, 1990). Heritability estimates range from 0.17 to 0.25.

As identified above, differences between breeds in seasonal breeding activity have been linked to seasonal shifts in both LH levels and sensitivity to oestradiol negative feedback. Indeed, successful selection for an earlier onset of the breeding season has been demonstrated (Smith *et al.*, 1992) and has produced animals with a reduced sensitivity to day-length-controlled negative oestradiol feedback, similar to that displayed in other breeds with different seasonal breeding patterns. These studies also confirmed that selection for lambing date has a significant effect on the date of first oestrus (Smith *et al.*, 1995).

The combined effect of puberty, oestrous activity, seasonality and post-partum joining interval determines the fertility level of a flock at a particular time. This trait is normally measured as number of ewes pregnant per ewe joined to the ram and, as such, the expression involves contributions from the ram and the fetus(es). Under conditions of good management and nutrition, fertility is generally close to its maximum. Given that the trait has a low heritablility (generally less than 0.1 (Fogarty, 1995)), the attention of breeders is generally focused on achieving a high flock fertility by optimizing management and nutrition.

Ovulation rate

Polygenic control Ovulation rate is a vital component of prolificacy and reproductive efficiency in any flock, because it sets the upper limit to these traits in any one sexual cycle. Since the development of laparoscopy as the preferred method for assessing ovulation rate at any time in the sexual cycle (Thimonier and Mauléon, 1969), a great deal of information has been collected on this trait, for many breeds of sheep. Despite this, a full understanding of the underlying genetic control of ovulation rate in the sheep is still a major challenge. This is largely because ovarian follicle growth and development are an enormously complex and highly coordinated process.

As indicated above, under 'Oestrus expression, seasonality and fertility', development of the ovarian primordial follicle population begins early in fetal life and is complete by the time of birth. Initiation of follicle growth is triggered by an as yet unknown factor, and follicle growth is continuous for the remainder of the reproductive life of the ewe. Only a tiny proportion of primordial follicles complete growth to the ovulatory stage. In cattle, follicle growth and development during the oestrous cycle occur in a wave pattern, and a mechanism of functional dominance of the largest follicles leads to ovulation. Available evidence supports a similar pattern in sheep.

The release of ova is the culmination of a complex of processes and interactions, including a gonadotrophin/steroid feedback system, involving the hypothalamus, pituitary and ovary, inter- and intraovarian steroids and follicular proteins, other endocrine hormones and local ovarian factors. While the gonadotrophins provide the stimulatory drive for recruitment and development of follicles, other factors act in an inhibitory fashion and damp down the actions of FSH and LH (for review, see Gong and Webb, 1996).

In contrast with cattle, where there is little variation in ovulation rate, there is considerable variability between breeds of sheep in both the mean ovulation rate and the distribution of values. Breed means vary from as low as 1.2 in the Merino up to as high as 5.3, as observed in the Cambridge (see Bindon *et al.*, 1996). Much of our current understanding of genetic control of ovulation rate comes from comparative breed studies, especially between breeds of prolific sheep.

Heritability estimates for ovulation rate vary considerably, but in general are higher in the moderately and highly prolific sheep than comparable estimates for litter size. Estimates range from a low of 0.05 in young Merinos (Piper *et al.*, 1980) to 0.50 in Finn ewes in Ireland (Hanrahan and Quirke, 1985). Fogarty (1994) reviewed available estimates of heritability and calculated a weighted mean estimate of 0.21 from nine studies. This mean value was characterized by a high standard deviation (coefficient of variation (CV) appoximately 30%). The trait is also characterized by moderate intrayear and interyear repeatabilities (0.37 and 0.31, respectively (Fogarty, 1994)).

Major genes affecting ovulation rate Quantitative genetics can be described as the discipline that embraces the inheritance of traits that have a continuous (or even semicontinuous, i.e. categorical) distribution. Based on theory developed by Fisher (1918) and Wright (1921) and later extended by others, the classical model assumes that the expression of quantitative traits is in part due to genes and that it is the product of many genes of small effect. However, in a few special cases, genes with a major influence on the phenotype (major genes) have been discovered.

The discovery in 1980 (Piper and Bindon, 1982) that the extremely high prolificacy of the Booroola Merino was due to a major gene (or tightly linked group of genes) was the stimulus for a decade of intensive study aimed at identifying the mechanism causing the exceptional prolificacy in the Booroola and, by extension, that in other prolific breeds.

Segregation studies with Booroola flocks established that the effects of what is now called the *Fec*^{B} gene (COGNOSAG, 1989) are additive. Recently, the *Fec*^{B} locus has been mapped on to sheep chromosome 6 (Montgomery *et al.*, 1994). Each copy of the gene adds 1.0 to 1.5 ova and this translates into an increase in litter size of 0.75–1.0 lambs in breeds of low to moderate prolificacy (Piper and Bindon, 1987). The most likely mechanism through which the gene operates is reduced feedback products from the ovary resulting in increased levels of the gonadotrophin, FSH (Piper and Bindon, 1987). The exact action of the *Fec*^{B} gene will most probably be elucidated from molecular studies.

Major genes have also been identified in several other breeds of sheep. In the Romney, Javanese, Cambridge, Belclare, Thoka and Olkuska breeds, animals of extreme prolificacy have been identified (see Fahmy, 1996) and available data are suggestive of segregation of major genes influencing ovulation rate. In the Romney breed, the segregation of an X-linked major gene, the 'Inverdale' (*Fec*^{I}) gene, has been demonstrated (Davis *et al.*, 1991). One copy of this gene causes an increase in ovulation rate of 1.0 ovum and litter size by 0.6 lambs. However, homozygous individuals are sterile.

The Booroola gene has been transferred to many breeds throughout the world and, as experience has grown with introgressing this gene into populations of widely different natural prolificacy, strategies for identification and utilization of major genes are being further developed (Elsen *et al.*, 1994). Furthermore, when molecular-marker technologies are added to the increasingly sophisticated statistical methodologies, the efficiency with which major genes affecting reproduction rate are utilized should improve substantially.

Embryonic mortality

Together with ovulation rate, EM accounts for variation in litter size. It can influence the output of a breeding flock through the loss of one embryo of a pair and cause ewes to have extended intervals before they return to service during mating. Mortality primarily occurs in the first 4 weeks of pregnancy and up to 20–30% of fertilized ova can be lost (Edey, 1966; Quirke and Hanrahan, 1983).

The relative importance of EM to genetic variation in litter size has been the subject of much research and it is only recently that some consensus has been found in the literature. There is a negative linear relationship between embryo survival and number of ova shed and, as the mean ovulation rate in a population increases, variation in litter size due to embryo survival increases.

Estimates of heritability for EM (as a trait of the dam) are generally low and less than 0.07 (Elsen *et al.*, 1994) and suggest that genetic variation makes only a small contribution to the observed variation within a population. Despite evidence from ovum transfer studies that the genes of the conceptus only play a very small part in its own survival (Hanrahan, 1982), recent findings suggest that the genotype of the embryo can be important in determination of survival (Bodin *et al.*, 1994).

Litter size

Litter size, or lambs born per ewe lambing, is the trait that has been the subject of most of the attention in genetic improvement programmes and selection experiments. In part, this is because the trait is automatically measured when maternal pedigrees are allocated to lambs. However, as with ovulation rate, it does show large between-breed variability, which in part reflects the influence of the major genes influencing ovulation rate. In contrast with ovulation rate, estimates of heritability of litter size are less variable. Over 53 estimates of heritability, Fogarty (1995) calculated a standard deviation of 0.07 against the mean of 0.10. Litter size still displays considerable variability (CV = 36% (Fogarty, 1995)).

Lamb survival

Reproductive wastage due to neo- and postnatal lamb mortality is one of the major causes of inefficiency in sheep-breeding enterprises. In large-scale studies in Western Australia and New Zealand (reviewed by Knight, 1990), lamb mortality was identified as the second largest cause of reproductive wastage (next to failure of ewes to lamb or partial failure of multiple ovulation – both of which are likely to have been primarily due to early EM). Most lambs died at birth or within 3 days after birth, and the mortality was primarily affected by breed and age of ewe, sex, birth rank and birth weight of the lamb. Within birth rank, the most frequent causes of death were dystocia in single lambs and starvation and exposure in twin-born lambs. Birth rank and ewe age effects were mostly accounted for by birth weight.

The genotype of the lamb and the dam have both been shown to influence lamb survival, although, compared with the above environmental effects, these influences are small. Differences between breeds can be substantial (Hight and Jury, 1970; Atkins, 1980), with this often reflecting substantial differences between breeds in the survival of lambs born in multiple litters. Atkins (1987) also examined differences in lamb survival between crosses of Merino strains and found that direct additive genetic differences and genetic differences in maternal effects contributed about equally to overall variation between blood-lines in lamb survival. Heterosis for lamb survival was a significant source of variation, whether the crosses were between strains or between blood-lines within strains, and the lower mortality of crossbreds was found in lambs born as singles and multiples.

Male Reproduction

Biology of male reproduction

The male contributes 50% of the genes of the new generation. However, in terms of his influence on genetic change, because of differential mating ratios, the male has a much bigger influence than the female. Therefore, in a self-replacing breeding flock, many fewer, and hence superior, males are selected to be parents than are females. This feature of sheep reproduction also means that failure of fertility in one ram can have a significant impact on the overall reproductive efficiency of the flock.

Genetics of male reproduction

Testicular size and spermatogenesis

Given the problems of conducting genetic studies on rates of spermatogenesis or on spermatogenic efficiency, the more practical measurement of the size of the testes has been the trait most utilized in studies aimed at identifying the extent of genetic variation. The average circumference or diameter has been equally used as the measure of testicular size.

Unlike many female measures of reproductive performance, testicular size can be easily measured at any time during the life of the ram. Consequently, studies of this trait have been conducted over a wide range of ages. Fogarty (1995) has summarized the findings of 14 studies, which show an average heritability of 0.24 (range to 0.69), with an average CV for the trait of 12%. In a major study of genetic variation in testicular size at ages between 4 months and 19 months of age in Merinos, Purvis (1985) estimated the heritability to range between 0.22 and 0.75, with the average value being 0.52. Genetic correlations between the ten age-based measures of testicular size were all high and positive.

In several of the genetic studies of testicular size, investigators have reported the trait as a function of live weight. Because of close genetic relationships between live weight and testicular size measured at the same age, the live weight-adjusted estimates of heritability are generally lower than the unadjusted estimates.

Libido and serving capacity

The term libido was first used by Freud in 1894 to define the force by which sexual instinct is represented in the human mind (Wodzicka-Tomaszewska *et al.*, 1981). In the context of sheep breeding, it refers to the desire of a ram to mount and inseminate ewes that are displaying oestrus. Serving capacity is a term used to describe the results of an evaluation of the capacity to mount a ewe and achieve intromission. Associated with the concepts of libido and serving capacity is a courtship ritual, which involves behavioural interactions between male and female. As with the failure of spermatogenesis, the ram that has inadequate libido or poor mating dexterity can have a profound impact on flock reproductive efficiency.

Measurement of serving capacity has been focused on pen tests of a set duration, during which a single (or several) ram(s) are exposed to a group of ewes artificially brought into oestrus. The number of mounts and ejaculations for each ram is recorded. Variations on this technique have also been used. For example, non-oestrous ewes restrained in cradles have been utilized in some studies (Rival and Chenoweth, 1982).

The relationship between serving capacity of rams and flock reproductive performance has been found to be dependent upon the length of time over which the evaluation was conducted. In evaluations conducted over time periods of more than the length of one oestrous cycle, the number of ewes inseminated or pregnant has not been correlated with measures of serving capacity (Kilgour, 1980).

Between-breed differences in various measures of libido and serving capacity have been found, but, in general, the breeds have not been compared at the same stage of sexual maturity and, in several cases, may have reflected seasonal differences in sexual activity between breeds (Lindsay and Ellsmore, 1968; Kilgour and Winfield, 1974).

One major study of genetic variation in 840 young Merino rams (Purvis, 1985) showed some differences between Merino blood-lines and within strains, but there was no evidence of any within-flock genetic variation for a range of measures of serving capacity. These data were characterized by a large number of rams showing little or no libido, and different results might have been obtained with rams that had been exposed to previous heterosexual experiences. In this study, no relationship was found between serving capacity and testicular size.

Male Measures of Female Reproductive Performance

The finding by Land (1973) that the quantitative expression of male and female gonadal traits was genetically correlated stimulated a large number of studies on this subject, particularly in sheep and cattle. Subsequently, Walkley and Smith (1980) demonstrated that, if the male trait had a moderate heritability and a high genetic correlation with the female trait, gains could be achieved that were greater than direct selection on the female trait. Land's (1973) suggestion was based on the knowledge that germ-cell differentiation and production in both sexes was largely controlled by the same endocrine system. Support for this concept was provided by results from flocks selected for increased female reproduction rate or its components, which had shown a correlated response in testicular size (Hanrahan, 1982; Burfening and Tulley, 1982).

Estimates of within-flock relationships between testicular size and female reproductive performance further supported the hypothesis. In the Australian Merino, moderate genetic correlations between hogget testicular size and ovulation rate, litter size and lambs weaned were found (Atkins and Mortimer,

1987; Purvis *et al.*, 1988). These estimates, together with the high heritabilities of testicular size, indicated that a useful genetic response in female reproductive measures would result from selection on testicular size in Merino flocks.

Selection of Finn–Dorset flocks on the basis of index of testicular size adjusted for live weight (Land *et al.*, 1982; Lee and Haley, 1990) did not improve ovulation rate, as originally postulated. However, Purvis *et al.* (1988) using the findings of their study of Merino flocks demonstrated that the absence of a correlated response in ovulation rate from selection on testicular size adjusted for live weight could be expected on the basis of the genetic relationships involving live weight and testicular size/ovulation rate. Similarly, the decrease in the mature size of the females in the Finn–Dorset line selected for increased live weight-adjusted testicular size was predictable on the basis of parameter estimates from the Merino study.

Selection Experiments on Components of Reproduction Rate

Knowledge of heritability and phenotypic variance and choice of an appropriate selection differential and generation intervals allow prediction of response to selection to be made. When theoretical responses to selection for increased prolificacy are compared with those achieved in experimental selection flocks or commercial selection schemes and account is taken of the actual selection differential, the agreement is good. This is despite these flocks differing substantially in breed, environment, age and selection intensity. Analysis of responses achieved in selection experiments where litter size has been the criterion (Bradford, 1985; Morris, 1990) shows gains of 1.3–1.5% to be the average. Where ovulation rate has been measured, the response in litter size has been shown to be fully explained by increases in ovulation rate.

Several commercial selection flocks have been formed by screening large populations for individuals of extreme performance (e.g., Meyer and Clarke, 1982). These schemes have shown a large initial response to screening, then a lull and then an accelerating response. Such a pattern may be due to the accumulation of additive and favourable dominant alleles during the screening, the subsequent lull occurring as a result of the initial slow accumulation of recessive alleles and the subsequent increasing response reflecting the expression resulting from the increasing frequency of these recessives (Bradford, 1985).

Estimation of Genetic Parameters for Reproduction Rate

As indicated above, most reproductive traits in sheep populations have distributions that are non-continuous and/or non-normal in their distribution. Which methods are optimal for estimating genetic parameters and breeding values for components of reproduction rate have been a topic of much interest in the literature over the past 20 years. The potential problems with such data are reflected by a consideration of litter size. Within populations, litter size can fall into three to six categories and have a distribution where one or two classes account for 90% of the population. Mathematical algorithms for estimating genetic and phenotypic parameters and predicting genetic means have been described that are based on assumptions of ordered categories and concepts of threshold (Gianola and Foulley, 1983; Harville and Mee, 1984). Studies that have compared the utility of these statistically more rigorous approaches with others that 'incorrectly' assume linearity and normality have generally shown good agreement. For example, Jorhensen (1994) evaluated a non-linear estimation of genetic parameters and prediction of breeding values for litter size on data from the Danish sheep-recording scheme. There was generally good agreement between the two approaches. He suggested that best linear unbiased prediction (BLUP) should not be replaced by non-linear algorithms until it can be performed with an animal model.

Conclusions

As indicated throughout this chapter, the study of the genetics of reproductive traits in the sheep is challenging for several reasons. Not only do many of the traits show categorical distributions but many are sex-limited in their expression. There are large differences between breeds for many of the components of reproduction and this, coupled with the low heritability of many of these traits, has complicated across-breed and across-flock assessment of genetic merit. However, breeders and experimentalists have shown that, despite the low heritability of traits such as litter size, because of high coefficients of variation it is possible to make significant genetic progress through simple selection on phenotype.

The discovery of the Booroola *Fec*^{B} gene has led to a major study of the physiology and genetics of such genes. This has resulted in significant advances in the knowledge of how to find and utilize these genes and they have been introgressed into many breeds that are normally non-prolific. The explosion of research on molecular genetics will result in the identification of genes with large effect on other components of the reproductive system and allow more sophisticated use of major genes in commercial sheep breeding.

The challenge that remains for researchers and breeders alike is to find ways to successfully integrate genetic improvement of reproductive performance into enterprise-based breeding programmes that operate under pastoral conditions.

Acknowledgements

The authors would like to thank the editors for their patience and the staff of the Commonwealth Scientific and Industrial Research Organisation (CSIRO) Division of Animal Production for their assistance. Sincere thanks also to Dr J.P. Hanrahan for his helpful comments.

References

Atkins, K.D. (1980) *Lamb growth and survival. Australian Journal of Experimental Agriculture and Animal Husbandry* 20; 272–279.

Atkins, K.D. (1987). Breeding objectives for terminal sires of sheep. *Proceedings of the Australian Association of Animal Breeding and Genetics* vol. 6, 221–224.

Atkins, K.D. and Mortimer, S. (1987). *Genetic Improvement of Reproductive Rate in Merino Sheep*. Final Report K/1/1065, New South Wales Department of Agriculture, Trangie, NSW. 67 pp.

Bindon, B.M. and Piper, L.R. (1982) Physiological characteristics of high fecundity sheep and cattle. In: Barton, R.A. and Smith, W.C. (eds) *Proceedings of the World Congress on Sheep and Cattle Breeding*, Vol. I. Dunsmore Press, Palmerston North, New Zealand, pp. 315–331.

Bindon, B. M., Piper, L. R., and Hillard, M. A. (1996). Reproductive physiology and endocrinology of prolific sheep. In: Fahmy, M.H. (ed.) *Prolific Sheep*. CAB International, Wallingford, UK, pp. 453–470.

Bodin, L., Reinaud, P., Lajous, D., Poivey, J.P. and Charpigny, G. (1994) Relationship between embryonic interferon (oTP) and embryo survivial in sheep. In: *Proceedings of the 5th World Congress on Genetics Applied to Livestock Production*, Guelph, Canada, Vol. 19, 245–248.

Bradford, G.E. (1985) *Selection for litter size*. In: Land, R.B. and Robinson, D.W. (eds) *Genetics of Reproduction in Sheep*. Butterworths, London, pp. 3–18.

Bradford, G.E. and Inounu, I. (1996). *Prolific breeds of Indonesia*. In: Fahmy, M.H (ed.) *Prolific Sheep*. CAB International, Wallingford, UK, pp. 146–151.

Brooks, A. N., McNeilly, A. S. and Thomas, G.B. (1995) Role of GnRH in the ontogeny and regulation of the fetal hypothalamo-pituitary–gonadal axis in sheep. *Journal of Reproduction and Fertility* Suppl. 49, 163–175.

Burfening, P. J. and Tulley, D. (1982) Effect of selection for high and low prolificacy on libido and scrotal circumference. *Proceedings of the 2nd World Congress on Genetics Applied to Livestock Production*, Madrid, Spain, vol. 7, pp. 536–542.

Caldani, M., Antoine, M., Batailler, M. and Duittoz, A. (1995) Ontogeny of GnRH systems. *Journal of Reproduction and Fertility* Suppl. 49, pp. 147–162.

COGNOSAG, (1989). *Standardised Genetic Nomenclature for Sheep and Goats*. Lavoisier, Paris, pp. 49–52.

Courot, M., Hochereau de Reviers, M.T. and Ortavant, R. (1970) Spermatogenesis. In: Johnson, A.D., Gomes, W.R. and Van Demark, N.L. (eds) *The Testis*, 1st edn, Vol. 1. Academic Press, New York, pp. 339–432.

Darwin, C. (1859) *On the origin of species by means of natural selection or the preservation of favoured races in the struggle for life*. Murray, London.

Davis, G.H., McEwan, J.C., Fennessy, P.F., Dodds, K.G., and Farquhar, P.A. (1991) Evidence for the presence of a major gene influencing ovulation rate on the X-chromosome of sheep. *Biology of Reproduction* 44, 620–624.

Dyrmundsson, O.R. (1973) Puberty and early reproductive performance in sheep. I. Ewe lambs. *Animal Breeding Abstracts* 41(6), 273–289.

Edey, T.N. (1966) Nutritional stress and pre-implantation embryo mortality in Merino sheep. *Journal of Agricultural Science* (Cambridge) 67, 287–293.

Elsen, J.M., Bodin, L., Francois, D., Poivy, J.P. and Teyssier, J. (1994) Genetic improvement of litter size in sheep. In: *Proceedings of the 5th World Congress on Genetics Applied to Livestock Production*, Guelph, Canada, Vol. 19, pp. 237–243.

Fahmy, M.H. (1990) Development of DLS breed of sheep: genetic and phenotypic parameters of date of lambing and litter size. *Canadian Journal of Animal Science* 70, 771–778.

Fahmy, M.H. (ed.) (1996) *Prolific Sheep*. CAB International, Oxford. UK.

Fisher, R.A. (1918) The correlation between relatives on the supposition of Mendelian inheritence. *Transactions of the Royal Society, Edinburgh* 52, 399–433.

Fogarty, N.M. (1984) Breeding for reproductive performance. In: Lindsay, D.R. and Pearce, D.J. (eds) *Reproduction in Sheep*. Australian Academy of Science and Australian Wool Corporation, Canberra, pp. 226–233.

Fogarty, N.M. (1995) Genetic parameters for live weight, fat and muscle measurements, wool production and reproduction in sheep: a review. *Animal Breeding Abstracts* 63(3), pp. 143.

Gianola, D. and Foulley, J.L. (1983) Sire evaluation for ordered categorical data with a threshold model. *Genetic Selection and Evolution* 15; 201–224.

Gong, J.G. and Webb, R. (1996) Control of ovarian follicle development in domestic ruminants: its manipulation to increase ovulation rate and improve reproductive performance. *Animal Breeding Abstracts* 64(3); 195–204.

Greef, J.C., Bouwer, L. and Hofmeyr, J.H. (1995) Biological efficiency of meat and wool production of seven sheep genotypes. *Animal Science* 61; 259–264.

Hanrahan, J. P. (1982) Selection for increased ovulation rate, litter size, and embryo survival. In: *Proceedings of the 2nd World Conference on Genetics Applied to Livestock Production*, Madrid, Spain, Vol. 5, pp. 204–309.

Hanrahan, J. P. and Quirke, J.F. (1985) Contribution of variation in ovulation rate and embryo survival to breed variation in litter size. In: Land, R.B. and Robinson, D.W. (eds) *Genetics of Reproduction in Sheep*, Butterworths, London, pp. 193–202.

Hanrahan, J.P. and Quirke, J.F. (1986) Breeding season and multiple births in small ruminants. In: *Proceedings of the 3rd World Congress on Genetics Applied to Livestock Production, Nebraska*, USA, Vol. 11, pp. 30–45.

Hight, G.K. and Jury, K.E. (1970). The influence of age, flock, and year, on some components of reproduction rate in Romney and Border Leicester × Romney ewes. *New Zealand Journal of Agricultural Research* 13, 641–659.

Harville, D.A. and Mee, R.W. (1984) A mixed-model procedure for analysing ordered categorical data. *Biometrics* 40, 393–408.

Jeffreys, A.J., Wilson, V. and Thein, S.L. (1985) Hypervariable 'minisatellite' regions in human DNA. *Nature* 314(7), 67–73.

Jenkin, G., McFarlane, J. and deKretser, D.M. (1995) Inhibin and activin in embryonic and fetal development in ruminants. *Journal of Reproduction and Fertility* Suppl. 49, 177–186.

Jorgensen, J.N. (1994) Estimation of genetic parameters for litter size in sheep using a non-linear method. *Acta Agriculturae Scandinavica. A. Animal Science* 44, 8–11.

Kilgour, R. and Winfield, C.G. (1974) Sheep mating behaviour. *New Zealand Journal of Agriculture* 128(4), 17–23.

Kilgour, R.J. (1980). The assessment and significance of sexual drive in the ram. In: Wodzicka-Tomaszewska, M., Edey, T.N. and Lynch, J.J. (eds) *Reviews in Rural Science. 4. Behaviour in Relation to Reproduction, Management and Welfare*. New England University Press, Armidale, NSW, Australia, pp. 43–46.

Knight, T. (1990) Reproductive wastage, a guide for fundamental research: a New Zealand perspective. In: Oldham, C.M., Martin, G.B. and Purvis, I.W. (eds) *Reproductive Physiology of Merino Sheep – Concepts and Consequences*. University of Western Australia, Perth, Australia, pp. 11–21.

Land, R.B. (1973) The expression of female sex-limited characters in the male. *Nature, London* 241, 208–209.

Land, R.B., Atkins, K.D. and Roberts, R.C. (1982) Genetic improvement of reproductive performance. In: Haresign, W. (ed.) *Sheep Production*. Butterworths, London, pp. 515–535.

Large, R.V. (1970) The biological efficiency of meat production in sheep. *Animal Production* 12, 393–401.

Lee, G.J. and Haley, C.S. (1990) Bodyweight adjusted testicular size as a selection criterion to improve production efficiency in sheep. In: *Proceedings of the 4th World Congress on Genetics Applied to Livestock Production, Edinburgh, UK, July 1990*, Vol. 16, pp. 370–373.

Liggins, G.C. and Kennedy, P.C. (1968) Effects of electrocoagulation of the foetal lamb hypophysis on growth and development. *Journal of Endocrinology* 40, 371–381.

Lindsay, D.R. and Ellsmore, J. (1968) The effect of breed, season and competition on mating behaviour in rams. *Australian Journal of Experimental Agriculture and Animal Husbandry* 8, 649–652.

McNatty, K.P., Smith, P., Hudson, N.L., Heath, D.A., Tisdall, D.J., O, W.-S., and Braw-Tal, R. (1995) Development of the sheep ovary during fetal and early neonatal life and the effect of fecundity genes. *Journal of Reproduction and Fertility* Suppl. 49, 123–135.

Martin, G.B. and Thomas, G.B. (1990). Roles of communication between the hypothalamus, pituitary gland and ovary. In: Oldham, C.M., Martin, G.B. and Purvis, I.W. (eds) *Reproductive Physiology of Merino Sheep–Concepts and Consequences.* University of Western Australia, pp. 23–40.

Matwijiw, I. and Faiman, C. (1989) Control of gonadotrophin secretion in the ovine fetus. II. A sex difference in pulsatile luteinizing hormone secretion after castration. *Endocrinology* 124, 1352–1358.

Matwijiw, I. and Faiman, C. (1991) Control of gonadotrophin secretion in the ovine fetus. IV. Male specific entrainment of hypothalamic control of luteinizing hormone secretion by testosterone in the female ovine fetus. *Endocrinology* 129, 1447–1451.

Mauléon, P. (1978) Ovarian development in young mammals. In: Crighton, D.B., Haynes, N.B., Foxcroft, G.R. and Lamming, G.E. (eds) *Control of Reproduction.* Butterworths, London, pp. 141–158.

Mesiano, S., Hart, C.S, Heyer, B.W., Kaplan, S.L. and Grumbach, M.M. (1991) Hormone ontogeny in the ovine fetus XXVI. A sex difference in the effect of castration on the hypothalamic–pituitary gonadotrophin unit in the ovine fetus. *Endocrinology* 129, 3073–3079.

Messaoud-Toumi, L.H., Taragnat, C. and Durand, P. (1993) Heterogeneity in the storage of gonadotropins in the ovine fetus and evidence of luteinizing hormone–follicle stimulating hormone cells in the fetal pituitary. *Biology of Reproduction* 48, 1239–1245.

Meyer, H.H. and Clarke, J.N. (1982) Effect of ewe ovulation rate and uterine efficiency on breed and strain variance in litter size. *Proceedings of the New Zealand Society of Animal Production* 42, 33–35.

Montgomery, G.W., Penty, J.M., Lord, E.A., Henry, H.M., Cambridge, L.M. and Broad, T.E. (1994) Mapping production traits in farm animals: the Booroola (FecB) locus. In: *Proceedings of the 5th World Congress on Genetics Applied to Livestock Production*, Guelph, Canada, Vol. 21, 17–20.

Morris, C. (1990) Theoretical and realised responses to selection for reproduction rate. In: *Proceedings of the 4th World Congress on Genetics Applied to Livestock*

Production, Edinburgh, UK, Vol. 16, pp. 309–318.

Phillips, D.J., Smith, P.R., Heath, D.A., Condell, L.A. and McNatty, K.P. (1992) Bioactive and immunoreactive FSH and immunoreactive inhibin concentrations in the ovine fetus. *Journal of Endocrinology* 34, 287–295.

Piper, L. R. and Bindon, B.M. (1982) Genetic segregation for fecundity in Booroola Merino sheep. In: Barton, R.A. and Smith, W.C. (eds) *Proceedings of the World Congress on Sheep and Beef Cattle Breeding*. The Dunmore Press, Palmerston North, vol. 1, pp. 395–400.

Piper, L.R. and Bindon, B.M. (1987). Industry utilisation of the Booroola Merino in Australia. In: McGuirk, B.J. (ed). *Merino Improvement Programs in Australia. Proceedings of a National Symposium*. Australian Wool Corporation, Melbourne, pp. 279–282.

Piper L.R., Bindon, B.M., Atkins, K.D. and Rogan, I.M. (1980) Genetic variation in ovulation rate in Merino ewes aged 18 months. *Proceedings of the Australian Society of Animal Production* 13, 409–412.

Purvis, I.W. (1985) Genetics of male and female reproductive traits. PhD thesis, University of New England, Armidale, New South Wales, Australia.

Purvis, I.W., Piper, L.R., Edey, T.N. and Kilgour, R.J. (1988) The genetic relationship between ovulation rate and testicular diameter in a random-breeding Merino flock. *Livestock Production Science* 18, 35–54.

Quirke, J.F. and Hanrahan, J.P. (1983) Comparison of the survival of fertilised eggs from adult ewes in the uterii of adult ewes and ewe lambs. *Journal of Reproduction and Fertility* 68, 289-294.

Rival, M. and Chenoweth, P.J. (1982) Libido testing of rams. *Proceedings of the Australian Society of Animal Production* 134, 174–175.

Scaramuzzi, R.J., Adams, N.R., Baird, D.T. Campbell, B.K., Downing, J.A., Findlay, J.K., Henderson, K.M., Martin, G.B., McNatty, K.P., McNeilly, A.S. and Tsonis, C.G. (1993) A model for follicle selection and the determination of ovulation rate in the ewe. *Reproduction, Fertility and Development* 5, 459–478.

Smith, J.F., Johnson, D.L. and Reid, T.C. (1992) Genetic parameters and performance of flocks selected for advanced lambing date. *Proceedings of the New Zealand Society of Animal Production* 52, 129–131.

Smith, J.F., Briggs, R.M., Parr, J., Johnston, D.L. and Duganzich, D.M. (1995) Seasonal changes in LH profiles of ewes selected for and against an early lambing date. *Proceedings of the New Zealand Society of Animal Production* 55, 224–227.

Smith, P., O, W.-S., Hudson, N.L., Shaw, L., Heath, D.A., Condell, L., Phillips, D.J. and McNatty, K.P. (1993) Effects of Booroola gene (Fec^B) on body weight ovarian development and hormone concentrations during fetal life. *Journal of Reproduction and Fertility* 98, 45–54.

Thimonier, J and Mauléon, P. (1969) Variations saisonniers du comportement d'oestrus et des activités ovarienne et hypophysaire chez les ovins. *Annales de Biologie animale, Biochimie et Biophysique* 9, 233–250.

Thomas, G.B., McNeilly, A.S. and Brooks, A.N. (1995) Effect of suppression of pituitary–testicular function during fetal life with a GnRH agonist on reproductive development in ram lambs during the first 28 weeks of life. *Journal of Reproduction and Fertility* Suppl. 49, 533–537.

Turnbull, K.E., Braden, A.W.H., and Mattner, P.E. (1977) The pattern of follicular growth and atresia in the ovine ovary. *Australian Journal of Biological Sciences* 30, 229–241.

Walkley, J.R.W. and Smith, C. (1980) The use of physiological traits in genetic selection for litter size in sheep. *Journal of Reproduction and Fertility* 59, 83–88.

Wang, C.T. and Dickerson, G.E. (1991) Simulation of life cycle efficiency of lamb and wool production for genetic levels of component traits and alternative management options. *Journal of Animal Science* 69, 4324–4337.

Wilkins, J.F. and Fowler, D.G. (1984) Ultrasound imaging for litter size diagnosis in breeding flocks. In: Lindsay, D.R. and Pearce, D.T. (eds) *Reproduction in Sheep.* Australian Academy of Science and Australian Wool Corporation, Canberra, pp. 182–184.

Wodzicka-Tomaszewska, M., Kilgour, R. and Ryan, M. (1981) Libido in the larger farm animals. *Applied Animal Ethology* 7, 203–238.

Wright, S. (1921) Systems of mating. *Genetics*, 6, 111–178.

Modern Reproduction Technologies and Transgenics

14

I. WILMUT, K.H.S. CAMPBELL and L. YOUNG
Roslin Institute (Edinburgh), Roslin, Midlothian EH25 9PS, UK

Introduction 395
Embryo Production 396
  Surgical recovery and transfer 396
  Laparoscopic techniques 397
  Non-surgical techniques 397
  Embryo production *in vitro* 397
  Nuclear transfer 400
Genetic Manipulation 403
  Direct injection 403
  Gene targeting 404
  Potential applications 405
Embryo Storage 405
Abnormal Fetal Development 407
Commercial Application 408
Acknowledgements 408
Note in Proof 408
References 409

Introduction

New technologies in reproduction fall into two categories concerned with either the production of embryos or their manipulation as a means of introducing genetic change. As exploitation of both groups of techniques depends upon an ability to recover, store and transfer embryos to recipient females, this chapter will also review these associated procedures. Unfortunately, experience with the first techniques to be applied has shown that culture or manipulation of ruminant embryos sometimes leads to the birth of unusually large offspring, with an unexpectedly high incidence of perinatal death, and so possible causes of this condition are considered. Finally, the potential value of these new technologies in sheep breeding and production is reviewed. Attention is paid to the limitations of the present techniques and the need for further research. No attempt is made to provide a comprehensive list of references, however, the reader should be able to trace key papers from the reviews that are cited.

Embryo Production

Surgical recovery and transfer

The sheep was the first of the domestic mammals for which routine procedures were established for the surgical recovery of embryos from donor females that had been induced to superovulate (reviewed by Moore, 1982). These procedures have been used very widely in research (reviewed by Wilmut, 1982) and on a small scale for the commercial import/export of sheep as embryos. While they are reliable, these techniques are expensive and involve major surgery. Typically, the oestrous cycle of the donor is controlled by administration of progestogen for a period equal to the length of the luteal phase, before repeated injections of gonadotrophin with follicle-stimulating activity are given to promote development and ovulation of a larger number of oocytes. Fertilization may be achieved by natural mating or artificial insemination (AI). Several days after mating or AI, anaesthesia is induced and the reproductive tract exposed through a midventral incision to allow the introduction of catheters into either the oviduct or the uterus, depending upon the stage of embryo recovery (embryos enter the uterus on day 3). Sterile medium is then passed through the tract to flush out the embryos. In the reverse of this process, embryos are transferred into a uterine horn by insertion of a glass Pasteur pipette through a small hole made with a blunt needle.

This procedure may yield around 6–10 embryos per donor per recovery, although the response is inherently very variable, as a result of differences in the number of follicles able to respond to the gonadotrophin. Furthermore, not all oocytes ovulated are fertilized and/or recovered. This number may be influenced by breed, season, body condition and nutrition (Moore, 1982). The number of times that surgical recovery may be attempted is limited by welfare and legislative considerations and by the formation of adhesions. The conception rate after embryo transfer is also influenced by a great many factors, but should be 60% or more if embryos are transferred immediately after recovery into recipients that were in oestrus at the same time as the donor. In this case, a yield of around four to six lambs per recovery might be expected, with a great deal of variation.

New approaches to embryo recovery and transfer have developed less intrusive means of surgical manipulation. In addition, embryos may be produced by *in vitro* oocyte maturation and fertilization, with subsequent culture of the fertilized eggs or, alternatively, nuclear transfer. These three opportunities are considered in turn.

Laparoscopic techniques

Laparoscopic techniques were introduced for embryo recovery and transfer in order to reduce the extent of the surgical intrusion (McKelvey and Robinson, 1984; reviewed by Wallace, 1992). This has advantages in terms of animal welfare and increased numbers of embryo transfers possible per day. A much smaller incision is required than in conventional surgery, with only partial exposure of the reproductive tract, using grasping forceps. The objectives are to flush medium through the uterine horns for embryo recovery or to transfer embryos in essentially the same procedures as already described. Laparoscopic observation of the ovaries before transfer allows transfer into a horn adjacent to an ovulation. While this approach does allow recovery of an increased total number of embryos through repeated flushing, the average yield from a single recovery is less than during one midventral surgery (Wallace, 1992).

Non-surgical techniques

The ability to recover and transfer embryos by non-surgical passage through the cervix has created new opportunities in cattle breeding by permitting large, field-scale application of these techniques. In sheep, there are two problems to overcome: firstly, the canal through the cervix is extremely tortuous and, secondly, the rectum is too small to allow manual manipulation of the cervix from the rectum. Although some success has been described (Mylne *et al.*, 1992), this is not currently a reliable option in sheep.

Embryo production in vitro

Embryo production *in vitro* involves three steps: recovery and maturation of the oocyte (IVM), fertilization (IVF) and culture (IVC) of the resulting embryo to a stage at which it can be transferred into the uterine horn of a recipient female. Such methods of embryo production in culture are well established and extensively used in both research and commercial breeding in cattle (see Gordon, 1994, for an exhaustive review). Although the analogous techniques are available in sheep, they have received far less attention. The three steps will be considered in turn.

Oocyte maturation

Oocytes suitable for maturation may be obtained by aspirating fresh abattoir-derived ovaries, which requires no surgical intervention and has the potential to dramatically increase oocyte availability. However, the genetic background of such ovaries is often unknown. Alternatively, oocytes may be aspirated from donor ewe ovaries, using laparoscopy. In cattle, ultrasound-guided equipment has recently been manipulated per rectum to allow effective transvaginal aspiration of developing oocytes. This technique of 'ovum pick-up' (OPU), combined with IVF and embryo culture, allows production of more embryos from a donor female than is possible by conventional superovulation and non-surgical recovery (Kruip *et al.*, 1991). This technique is applied while the animal is sedated and under local anaesthetic. Because of the anatomical differences, it is difficult to envisage how an OPU procedure could be applied in sheep. Repeated laparoscopy may be capable of harvesting a comparable number of oocytes, but is more intrusive and stressful and is perhaps less acceptable for use as a routine breeding procedure.

Reliable IVM of oocytes harvested from ovaries was first achieved in sheep (Staigmiller and Moor, 1984) and it is this procedure which provides the basis for current methods in all domestic species. Oocytes are obtained by aspiration from Graafian follicles ranging from 2 to 5 mm in diameter, in order to maximize the number of oocytes for IVM. However, there is evidence that better embryo development and birth rates are obtained with oocytes from the larger (and perhaps more mature) follicles (see Gordon, 1994). Further research is required to identify those facets of oocyte development which are deficient in oocytes from smaller follicles and to find ways of promoting their maturation in culture.

Those oocytes which have dark, even cytoplasm and are contained within a multilayered mass of cumulus cells are cultured for approximately 24 h in a complex medium containing serum, gonadotrophic hormones and often oestrogen. At the end of this maturation period, which mimics approximately the interval in a ewe between the onset of the luteinizing hormone surge and ovulation, most of the cumulus cells are removed to facilitate fertilization using hyaluronidase.

Fertilization

Although procedures have been established for the capacitation of sperm from most bulls (see Gordon, 1994), it seems that the same procedures are less reliable in sheep. The objective is to be able to obtain, consistently, a number of capacitated sperm that give high fertilization rates, with a minimal incidence of polyspermic penetration. Motile sperm are enriched by 'swimming up' through either a layer of culture medium or, alternatively, a gradient of Percoll polymer (e.g. Walker *et al.*, 1994; Thompson *et al.*, 1995). In contrast to the cow, simple media such as synthetic oviduct fluid (SOF) are routinely used for sheep IVF.

Embryo culture

Embryo culture is required as part of many potential applications. Despite a considerable research effort and a great improvement in effectiveness of the techniques, there is still not a method capable of producing embryos at a comparable rate to *in vivo*. Reliable sheep-embryo development from zygote to the transferable, blastocyst stage in culture was first achieved by coculture with cells from the oviduct (Gandolfi and Moor, 1987). Subsequent studies showed that this ability is not limited to cells from the reproductive tract (e.g. buffalo-rat liver cells have been used extensively with bovine embryos (see Gordon, 1994)), suggesting that the embryotrophic factors provided by the cells are not specific to the uterine environment. Coculture methods utilize complex media, such as medium 199, supplemented with a serum source and maintained in the presence of 5% carbon dioxide in air (Watson *et al.*, 1994). It is suggested that the cocultured cells assist in removing toxic products, providing beneficial components such as growth factors, and lowering oxygen tension locally.

An alternative culture strategy has established effective chemically defined media, and a number of such media have been studied with either sheep or cattle embryos (reviewed by Trounson *et al.*, 1994). One of the defined media most frequently used with sheep embryos has a composition based upon sheep oviduct fluid and is termed 'synthetic oviduct fluid' (Tervit *et al.*, 1972). A number of modifications to the original composition have been introduced, including a reduction in the glucose concentration and the introduction of both essential and non-essential amino acids (Gardner *et al.*, 1994). In SOF cultures, the concentration of oxygen in the environment is lowered, compared with the 20% used with coculture systems, to between 5 and 7%. Bovine serum albumin is usually included as a protein source, but more defined surfactants, such as polyvinyl alcohol, are currently being tested and may be applied to sheep embryo culture in the future.

Embryo culture in simple chemically defined medium exploits the paracrine and autocrine influences of embryos in culture and depends upon optimization of the numbers of embryos within a culture of specified volume. It is believed that by these means an effective concentration of biologically active molecules, such as growth factors, is maintained, while also allowing for the removal of toxic waste products, such as ammonia, by periodic replenishment of the medium (Gardner *et al.*, 1994).

In some cases, SOF medium has been supplemented with serum. A variety of different sources have been used, including fetal calf, sheep and human serum (e.g. Walker *et al.*, 1992). While the serum may increase the proportion of fertilized eggs that form transferable embryos, it appears to perturb development and can result in the birth of unusually large lambs if it is present throughout the culture period (Thompson *et al.*, 1995). It seems likely that the serum provides biologically active molecules, which are present at an inappropriate concentration or at an inappropriate time in the culture period. It may be that serum would be more appropriate if added toward the end of the culture (see later discussion on large lambs), when the embryo would normally be in the uterus.

Use of defined media eliminates the risk of disease introduction from cocultured cells and animal sera, but is currently compromised by the impurity of available serum albumin and is inferior to coculture in producing freezable embryos. Further research is likely to circumvent these problems.

Yield of embryos

While these *in vitro* techniques are able to produce a small number of embryos from a donor ewe their practical application is limited by the difficulty of obtaining a significant number of oocytes. There are two ways by which this limitation might be overcome: being able to carry out repeated recovery in some way (which has already been discussed) or being able to harvest a significant number of early, primordial follicles. At the time of birth, the ovaries of a ewe lamb contain tens of thousands of primordial follicles. During a lifetime, only a very small proportion of these follicles will mature and ovulate and an even smaller fraction will become lambs. The dream of being able to support growth and maturation of a number of these oocytes may be realized at some time in the future, but it is probably many years away. At present, it can be achieved with a low efficiency from primordial follicles of the mouse, which require a relatively short period of culture. However, a period of several weeks is required for such development in sheep and appropriate *in vitro* culture (IVC) systems would be expensive and demanding.

Nuclear transfer

In principle, nuclear transfer may be used to produce large numbers of genetically identical embryos. The most commonly used method of nuclear transfer involves removal of the chromosomes from a recipient oocyte before the transfer of a nucleus from a donor blastomere. The donor cell is fused to the recipient oocyte and the activation of development initiated (reviewed by Campbell *et al.*, 1996a). The transferred nucleus determines most of the characteristics of the resulting offspring. In the sheep, recipient oocytes have usually been recovered from donor ewes, but there is no apparent reason why oocytes matured in culture could not be used as techniques improve. Several factors will be considered which influence development after nuclear transfer, including coordination of the cell cycle in donor and recipient cells, as well as the stage of development of the donor cells.

Coordination of cell cycle

The cell cycle can be divided into a series of discrete events. In the early embryo, by far the greatest part of each cycle is taken up by the process of deoxyribonucleic acid (DNA) replication (S phase). There are brief periods between the previous cleavage and the S phase (G1 phase) and after the completion of replication and mitosis (G2 phase). Mechanisms exist to ensure that, during a single cell cycle, all of the chromosomal DNA is replicated only once (reviewed by Murray and Hunt, 1993). The integrity of the nuclear membrane is central to this control. During mitosis the nuclear envelope is absent and exposure of the chromosomes to the cytoplasm allows access of 'licensing factors', which trigger DNA replication during the subsequent cell cycle. During a cell cycle, nuclear-membrane integrity is controlled by a cytoplasmic kinase activity, known as maturation/mitosis/meiosis promoting factor (MPF). It acts to induce a number of processes, including breakdown of the nuclear membrane and condensation of the chromosomes, resulting in mitosis and so allowing access of the licensing factors. The MPF activity increases during the G2 phase of the cell cycle, after the completion of DNA replication, and is maximal at metaphase of the mitotic division.

Similar mechanisms act during meiosis and fertilization. At the time of ovulation, the oocyte is at metaphase of the second meiotic division and has a high level of MPF activity. The arrival of the sperm at fertilization or the induction of parthenogenetic development causes a resumption of meiosis and a decline in MPF activity (see Campbell *et al.*, 1996a). The chromatin then decondenses, a nuclear membrane is assembled and the resulting pronucleus enters the S phase.

In embryos reconstructed by nuclear transfer, the fate of the transferred nucleus depends upon the level of MPF activity at the time of cell fusion (Campbell *et al.*, 1993). If a nucleus is transferred to an oocyte at metaphase II, the MPF acts to cause nuclear-membrane breakdown and chromosome condensation, regardless of the stage of the donor cell cycle. As the MPF activity decays, the nuclear membrane re-forms and DNA replication is initiated. In these circumstances, the only donor cell-cycle stage which is appropriate for transfer would be that requiring DNA replication, i.e. donors in the G1 phase. Transfer of nuclei in either the S phase or the G2 phase would be expected to lead to aneuploidy.

A contrasting situation applies if transfer is to an oocyte with a low level of MPF activity, as it is the nucleus that determines whether there should be DNA replication or not. If recipient oocytes are parthenogenetically activated and cultured to allow MPF activity to decay, nuclear-membrane breakdown and chromosome condensation are not observed. In these circumstances, the oocyte provides an appropriate environment for donor nuclei at any stage of the cell cycle and has been termed a 'universal recipient'.

These observations established two different approaches to nuclear transfer that are able to maintain normal ploidy: the universal recipient or the transfer of diploid nuclei to oocytes with a high level of MPF activity. As there are serious technical difficulties in the synchronization of embryo-derived blastomeres in the G1 phase of the cell cycle, the universal recipient has been used more extensively.

Stage of development of donor cells

Until recently, nuclear transfer in mammals has only been successful when blastomeres removed from early embryos are used as nuclear donors. In a comparison of species, an apparent relationship exists between the latest stage of development from which the donor nucleus is able to promote development and the onset of embryonic transcription. The later the onset of embryonic transcription, the later the developmental stage from which development can be obtained. Two possible explanations have been proposed to explain this (see Campbell and Wilmut, 1994): following the onset of transcription, cells become progressively committed to a particular lineage and their ability to redirect development after nuclear transfer is reduced; alternatively, in those species which initiate transcription at a later developmental stage, the transferred nucleus undergoes more mitotic divisions in the absence of transcription and greater reprogramming can occur. The coordination of cell-cycle phases of the donor nucleus and recipient cytoplasm has not only resulted in development from a greater proportion of embryos reconstructed by nuclear transfer (Campbell *et al.*, 1994) but has also supported development to term following blastomere transfer from later stages of embryo development than had been possible previously (see Campbell *et al.*, 1996a). Furthermore, using the 'universal recipient', a small number of lambs were born following nuclear transfer from sheep embryonic cells that had been in culture for up to three culture passages (Campbell *et al.*, 1996b), although no lambs were born following transfer from later culture passages of these cells.

The availability of an established embryonic cell line has allowed the use of synchronization techniques to obtain donor cells at specific cell-cycle stages. In subsequent experiments, a synchronized population of diploid cells from the established embryonic cell line was obtained by reducing the serum concentration in the culture medium for several days. In these circumstances, the cells entered a quiescent state, G0, from which they could be induced to re-enter the cell cycle and initiate DNA replication at the required time by the restoration of the serum concentration. Following nuclear transfer from these embryo-derived cells synchronized in G0, several live lambs were born (Campbell *et al.*, 1996b). This represents a significant advance, as these are the first animals to be born following nuclear transfer from an established cell line, but the reasons behind the success are not yet understood.

When an embryo is reconstructed by nuclear transfer, it might be anticipated that normal development depends upon changes in gene expression in the transferred nucleus to the pattern normally observed in early development. In fact, the first cell cycles of development are governed by maternally derived messenger ribonucleic acid (mRNA) and proteins produced and stored in the oocyte before ovulation (see Johnson, 1981). At a species-specific stage (8–16 cells in the sheep), transcription from the embryonic genome starts, the maternally derived mRNA is destroyed and the embryonic genome progressively assumes control of development. This would suggest that transcription from a transferred nucleus should cease, until reinitiation at the appropriate time. Entry of a cell into the G0 phase is associated with changes in chromatin structure and these may facilitate reprogramming of gene expression. This would suggest that other, untested cell lines able to enter the G0 phase would also be suitable nuclear donors. Alternatively, the current success of the sheep embryonic line may reflect unknown characteristics of this cell line. The sheep donor cells do not resemble murine embryonic stem cells, but rather exhibit characteristics of differentiated cells. Nevertheless, it is possible that they are relatively undifferentiated or particularly amenable to reprogramming. Clearly, these questions should be addressed to discover the limits to this new approach to embryo production.

Despite the fact that the birth of lambs using an established cell line for nuclear transfer represents a significant advance, the present techniques have several limitations. Firstly, the efficiency is low, as only approximately 1% of reconstructed embryos developed to become lambs and some died after birth. As the new procedure has only been applied to one population of cells, it is not possible to know if this inefficiency reflects the influence of the cell line, the synchronization of cells in G0 or an interaction between the two factors. Secondly, the lambs appeared to have some of the characteristics of the large offspring born previously after blastomere nuclear transfer (Wilson *et al.*, 1994). Although a formal statistical analysis was not possible, there appeared to be an increase in the length of gestation and in birth weight.

In the light of the present limitations, it is unlikely that this new procedure will be used in the short term for commercial embryo production. The high costs and the low efficiency would prohibit large-scale application. However, further research will enhance the procedure and overcome these limitations and, at that time, large-scale application will become practicable.

Genetic Manipulation

Genetic manipulation may have one of two objectives: the introduction of an additional gene or a change to an existing gene. Direct injection of DNA into a nucleus offers the only proved method for the introduction of genetic change in sheep and it is limited to the addition of a gene. However, the availability of methods for the production of lambs by nuclear transfer from an established cell line will provide new opportunities for the introduction of changes by site-specific targeting.

Direct injection

Although direct injection of a gene has been used for a variety of applications, it has several limitations, which probably arise from the mechanism of incorporation of the 'transgene'. Usually embryos are recovered at the pronuclear stage of development. While they are secured on a holding pipette, a fine-drawn pipette is introduced into one of the pronuclei to allow the introduction of several hundred copies of the transgene. Those embryos that survive the injection process are transferred into recipient ewes for development to term. Transfer may be a few hours after injection or after IVC to the blastocyst stage of development.

It is believed that the injection of fluid causes breakages in the chromosomes and that inadvertently the mechanisms which act to repair the chromosomes incorporate some of the injected DNA into the chromosomes. Often a number of copies of the gene are integrated in a tandem array. This interpretation would certainly account for many of the observed effects of gene transfer (as reviewed Wilmut and Clark, 1990). First, many of the embryos die and only around 1% of injected embryos survive to become transgenic lambs. Second, the site of integration is apparently random and presumably reflects the position of the original break. As a result, the transgene may be within an existing gene and cause a lethal mutation. Research in mice suggests that such a mutation has occurred in between 5 and 10% of lines. Breeders may wish to produce animals that are homozygous for the change in order to show that a lethal mutation has not been introduced. In view of the risk of unforeseen effects of the transgene, they may also wish to show that animals with the genetic modification offer an advantage over other animals in the population. Similarly, it is believed that unpredictable influences of the DNA neighbouring the site of incorporation account for the great variation in the level of expression of the same transgene in different lines of mice. Whereas the transgene in some sites may be accessible to the transcription machinery, in others, where the chromatin is completely closed, transcription might not occur at all.

Despite these limitations, direct injection has been used to create lines of sheep able to produce a protein in their milk that is needed for the treatment of human disease (Wright *et al.*, 1991). Regulatory elements from sheep milk-protein genes are combined with the coding sequences for the human protein to direct production of the protein to the mammary gland. There are a number of lines of sheep secreting several grams per litre of human protein (Carver *et al.*, 1993). Although the recombinant protein is very similar to authentic human protein, there are some differences in post-translational modification and it is not known if they will modify the response of the patient to the treatment.

Gene targeting

While cells are in culture, it is possible to use well-established genetic techniques to introduce precise genetic changes. This approach depends upon recombination between sequences of DNA that are identical in order to introduce different neighbouring regions to a chromosome. Gene targeting is used in mice to induce specific changes in embryonic stem cells before selected cells are aggregated with another embryo (Hooper, 1992). Sometimes, such cells retain the ability to contribute to all of the tissues of the resulting chimera, including the germ line. In this way the change may be introduced into offspring. By targeting changes into the regulatory or coding sequences, studies have been made of the role of specific gene products and of the regulation of gene expression, respectively. As embryonic stem cells have only been isolated from mouse embryos, this is the only species for which such an approach is possible at present.

The recent development of methods for nuclear transfer from an established cell line in sheep (Campbell *et al.*, 1996a) makes it possible to consider targeting precise genetic change in species other than mice. This would have the advantage of avoiding the chimeric generation. When offspring are derived following the aggregation of cells from two embryos, only one of which has the targeted change, the effect of genetic change cannot be measured until the transgene is transmitted to offspring, which would have the change in every cell. While the delay for such breeding in mice is only a matter of a few weeks, the comparable period in sheep would be several years. In contrast, if a lamb were derived by nuclear transfer from a line with the change, every cell in the lamb would have the desired change, saving considerable time and money. It remains to be confirmed that the present cell line and others from which lambs may be obtained are able to tolerate the procedures for targeting and selection.

Potential applications

Although gene transfer has reached commercial application in producing pharmaceutical proteins, there are no agricultural applications at present. In part, this may reflect the great cost of gene transfer by direct injection and the inadequacies of the procedure. In addition, there is only a limited understanding of the molecular regulation of livestock physiology. It is to be expected that, as more genes are identified by gene-mapping projects and the techniques for gene targeting are established, agricultural applications will be studied in more detail. While it has been natural that many of the first applications to be considered have been concerned with aspects of animal performance that are already the subject of genetic selection, including growth rate, in fact genetic modification may be more likely to contribute in areas that cannot be addressed in any other way, such as wool-fibre type or resistance to pests.

Gene-targeting techniques offer all of the opportunities of direct injection and a number of others. It would be possible to insert coding sequences adjacent to regulatory elements expected to direct tissue- and stage-specific expression as desired for a particular application. In this way, coding sequences for a human protein required for the treatment of disease might be inserted within a milk-protein gene to target expression to the mammary gland and this would be expected to give appropriate expression. It remains to be confirmed that this offers a significant advantage over direct injection.

Modification to existing genes may be considered at several different levels. The most frequent application in mice has been to disrupt gene function in order to clarify the role of the gene product, but it is less likely that this would have many practical applications in agriculture. A gene may act to establish a homeostatic 'feedback loop', and complete disruption of that function would create an imbalance. In this way, the transfer of unregulated genes producing growth hormone led to the birth of piglets with serious abnormalities (Pursel *et al.*, 1989). In contrast, a smaller increase in growth-hormone level, with expression regulated by modified regulatory elements, might provide a useful change in metabolism, with none of the attendant disadvantages seen previously. Similarly, changes to genes concerned with the regulation of reproduction might increase the number of ovulations or change the pattern of seasonal breeding (discussed by Wilmut *et al.*, 1992).

Embryo Storage

Although embryos may be stored for several hours in balanced salt solutions at a temperature between 5°C and 20°C, prolonged storage depends upon freezing and thawing. The response to freezing and thawing depends upon the rate of cooling, the rate of warming and the choice and concentration of cryoprotective agent (Rall, 1992). There are, broadly, three main approaches that provide an acceptable combination of treatments. In the original protocols, embryos are incubated with permeating agents in such a way as to allow them to reach equilibrium. Typically, 10% glycerol is added, at room temperature (c. 20°C), several minutes before cooling. The embryos are then cooled to a temperature 3-4°C below the freezing-point of the medium before ice crystals are induced to form by placing a pair of forceps chilled in liquid nitrogen against the side of the container. Slow cooling, at around 0.3°C min^{-1}, is then continued to around −60°C, before the sample is transferred to liquid nitrogen (−196°C) for long-term storage. The embryos are warmed slowly before the glycerol is removed by transfer of the embryos between media with progressively lower concentrations of glycerol and a final rinse in medium with no glycerol.

While this protocol allows recovery of a large proportion of frozen embryos, it takes several hours and requires expensive equipment and a specialist, with the necessary microscope, to examine the embryos before transfer. Research has continued to increase embryo survival after freezing, shorten the time required for cooling and warming and reduce the dependence upon specialist equipment.

Slow cooling is necessary in order to allow water to leave the cells, and the maximum rate of cooling depends upon the total amount of water in the cell, among other factors. Incorporation in the medium of a molecule such a sucrose, which does not enter the cells, offers several advantages. Firstly, it reduces the time required for cooling, either by increasing the temperature from which the embryo may be transferred into liquid nitrogen or by increasing the maximum rate of cooling which is acceptable. Secondly, it reduces the amount of glycerol entering the cells and so can allow direct transfer of the embryo to a recipient ewe. Direct transfer of embryos frozen and thawed by the original freezing protocol is followed by rapid entry of water in an attempt to equalize the osmotic pressure inside and outside the cells. This rapid surge causes death of the embryos and must be avoided by serial dilution, as described above. Rapid cooling and thawing, using sucrose during dilution, is perhaps the most commonly applied procedure with ruminant embryos at present.

The third protocol involves a radically different approach. If concentrated solutions of cryoprotective agents are cooled very rapidly, at thousands of degrees per minute, they may form glasses, without formation of ice crystals. This is known as vitrification. Success with this approach depends upon accurate observance of the protocol, particularly those aspects designed to allow entry of cryoprotective agent. Typically, cryoprotective agent is added in stages. The first stage, at room temperature (or above), allows initial entry of some agent, before transfer to very high concentrations (up to 50%) at a lower temperature to prevent equilibration before the very rapid cooling stage. The effect of these high concentrations is to cause the withdrawal of water in order to restore osmotic equilibrium, and to create a mixture capable of forming a glass. Samples cooled in this way must be warmed quickly, to prevent formation of ice crystals, and transferred directly before more cryoprotective agent is able to enter the cells. Although vitrification has the appeal of simplicity, it is more difficult to apply and has not yet become the routine procedure of choice.

In using a procedure for embryo storage, it is essential to rigidly follow the established protocol. Under these circumstances, it is quite possible to store embryos for more efficient use of embryo-production technologies and synchronized recipients, for international trade and for research purposes.

Abnormal Fetal Development

Many of the protocols used for culture and manipulation of ruminant embryos have been associated with the birth of unusually large offspring, which also have a high incidence of sudden, unexplained perinatal death (reviewed by Walker *et al.*, 1996). As the scale of the increased birth weight is sometimes sufficient to threaten the life of the young and the mother, it is essential that the factors behind the change are identified and eliminated before there is large-scale application of these new reproduction technologies. While there are reports of large lambs and calves, there is no evidence of this effect in pigs, mice or humans after embryo manipulation, although the basis of this difference between species is not known.

The manipulations that lead to the production of large offspring include nuclear transfer (Wilson *et al.*, 1994), embryo coculture (Farin and Farin, 1995), culture in simple medium with serum added for the 5 days of culture (Thompson *et al.*, 1995; Walker *et al.*, 1996) or exposure of the embryo to a uterus at an inappropriate stage of the oestrous cycle by asynchronous embryo transfer (Wilmut and Sales, 1981) or after modification of the maternal hormonal environment (Kleeman *et al.*, 1994). However, the treatments are not as distinct as might be expected. In one study, when large calves developed from embryos produced by nuclear transfer, these embryos were cultured in a ligated sheep oviduct, with no regard to synchrony of the oestrous cycle (Bondioli, 1993). Thus, it is not possible to know to what extent the perturbation of development occurred as a result of nuclear transfer or culture.

The changes due to exposure to an inappropriate environment may all be induced by biologically active factors, such as growth factors, either present at an inappropriate time or an inappropriate concentration. In this way, temporary asynchronous transfer or elevated maternal progesterone may expose the embryos to unusual uterine secretions. The abnormal effects of culture have been associated with the presence of serum, as lambs born after embryo culture in serum-free medium had a normal birth weight (Thompson *et al.*, 1995). The same factor(s) may be responsible for perturbations caused in culture with serum and by exposure to the reproductive tract at an inappropriate stage of the cycle. It is interesting to note that some of the lambs with normal birth weight resulting from serum-free culture developed from oocytes matured *in vitro*, suggesting that maturation of oocytes according to the protocol used did not perturb development (Thompson *et al.*, 1995). This observation remains to be confirmed and the factor in serum which causes the changes to be defined.

The simplest hypothesis is that all of the factors which perturb development act through the same mechanism in the embryo, but this may not be the case. A disturbance of imprinted genes concerned with the regulation of fetal growth is one candidate mechanism. There is a clear need to understand the mechanism and to define the factors in the reproductive tract that influence embryo development. This will also assist in the optimization of embryo development in culture, while also eliminating the perturbations to development. Unfortunately, this is an intractable problem because of the difficulty of assessing the effect of a factor on embryo development in culture without transfer of each embryo and determination of embryo survival and fetal weight.

Commercial Application

Application of these techniques is likely to evolve over a period of years. The technical complexity of the procedures and the cost will limit their application to specific, specialist purposes in a small number of breeding centres. Embryo recovery and transfer within élite selection flocks have the potential to increase the rate of genetic improvement and are being applied in sheep as in cattle. In comparison with dairy cattle, dissemination of the improvement from beef herds and sheep flocks is handicapped by difficulties in the large-scale application of AI or non-surgical embryo transfer. Widespread sale of rams is likely to provide the most important means of dissemination of genetic gain from selection flocks.

The means of introducing genetic change using nuclear transfer will be used first by biotechnology companies, as it is easier to justify their costs and limitations for such applications. The experience gained in these projects will increase their efficiency and overcome the problems associated with the present protocols and ultimately allow their use for agricultural purposes. Similarly, identification of genes by mapping projects will provide new opportunities. Experience with the introduction of AI and embryo transfer in cattle breeding suggests that it is several years after a procedure becomes available before breeders appreciate how best it should be used. Opportunities for the use of the new reproduction technologies remain to be appreciated.

Acknowledgements

The research which provided the background for this review was supported by the Biotechnology and Biological Sciences Research Council, the Ministry of Agriculture Fisheries and Food, the Scottish Office Agriculture and Fisheries Department and the European Economic Community (EEC).

Note in proof

After submission of this chapter Wilmut *et al.* (1997) published results of their latest experiments. They have shown that nuclear transfer from differential fetal and adult cells provides normal development in sheep. The lamb born after nuclear transfer is the first mammal to develop from a cell derived from adult tissue.

References

Bondioli, K.R. (1993) Nuclear transfer in cattle. *Molecular Reproduction and Development* 36, 274–275.

Campbell, K. and Wilmut, I. (1994) Recent advances on *in vitro* culture and cloning of ungulate embryos. In: Smith, C., Gavora, J.S., Benkel, B., Chesnais, J., Fairfull, W., Gibson, J.P., Kennedy, B. and Burnside, E.B. (eds) *Proceedings of the Fifth World Congress on Genetics as Applied to Livestock*, Vol. 20, International Committee for World Congresses on Genetics Applied to Livestock Production, Guelph, Canada, pp. 180–187.

Campbell, K.H.S., Ritchie, W.A.R. and Wilmut, I. (1993) Nuclear cytoplasmic interactions during the first cell cycle of nuclear transfer reconstructed bovine embryos. implications for DNA replication and development. *Biology of Reproduction* 49, 933–942.

Campbell, K.H.S., Loi, P., Cappai, P. and Wilmut, I. (1994) Improved development of ovine nuclear transfer embryos reconstructed during the presumptive S-phase of enucleated activated oocytes. *Biology of Reproduction* 50, 1385–1393.

Campbell, K.H.S., Loi, P., Otaegui, P.J. and Wilmut, I. (1996a) Cell cycle co-ordination in embryo cloning by nuclear transfer. *Reviews in Reproduction* 1, 40–46.

Campbell, K.H.S., McWhir, J. Ritchie, W.A. and Wilmut, I. (1996b) Sheep cloned by nuclear transfer from a cultured cell line. *Nature* 380, 64–66.

Carver, A.S., Dalrymple, M.A., Wright, G., Cottom, D.S., Reeves, D.B., Gibson, Y.H., Keenan, J.L., Barrass, J.D., Scott, A.R., Colman, A. and Garner, I. (1993) Transgenic livestock as bioreactors: stable expression of human alpha-1-antitrypsin by a flock of sheep. *Biotechnology* 11, 1263–1270.

Farin, P.W. and Farin, C.E. (1995) Transfer of bovine embryos produced *in vivo* or *in vitro*: survival and foetal development. *Biology of Reproduction* 52, 676–682.

Gandolfi, F. and Moor, R.M. (1987) Stimulation of early embryonic development in the sheep by co-culture with oviduct epithelial cells. *Journal of Reproduction and Fertility* 81, 23–28.

Gardner, D.K., Lane, M., Spitzer, A. and Blatt, P. (1994) Enhanced rates of cleavage and development for sheep zygotes cultured to the blastocyst stage *in vitro* in the absence of serum and somatic cells: amino acids, vitamins, and culturing embryos in groups stimulate development. *Biology of Reproduction* 50, 390–400

Gordon, I. (1994) *Laboratory Production of Cattle Embryos.* CAB International, Wallingford, Oxfordshire, UK, 640 pp.

Hooper, M.L. (1992) Embryonal stem cells. In: Evans, H.J. (ed.) *Introducing Planned Changes into the Germline*. Harwood Academic Publishers, Switzerland.

Johnson, M.H. (1981) The molecular and cellular basis of preimplantation development. *Biological Reviews* 56, 463–498.

Kleeman, D.O., Walker, S.K. and Seamark, R.F. (1994) Enhanced fetal growth in sheep administered progesterone during the first three days of pregnancy. *Journal of Reproduction and Fertility* 102, 411–417.

Kruip, T.A.M., Pieterse, M.C., van Beneden, T.A.M., Vos, P.L.A.M., Wurth, Y.A. and Teverne, M.A.M. (1991) A new method for bovine embryo production: a potential alternative to superovulation. *Veterinary Record* 128, 208–210.

McKelvey, W.A.C. and Robinson, J.J. (1984) Normal lambs born following transfer of embryos by laparoscopy. *Veterinary Record* 115, 230.

Moore, N.W. (1982) Egg transfer in sheep and goat. In: Adams, C.E. (ed.) *Mammalian Egg Transfer.* CRC Press, Boca Raton, Florida, pp. 119–133.

Murray, A. and Hunt, T. (1993) *The Cell Cycle: an Introduction*. Oxford University Press, Oxford. 251 pp.

Mylne, M.J.A., McKelvey, W.A.C., Fernie, K. and Matthews, K. (1992) Use of a transcervical technique for embryo recovery in sheep. *Veterinary Record* 130, 450–451.

Pursel, V.G., Pinkert, C.A., Miller, K.F., Bolt, D.J., Campbell, R.G., Palmiter, R.D., Binster, R.L. and Hammer, R.E. (1989) Genetic engineering of livestock. *Science* 244, 1281–1288.

Rall, W.F. (1992) Cryopreservation of oocytes and embryos: methods and applications. *Animal Reproduction Science* 28, 237–246.

Staigmiller, R.B. and Moor, R.M. (1984) Effect of follicle cells on the maturation and developmental competence of ovine oocytes matured outside the follicle. *Gamete Research* 9 (2), 221–229.

Tervit, H.R., Whittingham, G. and Rowson, L.E.A. (1972) Successful culture *in vitro* of sheep and cattle ova. *Journal of Reproduction and Development* 30, 493–497.

Thompson, J.G., Gardner, D.K., Pugh, P.A., McMillan, W.H. and Tervit, H.R. (1995) Lamb birth weight is affected by culture system utilised during *in vitro* pre-elongation development of ovine embryos. *Biology of Reproduction* 53, 1385–1391.

Trounson, A.O., Pushett, D., Maclellan, L.J., Lewis, I. and Gardner, D.K. (1994) Current status of IVM/IVF and embryo culture in humans and farm animals. *Theriogenology* 41, 57–66.

Walker, S.K., Hear, T.M. and Seamark, R.F. (1992) *In vitro* culture of sheep embryos without co-culture: successes and perspectives. *Theriogenology* 37, 111–125.

Walker, S.K., Hill, J.L., Bee, C.A. and Warnes, D.M. (1994) Improving the rate of production of sheep embryos using *in vitro* maturation and fertilisation. *Theriogenology* 41, 330.

Walker, S.K., Hartwich, K.M. and Seamark, R.F. (1996) The production of unusually large offspring following embryo manipulation: concepts and challenges. *Theriogenology* 45, 111–120.

Wallace, J.M. (1992) Articifical insemination: an embryo transfer. In: Speey, A.W. (ed.) *Progress in Sheep and Goat Research*. CAB International, Wallingford, Oxfordshire. UK, pp. 1–24.

Watson, A.J., Watson, P.H., Warnes, D., Walker, S.K., Armstrong, D.T. and Seamark, R.F. (1994) Preimplantation development of *in vitro*-matured and *in vitro*-fertilized ovine zygotes: comparison between coculture on oviduct epithelial cell monolayers and culture under low oxygen atmosphere. *Biology of Reproduction* 50, 715–724.

Wilmut, I. (1982) Applications of egg transfer to animal production, breeding and research. In: Adams, C.E. (ed.) *Mammalian Egg Transfer*. CRC Press, Boca Raton, Florida, pp. 211–230.

Wilmut, I. and Clark, A.J. (1990) Basic techiques for transgenesis. *Journal of Reproduction and Fertility* Suppl. 43, 265–275.

Wilmut, I. and Sales, D.I. (1981) Effect of an asynchronous environment on embryonic development in the sheep. *Journal of Reproduction and Fertility* 61, 179–184.

Wilmut, I., Haley, C.S., Simons, J.P. and Webb, R. (1992) The potential role of molecular genetic manipulation in the improvement of reproductive performance. *Journal of Reproduction and Fertility* Suppl. 45 157–173.

Wilmut, I., Schnieke, A.E., McWhir, J., Kind, A.J. and Campbell, K.H.S. (1997) Viable offspring derived from fetal and adult mammalian cells. *Nature* 385, 810–813.

Wilson, J.M., Williams, J.D., Bondioli, K.R., Looney, C.R., Westhusin, M.E. and McCalla, D.F. (1994) Comparison of birth weight and growth chracteristics of bovine calves produced by nuclear transfer (cloning), embryo transfer and natural mating. *Animal Reproduction Science* 38, 73–83.

Wright, G., Carver, A. Cottom, D., Reeves, D., Scott, A., Simons, P., Wilmut, I., Garner, I. and Colman, A. (1991) High-level expression of active human alpha-1-antitrypsin in the milk of transgenic sheep. *Biotechnology*9, 830–834.

Developmental Genetics 15

N.E. COCKETT
Utah State University, Department of Animal, Dairy and Veterinary Sciences, Logan, UT 84322-4700, USA

Developmental Stages of the Embryo 413
Placental and Maternal Gene Expression Influencing the Embryo 416
  Maternal recognition of pregnancy 416
  Placental lactogens 417
  Peptide growth factors 418
  Pregnancy-associated glycoproteins 420
Gene Expression by the Embryo 421
  Genome activation 421
  Plasminogen 422
  Peptide growth factors 423
  Immune factors 425
  Sex determination 427
  Inactivation of the X chromosome 428
  Imprinted genes 429
Conclusions 431
References 431

Developmental Stages of the Embryo

Fertilization between the ovine sperm and oocyte results in the formation of a large cell with two pronuclei. During the preimplantation stage, this single cell undergoes cell divisions to form a zygote (one to four cells), a morula (24–40 cells) and, eventually, a blastocyst. The ovine preimplantation stage lasts through day 15 postfertilization (Table 15.1). The cells of the zygote are quite large, with a low nuclear-to-cytoplasmic ratio, but each subsequent cleavage step creates daughter cells (blastomeres) with ratios more similar to those in somatic cells. Mitotic division during each cleavage ensures that every blastomere has the proper chromosome number (in sheep $2n = 54$). Because mammalian zygotes contain little yolk, the zygote must rely on maternal secretions from the uterus to obtain nutrients. There is a low but constant metabolic need during early cleavage, but a

Table 15.1. Time of selected events in ovine embryonic and fetal development (adapted from Bazer *et al.*, 1987, and Jainudeen and Hafez, 1987).

Embryonic stage	Days (postfertilization)
Two-cell	1.0
Four-cell	1.3
Eight-cell	1.5
Genome activation*	1.5–2.0
Transport to uterus	2.7–3.0
Blastocyst	6–7
Hatching	7–8
Blastocyst elongation	11–16
Maternal recognition of pregnancy	12
Initial placentation	15
Primitive-streak formation	14
Open neural tube	15–21
Differentiation of the first somite	17
Apparent heart beat	20
Closed neural tube	21–28
Visible forelimb and hindlimb buds	28–35
Differentiation of digits	35–42
Differentiation of nostril and eyes	42–49
Hair follicles appear	42–49
Eyelids close	49–56
Tooth eruption	98–105
Hair covering body	119–126
Birth	145–155

*Occurs during the 8–16-cell stage.

dramatic rise occurs in the later shift from morula to blastocyst.

Compaction of the zygotic blastomeres occurs during the morula stage and results in a solid sphere of cells packed within the zona pellucida; the compaction process involves several gene products, including uvomorulin (Watson *et al.*, 1990), gap-junction proteins (Kidder *et al.*, 1987), tight-junction protein (Fleming and Hay, 1991) and components of the membrane cortical cytoskeleton (Fleming and Johnson, 1988). Cell polarization is also initiated during this stage and appears essential for tissue differentiation, which is initiated prior to the blastocyst stage.

Around day 6, the ovine blastocyst begins to form, with its characteristic fluid-filled blastocoelic cavity. The process of cavitation and its regulation has been recently reviewed by Watson (1992). During this process, the blastomeres continue to differentiate and establish two distinct regions: the inner cell mass (ICM) and the trophectoderm, or extraembryonic tissue. The ICM will later develop into three primary germ layers of the embryo (ectoderm, mesoderm and endoderm) during gastrulation. The trophectoderm, composed of squamous epithelium covered with dense microvilli, will contribute to the placental structures. During cavitation, a transtrophectoderm sodium ion (Na^+) gradient is established to actively transport Na^+ out of the trophectoderm cells into the blastocoelic cavity, which then drives the osmotic accumulation of water across the epithelium into the blastocoelic cavity (Watson, 1992). The transport of Na^+ is regulated by sodium–potassium adenosine triphosphatase (Na^+/K^+-ATPase) located in the basolateral plasma membrane of the trophectoderm. To prevent leakage of blastocoelic fluids, a permeability seal within the tight junctions is established. It is apparent that growth factors regulate cavitation, but the exact process is not yet elucidated. One suggestion is that the binding of the epidermal growth factor receptor (EGFR) by its ligand transforming growth factor alpha (TGF-α) can result in an increase of intracellular cyclic adenosine monophosphate (cAMP) levels, which in turn activates Na^+/K^+-ATPase (Rossier *et al.*, 1987). However, activation of other genes occurs once ligand-receptor binding is established, so many other possible regulation points exist.

Eventually, the blastocyst 'hatches' from the zona pellucida. The stimulus for hatching in sheep is not known, but in cattle it is due to expansion and contractions of the blastocyst, which rupture the zona because of distension (Bazer *et al.*, 1987). The expansion is due to cellular hyperplasia and fluid accumulation in the blastocoele and may be induced by uterine progesterone.

Following hatching, the ovine blastocyst expands to a diameter greater than .1 mm and the trophectoderm differentiates into a bilaminar layer, composed of the outer epithelial trophectoderm and the inner primitive endoderm. The blastocyst then undergoes considerable growth and, beginning on day 11, over the next 4–5 days elongates to a length of several centimetres. Most of this growth is due to hyperplasia of the trophectoderm and endoderm; the ICM is still small in comparison.

Movement of the embryo from the lower part of the ampulla of the oviduct, where fertilization takes place, to its final attachment site in the uterus occurs during the preimplantation stage. If multiple ovulations occur on the same ovary, intrauterine migration of embryos will also occur during elongation (Bazer *et al.*, 1987).

Implantation of the embryo into the uterus is associated with the development of a placenta. On day 15, a temporary connection is formed by extension of papillae from the trophectoderm into the lumen of uterine glands. A more permanent structure – the placentomes – is then formed between days 21 and 28, as fetal cotyledons fuse with maternal caruncles. In sheep, about 90–100 placentomes are evenly distributed between the pregnant and non-pregnant horns. About day 17, fusion of migrating trophoblast binucleated cells and cells of the uterine epithelium occurs, forming a more or less syncytial layer between the chorion and the maternal endometrium. Secretory granules within the resulting trinucleated cells are discharged directly towards the maternal tissue. Trophoblast binucleated cells involved in this fusion have probably arisen from mononucleated cells that have undergone mitotic division without cytokinesis. Thus, the ovine placenta is classified as synepitheliochorial (Wooding, 1992) and the tissues forming the placental barrier in sheep are maternal endothelium, connective tissue, syncytial layer (maternal epithelium and trophectoderm), fetal connective tissue and fetal epithelium.

During the embryonic stage (days 1–34 in sheep), the embryo undergoes rapid growth and development. The major tissues, organs and systems are established and prominent features of the external body are recognizable. The fetal period in sheep extends from day 35 until birth, which occurs about day 145–155. Timing of several important development steps in the ovine embryo and foetus are given in Table 15.1.

Placental and Maternal Gene Expression Influencing the Embryo

The placenta serves as an active endocrine organ, capable of synthesizing and secreting a broad range of factors, including hormones and growth factors. These factors allow the placenta to mediate and modulate the maternal environment, thereby allowing normal fetal development. Some of these hormones are exclusive to the placenta, including conceptus interferons (IFNs) and placental lactogens (PLs).

Maternal recognition of pregnancy

Maternal recognition of pregnancy occurs at day 12 in sheep, preceding attachment of the embryo to the uterine wall. At this time, the conceptus releases an antiluteolytic substance, which prolongs the life of the corpus luteum (CL) and ensures the release of progesterone, necessary for development of the conceptus and placental attachment (Bazer *et al.*, 1991). If this signal is not received during day 13 or 14, the CL regresses, due to the luteolytic action of prostaglandin $F_{2\alpha}$ ($PGF_{2\alpha}$), which is released in pulses from the uterine endometrium late in the oestrous cycle. The presence of an antiluteolytic factor was first demonstrated in sheep by Rowson and Moor (1967); they were able to extend the life of the CL of cyclic ewes for more than 1 month by injecting day 14–15 conceptus homogenates into the uterus. However, day 25 conceptus homogenates did not extend the life of the CL. This antiluteolytic substance was later purified by Godkin *et al.* (1982); in their study, an acidic protein of 21,000 daltons (Da) was secreted in large amounts by day 13–21 blastocysts into the culture medium. Researchers have referred to this antiluteolytic substance as trophoblastin (Martal *et al.*, 1979), Protein X (Godkin *et al.*, 1982), ovine trophoblast protein 1 (oTP-1) (Imakawa *et al.*, 1987; Stewart *et al.*, 1987) or IFN-τ (Roberts *et al.*, 1996). Low levels of the antiluteolytic factor are detectable at time of hatching, with massive quantities secreted by the mononucleated cells of the trophectoderm during elongation, up to 100 mg day^{-1} during peak production on days 13–14. Once synthesized, the product is released into the uterine lumen (reviewed by Bazer *et al.*, 1994; Roberts and Anthony, 1994). Its presence clearly suppresses the pulsating release of $PGF_{2\alpha}$, possibly by downregulating oestrogen receptors within the uterine endometrium, which in turn prevents increases in oxytocin receptors. Based on studies in which oTP-1 alone did not support a pseudopregnancy, it is apparent that oTP-1 does not act alone.

The sequence of nine oTP-1 complementary deoxyribonucleic acids (cDNAs) has been determined (Imakawa *et al.*, 1987; Stewart *et al.*, 1987; Charlier *et al.*, 1989; Nephew *et al.*, 1993); comparisons of these sequences indicate that several distinct and yet functional genes exist. The consensus sequence contains an open reading frame (ORF) of 595 bp, coding for a 23 amino acid signal peptide, followed by a 172 amino acid mature protein. These genes are differentially transcribed during early pregnancy (Nephew *et al.*, 1993) and may respond to different uterine stimulatory factors. Although significant expression is finished by day 21, one transcript was still detectable after day 45 (Nephew *et al.*, 1993).

Ovine trophoblast protein seems to react with the same type I receptor as other IFNs (Roberts *et al.*, 1992). It has been previously demonstrated that the binding of IFN with its receptor causes a cytoplasmic activation of a tyrosine kinase, *tyk-2*, which phosphorylates a cytoplasmic transcription factor, ISGF-2. This factor then migrates into the nucleus and activates the appropriate IFN-responsive genes (Schindler *et al.*, 1992). At this point, it is not known which genes oTP-1 ultimately activates upon binding with its receptor.

Placental lactogens

The first description of PLs in sheep was reported by Martal and Djiane (1975). Over the next decade, a number of studies characterized this placental hormone and its mechanism of action (reviewed by Anthony *et al.*, 1995). Placental lactogen is a non-glycosylated polypeptide with an apparent molecular weight of 22,000 Da (Warren *et al.*, 1990). The mature protein is 198 amino acids, generated from a 236 amino acid precursor, with 67%, 48% and 27% overall homology with bovine PL, ovine prolactin (PRL) and ovine growth hormone (GH), respectively. The protein is encoded by a single-copy gene, which contains consensus regulatory sequences for two activator proteins, AP-1 and AP-2, and GHF-1/Pit-1, as well as the thyroid hormone response element (Liang and Anthony, 1993); however, investigation of the regulation of expression has not yet been reported.

Kappes *et al.* (1992) detected ovine PL messenger ribonucleic acid (mRNA) in preparations of chorionic binucleated cells, beginning on day 60. Transcript levels peaked on day 120 and then plateaued until parturition. The increased expression is probably due to an increase in PL-secreting binucleated cells, rather than an increase in production from individual cells (Kappes *et al.*, 1992). Similar patterns of PL concentrations in maternal and fetal sera were found; however, peaks occurred on day 135 and day 90, respectively, suggesting separate regulation and methods of secretion for the two sources. Cotyledon mRNA levels and maternal serum levels of PL were positively correlated with fetal weight, suggesting an involvement of PL in regulating fetal growth and metabolism, possibly by influencing production of insulin-like growth factors (IGFs) by the fetus (Anthony *et al.*, 1995).

The identity of a specific receptor for PL in maternal and fetal sheep liver is still not resolved (reviewed by Anthony *et al.*, 1995); certain properties suggest that the PL receptor is different from the GH and PRL receptors, while other evidence suggests that the fetal PL receptor is a structural variant of the GH receptor. Regardless, several characteristics of the receptor are known. It has a high-affinity binding site for PL (K_d (dissociation constant) approximately 0.27 nM) with an apparent molecular weight of 38,000–44,000 Da (Freemark *et al.*, 1987). Recent evidence strongly suggests that the ovine PL receptor is a variant of the GH receptor (Freemark, 1991; Breier *et al.*, 1994; Pratt and Anthony, 1995). Although this variant GH receptor is present in the fetus, there is no specific GH binding in the fetal liver until term (Gluckman *et al.*, 1983).

Peptide growth factors

It has been demonstrated that a number of peptide growth factors are critical to the initiation, establishment and maintenance of pregnancy, as well as the coordination of the uterine and conceptus development during early pregnancy (Table 15.2). The key components of these pathways are IGF-I and IGF-II. These factors are regulated by stage of gestation and, for IGF-I, by oestrogen and progesterone (Paria and Dey, 1990). Insulin-like growth factor I and IGF-II are expressed by the uterine endometrium and secreted into the uterine lumen. In a study by Ko *et al.* (1991), the levels of IGF-I within uterine luminal fluids did not differ in pregnant and cyclic ewes from days 10, 12 and 14 but there was a decrease in IGF-I levels on day 16 in the pregnant ewes, unlike the sharp increase in cyclic ewes at day 16. In contrast, IGF-II levels within the uterine luminal fluids were very similar in pregnant and cyclic ewes. In both groups of females, IGF-II peaked on day 14, with a ten-fold greater concentration than IGF-I on day 14, and then decreased on day 16. The timing of peak IGF-II concentration on day 14 corresponds to the production of oTP-1 by the ovine conceptus (Farin *et al.*, 1989). Because of the overlap in timing of these two factors, Simmen *et al.* (1993) suggest that IGF-II may mediate oTP-1 expression. In support of this, Ko *et al.* (1991) demonstrated that simultaneous addition of IGF-I and IGF-II to cultured day 13 conceptuses increased oTP-1 secretion, while neither IGF-I nor IGF-II alone had any effect. It is apparent that these growth factors mediate the uterine response to the oTP-1 signal and may also maintain oTP-1 secretion by the conceptus, thus fulfilling both autocrine and paracrine functions. It is also possible that these factors influence the developmental rate of the embryo. Using a uterine epithelial cell line, HRE-H9, Simmen *et al.* (1993) showed that the addition of IGF-I and EGF enhanced protein synthesis but reduced protein secretion. In contrast, TGF-β_1 had no effect on protein synthesis but increased protein secretion. It should be noted that ovine primary oviduct cell cultures did not produce EGF, insulin or nerve growth factor (NGF) but did produce IGF-I and TGF-β_1, as well as IGF-II,

Table 15.2. Molecular aspects of implantation and potential involvement of peptide growth factors*. Reproduced with permission from Simmen and Simmen (1991).

	HBGFs	TGF-βs	TGF-α	IGF-I	IGF-II	IGFBPs	PDGF	CSF-I	EGF
Early embryogenesis	X†	X	X	X	X	X	X		
Trophoblast growth	X	X	X	X	X	X	X	X	X
Endometrial-cell proliferation, differentiation	X	X	X	X	X	X			X
Modulation of extracellular matrix	X	X							
Conceptus invasiveness					X	X	X		
Blood island formation	X								
Angiogenesis	X		X						X
Cell migration	X	X					X		
Cell adhesion			X					X	X
Regulation of steroidogenesis	X			X		X			
Protection of fetal allograft		X						X	
Stromal–epithelial		X		X					X

*References for each factor can be found within Simmen and Simmen (1991).

†X represents postulated functions for the indicated growth factors.

HBGF, heparin-binding growth factor; TGF, transforming growth factor; IGF, insulin-like growth factor; IGFBP, IGF binding protein; PDGF, platelet-derived growth factor; CSF, colony stimulating factor; EGF, epidermal growth factor.

TGF-α, IGF-I receptor, insulin receptor and basic fibroblast growth factor (bFGF) (Watson *et al.*, 1994). Thus, it is likely that IGF-I and TGF-β_1 play a role in this modulation, while EGF does not.

Receptors can regulate the mode of action of peptide growth factors. Insulin-like growth factors bind to at least two types of IGF receptors, as well as the insulin receptor, with different affinities (reviewed by Heyner *et al.*, 1993). The insulin and IGF-I receptors are structurally very similar; both possess a tyrosine kinase domain, which is activated following ligand binding. The IGF-II receptor does not have a tyrosine kinase domain; rather, it is identical to the mannose-6-phosphate (Man-6-P) receptor, although IGF-II and Man-6-P bind to different receptor domains. Insulin binds with high affinity to the insulin receptor and with lower affinity to the IGF-I receptor, while IGF-I binds with high affinity to the IGF-I receptor and with lower affinity to the insulin receptor. The IGF-II receptor binds IGF-II with high affinity and IGF-I with low affinity. As mentioned, all three of these receptors are produced by ovine primary oviductal cell cultures (Watson *et al.*, 1994).

Insulin-like growth factor-binding proteins (IGFBPs) play a major role in delivery of IGFs to target tissues and also modulate biological action of the IGFs by their capacity to bind the ligands with equal affinity to the receptors. There have been six IGFBPs identified so far, designated IGFBP-1 to IGFBP-6 (Shimasaki and Ling, 1991). The most abundant IGFBP in fetal and postnatal lambs is IGFBP-3 (Lord *et al.*, 1991), but IGFBP-2 binds with a higher affinity to IGF-II. The mode of action appears to be different for the two binding proteins: IGFBP-2 can pass freely between the blood and tissue fluid compartments, whereas IGFBP-3 must be cleaved into smaller subunits to enter extracellular fluids. Therefore, IGFBP-3 is important as an IGF carrier in the circulation, while IGFBP-2 may be important as a transporter between blood and tissue. In addition, IGFBP-1 concentration parallels IGF-II concentrations in the ovine uterine endometrium (Waites *et al.*, 1990).

Pregnancy-associated glycoproteins

During the last 15 years, a number of pregnancy-associated glycoproteins (PAGs) have been isolated from bovine and ovine placental tissue. These glycoproteins are secreted by binucleated cells of the trophectoderm. Expression occurs in a relatively constant pattern, beginning at time of implantation (day 15) and continuing throughout gestation, and can be detected in maternal serum by day 20 (Roberts *et al.*, 1996). The first sequence of ovine PAG was obtained by Xie *et al.* (1991), using day 15 conceptus and day 100 cotyledon cDNA libraries. A nucleotide similarity of about 50% was determined between ovine PAG and the aspartic proteinase family, which includes pepsin. However, the ovine PAG product is apparently non-reactive, as it is lacking one of two active aspartic acids (at position 215) that are essential for catalysis. There is a suggestion that bovine PAG is also non-reactive, indicated by the presence of a relatively bulky methyl side-chain on the alanine at position 34, which would also cause loss of enzymatic activity. However, both proteins have a deep substrate-binding cleft, which is apparently functional. The authors speculate that these proteins are capable of binding a peptide chain and may, in this way, compete with the major histocompatibility complex (MHC) for processed antigen being presented for immune recognition by T cells. Thus, these proteins could prevent maternal recognition of foreign epitopes on the fetus (Roberts *et al.*, 1996).

Another set of at least ten PAG proteins were identified using a monoclonal antibody, SBU-3, raised against trophoblast microvilli isolated from midgestation ovine cotyledons (Atkinson *et al.*, 1993). The SBU-3 antibody recognizes an *N*-glycolylneuraminic acid-containing carbohydrate epitope. Sequence identities of three of these PAGs that bind phytohaemagglutinin were about 78% with the previously described ovine PAG (Xie *et al.*, 1991), and the proteins from this study did not exhibit proteolytic activity on haemoglobin. It has not been determined whether these SBU-3 proteins are differentially expressed during pregnancy.

Gene Expression by the Embryo

Genome activation

> The making of a blastocyst is a complex and delicate procedure based upon the unfolding of a genetic program in a suitable environment. The genetic program basically consists of the ordinate activation of a hierarchy of regulatory genes which, in their turn, activate sets of subordinate genes which will determine cell fate and differentation.
>
> (Kidder, 1993)

Once fertilization occurs, the embryo has its complete genetic composition, which it will carry for the rest of its life; however, the ovine embryo's genome is transcriptionally inactive until the 8–16-cell stage (Crosby *et al.*, 1988). Therefore, the embryo is dependent on maternally inherited components stored within the oocyte for coordination of the first cleavage divisions. In the Crosby *et al.* (1988) study investigating the transition from maternal to embryonic transcriptional control, an RNA polymerase II and III inhibitor (α-amanitin) was added to cultures of various-aged sheep embryos. There was no effect on cleavage or protein synthesis during the first three cell cycles but the inhibitor blocked cleavage and protein synthesis thereafter. Regardless of species, several key steps must occur during the transition time from maternal to embryonic control (Telford *et al.*, 1990). First, there must be decay of maternally derived transcripts within the embryonic cells. In mice, there is a major turnover in maternal mRNA at the same time that embryonic mRNA is made. However, there is a more gradual decay of ribosomal RNA, ribosomes and small nuclear RNA, with up to 75% of the total remaining during this interval (Piko and Clegg, 1982). Thus, transcripts produced by the embryo soon after the switch are processed using ribosomal products of maternal origin. Second, synthesis of new mRNA from the embryonic genome occurs, with concurrent appearance of embryonic transcription products. Third, there are significant changes in specific protein synthesis. Crosby *et al.* (1988) showed that total protein synthesis was consistently high during the first two ovine cleavage divisions, dropped by 95% in the third division, remained low in the fourth and increased again in the fifth division. Interestingly, the timing of embryonic genome activation corresponds to the 'block' stage, at which time cultured embryos are very sensitive to culture conditions. However, cultured embryos can be rescued by coculture with oviduct cells (Gandolfi and Moor, 1987), suggesting that oviductal secretions modify the initiation of the transcriptional switch.

Plasminogen

Work by Menino *et al.* (1989) has demonstrated that plasminogen activator (PA), a serine protease that converts uterine plasminogen to the active enzyme plasmin, is produced by the embryo during early gestation. The resulting plasmin has trypsin-like specificity and can degrade fibrin, as well as a variety of other proteins, including components of basement membrane and extracellular matrix (Werb *et al.*, 1980). In sheep embryos, levels of PA are low until the morula stage, increase during morula–blastocyst transition and remain elevated through blastocoelic expansion and hatching (Menino *et al.*, 1989). The authors suggested that the resulting plasmin may cause a change in zona pellucida solubility and therefore aid in the hatching process. However, in this study, solubility of the zona pellucida was altered in the absence of plasminogen, and they concluded that other factors, e.g. physical distension of the zona pellucida by the expanding blastocoele, also contribute to hatching. In a later study (Menino and Williams, 1994), the authors used ouabain, an inhibitor of Na^+/K^+-ATPase and cavitation, to study associations between PA production and cavitation by the blastocyst. Interestingly, ouabain effectively suppressed not only blastocoelic expansion but also PA production. Thus, it is possible that cavitation is related to PA release, based on the timing of PA production, but the ouabain may have compromised the embryo's viability and therefore PA production as well. This group further characterized the type of PA produced by preimplantation sheep embryos and demonstrated that it is a urokinase-type PA, similar to the enzyme produced by embryos having invasive implantation, such as the mouse (Bartlett and Menino, 1993). However, a specific PA inhibitor (PAI-2) was also detected in preimplantation sheep embryos; this may function to regulate the proteolysis sustained by PA production. In a subsequent study, Bartlett and Menino (1995) evaluated PA production by migrating endodermal cells derived from sheep ICMs cultured on matrices of collagen IV, laminin or fibronectin. Outgrowth of endodermal cells was greatest on fibronectin and also fibronectin stimulated PA production. Addition of PAI-2 to the cultures eliminated PA activity and altered the pattern of endodermal outgrowth but did not inhibit cell migration. The authors concluded that PA may participate in cell movements in the early embryo, but that cell migration was not dependent on PA.

In an extensive study of growth-factor gene expression in the ovine fetus, Watson *et al.* (1994) used reverse transcription polymerase chain reaction (RT-PCR) on RNA preparations derived from several stages of cultured embryos, including one-cell zygotes, two- to eight-cell embryos, morulae and blastocyts, as well as primary oviduct epithelial-cell cultures. Transcripts for IGF-I, IGF-II, IGF-I receptor, TGF-α, TGF-β_1 and the insulin receptor were detected from the one-cell to the blastocyst stage. Because embryonic genome activation occurs at the 8–16-cell stage (Crosby *et al.*, 1988), it is likely that these transcripts are products of both maternal and embryonic origin. The transcript for bFGF decreased in relative abundance over the studied embryonic ages, suggesting that it may be of maternal origin only. As expected, transcript for oTP-1 was detected only in blastocyst samples. No transcript for IGF-II receptor, EGF, insulin or NGF genes was detectable in any of the tested embryos. However, the authors caution that the IGF-II receptor primers were based on heterologous sequences and may not have been able to amplify the appropriate transcript. They suggest further clarification of IGF-II receptor expression should be done upon publication of ovine sequences for this gene.

Epidermal growth-factor receptors were identified on the ovine trophoblast as early as day 15 (Gharib-Hamrouche *et al.*, 1993). Both EGF and TGF-α bound these receptors with the same affinity. However, it should be noted that TGF-α transcript is produced by the embryo from one-cell stage to day 30, while EGF is not produced by the pre- to peri-implantation embryo, the trophoblast or the uterus (Watson, 1992; Gharib-Hamrouche *et al.*, 1995). Appearance of EGFR might mirror TGF-α expression; however, day 15 embryos were the earliest examined in this study.

Expression of growth factors has also been examined in older ovine fetuses and neonates. In an extensive study on expression of the IGF-II and IGFBP-2 genes, Delhanty and Han (1993) examined transcripts in several tissues of the day 50 ovine fetus up to term, postnatal lambs at 2 days, 4 weeks and 9 weeks, and adult sheep. Relative mRNA levels of IGF-II and IGFBP-2 in liver, kidney, adrenal, skeletal muscle and choroid plexus for these ages are shown in Fig. 15.1. In general, IGF-II mRNA levels were high in fetal tissues from early gestational ages and decreased with maturation. In some tissues (kidney, adrenal and choroid plexus), relative abundance and patterns for IGFBP-2 mRNA were similar to those for IGF-II. In other tissues, such as skeletal muscle, IGFBP-2 mRNA levels fell sharply to barely detectable levels after day 99, while IGF-II mRNA levels decreased gradually over development. The pattern of both transcripts was quite different in liver; IGF-II levels peaked on day 100–119 and then gradually decreased with maturation, while IGFBP-2 increased over gestation and was highest in postnatal and adult animals. Plasma IGF-II concentrations were similar to mRNA levels within most tissues, while plasma IGFBP-2 concentrations were biphasic – low at day 50, peaking at late gestation and then declining gradually after birth until undetectable in adults. In contrast, IGFBP-3 was the most abundant binding protein in fetal plasma throughout

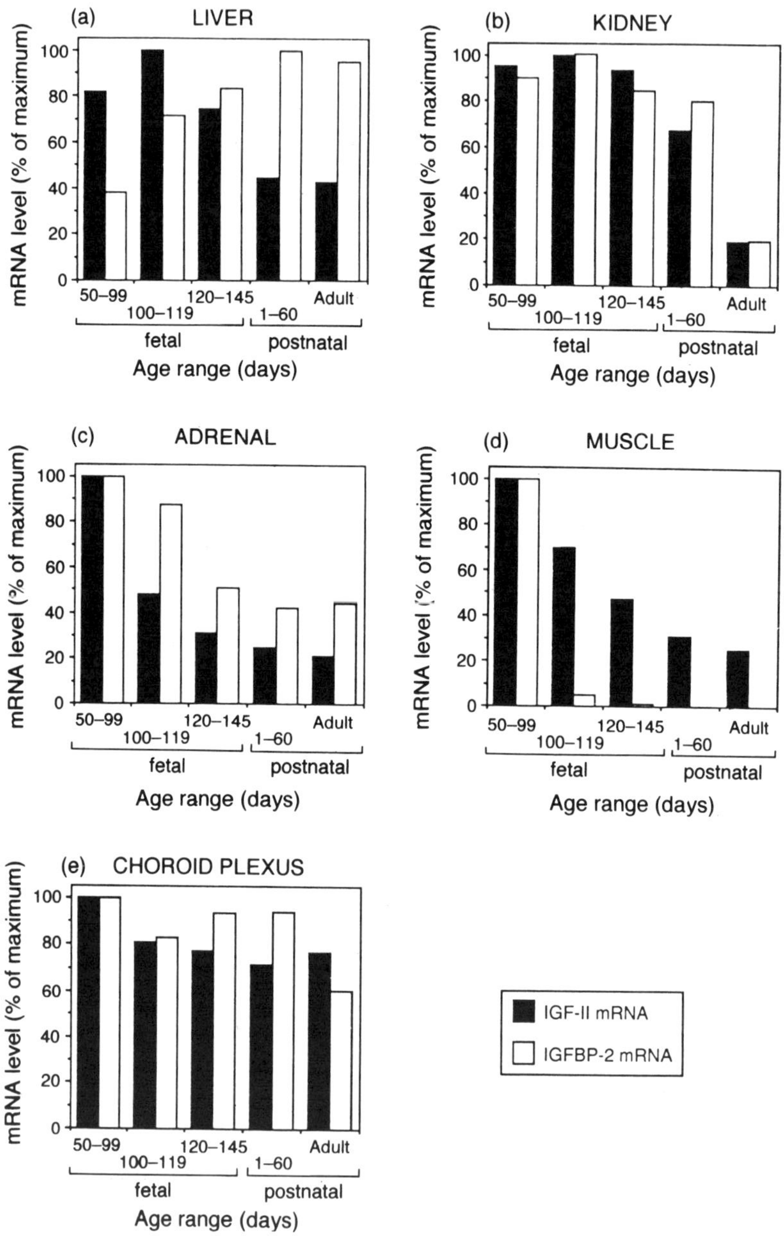

Fig. 15.1. The relationship between IGF-II and IGFBP-2 mRNA expression during development of the (a) liver, (b) kidney, (c) adrenal, (d) skeletal muscle and (e) choroid plexus (reproduced with permission from Delhanty and Han, 1993).

gestation, as well as during postnatal and adult periods.

There has also been emphasis on expression of the IGFs after placental restriction. McLellan *et al.* (1992) induced hypoxaemia in ovine day 115–120 fetuses by restricting uterine blood flow for 24 h and then measured IGF-I, IGF-II, IGFBP-1 and IGFBP-2 levels in the fetuses. Plasma concentrations and liver mRNA levels of IGF-I and IGF-II were lower in hypoxic fetuses than in controls, but only IGF-I mRNA levels in the liver were significantly different. In the hypoxic lambs, plasma concentrations of IGFBP-1 were 80% higher and mRNA levels in the liver were tenfold higher, whereas plasma concentrations of IGFBP-2 were 50% lower and mRNA levels in the liver and kidney were 60% lower. The levels of IGFBP-3 and IGFBP-4 did not differ between the two groups. The authors suggest that the greater levels of IGFBP-1 in hypoxic fetuses may be a mechanism for reducing growth under unfavourable conditions. Kind *et al.* (1995) continued these studies by examining IGF-I and IGF-II production in the day 121 fetus after chronic placental restriction, caused by removal of caruncles from the uterus. There was a reduction in IGF-I mRNA in fetal muscle, lung and kidney, while there was no change in IGF-I mRNA levels in placenta, heart or liver. The amount of mRNA in liver, muscle and kidney correlated with blood levels, suggesting that serum IGF-I originates from various tissue sites, and serum IGF-I levels correlated positively with fetal size. In agreement with the McLellan *et al.* (1992) study, restriction of placental growth did not alter IGF-II blood concentration or mRNA production in the examined tissues.

Immune factors

In ruminant animals, there is no transfer of maternal immune factors across the placenta to the developing fetus. In addition, the fetus develops only a rudimentary immune system during gestation. Thus, the immature immune system of the neonate does not adequately prevent disease in the first few weeks after birth. However, because the intestinal epithelia remain permeable to large protein molecules until approximately 24 h post-partum, the neonate can absorb immunoglobulins that are present in colostrum (Brandon, 1976). Therefore, the subsequent health and survival of young lambs is dependent on their success in obtaining these maternal immunoglobulins (Campbell, 1974; Halliday, 1974).

Development of the rudimentary immune response in the ovine fetus occurs in a predictable, sequential manner. Osburn (1988) reported the first appearance of lymphocytes in blood around day 32, in plasma on day 39, in the thymus on day 41, in lymph nodes on day 44–45 and in the spleen on day 56, and the first appearance of serum immunoglobulins on day 56. Maddox *et al.* (1987a, b) used a panel of monoclonal antibodies to investigate the appearance of lymphocyte-associated cell antigens in the thymus, in the spleen and throughout the body (Table 15.3). All subsets of lymphocytes except those with surface immunoglobulins (sIg^+) appeared in the thymus before the spleen.

Table 15.3. First appearance of cellular antigens within the ovine embryo.

Antigen	Gestational age (days)		
	Thymus†	Spleen‡	Body†
LCA	35	42–44	35
CD5$^+$	35	43–44	33
CD4$^+$	35–38	50–55	35
CD8$^+$	35	43–44	35
T19$^+$	50–58	57	50–58
MHC I$^+$	33	45–47§	19§
MHC II$^+$	35	45–47	25
sIg$^+$	50	45–50	39

*Reactivity: LCA = all leucocytes; CD5 = all T lymphocytes; CD4 = helper T cells; CD8 = cytotoxic or suppressor T cells; and T19 = T lymphocytes expressing the δ and γ chains of the T-cell receptor but CD4$^-$ and CD8$^-$ (Mackay *et al.*, 1987; Kimpton *et al.*, 1989).
†Adapted from Maddox *et al.* (1987a).
‡Adapted from Maddox *et al.* (1987b).
§Earliest age that was examined.

Recirculation of lymphocytes between blood and lymph, through the spleen and lymph nodes, starts at least by day 75 in fetal sheep (Kimpton *et al.*, 1989). The number of lymphocytes in the recirculating pool increases throughout gestation, as new cells are added to the blood, lymph and lymphoid tissues. The pool usually contains less than 5% B cells and appears to be thymus-dependent. The lymph node of a fetal sheep has a much higher lymphocyte output g^{-1} h^{-1} than the adult, with output of T-cell subsets being higher in the fetus but outputs being lower for B cells and MHC class II$^+$ cells. By cannulating the prescapular lymph node efferent ducts and collecting blood from the jugular vein of day 120–145 fetal lambs, Kimpton *et al.* (1989) were able to estimate the percentage of lymphocyte subsets in lymph and blood, respectively. They found that the percentage of lymphocytes that were CD4$^+$ was higher in lymph than in blood (54 and 29%, respectively), whereas the percentage of T19$^+$ cells was higher in blood than in lymph (24 and 9%, respectively). The percentages of CD8$^+$, MHC class II$^+$ and sIg$^+$ cells were the same in lymph and blood. Interestingly, the percentages of sIg$^+$ and MHC II$^+$ cells were much less than adult levels, with the percentage of B cells being five times higher in the lymph of adult animals compared with the fetal lambs. The authors suggest that a postnatal expansion of B cells is due in part to the rapid development of ileal Peyer's patches after birth.

Complement components are present in fetal lamb serum by day 123 (Rice and Silverstein, 1964). In calves, complement activity increases over gestation but reaches only 50% of adult activity at time of birth (Culbertson, 1976). In contrast, proliferative responses of blood mononuclear cells to mitogens and alloantigens occurred in day 120–150 fetal sheep, with responses comparable to those in the adult (Raghunathan *et al.*, 1984). Also, fetal sheep at days 80 and 117 were able to reject skin allografts (Osburn *et al.*, 1982), indicating a competent cell-mediated immunity.

In utero, the fetus is usually protected from infection by the placental barrier. However, using experimental challenges in the ovine fetus, it has been demonstrated that an antibody response was raised against the ϕX176 bacteriophage by day 41, ferritin at day 56 and flagellin and chicken red blood cells at day 58 (Silverstein *et al.*, 1963). Thus, a gradual progression of immune response occurs in the fetal sheep, probably in coordination with maturation of appropriate lymphocytic cells.

Sex determination

Primary sex differentiation is the developmental process in which the bipotential embryonic gonad differentiates into a testis or an ovary (Müller, 1992). In eutherian mammals, testis development is the key event that determines the sexual fate of the embryo. This development is triggered by a gene on the Y chromosome coding for a testis determining factor; in the absence of this factor, the gonad develops into an ovary. The testis determining factor probably induces the differentiation of support-cell lineages into Sertoli cells. The Sertoli cells then secrete anti-Müllerian hormone, which is responsible for regression of the female müllerian ducts, and also seem to trigger a cascade of events that cause male differentiation of other cell lineages, such as the Leydig cells. The Leydig cells secrete testosterone, which stimulates development of the male reproductive tract. In combination, these factors alter development from feminine to masculine characteristics. The secondary sexual characteristics (i.e. phenotypic sex) appear to be regulated not by direct gene action, but by sex hormones produced by the gonads.

Strong evidence implicates the sex-determining region Y gene (*SRY*) as being responsible for the testis-determining factor (recently reviewed by Koopman, 1995; Lovell-Badge and Hacker, 1995). The actual mechanism by which the SRY protein regulates Sertoli-cell differentiation is unknown; however, analysis of the *SRY* sequence indicates that it encodes a 79 amino acid HMG box that confers DNA-binding ability, suggesting that *SRY* is a transcription factor (Sinclair *et al.*, 1990). Sequence identities of the HMG box of the sheep *SRY* with the corresponding regions of human and mouse were 83.4 and 72.8%, respectively (Pomp *et al.*, 1995). An *SRY* transcript is found in the mouse fetus only between days 10.5 and 12.5, exclusively in the genital ridge (Koopman *et al.*, 1990). This corresponds to the presence of testis cords within the ridge, due to the differentiation and alignment of Sertoli cells. The *SRY* is also expressed in the adult testis of human, mouse and marsupial (Koopman *et al.*, 1990; Sinclair *et al.*, 1990; Foster *et al.*, 1992).

Previously, a male-specific histocompatibility (H-Y) antigen, detected by rejection of male skin grafts in female recipients, was suggested to be the testis-determining factor (Wachtel *et al.*, 1975). Subsequently, it was demonstrated that the H-Y antigen could not be the determining factor, because the expression of the H-Y antigen is not required for testis development and the presence of the H-Y antigen does not ensure that an XX gonad will undergo testicular development (McLaren *et al.*, 1984). Another protein, the serologically detectable male antigen (SDM), was identified, using antibodies produced in female rodents after inoculation with male cells (Goldberg *et al.*, 1971). Müller and Wachtel (1991) later showed that SDM is secreted by Sertoli cells and is identical to anti-Müllerian hormone.

The gene encoding a zinc finger protein (ZFY) was also briefly considered to be a candidate for the testis determining factor (Page *et al.*, 1987). This gene has a homologue on the X chromosome called ZFX, and both genes may code for some type of transcription activator; however, the actual function of this gene is not yet known. It was eliminated as the testis-determining factor, because neither ZFY nor ZFX is found on the sex chromosomes of marsupials (Sinclair *et al.*, 1988), expression of ZFY in mice occurs in the testis germ cells, which are not required for testis development (Koopman *et al.*, 1989), and the ZFY locus in humans does not map to the same position as the testis determining region (Müller, 1992).

Inactivation of the X chromosome

Male mammals have one X chromosome and one Y chromosome, while female mammals have two X chromosomes. To equalize dosage of X chromosome genes between the two sexes, only one of the female's X chromosomes is transcriptionally active (Lyon, 1961). The other inactivated X chromosome assumes a characteristic condensed structure, asynchronously replicates its DNA and, generally, is not transcribed into mRNA, although exceptions have been noted by Brown *et al.* (1992). Inactivation of the maternal (X^M) or paternal (X^P) X chromosome is usually random in female embryonic ectoderm, so that the embryo and resulting adult are mosaic for expression. This random inactivation is complete in most cell lineages by day 6.5 in the mouse (Grant and Chapman, 1988), but the corresponding time in sheep is not yet known. Using gene targeting in mouse embryo stem cells and *in vivo* chimeras, it has been clearly demonstrated that the X-active specific transcript (*Xist*) locus is responsible for regulation of inactivation (Penny *et al.*, 1996). In mouse and human, the gene is expressed exclusively by the inactive X chromosome and produces a large RNA transcript of 15–17 kb, which remains within the nucleus, associated with the inactive X chromosome, but because of no apparent ORF and no association with polysomes, the gene appears to lack protein coding potential (Brockdorff *et al.*, 1992; Brown *et al.*, 1992). The nucleotide sequences in mouse (Brockdorff *et al.*, 1992) and human (Brown *et al.*, 1992) share a 70% nucleotide similarity; the corresponding gene in sheep has not yet been isolated.

Interestingly, while random inactivation occurs within embryonic cell lineages, only the X^M is active within the trophectoderm and yolk-sac endoderm (reviewed by Chapman, 1986). This suggests that distinctive mechanisms of inactivation occur within these different cell lineages; however, methylation of CpG-rich islands is associated with the inactivated X chromosome, regardless of tissue (Norris *et al.*, 1991).

Studies examining the inactivation of the X chromosome have used cytogenetic, biochemical and genetic approaches. Cytogenetic approaches for detecting the inactive or allocyclic chromosome include differential staining of condensed heterochromatin in interphase cells, differential staining after hypotonic potassium chloride treatment of metaphase chromosomes or asynchronous replication of the allocyclic chromosome. Timing of inactivation can also be detected using biochemical techniques; with this approach, dosage differences between single male and female embryos are examined for X chromosome gene products, such as α-galactosidase or hypoxanthine phosphoribosyl transferase (HPRT). These studies require microquantitative assays, because of the limited tissue source, and are limited by the sensitivity of the assay in detecting the twofold dosage in females with two active X chromosomes versus the onefold dosage in males or in females with an inactive X chromosome. Genetic methods have utilized allelic variants in the X^M and X^P chromosomes; heterozygous patterns of mRNA transcripts are then examined within a single blastomere. However, the presence of two transcripts does not necessarily indicate that both X chromosomes are active; a heterozygous pattern of transcripts might also indicate a single embryonic copy and a maternal copy.

At some point during oogenesis in the female, the inactivated chromosome must be reactivated. This reactivation appears to occur during meiotic prophase. Both chromosomes remain active through ovulation and fertilization and at least to the 32-cell stage. Thus, while premeiotic germ cells have a single active X chromosome, the early cleavage-stage blastomeres have two active X chromosomes, until inactivation occurs at the 32-cell stage.

Imprinted genes

Parental imprinting can be defined as the parent-of-origin epigenetic modification of otherwise identical DNA sequences. The epigenetic modifications lead to differential expression patterns of the paternal versus maternal alleles at the imprinted locus. For all imprinted genes characterized so far, the differential expression pattern causes transcriptional silencing of either the paternal or the maternal allele, resulting in functional hemizygosity at the locus (for recent reviews of imprinting, see Efstratiadis, 1994; Tycko, 1994; Ruvinsky, 1995).

The total number of imprinted genes is essentially unknown, but studies using intercross matings of mice heterozygous for balanced translocations suggest that there are at least 15 established imprinted chromosomal regions. In addition, restriction-landmark genome scanning suggests that there might be as many as 100 loci characterized by parent-of-origin-specific methylation of associated CpG islands in the mouse. Information on at least nine imprinted autosomal genes in mammals has been published (reviewed by Ruvinsky, 1995). These include six paternally expressed genes – IGF-II, Snrpn, Znf127, U2afbinding protein-Sp2-splicing factor, insulin 1 and insulin 2 – and three maternally expressed genes – IGF-II receptor, H19 and Wilm's tumour supressor gene (WT1). For some of these genes, expression of the genes is altered as early as the two-cell stage, whereas, in other cases, the lack of expression is not apparent until after puberty. There is evidence that modifier genes influence the time and place of imprinting.

The actual mechanism of imprinting is essentially unknown. It is likely, however, that DNA methylation plays a role in at least the maintenance of imprinting, as most of the known imprinted genes exhibit parent-of-origin-specific methylation patterns. Also, silencing of imprinted genes is perturbed in mice that are homozygous for targeted mutations in the DNA methyltransferase gene.

Recent evidence suggests that the *callipyge* gene, responsible for pronounced muscle hypertrophy in the hindquarters of sheep, is controlled by a novel form of imprinting, referred to as polar overdominance (Cockett *et al.*, 1996). This gene has been previously mapped to ovine chromosome 18 (Cockett *et al.*, 1994). Using paternal half-sib families segregating for the muscular hypertrophy phenotype, the existence of two alleles, *CLPG* and *clpg*, is suggested, with *CLPG/clpg* animals being heavy-muscled and *clpg/clpg* animals being normal in appearance. However, when heterozygous callipyge females (*CLPG/clpg*) were mated to normal rams (*clpg/clpg*), none of the 35 resulting offspring exhibited the callipyge phenotype. All offspring were genotyped for the DNA markers flanking the *Callipyge* locus and showed a normal 1:1 segregation ratio of the corresponding chromosomal segment. In crosses between heterozygous parents (*CLPG/clpg*), only 29% (15/51) callipyge offspring were produced. These offspring were also genotyped with the flanking markers; as before, lambs that inherited the maternal *CLPG* and paternal *clpg* alleles were normal in appearance. In addition, animals homozygous for the *callipyge* mutation (*CLPG/CLPG*) did not express the trait. Thus, animals only express the callipyge phenotype if they have inherited the *CLPG* mutation from their sire, not if it is transmitted by the dam, and the inactivation of the maternal *CLPG* allele dominates the activation of the paternal allele. Characterization of the causal gene is needed to explain this novel imprinting mechanism.

The importance of imprinting in mammalian development is illustrated by the improper development of embryos derived from pronuclear transfer of two maternal or two paternal nuclei. Evolutionary forces leading to the establishment of imprinting remain elusive. Several hypotheses have been formulated, including the prevention of parthenogenesis, prevention of parent–offspring competition for the limited nutrient supply during gestation, control of the growth of the placenta and protection against trophoblastic diseases.

Conclusions

In addition to the genes discussed in the previous sections, expression of a number of other genes has been examined in the ovine embryo and/or fetus. Unfortunately, it is not possible to review these studies in the remaining space. Suffice to say, a vast number of genes must be expressed at the appropriate time and at the proper level to ensure the development of a one-cell zygote into a viable neonate lamb. As many experiments have demonstrated, products of these genes are interwined in a cascade of events. While many of the conclusions about these events have been gleaned from research conducted in heterologous systems, the ovine species is unique in many ways, thereby justifying the continuation of experiments in this species.

References

Anthony, R.V., Pratt, S.L., Liang, R. and Holland, M.D. (1995) Placental–fetal hormonal interactions: impact on fetal growth. *Journal of Animal Science* 73, 1861–1871.

Atkinson, Y.H., Gogolin-Ewens, K.J., Hounsell, E.F., Davies, M.J., Brandon, M.R. and Seamark, R.F. (1993) Characterization of placentation-specific binucleate cell glycoproteins possessing a novel carbohydrate: evidence for a new family of pregnancy-associated molecules. *Journal of Biological Chemistry* 268, 26679–26685.

Bartlett, S.E. and Menino, A.R., Jr (1993) Partial characterization of the plasminogen activator produced by ovine embryos *in vitro*. *Biology of Reproduction* 49, 381–386.

Bartlett, S.E. and Menino, A.R., Jr (1995) Evaluation of extracellular matrices and the plasminogen activator system in sheep inner cell mass and trophectodermal outgrowth *in vitro*. *Biology of Reproduction* 52, 1436–1445.

Bazer, F.W., Geisert, R.D. and Zavy, M.T. (1987) Fertilization, cleavage and implantation. In: Hafez, E.S.E. (ed.) *Reproduction in Farm Animals*, 5th edn. Lea and Febiger, Philadelphia, pp. 210–228.

Bazer, F.W., Thatcher, W.W., Hansen, P.J., Mirando, J.A., Ott, T.L. and Plante, C. (1991) Physiological mechanisms of pregnancy recognition in ruminants. *Journal of Reproduction and Fertility* Suppl. 43, 39–47.

Bazer, F.W., Ott, T.L. and Spencer, T.E. (1994) Pregnancy recognition in ruminants, pigs and horses: signals from the trophoblast. *Theriogenology* 41, 79–94.

Brandon, M.R. (1976) Selective transfer and catabolism of IgG in the ruminant. In: Hemmings, W.A. (ed.) *Maternofetal Transmission of Immunoglobulins.* Cambridge University Press, Cambridge, pp. 437–448.

Breier, B.H., Funk, B., Surus, A., Ambler, G.R., Wells, C.A., Waters, M.J. and Gluckman, P.D. (1994) Characterization of ovine growth hormone (oGH) and ovine placental lactogen (oPL) binding to fetal and adult hepatic tissue in sheep: evidence that oGH and oPL interact with a common receptor. *Endocrinology* 135, 919–928.

Brockdorff, N., Ashworth, A., Kay, G.F., McCabe, V.M., Norris, D.P., Cooper, P.J., Swift, W. and Rastan, S. (1992) The product of the mouse *Xist* gene is a 15 kb inactive X-specific transcript containing no conserved ORF and located in the nucleus. *Cell* 71, 515–526.

Brown, C.J., Hendrich, B.D., Rupert, J.L., Lafreniere, R.G., Xing, Y., Lawrence, J. and

Willard, H.F. (1992) The human *XIST* gene: analysis of a 17 kb inactive X-specific RNA that contains conserved repeats and is highly localized within the nucleus. *Cell* 71, 527–542.

Campbell, S.G. (1974) Experimental colostrum deprivation in lambs. *British Veterinary Journal* 130, 538–542.

Chapman, V.M. (1986) X-chromosome regulation in oogenesis and early mammalian development. In: Rossant, J. and Pedersen, R.A. (eds) *Experimental Approaches to Mammalian Embryonic Development.* Cambridge University Press, Cambridge, pp. 365–398.

Charlier, M., Hue, D., Martal, J. and Gaye, P. (1989) Cloning and expression of cDNA encoding ovine trophoblastin: its identity with a class-II alpha interferon. *Gene* 77, 344–348.

Cockett, N.E., Jackson, S.P. Shay, T.L., Nielsen, D., Steele, M.R., Green, R.D. and Georges, M. (1994) Chromosomal localization of the *callipyge* gene in sheep (*Ovis aries*) using bovine DNA markers. *Proceedings of the National Academy of Sciences of the USA* 91, 3019–3023.

Cockett, N.E, Jackson, S.P., Shay, T.L., Farnir, F., Berghmans, S., Snowder, G.D., Nielsen, D. and Georges, M. (1996) Polar overdominance at the ovine *callipyge* locus. *Science* 273, 236–238.

Crosby, I.M., Gandolfi, F. and Moore, R.M. (1988) Control of protein synthesis during early cleavage of sheep embryos. *Journal of Reproduction and Fertility* 82, 769–775.

Culbertson, M.R. (1976) Bovine hemostatic and complement bacterial endotoxin. PhD thesis, University of California, Davis, USA.

Delhanty, P.J.D. and Han, V.K.M. (1993) The expression of insulin-like growth factor (IGF)-binding protein-2 and IGF-II genes in the tissues of the developing ovine foetus. *Endocrinology* 132, 41–52.

Efstratiadis, A. (1994) Parental imprinting of autosomal mammalian genes. *Current Opinion in Genetics and Development* 4, 265–280.

Farin, C.E., Imakawa, K. and Roberts, R.M. (1989) *In situ* localization of mRNA for the interferon, ovine trophoblast protein-I, during early embryonic development of the sheep. *Molecular Endocrinology* 3, 1099–1107.

Fleming, T.P. and Hay, J.J. (1991) Tissue specific control of expression of the tight junction polypeptide ZO-1 in the mouse early embryo. *Development* 113, 295–304.

Fleming, T.P. and Johnson, M.H. (1988) From egg to epithelium. *Annual Review of Cell Biology* 4, 459–485.

Foster, J.W., Brennan, F.E., Hampikian, G.K., Goodfellow, P.N., Sinclair, A.H., Lovell-Badge, R., Selwood, L., Renfree, M.B., Cooper, D.W. and Graves, J.A.M. (1992) The human sex determining gene *SRY* detects homologous sequences on the marsupial Y chromosome. *Nature (London)* 359, 531–533.

Freemark, M. (1991) Immunoblot analysis of growth hormone and placental lactogen (PL) binding sites in fetal and maternal sheep liver. In: *The Endocrine Society 73rd Annual Meeting Program and Abstracts*, p. 418.

Freemark, M., Comer, M., Korner, G. and Handwerger, S. (1987) A unique placental lactogen receptor: implications for fetal growth. *Endocrinology* 120, 1865–1872.

Gandolfi, F. and Moor, R.M. (1987) Stimulation of early embryonic development in the sheep by co-culture with oviduct epithelial cells. *Journal of Reproduction and Fertility* 81, 23–28.

Gharib-Hamrouche, N., Chêne, N., Guillomot, M. and Martal, J. (1993) Localization and characterization of EGF/TGF-μ receptors on peri-implantation trophoblast in sheep. *Journal of Reproduction and Fertility* 98, 385–392.

Gharib-Hamrouche, N., Chêne N. and Martal, J. (1995) Comparative expression of TGF-α and EGF genes in the ovine conceptus and uterine endometrium in the peri-implantation period. *Reproduction, Nutrition and Development* 35, 291–303.

Gluckman, P.D., Bulter, J.H. and Elliott, T.B. (1983) The ontogeny of somatotropic binding sites in ovine hepatic membranes. *Endocrinology* 112, 1607–1612.

Godkin, J.D., Bazer, F.W., Moffat, J., Sessions, F. and Roberts, R.M. (1982) Purification and properties of a major, low molecular weight protein released by the trophoblast of sheep blastocysts on day 13–21. *Journal of Reproduction and Fertility* 65, 141–150.

Goldberg, E.H., Boyse, E.A., Bennett, D., Scheid, M. and Carswell, A.E. (1971) Serological demonstration of H-Y (male) antigen on mouse sperm. *Nature (London)* 232, 478–480.

Grant, S.G. and Chapman, V.M. (1988) Mechanisms of X-chromosome regulation. *Annual Review of Genetics* 22, 199–233.

Halliday, R. (1974) Variations in immunoglobulin concentrations in Merino and Scottish Blackface lambs. *Animal Production* 19, 301–308.

Heyner, S., Shah, N., Smith, R.M., Watson, A.J. and Schultz, G.A. (1993) The role of growth factors in embryo production. *Theriogenology* 39, 151–161.

Imakawa, K., Anthony, R.V., Kazemi, M., Marotti, K.R., Polities, H.G. and Roberts, R.M. (1987) Interferon like sequence of ovine trophoblast protein secreted by embryonic trophectoderm. *Nature (London)* 330, 377–379.

Jainudeen, M.R. and Hafez, E.S.E. (1987) Gestation, prenatal physiology and parturition. In: Hafez, E.S.E. (ed.) *Reproduction in Farm Animals*, 5th edn. Lea and Febiger, Philadelphia, pp. 229–259.

Kappes, S.M., Warren, W.C., Pratt, S.L., Liang, R. and Anthony, R.V. (1992) Quantification and cellular localization of ovine placental lactogen messenger ribonucleic acid expression during mid- and late gestation. *Endocrinology* 131, 2829–2838.

Kidder, G.M. (1993) Genes involved in cleavage, compaction and blastocyst formation. In: Gwatkin, R.B.L. (ed.) *Genes in Mammalian Reproduction*. Wiley-Liss, New York, pp. 45–71.

Kidder, G.M., Rains, J. and McKeon, J. (1987) Gap junction assembly in the preimplantation mouse conceptus is independent of microtubules, microfilaments, cell flattening, and cytokinesis. *Proceedings of the National Academy of Science of the USA* 84, 3718-3722.

Kimpton, W.G., Washington, E.A. and Cahill, R.N.P. (1989) Recirculation of lymphocytes subsets ($CD5^+$, $CD4^+$, $CD8^+$, $T19^+$ and B cells) through fetal lymph nodes. *Immunology* 68, 575-579.

Kind, K.I., Owens, J.A., Robinson, J.S., Quinn, K.J., Grant, P.A., Walton, P.E., Gilmour, R.S. and Owens, P.C. (1995) Effect of restriction of placental growth on expression of IGFs in fetal sheep: relationship to fetal growth, circulating IGFs and binding proteins. *Journal of Endocrinology* 146, 23-34.

Ko, Y., Lee, C.Y., Ott, T.L., Davis, M.A., Simmen, R.C.M., Bazer, F.W. and Simmen, F.A. (1991) Insulin-like growth factors in sheep uterine fluids: concentrations and relationship to ovine trophoblast protein-1 production during early pregnancy. *Biology of Reproduction* 45, 135-142.

Koopman, P. (1995) The molecular biology of *SRY* and its role in sex determination in mammals. *Reproduction, Fertility and Development* 7, 713-722.

Koopman, P., Gubbay, J., Colligon, J. and Lovell-Badge, R. (1989) *Zfy* gene expression patterns are not compatible with a primary role in mouse sex determination. *Nature (London)* 342, 940-942.

Koopman, P., Munsterberg, A., Capel, B., Vivian, N. and Lovell-Badge, R. (1990) Expression of a candidate sex-determining gene during mouse testis differentiation. *Nature (London)* 342, 940-942.

Liang, R. and Anthony, R.V. (1993) Characterization of the gene encoding ovine placental lactogen. *Biology of Reproduction* 48 (Suppl. 1), 177.

Lord, A.P.D., Martin, A.A., Walton, P.E., Ballard, F.J. and Read, L.C. (1991) Insulin-like growth factor binding proteins in tissue fluids from lambs. *Journal of Endocrinology* 129, 59-68.

Lovell-Badge, R. and Hacker, A. (1995) The molecular genetics of *Sry* and its role in mammalian sèx determination. *Philosophical Transactions of the Royal Society of London, Series B, Biological Sciences* 350, 205-214.

Lyon, M.F. (1961) Gene action in the X-chromosome of the mouse (*Mus musculus* L.) *Nature (London)* 190, 372-373.

Mackay, C.R., Maddox, J.F. and Brandon, M.R. (1987) Lymphocyte antigens of sheep: identification and characterization using a panel of monoclonal antibodies. *Veterinary Immunology and Immunopathology* 17, 79-89.

McLaren, A., Simpson, E., Tomonari, K., Chandler, P. and Hogg, H. (1984) Male sexual differentiation in mice lacking H-Y antigen. *Nature (London)* 312, 552-555.

McLellan, K.C., Hooper, S.B., Bocking, A.D., Delhanty, P.J.D., Phillips, I.D., Hill, D.J. and Han, V.K.M. (1992) Prolonged hypoxia induced by the reduction of maternal uterine blood flow alters insulin-like growth factor-binding protein-1 (IGFBP-1) and IGFBP-2 gene expression in the ovine foetus. *Endocrinology* 131, 1619-1628.

Maddox, J.F., Mackay, C.R. and Brandon, M.R. (1987a) Ontogeny of ovine lymphocytes I. An immunohistological study on the development of T lymphocytes in the sheep embryo and fetal thymus. *Immunology* 62, 97-105.

Maddox, J.F., Mackay, C.R. and Brandon, M.R. (1987b) Ontogeny of ovine lymphocytes II. An immunohistological study on the development of T lymphocytes in the sheep fetal spleen. *Immunology* 62, 107-112.

Martal, J. and Djiane, J. (1975) Purification of lactogenic hormone in sheep placenta. *Biochemical and Biophysical Research Communications* 65, 770-778.

Martal, J., Lacroix, C., Loudes, C., Saunier, M. and Wintenberger-Torres, S. (1979) Trophoblastin, an antiluteolytic protein present in early pregnancy in sheep. *Journal of Reproduction and Fertility* 56, 63-72.

Menino, A.R., Jr and Williams, J.S. (1994) Suppression of plasminogen activator production in sheep embryos *in vitro* after treatment with cycloheximide or ouabain. *Journal of Reproduction and Fertility* 102, 65-71.

Menino, A.R., Jr, Dyk, A.R., Gardiner, C.S., Grobner, M.A., Kaaekuahiwi, M.A. and Williams, J.S. (1989) The effects of plasminogen on *in vitro* ovine embryo development. *Biology of Reproduction* 41, 899-905.

Müller, U. (1992) The Y chromosome and its role in primary sex differentiation. *Journal of Animal Science* 70 (Suppl. 2), 1-7.

Müller, U. and Wachtel, S.S. (1991) Are testis-secreted H-Y serological antigen and Müllerian inhibiting substance the same? *American Journal of Human Genetics* 49 (Suppl.), 22.

Nephew, K.P., Whaley, A.E., Christenson, R.K. and Imakawa, K. (1993) Differential expression of distinct mRNAs for ovine trophoblast protein-1 and related sheep type 1 interferons. *Biology of Reproduction* 48, 768-778.

Norris, D.P., Brockdorff, N. and Rastan, S. (1991) Methylation status of CpG-rich islands on active and inactive mouse X chromosomes. *Mammalian Genome* 1, 78–83.

Osburn, B.I. (1988) Ontogeny of host defense systems and congenital infections. *Progressive Clinical and Biological Research* 281, 15–32.

Osburn, B.I., MacLachlan, N.J. and Terrell, T.G. (1982) Ontogeny of the immune system. *Journal of the American Veterinary Medical Association* 181, 1049–1052.

Page, D.C., Mosher, R., Simpson, E.M., Fisher, E.M.C., Mardon, G., Pollack, J., McGillivray, B., de la Chapelle, A. and Brown, L.G. (1987) The sex-determining region of the human Y chromosome encodes a finger protein. *Cell* 51, 1091–1104.

Paria, B.C. and Dey, S.K. (1990) Preimplantation embryo development *in vitro*: cooperative interactions among embryos and role of growth factors. *Proceedings of the National Academy of Science of the USA* 87, 4756–4760.

Penny, G.D., Kay, G.F., Sheardown, S.A., Rastan, S. and Brockdorff, N. (1996) Requirement for *Xist* in X chromosome inactivation. *Nature (London)* 379, 131–137.

Piko, L. and Clegg, K.B. (1982) Quantitative changes in total RNA, total poly(A) and ribosomes in early mouse embryos. *Developmental Biology* 89, 362–378.

Pomp, D., Good, B.A., Geisert, R.D., Corbin, C.J. and Conley, A.J. (1995) Sex identification in mammals with polymerase chain reaction and its use to examine sex effects on diameter of day-10 or -11 pig embryos. *Journal of Animal Science* 73, 1408–1415.

Pratt, S.L. and Anthony, R.V. (1995) The growth hormone receptor messenger ribonucleic acid present in ovine fetal liver is a variant form. *Endocrinology* 136, 2150–2155.

Raghunathan, R., Miller, M.E., Wuest, C. and Faust, J. (1984) Ontogeny of the immune system: fetal lamb as a model. *Pediatric Research* 18, 451-456.

Rice, C.E. and Silverstein, A.M. (1964) Hemolytic complement activity of sera of foetal and newborn lambs. *Canadian Journal of Comparative Medicine* 28, 34–37.

Roberts, R.M. and Anthony, R.V. (1994) Molecular biology of trophectoderm and placental hormones. In: Findlay, J.K. (ed.) *Molecular Biology of the Female Reproductive System*. Academic Press, San Diego, pp. 395–440.

Roberts, R.M., Cross, J.C. and Leaman, D.W. (1992) Unique features of the trophoblast interferons. *Pharmacological Therapy* 51, 329–345.

Roberts, R.M., Xie, S. and Mathialagan, N. (1996) Maternal recognition of pregnancy. *Biology of Reproduction* 54, 294–302.

Rossier, B.D., Geering, K. and Kraehenbuhl, J.P. (1987) Regulation of the sodium pump: how and why? *Trends in Biochemical Science* 12, 483–487.

Rowson, L.E.A. and Moor, R.M. (1967) The influence of embryonic tissue homogenate infused into the uterus on the life span of the corpus luteum in sheep. *Journal of Reproduction and Fertility* 13, 511–516.

Ruvinsky, A. (1995) Gametic imprinting in mammals: phenomenon and possible applications. *AgBiotech News and Information* 7, 69N–76N.

Schindler, C., Shuai, K., Prezioso, V.R. and Darnell, J.E. (1992) Interferon-dependent tyrosine phosphorylation of a latent cytoplasmic transcription factor. *Science* 257, 809–813.

Shimasaki, S. and Ling, N. (1991) Identification and molecular characterization of insulin-like growth factor binding proteins (IGFBP-1, -2, -3, -4, -5 and -6). *Progress in Growth Factor Research* 3, 243–266.

Silverstein, A.M., Uhr, J., Kraner, K.L. and Lukes, R.J. (1963) Fetal response to antigenic stimulus II. Antibody production by the fetal lamb. *Journal of Experimental Medicine* 117, 799–805.

Simmen, F.A. and Simmen, R.C.M. (1991) Peptide growth factors and proto-oncogenes in mammalian conceptus development. *Biology of Reproduction* 44, 1–5.

Simmen, R.C.M., Ko, Y. and Simmen, F.A. (1993) Insulin-like growth factors and blastocyst development. *Theriogenology* 39, 163–175.

Sinclair, A.H., Foster, J.W., Spencer, J.A., Page, D.C., Palmer, M., Goodfellow, P.N. and Marshall Graves, J.A. (1988) Sequences homologous to *ZFY*, a candidate human sex-determining gene, are autosomal in marsupials. *Nature (London)* 336, 780–783.

Sinclair, A.H., Berta, P., Palmer, M.S., Hawkins, J.R., Griffiths, B.L., Smith, M.J., Foster, J.W., Frischauf, A.-M., Lovell-Badge, R. and Goodfellow, P.N. (1990) A gene from the human sex-determining region encodes a protein with homology to a conserved DNA-binding motif. *Nature (London)* 346, 240–245.

Stewart, H.J., McCann, S.H.E., Barker, P.J., Lee, K.E., Lamming, G.E. and Flint, A.P.F. (1987) Interferon sequence homology and receptor binding activity of ovine trophoblast antiluteolytic protein. *Journal of Endocrinology* 115, R13–R15.

Telford, N.A., Watson, A.J. and Schultz, G.A. (1990) Transition from maternal to embryonic control in early mammalian development: a comparison of several species. *Molecular Reproduction and Development* 26, 90–100.

Tycko, B. (1994) Genomic imprinting: mechanism and role in human pathology. *American Journal of Pathology* 144, 431–443.

Wachtel, S.S., Ohno, S., Koo, G.C. and Boyse, E.A. (1975) Possible role for H-Y antigen in the primary determination of sex. *Nature (London)* 257, 235–236.

Waites, G.T., Whyte, A. and Bell, S.C. (1990) Localization of antigen defined by a monoclonal antibody to human 32-kDa insulin-like growth factor-binding protein in the sheep uterus at preimplantation stages of pregnancy. *Placenta* 11, 233–240.

Warren, W.C., Liang, R., Krivi, G.G., Siegel, N.R. and Anthony, R.V. (1990) Purification and structural characterization of ovine placental lactogen. *Journal of Endocrinology* 126, 141–149.

Watson, A.J. (1992) The cell biology of blastocyst development. *Molecular Reproduction and Development* 33, 492–504.

Watson, A.J., Damsky, C.H. and Kidder, G.M. (1990) Differentiation of an epithelium: factors affecting the polarized distribution of Na/K-ATPase in mouse trophectoderm. *Developmental Biology* 11, 41–48.

Watson, A.J., Watson, P.H., Arcellana-Panlilio, M., Warnes, D, Walker, S.K., Schultz, G.A., Armstrong, D.T. and Seamark, R.F. (1994) A growth factor phenotype map for ovine preimplantation development. *Biology of Reproduction* 50, 725–733.

Werb, Z., Banda, M.J. and Jones, P.A. (1980) Degradation of connective tissue matrices by macrophages. I. Proteolysis of elastin, glycoproteins and collagen by proteinases isolated from macrophage. *Journal of Experimental Medicine* 152, 1340–1357.

Wooding, F.B.P. (1992) Current topic: the synepitheliochorial placenta of ruminants: binucleate cell fusions and hormone production. *Placenta* 13, 101–113.

Xie, S., Low, B.G., Nagel, R.J., Kramer, K.K., Anthony, R.V., Zoli, A.P., Beckers J. and Roberts, R.M. (1991) Identification of the major pregnancy-specific antigens of cattle and sheep as inactive members of the aspartic proteinase family. *Proceedings of the National Academy of Sciences of the USA* 88, 10247–10251.

Genetic Resources and Conservation

16

R.W. PONZONI
South Australian Research and Development Institute, PO Box 397, Adelaide, SA 5001, Australia

Introduction 438
The Current Situation 439
Justification for the Conservation of Sheep Genetic Resources 442
The economic justification 443
The scientific justification 443
The cultural and historical justification 444
Which Breeds should be Targeted for Genetic Conservation? 445
Economic considerations 446
Scientific considerations 446
Cultural and historical considerations 447
Making a final choice 447
Methods of Conservation 449
Live-animal conservation 449
Cryopreservation of gametes and embryos 450
Choice of method 452
Genetic and Statistical Considerations 453
Conceptual framework 454
Effective population size 455
Relation between inbreeding, loss of genetic variance and effective population size 457
Real sheep populations 458
Frozen semen and embryos 458
Genetic Conservation Efforts at Different Levels 459
International programmes 460
Regional efforts 461
National efforts 462
Conclusions 464
Acknowledgements 467
References 467

Introduction

In broad terms, conservation is the sustainable use of natural resources, such as soils, water, plants, animals and minerals. The natural resources of any area constitute its basic assets. Wasteful (or no) use of those resources implies a loss, measurable in economic, scientific, social, cultural or aesthetic terms. Livestock are an animal resource that has served mankind for several millennia. Sheep in particular have been domesticated at least since 9000 BC (Ryder, 1984). There are currently about 1087 million sheep worldwide (FAO, 1994). The species is present in all continents, and may be categorized into more than 900 breeds (Scherf, 1995). It is estimated that, for at least 119 (approximately 13%) of these breeds, the number of animals is too low to ensure their long-term survival. Moreover, Hall and Ruane (1993) estimate that, in the past 100 years, 148 sheep breeds have become extinct (i.e. such breeds can no longer be found or easily re-created). Reviews on endangered sheep breeds have been conducted by Ryder (1979), Lauvergne (1990) and Ponzoni (1992b).

Historically, conservation of sheep genetic resources has not been a major issue. The enormous variation observed today among sheep breeds is the combined effect of artificial selection imposed by humans, natural selection forces specific to the environment in which each breed developed and random processes of genetic drift (due to finite population size). Sheep breeders and geneticists have devoted a considerable amount of attention to the accurate identification of 'superior' individuals and to their intensive propagation. They have not hesitated to substitute one breed (or strain) with another one, in the pursuit of greater profit or personal satisfaction. The development of reproductive technologies, such as artificial insemination and embryo transfer (often used in conjunction with multiple ovulation), have made the replacement of one sheep population by another one easier and faster, thus increasing the probability of putting a greater number of breeds (or strains) at risk. Change and loss have always occurred in sheep populations, but it is the realization that a substantial number of breeds are currently at risk, coupled with the wider use of reproductive technologies, which have the potential for greatly increasing the rate at which breeds disappear, that makes conservation of sheep genetic resources an issue.

There are several factors that can result in a numerical reduction of a breed (or strain) of sheep, to the point of placing it at risk. Most factors are associated with economic pressures, and include breed replacement due to indiscriminate crossbreeding with an 'improved' breed (or strain), excessive emphasis on one or a few traits instead of focusing on overall productivity, and changes in the production environment. Natural (e.g. floods, droughts) and man-made (e.g. land clearing, social unrest, wars) disasters can also result in the loss of numerically weak breeds.

In this chapter, an argument in favour of the conservation of sheep genetic resources is developed by briefly describing the present situation, outlining a justification for the conservation of sheep genetic resources, discussing which breeds should be targeted, examining possible methods of conservation, looking at genetic and statistical considerations with regard to conservation programmes and pointing to some outstanding international and national efforts in the area. Finally, the essential aspects to be considered in a national programme of conservation of animal genetic resources are discussed.

The Current Situation

Breeds are the main basis of sheep genetic conservation. A sheep breed is a homogeneous group with definable and identifiable external characteristics that enable it to be separated by visual appraisal from other similarly defined groups. Alternatively, it may be a homogeneous group for which geographical separation from phenotypically similar groups has led to the general acceptance of its individual identity. These definitions of the term 'breed' are adaptations from more general ones given by the Food and Agricultural Organization (FAO) (Scherf, 1995). Breeds have evolved in the context of geographical, cultural and productive requirements, and the term 'breed' should not necessarily be interpreted as technical, often having a predominantly cultural connotation. In principle, valuable sheep genes could be conserved by establishing large random mating multibreed populations, without reference to any particular breed (Hall and Bradley, 1995). However, these authors point out that this approach is not generally favoured, because it would be difficult to evaluate and retrieve the genes. Furthermore, individual breeds are likely to have unique combinations of genes, as a result of adaptation to different environments, and these combinations would be broken up (and very difficult to re-create) in multibreed populations. Sheep with valuable traits resulting from such gene combinations will only be amenable to evaluation and use if bred as discrete populations. Therefore, the conservation of sheep genetic resources is likely to remain primarily focused on the maintenance of breeds. The exception to this occurs when a gene of major effect and interest is detected. In such cases, the aim may be to preserve the gene so that it can be introduced into whichever breed it is deemed desirable.

Figure 16.1 shows the distribution of sheep numbers and breeds in the world, by regions, as defined in Scherf (1995). The trends observed should be interpreted with caution, because the number of breeds in some regions (notably Africa) may be underestimated, due to lack of characterization of some populations. In contrast, a much greater fraction of the true number of breeds is likely to have been identified in Europe. In spite of these limitations, the values given by Scherf (1995) broadly agree with those from other sources (e.g. Hall and Ruane, 1993), and Fig. 16.1 indicates that Europe and the Asian–Pacific regions are the richest in terms of sheep-breed diversity, jointly accounting for more than 70% of the world's documented breeds. Latin America and the Caribbean and North America are the regions with the smallest number of sheep breeds. Note that, according to FAO's regionalization, the countries in northern Africa are included in the Near East, due to agroecological and

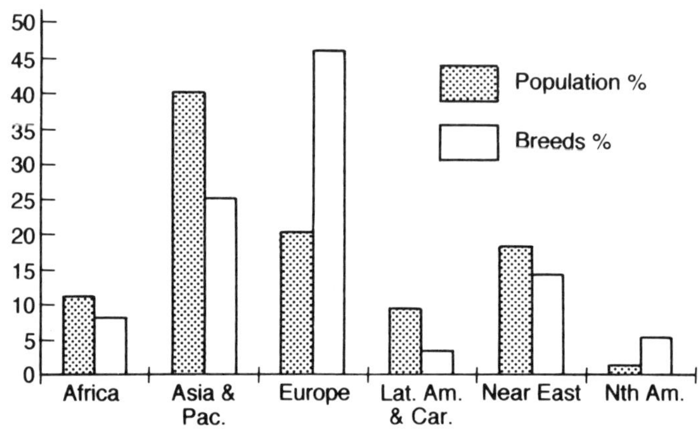

Fig. 16.1. Percentage of the world's sheep population and sheep breeds by region (from Scherf, 1995).

climatic similarities. Cyprus, Israel and Turkey are geographically in Asia but are included by FAO in Europe. Some of the republics of the former USSR are considered under Europe, whereas others are included in the Near East. Countries included in the Near East region share a similar culture and often language. Details of the countries included in each region can be found in Scherf (1995).

The degree of risk for a breed depends on several factors, and criteria have been established to categorize livestock populations in this respect (Table 16.1). The categories are, of course, somewhat arbitrary, but they serve the purpose of grouping and drawing attention to populations at greater risk. The categories 'critical' and 'endangered' are further divided into 'maintained' and 'not maintained' (at-risk) populations. 'Maintained' populations are those for which active conservation programmes are in place, or which are kept by research institutions or private organizations (Scherf, 1995).

Figure 16.2 shows the percentage of sheep breeds at 'critical' plus 'endangered' levels and maintained (not at risk) and 'critical' plus 'endangered' not maintained (at risk), by region. Again, the trends observed have to be treated with caution. The estimate of breeds at risk for Africa, Asia and the Pacific, Latin America and the Caribbean and the Near East are considered to be gross understatements of the true situation (Scherf, 1995). Information from those areas in current databases is limited or absent. This is particularly so for breeds possibly at greatest risk, due to the likelihood of biased reporting of the more common breeds. No active sheep-conservation programmes are recorded in these regions. Sheep breeds are better documented in Europe and North America. There is a relatively high fraction of breeds at risk in these regions, the actual numbers being 49 and 16, respectively. Europe has, by far, the greatest number (33) of active sheep-conservation programmes.

Table 16.1. Criteria for determining breeds at risk (from Scherf, 1995).

Risk category	Population characteristics
Extinct	No longer possible to re-create population
Critical	Total number of breeding females less than 100, or total number of breeding males less than or equal to five, or overall population size slightly above 100 but decreasing and less than 80% of females bred pure
Endangered	Total number of breeding females between 100 and 1000, or total number of breeding males between six and 20, or overall population slightly below 100, increasing and more than 80% of females bred pure, or overall population size slightly above 1000 but decreasing and less than 80% of females bred pure
Not at risk	Total number of breeding females and males greater than 1000 and 20, respectively, or overall population size approaching 1000, increasing and close to 100% of females bred pure

Despite the limitations of the currently available information, one may conclude that, in both developed and developing regions, there is a problem of loss of sheep-breed diversity. The reasons for the erosion of the breed diversity, however, are not necessarily the same in both instances (Barker, 1994). In the developed regions, sheep breeding has been primarily driven by market forces (competition among breeds and breeders), which have frequently led to the replacement of some breeds by others. Breeds less suited to meet market demands gradually decrease in numbers, become increasingly vulnerable and may eventually become extinct. However, market forces only consider present and near-future scenarios, and long-term economic predictions are characterized by a very low accuracy. Breeds or strains currently at risk could be of

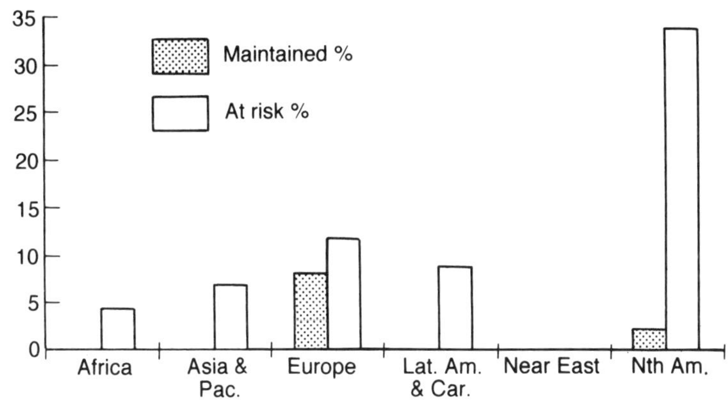

Fig. 16.2. Percentage of sheep breeds maintained and at risk (critical and endangered levels combined), by region (from Scherf, 1995).

value in the more distant future. So, in the developed regions, the desire to contain breed losses may be perceived to a large extent as a matter of insurance.

The situation in developing regions is often different. Breeds as such are not always clearly defined and recognized. There may be several geographically separated populations, sharing many common attributes but having some differences. Sometimes breed names change without a change in appearance, and vice versa. While some of these breeds or strains are clearly under threat and have been reduced in numbers, others, though still numerically strong, are far from safe. Typically, indigenous sheep, particularly those from tropical and subtropical regions, are well adapted to their environment. They are shepherded in low input production systems that are part of the tradition and way of life of the people in question. Production parameters for these populations are often lower than for their counterparts in developed regions. Although such production figures are not strictly comparable between regions, they have frequently been used as a basis of unrealistic expectations in developing regions, and have led to indiscriminate crossbreeding of local populations with 'improved' breeds. The F_1 crossbreds have usually performed well, thus providing a further incentive to the replacement of the local population. However, beyond the F_1, the outcome has most often been less desirable, due to the poor adaptation of the 'improved' breed to the new environment. Coupled with an inability to further develop the F_1, such crossing programmes have mainly been a failure from the production point of view and have resulted in a 'dilution' of the genes of the local population. The imperative for developing regions, often badly needing increases in livestock yields, is to salvage indigenous, well-adapted breeds, and to genetically improve them in a manner compatible with the environment in which they are supposed to perform. A recent study in Peru (Burfening and Carpio, 1995) illustrates how a comprehensive analysis of productivity may show that 'improved' breeds are not necessarily better than the native (Criollo) breed, which has superior ewe reproductive rate and lamb survival.

Justification for the Conservation of Sheep Genetic Resources

In the previous section, arguments in favour of conservation in developed and in developing regions were put forward, namely, as insurance in the former and as a means of realistically increasing livestock production in the latter. In this section, further elaboration of these arguments is presented. Also, several other justifications that have been given for the conservation of livestock breeds or strains are presented, drawing upon considerations made by a number of authors who have treated the subject earlier (Jewell, 1971; Smith, 1984a,b; Hall, 1990, 1993; National Research Council, 1993; Wiener, 1994).

The replacement of less competitive breeds or strains by others has been traditionally considered a normal part of the livestock improvement process. Natural selection favours animals that are best suited to particular environments, whereas artificial selection shifts the process in the pursuit of some sort of benefit to humans. The latter force has probably been the main determining factor in the creation of the enormous sheep-breed diversity currently observed (Hall, 1993). Aided by powerful reproductive technologies, between- and within-breed artificial selection can seriously reduce the genetic variation available in the species. Because sheep are mainly kept in relatively uncontrolled environments (in contrast to pigs and poultry), they are highly dependent on genetic variation to adapt to climatic conditions, nutritional levels and disease challenges. Furthermore, unpredictable factors could change current demand for sheep products and alter production systems. The main reasons for conserving sheep genetic resources may be grouped into economic, scientific and cultural and historical, and they are discussed in some detail below.

The economic justification

This is essentially a practical consideration, based on the assumption that preserved breeds or strains may contribute to improvements in efficiency and/ or quality of production in commercial stocks some time in the future. It is a reasonable justification, because it is well known that production and marketing environments change over time. The economic benefits of conservation, however, are very difficult to ascertain, because the nature and magnitude of future change are not known. For instance, the benefit derived from conserving a rare breed resistant to a particular disease will depend on the future incidence of the disease and on the availability of alternative control methods.

Nevertheless, Smith (1984a) concludes that, even with a very low probability of use of the conserved stock, conservation programmes can be economically justified. In the economic context, generally, conservation may be compared to a gamble with a high probability of no pay-off and a low probability of a high pay-off. In developed regions, where one may assume market forces are operating and breeds currently being used are improving genetically, radical changes in the production and marketing systems would be required for a displaced breed to make its way back into commercial flocks. Hence, the probability of pay-off in developed regions is likely to be very low. In contrast, in developing regions, detailed evaluations of locally adapted breeds are infrequently available and often breed replacement is based on partial analyses of productivity. Under such circumstances, the probability of a high pay-off is much greater than in developed regions. Smith's (1984a) detailed study provides an excellent framework for the examination of cost–benefit relations in conservation programmes. However, the uncertainty associated with our inability to predict the future will almost always remain.

The scientific justification

Genetic research and development depend on the availability of genetic variation. Maintaining as much genetic variation as possible is likely to be useful in many scientific investigations. Scientists often use genetically different breeds, strains or selection lines to elucidate underlying mechanisms responsible for nutritional requirements, reproduction, growth, carcass characteristics, wool production, disease resistance, behaviour and other biological aspects of sheep. For instance, it is known that several sheep breeds throughout the world have an outstanding reproductive rate (e.g. Booroola Merino, Finnish Landrace, Romanov). Research has shown that in the Booroola Merino there is a single gene, the Booroola fecundity gene (*FecB*), with a major effect on ovulation rate, whereas this does not appear to be the case with the Finnish Landrace and the Romanov breeds (Piper *et al.*, 1988). This finding is of practical interest, because it influences the way in which the Booroola Merino, on the one hand, and Finnish Landrace and Romanov, on the other, might be used in industry. Many scientists would argue that it is also a finding of scientific interest in its own right, because it is a step in the direction of explaining the underlying mechanisms determining high reproductive rate in different breeds, irrespective of its immediate application or its monetary worth to the sheep industry.

There are several breeds that are not numerically strong but have special features worth investigating from a scientific point of view. There are probably many more, in both developed and developing regions, that have features we do not know about and that could be even more valuable. Much potential for knowledge gain could be lost if sheep-breed and strain diversity were seriously reduced. There is little doubt that diverse sheep genotypes will be very useful as gene-mapping work progresses and scientists identify and manipulate individual genes involved in the various biological processes associated with sheep production.

In the course of their work, geneticists often create selection lines within a breed or strain (Robertson, 1980). These are populations usually intensely selected, each one for a single trait (or a combination of traits). After a few generations, the lines often diverge, that is, they become genetically different, thus providing, among other things, valuable material for basic genetic and physiological research. Unfortunately, producers frequently perceive these lines as being of very limited value, because they are not commercially competitive or aesthetically appealing. Funding agencies have been particularly insensitive to this type of population, not hesitating to discontinue support for the maintenance of selection lines and sometimes systematically refusing to fund new ones. The lack of commitment to the maintenance of this source of genetic variation contrasts sharply (and is difficult to reconcile) with funding agencies' inclination to finance recombinant deoxyribonucleic acid (DNA) work that could result in 'new' or 'useful' genetic variation.

The cultural and historical justification

There are cultural and historical considerations that have a role in the conservation of sheep breeds and strains. The preservation of geographical sites, buildings, works of art and museum collections of a diversity of objects is widely accepted. The same approach may be taken with regard to the conservation of some sheep populations, which can provide visual evidence of a nation's or region's heritage. In France, for instance, a small (about 120 ewes and 15 rams) flock of Rambouillet Merinos is kept at the Bergerie Nationale (National Sheep Farm) de Rambouillet, where they have been bred for more than 200 years (Anonymous, 1980). Rambouillet Merinos played an important role in the creation and improvement of other wool breeds throughout the world. The flock maintained at the Bergerie Nationale de Rambouillet is reputed to be an almost pure descendant of the original flock established in 1786.

In some cases a particular breed of sheep may be associated with a people's traditions. Henson (1992) reports work with the Navajo Indians (USA) attempting to rescue all that is left of their ancient sheep breed, the Navajo Churro, which originates from the Spanish Churro (Mason, 1996). The breed is light-boned with clean legs and face, and was well adapted to extremes of heat and cold of the Navajo territory. The fleece had a dense, fine undercoat, with long, coarse hairs growing through it. The Navajo people used the wool to weave beautiful rugs of great cultural importance, in traditional patterns. Unfortunately, crossing with other breeds to 'improve' meat production has made pure Navajo Churro sheep very rare. Utah State University has been actively helping to locate 'pockets' of the breed, multiplying sheep numbers and supplying weaving families with the wool they need. Although the rugs woven by the Navajo people have monetary value if offered in the market, the worth of the Navajo Churro breed to the Navajo people goes beyond what is usually taken into consideration by conventional accounting procedures.

There are instances in which the way of life of a particular people is associated with a certain breed. Desert Sudanese sheep are kept by nomadic people in Sudan, where six different varieties of this breed type are recognized (Devendra and McLeroy, 1982). The Abbala (camel-breeding, nomadic) people are reputed to have the best strain or variety of Desert Sudanese sheep. Their sheep are run with camels and are well adapted to the dry climate. Management is rigorous, including controlled mating and lambing and a dry-season watering interval of 3–5 days. It is easy to think of many sheep breeds that would be unsuitable for the production system and way of life of the Abbala people. Desert Sudanese sheep are not currently at risk. However, any ill-conceived programme of genetic 'improvement' (of which there are many examples in developing regions) would not only threaten the breed, but would also be likely to undermine an established way of life.

Which Breeds Should be Targeted for Genetic Conservation?

From the point of view of a conservation biologist, it may be argued that every breed currently in existence should be allowed to survive. Sheep-breed diversity is part of the world's general biodiversity, of which we are an integral part and which we have no right to destroy. Furthermore, it may be argued that we have the duty to preserve current biodiversity for the benefit of future generations.

However, in a world dominated by short-term economic considerations, with 119 sheep breeds identified as having population sizes below 'safe' levels (Table 16.1) and limited funds available for sheep genetic-conservation projects, it is highly unlikely that programmes for every threatened breed will ever be established. Several authors (e.g. Brem *et al.*, 1989; Henson, 1992; Hall, 1993; National Research Council, 1993; Barker, 1994; Hall and Bradley, 1995; Hammond and Leitch, 1995) have discussed and tried to establish criteria of breed eligibility and priorities for conservation programmes. The treatment of the subject that follows attempts to draw together the considerations of the above-mentioned authors and generally parallels the kind of justifications for the conservation of genetic resources put forward in the previous section.

Economic considerations

On economic grounds, a breed considered to be at risk according to the criteria described in Table 16.1 would be a candidate for a conservation programme if it had one or more of the following attributes.

1. One, several or a particular combination of traits of economic importance.
2. One, several or a particular combination of traits currently of little or no economic importance, but which could become important in the future.
3. Presence at a relatively high frequency of a gene with a major effect on a trait(s) of economic importance.
4. Presence at a relatively high frequency of a gene with a major effect on a trait(s) currently of little or no economic importance, but which could become important in the future.
5. Trait levels in the breed not necessarily outstanding, but high levels of heterosis expected when crossed to some other breeds.

The great weakness with economic criteria as they apply to the choice of breeds for conservation programmes is that future economic situations are extremely difficult to predict with any degree of certainty. Although Smith (1984a) indicates that even with low probability levels of future breed usage, genetic conservation programmes can be profitable, it will always be very difficult to persuade funding agencies to invest in the maintenance of breeds that have been (or are currently being) displaced by more profitable stocks. Admittedly, if funding agencies broadened their views and became genuinely concerned with long-term issues of agricultural and social development, winning the case for breeds that could be valuable in some more or less distant future could become easier.

Scientific considerations

On scientific grounds, a breed, strain or selection line considered to be at risk would be a candidate for a conservation programme if it had unique or very rare characteristics, not generally found in other populations. Note that to be of scientific interest the traits in question do not have to be of economic importance. They could be aspects of behaviour, a special ability to cope with a certain diet or a peculiar colour pattern. If a choice of more than one population with the same features were available, preference should be given to the one that had suffered the least contamination through crossing with other breeds. The latter consideration may be particularly important with local populations in some developing regions, where often indigenous breeds have been indiscriminately crossed with introduced breeds.

Cultural and historical considerations

On cultural and historical grounds, a breed considered to be at risk would be a candidate for a conservation programme if it had one or more of the following attributes.

1. Special national or regional significance.
2. Advantages from the standpoint of environmental care or ecological equilibrium, where the preservation of fragile environments is concerned.
3. Recreational value and attractiveness due to special or unusual form, colour, wool type or behaviour.

The importance of cultural and historical considerations should not be underestimated. Animal breed exhibits can be a focal point at fairs and other special occasions. There are great cultural benefits in using livestock breeds to illustrate a nation's or a region's agricultural development. Livestock breeds are sometimes kept in national parks, which can be a great tourist attraction. Furthermore, specific breeds are sometimes an integral part of a way of life and provide special products to a diversity of people around the world.

Making a final choice

Making a decision about which breeds will be the target of conservation programmes and which ones will not is unlikely to be an easy task. Often, one finds that some of the breeds at greatest risk are among those we know least about. Under such circumstances, making logical decisions can be impossible. In some instances, breeds may be eligible under more than one criterion. For example, a breed may be of scientific interest because of some special trait, which could be thought of as of potential economic worth in some more or less distant future, and it could also have some cultural and historical merit. Ancestral Merino strains could fall into this category.

In contrast, there may be several breeds or vaguely defined populations which could all be candidates for conservation for the same (or a very similar) reason. Typically, this situation is likely to be encountered in developing regions, where often populations that look and perform in a comparable manner have different local names, whereas populations that appear to be different may have the same name. The situation may also occur in developed regions, but less frequently. Under such circumstances, it is important to ascertain whether populations that look the same are actually the same (or very similar) genetically, and whether populations that look a little different are actually different genetically. The documented history of the populations, as well as the knowledge of experts and local people, can all be useful in tracing the origin, development, distribution and possible migration among the populations in question. The value of what might be deemed anecdotal and non-scientific information should not be underestimated.

There are, of course, more scientific methods of studying the genetic relationship among different breeds, and these have been discussed by Barker (1994), Hall and Bradley (1995) and Hammond and Leitch (1995). The FAO has been developing an initiative (Global Project for the Maintenance of Domestic Animal Genetic Diversity (Hammond and Leitch, 1995)), which would utilize microsatellite technology, initially focusing on 14 species (including sheep) which account for more than 90% of food and agriculture globally. The primary objective of the project is to increase the cost-effectiveness of the conservation effort by establishing the genetic relationship among the breeds of the 14 chosen domestic-animal species (Barker *et al.*, 1993). Genetic markers, such as microsatellites, can be used in the calculation of the genetic distance between breeds. The genetic distance between two populations, say P_1 and P_2, is a measure of the genetic differences between the populations (i.e. characterized by allele frequencies at a number of loci, it is the probability that an allele chosen at random will be present in P_1 but not in P_2). When available, it is suggested that information on the genetic distance between populations should be used as an indication of their 'genetic uniqueness', and breeds that are more distant from other breeds should be given a higher priority in conservation programmes.

While it is true that isolated populations will tend to accumulate different alleles as a consequence of mutation and genetic drift, populations will also differ because of natural and artificial selection pressure imposed upon them. However, loci responsible for differences due to natural and artificial selection are infrequently available for genetic-distance analyses. Therefore, genetic distance cannot account for the effects of natural selection on adaptation or general fitness, or for the response to artificial selection on morphological and on economically important traits. This constitutes a serious limitation of the technique in the context of making breed choices for conservation programmes. Based on genetic-distance analysis, one could declare that a group of breeds is basically the same genetically, when in reality the breeds might differ substantially in adaptation and production traits, highly decisive factors when making economic and scientific decisions about conservation. It may be fair to state that genetic-distance measures can be used to demonstrate that populations that look and perform similarly are actually different, but cannot be used as evidence that populations that look and perform differently are the same genetically.

Barker *et al.* (1993) correctly point out that genetic-distance information should only be used as an initial guide to population structure and breed differentiation in making conservation decisions. Final decisions must also take into account any available information on aspects such as traits of economic value, specific adaptive features, presence of unique genes or phenotypes and local or regional significance of the breed. A risk is that given the effort and investment involved in a programme to measure the genetic distance among livestock breeds and the scientific aura likely to surround it, unscrupulous administrators and decision-makers could misuse the information to legitimize a hostile attitude towards breeds shown to be related to another one that has been chosen to be maintained in a conservation programme.

Furthermore, when the prime motive for conservation is of a cultural or historical nature, it will be little consolation to the people witnessing the disappearance of a breed they cherish to know that in some other region or nation a population deemed closely related to their own is being maintained.

Methods of Conservation

The subject of conservation methods for livestock has been thoroughly examined and discussed by several authors (Brem *et al.*, 1989; Henson, 1992; Mukherjee, 1992; National Research Council, 1993). In principle, there are three main methods of conserving livestock: maintaining live populations, cryopreserving gametes or embryos and establishing stores of the basic genetic material, DNA. In this section only the first two methods are discussed in detail. The DNA may be stored in gene libraries, and as a complementary measure to other conservation methods. Gene libraries may become more important in the future, but at present storing DNA is not a practical method of sheep conservation and usage. At present, it is not possible to regenerate viable individuals from isolated DNA.

Live-animal conservation

Live-animal conservation is also known as *in situ* conservation. It involves the maintenance of live populations in a 'natural' environment, which could be the population's original environment or one resembling it as much as possible. Live-animal conservation programmes have a number of desirable features. For instance, while attention to details of sampling and design is required (see below, under 'Genetic and Statistical Considerations'), once the population has been established, its management does not require any complicated technology and it may be run like any other flock in the area. The animals that are being maintained can be recorded and characterized for all traits considered important or of potential interest. At all times, they can be seen and become a focus of attraction to farmers, students or visitors, thus contributing to continued interest in the conservation programme. If the flock is sufficiently large there may be sufficient offtake in the form of wool, surplus sheep and skins for the flock to be profitable or at least self-funding. Again, flock size permitting, it may be possible (and tempting) to do some selection and genetically improve some traits and/or select against 'defects'. However, care should be exercised to avoid excessive emphasis on individual traits, which could result in a genetic deterioration in the attributes for which the breed was selected in the first place.

An additional advantage often attributed to live-animal conservation programmes is that, by maintaining the population in an environment as close to 'natural' as possible, it will continue to adapt and evolve. Note, however, that maintained populations are often small and, to avoid problems with inbreeding and genetic drift, restrictions are imposed on the contribution of each family to the next generation and on the genetic relation between mates. Such restrictions (i.e. minimizing inbreeding and genetic drift) interfere with natural selection and are not conducive to simultaneous evolutionary change (Franklin, 1980). They defer the elimination of genes (or gene combinations) responsible for lower fitness among some individuals. Disallowing evolutionary change should not be viewed as a serious problem if, by so doing, one is conserving genetic variation for future evolutionary change.

There are some shortcomings associated with live-animal conservation programmes. Where land resources are limited and labour is very costly, the programme can be perceived as being too expensive or simply not possible. After it has started, sudden withdrawal of funding can terminate the effort. Even when precautions are taken, inbreeding and genetic drift are inevitable in the long term in small populations, so, in time, changes are likely to occur. If the population is maintained as a single flock, a disease outbreak could eliminate the population. The problem would be less serious if several populations were maintained at different locations. Live animals are more expensive to transfer internationally and are subject to more disease-control restrictions than frozen semen or embryos, but this should not be viewed as a limitation, since the live animals can be the source of semen and embryos, which may be more easily transferred.

Cryopreservation of gametes and embryos

Cryopreservation is one of the forms of what is termed *ex situ* conservation. Generally, *ex situ* methods involve the preservation of animals (or parts of them) in a situation removed from their 'normal' habitat. In addition to cryogenic preservation of gametes and embryos, they include other methods, such as preservation of blood or other tissue and of DNA material and maintenance of captive live animals in a situation far removed from their original environment.

With sheep, the technologies of cryopreservation of semen and of embryos have now reached the stage of being routinely used in research and breeding programmes, whereas the status of the technology for the preservation of frozen ova is not yet as advanced. So, in practice, cryogenic preservation of sheep may currently involve preservation of semen or embryos but not of ova.

Cryopreservation methods have a number of desirable features, of which the alleged relatively low cost is the one most frequently advanced. Smith (1984a) showed that, for the assumptions he made, initial costs are greater for frozen semen and embryos than for the establishment of a live-animal population, but that subsequently the opposite is true. Live-animal populations require recurrent input every year, whereas frozen semen or embryos have very low maintenance costs. For a 20-year period of conservation, the cost of maintaining a live-sheep population is the same as that of maintaining frozen embryos. After 5 years, frozen semen becomes the least expensive conservation method. Of course, as Smith (1984a) suggests, with improvement in the technologies involved, the relative costs are likely to vary and, before making decisions, each case should be examined in its own right and in the light of current technology status and resources available.

Space and labour requirements of frozen semen and embryos are low. A very large number of samples can be stored in little space at a single facility, under the care of a few technicians. When land is a limiting factor, cryopreservation offers an alternative. Furthermore, from a genetic point of view, cryopreserved populations do not suffer continuous losses due to genetic drift or selection. Genetically, the population remains intact from the time of sampling until the time of use, assuming, as one may, that there are no long-term effects associated with the freezing technology. As noted earlier, disease control and between-country transfers of frozen semen or embryos are likely. to be considerably easier than for whole live animals.

Funding agencies are particularly attracted to cryopreservation and *ex situ* methods generally. This seems to be part of a more general preference for options requiring more sophisticated technology, to the detriment of those using simpler techniques, somewhat irrespective of the ultimate outcome. It should be realized, however, that, as far as sheep are concerned, cryopreservation methods have a number of important shortcomings. The technology required to collect and freeze semen and embryos can be readily transferred throughout the world, but the process can be expensive for countries in which it is not yet established. Continued reliance on expertise from developed regions will perpetuate dependency and control of resources in developing regions by developed nations. When the technology is not well established, the ever-present risk of loss due to accident, power failure or human error will be much greater than in laboratories where it is routine practice. While not very expensive to run, they have specialized annual maintenance requirements (e.g. liquid nitrogen), the acquisition of which can pose problems in developing regions. Preserved semen and embryos only generate expenses, no income, so, when funds are scarce, administrators will often be inclined to consider them a luxury and suggest that they be reduced in size or eliminated altogether.

From a genetic standpoint, cryopreservation methods are highly effective for the conservation of known genes from a particular population. Even relatively small samples will include all but the rarest genes (see below, under 'Genetic and Statistical Considerations'). However, for the preservation of whole breeds, where relative gene frequencies are important, sampling considerations are important. If the semen or embryo sample is not properly obtained or is too small, there could be a shift away from the gene frequencies in the original population. While a population is cryopreserved, it cannot be evaluated for economically important traits or scientifically studied. Of course, it cannot be seen and enjoyed by the public and, because sometimes what is not seen is more easily forgotten, there is a greater risk that changing administrators and decision-makers of funding agencies will deem it unimportant or irrelevant. Because a frozen population will not be subject to artificial and/or natural selection, it could be continuously losing ground in terms of commercial value in relation to breeds currently used. Note also that if semen is preserved from a population, there will be a period of upgrading prior to the re-establishment of the breed, and this will be costly and time-consuming.

A live population is self-renewing, whereas a store of frozen semen or embryos is not. Therefore, repeated access to the latter will eventually deplete the store unless steps are taken to replenish it. A way of avoiding eventual depletion of semen and embryo stores would be to make access conditional on an undertaking by the user to replenish the store. In some circumstances this may be relatively easy (e.g. if the breed still exists as a live population), whereas in other situations it may be difficult to implement (e.g. when only semen samples exist and the breed is extinct).

Choice of method

In general, the two methods should not be viewed as mutually exclusive, but as complementary. For instance, a live-animal conservation programme of small size could accumulate undesirable recessive genes over time. If the frequency of such genes became high, they would be very difficult to eliminate. Reactivation of frozen semen or embryos would help remedy the problem. Similarly, a semen or embryo bank can be destroyed by accident, in which case it could be re-established if a live population of the breed were available.

The discussion of choice of method for a conservation programme is best conducted in the context of a strategy for the different risk categories defined in Table 16.1. A summary of the main features of live-animal and cryopreservation methods is presented in Table 16.2. No individual procedure is best in every respect. However, their relative importance varies with current population size and risk status (Table 16.1). When a population is categorized as ‘not at risk’, the best means of conservation is by usage, that is, utilizing the breed commercially and improving it by developing for it an appropriate breeding objective, establishing a performance-recording and genetic-evaluation system and providing guidelines for the use of the selected individuals. Great care should be exercised to ensure that the ‘improvement programme’ does not place excessive emphasis on extreme values for some production traits at the expense of others that originally made the breed attractive (e.g. adaptation to an adverse environment, resistance to disease). If a breed is being effectively used commercially, little or no effort will be required to conserve it. Freezing semen and/or embryos may be part of the overall genetic-improvement programme, but is not as important as a conservation measure for populations not at risk as it is for those that are at risk. Conservation by usage is especially relevant to

Table 16.2. Main features of live-animal and cryopreservation methods.

	Conservation method	
Feature	Live animals	Cryopreservation
Technology requirements	Low	Medium to high
Costs		
Initial	Medium to high	Low to medium
Maintenance	Low to medium	Low
Genetic considerations		
Initial drift	Depends on sampling	Depends on sampling
Annual drift	Yes	No
Breed characterization	Possible	Not possible
Selection for use	Possible	Not possible
Evolutionary adaptation	Very little	None
Access		
Public visibility	Yes	No
Educational value	Yes	No
International transfer	Difficult	Easier
Risk	Low to moderate	Moderate, but depends greatly on skills and facilities to apply the technology

some indigenous breeds in developing regions.

For populations classified as 'critical' or as 'endangered' (Table 16.1), the most appropriate course of action would be to establish a live-animal conservation programme, if possible replicated in more than one location, and use cryopreservation of semen and/or embryos as a form of insurance against possible problems with the live-animal part of the programme. Freezing embryos and semen from every sire available is highly recommended. Expansion of the population in size would be highly desirable and should be a high priority, particularly for breeds at a 'critical' level.

When a population is nearing extinction, semen should be preserved from every available sire, as well as embryos representing as many sires and dams as possible. This may be the only option left to save the population from complete extinction. The live-animal component of the conservation effort could consist in an upgrading programme, using dams from another breed.

Genetic and Statistical Considerations

Livestock-conservation programmes virtually always face problems associated with finite population size. Such problems are not exclusive to genetic conservation efforts, and they are also present in the usual livestock genetic improvement programmes. Nicholas (1980), Ponzoni (1992a), James (1993) and Aggrey *et al.* (1995) deal with various aspects of finite population size in genetic-improvement programmes, and their considerations are particularly relevant to situations in which the breed in question is not at risk and conservation by usage is in operation (or about to be implemented).

Note, however, that the number of breeds for which conservation-by-usage programmes can be undertaken will be limited and that there will be many more breeds currently considered not commercially viable, but worth conserving for one or more of the justifications presented above, under 'Which Breeds should be Targeted for Genetic Conservation?'. Therefore, one frequently finds situations in which populations are at risk. The question then is: how does one go about maintaining them in such a way that they retain as much as possible of the original genetic variation and that they remain viable in the long term? Good intentions and funds are, of course, required for a successful conservation programme, but the effort can be negated by poor sampling and design. Insufficient attention to genetic and statistical aspects can result in radical changes in the characteristics of the population one is trying to conserve and in an unacceptable reduction of its long-term viability.

Rochambeau *et al.* (1979), Notter and Foose (1987) and Henson (1992) provide very detailed treatments of the genetic consequences of small population size in animal populations. Other, more specific, studies have examined the effect of population subdivision on effective population size (Chesser *et al.*, 1993; Nei and Takahata, 1993) and the consequences of multiple paternity (Sugg and Chesser, 1994) and of using individuals of unknown ancestry in a breeding programme (Willis, 1993). The Animal Genetic Resources Group of the Animal Production and Health Division of the FAO (Rome, Italy) is currently preparing publications that will serve as manuals and user guides to livestock conservation programmes (cited in Scherf, 1995). Such manuals could prove extremely useful for technical and field personnel in charge of conservation programmes. They could provide necessary guidelines for those with limited training in quantitative genetics, and also they might help to neutralize administrators' attempts to reduce flock (or cryogenic sample) size below acceptable limits.

A thorough treatment of all genetic and statistical aspects of genetic conservation of livestock would be lengthy and is beyond the scope of this chapter. The publications referred to above should be consulted when looking for in-depth information on the subject. The more modest aim here is to describe the main consequences of finite population size in relation to conservation programmes, and to show how some of the 'minimum' or 'acceptable' numbers of animals sometimes quoted are arrived at.

Conceptual framework

A population of finite size cannot indefinitely avoid inbreeding. The number of ancestors of new individuals doubles each generation, and will at some time exceed the number of animals then breeding (James, 1993). So, for any two members of a population, a common ancestor would be found if pedigrees were traced back far enough. In livestock-conservation programmes, ensuring that inbreeding does not accumulate rapidly is very important.

Another (and related) aspect of finite population size is the random drift of gene frequencies, which arises from two sources: (i) unequal contribution of gametes to the next generation by different individuals; and (ii) unequal contribution to gametes by the two alleles present in heterozygotes. In other words, by chance some individuals and alleles may be overrepresented in the next generation, whereas others may be underrepresented. The first source can be controlled by ensuring that all parents of one generation contribute equally to the next, which is possible if the number of male parents equals the number of female parents. However, in genetic-improvement programmes and also most frequently in conservation programmes, the number of females exceeds the number of males. The second source cannot be controlled in sheep, because it would require having a collection of completely inbred lines, which is not possible with this species.

In populations that are not undergoing selection, inbreeding and genetic drift are manifestations of the same phenomenon. James (1993) gives a simple example illustrating how drift occurs first and is then followed by inbreeding. Assume that a line is begun by sampling two individuals (one male and one female, unrelated) from a large population. The gene frequency in these two individuals will drift away from that in the base population. When the two individuals are mated, their progeny will not be inbred in the first generation. However, because all first-generation progeny are full sibs, inbreeding cannot be avoided in the next generation. In a population maintained using a constant mating system, inbreeding and drift increase at the same rate, so measuring one allows calculation of the other one.

Effective population size

Central in discussions and considerations about genetic drift and inbreeding is the concept of effective population size (N_e), which should be distinguished from the census size (N). The effective population size takes into account the fact that, in sheep flocks and other livestock populations, mating conditions are far from ideal in terms of factors such as relative number of males and females, progeny contributed to the next generation by each individual and population size over time. By definition, N_e is the number of individuals that would give rise to the same inbreeding or genetic drift if they bred in the manner of the idealized population. Thus, by converting N to N_e one can apply formulae devised for idealized conditions.

There are many factors that can affect N_e, and some of those relevant in sheep-conservation programmes are examined in what follows.

Different number of males and females

Typically, in sheep flocks the number of females is different from the number of males. The effective population size is then calculated as:

$$N_e = (4N_m N_f)/(N_m + N_f)$$

where N_m and N_f are the actual number of males and females, respectively. Effective population size is primarily controlled by the less numerous sex – in sheep, most frequently the male. For instance, in a flock consisting of five rams and 100 ewes, the census number is 105, but the effective population size is 19. Figure 16.3 shows N_e values for different male-to-female ratios. It is clear that similar numbers of females give rise to very different effective population sizes, depending on the number of males.

Variance in progeny numbers

In an ideal population, each individual is assumed to be equally likely to contribute progeny to the next generation, and then the mean and variance of family size are equal to two. Variation in family size is probably the most important deviation from the breeding system in an idealized population. It can be shown that in a population of constant size (mean family of two), the effective population size is:

$$N_e = 4N/(2 + \sigma^2)$$

where σ^2 is the variance in progeny number. It can be seen that, when σ^2 equals two, the effective population size equals the census size. By controlled breeding, the variance of family size can be reduced to values smaller than two, in which case N_e becomes greater than N. In the extreme case, when σ^2 equals zero, the effective population size is twice the census number. This may be important in conservation programmes, where costs and space are limiting factors. A breeding system resulting in a variance of progeny number equal to zero will give the lowest possible rate of inbreeding for a given number of breeding individuals. Note that, in the long term, the mating pattern adopted also matters.

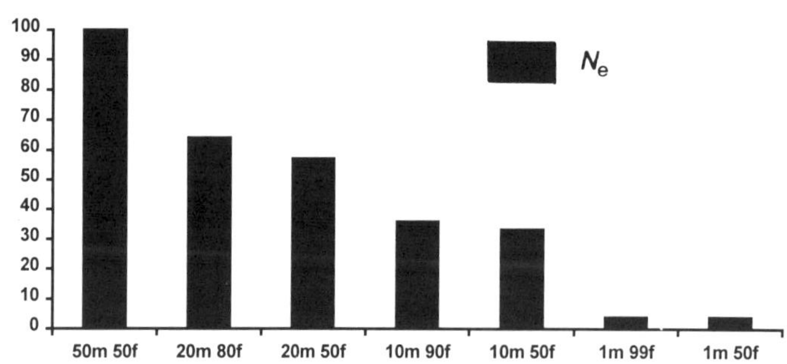

Fig. 16.3. Effective population size (N_e) resulting from different male (m) to female (f) ratios.

Fluctuations in population size

A number of factors can induce fluctuations in population size in a conservation programme (e.g. natural disasters such as drought, need to expand or contract population). When population size varies from one generation to the next, the effective population size is given by:

$$N_e = t/(1/N_1 + 1/N_2 + \ldots + 1/N_t)$$

where $N_1, N_2, \ldots, N_t$ are the population sizes in generations 1 to t, respectively. Effective population size is greatly reduced by periods of small population size. For instance, if a population has size 100 in four generations and size 10 in one generation, N_e is approximately equal to 36, which is closer to 10 than to 100. This too has implications for sheep-conservation programmes, because occasional reductions of population size have long-term consequences, which are not necessarily remedied by later expansions.

Relation between inbreeding, loss of genetic variance and effective population size

The increase of the coefficient of inbreeding and the reduction of the genetic variance in a population of finite size can be expressed as a function of N_e. The inbreeding coefficient at generation t for a population of effective size N_e is given by:

$$F_t = 1 - (1 - 1/2N_e)^t$$

Similarly, the additive genetic variance at generation t is:

$$G_t = G_0 (1 - 1/2N_e)^t$$

where G_0 is the original amount of additive genetic variance. So, in populations of finite size the inbreeding coefficient increases and the additive genetic variance decreases by an amount $1/2N_e$ each generation.

Inbreeding has deleterious effects on survival and reproduction and it also affects production traits, such as fleece weight and growth rate (Lamberson and Thomas, 1984). Natural and artificial selection will tend to counter the negative effects of inbreeding, and animal breeders often tolerate inbreeding levels as high as 1% per generation. In conservation programmes, one may be forced to start with a relatively small group of animals of unknown parentage, but probably related to each other. To be 'on the safe side', one may decide to limit further increases in inbreeding to 0.5% per generation, which would require an N_e value of 100. This can be achieved, for example, with 50 males and 50 females per generation (Fig. 16.3) or with 30 males and 150 females.

The additive genetic variance is reduced each generation by $1/2N_e$ to a fraction $(1 - 1/2N_e)$. After ten generations, with N_e equal to 100, the population would still retain 95% of the additive genetic variance present in the population at the beginning of the programme. These results show that N_e values that are satisfactory to keep inbreeding low are also satisfactory in maintaining a large fraction of the available additive genetic variance over many generations.

Real sheep populations

In sheep populations, females remain in the breeding flock for several years, and males often do so too. This means that at any particular time there are progeny and parents of different generations in the flock. This overlap of generations poses no problem in principle for the calculation of N_e, increase of the inbreeding coefficient or decrease of the additive genetic variance. Turner and Young (1969), Hill (1972, 1974, 1979), Nicholas (1987) and James (1993) deal with the problem of overlapping generations and provide equations derived under less restrictive assumptions than those given in the section above. Garcia (1991) and Henson (1992) provide guidelines on appropriate age structures and replacement rates for domestic livestock conservation programmes.

Frozen semen and embryos

In addition to genes of already known effect, it is likely that other interesting and useful genes not yet identified will be detected some time in the future. The broader the genetic base of the stock stored, the greater the chance of finding such genes. When cryopreservation is practised, there is no further drift and inbreeding, apart from that involved in initially establishing the sample. The creation of the sample of semen or of embryos is equivalent to imposing a one-off bottleneck on the population in question. The sample may be preserved for a period much greater than the time between generations before it is used. In the context of conservation of genetic resources, a sample of semen or of embryos should fulfil the following requirements.

1. It should include as high as possible a fraction of the alleles present in the original population.
2. It should be large enough to enable the frozen semen or embryos to be used in supporting or re-establishing the breed at risk in question.

The probability that a gene occurring in a population is not included in a store depends on the gene frequency and on the size and form of the store (Smith, 1984b). For semen stores, the probability that an allele with frequency p in the original population is not in a semen sample of ten or more doses of semen from N males may be calculated as:

$$(1 - p)^{2N}$$

assuming that the males are unrelated. Notter and Foose (1987) show that a sample of semen representing 45 sires has a probability of 0.01 of excluding a gene with frequency 0.05 in the original population. This would represent a very good sample, and it generally means that achievable sample sizes should be able to capture most of the alleles, even those at relatively low frequencies, in the original population.

In the case of an embryo store of more than ten embryos per mating, the probability that none of the embryos contain an allele in the parent population may be calculated as:

$$(1 - p)^{4N}$$

where N is now the number of pair matings and the pairs are assumed to be unrelated. With values for N of the order of 15–20, these probabilities become very small, unless the gene in question is at a very low frequency. Thus, the risk that a gene present in the original population is lost can be made very low with achievable embryo numbers.

Note that important assumptions involved in these results are that the individuals represent a random sample of the original population and that they are non-inbred and unrelated to each other. In populations already at risk, meeting these assumptions may be difficult. Sampling from a population at risk requires very careful consideration if one is to ensure that a high fraction of the remaining genes is included in the sample.

Note also that in the above considerations no allowance was made for losses or inefficiencies associated with the technology of freezing semen or embryos. For instance, the semen of some individuals may be less able to withstand freezing, and not all frozen embryos will necessarily result in viable individuals and produce the desired number of progeny. Numbers will have to be adjusted (increased) according to the current 'state of the art' of the technology. Well-equipped laboratories where the technology has been applied for some time are likely to have greater rates of success than those that are less well equipped and in which the technology has only recently been introduced.

Genetic Conservation Efforts at Different Levels

Alderson (1990), Henson (1992), National Research Council (1993) and Heywood (1995) give accounts of conservation efforts being made at different levels (e.g. international, national), including those specifically related to sheep. Jointly, these publications provide a comprehensive inventory of current activities in the field. The purpose of this section is to review briefly the present situation, emphasizing the range of approaches adopted at different levels and under a variety of circumstances.

Changing patterns in livestock breeding have been taking place for centuries, but it is only relatively recently that the need to prevent loss of livestock genetic diversity through breed replacement has been recognized. The development of programmes of conservation of genetic resources for crops and pastures has advanced at a much faster rate than for domestic livestock, aided by a number of advantageous characteristics of plants over animals, such as lower cost per unit, greater reproductive rate and ease of collection and storage. Note also that, traditionally, plant breeders have made use of conserved stocks in genetic improvement programmes, whereas animal breeders have not. Nevertheless, the inclusion of a major (and very well attended) symposium on livestock conservation at the latest World Congress on Genetics Applied to

Livestock Production (Guelph, Canada, 1994) and of a chapter on breed conservation in a recently published book on animal breeding (Wiener, 1994) and the publication of a book specifically documenting the characteristics of some livestock breeds that fall outside the mainstream of current popularity (National Research Council, 1991) are all indicators of an increased awareness of the problem.

International programmes

The FAO has had an interest in this subject since 1946, when it was given the task of conducting work on the evaluation and conservation of both plant and animal genetic resources (see historical account by Beilharz, 1983). An expert consultation in 1992 reaffirmed FAO's global leadership role in the management of animal genetic resources and the development and implementation of a global programme was recommended. The goal of the programme is to overcome the erosion and to enable better development of animal genetic resources (Scherf, 1995). The strategy for such an undertaking includes the description and monitoring of existing animal genetic resources, the characterization of breeds at the molecular level to establish the relationship among them, the development of a global database, the provision of guidelines for live-animal conservation and cryopreservation, training in all aspects of conservation genetics and communicating to the world community the importance of looking after animal genetic resources.

An extremely valuable outcome of the FAO global programme has been the publication of a World Watch List (WWL) for domestic animal diversity (Scherf, 1995). The aim of the WWL is to communicate the state of farm-animal resources, and in particular to provide a catalogue of breeds at risk for 28 mammalian and avian species. It constitutes a key document for all those somehow involved in livestock conservation, and it should be very valuable as an educational tool on the status of domestic animal breeds and their sustainable use. In such an ambitious document, it is not surprising that one can find a few loose ends and inaccuracies. For instance, I see little point in listing the Booroola Leicester as a sheep breed at risk in Australia, when the breed is the result of incorporating *FecB* into the Border Leicester and neither the gene nor the latter breed is at risk. Furthermore, the numbers of Booroola Leicester are more likely to be low due to the early stage of the breed's development than to its replacement by other stock. This and some other inaccuracies do not detract from the value of the publication or the merit of those responsible for its preparation. When inaccuracies occur, the problem appears to be more a reflection of the poor quality of the information supplied by the national coordinators, rather than any fault on the part of those collating it. Greater rigour and commitment by those supplying country-specific information should greatly increase the value of the WWL.

Although the FAO has taken a leading role in the international scene of livestock conservation, there are many other government and non-government organizations with an international scope and activities in this field (e.g. the United Nations Environment Programme (UNEP) and the World Bank; see National Research Council (1993) for a comprehensive list). Overlapping of aims is not uncommon, and there seems to be room for improvement in the efficiency with which international activities are carried out by some further coordination. For instance, the report prepared for UNEP by Heywood (1995) includes some information that is complementary to that contained in the FAO's WWL. It has an extensive table (pp. 583–587) listing the number of indigenous breeds of the main mammalian livestock species involved in some sort of conservation programme, by country, but prepared independently of the information in Scherf (1995).

Regional efforts

A regional approach to the conservation of genetic resources has much to offer in terms of helping to rationalize the effort. Political boundaries often cut across ecological, historical and cultural boundaries and, if feasible, a regional focus may add strength and logic to the overall programme. Adalsteinsson (1994) describes the conservation activities with farm animals conducted by the Nordic countries (Denmark, Finland, Iceland, Norway and Sweden). In 1979, a working group appointed by the Nordic Council of Ministers (NMR) recommended the establishment of a Nordic gene bank for farm animals. This led to the initiation of some national conservation activities that saved from extinction an old Icelandic chicken population, as well as some Finnish, Danish and Norwegian cattle breeds. A formalized project (Nordic Gene Bank Cooperation – Farm Animals) began in 1984, when a steering group was appointed, with one member per country. The steering group meets twice a year and there is a secretary in charge of overseeing matters between meetings. Each country has its own Conservation Committee, which is responsible for the conservation programmes within the country. Funds for meetings and for some specific conservation measures are provided by the national Ministries of Agriculture, but private initiative also plays a role. Adalsteinsson (1994) describes the programme in each of the participating countries. Overall, 103 breeds of 13 animal species have been identified. Farmers maintaining a conserved breed may receive a form of subsidy, which is supposed to compensate for the effort and loss of revenue of using a less profitable breed. Granting of subsidies is perceived as a less expensive alternative form of conservation than setting up an independent flock or herd. Other conserved populations may be kept by teaching organizations and agricultural museums. The Nordic programme has suffered some losses of cryopreserved material, which highlights the importance of considering live-animal conservation and cryopreservation as complementary methods, rather than as alternatives.

The NMR has issued a number of publications and reports on domestic-animal breeds, as well as on details of the activities carried out by the programme. The Nordic experience provides an example of regional integration and coordination of efforts in the conservation of domestic animal diversity. If logic prevailed, one would think that activities along similar lines could be implemented in other regions, such as the Caribbean, southern South America or south west Asia.

National efforts

The relative importance of animal production from an environmental, social, cultural and economic point of view differs between countries. Their perceived needs for conservation also differ, and each country should determine which of its livestock populations are in need and worthy of conservation efforts. International and regional activities can play a crucial role in providing an overall framework and logical guidelines to arrest the erosion of domestic-animal diversity and to increase the efficiency of conservation. However, national efforts are likely to remain the basic building-blocks upon which regionally or globally conceived programmes will depend.

A number of programmes at a 'country' level have been established, and they are reviewed in the above listed references. Not surprisingly, these have been more common in developed than in developing regions, where they occur mainly among larger countries with good infrastructure and technical 'know-how', such as Argentina, Brazil and India. In this section, two examples (one government-funded and the other privately funded) of livestock-conservation programmes at a country level are given.

A government funded programme in Hungary (account based on National Research Council, 1993)

Efforts to preserve old breeds in Hungary date back to the 1950s, when attempts were made to save the remaining animals of some traditional breeds. In 1973 the Hungarian Ministry of Agriculture and Food gave the Institute for Animal Breeding and Feed Control responsibility for maintaining endangered breeds. The programme includes activities with cattle, geese, pig and sheep populations. With financial support from the government, state and cooperative farms have been maintaining non-commercial populations of livestock. There has been research based on these populations in areas such as the evolution and genetic structure of the breeds and evaluation under present production conditions. In the case of sheep, for instance, the Racka breed has been a focus of attention. The Racka belongs to the Zackel group of sheep breeds, of which there are populations in several south-east European countries (Mason, 1996). It is a dairy and meat sheep, coarse-woolled, usually white, with light brown face and legs, and long, vertically growing corkscrew horns.

The state also runs Hortobagy National Park, which may be cited as an example of a national park designed as a conservation area to preserve rare breeds. The conservation programme at the park includes the protection of indigenous plants and wildlife, as well as the perpetuation of the regional skills and traditions of farmers and their traditional breeds. The conservation programme has its own state-funded budget, land at the park and maintenance staff. Close links exist with university personnel, who conduct the breeding programme and use the animals for research purposes. The park derives some of its income from fees paid by the visiting public.

A privately funded programme in the UK (account based on Alderson, 1990; National Research Council, 1993)

The Rare Breeds Survival Trust (RBST) is one of the best known non-government organizations in the field of livestock conservation. It is privately funded and managed as a charitable organization. Its aims are to ensure the survival of rare farm breeds that are part of Britain's agricultural heritage, and which constitute a source of genetic variation for future generations of livestock breeders. Since its creation in 1973, its work has included emergency actions and the establishment of conservation, evaluation and support programmes.

The RBST has about 8500 members, of which about 1000 are owners and breeders of small groups of animals of rare breeds. Financial support comes entirely from private sources (membership contributions, donations, sponsorship, corporate grants). It operates as an umbrella organization, offering networking services to coordinate the voluntary initiatives of breeders of rare breeds, including farm parks. Farm parks are independent centres owned and operated by individuals, landowning charities or city councils. They are recognized by the RBST in an approved scheme, which identifies those farm parks genuinely involved in the conservation of rare breeds. Such parks have relatively large breeding units of rare breeds and are open to the public as educational and visitor centres. Farm parks have proved to be an effective means of stimulating public interest in and support of conservation activities.

Other major activities of the RBST include the definition of criteria to classify a breed as rare, the payment of subsidies to farmers keeping rare breeds, recording pedigrees and monitoring inbreeding. The RBST staff conduct workshops and livestock sales, provide advice on breeding matters and publish a monthly magazine. The RBST owns the island of Linga Holm (off the north coast of Scotland), where a flock of the extraordinary North Ronaldsay breed is kept. Adapted to life on rocky shores, North Ronaldsay sheep show a number of remarkable features, of which their ability to survive on a diet of seaweed is probably the best known. Also, their feeding behaviour does not follow a diurnal (sunrise to sunset) pattern, as in other breeds, but follows the tidal cycle instead. The breed is very salt tolerant and capable of living on a diet so low in copper that it would kill other breeds.

The RBST provides a magnificent example of how livestock conservation can be supported at a national level, while at the same time placing emphasis on increasing public awareness about the issues involved. The success of the activities undertaken by the RBST can be measured by results. Since its creation in 1973, no British breeds of livestock have become extinct and the population size of almost all endangered breeds of farm livestock in Great Britain has increased.

Conclusions

Unlike many endangered wild species, sheep and other major livestock are far from being threatened with extinction. However, conservation in sheep and other livestock is important, because breeds and native or local populations are major reservoirs of genetic diversity within a species. Thus, breed losses generally imply a reduction of genetic diversity. In developed regions breed losses mainly occur due to competition from other breeds that are perceived as more profitable. In developing regions, however, the reasons are often less logical and include ill-conceived agricultural development and/or genetic improvement programmes. The extent to which the losses will actually reduce genetic variation within a species is neither precisely known nor easily measured. In developed regions a greater amount of knowledge has been gained about the status and trends of livestock breeds than in developing regions. The latter often lack well-characterized breeds or census data and, unfortunately, that situation can be conducive to genetic erosion.

Traditionally, sheep genetic improvement programmes have relied on existing genetic variation (between and within populations). The paradox is that the standard procedures to achieve improvement (i.e. breed replacement, crossbreeding and selection in a hierarchical structure within a breed) can, if unchecked, have devastating effects on genetic deversity, thus undermining the very basis of future improvement.

The conflict may be further illustrated by examining the generalized views and perceptions of geneticists and practising animal breeders and comparing them with those of conservation geneticists. Geneticists working with livestock have been trained as 'animal improvers'. Their methods include breed replacement, crossbreeding, selection within populations and dissemination of genetic gains. They use them effectively, neither looking back nor too far ahead. Practising livestock breeders worldwide are under pressure, endeavouring to remain economically viable. They do not look favourably upon less productive strains or breeds, and they are often willing to combine their traditional practices with recommendations made by geneticists, in the pursuit of short- and medium-term economic gains.

In this kind of scenario, the point of view of the conservation geneticist is unlikely to be readily accepted. For the conservation geneticist, breed diversity is of value due to potential economic worth sometime in the future, scientific interest or cultural and historical connotations, and this may be quite independent of commercial merit according to current accounting systems. From his (or her) standpoint, genetic improvement has a very high price, namely, a reduction in genetic variability and a sacrifice of long-term advantages, in exchange for the overriding short- and medium-term benefit of desired levels of relatively uniform and predictable performance. The conservation geneticist would prefer to allow genetic diversity to continue to evolve on a large scale, in order to ensure that genetic resources will be available as building-blocks for future agricultural, scientific and cultural developments.

These contrasting views mean that an outcome perceived as an achievement by one sector could be considered a major disaster by the other one. For instance, an effective breed-replacement programme would be considered a success by the conventional geneticists and breeders involved, but it would be counted as a loss by the conservation geneticist if the replaced breed had unique attributes (productive, scientific, cultural or historical) not found in other populations. In contrast, payment of a subsidy to farmers to encourage them to maintain a local breed would be perceived as an impediment to 'progress' by conventional geneticists and breeders. There are, of course, less extreme, conciliatory views, which I believe are gaining supporters. Nevertheless, by looking at such extremes one can better gauge the scope of the problem.

At a time when rapid change and 'progress' are frequently presented and imposed as inevitable, it is hard to be very optimistic about the fate of threatened breeds in some regions. Policies developed at international and national levels can greatly influence the course of sheep and other livestock genetic diversity in the coming years. The FAO's development of an integrated programme for the global conservation of domestic-animal genetic resources will, if followed up by implementation in the field at national and regional levels, play a vital role. Country programmes are likely to constitute the basic units upon which endeavours at a higher level will build. In order to be successful, such programmes must have a well-defined overall strategy. They should address a number of key issues, which, if neglected, could seriously detract from the value of the programme. Due attention to the following aspects appears essential.

Inventory and characterization of sheep breeds

This is the logical starting-point. The aims should be to have a complete list of existing breeds, strains or populations, to establish their status (as defined in Table 16.1) and to compile as much information as possible regarding the breeds' attributes.

Creation of a database

This step follows from the previous one. All the information should be safely stored in a manner that makes it readily accessible whenever judgements and/or decisions have to be made.

Education

Conservation efforts should be accompanied by vigorous education campaigns. Often, reference is made to and emphasis placed on the need to include conservation genetics in animal-breeding courses at the university level. While this is, no doubt, a very important aspect, education of conventional geneticists, practising breeders, administrators and politicians may be just as important. Well-meaning people may, unknowingly, make decisions that are hostile to the conservation of genetic resources, simply because of lack of information or incomplete understanding of the problem. Such people will only become sensitive to the point of view of conservation geneticists if they fully understand the issues involved.

Liaison with genetic improvement programmes
Where genetic-evaluation and performance-recording services exist, attempts should be made to develop strong links with them, in order to ensure that, when appropriate, they address conservation issues. The ultimate aim should be to make genetic conservation an integral part of genetic improvement efforts.

Liaison with funding bodies
One may find that agencies funding livestock research and development (especially those that are industry-based) do not have genetic conservation on their agenda. In such cases, communication and shared experiences may contribute to an improved perception and provide some opportunity for future collaboration. The dialogue may be used to introduce notions likely to initially shock the funding body's representatives, such as that, in genetic conservation, arguments other than short-term economics are used. A less ambitious and more achievable objective may be to get the funding body to accept that genetic conservation is an issue and that it is likely to become more important in the future.

Genetic-impact statements
The National Research Council (1993, p. 20) and Hall and Bradley (1995) suggest that designers of livestock-improvement programmes should consider the potential impact on genetic diversity. This would be particularly important in developing regions, where a development programme could include the introduction of an 'improved' exotic breed, which could displace a local native population. The genetic impact statement would define the projected effects of the development on the number and genetic 'purity' of animals of the native breeds in the vicinity. Minimum numbers of native females to be mated pure could be established in order to avoid the erosion of the local gene pool. Structured crossbreeding programmes could be implemented to capture any advantage the introduced breed might have, without leading the native population to extinction. The notion of genetic-impact statements should not be entirely foreign to sheep producers. It is akin to the complaints sometimes made by powerful breed societies when they feel that the introduction and/or dissemination of another breed might bring about serious competition for a share of the market and/or be the cause of disrepute, due to the offer of a lower-quality or contaminated product (e.g. pigmented fibres in high-quality white wool).

Initiation of conservation programme(s)
Finally, breeds, strains or populations considered at risk and unique according to the criteria given above should be preserved. The best combination of the live-animal and the cryopreservation components will depend on the specific circumstances.

Achieving a balance between the pressure to increase productivity and addressing conservation issues will not be an easy task. However, progress made in the above areas in many nations worldwide would greatly contribute to the maintenance of breed diversity. This should assist in meeting the challenges of future production and marketing environments, in broadening the scope of scientific research and in enhancing our cultural and historical appreciation of the role of sheep (and livestock generally) in the way of life of many people.

Acknowledgements

I wish to thank Drs K. Hammond, Helen Leitch and Beate Scherf of the Animal Genetic Resources group of the FAO, Rome, for providing information on their global programme on the management of domestic-animal diversity. Dr Stephen Hall of the Department of Clinical Veterinary Medicine, Cambridge, was kind enough to send me a large number of papers on the subject of livestock conservation, which I found extremely useful. Dr John James read a version of the chapter and made many very valuable suggestions.

References

Adalsteinsson, S. (1994) Conservation activities in the Nordic countries: farm animals. In: Adalsteinsson, S. (ed.) *Genetic Resources in Farm Animals and Plants – Report from Research Symposium, 27–29 May, 1994, As, Norway.* Nordic Council of Ministers, Copenhagen, Denmark, pp. 63–70.

Aggrey, S.E., Lin, C.Y. and Cheng, K.M. (1995) Size of breeding populations required for selection programs. *Theoretical and Applied Genetics* 91, 553–556.

Alderson, L. (ed.) (1990) *Genetic Conservation of Domestic Livestock.* CAB International, Wallingford, UK, 242 pp.

Anonymous (1980) Livestock from France. *Bulletin de l'Élévage français* Special No., Spring, pp. 45.

Barker, J.S.F., Bradley, D.G., Fries, R., Hill, W.G., Nei, M. and Wayne, R.K. (1993) *An Integrated Global Programme to Establish the Genetic Relationships Among the Breeds of Each Domestic Animal Species.* Report of a working group for the Animal Production and Health Division, FAO, Rome, Italy, 32 pp.

Barker, J.S.F. (1994) Animal breeding and conservation genetics. In: Loeschcke, V., Tomiuk, J. and Jain, S.K. (eds) *Conservation Genetics.* Birkhauser Verlag, Basle, Swizerland, pp. 381–395.

Beilharz, R.G. (1983) Conservation of animal genetic resources. In: Peel, L. and Tribe, D.E. (eds) *Domestication, Conservation and Use of Animal Resources.* Elsevier Science Publishers, Amsterdam, The Netherlands, pp. 93–105.

Brem, G., Brenig, B., Muller, M. and Springmann, K. (1989) *Ex situ* cryoconservation of genomes and genes of endangered cattle breeds by means of modern biotechnological methods. *FAO Animal Production and Health Paper* 76, 123 pp.

Burfening, P.J. and Carpio, M. (1995) Improving Criollo sheep in Peru through crossbreeding. *Small Ruminant Research* 17, 31–35.

Chesser, R.K., Rhodes, O.E., Sugg, D.W. and Schnabel, A. (1993) Effective sizes for subdivided populations. *Genetics* 135, 1221–1232.

Devendra, C. and McLeroy, G.B. (1982) *Goat and Sheep Production in the Tropics*, 1st edn. Longman Scientific and Technical, Marlow, Essex, England, 271 pp.

FAO (1994) *FAO Yearbook*, Vol. 48: *Production*. FAO Statistics Series no. 125, FAO, Rome.

Franklin, I.R. (1980) Evolutionary change in small populations. In: Soule, M.E. and Wilcox, B.A. (eds) *Conservation Biology: an Evolutionary–Ecological Perspective*. Sinauer Associates, Sunderland, Massachusetts, USA, pp. 135–149.

Garcia, C. (1991) Diseno de un plan de conservacion de las razas vacunas autoctonas de Galicia. *Informacion Tecnica Economica Agraria* 87A, 268–274.

Hall, S.J.G. (1990) Genetic conservation of domestic livestock. *Oxford Reviews of Reproductive Biology* 12, 289–318.

Hall, S.J.G. (1993) Why are there so many breeds of livestock? In: Clason, A., Payne, S. and Uerpmann, H.-P. (eds) *Skeletons in Her Cupboard – Festschrift for Juliet Clutton-Brock*. Oxbow Monograph 34, Oxbow Books, Oxford, UK, pp. 99–107.

Hall, S.J.G. and Bradley, D.G. (1995) Conserving livestock breed biodiversity. *Trends in Ecology and Evolution* 10, 267–270.

Hall, S.J.G. and Ruane, J. (1993) Livestock breeds and their conservation: a global overview. *Conservation Biology* 7, 815–825.

Hammond, K. and Leitch, H.W. (1995) The FAO global programme for management of farm animal genetic resources. Paper presented at Biotechnology's Role in the Genetic Improvement of Farm Animals, Beltsville, Maryland, USA, 14–17 May 1995, 24 pp.

Henson, E.L. (1992) *In situ* conservation of livestock and poultry. *FAO Animal Production and Health Paper* 99, 112 pp.

Heywood, V.H. (ed.) (1995) *Global Biodiversity Assessment*. Cambridge University Press, Cambridge, UK, 1140 pp.

Hill, W.G. (1972) Effective size of populations with overlapping generations. *Population Biology* 3, 278–289.

Hill, W.G. (1974) Prediction and evaluation of response to selection with overlapping generations. *Animal Production* 18, 117–139.

Hill, W.G. (1979) A note on effective population size with overlapping generations. *Genetics* 92, 317–322.

James, J.W. (1993) Risks of inbreeding, drift and loss of genetic variation. In: *Short Course in Animal Breeding – Design of Livestock Breeding Programs*. Animal Genetics and Breeding Unit, University of New England, Armidale, New South Wales, Australia, pp. 169–179.

Jewell, P.A. (1971) The case for the preservation of rare breeds of domestic livestock. *Veterinary Record* 89, 524–527.

Lamberson, W.R. and Thomas, D.L. (1984) Effects of inbreeding in sheep: a review. *Animal Breeding Abstracts* 52, 287–297.

Lauvergne, J.J. (1990) Endangered breeds of sheep. *FAO Animal Production and Health Paper* 80, 235–242.

Mason, I.L. (1996) *World Dictionary of Livestock Breeds, Types and Varieties*, 4th edn. CAB International, Wallingford, Oxfordshire, UK.

Mukherjee, T.K. (1992) *Ex situ* conservation of animal genetic resources. In: *Proceedings of the 6th AAAP Animal Science Congress, Bangkok, Thailand*, Vol. 1. Animal Production and Rural Development Department, Bangkok, pp. 115–130.

National Research Council (1991) *Microlivestock: Little-known Small Animals with a*

Promising Economic Future. Board on Science and Technology for International Development, National Academy Press, Washington, DC, USA, 449 pp.

National Research Council (1993) *Managing Global Genetic Resources: Livestock.* Committee on Managing Global Genetic Resources: Agricultural Imperatives, National Academy Press, Washington, DC, USA, 276 pp.

Nei, M. and Takahata, N. (1993) Effective population size, genetic diversity, and coalescence time in subdivided populations. *Journal of Molecular Evolution* 37, 240–244.

Nicholas, F.W. (1980) Size of population required for artificial selection. *Genetical Research* 35, 85–105.

Nicholas, F.W. (1987) *Veterinary Genetics.* Clarendon Press, Oxford, UK, 580 pp.

Notter, D.R. and Foose, T.J. (1987) *Concepts and Strategies to Maintain Domestic and Wild Animal Germplasms.* Commissioned paper in support of Technologies to Maintain Biological Diversity, Office of Technology Assessment, Washington DC, USA, 103 pp.

Piper, L.R., Bindon, B.M., Davis, G.H. and Elsen, J.M. (1988) Control of litter size: major genes and industry utilization of the Booroola F gene. In: *Proceedings of the 3rd World Congress on Sheep and Cattle Breeding, Paris, France,* Vol. 2. INRA, Paris, France, pp. 589–609.

Ponzoni, R.W. (1992a) Genetic improvement of hair sheep in the tropics. FAO Animal Production and Health Paper 101, 168 pp.

Ponzoni, R.W. (1992b) A global review of the genetic resources of sheep and goat breeds. *FAO Animal Production and Health Paper* 104, 157–175.

Robertson, A. (ed.) (1980) *Selection Experiments in Laboratory and Domestic Animals.* Commonwealth Agricultural Bureaux, Slough, UK, 245 pp.

Rochambeau, H. de, Chevalet, C. and Malafosse, A. (1979) *Le Controle de la consanguinité dans les petites populations.* Bulletin Technique du Departement de Génétique Animale no. 31, Institut National de la Recherche Agronomique, France, 122 pp.

Ryder, M.L. (1979) Preserving rare breeds of sheep and goats. *Span* 22, 11–13.

Ryder, M.L. (1984) Sheep. In: I.L. Mason (ed.) *Evolution of Domesticated Animals.* Longman, New York, USA, pp. 63–85.

Scherf, B.D. (ed.) (1995) *World Watch List for Domestic Animal Diversity,* 2nd edn. FAO, Rome, 769 pp.

Smith, C. (1984a) Estimated costs of genetic conservation in farm animals. *FAO Animal Production and Health Paper* 44(1), 21–30.

Smith, C. (1984b) Genetic aspects of conservation in farm animals. *FAO Animal Production and Health Paper* 44(1), 31–41.

Sugg, D.W. and Chesser, R.K. (1994) Effective population sizes with multiple paternity. *Genetics* 137, 1147–1155.

Turner, H.N. and Young, S.S.Y. (1969) *Quantitative Genetics in Sheep Breeding.* Cornell University Press, Ithaca, New York, USA, 332 pp.

Wiener, G. (1994) *Animal Breeding,* 1st edn. Macmillan, London, UK, 208 pp.

Willis, K. (1993) Use of animals with unknown ancestries in scientifically managed breeding programs. *Zoo Biology* 12, 161–172.

Genetic Improvement of Wool Production

17

K.D. ATKINS
New South Wales Agriculture, Orange Agricultural Institute, Forest Road, Orange, NSW 2800, Australia

Introduction 471
Traits of Importance 472
Raw-wool traits and their relationship with processing performance 472
Economic value of raw-wool traits 474
Genetic Parameters for Wool Traits 479
Fleece weight 480
Yield 483
Average fibre diameter 483
Vegetable matter 484
Staple strength 484
Staple length 484
Colour 485
Style 485
Fibre-diameter variability 485
Realized Responses to Selection 485
Consequences of selection for fleece weight 486
Consequences of selection for wool quality 491
Breeding Programmes for Improved Fleece Value 493
Potential genetic gains 493
Breeding in practice 496
Concluding Remarks 497
References 498

Introduction

Wool is not a homogeneous product but varies enormously among breeds in its characteristics and therefore its end use. Merino wools are relatively fine, soft and white and are primarily processed through the worsted manufacturing process to produce light- and medium-weight apparel fabric. The wools from the majority of British breeds and crosses with finer types are coarser, mainly white and processed through both the worsted and the woollen manufacturing systems to produce heavyweight apparel fabric and fabric for upholstery and other interior textiles. The carpet-type wools from speciality carpet-wool breeds, as well as 'hairy' sheep breeds, are generally coarse, medullated and sometimes pigmented. This wool is most frequently used in floor coverings and other furnishing fabrics and fillers. It is also true that the contribution of wool to the income of sheep producers varies considerably. For example, wool may contribute as much as 80% to a Merino producer's income but only 10–30% of a lamb producer's income.

For these reasons, this chapter is restricted to the apparel-wool sector, largely reflecting the Merino and crossbred Merino types. Such wools dominate world wool production and form the vast majority of the total value of the traded wool in the world. However, the principles of wool genetics are applicable over the broad spectrum, although the relative importance of wool within comprehensive breeding programmes will vary according to wool's relative contribution to income. A review by Ponzoni *et al.* (1990) provides useful information on the relative importance of quality traits across the broad range of wool types.

Genetic improvement of raw wool is aimed at:

- increasing the production of fibre from a fixed resource (usually land) to give a greater economic return per hectare; and/or
- improving the quality of the fibre so that it can be processed into superior end-products, thus attracting a higher unit value for the wool produced.

The development of efficient breeding programmes relies on the following steps, which are covered in detail within the chapter:

- Identifying the appropriate traits that should be improved and attaching a relative economic value that will dictate the emphasis to be applied to each within the breeding programme.
- A precise knowledge of the genetic parameters (heritabilities and correlations), so that an effective prediction of response to selection can be made and so that animals may be appropriately evaluated for their potential contribution to the breeding programme.
- Evaluating selection strategies that will lead to the most cost-effective means for achieving progress in the breeding objective.

Traits of Importance

Raw-wool traits and their relationship with processing performance

The transformation of raw wool into the finished product involves a long pipeline of successive processes. The length and interdependence of successive stages in the wool pipeline demand an objective means of specifying the key characteristics of raw wool that might influence its ability to be processed. 'The ultimate objective of this system of specification is to allow the prediction of processing performance, and intermediate and final product quality from the original raw wool' (Rogan, 1988).

The processing pipeline can be reasonably reduced to four major steps: (i) top-making; (ii) spinning; (iii)weaving/knitting; and (iv) garment manufacture.

Top-making

The raw wool is cleaned by removing dust, grease, salts and other skin products by scouring the wool. Vegetable matter (VM) is also removed, either physically (by carding and combing the fibres) or, in cases when the content is very high, chemically, by the process of carbonizing. In the case of wools destined for the apparel trade, the end-product of this process is the top, a thick, loose rope of clean, aligned fibres. The success of the top-making operation is determined by:

- the yield of top (or, inversely, the level of waste or noil);
- fibre length in the top, or hauteur;
- variation in hauteur;
- freedom from neps and contaminants.

The raw-wool characteristics having the greatest impact on top-making are the clean scoured yield, the staple length and the likelihood of the staples breaking during top-making, as determined by the strength of the staples and the position along the staples that the breakage will occur. The task of removing significant amounts of VM may also result in additional breakages of fibres (Whiteley, 1994). Fibre diameter is an issue, since the top is usually contracted at a specified diameter by the spinner. Using the raw wool measurements of yield, length, strength, position of break and fibre diameter in a prediction equation has been shown to account for about 80% or more of the variation in hauteur among sale lots of wool (Andrews *et al.*, 1985).

Some of the so-called 'style' components of wool may also be of importance. For example, weathering of the tip will lead to greater wastage, while good crimp definition (that is, the clarity of crimp along the staple, not its frequency) will also increase hauteur and reduce waste (Stevens and Crowe, 1994).

Spinning

The wool top is converted to worsted yarn by a process of drawing a relatively small number of fibres to form a web and then applying a twist to this web. The number of fibres in the yarn cross-section and their diameter will determine the 'thickness' of the yarn and the ultimate weight of the resulting fabric. Since there is variation in the number of fibres in the cross-section along the yarn, there are lower limits to the average number of fibres that can be included in the cross-section without causing unreasonable compromises in the strength and evenness of the yarn and its likelihood to break during spinning and weaving (Lamb and Yang, 1994). Additionally, the speed of the spinning process has increased enormously, to further limiting reductions in fibre number in the cross-section. The net result is that fibre diameter is of overwhelming importance in determining the fabric weight and the evenness of the yarn achieved from spinning. The consistent move of processors towards lighter-weight fabrics, in response to consumer demands, has highlighted the requirement and demand for finer fibres to achieve these fabrics.

Fibre-diameter variation will also influence the number of fibres in the yarn cross-section: wider variation leads to an increase in the number of fibres needed to achieve the same yarn uniformity and strength. In relative terms, a reduction in the coefficient of variation of the diameter of 5% (where the mean of sale lots is about 22–23% (Naylor *et al.*, 1995)) is equivalent to a reduction of 1 micrometre (μm) in average diameter in achieving the equivalent degree of yarn uniformity (de Groot, 1992). Other characteristics, such as length, crimp definition and strength, have some influence on fabric quality during the spinning process, although extreme variations in these traits have been removed or substantially reduced as a result of blending during topmaking (Whiteley, 1994).

Weaving/knitting
The spun yarn is then either woven or knitted to produce the fabric or cloth. Different weave and fabric constructions will alter both the weight and the appearance of the final product. Fabric dyeing and finishing and finally garment manufacture are further processes required before the product reaches the consumer. Late-stage processing is heavily dependent on yarn quality which is dominated by average fibre diameter. Wool colour and contamination are two specific traits that can interfere with the dyeing process.

Garment manufacture
Finally, the end-product is produced in line with the demands of consumers for garments that are easy to care for, soft, free from skin irritation and lightweight, and that retain their appearance with wear (Whiteley, 1994). The trends in the last 10–15 years towards lighter-weight fabrics, improved softness and better fabric drape all point to the use of wools of finer diameter to satisfy consumer demands.

The relative significance of wool characteristics over the various processing stages can be summarized according to Table 17.1, adapted from Whiteley (1994). Average fibre diameter is the dominant characteristic at all stages of the processing chain. A wide range of other characteristics have some influence, although the number of traits and their importance decrease after top-making.

Economic value of raw-wool traits

Since most of the traits that determine processing value are measured prior to sale, analysis of wool sales (as routinely undertaken in Australia by the International Wool Secretariat (IWS, 1996 *et sqq.*)) provides an assessment of the relative value of these traits. Over the period from 1991 to 1995, the relative contribution of the major traits has shown the expected dominance of fibre diameter in determining wool value (Fig. 17.1).

Table 17.1. Influence of raw-wool characteristics on apparel processing.

Raw-wool characteristics	Scouring and top-making	Spinning	Weaving	Garment manufacture
Fibre diameter	***	*****	***	***
Yield				
Wax, suint and dirt	****	*	–	–
Vegetable matter	*	*	–	*
Strength	**	*	*	–
Length	***	**	*	–
Style	**	*	–	–
Fibre crimp	*	*	–	–
Colour	*	–	–	**
Diameter variability	–	*	*	*

Fibre diameter

Average fibre diameter is measured in micrometres by a range of procedures. Almost all wool sold has a presale test of diameter available and, within a sale, this measure dominates the value of the wool. The majority of Merino wool lies within the range of 18–25 μm but the effect of diameter on price is distinctly and steeply curved, with a relatively greater increase in price with finer diameter. A useful measure of the relative value of fibre diameter is the premium associated with a reduction in fibre diameter of 1 μm at a particular point on the distribution (termed micron premium (MP)). So the MP for 22 μm wool is the

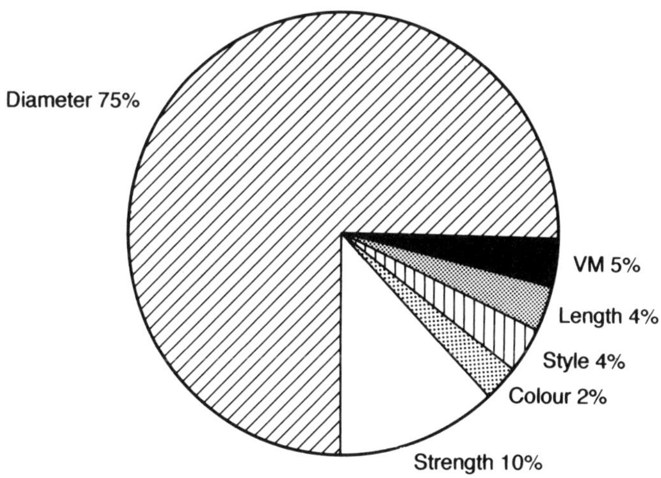

Fig. 17.1. Relative influence of raw-wool traits on price.

relative increase in value (as a percentage) of 21 µm wool over 22 µm wool.

The MP for each micron category of Merino wool over the period 1985–1995 compared with the previous 10-year period is shown in Fig. 17.2. The non-linear trend in the premium is apparent in the current market and these premiums are substantially higher than those operating 10–20 years ago. The demands by consumers for lighter-weight, soft-handling fabrics have clearly sent a strong signal to the wool producer through the market.

For purposes of comparison with the other measures of wool quality, the impact of fibre diameter on price over the period 1991–1995 has been estimated for fine (19–20 µm), medium (21–22 µm) and broad wools (23–24 µm) in Table 17.2. The tabled values are the premiums for a 1 µm change in diameter at the midpoint of these descriptive classifications. The final column in Table 17.2 takes the economic values and scales them according to the likely within-flock standard deviation, providing a more appropriate comparison between traits for the relative impact of variation in quality on price.

Vegetable matter

The removal of VM contamination can be a major cost to processors. The VM content in wool is measured presale and expressed as a percentage (dry weight) of the clean wool content. Discounts received per percentage increase in VM are greater for wools of fine diameter than among medium and broad wools (Table 17.2). Although only the linear component of the regression of price on VM is shown in Table 17.2, there is evidence of non-linearity with wools of VM content of 1–3% attracting a lesser discount than the more heavily contaminated wools.

Staple strength

Traditionally, staple strength was assessed by the 'finger-flick' test, described by Rogan (1988). Since 1985, the optional inclusion of presale measurement of staple strength has resulted in the majority of wool being presented for sale with measured strength. The measure is the average force required to break staples of a given thickness, expressed as newtons per kilotex (N $ktex^{-1}$). The critical

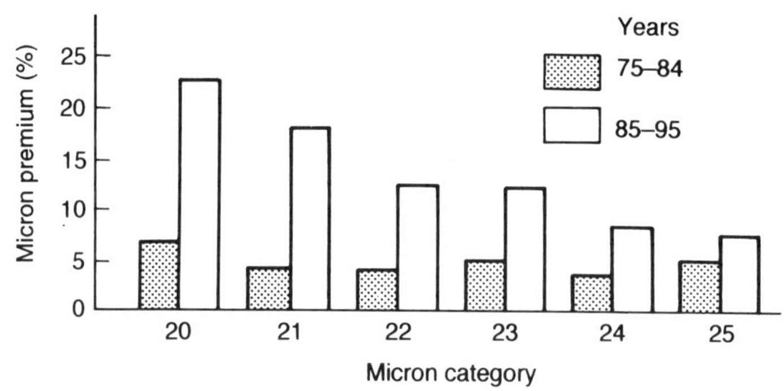

Fig. 17.2. Micron premiums for diameter categories (20–25 µm) between 1975 and 1995.

Table 17.2. Relative economic values of wool-quality traits.

	Fine wool (19–20 μm)	Medium wool (21–22 μm)	Broad wool (23–24 μm)	Relative value per standard deviation
Fibre diameter (% m^{-1})	21.0%	12.2%	5.3%	10% to 30%
Vegetable matter (% %VM^{-1})	–2.2%	–1.3%	–1.1%	–1% to –3%
Staple strength (% N^{-1})				
< 30–35 N ktex^{-1}	–1.2%	–0.9%	–0.6%	–4% to –9%
> 30–35 N ktex^{-1}	+0.10%	+0.06%	+0.04%	+0.5% to 1%
Staple length (% mm^{-1})				
below 90–95 mm	–0.25%	–0.20%	–0.12%	–1% to –2%
above 90–95 mm	–0.06%	–0.02%	0	0% to –1%
Style (% per grade)	3.5%	2.3%	1.0%	1% to 2%
Colour (% per grade)	3.5%	3.0%	2.9%	1% to 2%
Diameter variation (% %cvFD^{-1})	–	–	–	3% to 10% (see text)

minimum strength for sound (or non-tender) wool is in the range 30–35 N ktex^{-1}, depending on diameter. Below this critical level, significant discounts are applied, while above the threshold very small premiums are paid. The discounts and premiums for staple strength over the period 1991–1995 (Table 17.2) were essentially linear below and above the critical threshold, but of very different magnitude. Further, the discounts for tender wools are significantly higher for wools of finer diameter.

Staple length

Optional inclusion of presale measurement of staple length was commenced at the same time as that of staple strength, again replacing a subjective assessment of length. The measure is the average length of the unstretched staple in millimetres (mm). Analyses of prices received also shows a critical length of about 90–95 mm, depending on diameter, below which significant discounts are applied (Table 17.2). The discounts for shorter staple lengths are again higher in wools of finer diameter. Above the critical length, premiums are small and variable, with some evidence of a slight discount for long staples in finer types.

Style

Wools are currently graded subjectively prior to sale into a number of style grades. The grading is done by experienced assessors, using a variety of traits (Jackson and Rottenbury, 1994), including:

- crimp definition and frequency;
- staple tip and shape;

- staple colour;
- weathering;
- handle;
- dust contamination.

Most of these components, except perhaps handle, can be measured using image analysis, and procedures are in place for the objective measurement of style. Without information on the components, it is difficult to determine which of the style components may have a greater or lesser impact on price, although crimp definition and weathering have been identified as traits of some importance in processing (Stevens and Crowe, 1994). Based on prices received according to style grade, premiums and discounts for different styles are identifiable (Table 17.2). However, in valuing style, it is important to note the distribution of sale wools according to style. Over the same period (1991–1995), 5–10% of Australian Merino wool was classified as average or inferior top-making style, the vast majority (65%) as good top-making style, about 25–30% as best top-making style and less than 1% as spinners or better style. The premiums between adjacent grades were approximately linear between good, best and spinners styles, with the premiums for improved style greater for wools of finer diameter.

Colour

From a processing viewpoint, wool colour after scouring is the most important trait, as it can limit the colour range and brightness of the dyed fabric (Whiteley, 1994). The relationship between greasy wool colour and scoured colour of the top is not particularly strong (Clark and Whiteley, 1977; Pattinson and Whiteley, 1984), but assessed colour of the greasy sample is still the basis on which most wool is traded. The discounts that apply to discoloured wool (Table 17.2) are moderate in size and slightly greater for wools of finer diameter. An objective measure of scoured-wool colour based on light-reflectance properties is available (Clark and Whiteley, 1978) and is becoming more frequently used presale. Market signals based on measured colour may give a more powerful stimulus for breeding programmes in the future.

Fibre-diameter variability

Although diameter variability is not currently measured at sale, its economic value for processing can be directly inferred from its mathematical equivalence with average fibre diameter for spinning (1 μm average diameter equates to 5% coefficient of variation of diameter). On this basis, it is included in Table 17.2 for comparison with the other quality traits.

Other traits

There are a number of other traits that may have some impact on processing, but either their relationship with raw-wool traits is unclear or market signals are impossible to gauge. For example, variability in staple strength and variability in staple length have been shown to have some impact on late-stage processing, but the level of variation within the yarn is more likely to be determined by factors influencing fibre breakage in carding and combing than by inherent variability in the fleece (Bell, 1987).

Resistance to compression, which is highly related to crimp frequency (Watson *et al.*, 1977), is of contentious value to processors of apparel wool. It would seem to be associated with less even and weaker yarn (Hunter *et al.*, 1984) and higher levels of waste in top-making (Kurdo *et al.*, 1986). If measures of resistance to compression were available at sale, it might attract a negative economic value. Traits such as resistance to compression, the components of style and diameter variability will only receive clear market signals when they are included in sale catalogues. Without such market information, there is always a danger of overemphasizing their individual importance in breeding programmes.

Suggested traits such as fibre circularity or alignment of scales are at present of such little theoretical and practical significance that they do not warrant serious attention.

In summary, the relative economic value of fibre diameter dominates that of other wool-quality traits. Of the remainder, staple strength in wools of less than 30–35 N $ktex^{-1}$ and diameter variability are the next most important traits. However, it bears repeating that the value placed on reduced diameter variability is driven solely by its relationship with spinning performance through average fibre diameter and is not currently reflected in true market value. The remaining traits – VM, staple length, style and colour – are traits of relatively minor economic importance.

Genetic Parameters for Wool Traits

To make effective genetic progress, we need estimates of the level of phenotypic variation, heritabilities and correlations, both phenotypic and genetic, among the traits we wish to improve. The task of the geneticist is to combine such estimates into selection strategies that will give a precise prediction of likely genetic progress and provide a basis for genetically evaluating animals as potential parents. For the latter purpose, adjustments for identifiable sources of environmental factors (such as birth type, age of animal or age of dam) are also required to improve the genetic model.

Numerous reports and reviews are available which detail estimates of these parameters from research flocks. The approach taken here has been to estimate these genetic parameters from an analysis of published estimates from Australian Merino populations. The data were taken from published reports and a weighting factor (the inverse of the estimated sampling variance) used in a weighted analysis of variance. The models fitted to these data included factors such as:

Age	Yearling (to 12 months of age)
	Hogget (13–20 months of age)
	Adult
Strain	Fine, medium, broad
Method of analysis	Dam/offspring
	Paternal half-sib
	Animal-model restricted maximum likelihood (REML)

Table 17.3 contains weighted-average estimates of phenotypic standard deviation and environmental effects while Table 17.4 shows the estimates of heritability, genetic correlations and phenotypic correlations among selected traits of importance. The sources of the information included in the analyses are shown at the foot of each table. Strain effects on parameters were generally of little significance and so the estimates presented here have been combined across strains.

Fleece weight

The greasy fleece, as shorn, contains wax, suint, dust, VM and other impurities, as well as fibre. However, the genetic correlation between greasy- and clean-fleece weight is very high and the heritabilities of both traits are little different from each other.

Age at shearing has a significant impact on heritability, increasing from about 0.35 in yearling animals up to about 0.45 in adults. These heritabilities have been estimated generally after fitting fixed environmental effects, such as birth type, birth date, age of dam, etc. These environmental factors diminish in importance with age (Table 17.3), suggesting that the heritability of yearling fleece weight, in particular, would be even lower if such effects were ignored. The genetic correlations between fleece weight at the various ages are very high but not unity. The impact of these age-related genetic parameters has been considered in some detail by Atkins (1990), showing that repeat information on fleece weight can add significantly to selection accuracy. For example, including a second test shearing at 16–18 months of age for selected animals following a yearling fleece weight can lead to substantial genetic benefits over a single-stage yearling selection policy and marginal benefits over a single-stage hogget-selection policy.

Maternal effects have only recently been investigated. The reports of Mortimer and Atkins (1994), Hickson *et al.* (1995a) and Torshizi *et al.* (1995) all indicate that small but significant maternal genetic effects probably exist in Merinos at both yearling and hogget ages. The consistently higher estimates of heritability for fleece weight derived from dam/offspring compared with both REML and paternal half-sib analyses also point to the existence of maternal genetic effects.

Genetic correlations between fleece weight and quality traits are generally modest. The most important correlation is the low to moderate positive

Table 17.3. Phenotypic variation in and environmental effects on measures of wool production and wool quality.

Trait	Code	Coefficient of variation (cv) (%)	Birth type (single–twin %)	Age of dam (adult–maiden %)	Age (30-day difference %)
Greasy-fleece weight (gfw)	y-gfw	14	+7%	+3%	+9%
	h-gfw	14	+5%	+2%	+4%
	a-gfw	14			
Clean-fleece weight (cfw)	y-cfw	15	+9%	+2%	+5%
	h-cfw	15	+5%	+1%	+4%
	a-cfw	16			
Fibre diameter (fd)	y-fd	7	–1%	+2%	+3%
	h-fd	7	–1%	+1%	+1%
	a-fd	8			
Yield	yie	7	0%	0%	0%
Vegetable matter	vm	75	+11%	+4%	+9%
Staple strength	str	30	–	–	–
Staple length	len	10	0%	0%	+1%
Colour	col	(score*)			
Cv of fibre diameter	cv	12	–3%	–1%	–
Dust (style)	dust	25	0%	+1%	+1%
Crimp definition (style)	char	(score*)			

*Higher scores are associated with better colour or better crimp definition.

y-, yearling (8–12 months of age); h-, hogget (13–20 months of age); a-, adult (> 20 months of age).

Sources: Morley (1955, 1956); Young *et al.* (1960); Beattie (1961); Pattie (1965); Lax and Brown (1967); Mullaney *et al.* (1969); Watson *et al.* (1977); Gregory and Ponzoni (1981); Walkley *et al.* (1987); Mortimer and Atkins (1989, 1993); Atkins (1990); James *et al.* (1990); Lewer *et al.* (1992); Taylor and Atkins (1992); Greeff *et al.* (1995); Ponzoni *et al.* (1995); Rogan *et al.* (1995); Swan *et al.* (1995).

Table 17.4. Heritabilities (on diagonal), genetic correlations (below diagonal) and phenotypic correlations (above diagonal) among wool-production and wool-quality traits.

	y-gfw	h-gfw	a-gfw	y-cfw	h-cfw	a-cfw	y-fd	h-fd	a-fd	yie	str	len	col	cv	dust	char
y-gfw	**0.36**	0.55	0.40	0.85	0.45	0.35	0.25	0.15	0.20	–0.05	0.10	0.30	–0.15	0.00	–0.05	0.00
h-gfw	0.70	**0.40**	0.60	0.45	0.85	0.50	0.15	0.25	0.20	–0.05	0.10	0.30	–0.15	0.00	–0.05	0.00
a-gfw	0.55	0.75	**0.45**	0.35	0.50	0.85	0.20	0.20	0.30	–0.05	0.10	0.25	–0.15	0.00	–0.05	0.00
y-cfw	0.80	0.55	0.45	**0.34**	0.55	0.40	0.25	0.15	0.20	0.40	0.10	0.40	0.00	0.00	0.00	0.00
h-cfw	0.55	0.80	0.60	0.70	**0.38**	0.60	0.15	0.25	0.20	0.40	0.10	0.40	0.00	0.00	0.00	0.00
a-cfw	0.45	0.60	0.80	0.55	0.75	**0.43**	0.20	0.20	0.30	0.40	0.10	0.35	0.00	0.00	0.00	0.00
y-fd	0.15	0.20	0.30	0.15	0.20	0.30	**0.45**	0.70	0.60	0.00	0.20	0.15	–0.10	–0.10	0.05	–0.10
h-fd	0.15	0.20	0.30	0.15	0.20	0.30	0.90	**0.50**	0.80	0.00	0.25	0.15	–0.10	–0.10	0.05	–0.10
a-fd	0.15	0.20	0.30	0.15	0.20	0.30	0.80	0.90	**0.50**	0.00	0.20	0.10	–0.10	–0.10	0.05	–0.10
yie	–0.25	–0.15	–0.15	0.25	0.30	0.30	0.00	0.00	0.00	**0.50**	0.20	0.25	0.35	–0.05	0.05	0.05
str	–0.15	–0.15	–0.15	0.10	0.10	0.10	0.10	0.30	0.30	0.20	**0.30**	0.05	0.00*	–0.40	0.00*	0.00*
len	0.20	0.15	0.10	0.35	0.30	0.20	–0.10	0.10	0.10	0.25	–0.05	**0.40**	0.15	–0.15	0.30	0.05
col	–0.10	–0.10	–0.10	0.15	0.15	0.15	–0.15	–0.15	–0.15	0.55	0.00*	0.00	**0.35**	0.00	–0.10	0.10
cv	0.20	0.10	0.10	0.20	0.10	0.10	–0.05	–0.10	–0.15	–0.05	–0.50	–0.05	–0.10	**0.40**	0.00	–0.20
dust	0.10	0.10	0.10	0.25	0.25	0.25	0.15	0.15	0.15	0.10	0.00*	0.55	–0.05	–0.15	**0.20**	–0.05
char	0.00	0.00	0.00	0.10	0.10	0.10	–0.10	–0.10	–0.10	0.15	0.00*	0.00	0.20	–0.15	0.05	**0.40**

*No information available, assumed zero.

For code, see Table 17.3. Sources as for Table 17.3, plus: Beattie (1962); Brown and Turner (1968); Mullaney *et al.* (1970); Gregory (1982a,b); Purvis and Piper (1992); Lewer *et al.* (1994); Hickson *et al.* (1995a); Torshizi *et al.* (1995).

correlation with fibre diameter. This antagonistic relationship is expected to result in an increase in diameter with selection on fleece weight alone, but the modest magnitude of the correlation still allows fibre diameter to be maintained at or reduced from its current level, while allowing genetic improvements in fleece weight. Selection for either greasy- or clean-fleece weight is likely to result in a desirable increase in staple length and a small undesirable increase in the coefficient of variation of the diameter. Selection for greasy-fleece weight is likely to lead to a slight reduction in both yield and staple strength, while selection for clean-fleece weight might be expected to result in modest increases in yield, a slight improvement in staple strength and an undesirable increase in dust penetration and deterioration in crimp definition.

Yield

Although the production environment will directly influence the content of the non-wool components of the fleece, the heritability of yield within a flock is very high (0.5). Selection for higher yield is likely to lead to modest improvements in clean-fleece weight, staple length, staple strength and whiteness of wool. Despite the favourable relationships with economically important traits, many breeders are reluctant to allow yield to increase, fearing that high yields will leave the animal exposed to weathering damage and increased likelihood of dust penetration. The genetic correlations (Table 17.4) would suggest that such fears are probably unfounded and that an increase in yield would be a desirable outcome.

Average fibre diameter

The heritability of average fibre diameter is high. Age effects on heritability are small and the genetic correlations between measures of diameter at different ages are closer to unity than those for fleece weight (Table 17.4). The pattern of a high, age-independent heritability is further supported by the relatively small influence of specific environmental effects and the almost total lack of evidence for maternal genetic effects (Mortimer and Atkins, 1994; Hickson *et al.*, 1995a; Torshizi *et al.*, 1995). Although the genetic correlations among ages are very high, Atkins (1990) and Hickson *et al.* (1995b) showed that there was genetic variation among animals in their fibre-diameter response with age. This trait, termed diameter stability, has a heritability of 0.2–0.3 (depending on the age interval involved), but appears poorly correlated with other production traits or potential indirect selection criteria (Cottle *et al.*, 1995; Hickson *et al.*, 1995b).

Apart from the moderately unfavourable genetic correlation with fleece weight, selection to reduce fibre diameter is likely to lead to a modest reduction in staple strength and a slight increase in the coefficient of variation of the diameter. However, reduced fibre diameter would lead to improved wool colour and lowered dust penetration. Although not shown in Table 17.4, the strongest relationship with fibre diameter, both phenotypically and genetically, is that between reduced fibre diameter and increased follicle density (–0.6 to –0.7).

Vegetable matter

The only report of the heritability of VM was an estimate of 0.06 ± 0.04 by Mortimer and Atkins (1993), who concluded that selection for reduced VM would be an unrewarding avenue to pursue.

Staple strength

Staple strength appears to be moderately heritable (Table 17.4), but, when combined with its high phenotypic variance (Table 17.3), it becomes a very attractive trait for genetic modification. Direct measurement of staple strength is relatively expensive, but Lewer and Li (1994) showed that the coefficient of variation of the diameter was a strongly correlated indirect measure of staple strength, which could be applied at a fraction of the cost. A subsequent report by Greeff *et al.* (1995), from a wide range of Merino populations, has confirmed the initial estimates of Lewer and Li (1994), such that indirect selection on diameter variability is likely to be 60% as effective as direct selection. The reservations of Ponzoni *et al.* (1990) that staple strength in young animals may not be exactly the same trait as predisposition to tender wool in adult ewes still highlight the need for further estimates of genetic parameters across age classes of animals, particularly in the presence of the physiological stress of reproduction. Equally, we need to ensure that increasing the commonly high levels of staple strength in young stud rams will improve the staple strength of their progeny under less favourable commercial environments.

A moderately positive genetic correlation between staple strength and fibre diameter suggests that retaining strength while reducing diameter or preventing upward drift in fibre diameter while selecting for increased staple strength may need particular attention.

Staple length

The small and non-linear economic value of staple length is unlikely to lead to a strong incentive for including staple length as a trait in its own right within an economic breeding objective for Merinos. The heritability of the trait is high and genetic correlations with fleece weight would indicate that staple length will increase as a correlated response to selection. Genetic correlations between staple length and other wool quality measures are generally close to zero with the exception of the strong positive (undesirable) correlation with dust penetration.

Colour

Despite the only moderate relationship between greasy and scoured colour mentioned earlier among sale lots, the only genetic study of the two traits reported a very high genetic relationship (James *et al.*, 1990). Further information on this relationship and associations between wool colour and production traits are required before we can have sufficient confidence to include either greasy or scoured colour into breeding programmes. The genetic parameters reported in Table 17.4 are for visually scored greasy colour. A moderate heritability of greasy colour and desirable associations with increased clean-fleece weight (largely through increased yield) and reduced fibre diameter are indicated by the parameters.

Style

At present, information on genetic parameters for the components of style come from subjective scores. The most detailed and complete information of this type was recently reported by Lax *et al.* (1995), who foreshadowed the genetic analysis of objective measurements of style and relationships between young- and adult-animal assessments. The study of Lax *et al.* (1995) reported simultaneous assessment of five style components, each with a moderate heritability (0.2–0.35) and low to moderate positive correlations among the components. In Table 17.4, estimated parameters from all sources have been presented for dust penetration, crimp definition and greasy colour only. With such parameters available and the prospect of even more precise estimates shortly, the major requirement for breeding programmes is better information on the likely economic impact for each of the style components.

Fibre-diameter variability

Variation in diameter can be expressed as a standard deviation, coefficient of variation or any number of proportions of the distribution (deciles, quartiles, etc.), but the coefficient of variation is the expression most commonly used. The coefficient of variation of diameter is a trait of moderate heritability, having low genetic correlations with other traits, apart from the high genetic and phenotypic association with staple strength.

Realized Responses to Selection

Estimated genetic parameters have given us a relatively precise prediction for expected responses to selection. Strictly, though, such estimates are population-specific and are relevant to one generation of selection, despite their frequent use across a wide range of populations and for many rounds of selection. Selection flocks, where selection has been based on a single trait or a restricted range of traits, have been used widely in the Merino. The major aim of such flocks was to check the prediction that selection response would be effective (Morley, 1951), but important secondary aims were to improve the understanding of wool and skin biology (McGuirk, 1980) and to investigate the physiological consequences of selection on production traits.

Consequences of selection for fleece weight

Direct responses to selection

McGuirk (1980) and Rogan (1984) reviewed the responses to selection in a number of experiments where fleece weight was either the sole selection criterion or the dominant selection criterion in a multitrait objective. Based on a heritability of 0.4 for hogget clean fleece weight, the expected responses in all flocks were between 1.0 and 2.0% per year, accounting for variation in selection intensity (through population size and structure) and the number of traits in the objective. The achieved responses to selection across experiments (Table 17.5) varied considerably, with more rapid gains achieved in the single character selection lines (experiments 1 and 2) than in most of the multitrait lines (experiments 3–6). While the selection responses might be described as qualitatively in agreement with prediction, marked variation across years within experiments was frequently encountered. This can be seen in the variation between estimates for various time periods in Table 17.5.

A clearer demonstration of this variability can be seen in the pattern of response from the divergent single-character selection lines for fleece weight (experiment 1 in Table 17.5), which have been maintained by NSW Agriculture at Trangie for more than ten generations of selection, as shown in Fig. 17.3. These lines were selected for high and low clean fleece weight measured at about 15 months of age and the selection responses measured as deviations from a randomly selected control flock (described by Pattie and Barlow, 1974). The pattern of response was very consistent in the downward direction at a rate of 2.1% per year. However, responses in the upward direction have been lower and more variable, with an average response of little more than 1% per year. After accounting for linkage disequilibrium, inbreeding and actual selection intensities, Atkins (1988) reported that the response realized in the downward selection line was consistent with a base population heritability of 0.49, while in the upward selection line the realized responses were equivalent to a base population heritability of only 0.24.

At least part of this uneven pattern of response across years can be attributed to genotype × environment interactions or, specifically, to a genotype × level-of-nutrition interaction. A number of studies have recorded the wool-growth rates of sheep from the high and low selected lines in comparison with their control line under defined levels of nutrition in pens (Ahmed *et al.*, 1963; Williams and Winston, 1965; Williams, 1966; Saville and Robards, 1972; Robards *et al.*, 1974). These levels of nutrition have varied from

Table 17.5. Responses in selection experiments where selection has been solely or principally for fleece weight.

Experiment (organization)	Selection criteria	Annual flock size (breeding ewes)	Period of response	Annual response in fleece weight (%)	Reference
1 (NSWAg)	High fleece weight	100	1952–81	+1.2	Rogan (1984)
	Low fleece weight	100	1952–81	−2.1	Atkins (1988)
2 (CSIRO)	High fleece weight	32–50	1954–76	+1.6	McGuirk (1983)
	Low fleece weight	32–50	1954–76	−3.1	Davis and McGuirk (1987)
3 (NSWAg)	High fleece weight, maintaining crimp frequency, skin wrinkle and face cover	200	1947–67	+1.0	McGuirk (1983)
4 (CSIRO)	High fleece weight, maintaining fibre diameter and skin wrinkle	200–600	1950–74	+1.1	McGuirk (1983)
			(1950–59)	(+2.2)	Turner *et al.* (1968)
			(1966–74)	(+2.6)	Turner and Jackson (1978)
5 (CSIRO)	High fleece weight, maintaining crimp frequency and skin wrinkle	100–250	1950–74	+0.9	McGuirk (1983)
			(1950–59)	(+2.2)	Turner *et al.* (1968)
			(1966–74)	(+1.1)	Turner and Jackson (1978)
6 (Roseworthy)	High fleece weight, after visual culling	200	1953–76		Mann *et al.* (1980)
			(1968–76)	(+2.7)	

NSWAg, New South Wales Agriculture, Trangie Agricultural Research Centre, New South Wales; CSIRO, Commonwealth Scientific and Industrial Research Organisation, Cunnamulla, Queensland and Armidale, New South Wales; Roseworthy, Roseworthy Agricultural College, South Australia.

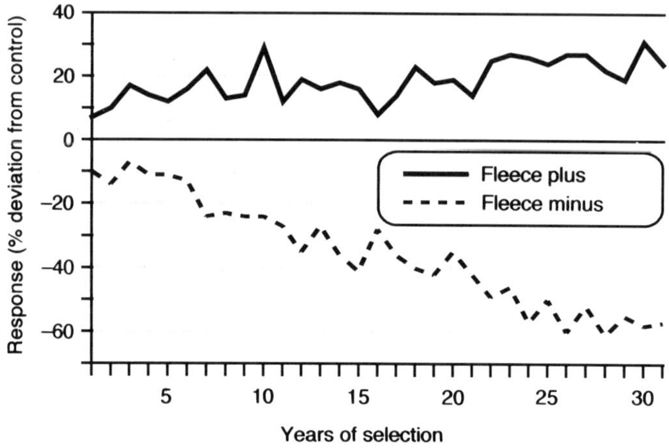

Fig. 17.3. Response to selection for high and low clean-fleece weight.

below maintenance to *ad libitum* with varying diet quality. Figure 17.4 shows the daily wool growth rates of the selected lines in comparison with the wool-growth rates of the randomly selected control line from a number of these animal-house experiments. As nutritional level increased, assessed by the production level of the control line, the differences between the selected lines also increased in both absolute and relative terms. Williams *et al.* (1972) showed

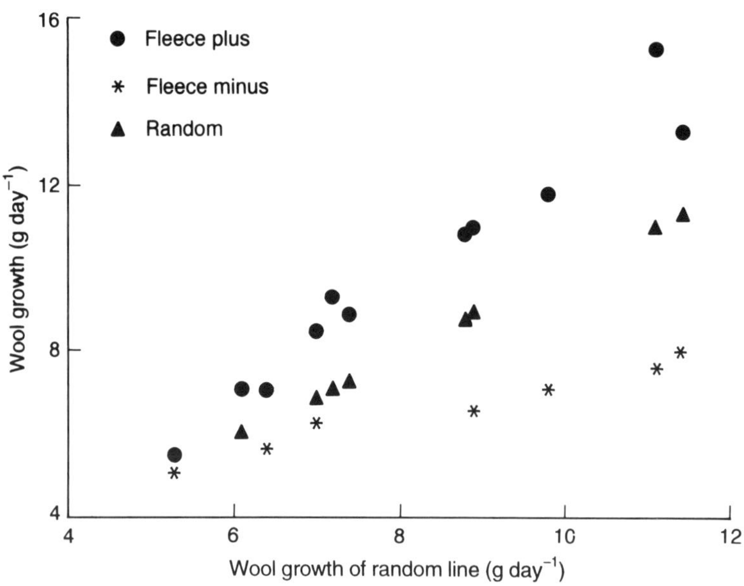

Fig. 17.4. Wool growth of high- and low-fleece-weight lines in response to level of nutrition.

Table 17.6. Response of high- and low-fleece-weight lines to amino acid supplementation (from Williams *et al.*, 1972).

Selection line	Treatment	Wool growth (g cm^{-2} day^{-1})	Response (%)
High fleece weight	Standard diet	743	–
	+ cystine	1114	+ 50
	+ methionine	1190	+ 60
Low fleece weight	Standard diet	639	–
	+ cystine	715	+ 12
	+ methionine	757	+ 18

that this genotype × nutritional-level interaction is probably related to the availability of sulphur-containing amino acids reaching the abomasum. Their results from the effect of abomasal infusion of either cystine or methionine on wool growth in the high and low selected lines (Table 17.6) show the much greater sensitivity of the high-fleece-weight line to supplementation.

Thus, selection of the Merino for increased fleece weight has produced an animal that will selectively respond to increased levels of nutrition, reflecting the higher availability of sulphur-containing amino acids reaching the small intestine. Indeed, Atkins (1988) demonstrated higher estimates of annual response in good nutrition years for both the high- and low-fleece-weight lines. However, variation between seasons within a year will also have a large impact on the superiority of the selected lines in a particular year.

Correlated responses to selection

Realized correlated changes to selection on fleece weight in a range of wool and skin traits have been reported by Barlow (1974) and Turner *et al.* (1970). The observed responses were generally consistent with expectations based on estimated genetic correlations. Response in fleece weight arose primarily from response in wool production per unit area of skin, reflecting positive genetic relationships between clean-fleece weight and yield, fibre density, staple length and fibre diameter. Responses in skin surface area, through either body size or skin wrinkling, were generally small and inconsistent. One feature of the correlated responses in the components of fleece weight noted by Barlow (1974) was the strong negative association between response in follicle density and fibre diameter within selected flocks. Large positive responses in follicle density within a year were associated with large negative responses in fibre diameter and vice versa. This would indicate that, since follicle density is largely determined *in utero* and in early life, suppression of genetic potential for density will be compensated for by later increases in fibre cross-sectional area. Alternatively, a favourable environment for follicle development will tend to suppress the potential for later expression of high fibre volume.

These observations on the correlated responses to selection in the components of fleece weight only partially fit our expectation. By reversing the process and observing responses in fleece weight to selection on the individual components, Davis and McGuirk (1987) concluded that selection on each component was not effective in increasing fleece weight. It would appear that selection for one component resulted in compensatory changes in other components, such that clean-fleece weight did not respond. For example, selection for high follicle density resulted in reduced fibre diameter and selection for more skin wrinkle resulted in shorter fibre length, with no response in clean-fleece weight in either case (Turner *et al.*, 1970). This apparent repartitioning of the contribution of the components, with no net increase in fleece weight, may be due to non-linearity in the genetic relationships among the components (Atkins, 1988). Whatever the explanation, there remains some doubt over the robustness of the genetic model involving the components of total fleece weight.

Correlated responses among wool quality traits have commonly confirmed expectations. The most significant responses were the correlated increases in fibre diameter, which resulted in an average 2 μm difference between the 'fleece plus' and its 'control' line, staple length and yield (Barlow, 1974). Slight increases were also observed by Taylor (1993) in crimp definition and staple strength, which were in line with expectations. However, the latter author noted no substantial correlated response in colour, dust penetration or diameter variability between the 'fleece plus' and 'control' animals.

Effects on food intake and efficiency of wool production

Feed is often limiting under pastoral conditions, so that the impact of selection for fleece weight on efficiency of production has been extensively studied. Comparisons of lines selected for high fleece weight, with either control or low-fleece-weight lines, have indicated little or no differences, either in voluntary food intake in pens (Williams, 1979) or in digestive function (Piper and Dolling, 1969). Williams (1979) also argued that selection for fleece weight had not resulted in any noticeable change in energy requirements either to maintain or to change live weight. Thus, the lack of correlated response in live weight to selection on fleece weight (Barlow, 1974) implies little if any change in food intake to selection on fleece weight, at least in pens.

Since differences in intake between selected lines are small, most of the genetic response in fleece weight has been associated with genetic differences in efficiency of conversion of feed to wool. Robards *et al.* (1974) summarized the results of pen-efficiency studies involving the Trangie high-fleece-weight line and its control. Efficiency of conversion of feed to wool was measured at an intake that maintains live weight, termed net efficiency. The results indicate that the selected line had a superiority over the control line of similar magnitude for both wool growth and net efficiency of wool growth. In addition, there was an apparent improvement with time in efficiency as a correlated response to selection.

Information on food intake and efficiency of wool production of sheep at pasture is more limited. Hamilton and Langlands (1969) found a difference of 10–15% in pasture intake between sheep selected for high and low fleece weight.

They also reported that the 'fleece plus' animals grew more than twice as much wool as the 'fleece minus' in the same experiment, so that the high-fleece-weight animals were still substantially more efficient in their utilization of pasture for wool growth. Support has been reported by Lee *et al.* (1995), who showed a positive genetic correlation between intake and wool growth in grazing adult ewes. The genetic correlation between wool growth and efficiency of wool growth was still high in their study, but less than unity. It is possible that selection for increased fleece weight has influenced grazing behaviour, allowing higher pasture intake and intake per unit live weight (Lee *et al.*, 1995).

Biochemical and physiological consequences of selection

Selection for increased fleece weight has not only changed the amount of wool produced but also resulted in a number of changes in the chemical and physical properties of the wool and morphological changes in skin tissue.

Selection for increased fleece weight has led to a lower sulphur content of wool (Piper and Dolling, 1966), due to a lower proportion of high-sulphur proteins (Campbell *et al.*, 1972). Apart from leading directly to a reduction in resistance to compression, the lower sulphur content may alter the ability of such wools to take and retain set, although chemical treatment apparently allows this feature to be overcome during processing (McGuirk, 1980).

Visual differences between high- and low-fleece-weight lines in the pattern of follicle arrangement in the skin have been described by Nay (1966, 1970). The follicles of sheep from high-fleece-weight lines tend to be deeper in the skin and straighter and to have a better developed vascular supply. Jackson *et al.* (1975) showed that the follicle characteristics of density, depth and degree of curvature together accounted for 83% of the genetic variation in fleece weight. The implications of these observations are not obvious, given that selection for increased follicle depth was as ineffective in increasing fleece weight as selection for the individual components of fleece weight, including density (Davis and McGuirk, 1987). However, these results demonstrate that structural differences in the skin are apparently associated with the effectiveness and capacity of the skin and its follicles to produce fibre. Further discussion on the possible interaction of structural form and physiological function with genetic capacity for wool growth can be found in Williams (1987) and Black (1987).

Consequences of selection for wool quality

In contrast to the wealth of knowledge on realized responses to selection for fleece weight, selection lines for wool-quality traits of economic importance are relatively few. Companion pairs of flocks to the fleece weight selection lines of experiment 2 in Table 17.5 were those selected for high and low fibre diameter, high and low staple length and high and low yield, as reported by Turner *et al.* (1970). Direct and correlated responses to selection in these lines and those selected for high and low clean fleece weight over the period 1954–1964 are shown in Table 17.7 (from Turner *et al.*, 1970). Also included in Table 17.7 are

Table 17.7. Direct and correlated responses to selection, expressed as percentage deviations from the control line (from Turner *et al.*, 1970).

Selection trait	Line	Greasy-fleece weight	Clean-fleece weight	Follicle density	Fibre diameter	Staple length	Yield
Clean-fleece	High	+15.2	+18.2	+13.7	−1.0	+7.0	+2.7
weight	Low	−14.9	−14.6	+0.4	−3.1	−4.6	−0.1
Follicle density	High	−1.5	−0.4	+31.0	−6.1	−4.5	+1.2
	Low	−2.3	−7.3	−15.0	+5.4	−3.3	−4.8
Fibre diameter	High	+2.2	+1.1	−14.4	+13.4	−3.3	−1.0
	Low	−6.3	−5.4	+35.5	−13.6	−3.5	+1.0
Staple length	High	+2.7	+5.4	+16.3	−6.0	+12.3	+2.8
	Low	−5.4	−16.1	+4.6	−0.2	−20.1	−11.4
Yield	High	−4.2	+6.0	+7.9	−1.0	+6.7	+10.6
	Low	+7.4	−6.3	−1.1	+1.9	−7.1	−12.4
Significant difference ($P = 0.05$)		2.4	2.7	3.3	1.4	2.0	1.4

the responses in the lines selected for high and low follicle density, which was referred to earlier as the character having the strongest genetic correlation with fibre diameter.

Despite the very small size of these selection lines, the results do provide some confirmation of the estimated genetic parameters:

- Selection on fibre diameter showed substantial change in diameter, a positive genetic correlation with fleece weight and little change in either yield or staple length.
- Selection on staple length showed greater response in length in the downward direction, positive genetic correlations with yield and clean-fleece weight and some ambiguity with respect to diameter changes.
- Selection on yield produced substantial responses in yield, a negative genetic correlation with greasy-fleece weight and positive genetic correlations with both clean-fleece weight and staple length.

The results in Table 17.7, particularly those involving follicle density and, to a lesser extent, staple length, provide some evidence of the non-linearity among the components of fleece weight alluded to earlier. All responses in density demonstrate asymmetry and dictate caution in using measures of follicle density as indirect traits for selection. Selection lines established later by the Commonwealth Scientific and Industrial Research Organisation (CSIRO) for primary-follicle density and secondary/primary follicle ratio (Rendel and Nay, 1978) and another for total follicle number (Davis and McGuirk, 1987) all resulted in a slight to significant reduction in fleece weight.

Breeding Programmes for Improved Fleece Value

Potential genetic gains

We have seen that the two dominant traits in a breeding objective for improved fleece value will be the weight of wool produced and the fibre diameter of that wool. The actual value to be attached to fibre diameter relative to fleece weight can vary enormously, according to the type of wool produced and the specific market period of interest. Here we wish to examine the likely genetic gains possible for a range of scenarios and investigate the impact on other traits of economic importance.

Firstly, a selection index has been shown to be the most efficient means of combining traits to maximize response. For an index, we need to derive a breeding objective; the process involved is described in detail by Ponzoni (1986). This requires some basic assumptions on production levels, apart from knowledge of genetic and economic parameters. Secondly, in order to make an effective prediction of potential rates of gain, we need to make some assumptions on flock size and structure. None of the assumptions involved are too critical to the outcome, but the following values have been assumed.

Flock size	1000 breeding ewes	
Net reproduction rate	80%	
Annual death rate	5%	
	Males	*Females*
Age-groups	1	4
Age at first lambing	2 years	2 years
Mating ratio	1 ram:40 ewes	
Fleece weight	4 kg clean	
Fibre diameter	21 μm	

Simple objective: fleece weight + diameter

Expected selection responses over 10 years are shown in Fig. 17.5, with all selection emphasis placed on an index of clean-fleece weight and fibre diameter (both measured at hogget age) over a range of MP from 0% (zero economic value for fibre diameter) up to 20%. Genetic responses in the component traits are particularly sensitive to changes in MP over a relatively narrow range – from 2–3%, where fibre diameter would be maintained at its current level and most improvement achieved through fleece weight, to 10% MP or more, where most improvement would come from reduced fibre diameter.

The remaining wool quality traits were included in the model, but with zero economic values, to monitor expected changes to selection on fleece weight and diameter. The responses in the quality traits over the range of MP were:

Yield	Change of between +2.6% (0% MP) to +1.2% units (20% MP)
Strength	Responses (Fig. 17.5) show a slight positive response at very low MP to a decline of up to 4 N $ktex^{-1}$ at

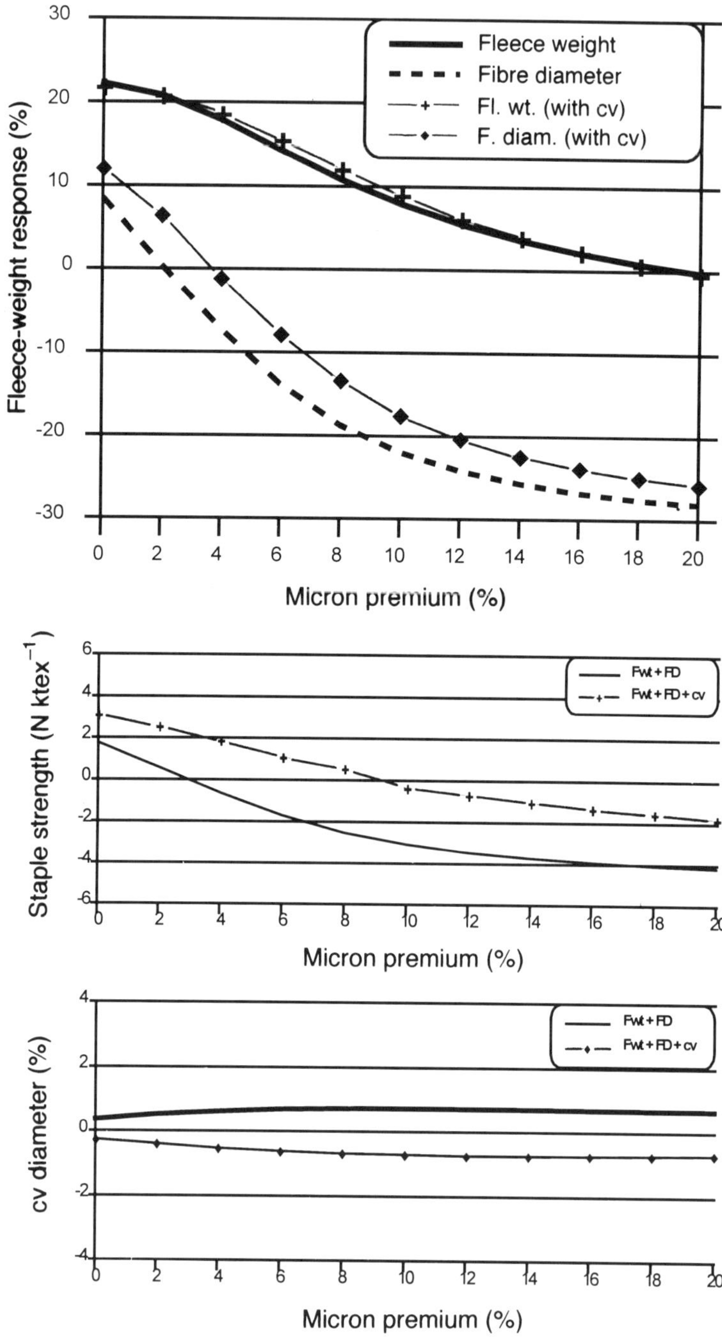

Fig. 17.5. Potential rates of response. cv, coefficient of variation.

	high MP
Length	Positive correlated changes of between 4.6 mm (0% MP) and 0.3 mm (20% MP)
Colour	Consistent small improvement in colour of 0.1–0.2 of a subjective grade
Crimp definition	Consistent small improvement in crimp definition of about 0.1 of a subjective grade
Dust penetration	Small increases in dust penetration at low MP ranging to a small decline in dust penetration above 10% MP
Diameter variability	Coefficient of variation of diameter showed an increase of between 0.5 and 1.0% units over the full range of MP (Fig. 17.5)

In summary, then, selection for fleece weight and fibre diameter would be expected to make a positive contribution to improved staple length, wool colour and crimp definition, but with some significant decline in staple strength (at a high emphasis on fibre diameter) and increase in dust penetration (at a low emphasis on diameter).

Expanded objective: fleece weight + diameter + wool quality

In the second model, the additional wool quality traits were assigned an economic value in keeping with the maximum values from Table 17.2 for medium wools. After fibre diameter, staple strength and diameter variability were the two most important wool-quality traits and so the coefficient of variation of the diameter was included as a selection trait (together with fleece weight and average diameter) and the potential responses re-estimated (Fig. 17.5). The effect of the expanded objective and expanded selection traits was relatively minor for the dominant traits in the objective but had a demonstrable impact on those traits of secondary importance. Specifically, the impact on potential gains was:

- little change in fleece weight response;
- reduced response in fibre diameter by up to 0.5 μm over 10 years;
- greater response (or lesser reduction) in staple strength, by about 2 N $ktex^{-1}$, although staple strength was still predicted to decline in indexes above 10% MP;
- consistent reduction in coefficient of variation of diameter across all MP.

We can conclude that the breeding programme can be adjusted to incorporate additional traits as required. Of course, the appropriate economic value for all traits in the objective will depend on current production levels and assessments by individual breeders of future market conditions. Our genetic prediction, though, has been shown to be robust for a range of traits under a wide range of market scenarios.

Breeding in practice

Given the potential genetic gains that can be made in fleece value, it is reasonable to ask how much of this potential is currently being captured in the industry's breeding programmes. It is not possible to provide an answer for the whole Merino industry, but information derived from about 50 flocks that account for 10% of the Australian stud Merino flock does provide some indication (Atkins and Casey, 1994). These flocks do not represent a random sample of studs but give an indication of the genetic gains likely to be achieved in an important, and probably representative, sector of the breeding industry.

Breeding objectives

Personalized breeding objectives have been developed for many of these ram breeding flocks, using the OBJECT software described by Atkins *et al.* (1994). The average objective corresponds with an MP of 5.5%, with a range from 2% (fibre diameter maintained and most emphasis on fleece weight) to over 10% (major emphasis on reduced fibre diameter and some emphasis on fleece weight). Although most of the personalized objectives lie in a relatively narrow, but genetically sensitive, band of MP, the process of establishing a personalized breeding objective has a number of inherent advantages for the subsequent operation of the breeding programme (Atkins and Casey, 1994). These include the cushioning of market volatility, examining the effect of alternative indexes, understanding the impact of the objective on potential selection candidates and recognizing when the objective might need to be altered.

Selection traits

A range of traits recorded and ages of measurement among breeding programmes make generalizations difficult. However, measured clean-fleece weight and mean and variability of diameter among hogget rams and measured greasy-fleece weight with mean and variability of diameter among hogget ewes gave the closest approximation to the achieved correlation between the average index and objective.

Selection efficiency

In Table 17.8, the sample of industry flocks recorded are assessed in terms of their potential selection intensity (numbers selected of total numbers available), effective selection intensity (numbers selected of total numbers measured), generation interval and selection emphasis on the index, separately for males and females. Selection intensities among females were both lower and more variable than those among males. Female selection intensity was substantially higher in flocks with an open-nucleus structure of two or more tiers of females, leading to a higher variability among flocks. The loss of selection intensity before measurement was relatively greater in females, and the difference between potential and effective intensities represented a reduction in available numbers of about 20% in both males and females. While fleece weight and fibre diameter are the dominant traits in the breeding objectives, other traits, such as body weight, reproduction rate, disease resistance and additional wool-quality traits, can be of direct economic importance and are often included in a formal breeding objective (Ponzoni, 1986). There are further traits, such as those related to conformation and protection from the environment, that are visually assessed and account for some of the available selection differential, even when they may have no correlation with performance (Lewer and MacLeod, 1990). An indirect measure of the impact of these additional traits is the selection emphasis, defined as the percentage of the total selection differential applied to the measured traits in the index of a nominated objective. Predicted selection emphasis (Table 17.7) indicated that only 59% of the effective selection intensity available for males was used on index traits and even less among females.

Likely genetic gains

Using the information above and our knowledge of genetic parameters results in the predicted gains within commercial breeding programmes relative to the potential gains nominated previously (Fig. 17.6). Less than 50% of the potential gains in both fleece weight and fibre diameter would be achieved, on average, in these commercial breeding programmes, although the variation across flocks is substantial (Table 17.8). With less than 60% of the available emphasis for selection being applied to measured traits in males and even less in females, there are clear deficiencies in the effectiveness of selection. Further, the average male-generation interval is unreasonably long and apparently not supported by effective across-age selection among sires (Atkins and Casey, 1994). Finally, selection among females appears to be inefficiently exploited, in terms of both selection intensity and accuracy.

Concluding Remarks

The general principles followed in this chapter are those broadly used in all animal breeding applications.

1. Identifying the components of the breeding objective for improved value of

Table 17.8. Measures of selection intensity and accuracy among Merino breeding programmes, expressed as mean and coefficient of variation (cv).

	Male		Female	
	Mean	cv (%)	Mean	cv (%)
Selection intensity (standardized selection differential)				
Potential	2.47	7	0.98	23
Effective	2.32	13	0.79	31
Generation interval (years)	3.37	19	3.71	11
Selection emphasis (%)	59	18	44	32

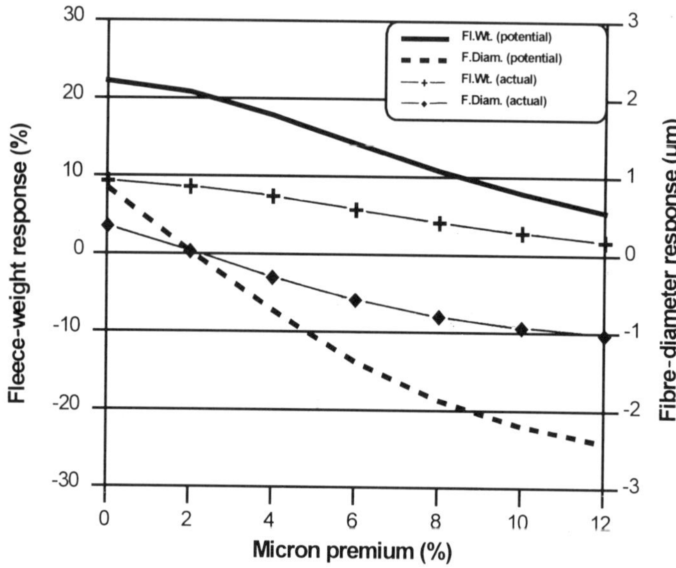

Fig. 17.6. Likely genetic gains in Merino breeding programmes.

wool production highlighted the relative importance of the weight of wool and its mean fibre diameter. A number of other wool-quality traits contribute to total value, but their current economic value and likely influence on wool processing or the properties of the woollen garment are relatively small.

2. Genetic parameters for most traits associated with wool production and quality have been precisely estimated from a large number of studies. Our predictions of likely responses are therefore precise and, at least for fleece weight, have been well supported by observed responses in selection lines. Realized responses to selection in wool-quality traits have been less extensively studied, although research currently in progress includes single trait selection lines for staple strength and multitrait-selection index lines for varying combinations of fleece weight and fibre diameter.

3. The clarity of the economic signals and the simplicity and relatively high heritability of selection traits leads to predictions of rapid potential rates of progress for the industry's breeding programmes. That only a fraction of this potential is probably being achieved in commercial breeding programmes highlights the continuing need for scientists to contribute to the design and monitoring of the breeding programme as a whole.

References

Ahmed, W., Dun, R.B. and Winston, R.W. (1963) The efficiency of conversion of feed to wool in Merino flocks selected for and against fleece weight. *Australian Journal of Experimental Agriculture and Animal Husbandry* 3, 269–275.

Andrews, M.W., Douglas, S.A.S., Ireland, A.H.M., Quirk, R.J. and Rottenbury, R.A. (1985) *Final Report by TEAM Project Management Committee to the Australian Wool Corporation*, Australian Wool Corporation, Melbourne, 50 pp.

Atkins, K.D. (1988) Consequences of selection for wool production and implications for Merino improvement programmes. In: *Proceedings of 3rd World Congress on Sheep and Beef Cattle Breeding, Paris*, Vol. 2. Institute National de la Recherche Agronomique (INRA), Paris, pp. 409–428.

Atkins, K.D. (1990) Incorporating parameters for lifetime productivity into breeding objectives for sheep. In: *Proceeding of 4th World Congress on Genetics Applied to Livestock Production, Edinburgh*, Vol. 15. pp. 17–26.

Atkins, K.D. and Casey, A.E. (1994) Recent developments in wool sheep breeding. In: *Proceeding of 5th World Congress on Genetics Applied to Livestock Production, Guelph, Canada*, Vol. 18.pp. 23–30.

Atkins, K.D., Semple, S.J. and Casey, A.E. (1994) OBJECT – Personalised breeding objectives for Merinos. In: *Proceeding of 5th World Congress on Genetics Applied to Livestock Production, Guelph, Canada*, Vol. 22. pp. 79–80.

Barlow, R. (1974) Selection for clean fleece weight in Merino sheep. II. Correlated responses to selection. *Australian Journal of Agricultural Research* 25, 973–994.

Beattie, A.W. (1961) Relationships among productive characters of Merino sheep in north-western Queensland. 1. Estimates of phenotypic parameters. *Queensland Journal of Agricultural Science* 18, 437–445.

Beattie, A.W. (1962) Relationships among productive characters of Merino sheep in north-western Queensland. 2. Estimates of genetic parameters, with particular reference to selection for wool weight and crimp frequency. *Queensland Journal of Agricultural Science* 19, 17–26.

Bell, P.J.M. (1987) Topmaking today. *Wool Technology and Sheep Breeding* 36(2), 101–108.

Black, J.L. (1987) Mechanisms controlling the rate of growth, composition and morphology of wool. In: McGuirk, B.J. (ed.) *Merino Improvement Programs in Australia*. AWC, Melbourne, pp. 189–206.

Brown, G.H. and Turner, H.N. (1968) Response to selection in Australian Merino sheep. II. Estimates of phenotypic and genetic parameters for some production in Merino ewes and an analysis of the possible effects of selection on them. *Australian Journal of Agricultural Research* 19, 303–322.

Campbell, M.E., Whiteley, K.J. and Gillespie, J.M. (1972) Compositional studies of high- and low-crimp wools. *Australian Journal of Biological Sciences* 25, 977–987.

Clark, M.J. and Whiteley, K.J. (1977) Some aspects of the measurement of colour in raw wool. *Journal of the Textile Institute* 69(5), 121–130.

Clark, M.J. and Whiteley, K.J. (1978) Some observations on the colour of Australian wool. *Wool Technology and Sheep Breeding* 25(1), 5–9.

Cottle, D.J., Russell, B.C. and Atkins, K.D. (1995) Do Merino hoggets with 'positive micron' have a propensity for 'micron blowout'? *Proceedings of Australian. Association of Animal Breeding and Genetics* 11, 525–528.

Davis, G.P. and McGuirk, B.J. (1987) Genetic relationships between clean wool weight, its components and related skin characters. In: McGuirk, B.J. (ed.) *Merino Improvement Programs in Australia*. AWC, Melbourne, pp. 189–206.

de Groot, G.J.J.B. (1992) The effect of coefficient of variation of fibre diameter in wool tops on yarn and fabric properties. *Wool Technology and Sheep Breeding* 40, 60–64.

Greeff, J.C., Lewer, R.P., Ponzoni, R.W. and Purvis, I. (1995) Staple strength: progress towards elucidating its place in Merino breeding. *Proceedings of Australian*

Association of Animal Breeding and Genetics 11, 595–601.

Gregory, I.P. (1982a) Genetic studies of South Australian Merino sheep. III. Heritabilities of various wool and body traits. *Australian Journal of Agricultural Research* 33, 355–362.

Gregory, I.P. (1982b) Genetic studies of South Australian Merino sheep. IV. Genetic, phenotypic and environmental correlations between various wool and body traits. *Australian Journal of Agricultural Research* 33, 363–373.

Gregory, I.P. and Ponzoni, R.W. (1981) Genetic studies of South Australian Merino sheep. II. Environmental effects on wool and body traits at 15–16 months of age. *Australian Journal of Agricultural Research* 32, 657–667.

Hamilton, B.A. and Langlands, J.P. (1969) Efficiency of wool production of grazing sheep. 1. Differences between Merino sheep selected for high and low fleece weight. *Australian Journal of Experimental Agriculture and Animal Husbandry* 9, 249–253.

Hickson, J.D., Swan, A.A., Kinghorn, B.P. and Piper, L.R. (1995a) Maternal effects at different ages in Merino sheep. *Proceedings of Australian Association of Animal Breeding and Genetics* Adelaide 11, 416–420.

Hickson, J.D., Kinghorn, B.P., Piper, L.R. and Swan, A.A. (1995b) Micron blowout – using available genetic variation. *Proceedings of Australian, Association of Animal Breeding and Genetics* 11, 529–533.

Hunter, L., Turpie, D.W.F. and Gee, E. (1984) The effect of wool fibre properties and breed of sheep on worsted processing performance and on yarn and fabric properties. In: *Proceedings of 3rd World Congress on Sheep and Beef Cattle Breeding, Pretoria*. South African Stud Book and Livestock Improvement Association, Bloemfontein, pp. 703–710.

International Wool Secretariat (1996 *et sqq.*) *Wool Premiums and Discounts: Analysis of Wool Sold at Auction*. Quarterly series published by IWS and Wool International, Melbourne.

Jackson, N. and Rottenbury, R.A. (1994) *Style Metrology*. Specification of Australian Wool and its Implications for Marketing and Processing, CSIRO and IWS, Sydney, 14 pp.

Jackson, N., Nay, T. and Turner, H.N. (1975) Response to selection in Australian Merino sheep. VII. Phenotypic and genetic parameters for some wool follicle characteristics and their correlation with wool and body traits. *Australian Journal of Agricultural Research* 26, 937–957.

James, P.J., Ponzoni, R.W., Walkley, J.R.W. and Whiteley, K.J. (1990) Genetic parameters for wool production and quality traits in South Australian Merinos of the Collinsville family group. *Australian Journal of Agricultural Research* 41, 583–594.

Kurdo, K.O., Whiteley, K.J. and Smith, L.J. (1986) The influence of resistance to compression on the processing performance of superfine wools. Part 1: Topmaking. *Journal of Textile Institute* 77(2), 104–118.

Lamb, P.R. and Yang, S. (1994) *The Effect of Wool Properties on Spinning Performance and Yarn Properties*. Specification of Australian Wool and its Implications for Marketing and Processing, CSIRO and IWS, Sydney, 11 pp.

Lax, J. and Brown, G.H. (1967) The effects of inbreeding, maternal handicap, and range in age on 10 fleece characteristics in Merino rams and ewes. *Australian Journal of Agricultural Research* 18, 689–706.

Lax, J., Swan, A.A. and Purvis, I.W. (1995) Genetic and phenotypic relationships involving subjectively assessed 'style' traits in hoggets from the CSIRO fine wool flock. *Proceedings of Australian Association of Animal Breeding and Genetics* 11, 521–524.

Lee, G.J., Atkins, K.D. and Mortimer, S.I. (1995) Variation between Merino ewes in pasture intake. 2. Within-flock genetic parameters for intake and some production traits. *Livestock Production Science* 41, 143–150.

Lewer, R.P. and Li, Y. (1994) Some aspects of selection for staple strength. *Wool Technology and Sheep Breeding*. 42, 103–111.

Lewer, R.P. and McLeod, I.M. (1990) One- and two-stage selection indexes for Australian Merino sheep, incorporating visual selection criteria. In: *Proceedings of 4th World Congress on Genetics Applied to Livestock Production, Edinburgh*, Vol. 15. pp. 61–64.

Lewer, R.P., Woolaston, R.R. and Howe, R.R. (1992) Studies on Western Australian Merino sheep. I. Stud, strain and environmental effects on hogget performance. *Australian Journal of Agricultural Research* 43, 1381–1397.

Lewer, R.P., Woolaston, R.R. and Howe, R.R. (1994) Studies on Western Australian Merino sheep. II. Genetic and phenotypic parameter estimates for objectively measured traits on ram and ewe hoggets using different model types. *Australian Journal of Agricultural Research* 45, 829–840.

McGuirk, B.J. (1980) Selection for wool production in Merino sheep. In: Robertson, A. (ed.) *Selection Experiments in Laboratory and Domestic Animals*. Commonwealth Agricultural Bureaux, Slough, UK, pp. 176–197.

McGuirk, B.J. (1983) Selection for wool production. In: Haresign, W. (ed.) *Sheep Production*. Butterworths, London, pp. 545–566.

Mann, T.L.J., Ponzoni, R.W. and Polkinghorne, R.W. (1980) Response to partial selection on clean fleece weight in South Australian strong-wool Merino sheep. V. Estimation of genetic gain by contemporary comparison of progeny of sires two years apart in age. *Australian Journal of Agricultural Research* 31, 1029–1035.

Morley, F.H.W. (1951) Selection for economic characters in Australian Merino sheep. I. Estimates of phenotypic and genetic parameters. *Scientific Bulletin of Department of Agriculture, NSW* 73.

Morley, F.H.W. (1955) Selection for economic characters in Australian Merino sheep. V. Further estimates of phenotypic and genetic parameters. *Australian Journal of Agricultural Research* 6, 77–90.

Morley, F.H.W. (1956) Selection for economic characters in Australian Merino sheep. VI. Inheritance and interrelationships of some subjectively graded characteristics. *Australian Journal of Agricultural Research* 7, 873–881.

Mortimer, S.I. and Atkins, K.D. (1989) Genetic evaluation of production traits between and within flocks of Merino sheep. I. Hogget fleece weights, body weight and wool quality. *Australian Journal of Agricultural Research* 40, 433–443.

Mortimer, S.I. and Atkins, K.D. (1993) Genetic evaluation of production traits between and within flocks of Merino sheep. II. Component traits of the hogget fleece. *Australian Journal of Agricultural Research* 44, 1523–1539.

Mortimer, S.I. and Atkins, K.D. (1994) Direct and maternal additive genetic effects on wool production of Merino sheep. In: *Proceedings of 5th World Congress on Genetics Applied to Livestock Production, Guelph, Canada*, Vol. 18. pp. 103–106.

Mullaney, P.D., Brown, G.H., Young, S.S.Y. and Hyland, P.G. (1969) Genetic and phenotypic parameters in fine-wool Merino, Corriedale and Polwarth sheep. I. Influence of various factors on production. *Australian Journal of Agricultural Research* 20, 1161–1176.

Mullaney, P.D., Brown, G.H., Young, S.S.Y. and Hyland, P.G. (1970) Genetic and phenotypic parameters in fine–wool Merino, Corriedale and Polwarth sheep. II.

Phenotypic and genetic correlations, heritability and repeatability. *Australian Journal of Agricultural Research* 21, 527–540.

Nay, T. (1966) Wool follicle arrangement and vascular pattern in the Australian Merino. *Australian Journal of Agricultural Research* 17, 797–805.

Nay, T. (1970) Follicle characteristics in a group of Merino sheep selected up and down for fleece weight. *Australian Journal of Agricultural Research* 21, 951–954.

Naylor, G.R.S., Phillips, D.G. and Veitch, C.J. (1995) The relative importance of mean diameter and coefficient of variation of sale lots in determining the potential skin comfort of wool fabrics. *Wool Technology and Sheep Breeding* 43, 69–82.

Pattie, W.A. (1965) Selection for weaning weight in Merino sheep. 2. Correlated responses in other production characters. *Australian Journal of Experimental Agriculture and Animal Husbandry* 5, 361–368.

Pattie, W.A. and Barlow, R. (1974) Selection for clean fleece weight in Merino sheep. I. Direct response to selection. *Australian Journal of Agricultural Research* 29, 1077–1086.

Pattinson, R.D. and Whiteley, K.J. (1984) Appraisal and measurement of the colour of Australian wool and the role of colour in sale by description. *Wool Technology and Sheep Breeding* 32, 181–189.

Piper, L.R. and Dolling, C.H.S. (1966) Variation in the sulphur content of wool of Merino sheep associated with genetic differences in wool-producing capacity. *Australian Journal of Biological Sciences* 19, 1179–1182.

Piper, L.R. and Dolling, C.H.S. (1969) Efficiency of conversion of food to wool. V. Comparison of the apparent digestibility of sheep selected for high clean wool weight with that of sheep from a random control group. *Australian Journal of Agricultural Research* 20, 579–587.

Ponzoni, R.W. (1986) A profit equation for the definition of the breeding objective of Australian Merino sheep. *Journal of Animal Breeding and Genetics* 103, 342–359.

Ponzoni, R.W., Rogan, I.M. and James, P.J. (1990) Genetic improvement of apparel and carpet wool production. In: *Proceedings of 4th World Congress on Genetics Applied to Livestock Production, Edinburgh*, Vol. 15. pp. 149–166.

Ponzoni, R.W., Grimson, R.J., Jaensch, K.S., Smith, D.H., Gifford, D.R., Ancell, P.M.C., Walkley, J.R.W. and Hynd, P.I. (1995) The Turretfield sheep breeding project: messages on phenotypic and genetic parameters for South Australian Merino sheep. *Proceedings of Australian Association of Animal Breeding and Genetics* 11, 303–313.

Purvis, I.W. and Piper, L.R. (1992) The development of an ultra-fine strain of Merino sheep. *Proceedings of Australian Association of Animal Breeding and Genetics* 10, 109–113.

Rendel, J.M. and Nay, T. (1978) Selection for high and low ratio and high and low primary density in Merino sheep. *Australian Journal of Biological Sciences* 29, 1077–1086.

Robards, G.E., Williams, A.J. and Hunt, M.H.R. (1974) Selection for crimp frequency in the wool of Merino sheep. 2. Efficiency of conversion of food to wool. *Australian Journal of Experimental Agriculture and Animal Husbandry* 14, 441–448.

Rogan, I.M. (1984) Selection for wool production. In: *Proceedings of 2nd World. Congress on Sheep and Beef Cattle Breeding, Pretoria*. South African Stud Book and Livestock Improvement Association, Bloemfontein, pp. 367–378.

Rogan, I.M. (1988) Genetic variation and covariation in wool characteristics related to processing performance and their economic significance. In: *Proceedings of 3rd World Congress on Sheep and Beef Cattle Breeding, Paris*, Vol. 2. Institute National de la Recherche Agronomique (INRA), Paris, pp. 383–407.

Rogan, I.M., Atkins, K.D. and Coelli, K.A. (1995) Shearing Merino ram weaners – is it worth it? *Wool Technology and Sheep Breeding* 43, 56–68.

Saville, D.G. and Robards, G.E. (1972) Efficiency of conversion of food to wool in selected and unselected Merino types. *Australian Journal of Agricultural Research* 23, 117–130.

Stevens, D. and Crowe, D.W. (1994) *Style and Processing Effects.* Specification of Australian Wool and its Implications for Marketing and Processing, CSIRO and IWS, Sydney, 12 pp.

Swan, A.A., Lax, J. and Purvis, I.W. (1995) Genetic variation in objectively measured wool traits in CSIRO's fine wool flock. *Proceedings of Australian Association of Animal Breeding and Genetics* 11, 516–520.

Taylor, P.J. (1993) *The Effect on Wool Quality of Long Term Selection For Increased Fleece Weight: Final Report to the Australian Wool Corporation.* NSW Agriculture, Trangie, 15 pp.

Taylor, P. and Atkins, K. (1992) Merino breeding programs and fibre diameter variability. *Wool Technology and Sheep Breeding.* 40, 97–101.

Torshizi, R.V., Raadsman, H.W. and Nicholas, F.W. (1995) An investigation of the potential for early (indirect) selection in Australian Merino sheep. *Proceedings of Australian Association of Animal Breeding and Genetics* 11, 314–317.

Turner, H.N. and Jackson, N. (1978) Response to selection in Australian Merino sheep. VIII. Further results on selection for high clean wool weight with attention to quality. *Australian Journal of Agricultural Research* 29, 615–629.

Turner, H.N., Dolling, C.H.S. and Kennedy, J.F. (1968) Response to selection in Australian Merino sheep. I. Selection for high clean wool weight, with a ceiling on fibre diameter and degree of skin wrinkle: response in wool and body characteristics. *Australian Journal of Agricultural Research* 19, 79–112.

Turner, H.N., Brooker, M.G. and Dolling, C.H.S. (1970) Response to selection in Australian Merino sheep. III. Single character selection for high and low values of wool weight and its components. *Australian Journal of Agricultural Research* 21, 955–984.

Walkley, J.R.W., Ponzoni, R.W. and Dolling, C.H.S. (1987) Phenotypic and genetic parameters for lamb and hogget traits in a flock of South Australian Merino sheep. *Australian Journal of Agricultural Research* 27, 205–210.

Watson, N., Jackson, N. and Whiteley, K.J. (1977) Inheritance of the resistance to compression property of Australian Merino wool and its genetic correlation with follicle curvature and various wool and body characters. *Australian Journal of Agricultural Research* 28, 1083–1094.

Whiteley, K.J. (1994) The influence of wool fibre characteristics on processing and garment performance. In: Azzarini, M. and Cardellino (eds) *Proceedings of IVth World Merino Congress, Montevideo, Uruguay.* SUL, Montevideo, pp. 209–227.

Williams, A.J. (1966) The efficiency of conversion of feed to wool during limited and unlimited feeding of flocks selected on clean fleece weight. *Australian Journal of Experimental Agriculture and Animal Husbandry* 6, 90–95.

Williams, A.J. (1979) Speculations on the biological mechanisms responsible for genetic variation in the rate of wool growth. In: Black, J.L. and Reis, P.J. (eds) *Physiological and Environmental Limitations to Wool Growth.* UNE, Armidale, pp. 337–354.

Williams, A.J. (1987) Physiological consequences of selection for increased fleece weight. In: McGuirk, B.J. (ed.) *Merino Improvement Programs in Australia.* AWC, Melbourne, pp. 481–494.

Williams, A.J. and Winston, R.J. (1965) Relative efficiencies of conversion of feed to wool at three levels of nutrition in flocks genetically different in wool production. *Australian Journal of Experimental Agriculture and Animal Husbandry* 5, 390–395.

Williams, A.J., Robards, G.E. and Saville, D.G. (1972) Metabolism of cystine by Merino sheep genetically different in wool production. II. The responses in wool growth to abomasal infusions of l-cystine or dl-methionine. *Australian Journal of Biological Science* 25, 1269–1276.

Young, S.S.Y., Turner, H.N. and Dolling, C.H.S. (1960) Comparison of estimates of repeatability and heritability for some production traits in Merino rams and ewes. II. Heritability. *Australian Journal of Agricultural Research* 11, 604–617.

Genetics of Lamb and Meat Production 18

R.G. BANKS
National LAMBPLAN Coordinator, Department of Animal Science, University of New England, NSW 2351, Australia

Introduction 505
Production Systems and Breeding Objectives 507
Breed Evaluations 509
Potential for Genetic Change 510
Genetic parameters 510
Results of selection experiments 511
Major genes 512
Implementation 513
Genetic information systems 514
Organized breeding schemes 515
Genetic trends 516
Developed and developing countries 517
Prospects 517
References 518

Introduction

Sheep meats have formed a component of many human diets worldwide for a considerable period, and in consequence production systems have evolved based around the capacity of sheep to produce food and fibre from low-quality/roughage-based diets (Ryder, 1983). Through historic and prehistoric time, genetic differences have been sampled and exploited, providing substantial within- and between-breed genetic variation among sheep populations around the world. This variation includes that in traits affecting lamb and meat production, and provides potential for genetic improvement of both cost-of-production and quality traits, within and across the diverse range of production systems in which sheep are used worldwide. The traits contributing to income and cost from sheep enterprises can be conveniently grouped into five categories: reproduction traits, wool traits, growth and carcass traits, disease traits and feed-intake and conversion traits. These components of animal performance can be categorized according to whether they are expressed primarily in the growing animal – 'growth phase' – or in the reproductive female – 'maternal phase' (Table 18.1).

Within each category there will be a number of traits for which genetic variation exists or can be expected to exist, and hence which can be manipulated by genetic selection. This chapter focuses upon the genetic variation existing in live-animal traits, excluding milk as a direct product, 'fine' aspects of carcass composition and meat quality and disease-resistance/resilience traits. The existence and utilization of genetic variation within these traits is considered under four headings:

- production systems and breeding objectives;
- breed/genotype evaluations;
- genetic parameters and scope for genetic change by selection;
- organization and implementation of information and improvement systems.

This structure reflects a suggested formal approach to breeding-programme design (Cunningham, 1979; Harris *et al.*, 1984), chosen here both for its logical flow and because it reflects a bias towards rational management of genetic variation for current and future needs. There is also some concentration on Australasian systems, reflecting both the author's experience and the diversity of production systems in those environments, with sheep being grown for essentially all major products in both single- and multipurpose production systems. Finally, no attempt is made at an exhaustive review of the literature under any of these headings; excellent reviews exist for each, and the approach here is to note relevant specific reviews and conference sessions and to highlight implications for utilization of existing genetic variation. It will become clear that existing knowledge provides an excellent basis for such utilization in a range of environments and production systems.

Table 18.1. Traits contributing to income and cost in sheep enterprises.

Trait	Maternal phase	Growth phase
Number of lambs born/weaned	+	
Milk production/mothering ability	+	
Fleece weight	+	+
Fibre diameter/fleece quality	+	
Mature weight	+	
Live weight at sale/growth rate		+
Carcass fatness		+
Carcass conformation		+
Disease resistance/resilience	+	+
Feed intake and conversion	+	+

Production Systems and Breeding Objectives

Formal breeding-programme design begins with definition of the production system, leading to definition of a breeding objective as the marginal effects on profit of unit changes in the production level of each trait. Useful reviews of theoretical and practical issues in breeding objectives for sheep production systems are given by Ponzoni (1986a), James (1986) and Bradford and Meyer (1986). More detailed studies include: for Merinos, e.g. Ponzoni (1986b); for dual-purpose sheep, Fogarty (1987), Fogarty and Gilmour (1993), Stafford and Walkley (1979); for terminal sires, e.g. Atkins (1987). More recent studies with increasing focus on the economic value of reducing carcass fat include Waldron *et al.* (1991), Simm and Dingwall (1991), Binnie and Clarke (1992) and Bennett (1990). A number of significant challenges have been tackled in these studies:

- Recognizing differences in breeding objectives for different breeds/ genotypes within structured crossing systems: for example, Atkins (1987) and Stafford and Walkley (1979) for terminal sire and crossing breeds in the Australian sheep industry. While Smith (1965) concluded that specialized sire and dam lines for sheep meat production would probably be of little value where sheep perform meat- and wool-production roles within a region and particularly where there is any differentiation among commercial enterprises in degree of specialization for wool production, distinct sire, dam and wool breeds or genotypes are usually maintained and used. These will have different optimal objectives and hence usually different selection criteria.
- Coping with changing market behaviour over time: for example, Ponzoni (1986b) work in wool sheep has addressed changing price effects of fibre diameter in Merinos, and most studies through the 1980s have placed more emphasis on price effects of carcass fatness. The particular challenge addressed in this respect by several authors, is the genetic antagonism between increased growth rate/carcass weight and carcass fatness, and the importance of obtaining as precise estimates as possible for the economic value of carcass fat relative to lean in view of this antagonism.
- Directly accounting for feed intake in the objective function: both Ponzoni (1986b) and Atkins (1987) have addressed this issue by estimating mature ewe feed intake and its relationship to ewe body weight and feed costs.
- Estimating price effects in situations of poor information and information flow: for example, Waldron *et al.* (1991), Simm and Dingwall (1991). Sheep production systems are often characterized by poor flow of price signals from consumer to breeder, and this has hampered both estimation of realistic price effects and effective application of breeding programmes to meet estimated breeding objectives.
- Combining these uncertainties: Fogarty and Gilmour (1993) explicitly addressed price variation and parameter uncertainty in their examination of comprehensive breeding objectives for dual-purpose sheep in Australia producing medium-micron apparel wool.

One area requiring further study is the combined effect of short- and long-term fluctuations in relative prices for biologically separate components of overall enterprise profit such as 'meat' and 'wool'. Observation in Australia, at least, suggests that, where such fluctuations are combined with poor price signals, breeding objectives in practice place low or zero value on change in most or all traits: breeders are effectively using utility values incorporating price variation as the effective economic values, with some real weighting against price variation. Given that essentially all sheep breeding programmes have multitrait goals, based on the multiple-product nature of sheep enterprises, this may be a widespread problem.

The breeding objective studies noted (and others) also reflect the diversity of sheep production systems and hence genotypes and of formal or informal breeding objectives. Further, breeding objectives have tended to evolve from a focus on costs of production – reproduction and early growth rate – towards greater emphasis on carcass composition, as markets in, at least, Western countries have shown clearer and clearer preference for leaner meat (Hopkins, 1988).

This significant change in breeding objective focus, together with the price-fluctuation concern noted above, and in the absence of effective breed, national or even international evaluation systems (such as exist widely for dairy cattle), appears to have contributed to typically low rates of genetic change in sheep populations (Smith, 1984). Thus, while considerable attention has been devoted to the problems of definition of breeding objectives for sheep, factors to some degree outside the formal scope of the production systems under study appear to have hindered genetic progress towards the breeding objectives.

Further, even for relatively similar production systems and genotypes there may still be regional or national differences in breeding objectives, reflecting in part the 'portfolio' of breeds that is available; for example, Merinos are the predominant sheep breed in both South Africa and Australia, but in the former country there has been more effective development of a dual-purpose or mutton-type Merino. In Australia this role has been widely filled by Border Leicester–Merino cross females, with the result that breeding objectives for Merinos in Australia (in both theory and practice) have placed lower emphasis on reproduction, growth and carcass traits than might otherwise have been the case. Extending this point, there is, at least in Western countries, a clear consumer distinction between meat from young sheep and that from surplus older sheep.

Overall, formal breeding objective studies for sheep have successfully tackled the most direct challenges outlined above. A satisfactory framework exists for developing workable breeding objectives for a wide range of sheep production systems. These objectives may be applied either formally in economic indexes (e.g. Simm and Dingwall, 1991) or through 'desired-gains' index (e.g. Banks, 1994), as appropriate to the maturity of markets and available information systems. More subtle problems affecting 'realized breeding objectives' remain, exacerbated by the multitrait nature of sheep enterprises. Possible solutions to this complex problem will be discussed below, under 'Implementation'.

Breed Evaluations

Given a breeding objective, rational utilization of genetic resources proceeds through evaluation of available genetic material. This is usually treated in two stages: at the breed or strain level, followed by within-breed evaluation (although between- and within-breed information can be used simultaneously, e.g. Kinghorn, 1986). Breed evaluations are based on measuring the mean performance of groups of animals with known proportions of genes from particular breeds or genotypes. Most commonly, first-cross progeny groups are compared – animals sired by different breeds and born to a common dam genotype. Where maternal-phase traits are of interest, more complex designs are usually necessary, both to obtain data on maternal genetic and heterotic effects and to minimize or avoid differences in maternal performance due to rearing environment.

As would be expected from the enormous diversity of sheep production systems, breeding objectives and breeds/genotypes, breed evaluations varying widely in scale and basis for evaluation abound in the literature. Baker (1986), Young *et al.* (1986) and Hohenboken (1986) together provide an overview of breed evaluation studies in sheep to that time, and in particular the problems of interpretation of results, where heterosis may be expected to be important (either in evaluation results or in later exploitation), and of determining useful/suitable bases for comparison and ranking. The latter issues are of particular importance in sheep, given (as noted above) the multitrait nature of sheep enterprises and therefore the higher likelihood of intermittent re-ranking as relative product prices change, compared with single-trait enterprises and species.

As with breeding-objectives studies, there has been a tendency, at least in Western countries, for breed evaluations to evolve from a focus on dual- or triple-purpose (wool, reproduction and lamb growth) genotypes towards greater focus on carcass composition. Thus, during the 1960s and 1970s much attention was focused on Finnish Landrace and Romanov sheep, while more recently breed evaluations in several countries have included the Texel breed, stimulated by clearer recognition of consumer aversion to dietary fat (Wolf *et al.*, 1980; Clarke and Kirton, 1990; Leymaster and Jenkins, 1993).

Even here, the problems of determining the bases for comparison do not disappear. Results from trials where Texels have been included have, for example, consistently shown advantages in some aspects of carcass composition, but typically in trade-offs in growth rate, such that lean growth rates per day may not differ significantly from other breeds evaluated. This raises a potential conflict, particularly in seasonal pasture-based systems, where fast growth rate is advantageous: benefits to the retailer in carcass value may have to be traded off against costs to the commercial producer due to slower lamb growth. Conversely, high growth potential genotypes, such as Oxford Downs and American Suffolks, may bring carcass disadvantages, particularly where markets have developed around animals slaughtered at intermediate carcass weights

(18–20 kg). Where trading systems are inefficient, such trade-offs may well lead to considerable suboptimal use of new genotypes in both the short and the long term.

Finally, as industry-wide genetic-evaluation procedures evolve, there is a strong likelihood that formal breed evaluations will decline in importance. More extensive genetic evaluations using better-defined genetic models support the tendency for progressive breeders to combine genetic material from more than one source/breed. While this development places greater and greater reliance on field data, it has the operational advantage that genetic evaluations are truly current, rather than representing historical 'snapshots' of a range of genotypes. For highly commercially orientated breeds and breeding programmes this is unlikely to cause any problems, but it almost invariably increases the distinction between currently viable breeds/genotypes and those that have lost favour or have only minority status for any reason.

Potential for Genetic Change

Genetic parameters

Fogarty (1995) provides an extensive review of genetic parameters for sheep production traits. Traits are summarized in four groups – live weights, wool weight and fibre diameter, body composition and reproduction – and estimates reported with method of estimation and breed identity where available. The reviewed estimates are summarized as weighted means of heritabilities, repeatabilities and genetic and phenotypic correlations. Tables 4 and 5 in Fogarty (1995) provide a simple summary of these genetic parameters, and clearly show that useful genetic variation exists for the range of traits reviewed.

Fogarty (1995) also notes the recent use of restricted maximum likelihood (REML) methods (Meyer, 1989), including those fitting maternal genetic (m2) and common environmental (c2) effects. Such models can provide a more detailed specification of the available genetic variation (Mercer *et al.*, 1994; Mortimer and Atkins, 1994; Nasholm and Danell, 1994; Swan and Hickson, 1994). Typically, maternal and common environmental effects comprise a significant proportion of total genetic variance for early weights and decline in importance as animals grow towards mature weights. Separation of direct from maternal and common environmental genetic effects will be beneficial in breeding programmes for terminal-sire and dam/maternal breeds.

Bennett's (1990) review of selection for growth and carcass composition included detailed discussion of relationships among carcass composition measures such as different sites for fat measurement. Genetic correlations were generally high but did not reach 1. This leads to discussion of the most appropriate sites for use as selection criteria, with choices likely to focus on measurement ease and value for predicting measures of total body composition. This discussion reflects a trend towards genetic evaluation using a range of new measurement technologies (Simm, 1987; Ball and Thompson, 1995; Jopson *et al.*, 1995; Kirton *et al.*, 1995), which allow more detailed analysis of genetic variation in body composition measures and which are now being applied to detailed analysis of feed utilization and energy balance by the sheep. As this work continues, it will add to the very limited number of studies directly concerned with genetic variation in feed intake and utilization (Lee *et al.*, 1990; Vangen and Thompson, 1992).

Mercer *et al.* (1994), in discussing genetic parameters for four sire breeds, commented that 'underlying biology of the various breeds is much the same, even though there may be differences in age at maturity'. Allowing for the apparent effects of inbreeding in some cases (Brash *et al.*, 1994), it seems reasonable to extend this comment over a wider range of breeds and traits in sheep. Against this, application of more powerful measurement technologies, which allow more detailed analysis of carcass composition, seems likely to highlight differences between genotypes in both trait means and patterns of genetic (co)variance at this level.

Most importantly, while common patterns are revealed in the broad production traits reviewed, software tools for parameter estimation from research and field data are now more widely available, providing a sound basis for the development of breeding programmes in essentially any sheep population.

Results of selection experiments

Parameter estimates provide one form of evidence of potential for genetic change. A second form is provided by selection experiments. Research activity in this area has focused on three broad areas:

- Effects of selection for direct measures of female reproductive performance (e.g. Hanrahan, 1984; Bradford, 1985).
- Effects of selection for indirect measures of reproductive performance, such as testicular diameter in males (e.g. Ricordeau *et al.*, 1986; Haley *et al.*, 1990; Matos and Thomas, 1992).
- Effects of selection for aspects of carcass composition, particularly subcutaneous fat depths (e.g. McEwan et al., 1989; Cameron, 1992; Cameron and Bracken, 1992; Bishop, 1993; Solis-Ramirez *et al.*, 1993; Ball and Thompson, 1995; and reviews by Bennett, 1990; Simm, 1994).

It is beyond the scope of this chapter to review this topic in any detail, but some key features can be noted:

- Selection applied has usually been effective in generating divergence, with realized heritabilities broadly in line with prior population estimates.
- Correlated responses have been harder to predict, reflecting a combination of limited (and often poor) estimates of genetic correlations and the often less obvious basis for predicting such correlated effects from prior understanding of sheep biology.

The first result – general effectiveness of selection in generating responses in the trait directly selected for – is reassuring, as is the confirmation that genetic divergence between selection lines is inherited in purebred and crossbred progeny. These results support the conclusion, from genetic-parameter estimations, that usable genetic variation exists for production traits in sheep. Where particular breeds are used predominantly in crossing, it is important to estimate effects of selection (and so of genetic differences) in commercial crossbred progeny and, if possible, over the range of likely feeding regimes and slaughter points for lambs. Such studies are limited in sheep (Lewis *et al.*, 1994; Hall *et al.*, 1995), but will become important for fine-tuning selection indexes for use in grandparental and/or parental flock-breeding programmes.

Regarding correlated responses, it is more realistic to view these as helping to develop our understanding of genetic and phenotypic relationships between measured traits of sheep and as almost inevitable, given the often simple nature of selection criteria applied. Indeed, particularly in the area of body composition, selection experiments comprise the principal method by which selection criteria for application in commercial breeding programmes will be identified and refined to meet more complex breeding objectives (Cameron, 1992; Cameron and Cienfuegos-Rivas, 1994; Oddy *et al.*, 1995; Bishop *et al.*, 1995; Kirton *et al.*, 1995).

Major genes

Work aimed at identifying and characterizing major genes has been a feature of sheep breeding research since the nature of the Booroola fecundity gene (*FecB*) was elucidated (Piper and Bindon, 1982). Similar genes affecting ovulation rate have been reported (Owen *et al.*, 1990) and attempts made to exploit such variation both in the 'source' populations and by introgression in other breeds and flocks (King *et al.*, 1994; Southey *et al.*, 1994). During this time also, statistical techniques able to distinguish variation due to major genes from background genetic variation have developed considerably (Hoeschele, 1988; Kinghorn and Kerr, 1995). Application of these techniques to the genetic analysis of selection lines has begun, and recent evidence suggests the existence of major genes for other production traits, including body weight and disease resistance (R.J. Kerr, Armidale, 1996, personal communication).

The *callipyge* gene – a major gene with significant effects on carcass composition – was identified originally in an American Dorset flock selected with emphasis on leg conformation (Snowder *et al.*, 1994; Cockett *et al.*, 1995). This gene maps to the interval on chromosome 18 spanned by the markers CSSM18 and TGLA122, and exhibits inheritance suggesting dominant negative imprinting. More recently, a major gene with apparently smaller and more variable effects on muscle dimensions and carcass composition has been detected in an Australian Poll Dorset flock (the Carwell flock) and tentatively mapped to the same region on chromosome 18. One pleiotropic effect of the *callipyge* gene is increased toughness in some muscles (Carpenter and Solomon,

1995; Meyer *et al.*, 1995). No direct or recent pedigree links are known between the American and Australian flocks in which *callipyge* was detected, pointing to the possibility that this allele may be segregating in other breeds/populations.

Both *FecB* and *callipyge* appear to have been concentrated in commercial breeding flocks under relatively simple selection regimes. Widespread application of best linear unbiased prediction (BLUP) procedures for genetic evaluation, which will efficiently select families in which major genes are segregating (Meuwissen and Goddard, 1996), together with newer techniques for major-gene quantative-trait-locus (QTL) detection and the beginnings of use of the ovine genome map, suggests that many more major genes may be detected in the next decade. Experience with both *FecB* and *callipyge* suggests that this will raise several challenges for sheep breeding researchers and programme managers, including:

- developing cost-effective strategies for rapid assessment of pleiotropic effects of such genes, most probably in conjunction with commercial exploitation;
- developing optimal strategies for reporting major-gene/QTL status for traits in genetic evaluations;
- developing programme design strategies that simultaneously optimize power of detection of major genes/QTLs and have the capacity to exploit such variation rapidly.

Detectable alleles with large effects on commercially important traits seem likely to be a central component of utilizing genetic variation in future, rendering the distinction applied here (genetic parameters for additive and non-additive models, and major genes) largely meaningless. The range of traits subject to genetic management in sheep, together with the enormous diversity contained within and between breeds, points to considerable enrichment of our models and understanding of this genetic variation through the blending of major-gene/QTL approaches with more traditional analysis of genetic variation.

Implementation

This brief overview highlights the existence of useful genetic variation in a wide range of traits relevant to lamb production and a well-developed foundation for developing and applying breeding objectives. Against this, Smith (1984) suggested that rates of genetic change in sheep populations had been lower than those achievable up to that time. In this section, key components of improvement systems for sheep meat production will be discussed and areas for likely improvement in utilization of knowledge of genetic variation in sheep identified.

Genetic information systems

Definition of breeding objectives and knowledge of the relevant genetic parameters together form the foundations for genetic information systems in any livestock-breeding programme. Details of design and management of such systems will differ between production systems – affecting traits measured and reported and indexing approaches – and between countries, reflecting differences in levels and mechanisms of public and private investment in genetic improvement. Such design details must clearly be addressed by system management, with genetic and financial cost-effectiveness as key design criteria.

Some general features and issues for sheep genetic information systems are apparent, and they influence the effectiveness of sheep-improvement programmes:

- Sheep in most countries have a much lower value per head than other livestock species (such as cattle and pigs). This frequently means that measurement and recording are limited, even within stud or nucleus tiers of multi-tier systems. This problem is exacerbated if, as is often the case, commercial sheep industries have poor and/or distorted price signals.
- Where commercial industries are based on structured crossing programmes, as in Australia and the UK, there are often considerable operational challenges in ensuring that information reported for different breeds supports both breed specificity of breeding objective and a level of uniformity or consistency of format that maximizes understanding through all appropriate sectors of the industry.
- Depending on the country, the production system and the breed role, there may be enormous diversity of flock-size and trait combinations in data (compared with, for instance, dairy cattle). This suggests that sheep genetic evaluation is more likely to benefit significantly from the increasing power of personel computers (PCs) and the rapid development of user-friendly prediction software (Groenveld *et al.*, 1990; Gilmour, 1994) than that in other livestock species.

These features have together delayed the application of BLUP systems for sheep genetic evaluations in most countries until later than other species, but most countries with significant sheep populations now have BLUP-based information systems in place. In most cases, field data are submitted by breeders for central processing, with records typically including a weaning weight, records of lambing performance for ewes, later weight(s) and a wool weight. Prediction models now typically separate direct and maternal effects for at least weaning weight. Some systems provide breeding values for fat depth and eye muscle, where scan data are available. Systems vary in approaches taken to combining traits into selection indexes; for example, in both New Zealand and the UK lean growth indexes are provided, combining weight, fat-depth and eye-muscle-depth breeding values (Simm and Dingwall, 1989), while in Australia LAMBPLAN provides a series of 'desired gains' indexes for terminal-sire and dual-purpose breeds (Banks, 1994). Most national systems are moving towards or have implemented across-flock genetic evaluations, typically relying on sire referencing to establish or strengthen genetic links between flocks. Both theory (Smith and Banos, 1991) and experience in other livestock suggest that across-flock evaluation will lead to significant acceleration of genetic progress.

Scope for international genetic evaluations in sheep has been limited (in comparison with dairy cattle, for example) by the absence of any significant trade in semen and by very tight quarantine regulations for many countries. Strong pedigree links exist between Australia and New Zealand for a number of terminal sire and dual-purpose breeds, and so combined genetic evaluations may be possible for those two countries. More general development of such evaluations, as has happened with dairy cattle, is likely to remain limited by quarantine restrictions and lower individual sire commercial values than are the case for dairy bulls.

Organized breeding schemes

Partly motivated by the high relative costs of measurement in sheep and partly to overcome problems raised by either very small or very large flock sizes, there has been considerable focus on cooperative breeding structures in sheep breeding. These have taken three main forms: group breeding schemes, involving a central nucleus flock supplying sires to a number of 'base' flocks (e.g. Anderson and Curran, 1990), sire-referencing schemes, where reference or link sires are identified for use via artificial insemination (AI) in a number of flocks (Guy and Croston, 1994), and ram circles, which are effectively a precursor to sire referencing and do not rely so much on use of AI (Gjedrem, 1969). In France, improvement schemes for several breeds are highly structured and involve a combination of on-farm central testing and organized progeny testing.

Positive genetic trends produced in such schemes are encouraging wider involvement by breeders and increased interest in full exploitation of BLUP technology, intense selection based on across-flock evaluation and cooperative progeny testing of sires. These in turn offer considerable potential for acceleration of genetic trends. An example is provided by the major terminal-sire breed in Australia, the Poll Dorset, in which, over the period 1989–1996, there has been development from limited use of LAMBPLAN, through establishing a central nucleus flock to evaluate and apply computed tomography (CT) technology, to the initiation of sire referencing across the breed based on young (10–12-month-old) performance-tested sires (Fig. 18.1). In countries without statutory requirements for ram registration, cooperative nucleus schemes and/or sire referencing seem likely to become the basis of wide-scale breed improvement programmes.

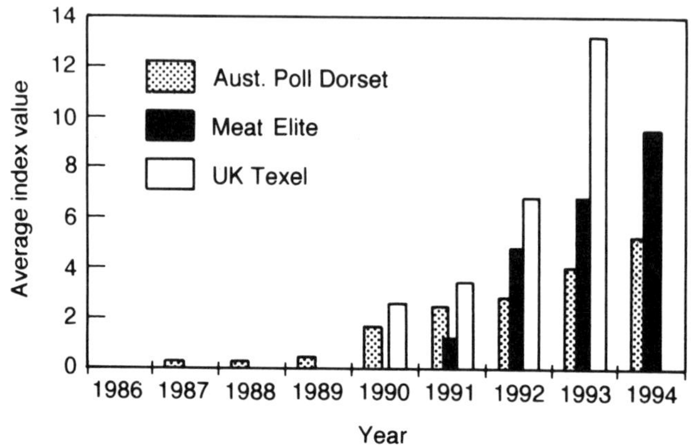

Fig. 18.1. Genetic trends in terminal-sire breeding: Australia and UK. The LAMBPLAN 60:20:20 index is designed to achieve genetic improvement in growth rate, leanness and muscling in the ratio 3:1:1 and to have an index standard deviation of 10 units.

Genetic trends

As noted previously, ample genetic variation exists for genetic improvement of production traits in all sheep breeds and populations studied. Genetic information systems are in place to provide accurate and timely breeding values, and systems have been developed for exploiting genetic resources available within whole breeds. Systems based on BLUP are now available to evaluate genetic progress within large sheep populations. Typically, positive trends are being reported for weight traits and, more recently, for fat depth and muscle depth. Figure 18.1 includes genetic trends for three populations: the Australian Poll Dorset breed, the Meat Elite nucleus flock within that breed and the UK Texel Sire Referencing Scheme (the Meat Elite flock was established as a nucleus within the Poll Dorset breed to simultaneously apply and evaluate CT technology through an élite nucleus flock). For simplicity, trends for weight, fat and muscle have been combined into the 60 : 20 : 20 index available through LAMBPLAN (Banks, 1994). Note that the UK scheme uses and reports its own lean growth index (Guy and Croston, 1994).

As well as providing clear evidence of positive genetic trends, these examples illustrate the capacity for improved genetic trends in well-run nucleus flocks (the Meat Elite example) and through sire referencing (the UK Texel scheme). Further, both the Poll Dorset breed in Australia and the UK Texel scheme show signs of acceleration in genetic trends.

Developed and developing countries

Evidence concerning genetic trends suggests that in countries with national sheep evaluation programmes and using BLUP technology, factors that had in the past limited wide-scale improvement are being overcome. In all cases, this development has relied on a substantial infrastructure for research and advisory support and a commercial market for at least some sheep products. In many developing countries, some or all these elements are missing, 'flock' sizes are often extremely small, control over matings is limited, and breeding goals either are very loosely defined or include traits or aspects of performance that are not easily measured (Hodges, 1992). In these circumstances initiation of nucleus flocks using institutional and international resources has advantages and provides a means of simultaneously addressing genetic improvement and conservation issues (Hammond, 1994).

Evidence for genetic variation in production traits in a wide range of breeds and production systems in developed and developing countries, together with increasing availability of software tools for variance component estimation and breeding-value prediction, means that, if the non-genetic aspects of implementing breeding programmes can be overcome, satisfactory rates of genetic improvement should be achieved. Capturing value from such programmes and maintaining the necessary management and scientific resources remain a far greater challenge.

Prospects

In reviewing the application of animal breeding theory to sheep breeding for meat and lamb production it is hard to avoid the conclusion that, despite a sound knowledge base on genetic resources and means to utilize them, breeding programmes have, until recently, not been very effective in generating industry-wide genetic improvement and have not helped overcome problems of declining consumption and competitiveness of sheep meats. Technical developments in breeding-value prediction and breeding-programme organization appear to be rapidly improving the effectiveness of meat-sheep breeding programmes in a number of Western countries. Where such genetic improvements are linked to improved commercial production systems and there is a close integration between the breeding, production and processing/retailing sectors of the sheep-meat industries, there appears to be considerable scope for real improvement in consumer appeal through both reductions in carcass fatness and greater price competitiveness from more efficient production. Improved harvesting of the value of genetic improvement in sheep-meat industries may therefore be stimulated by, and in turn encourage, significant acceleration of genetic progress for sheep-meat and lamb production traits and improved utilization of both sheep genetic resources and our knowledge of them.

References

Anderson, S. and Curran, M.K. (1990) Selection and response within the nucleus of a group breeding scheme. *Animal Production* 51, 593–599.

Atkins, K.D. (1987) Breeding objectives for terminal sire breeds of sheep. *Proceedings of the Australian Association of Animal Breeding and Genetics* 6, 221–224.

Baker, R.L. (1986) Evaluation and utilization of breed resources – sheep and goats. In: *Proceedings of the 3rd World Congress of Genetics Applied to Livestock Production* Lincoln, Nebraska, USA, Vol 9. pp. 495–496.

Ball, A.J. and Thompson, J.M. (1995) The effect of selection for differences in ultrasonic backfat depth on the feed utilisation for maintenance and biological efficiency in sheep. *Proceedings of the Australian Association of Animal Breeding and Genetics* 11, 403–407.

Banks, R.G. (1994) LAMBPLAN – Genetic evaluation for the Australian lamb industry. In: *Proceedings of the 5th World Congress of Genetics Applied to Livestock Production* Guelph, Ontario, Canada, Vol 18. pp. 15–18.

Bennett, G.L. (1990) Selection for growth and carcass composition in sheep. In: *Proceedings of the 4th World Congress of Genetics Applied to Livestock Production* Edinburgh, UK, Vol 15. pp. 27–36.

Binnie, D.B. and Clarke, J.N. (1992) Estimation of relative economic values for sheep breeding programmes. *Proceedings of the New Zealand Branch of the Australian Economics Society* 71, 86–190.

Bishop, S.C. (1993) Selection for predicted carcass lean content in Scottish Blackface sheep. *Animal Production* 56, 379–386.

Bishop, S.C., Cameron, N.D., Speake, B.K., Bracken, J., Ratchford, I.A.J. and Noble, R.C. (1995) Responses in adipocyte dimensions to divergent selection for predicted carcass lean content in sheep. *Animal Science* 60, 215–222.

Bradford, G.E. (1985) Selection for litter size. In: Land, R.B. and Robinson, D.W. (eds) *Genetics of Reproduction in Sheep*. Butterworths, London, pp. 3–18.

Bradford, G.E. and Meyer, H.H. (1986) Economic evaluation of breeding objectives for sheep and goats: practical considerations and examples. In: *Proceedings of the 3rd World Congress of Genetics Applied to Livestock Production* Lincoln, Nebraska, USA, Vol 9. 479–492.

Brash, L.D., Fogarty, N.M., Barwick, S.A. and Gilmour, A.R. (1994) Genetic parameters for Australian maternal and dual-purpose meatsheep breeds. I. Liveweight, wool production and reproduction in Border Leicester and related types. *Australian Journal of Agricultural Research* 45, 459–468.

Cameron, N.D. (1992) Correlated physiological responses to selection for carcass lean content in sheep. *Livestock Production Science* 30, 53–68.

Cameron, N.D. and Bracken, J. (1992) Selection for carcass lean content in a terminal sire breed of sheep. *Animal Production* 54, 367–377.

Cameron, N.D. and Cienfuegos-Rivas, E. (1994) Genetic and phenotypic relationships between physiological traits and performance traits in sheep. *Génétique, Sélection et Evolution*. 26: 137–150.

Carpenter, C.E. and Solomon, M.B. (1995) Post-mortem processing procedures to tenderise Callipyge lamb. *Proceedings of the International Congress of Meat Science and Technology* San Antonio, Texas, USA, 41, 620–621.

Clarke, J.N. and Kirton, A.H. (1990) How good are Texels? *Proceedings of the Ruakura Farmers' Conference* 42, 163–172.

Cockett, N.E., Logan, S.P, Jackson, S.P., Ludbrook, G.D. and Snowder, G.D. (1995) Genomic imprinting: a possible explanation for the non-Mendelian inheritance pattern of the callipyge phenotype (muscle hypertrophy) in sheep. *Journal of Animal Science* 73 (Suppl. 1); 7.

Cunningham, E.P. (1979) Quantitative genetic theory and livestock improvement. *Proceedings of the Australian Association of Animal Breeding and Genetics* 1, 8–16.

Fogarty, N.M. (1987) Breeding objectives for sheep: prime lamb dams. *Proceedings of the Australian Association of Animal Breeding and Genetics* 6, 217–220.

Fogarty, N.M. (1995) Genetic parameters for live weight, fat and muscle measurements, wool production and reproduction in sheep: a review. *Animal Breeding Abstracts* 63, 101–142.

Fogarty, N.M. and Gilmour, A.R. (1993). Sensitivity of breeding objectives to prices and genetic parameters in Australian Corriedale and Polwarth dual-purpose sheep *Australian Journal of Experimental Agriculture* 33, 259–268.

Gilmour, A.R. (1994). BVEST – multi-trait animal model breeding value software. In: *Proceedings of the 5th World Congress of Genetics Applied to Livestock Production* Guelph, Ontario, Canada, Vol 22. pp. 57–58.

Gjedrem, T. (1969) Some attempts to increase the efficiency of sheep selection. *Acta Agriculturae Scandinavica.* 19; 116–126.

Groenveld, E., Kovac, M. and Wang, T. (1990). PEST, a general purpose BLUP package for multivariate prediction and estimation. In: *Proceedings of the 4th World Congress of Genetics Applied to Livestock Production* Edinburgh, UK, Vol 13. pp. 319–331.

Guy, D.R. and Croston, D. (1994) UK experience and progress with sheep sire referencing schemes. In: *Proceedings of the 5th World Congress of Genetics Applied to Livestock Production* Guelph, Ontario, Canada, Vol 18. pp. 55–58.

Haley, C.S., Lee, G.J., Ritchie, M. and Land, R.B. (1990) Direct response in males and correlated responses for reproduction in females to selection for testicular size adjusted for body weight in young male lambs. *Journal of Reproduction and Fertility* 89, 383–396.

Hall, D.G., Luff, A.F., Fogarty, N.M. and Holst, P.J. (1995) LAMBPLAN and weight and fat responses in second cross lambs. *Proceedings of the Australian Association of Animal Breeding and Genetics* 11, 185–189.

Hammond, K. (1994) Conservation of domestic animal diversity: global overview. In: *Proceedings of the 5th World Congress of Genetics Applied to Livestock Production* Guelph, Ontario, Canada, Vol 18. pp. 55–58.

Hanrahan, J.P. (1984) Results of selection for litter size and ovulation rate in sheep. In: *Proceedings of the 2nd World Congress of Sheep and Beef Cattle Breeding.* Pretoria, South Africa. pp. 483–493.

Harris, D.L., Stewart, T.S., and Arboleda, C.R. (1984) Animal Breeding Programs: Systematic Approach to their Design. AAT-NC-8. Peoria, Illinois, USA.

Hodges, J. (1992) Issues and Financial Requirements of the Programmes for the Conservation of Indigenous Livestock Breeds. *The Management of Global Animal Genetic Resources* 104, FAO, Rome.

Hoeschele, I. (1988) Statistical techniques for detection of major genes in animal breeding data. *Theoretical and Applied Genetics* 76, 311–319.

Hohenboken, W.D. (1986) Costs and benefits of breed utilization strategies in sheep. In: *Proceedings of the 3rd World Congress of Genetics Applied to Livestock Production*, Lincoln, Nebraska, USA, Vol.9. pp. 509–523.

Hopkins, D.L. (1988) Consumer attitudes toward lamb and beef cuts of different fatness.

Proceedings of the Australian Society of Animal Production 17, 420.

James, J. (1986) Economic evaluation of breeding objectives in sheep and goats – general considerations. In: *Proceedings of the 3rd World Congress of Genetics Applied to Livestock Production*, Vol. 9. pp. 470–478.

Jopson, N.B., McEwan, J.C., Dodds, K.G. and Young, M.J. (1995) Economic benefits of including computed tomography measurements in sheep breeding programmes. *Proceedings of the Australian Association of Animal Breeding and Genetics* 11, 194–197.

King, J.W.B., Russel, A.J.F., Beck, N.F.G., Williams, A.F.J. and Williams, S.P. (1994) Use of the Thoka gene from Icelandic sheep for the improvement of fecundity. In: *Proceedings of the 5th World Congress of Genetics Applied to Livestock Production*, Guelph, Ontario, Canada, Vol 18. pp. 155–158.

Kinghorn, B.P. (1986) Mating plans for selection across breeds. In: *Proceedings of the 3rd World Congress of Genetics Applied to Livestock Production*, Lincoln, Nebraska, USA, Vol. 12. pp. 233–244.

Kinghorn, B.P. and Kerr, R.J. (1995) Use of segregation analysis to help estimate genotype effects at known major loci. *Proceedings of the Australian Association of Animal Breeding and Genetics* 11, 271–274.

Kirton, A.H., Carter, A.H., Clarke, J.N., Sinclair, D.P., Mercer, G.J.K. and Duganzich, D.M. (1995) A comparison between 15 ram breeds for export lamb production. I. Liveweights, body components, carcass measurements, and composition. *New Zealand Journal of Agricultural Research* 38, 347–360.

Lee, G.J., Atkins, K.D. and Mortimer, S.L. (1990) Genetic and phenotypic components of feed intake by grazing ewes In: *Proceedings of the 4th World Congress of Genetics Applied to Livestock Production*, Edinburgh, UK, Vol. 15. pp. 53–56.

Lewis, R.M., Simm, G. and Murphy, S.V. (1994) Selection for lean growth in Suffolk sires translates into leaner carcasses in their crossbred progeny. In: *Proceedings of the 5th World Congress of Genetics Applied to Livestock Production*, Guelph, Ontario, Canada, Vol. 18. pp. 63–66.

Leymaster, K.A. and Jenkins, T.G. (1993) Comparison of Texel- and Suffolk-sired crossbred lambs for survival, growth and compositional traits. *Journal of Animal Science* 71, 859–869.

Matos, C.A.P. and Thomas, D.L. (1992) Physiology and genetics of testicular size in sheep: a review. *Livestock Production Science* 32, 1–30.

McEwan, J.C., Clarke, J.N., Knowler, M.A. and Wheeler, M. (1989) Ultrasonic fat depths in Romney lambs and hoggets from lines selected for different production traits. *Proceedings of the New Zealand Society of Animal Production* 19, 113–120.

Mercer, J.T., Brotherstone, S., Bradfield, M.J. and Guy, D.R. (1994) Estimation of genetic parameters for use in sheep sire reference schemes. In: *Proceedings of the 5th World Congress of Genetics Applied to Livestock Production*, Guelph, Ontario, Canada, Vol. 18. pp. 39–42.

Meuwissen, T.H.E. and Goddard, M.E. (1996) The use of marker haplotypes in animal breeding schemes. *Génétique, Sélection et Evolution* 28 (161–176).

Meyer, H.H., Busboom, J.R., Bahader, A. and Thompson, J.M. (1995) Carcass and cooked meat characteristics of lambs sired by Dorset rams heterozygous for the callipyge gene, Suffolk rams and Texel rams. *Journal of Animal Science* 73 (Suppl. 1), 297.

Meyer, K. (1989) Restricted maximum likelihood to estimate variance components for animal models with several random effects using a derivative-free algorithm. *Génétique, Sélection et Evolution* 21, 317–340.

Mortimer, S.L. and Atkins, K.D. (1994) Direct additive and maternal effects on wool production of Merino sheep. In: *Proceedings of the 5th World Congress of Genetics Applied to Livestock Production*, Guelph, Ontario, Canada, Vol. 18. pp. 103–106.

Nasholm, A. and Danell, O. (1994). Maternal effects on lamb weights. In: *Proceedings of the 5th World Congress of Genetics Applied to Livestock Production*, Guelph, Ontario, Canada, 18. pp. 163–166.

Oddy, V.H., Speck, P.A., Warren, H.M., and Wynn, P.C. (1995) Protein metabolism in lambs from lines divergently selected for weaning weight. *Journal of Agricultural Science* (Cambridge) 124, 129–137.

Owen, J.B., Whitaker, C.J., Axford, R.F.E. and Ap Dewi, I. (1990) Expected consequences of the segregation of a major gene in a sheep population in relation to observations on the ovulation rate of a flock of Cambridge sheep. *Animal Production* 51, 277–282.

Piper, L.R. and Bindon, B.M. (1982) Genetic segregation for fecundity in Booroola Merino sheep. In: *Proceedings of the 1st World Congress of Sheep and Beef Cattle Breeding*, Palmerston North, New Zealand, Vol. 1. pp. 395–400.

Ponzoni, R. (1986a) Economic evaluation of breeding objectives in sheep and goats – summary and commentary. In: *Proceedings of the 3rd World Congress of Genetics Applied to Livestock Production*, Lincoln, Nebraska, USA, Vol. 9. pp. 465–469.

Ponzoni, R.W. (1986b) A profit equation for the definition of the breeding objective of Australian Merino sheep. *Journal of Animal Breeding and Genetics* 103, 342–357.

Ricordeau, G., Poivey, J.P., Bodin, L., Barrillet, F. , Boussely, M. and Mollaret, R. (1986) Importance of testicular measurements of young males tested on individual performance for improvement of their daughters prolificacy: application to the selection scheme of the Lacaune milking breed. In: *Proceedings of the 3rd World Congress of Genetics Applied to Livestock Production*, Lincoln, Nebraska, USA, Vol. 11. pp. 72–77.

Ryder, M.L. (1983) *Sheep and Man*. Duckworth, London.

Simm, G. (1987). Carcass evaluation in sheep breeding programmes. In: Marai, I.F.M. and Owen, J.B. (eds) *New Techniques in Sheep Production*. Butterworths, UK, pp. 125–144.

Simm, G. (1994) Developments in improvement of meat sheep. In: *Proceedings of the 5th World Congress of Genetics Applied to Livestock Production*, Guelph, Ontario, Canada, Vol. 18. pp. 3–10.

Simm, G. and Dingwall, W.S. (1989) Selection indices for lean meat production in sheep. *Livestock Production Science* 21, 223–234.

Smith, C. (1965). The use of specialised sire and dam lines in selection for meat production. *Animal Production* 7, 337–344.

Smith, C. (1984) Rates of genetic change in farm livestock. *Research and Development in Agriculture* 1, 79–85.

Smith, C. and Banos, G. (1991) Selection within and across populations in livestock improvement. *Journal of Animal Science* 69, 2387–2394.

Snowder, G.D., Busboom, J.R., Cockett, N.E., Hendrix, F. and Mendenhall, V.T. (1994). Effect of the callipyge gene on lamb growth and carcass characteristics. In: *Proceedings of the 5th World Congress of Genetics Applied to Livestock Production* Guelph, Ontario, Canada, Vol. 18. pp. 51–54.

Solis-Ramirez, J., Blair, H.T. and Purchas, R.W. (1993) Direct and correlated responses to selection for high or low ultrasonic backfat depth in Southdown sheep. *New Zealand Journal of Agricultural Research* 36, 133–141.

Southey, B.R., Thomas, D.L., Gottfredson, R.G. and Elphick, J.E. (1994) Introgression of the gene for increased ovulation rate (FecB) from the Booroola Merino into a Rambouillet flock. In: *Proceedings of the 5th World Congress of Genetics Applied to Livestock Production*, Guelph, Ontario, Canada, Vol. 18. pp. 151–154.

Stafford, J.E. and Walkley, J.R.W. (1979) Breeding objectives and selection criteria for Australian prime lamb production. *Proceedings of the Australian Association of Animal Breeding and Genetics* 1, 337–353.

Swan, A.A. and Hickson, J.D. (1994). Maternal effects in Australian Merinos. In: *Proceedings of the 5th World Congress of Genetics Applied to Livestock Production*, Vol. 18. pp. 143–146.

Vangen, O. and Thompson, J.M. (1992). The use of CAT-scanning to measure maintenance efficiency in genetic lines of sheep. In: *Proceedings of the 43rd Annual Meeting of European Association of Animal Production*, Madrid, Spain.

Waldron, D.F., Clarke, J.N. and Rae, A.L. (1991) Analysis of lamb schedules and relative economic values of lean and fat. *Proceedings of the New Zealand Society for Animal Production* 51, 405–409.

Wolf, B.T., Smith, C., and Sales, D.I. (1980) Growth and carcass composition in the crossbred progeny of six terminal sire breeds of sheep. *Animal Production* 31, 307–313.

Young, L.D., Dickerson, G.E.,Ch'ang, T.S. and Evans, R. (1986) Heterosis retention in sheep crossbreeding. In: *Proceedings of the 3rd World Congress of Genetics Applied to Livestock Production*, Lincoln, Nebraska, USA, Vol. 9. pp. 497–508.

Genetics of Meat Quality 19

J.M. THOMPSON and A.J. BALL
Department of Animal Science, University of New England, Armidale, 2351 NSW, Australia

Introduction 523
Genetic Variation in Body Composition 524
Between-breed variation in composition 524
Between-breed variation in fat partitioning 526
Within-breed variation in composition 527
Response to Single-trait Selection 527
Callipyge-gene effects – composition 529
Genetic Variation in Visual Meat Quality 530
Colour 530
Intramuscular fat 530
Fatty-acid profile 531
Protein and amino acid concentrations 532
Genetic Variation in Eating Quality 532
Tenderness 532
pH 534
Flavour and odour 534
Conclusion 534
References 535

Introduction

Sheep-meat quality refers to both the composition of retail cuts (i.e. proportions of muscle, bone and fat) and the visual and eating, or sensory, qualities of the meat. Over recent years, results from a number of studies have highlighted the consumer preference for leaner cuts of lamb (Thatcher and Couchman, 1983; Hopkins *et al.*, 1985). As discussed in Chapter 18 and elsewhere in this volume, there is now good evidence that there has been genetic progress towards production of leaner lamb carcasses, due to economic signals against overfatness.

Visual aspects of meat quality include the colour of both fat and lean and the level of intramuscular fat in the lean. Eating quality refers to the sensory attributes of meat, including tenderness, juiciness and flavour of the product. As both visual meat quality and sensory attributes of the meat can generally only be measured in the slaughtered animal, there is limited opportunity to include eating-quality traits in a breeding programme.

This chapter will examine genetic variation in carcass composition, at both the between-breed and within-breed levels. Genetic variation in some measured meat-quality traits will be examined, along with the correlated responses that occur in meat quality with changes in body composition.

Genetic Variation in Body Composition

In most meat-sheep breeding schemes around the world, the major emphasis has been placed on economically important traits, such as growth rate and reproduction. Certainly, under payment systems that discriminate solely on the quantity of the output, there has been little justification for changing the emphasis of these breeding schemes to include traits associated with the quality of the product. However, this is beginning to change, with several countries around the world implementing payment systems based on carcass weight and a measure of fatness. Breeders have responded to these changes in a number of ways. Firstly, they have utilized those genotypes which have less total body fat or, alternatively, those breeds that have less fat in the economically important carcass depots and more in the internal fat depots. Secondly, breeders have implemented breeding schemes that have applied selection pressure on carcass traits such as fat depth and eye-muscle dimensions. A number of selection experiments have documented that rapid changes in composition are possible by selecting on live-animal estimates of carcass weight and muscle dimensions, although there has been little attention paid to other correlated responses that may occur as a consequence of selection pressure placed on specific carcass traits. New measurement technology offers the potential for more accurate measurement of carcass traits in the live animal and consequently improved genetic gains, although this will need to be balanced against the increased costs.

Between-breed variation in composition

Sheep breeds vary greatly in carcass composition when compared at the same weight and, to a lesser degree, at the same age (Butterfield, 1988). McClelland and Russel (1972) examined chemical body composition in two breeds of similar mature size and found that they had similar proportions of total chemical fat in the body when compared at the same empty body weight. From this result they suggested that, when animals of different mature size were slaughtered at the same stage of maturity, the proportion of fat in the body would be similar. In a subsequent experiment, McClelland *et al.* (1976) demonstrated that in four diverse genotypes (Suffolk, Soay, Oxford Down and Southdown), which varied greatly in mature size, differences in body composition due to both breed and sex were largely eliminated by comparison at the same stage of maturity of body weight. However, the authors were concerned that, as the range of maturity in their breeds was restricted between 40 and 70%, further experimentation should cover a wider range of maturity.

In an experiment by Butterfield *et al.* (1983), rams from large and small strains of Merino sheep were fed *ad libitum* a pelleted ration from weaning until maturity. In this experiment, mature animals were defined as 'those in which weekly body weight gains had become minimal and where muscle plus bone weight growth had apparently ceased'. As expected, the large and small strains plateaued at vastly different mature weights (115 and 90 kg for the large and small strains, respectively), although when the mature group of animals were slaughtered and their bodies dissected they were found to have similar proportions of muscle, bone and fat in the body. It was also of note that although the different tissues in the body were laid down at a differential rate, namely, muscle was average impetus, bone low impetus and fat high impetus, there was no difference between the strains in the maturing patterns for these tissues.

Based on these results, it is possible to build up a model of breed differences in body composition. The experiment of Butterfield *et al.* (1983) and others (Thompson *et al.*, 1985) showed that, although breeds may differ in mature size, they have a similar percentage composition end-point at maturity, and there was no breed effect on the rate at which body tissues matured relative to body weight. Therefore, in breed comparisons in which the end-point was comparison at the same body weight, large mature-size breeds would be less mature than small mature-size breeds and consequently have a less mature body composition, namely, a lower proportion of fat, a greater proportion of bone and a similar proportion of muscle. Similarly, if large and small mature-size breeds are compared at the same age, differences in stage of maturity will still exist, albeit much reduced. At the same age, large mature-size breeds will be slightly less mature than small mature-size breeds and this will be reflected in percentage composition, whereby large mature-size breeds will have a slightly lower percentage of fat, a slightly greater proportion of bone and a similar proportion of muscle. It follows that at the same stage of maturity there is little difference between breeds in body composition.

As proposed by Taylor (1987), most experiments that have compared a range of breeds have found that differences in body composition can generally be explained by the mature-weight/body-composition model. Deviations from the mature-weight/body-composition model may occur for a number of reasons. As noted by Wood *et al.* (1980), there are some real exceptions to this model, which are of particular interest. An outlier for sheep is the Texel breed, which appears to have less total body fat than expected for its mature size (Leymaster and Jenkins, 1993). Breeds that do not fit the general mature-size/body-composition relationship are interesting deviations that may have application in improved meat production systems. While other breeds may appear not to fit the mature-weight/body-composition model, this may be due to variation in fat partitioning, as discussed in the following section. As discussed by Thompson (1983), this is often the problem when the measurement of fat is restricted to one or a limited number of fat depots (e.g. subcutaneous fat depth) and may not necessarily reflect total body fat in the animal.

If genetic differences in body composition are simply a function of stage of maturity, in a straight-breeding enterprise increasing carcass leanness at the same slaughter weight by changing to a larger mature-size breed is unlikely to result in any improvement in biological efficiency of the sheep enterprise (Thompson, 1991). This is because the increase in lean-tissue output is effectively cancelled out by an increase in feed-maintenance costs of the larger females. If, however, a larger mature-size terminal-sire breed is used in a crossbreeding enterprise, biological efficiency is substantially increased by increasing mature size (Thompson, 1991).

Between-breed variation in fat partitioning

As discussed, sheep have a differential pattern of growth for muscle, bone and fat (Butterfield *et al.*, 1983). Similarly, there is a differential pattern of fat deposition relative to total fat (Butterfield *et al.*, 1985; Thompson *et al.*, 1987). On a chemical basis, fat contained in the bone marrow has been shown to have a very low growth impetus. Fat associated with the muscle and viscera components are the next group with a low growth impetus, followed by fat in the dissectible depots with a high growth impetus. Within the dissectible fat depots, there is also an order of fat deposition, with fat between the muscles (i.e. intermuscular fat) having the lowest growth impetus and subcutaneous fat having the highest growth impetus. The ranking of the other depots is not clearly defined (Butterfield *et al.*, 1985).

The order of chemical-fat deposition within the body has some interesting implications for meat-quality traits in sheep carcasses. The very low growth impetus of chemical fat is possibly a replacement effect when bone marrow ceases to be a site for red blood-cell production. It is estimated that this function ceases at an early stage of growth in the growing animal. That the pattern of deposition was similar for the lipid in the viscera and that in the musculature suggests that this fat is associated to some degree with protein turnover in the body. Given the importance that some markets place on the levels of intramuscular fat, particularly in beef, these results suggest that it has a different functional role in the body, which may provide some opportunity for manipulation independently of total body fat.

A number of studies have reported differences in fat partitioning between breeds of sheep. These differences appear to be associated with the maternal traits of the breed, particularly lactation. In an attempt to provide a physiological explanation for the differences in fat partitioning in sheep, Wood *et al.* (1980) proposed that those breeds with a high reproductive rate and high milk production require larger proportions of non-carcass fats for the maintenance of pregnancy and lactation. The effect of pregnancy on fat partitioning was demonstrated in a study by K. Saka, J.M. Thompson and G.N. Hinch (unpublished data), in which pregnant and non-pregnant Merino ewes were maintained at the same maternal body weight during late pregnancy. While there was no change in total body fat, the pregnant ewes had a lower proportion of fat in the internal depots (kidney and omental fat) and more in the subcutaneous depot than non-pregnant ewes. These results are consistent with pregnancy causing greater mobilization of the non-carcass depots. More recent results from Afonso (1992) show that lactation stress results in preferential mobilization of the internal fat depots. These results are consistent with between-breed differences that were related to milk-production potential (Geenty *et al.*, 1979; Butler-Hogg *et al.*, 1986).

Within-breed variation in composition

Even though there is considerable between-breed variation in body composition, genetic change within a breed is an option for manipulating carcass composition. Obviously, genetic change depends on selection for desirable carcass traits in the live animal. To date, ultrasonics has been the main technology that has been used to measure muscle and fat depth in the live animal. A number of studies have identified significant genetic variation in both ultrasonic fat and muscle depths. Simm (1992) reviewed the results of a number of selection studies and concluded that ultrasonic fat depth on live animals was heritable and had responded to selection, although Morris *et al.* (1997) suggested that these results were from relatively short-term selection experiments and that some of the published results should be treated with caution. Since the review of Simm (1992), there have been a number of studies that have estimated heritabilities and genetic correlations for muscle and fat dimensions in lamb and these are summarized in Table 19.1.

Response to Single-trait Selection

The literature contains a number of reports of experiments that have demonstrated a response to selection for low back-fat thickness in sheep. Cameron and Bracken (1992) reported the short-term selection response to use of an index containing live weight, real-time ultrasound (RTUS) fat depth and eye-muscle dimensions. This index, which was designed to change body composition without any change in live weight, resulted in symmetrical divergence. Slaughter data on crossbred progeny indicated that selection had resulted in similar rates of fat deposition but a different rate of lean growth (Cameron and Bracken, 1992).

More recently, results from two New Zealand selection experiments have been reported: one in the Southdown breed conducted at Massey University (Solis-Ramirez *et al.*, 1993) and one with Coopworths at Invermay (Morris *et al.*, 1996), which included a control line that provided the opportunity to study

Table 19.1. Heritabilities for carcass traits summarized from the literature (after the review of Simm, 1992).

Parameter	Heritability	SE	Source
Muscle depth	0.27	0.09	Conington *et al.* (1995)
	0.23	0.04	Pollott *et al.* (1994)
	0.15	0.06	Gilmour *et al.* (1994)
	0.46	0.22	Olesen and Husabo (1994)
	0.36	0.14	Bishop (1993)
Muscle width			
	0.32	0.05	Pollott *et al.* (1994)
Fat depth			
	0.16	0.06	Conington *et al.* (1995)
	0.31	0.05	Pollott *et al.* (1994)
	0.33	0.09	Gilmour *et al.* (1994)
	0.26	0.17	Olesen and Husabo (1994)
	0.39	0.13	Bishop (1993)
	0.68	0.09	Liboriussen (1995)
	0.33	0.07	Solis-Ramirez *et al.* (1993)
	0.28	0.08	Fogarty *et al.* (1994)
	0.38	0.02	Morris *et al.* (1996)

the symmetry of response to selection above and below the control line. At Massey, a 1.7 mm divergence was achieved between the high and low selection flocks after 8 years of selection (46% of the high mean (Solis-Ramirez *et al.*, 1993)). At Invermay, Morris *et al.* (1996) reported results from an experiment in which a 2.9 mm divergence in ultrasonic back fat was achieved in 14 years (the high back-fat flock response was 89% above that of the control flock and the low back-fat flock response was 36% below that of the control flock). This asymmetrical response was mainly because of declining variability in the low back-fat flock as the mean was reduced. Morris *et al.* (1996) suggested that the reduced variation in the low line resulted in a lower heritability for fat depth adjusted for weight when compared with the heritability for the high line. The net result was that the response in the high line was 2.25 times greater than in the low line. Results from these studies demonstrate that selection on ultrasonic data was indeed successful in changing muscle and fat composition of the carcass.

As a consequence of these experimental results and developments in measurement technology (particularly RTUS to measure fat and muscle depth), a number of performance recording schemes have been developed, which provide genetic evaluation of carcass traits in meat sheep. Future development of these genetic evaluation systems is likely to be constrained by the accuracy of measurement of the carcass traits. Computer-aided tomography (CAT) is an imaging technology, developed for medical applications, which potentially offers much more accurate measurement of carcass traits in the live animal.

Simm and Dingwall (1989) examined the potential gains that CAT traits could deliver if incorporated in selection indices. Jopson *et al.* (1995) estimated that CAT had the potential to increase response to selection by 70% relative to RTUS measurements. Given the greater accuracy, there is no doubt that CAT offers the potential for more accurate carcass measurement and therefore the potential to deliver increased gains in a breeding programme, although the use of CAT in commercial breeding programmes has been constrained by the very high capital and operational costs. Thompson *et al.* (1996) modelled the genetic gains that would accrue with the use of CAT technology for evaluation of carcass merit in a meat-sheep breeding nucleus supplying breeding stock to a commercial population. They evaluated a range of selection strategies incorporating varying proportions of candidates measured using CAT or RTUS and assessed the financial performance of these different strategies. They concluded that CAT technology could be profitably applied in large, well-integrated operations, but involved higher exposure to risk than RTUS, thereby limiting its likely use to well-organized large commercial or national programmes.

Ball and Thompson (1995) examined the implications of selection for increased leanness on biological efficiency at the enterprise level. Using realized responses from the high- and low-back-fat selection lines from Invermay, they estimated that selection for decreased back-fat depth would result in a net increase in biological efficiency in a straight-breeding enterprise. Selection for increased leanness was estimated to increase the cost of maintaining the ewe, although this is more than offset by an increase in the lean content of the lambs at the same slaughter weight.

Callipyge*-gene effects – composition*

The most notable major-gene effect on carcass composition and meat quality that has been identified for sheep is the *callipyge* gene. Callipyge sheep were first identified in an American Dorset flock selected with emphasis on leg conformation. Cockett *et al.* (1993) indicated that an unusual inheritance feature existed for the *callipyge* gene, suggesting that a dominant negative-imprinting gene may be responsible for the major effect on muscling and composition. It has pronounced effects on muscle development, with the major muscles of the leg, the loin and the shoulder being significantly greater in callipyge sheep when compared with controls (Koohmaraie *et al.*, 1995). Goodson *et al.* (1995) reported higher eye-muscle area (30–69%), similar slaughter and carcass weights, higher dressing percentage and lower carcass-fat thickness (24–45%) in the callipyge sheep. Koohmaraie *et al.* (1995) also reported increased eye-muscle areas (34%), lower carcass-fat depths and reduced marbling scores. Jackson and Green (1993) showed that the callipyge sheep had greater feed efficiency, which may be related to lung and liver weights and kidney and small-intestine mass (Koohmaraie *et al.*, 1995). The *Callipyge* locus has been mapped to the ovine chromosome 18. The callipyge phenotype is characterized by a non-Mendelian inheritance pattern (polar overdominance), where only the heterozygous individuals that inherit the gene from their sire express the phenotype (Cockett *et al.*, 1996).

Leroy *et al.* (1995) indicated that there were extreme animals present within the Belgian Texels that could be classified as double-muscled. These animals have a higher dressing percentage, lower total fat content, greater eye-muscle area, increased carcass conformation and higher-price returns per kilogram. Interestingly, plasmatic creatinine and creatine could be used to distinguish the double-muscled sheep from the controls (Leroy *et al.*, 1995).

Recently, another major gene for muscling has been identified in an Australian Poll Dorset flock (R.G. Banks and R. Kerr, unpublished data). The 'CARWELL' gene seems to have smaller effects than those of the *callipyge* gene, with an average effect of +3 mm in eye-muscle depth for one copy of the gene. At this point in time, the gene is believed to be maternally imprinted and there appears to be no effect of the gene on meat quality traits, particularly tenderness.

Genetic Variation in Visual Meat Quality

Colour

Meat colour is a function of the myoglobin content and light-scattering properties of the muscle. Studies in beef (Dikeman, 1994) report that myoglobin content is moderately to highly heritable. The extent of light scattering – and hence the lightness or darkness of meat – is dependent upon the myofibrillar volume. While no reports of heritability for the myoglobin content or light-scattering properties of muscle were found in lamb, Kuchtik *et al.* (1996) reported significant breed differences in reflectance values of joints. It is therefore likely that there is genetic variance in these traits and that it would respond to selection. Young *et al.* (1993b) indicated that cooked Merino meat was darker than cooked Coopworth meat and suggested that the darker colour was a result of higher iron content, present as myoglobin and other iron-containing proteins.

Legrand *et al.* (1995) suggested that breed differences may occur with regard to fat quality, especially fat colour. They showed that ram lambs sired by Texels had a greater acceptability-of-colour score for the subcutaneous fat when compared with lambs sired by Charollais. Crouse *et al.* (1981) also reported breed effects on fat colour, with Suffolk-sired lambs having yellower fat than lambs sired by Rambouillets.

Intramuscular fat

Intramuscular fat, or marbling, is generally not considered as important a meat-quality trait in sheep meat compared with beef. Even in beef, the relationship between intramuscular fat and eating quality can only be considered variable. At medium levels of intramuscular fat, the consensus from a number of US studies would be that a slight positive relationship exists with tenderness (Cross, 1977). However, at lower levels of intramuscular fat (marbling score 1 and 2) in beef, Australian data have failed to find a significant relationship with any attribute of eating quality (Hearnshaw *et al.*, 1995).

While the contribution of intramuscular fat to eating quality of sheep meat is still not quantified, it would be wise to heed the pig-breeding experience. In earlier studies, intramuscular-fat content in pig meat was not identified as a major contributor to eating quality and little emphasis was placed on changes that occurred in selection programmes which decreased back fat in pigs. It is now generally agreed that the decline in eating quality in the modern-day pig is associated with the decline in intramuscular-fat content (Wood and Cameron, 1994). In a number of European countries, large breeding programmes have been commenced to try to address the low levels of intramuscular fat.

Kempster (1989) raised concerns over the drop in eating quality of large, very lean, lamb carcasses. It may be timely for the lamb industry to investigate the importance of intramuscular fat in their product or, at the very least, to monitor the changes that occur with modern genotypes and production systems. A number of studies have indicated variable and often low correlations between fat depots in lambs (Olson *et al.*, 1976) and other species (e.g. pigs, Duneic *et al.*, 1961; cattle, Renand, 1995), which suggests that a decrease in dissectible fat depots may be possible without concomitant changes in the intramuscular depot.

Selection for a change in back-fat thickness has altered the intramuscular-fat content of the longissimus dorsi muscle. J.C. McEwan (unpublished data) indicated that the fat line from the Coopworth selection experiment (Morris *et al.*, 1996) had 21% more fat than the lean line. He indicated that most of this change could be accounted for by a correspondingly small but significant reduction in the water content of the longissimus dorsi muscle of the fat line. This change in fat partitioning is supported by data from the Massey lines, where, after adjustment to a constant total half-carcass dissectible-fat weight, the low back-fat line had 28% less intramuscular fat than the high back-fat line (Kadim *et al.*, 1989).

Fatty-acid profile

Cameron *et al.* (1994) examined the changes that occurred in the lipid content of the adipose-tissue and fatty-acid profiles of subcutaneous fat from lines of Texel–Oxford and Scottish Blackface rams that had been divergently selected for carcass lean content. They showed that, although back-fat depth responded to selection, there were no significant changes in the individual fatty-acid concentrations of subcutaneous fat or in the proportion of unsaturated fatty acids. They did, however, show that there were breed differences for the concentration of myristic acid (C14:0), with Scottish Blackface rams having higher concentrations than Texel–Oxford rams.

Cameron *et al.* (1994) showed that across breeds and lines the concentration of unsaturated fat in the subcutaneous depot was positively correlated with subcutaneous-fat depth. In the Texel–Oxford breed, selection for a reduction in back-fat thickness lowered the lipid content within the adipose tissue from the subcutaneous depot. In contrast, there was no effect within the Scottish Blackface breed (Cameron *et al.*, 1994). Cramer *et al.* (1970) observed that the subcutaneous fat from Merinos was more unsaturated than that from coarse-woolled Romneys. Wong *et al.* (1975) indicated that the proportions of volatile branch-chain fatty acids are important components of odours released during the cooking of both sheep and goat meat.

More recently, Webb and Casey (1995) showed that there were genetic differences in the fatty-acid composition of subcutaneous adipose tissue. After correction for differences in maturity, the concentrations of C16:0, C16:1 and C18:0 differed between South Australia mutton Merinos and Dorpers. These results indicate that selection could change lipid contents within the fat depots, although the effects are likely to differ between breeds. Breed influenced the proportions of myristic (C14:0), heptadecenoic (C17:1) and oleic (C18:1) acids; however, when compared at equivalent levels of fatness, the breed differences in the proportions of C17:1 and C18:1 in the subcutaneous tissue were neligible. These results indicate that selection could change lipid contents within the fat depots, although the effects are likely to differ between breeds.

Protein and amino acid concentrations

Swartvagherova *et al.* (1996) recently presented results of breed effects on both the protein concentration and relative proportions of essential and non-essential amino acids of the longissimus dorsi muscle. In a comparison of three breeds, they showed that there was no breed effect on protein concentration, while breed affected the relative proportions of all essential amino acids except lysine. Large breed differences existed in proline, and all other non-essential amino acids exhibited breed differences, except tyrosine and histidine. These results indicate that breed differences may be used to select for desirable improvements in the concentrations of both essential and non-essential amino acids.

Genetic Variation in Eating Quality

Tenderness

Tenderness of meat is a function of myofibrillar and connective-tissue toughness. The connective-tissue component of toughness is often referred to as ‘background toughness’. Connective-tissue content of muscle varies within the carcass, with muscles such as the longissimus dorsi having a low connective-tissue content compared with muscles such as semitendinosus. As the animal ages, there is little change in the total connective tissue content of muscles, although there is a large increase in the thermal stability. Myofibrillar toughness tends to be related to the extent of the chemical bonding between the actin and myosin filaments within the myofibril. Myofibrillar tenderness is affected by the rate of glycolysis post-slaughter. Fast-glycolysing muscle tends to have a lower shear force and is more tender. Therefore, a rapid attainment of a low-pH/high-temperature condition is thought to activate the proteolytic enzymes of the lysosome. The early post-mortem changes are a complex interaction of pH, temperature and glycoloysis, which is reflected in the wide variation in heritability estimates for meat tenderness (Kemp, 1994).

Young *et al.* (1993b) used a trained taste panel to examine the differences in consumer-assessed tenderness of the loins of terminal progeny from Merino dams put to six sire breeds. They showed that the pure Merinos had a higher overall tenderness score than Oxford Down, Suffolk, Poll Dorset and Texel cross lambs; there was no difference between lambs sired by Border Leicesters and any of the other breeds. In a different study reported in the paper of Young *et al.* (1993b), the tenderness of loins from pure Merinos and Coopworths did not differ. Young and Dobbie (1994) compared the intramuscular collagen content of a forequarter muscle (supraspinatus) and two hindquarter muscles (semimembranosus and biceps femoris) of straight-bred Romney lambs with lambs sired by a terminal sire (Texel). They showed that, between the ages of 100 and 215 days, there was no effect of breed on the collagen content of either the forequarter or hindquarter muscles. This suggests that use of terminal sires will have little effect on the tenderness of lamb meat.

The collagen component of two muscles (semimembranosus and biceps femoris) was compared by Young *et al.* (1993a) in sheep selected for weight as yearlings. There was no line effect on the collagen concentration or solubility or on the shear force of the cooked semimembranosus muscle. They showed that the collagen concentration was 14% higher in the high-body-weight line, although there was no line effect on collagen solubility. Even though a line effect was apparent for the biceps femoris, Young *et al.* (1993a) suggested that the difference would only equate to a change of ½ in a tenderness score (on a 9-point scale) and that this would be unimportant to consumers' perception of lamb tenderness. Overall, they concluded that selection for increased body weight at weaning is unlikely to have any effect on the tenderness of lamb.

Callipyge-*gene effects – tenderness*

Meat from longissimus dorsi of callipyge sheep was shown to have much higher shear-force values, even after ageing for 21 days (Goodson *et al.*, 1995; Koohmaraie *et al.*, 1995). Connective-tissue content of the longissimus and myrofibrillar tenderness were higher for callipyge sheep, which resulted in lower consumer acceptance for tenderness. Koohmaraie *et al.* (1995) have shown that similar pH and temperature declines occurred with ageing in callipyge and control phenotypes. The callipyge sheep show higher calpastatin and μ-calpain levels, but no difference in μ-calpain, increased satellite-cell proliferation and suggestions of a higher capacity for protein synthesis and a reduced rate of protein degradation (Koohmaraie *et al.*, 1995). Koohmaraie *et al.* (1995) indicated that the troponin-T, desmin, nebulin and titan rates of degradation at day 20 for the *callipyge* genotype are similar to rates found at day 1 for normal genotypes. This suggests that muscle from callipyge-phenotype sheep tenderizes only slightly over time.

pH

Young *et al.* (1993b) showed that the mean ultimate pH of meats from Coopworths was significantly lower than that for Merinos (5.77 and 6.16, respectively) and that pH means of meats from lambs sired by five breeds lay between the two. However, ultimate pH was a poor indicator of the quality attributes of aroma, flavour, tenderness, texture and juiciness and overall acceptability. Recently, Hopkins *et al.* (1996) indicated that there was a genotype effect on the ultimate pH, with Border Leicester × Merino and Merino lambs having a higher pH than lambs sired by Texels and Poll Dorsets. They indicated that breed effects may need to be considered when managing lambs prior to slaughter.

Flavour and odour

Crouse *et al.* (1981) showed that there was no difference in the flavour of cooked meat from Suffolk- and Rambouillet-sired lambs. Young *et al.* (1993b) examined the differences in the flavour and odour intensities of cooked lean and subcutaneous fat between Merino and Coopworth lambs. They showed that the sheep-meat flavour of the lean was stronger for the Coopworth and that the foreign flavour was stronger for the Merino. However, they indicated that the ultimate pH differed between the two breeds and may have confounded the results for flavour. Although a trained taste panel identified breed differences in the odour from the fat, the means were not significantly different for either sheep-meat or foreign odour. In the same study, Young *et al.* (1993b) showed that there was no effect of sire breed on aroma and flavour attributes as assessed by a trained taste panel.

Conclusion

Genetic variation in body composition exists at both the between-breed and the within-breed levels. At the between-breed level, the majority of the variation appears to be associated with variation in mature size. For straight-breeding enterprises, a change in mature size may necessarily be associated with a large change in biological efficiency at the flock level. Within-breed selection for changes in high or low back-fat and muscle depth have resulted in changes in the relative rates of deposition of lean and fat, which Ball and Thompson (1995) showed resulted in an increase in biological efficiency at the enterprise level.

Genetic variation also exists in meat-quality traits, although, as measurement is generally restricted to the slaughtered animal, utilization of this information in a breeding programme is more difficult. As many of the meat-quality traits are interrelated, it is important not only that the genetic variation be identified but also that the interrelationships between traits be understood, so that gains in one trait will not be offset by a decline in another trait.

The *callipyge* gene clearly offers potential for improvements in carcass composition, carcass conformation and the efficiency of lean deposition. However, there are clear reductions in overall consumer acceptability of meat produced from callipyge sheep. It remains to be seen whether techniques for tenderization of the meat from callipyge sheep (Carpenter and Solomon, 1995; Solomon *et al.*, 1995) will enable producers to increase production, while maintaining customer satisfaction, through inclusion of the *callipyge* gene into commercial lamb populations.

References

Afonso, M.J.J. (1992) Changes in body composition in sheep during growth, pregnancy and lactation. PhD thesis, University of New England.

Ball, A.J. and Thompson, J.M. (1995) The effect of selection for differences in ultrasonic backfat depth on feed utilization for maintenance and biological efficiency in sheep. *Proceedings of the Australian Society of Animal Breeding and Genetics* 11, 403–407.

Bishop, S.C. (1993) Selection for predicted carcass lean content in Scottish Blackface sheep. *Animal Production* 56(3), 379–386

Butler-Hogg, B.W., Whelehan, O.P. and Mummery P. (1986) Carcass quality of dairy sheep. *Animal Production* 42, 461 (abstract)

Butterfield, R.M. (1988) *New Concepts in Sheep Growth*. Griffin Press, Netley, South Australia.

Butterfield, R.M., Griffiths, D.A., Thompson, J.M., Zamora, J. and James, A.M. (1983) Changes in body composition relative to weight and maturity in large and small strains of Australian Merino rams. 1. Muscle, bone and fat. *Animal Production* 36, 29–37.

Butterfield, R.M., Thompson, J.M. and Reddacliff, K.J. (1985) Changes in body composition relative to weight and maturity of Australian Dorset Horn rams and wethers. 3. Fat partitioning. *Animal Production* 40, 129–134.

Cameron, N.D. and Bracken, J. (1992) Selection for carcass lean content in a terminal sire breed of sheep. *Animal Production* 54, 367–377.

Cameron, N.D., Bishop, S.C., Speake, B.K., Bracken, J. and Noble, R.C. (1994) Lipid composition and metabolism of subcutaneous fat in sheep divergently selected for carcass lean content. *Animal Production* 58, 237–242.

Carpenter, C.E. and Solomon, M. (1995) Post-mortem processing procedures to tenderise Callipyge lamb. In: *Proceedings of the 41st Annual International Congress of Meat Science and Technology, San Antonio, USA*, Vol. 11. pp. 620–621.

Cockett, N.E., Jackson, S.P., Shay, T.L., Nielsen, D. Moore, S.S., Steele, M.R., Barendise, W., Greene, R.D. and Georges, M. (1993) Chromosomal localisation of the *Callipyge* gene in sheep (*Ovis aries*) using bovine DNA markers. *Proceedings of the National Academy of Sciences USA* 91, 3019–3023.

Cockett, N.E., Jackson, S.P. Shay, T.L., Farnir, F., Berghmans, S., Snowder, G.D., Nielsen, D.M. and Georges, M. (1996) Polar overdominance at the ovine Callipyge locus. *Science* 273(5272), 236–238

Connington, J., Bishop, S.C., Waterhouse, A. and Simm, G. (1995) Genetic analysis of early growth and ultrasound measurements in hill sheep. *Animal Production* 61(1), 85–95.

Cramer, D.A., Pruett, J.B., Kattnig, R.M. and Schwartz, W.C. (1970) Comparing breeds of sheep. II Carcass characteristics. *Proceedings of the Western Section of the American Society of Animal Science* 21, 267a–271a.

Cross, H.R. (1977) Effects of amount, distribution and texture of marbling on cooking properties of beef longissimus. *Journal of Food Science* 42, 182–185.

Crouse, J.D., Busboom, J.R., Field, R.A. and Ferrell, C.L. (1981) The effect of breed, diet, sex, location and slaughter weight on lamb growth, carcass composition and meat flavour. *Journal of Animal Science* 53, 376–386.

Dikeman, M.E (1994) Genetics of meat quality. In: *Proceedings of the 5th World Congress on Genetics Applied to Livestock Production*, Guelph, Ontario, Canada, Vol. 19. pp. 437–438.

Duneic, H. Keilanowski, J. and Osinka, Z. (1961) Heritability of chemical fat content in the loin muscles of baconers. *Animal Production* 3, 195–198.

Fogarty, N.M., Brash, L.D. and Gilmour, A.R. (1994) Genetic parameters for reproduction and lamb production and their components and liveweight, fat depth and wool production in Hyfer sheep. *Australian Journal of Agricultural Research* 45, 443–457.

Geenty, K.G., Clarke, J.N. and Jury, K.E. (1979) Carcass growth and development of Romney, Corriedale, Dorset and crossbred sheep. New Zealand Journal of Agricultural Research 22, 23–32.

Gilmour, A.R., Luff, A.F., Fogarty, N.M. and Banks, R. (1994) Genetic parameters for ultrasound fat depth and eye muscle measurements in live Poll Dorset sheep. *Australian Journal of Agricultural Research* 45, 1281–1291.

Goodson, K.J, Miller, R.K. and Savell, J.W. (1995) Carcass characteristics, muscle components and palatability assessments of meat from Callipyge versus normal lambs. In: *Proceedings of the 41st Annual International Congress of Meat Science and Technology*, Vol. 11. pp. 624–625.

Hearnshaw, H., Shorthose, W.R., Melville, G., Rymill, S.R., Thompson, J.M., Arthur, P. and Stephenson, P.D. (1995) Are carcass grades a useful indication of consumer assessment of eating quality of beef. In: *Meat '95: The Australian Meat Industry Research Conference, September 1995.*

Hopkins A.F., Congram, I.D. and Shorthose, W.R. (1985) *Australian Consumer Requirements for Lamb and Beef. 2. Consumer Preferences for Three Common Cuts of Lamb of Varying Weights and Fatness and Subject to Varying Chilling Rates.* Research Series, Report No 19, Livestock and Meat Authority of Queensland, Brisbane, Queensland, Australia.

Hopkins, D.L. Fogarty, N.M. and Menzies, D.J. (1996) Muscle pH of lamb genotypes. *Proceedings of the Australian Society of Animal Production* 21, 347.

Jackson, S.P and Green, R.D. (1993) Muscle trait inheritance, growth performance and feed efficiency of sheep exhibiting a muscle hypertrophy phenotype. *Journal of Animal Science* 71 (Suppl. 1), 241.

Jopson, N.B., McEwan, J.C. Dodds, K.G. and Young, M.J. (1995) Economic benefits of including computed tomography measurments in sheep breeding programmes. *Proceedings of the Australian Association of Animal Breeding and Genetics* 11, 194.

Kadim, I.T., Purchas, R.W., Rae, A.L. and Barton, R.A. (1989) Carcass characteristics of southdown rams from high and low backfat selection lines. *New Zealand Journal of Agricultural Research* 32, 181–191.

Kemp, R.A. (1994) Genetics of meat quality in cattle. In: *Proceedings of the 5th World Congress on Genetics Applied to Livestock Production*, Guelph, Ontario, Canada, Vol. 19. pp. 439–445.

Kempster A.J. (1989) Carcass and meat quality research to meet market needs. *Animal Production* 48, 483–496.

Koohmaraie, M., Shackelford, S.D., Wheeler, T.L., Lonergan, S.M. and Doumit, M.E. (1995) A muscle hypertrophy condition in lamb (Callipyge): characterisation of effects on muscle growth and meat quality traits. *Journal of Animal Science* 73, 3596–3607.

Kuchtik, J., Subrt, J. and Horak, F. (1996) Quality analysis of meat of slaughter lambs. *Zivocisna Vyroba* 41, 183–188.

Legrand, I., Denoyelle, C. and Quilichini, Y. (1995) Effects of breed and rationing concentrates on lambs subcutaneous adipose tissue quality. In: *Proceedings of the 41st Annual International Congress of Meat Science and Technology*, Vol. 11. pp. 116–117.

Leroy, P.L., Marcq, F., Charlier, C. and Farnir, F. (1995) Growth and carcass performance of purebred Texel rams tested in station. In: *Proceedings of the 40th Annual European Association of Animal Production, Dublin*. pp. 221–222.

Leymaster, K.A. and Jenkins, T.G. (1993) Comparison of Texel- and Suffolk-sired crossbred lambs for survival, growth and compositional traits. *Journal of Animal Science* 71, 859–869.

Liboriussen, T. (1995) Genetic parameters for daily gain, muscle area and fat thickness in sheep. In: *Proceedings of the 46th Annual European Association Animal Production, Prague*.

McClelland, H.T. and Russel, A.J.F. (1972) The distribution of body fat in Scottish Blackface and Finnish Landrace lambs. *Animal Production* 15, 301–306.

McClelland, H.T., Bonatti, B. and Taylor, St.C.S. (1976) Breed differences in body composition of equally mature sheep. *Animal Production* 23, 281–293.

Morris, C.A., McEwan J.C., Fennessy, P.F., Hickey, S.M., Bain, W.E. and Greer, G.J. (1997) Response to selection for low and high backfat depths in Coopworth sheep. *Animal Science* (in press).

Olesen, I. and Husabo, J.O. (1994) Effect of using ultrasound muscle depth and fat depth on the accuracy of predicted phenotypic and genetic values of carcass traits on live ram lambs. *Acta Agriculturae Scandinavica Section A* 44(2), 65–72.

Olson, L.W., Dickerson, G.E., Crouse, J.D. and Glimp, H.A. (1976) Selection criteria for intensive market lamb production, carcass and growth rate traits. *Journal of Animal Science* 43, 90–101.

Pollott, G.E., Guy, D.R. and Croston, D. (1994) Genetic parameters of lamb carcass characteristics at 3 end points – fat level, age and weight. *Animal Production* 58(1), 65–75.

Renand, G. (1995) Actual basis for genetic selection in beef. In: Ouali, A., Demeyer, D.I and Smulders, F.J.M. (eds) *Expression of Tissue Proteinases and Regulation of Protein Degradation as Related to Meat Quality*. European Consortium for Continuing Education in Advanced Meat Science and Technology, pp. 413–428.

Simm, G. (1992) Selection for lean meat production in sheep. In: Speedy, A.W. (ed.) *Progress in Sheep and Goat Research*. CAB International, Wallingford, Oxfordshire, pp. 193–215.

Solis-Ramirez, J., Blair, H.T. and Purchas, R.W. (1993) Direct and correlated responses to selection for high and low ultrasonic backfat depth in Southdown sheep. *New Zealand Journal of Agricultural Research* 36, 133–141.

Solomon, M.B., Long, J.B., Eastridge, J.S. and Carpenter, C.E. (1995) Tenderising Callipyge lamb with Hydrodyne® process. In: *Proceedings of the 41st Annual International Congress of Meat Science and Technology*, Vol. 11. pp. 622–623.

Swartvagherova, K., Ochodnicky, D., Margetinova, J., Palanska, O. and Ceresnakova, Z. (1996) Protein content and amino acid concentrations in m-longissimus dorsi of lambs of various genotypes. *Zivocisna Vyroba* 41, 87–89.

Taylor, St.C.S. (1987) An evaluation of genetic size scaling in breed and sex comparisons of growth, food efficiency and body composition. *Proceedings of the Australian Association of Animal Breeding and Genetics* 6, 1–12.

Thatcher, L.P. and Couchman, R.C. (1983) Determining consumer requirements for lamb loin chops – a preliminary study. *Review of Marketing and Agricultural Economics* 51, 167–177.

Thompson, J.M. (1983) Genetic variation in the total amount and partitioning of fat in the body of sheep. In: *Implications of Developments in Meat Science, Production and Marketing of Lamb Products: Proceedings of National Workshop, Orange Agricultural College, Orange, Australia.*

Thompson, J.M. (1991) Meat production. In: *Cottle, D.J. (ed.) Australian Sheep and Wool Handbook*. INKATA Press, Melbourne, pp. 243–251.

Thompson, J.M., Butterfield, R.M. and Perry, D. (1985) Food intake, growth, and body composition in Australian Merino sheep selected for high and low weaning weight. 2. Chemical and dissectible body composition. *Animal Production* 40, 64–70.

Thompson, J.M. , Butterfield, R.M. and Perry, D. (1987) Food intake, growth and body composition in Australian Merino sheep selected for high and low weaning weight. 4. Partitioning of dissected and chemical fat in the body. *Animal Production* 45, 49–60.

Thompson, J.M. Kinghorn, B.P. and Banks, R.G. (1996) Application of non-invasive techniques for carcass evaluation in breeding. In: *47th Meeting of the European Association for Animal Production, Lillehammer, Norway.*

Webb, E.C. and Casey, N.H. (1995) Genetic differences in fatty acid composition of subcutaneous adipose tissue in Dorper and SA Mutton Merino wethers at different liveweights. *Small Ruminant Research* 18, 81–88.

Wong, E., Johnson, C., Smith, M.E. and Braggins, T.J. (1975) The contribution of 4-methyloctanoic (hirinoic) acid to mutton and goat meat flavour. *New Zealand Journal of Agricultural Research* 18, 261–266.

Wood, J.D. and Cameron, N.D. (1994) Genetics of meat quality in pigs. In: *Proceedings of the 5th World Congress on Genetics Applied to Livestock Production*, Guelph, Ontario, Canada, Vol. 19: pp. 458–464.

Wood, J.D., MacFie, H.J.H., Pomeroy, R.W. and Twin, D.J. (1980) Carcass composition of four sheep breeds: the importance of type of breed and stage of maturity. *Animal Production* 30, 135–152.

Young, O.A. and Dobbie, S.L. (1994) Characteristics of intramuscular collagen in 2 sheep breeds. *New Zealand Journal of Agricultural Research* 37, 93–97.

Young, O.A., Hogg, B.W., Mortimer, B.J. and Waller, J.E. (1993a) Collagen in two muscles of sheep selected for weight as yearlings. *New Zealand Journal of Agricultural Research* 36, 143–150.

Young, O.A., Reid, D.H. and Scales, G.H. (1993b) Effect of breed and ultimate pH on the odour and flavour of sheep meat. *New Zealand Journal of Agricultural Research* 36, 363–370.

Genetics of Milk Production

20

F. BARILLET
INRA, Station d'Amélioration Génétique des Animaux, BP 27, 31326 Castanet-Tolosan Cedex, France

Introduction 539
Breeding Strategy 540
Evaluating breed resources 540
Comparisons between breeds 542
Breeding Objectives and Dairy-selection Criteria 543
Genetic parameters for dairy traits 544
Breeding objectives and dairy-selection criteria 544
Conception of Breeding Programmes for Dairy-sheep Populations 546
Pyramidal management of the population 546
Optimization of the breeding programme 547
Rationalization and simplification of milk-recording 548
Implementation of Breeding Schemes for Dairy Sheep in Europe 550
Survey of milk-recording 550
Use of artificial insemination 552
Genetic evaluation in dairy sheep 552
Published estimates of genetic trend for milk traits 554
Description of an Efficient Breeding Programme for a Dairy-sheep Population 556
The Lacaune breeding scheme 557
Phenotypic and genetic trends for selected and base populations 557
Secondary traits 558
Conclusion 559
References 560

Introduction

Dairy sheep have traditionally been farmed in Europe, especially in Mediterranean areas. Unlike cattle, which are exclusively milked after calving, dairy-sheep breeding is much more varied and complicated. The system of milking directly after lambing is usually only used in countries or areas where milking ewes is an exception. In Mediterranean countries, where there is a population of several million dairy sheep, normal husbandry systems include a suckling period (or a combined suckling-plus-milking period) of at least 1 month. Consequently the Mediterranean dairy-sheep systems have a dual purpose, with income from meat and milk, the milk being generally processed into high-quality cheese. Very often, there is a complete association between a region of production, a breed and the brand name of a cheese. In almost all cases, the production system is characterized by an out-of-season lambing period from October to January for adult ewes. But in some area or countries the practice of three lambings in 2 years exists also. Milking only begins after weaning in winter and ends in summer for the usual one out-of-season lambing per year. A wide range of local production systems (climate, feeding, etc.) may be observed, from intensive breeding to semi-extensive systems, with grazing even in winter and often transhumance in summer. In this type of situation, milk production is affected by other sheep-production aspects, such as meat, wool and even adaptation to local breeding conditions. Thus, in the Mediterranean basin, most sheep milk is produced by multipurpose local breeds with low or medium milk yields, which have been raised under traditional husbandry conditions (Boyazoglu and Flamant, 1990). However, some of the populations have evolved into dairy types, generally with a simultaneous adaptation of the breeding system to modern husbandry methods, using machine-milking, on-farm milk recording, artificial insemination (AI) and new feeding practices.

Breeding Strategy

The breeding strategy to improve dairy traits may involve either crossbreeding or purebreed selection programmes. While purebreed selection is the method generally used for the genetic improvement of dairy sheep in the Mediterranean basin, crossbreeding is also being used. However, farmers are generally not familiar with the various crossbreeding systems, and these different systems are often too complicated to manage in dairy production. Thus farmers rely only on the simplest scheme, which involves upgrading of the local populations to dairy-type sheep (Zervas *et al.*, 1975; Mavrogenis, 1995). Given the wide range of local production conditions and multiple-purpose systems, any definition of a breeding strategy needs to take into account not only milk-yield merit but also other traits, such as body size, sexual precocity, out-of-season lambing ability, prolificacy, growth rate, wool and adaptation to local breeding conditions (Barillet, 1990a).

Evaluating breed resources

Theoretically, it is desirable to have precise knowledge of the differences in average genetic levels between the breeds for all the aforementioned traits and the specific abilities of each candidate breed for the local breeding system under consideration. This would be a necessary first step, in order to be able to choose the best breeding strategy (crossbreeding or purebreed) for each local situation. But, in practice, there are many different local dairy-sheep breeds, with each breed generally being located in a regional area of production with its own specific management conditions (Gabina, 1997). This means that there is significant confusion between the average genetic merit of the considered breed and its breeding system, so that the comparison of the milk-yield levels between breeds (using only published results) is often less informative than expected. Naturally the same difficulty exists for the other traits described above. Taking only the average milk yield, it is necessary to observe some obvious and simple rules if any comparison of the recorded results is to be reliable.

- The application of International Committee for Animal Recording (ICAR) regulations, which recommend the recording and calculation of milk yield in the milking-only period (once the lambs have been weaned), if ewes are being milked after a suckling or suckling-plus-milking period (ICAR, 1992). Indeed, ewes suckling on a test day must not be controlled, because it is impossible to measure the individual milk yield of suckling ewes or suckling-plus-milked ewes simply and accurately (conditions essential for the large-scale application of milk recording on farms). Given that we shall consider the milking period only, the range of the relationships between total lactation, milk yield and daily milk during milking only are shown in Table 20.1 for different production levels. It should be noted that it is easy to reverse the ranking of two breeds if the published results correspond to milk yield at milking period for one breed and to an estimated total lactation for the other breed.
- Remembering that the average production of several thousand ewes is always more representative than the results of a hundred ewes (particularly if these ewes correspond to a biased sampling of élite females).

Because these obvious and simple regulations are not always applied, dairy-

Table 20.1. Milk yield at milking period only and total lactation for different levels of production.

Milking period only		Total lactation (range of estimated production in litres)		
		Length of the suckling period (days)		
Milk yield (l)	Range of daily milk yield (l)	0 (colostral phase)	30	45
100	0.6–1.0	100	120–150	130–160
150	0.8–1.2	150	180–220	190–230
200	0.9–1.4	200	240–280	260–300
250	1.1–1.6	250	300–350	320–370

sheep literature sometimes offers 'unrealistic information' on the differences in milk yield (and composition) for average genetic levels between breeds. Some authors tried to classify Mediterranean breeds according to their supposed average genetic level for milk yield. However, such classifications are not reliable enough, especially for the supposed medium-milk-yield breeds (in the range 120–180 l) for the reasons explained above. The only conclusion that can be reached is that in the Mediterranean basin some breeds have a good to high level of phenotypic means (in the range 200–300 l per lactation in the milking-only period), e.g. Awassi, Chios, Comisana, Lacaune and Sarda breeds. Furthermore, it is necessary to account for the existence or the lack of an efficient breeding programme for each candidate breed (see below, under 'Implementation of Breeding Schemes for Dairy Sheep in Europe'). Thus, considering these various problems, it is realistic to organize accurate comparisons between breeds in the same environment when deciding whether to keep the local breed or to import an improved breed.

Moreover, it is important to have a good idea of the expected genetic gain over 10 years in the case of direct selection for milk yield of the native population. The genetic coefficient of variation for milk yield (at milking period only) is between 10 and 15% (Barillet and Boichard, 1987). Thus the genetic standard deviation increases from about 10 l for an average level of 100 l to 30 l for a population with an average milk yield of 250 l. So the predicted genetic gain in 10 years may reach about 15–25 l in a population with a 100 l production level and 40–65 l in a population with a 250 l production level (Table 20.2). It should be noted that these figures represent an efficient breeding programme.

Comparisons between breeds

In the 1960s and 1970s, numerous comparisons between breeds were carried out in western Mediterranean countries (France, Greece, Italy, Spain) and also in Israel and Cyprus (Goot, 1966; Flamant and Ricordeau, 1969; Ricordeau and Flamant, 1969a,b; Zervas *et al.*, 1975; Kalaissakis *et al.*, 1977; Boyazoglu *et al.*, 1979; Mavrogenis, 1988). Attempts to improve local livestock production were

Table 20.2. Expected asymptotic gain for milk yield in 10 years according to the milk-yield level of the breed (in the case of milking after a 1-month suckling period).

Milking period only		
Milk yield (l)	Genetic standard deviation (l)	Expected genetic gain in 10 years (range in litres)
100	10–15	15–25
150	15–22	20–40
200	20–30	30–55
250	25–35	40–65

based on the importation of improved breeds (generally the East Friesian breed, in all countries, and Chios or Awassi breeds in eastern Mediterranean areas). It should be noted that Chios and Awassi are semifat-tailed and fat-tailed breeds, respectively, which is not a suitable characteristic for the meat market in western Mediterranean countries. This is one of the reason why the East Friesian breed was generally chosen for the comparisons carried out in Western European countries, expecting an improvement for both milk yield and prolificacy. But, in most cases, the crossbreeding results showed very disappointing performances for genotypes with more than 50% Friesian genes, which appeared to be poorly adapted to the Mediterranean conditions.

Consequently, upgrading of native sheep populations through reliance on improved breeds appeared in most cases to be too difficult to manage. In these conditions, the breeding strategy involved either the creation of synthetic lines through crossbreeding of local and imported productive breeds (FSL (Friesian × Sarda × Lacaune) in France, Frisarta in Greece, Assaf in Israel), to avoid having more than 50% of the genes coming from the imported breed, or through dairy selection of local breeds in their native area of production (Flamant and Barillet, 1982). The balance between cost/time and specific situations must therefore be taken into account for any final decision. In practice the genetic and economic comparisons carried out in the 1970s in Western Europe concluded most often that it was wiser to implement the dairy selection of the local breeds in their specific area and conditions of production. On the other hand, in Israel, the crossbreed Awassi × East Friesian, called the Assaf breed, has replaced the purebreed Awassi in many improved dairy flocks since the first importation of East Friesian breeding stock in 1955 (Epstein, 1985).

Thus, generally, in the western Mediterranean basin, efforts in research and extension have concentrated on the design and implementation of breeding programmes using local breeds in their specific area of production.

Breeding Objectives and Dairy-selection Criteria

As described above, the Mediterranean dairy-sheep systems have a dual purpose, with two-thirds (to three-quarters) income from milk and one-third (to one-quarter) from meat, especially for medium-milk-yield breeds with one out-of-season lambing period a year. Under these conditions, the market demand is to increase the milk-yield productivity of the ewes, seeing that often the production of ewe milk is stagnant, despite the increase in consumption of sheep cheese, because most of local breeds are low- or medium-milk-yield breeds. Consequently, regarding the weighted income from milk, the main goal is to improve dairy traits of the local breeds in their specific area of production.

Genetic parameters for dairy traits

When using the standard A_4 milk-recording test (monthly recording of the two daily milkings), genetic parameters for dairy traits (Table 20.3) are similar for sheep and cattle (Barillet, 1985, 1989; Barillet and Boichard, 1987). Milk (M), fat (F) and protein yields (P) have a moderate heritability (~0.30) and are highly correlated (~0.85–0.95). Fat (F%) and protein contents (P%) have a high heritability (~0.50–0.60 when measured at a standard A_4 testing) and are positively correlated (+0.7). Fat is more variable than protein, and the fat/protein ratio of genetic standard deviations reaches 1.3 for yields and 1.85 for content. The antagonism with M is stronger for P% (–0.4) than for F% (–0.3). Genetic correlations between F% and P% and between F and P are close to zero (slightly negative), except that the genetic correlation between F and F% is clearly positive (0.3). All these parameter estimates are in general agreement with those in the literature for dairy cattle (Maijala and Hanna, 1974; Boichard and Bonaïti, 1987).

Breeding objectives and dairy-selection criteria

As for dairy cattle, the breeding objectives are usually merged in the increase of milk income rather than in the progress of margin per ewe. Such a simplification is acceptable if the dairy selection keeps or increases the feed efficiency. Therefore, a divergent selection experiment has been in progress since 1988 in the Lacaune breed to study the components of feed efficiency according to genetic merit (Marie *et al.*, 1996).

The ewe milk is mainly processed into specific high quality cheeses known for their high ratio of fat to dry content. For this reason, useful yield (U) and content (U%), defined as a linear combination of F and P or F% and P%, may be chosen as the main and secondary dairy-selection criteria (Barillet, 1985; Barillet and Boichard, 1987). Since each content is weighted with the reciprocal of its standard deviation (1 for F% and 1.85 for P%), U% gives the same

Table 20.3. Genetic parameters for Lacaune-breed milk traits in first lactation (1487 Lacaune ewe lambs born from 176 rams) with A_4 milk-recording testing*: heritabilities on diagonal and genetic correlations below diagonal (from Barillet and Boichard, 1987).

Traits	Milk yield	Fat yield	Protein yield	Fat content	Protein content
Milk yield	**0.32**				
Fat yield	0.82	**0.29**			
Protein yield	0.92	0.91	**0.27**		
Fat content	–0.34	0.24	–0.05	**0.62**	
Protein content	–0.47	–0.05	–0.10	0.75	**0.53**

*Monthly sampling of the two daily milkings.

economic value to an increase of one genetic standard deviation in F% as in P%. So the genetic correlation of U% is the same with F% as with P% (~ 0.94). The main selection criterion (U) can be considered as cheese output, i.e. a dry matter yield (F + P), by the ewe. In the same way, the secondary criterion (U%) represents the cheese yield best predicted by the dry matter content (F% + P%).

The genetic correlation between U and U% is zero. Such a result is already known in dairy cattle (Alps and Averdunk, 1984), using different weightings (1–1.2) of protein relative to fat. Although selecting exclusively on U should maintain dry-matter content at a stable genetic level, it should not be advised for dairy ewes. According to the sign of the genetic correlation between P and both contents (null to slightly negative), this selection leads to a possible decrease in P% and to an almost certain increase in F% (Table 20.4). Accordingly, it seems preferable to use as selection criterion a combination of U and U% or of U and P%. The second option is rather satisfactory, when simplification of the milk-recording design is required (Barillet, 1990b). Finally, the global dairy criterion of selection is a linear combination of F, P and P%: F + k_1P + k_2P%, where F + k_1P = U and k_1 (1.85) and k_2 are constants. It can be pointed out that k_1 can be chosen in a large range of values (between 0 and ∞), in view of the high genetic correlation between F and P (~ 0.92) and the fat/protein-yields ratio of genetic standard deviations, which reaches 1.3. On the other hand, k_2 depends strongly on the payment for the chemical composition of the milk. In the French situation (Lacaune breed), k_1 = 1.85 and k_2 = 1/10 to 1/20, with U expressed in kg and P% expressed in g l^{-1}.

Nevertheless, such a criterion has to be used only when the breeding programme reaches the asymptotic annual genetic gain. In other words, at the starting period of the breeding programme (see next section), it is recommended to select on milk yield only, in order to reduce the costs of the implementation

Table 20.4. Predicted asymptotic genetic changes in 10 years according to the dairy criteria of selection: selection scheme with progeny test of rams on 40 daughters (for a 200 l production-level breed).

	Genetic changes in 10 years						
Criteria of selection	Milk yield (M) (l)	Fat yield (F) (kg)	Protein yield (P) (kg)	Dry-matter yield (F + P) (kg)	Fat content (F%) (g l^{-1})	Protein content (P%) (g l^{-1})	Dry-matter content (F% + P%) (g l^{-1})
Milk yield	53	2.746	2.253	4.999	–3	–2.2	–5.2
Useful yield (F + 1.85P)	48	3.138	2.375	5.513	+0.3	–0.3	0
Useful yield + 1/5 P%	35	2.746	2.155	4.901	+2.8	+1.9	+4.7

of the on-farm milk recording on a large scale, seeing, moreover, that the dry-matter content of low- or medium-milk-yield local breeds is generally high.

Conception of Breeding Programmes for Dairy-sheep Populations

Whatever the proposed programme, it should be kept in mind that an efficient breeding programme must ensure genetic gain and also its diffusion to the entire population (recorded and unrecorded flocks). In other words, our purpose is to develop an efficient breeding programme for a dairy-sheep population (300,000 to 3 million ewes) and not for a few solitary breeders.

Pyramidal management of the population

Whatever the dairy species, the most efficient breeding programmes need to progeny-test AI males born from assortative matings. Consequently, such a programme requires on-farm milk-recording, AI (eventually embryo transfer) and an accurate genetic evaluation (breeding value) of the animals. However, in practice, biological and economic constraints mean that these selection tools have to be adapted to sheep conditions (Barillet and Elsen, 1979; Flamant and Barillet, 1982).

Artificial insemination for sheep on a large scale (several thousand per day over a short period of 2–3 months a year) is most often carried out with fresh semen and induced oestrus. The diffusion potential of an AI ram is multiplied by 10–20 compared with a natural mating (NM) ram, but remains very limited compared with AI bulls (Barillet *et al.*, 1984). The official A_4 milk-recording (monthly recording of the two daily milkings) is the most popular method used in dairy-cattle recording. For dairy sheep, its cost per animal relative to the income is two to three times as much as for dairy cattle (Flamant and Poutous, 1970). Under these conditions, the development of a dairy-sheep breeding programme is suitable only in the framework of pyramidal management of the population, for both AI and NM rams, which is far removed from the dairy-cattle situation, based on only a few AI bulls; indeed, with such a pyramidal organization, the breeding tools required by the breeding programme (pedigree-recording, official milk recording, AI, progeny test, assortative mating) can be concentrated in an open selected nucleus (Elsen and Mocquot, 1974).

In pyramidal management of the population (Figure 20.1), the breeders are divided into two groups: the breeders of the nucleus flocks and the breeders for the commercial flocks of the base population (Barillet *et al.*, 1986). Official milk-recording is used only in the nucleus. Artificial insemination is primarily reserved for the nucleus flocks but may be used elsewhere. The breeding centres play a crucial role in such a scheme (Barillet and Elsen, 1979). These specific centres, managed by the breeders associations, have to raise the young males born from assortative matings in the nucleus, from weaning (at 1 month old) to

8 months of age. Thereafter, these males either join the AI centres for a subsequent progeny test or are sold to commercial flocks for NM. Note that the gene flow from the nucleus to the base population is based partly on AI and partly on these NM rams, born in the nucleus, raised in breeding centres and used in the base population. In spite of its good stability, the nucleus remains open. This means that each year some of the best breeders in the base population may join the nucleus (using the results of a simplified recording implemented in the commercial flocks).

Optimization of the breeding programme

The main results of several gene-flow studies, carried out for such a pyramidal breeding programme, using the Lacaune demographic and genetic parameters (Elsen and Mocquot, 1974; Barillet and Elsen, 1979; Vallerand and Elsen, 1979; Barillet *et al.*, 1988), can be summarized as follows for the nucleus flocks:

- The optimum size of the nucleus ranges between 10 and 20% of the population that is to be improved.
- The asymptotic annual genetic gain can be expected from 10 to 15 years after starting the breeding programme. It is 50% higher when AI is used, rather than simply relying on NM in the nucleus.
- Because proved rams have limited semen production, they cannot be as highly selected as in dairy cattle, and their superiority over young unproved

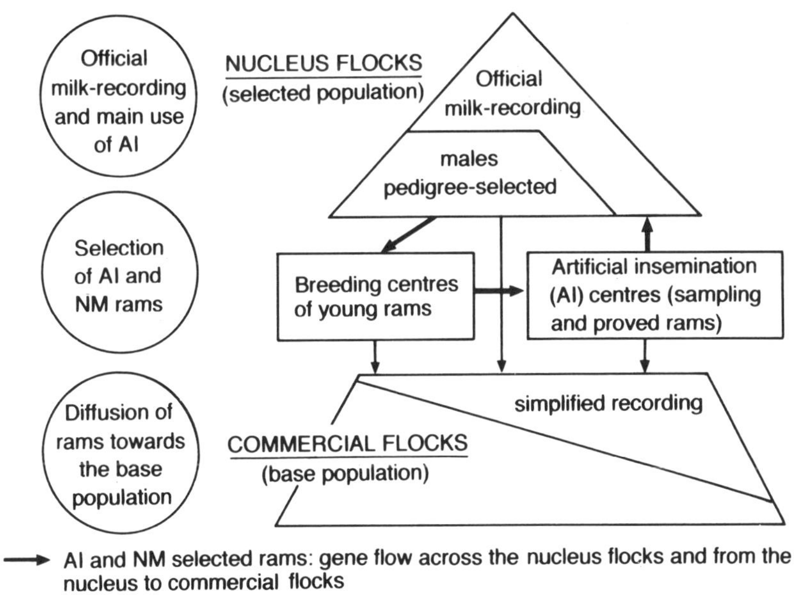

Fig. 20.1. Pyramidal management of the population.

rams is lower than in dairy cattle. Consequently, the best solution in order to maximize both the annual genetic gain and the discounted benefits is to progeny-test quite a large number of males in the nucleus: in a pure AI optimal situation, 40–50% of the female population should be mated with sampling rams.

Although such a programme is quite different from the usual dairy-cattle situation, it may provide the same expected annual genetic gain – for example, 0.20 genetic standard deviation per year, i.e. 2–2.5% of the mean population genetic level.

The main results of the same gene-flow studies for the commercial flocks (base population) are as follows:

- At the equilibrium, the annual genetic gain is the same in the nucleus and the base population.
- This equilibrium is reached more or less rapidly, depending on the diffusion rate of the rams from the nucleus to the base population.
- At the equilibrium, the lag between the nucleus and base populations may be limited to 5–10 years if the diffusion rate (*d*) reaches 100% (Fig. 20.2).

Rationalization and simplification of milk-recording

The rationalization and simplification of milk-recording is recommended because fixed costs are too high using the standard A_4 testing (monthly recording of the two daily milkings), which is considered to be the international reference in dairy cattle (ICAR nomenclature). Consequently, the official recording must be limited to those breeders who belong to the on-farm open nucleus, while a very simplified recording system can be implemented (unofficial recording) for commercial flocks. For instance, as practised in France (Cottier, 1972), a bimonthly recording can be carried out (for milk yield only)

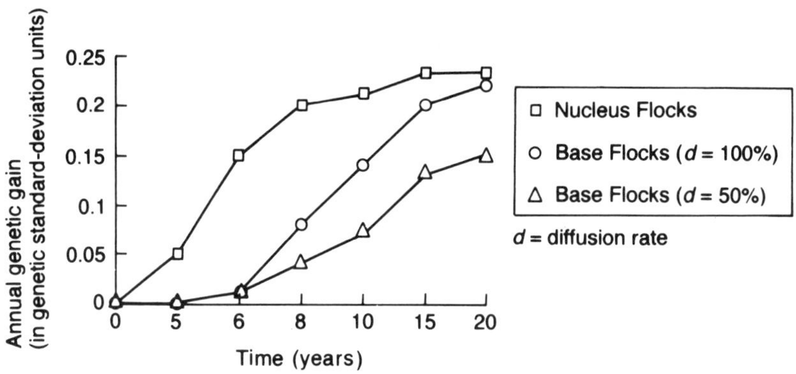

Fig. 20.2. Annual genetic gain for the nucleus and base flocks.

in order to rank ewes within the flock and to plan matings with rams selected from the nucleus. It is called simplified milk-recording in France (see Figure 20.1).

For the nucleus flocks, during the starting period of the breeding programme (at least the first 10–15 years), it is recommended to record only milk yield, in order to reduce the costs and implement the on-farm recording on a large scale. The first goal of this starting period is to reach the optimum size for the nucleus (10–20% of the whole population to be improved). Since during the starting period the genetic gain for milk yield is small, milk composition, which is expensive to collect, may be neglected (Barillet *et al.*, 1986). Logically, ICAR regulations for milk-recording in sheep describe qualitative tests as optional and recommend that it be implemented only in the framework of integrated selection schemes which are already efficient with respect to milk yield (ICAR, 1992).

Furthermore, to reduce the cost of the standard A_4 official milk-recording, simplified methods are suitable for achieving a large enough number of recorded animals to ensure a genetic gain for the entire sheep population (Flamant and Poutous, 1970; Barillet, 1985, 1990b; Maria and Gabina, 1992; Sanna *et al.*, 1995c): two simplified testings for milk yield (the alternating monthly test (AT) and corrected monthly test (AC) methods) are approved in ICAR regulations for sheep (ICAR, 1992). For these simplified methods, only one of the two daily milkings is recorded, alternating a monthly test for the AT method and a corrected monthly test for evening/morning differences for the AC method, using the total volume of milk produced by the whole flock for the two milkings concerned (bulk-tank weights).

In France, milk-yield recording is carried out by means of the AC method (always at the morning milking) and, for breeding programmes at the equilibrium, qualitative testing (milk composition) is also practised using the following simplified design (Fig. 20.3): three to four times during the first 4 test days for each ewe at morning milking only (Barillet, 1985).

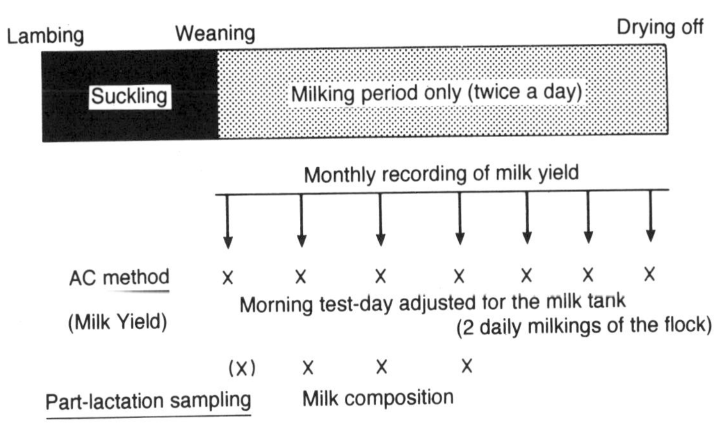

Fig. 20.3. Simplified recording design for milk composition in the nucleus flocks.

When using the simplified recording design for milk composition described in Fig. 20.3, milk contents determined from these reduced samplings (or partial sampling) are highly correlated (0.98) with contents determined from samples collected over the complete lactation period (A_4 method) but are less heritable (Barillet, 1985, 1990b): 0.35–0.45 for F% and P% with the reduced samplings (Fig. 20.3), compared with 0.50–0.60 for these traits with the A_4 method. Nevertheless, genetic parameters for dairy traits recorded with the A_4 or partial sampling are similar (except for heritability for F% and P%), as shown in Table 20.5 (Barillet and Boichard, 1994). The balance between cost and genetic efficiency is clearly in favour of the simplified recording design for milk composition.

Implementation of Breeding Schemes for Dairy Sheep in Europe

In the past, western Mediterranean breeding programmes for dairy sheep underwent a well-known evolution (as in dairy cattle) from on-station schemes to on-farm programmes. This means that nowadays in southern Europe several breeding programmes are being carried out on a large scale, using on-farm milk-recording and AI, combined with an accurate genetic evaluation of the animals.

Survey of milk-recording

In 1994 the ICAR sheep working group carried out a survey on milk-recording of dairy sheep in 31 member countries. From the 20 answers received (Table 20.6), the following can be shown (Barillet and Astruc, 1994):

- The ratio of recorded ewes relative to the whole population remains in most cases insufficient to enable the organization of efficient breeding programmes on a large scale for the whole population. Among the countries with large dairy sheep populations, the percentage of recorded ewes is less

Table 20.5. Genetic parameters for Lacaune-breed milk traits in first lactation (130,409 Lacaune ewe lambs born from 2670 rams) with a partial sampling of milk composition: heritabilities on diagonal and genetic correlations below diagonal (from Barillet and Boichard, 1994).

Traits*	Milk yield	Fat yield	Protein yield	Fat content	Protein content
Milk yield	**0.30**				
Fat yield	0.83	**0.28**			
Protein yield	0.91	0.89	**0.29**		
Fat content	–0.31	0.26	–0.06	**0.35**	
Protein content	–0.40	–0.04	–0.03	0.63	**0.46**

*With partial sampling for milk composition (see Fig. 20.3).

Table 20.6. Size of populations of dairy sheep, type of milk-recording and number of recorded ewes in the member countries of ICAR (from Barillet and Astruc, 1994).

		Official milk-recording		Other milk-recording	
Country	Size of population: total number of females	Number of recorded females (% population)	Number of recorded flocks (year)	Number of recorded females (% population)	Number of recorded flocks (year)
Belgium	1,000	–	–	95 (9.5%)	3 (1992)
France	1,235,000	259,318 (21.0%)	798 (1993)	539,005 (43.6%)	1,648 (1993)
Germany	59,421	1,460 (2.5%)	270 (1992)	–	–
Greece	10,108,000	54,700 (0.5%)	675 (1992)	–	–
Hungary	64,860	1,154 (1.8%)	21 (1993)	–	–
Israel	51,200	–	–	6,200 (12.1%)	6 (1992)
Italy	5,408,361	211,247 (3.9%)	1,911 (1993)	–	–
The Netherlands	–	–	3 flocks	–	–
Portugal	500,000	21,448 (4.3%)	460 (1992)	–	–
Slovenia	6,400	458 (7.2%)	24 (1993)	–	–
Spain	4,000,000	90,757 (2.2%)	384 (1992–93)	–	–
Sweden	(30 flocks)	–	–	–	–
Switzerland	3,250	–	–	340 (10.5%)	34 (1993)
Tunisia	210,000	2,200 (1.0%)	10 (1992)	–	–

than 10%, except for France, which makes extensive use of milk-recording (64.6% of recorded ewes). Countries with smaller dairy-sheep populations – Switzerland (10.5%) and Israel (12.1%) – have a percentage of recorded ewes up to 10%. If we now take a look at breeds (Table 20.7), among the large breeds (up to 10,000 recorded ewes), the following are closer to or higher than 10% of recorded ewes: Comisana (Italy), Corsica (France), Karagouniko (Greece), Lacaune (France), Latxa (Spain) and Manech (France). We have also to point out the large number of recorded ewes for the Sarda breed in Italy (Table 20.7), even if the ratio relative to the whole population is below 10%.

- The size of each on-farm nucleus (official milk-recording) appears to be higher in breeds for which the standard A_4 method has been replaced by the AC method (Lacaune, Manech and Corsica breeds in France) or the AT method (Latxa and Carranzana breeds in Spain). This survey confirms that milk-recording has to be simplified if it is to be implemented on a large scale in dairy sheep.

Use of artificial insemination

Although the dam–daughter selection pathway is not as negligible as in dairy cattle, because of a higher prolificacy in sheep, the most efficient pathways are those for the dam–son, the sire–son and the sire–daughter (Barillet and Elsen, 1979). Furthermore, in most cases, the average size of the dairy flocks ranges from 200 to 400 ewes. Thus, an efficient breeding programme cannot be organized on a strict within-flock basis, which allows in practice (accounting for the size of the flocks) managing only the dam–daughter and dam–son selection pathway. In other words, the breeding plan has to be carried out between flocks of the nucleus in order to be able to progeny-test rams, i.e. also to make efficient use of the sire–son and sire–daughter pathways. Under these conditions, AI is a powerful tool for creating genetic connections between flocks, thus enabling interflock organization of the breeding programme. If the impact of AI is moderate in the nucleus flocks (AI rate in the range of 10–20%), NM and AI have to be combined to carry out progeny tests of rams and to plan assortative matings. When the AI rate increases up to 40–50% in the nucleus flocks, it is possible to rely more on AI rams for the breeding scheme and the genetic gain can be 35% higher than for the previous situation. Finally, if the AI rate reaches 70–80% in the nucleus flocks, it also becomes possible to plan for a very efficient and fast diffusion of the genetic gain inside and outside the nucleus flocks.

The use of AI in the nucleus started in the early 1970s and late 1970s, respectively, for the Lacaune and Manech breeds in France, while in the dairy-sheep nucleus flocks of Greek, Italian and Spanish breeds (Karagouniko, Sarda, Latxa, Churra and Manchega) it started more recently, in the middle or late 1980s (Hanocq *et al.*, 1993; de la Fuente *et al.*, 1995; Georgoudis *et al.*, 1995; Jurado *et al.*, 1995; Sanna *et al.*, 1995a; Ugarte *et al.*, 1995). Thus, as shown in Table 20.7, French breeds (Lacaune and Manech) make more use of AI than the other breeds from western Mediterranean countries. However, for the above-mentioned breeds, the increase of use of AI in the late 1980s is really important.

Genetic evaluation in dairy sheep

Milk-recording in private flocks, the use of AI and an accurate evaluation of the animals are the main tools for constructing an efficient breeding programme. As

Table 20.7. Size of population, milk-recording and breeding programme for dairy-sheep breeds accounting for more than 10,000 recorded ewes (year 1993 or 1994).

			Milk recording			Breeding programme using artificial insemination (AI)		
Country	Breed	Total population	Recorded flocks	Recorded ewes	% Population recorded	AI year^{-1}	AI progeny-tested rams year^{-1}	Year of starting
France	Lacaune	725,000	376* 1501†	156,624* 511,110†	21.6* 70.5†	126,000* 256,000†	450	1968
France	Manech and Basco-Bearnaise	440,000	338* 147†	85,214* 33,833†	20.8* 8.3†	40,000* 15,000†	130	1975
France	Corsica	100,000	84	17,480	17.5	1,000		1993
Greece	Karagouniko	200,000	350	20,000	10	6,000	20	1986
Greece	Lesvos	175,000	130	10,800	6.2	–	–	–
Italy	Sarda	3,700,000	854	102,967	2.8	12,000	40	1986
Italy	Comisana	990,000	723	84,621	8.5	–	–	–
Portugal	Serra da Estrela	280,000	370	15,420	5.5	–	–	–
Spain	Latxa and Carranzana	440,000	208	60,875	18	11,500	50	1984
Spain	Churra	1,700,000	87	25,861	1.5	16,000	50	1986
Spain	Manchega	1,100,000	119	60,000	5.4	15,000	50	1986

*Nucleus flocks using official milk-recording (according to ICAR regulations) (see Fig. 20.1).
†Commercial flocks of the base population using simplified milk-recording.
Sources: Barillet and Astruc, 1994; de la Fuente *et al.*, 1995; Georgoudis *et al.*, 1995; Jurado *et al.*, 1995; Sanna *et al.*, 1995a; Ugarte *et al.*, 1995.

best linear unbiased prediction (BLUP) applied to an animal model (AM) is now considered to be the international reference, AM genetic evaluation systems have been implemented for dairy sheep in countries such as France, Italy and Spain since 1991 (Gabina and Barillet, 1991; Pagnacco *et al.*, 1991; Barillet *et al.*, 1992; Serrano *et al.*, 1993; Sanna *et al.*, 1994).

In theses countries, different organizations cooperate on the genetic evaluation of different breeds, as described in the first survey carried out in 1994 on this topic (Astruc *et al.*, 1994). For breeds involved in this survey, at the end of 1993, the size of the data set for genetic evaluation ranged from 18,194 lactations (Langhe breed) to 1,682,531 lactations (Lacaune breed). The number of males evaluated varied from 31 in the Corsica breed to 10,619 in the Lacaune breed, according to the population size for each breed and its breeding programme.

In all cases, evaluation is based on a repeatability AM, which always includes an additive genetic effect, a female permanent environmental effect and a main effect, such as the classical flock–year or flock–year–season effect, which is replaced by some authors by a flock–year–age or flock–year–parity effect. Other fixed effects may be included in the model. From the survey, it appears that, while all organizations are using the AM method, there are important differences in data editing and analytical models used for genetic evaluation of dairy sheep in France, Italy and Spain.

The previous evaluation was based on the analysis of lactation yields (at milking period only). These lactation yields are derived from elementary test-day information combined with Fleischmann's rules (ICAR, 1992). An alternative, which has been studied for the Lacaune breed in first lactation (Barillet and Boichard, 1994), is to analyse individual test-day records. In sheep, because cost constraints are much more limiting than in cattle, a simplified recording design will be developed for milk composition (Fig. 20.3), justifying further research towards a test-day evaluation (Baro *et al.*, 1994; Carta *et al.*, 1995).

Published estimates of genetic trend for milk traits

For Lacaune, Latxa, Manchega, Manech and Sarda breeds, estimates of genetic trend from AM evaluations have been published (Barillet *et al.*, 1994; Sanna *et al.*, 1994b, 1995b; Jurado *et al.*, 1995; Ugarte *et al.*, 1995). The main purpose of this review is to compare the efficiency of different breeding schemes according to the AI rate in the nucleus flocks (Tables 20.7 and 20.8).

Results are consistent with the organization of the nucleus flocks for each breed. Pyramidal management of the population and AI within the nucleus flocks are emphasized, because they make it possible to accurately progeny-test rams and apply high selection intensity. The AM evaluation confirmed the efficiency of the Lacaune breeding programme (Fig. 20.4), since the genetic gain for milk yield reached 5.7 l year^{-1} in the 1980s (Fig. 20.5), i.e. almost two-thirds of the phenotypic trend over the period analysed (Fig. 20.6). The genetic gain

Table 20.8. Published estimates of genetic trend for milk yield in dairy sheep (milk yield expressed as mature equivalent lactation).

			Annual genetic trend		
Country	Breed	Period	% Population mean	In litres	AI rate in the nucleus flocks (year 1993 or 1994) (%)
France	Lacaune	1981–1992	2.4	5.7	81
	Manech (Blond-faced)	1984–1992	2.0	2.1	44
Italy	Sarda	1990–1994	0.9	1.6	12
Spain	Latxa (Blond-faced)	1985–1993	1.2	1.1	18
	Latxa (Black-faced)	1985–1993	0.8	0.8	18
	Manchega	1986–1992		0.8	20

Sources: Barillet *et al.*, 1994; Sanna *et al.*, 1995b; Jurado *et al.*, 1995; Ugarte *et al.*, 1995.

reached about 2.4% year^{-1} of the population mean or 0.19 genetic standard deviation (Table 20.8). This result confirms that the genetic trend of a sheep population may be of the same magnitude as in dairy cattle, despite the limitation of AI in sheep.

This possibility was confirmed with the Manech (Blond-faced) example. The Manech breeding programme became really efficient in the 1980s, when the AI rate was large enough (40%) for AI progeny-testing of at least 100 males year^{-1} born from assortative mating in the nucleus flocks; the estimated genetic gain was 2.1 l year^{-1}, i.e. 2% of the selected population mean. For the Sarda breed, which already has a worthwhile milk-yield level, the genetic gain was only 0.4 l year^{-1} at the start of the AI breeding scheme in the 1980s, because of the 5% rate for AI in the nucleus flocks; only the dam–son selection pathway within the flock was really efficient (Sanna *et al.*, 1994b). Since 1990, for the Sarda breed, the yearly genetic trend for milk yield has increased to 1.6 l (0.9% of the population), now that the AI rate is over 10%, which emphasizes AI as the tool to carry out a cooperative breeding programme across flocks of the nucleus (Sanna *et al.*, 1995b). In the Latxa and Manchega breeds, progress is also beginning to be evident, with an annual genetic gain between 0.8 and 1.2% (0.8 to 1.1 l), consistent with the AI selection effort since the late 1980s (Jurado *et al.*, 1995; Ugarte *et al.*, 1995).

Description of an Efficient Breeding Programme for a Dairy-sheep Population

A close look at the Lacaune breed will be useful, as it has the oldest and most efficient dairy-sheep breeding scheme (see Fig. 20.4).

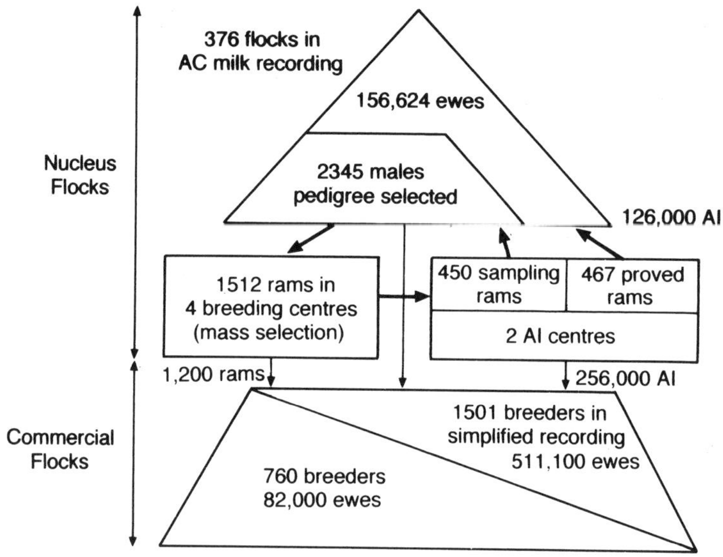

Fig. 20.4. Breeding scheme in the Lacaune breed (1993).

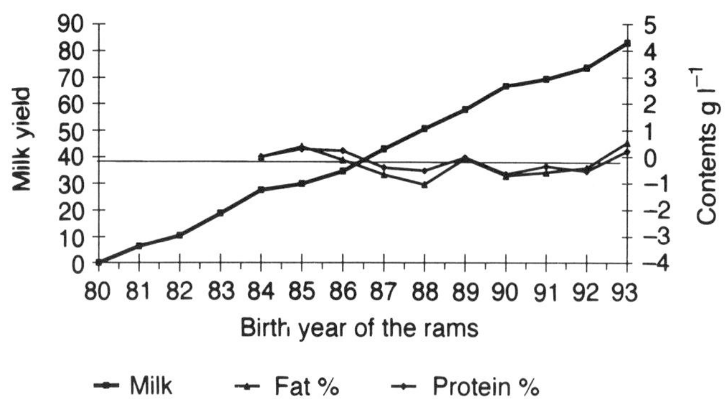

Fig. 20.5. Genetic trend for Lacaune rams (animal-model evaluation 1995).

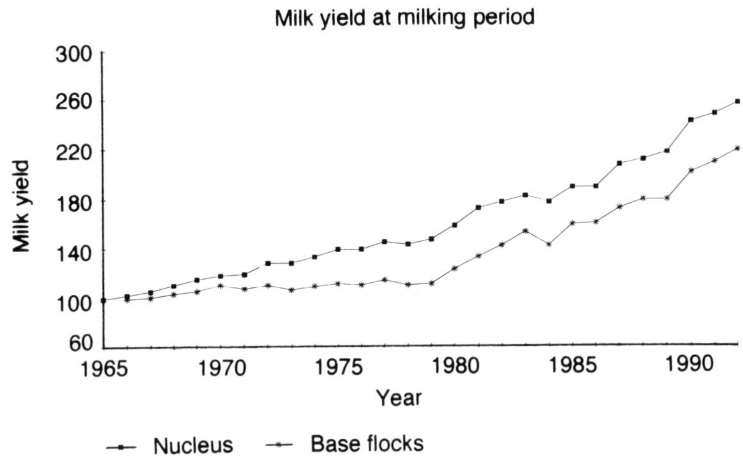

Fig. 20.6. Phenotypic trend for milk yield in the Lacaune nucleus flocks and commercial flocks.

The Lacaune breeding scheme

The whole population includes 750,000 ewes on 2637 farms, while in 1993 the nucleus included 156,624 ewes owned by 376 breeders. Recording of fat and protein has been generalized for all nucleus flocks since 1987. The rate of AI reached 80% in the nucleus in 1993. Of the ewes in the nucleus flocks, 126,000 were inseminated; 45% were used as mates of the 450 young sampling rams on progeny test, and the others were mated with 467 AI-proved rams, especially the best 120 élite rams. These élite rams were mated with the élite ewes of the nucleus, giving 2345 male lambs with a selected pedigree. About 1500 of these young rams born from assortative matings enter the breeding centres when they are 1 month old. Of these, 1000 are sold to commercial flocks. After some mass selection of the young rams in the breeding centres on type and growth-rate traits, the best 450 rams on the pedigree index for milk traits entered the AI centres at 8 months of age to be progeny-tested in the nucleus.

In the base population, about 40% of the adult ewes were inseminated with AI-proved rams (256,000 AI in 1993) and 1000 pedigree selected rams were sold from the nucleus to the base population. The diffusion rate (by AI plus NM rams) reached about 100% for commercial breeders practising simplified milk-recording (1501 flocks in 1993). In conclusion, with the pyramidal framework, AI was dominant in the nucleus flocks for the genetic-gain creation, while the diffusion to the base population combined AI and NM.

Phenotypic and genetic trends for selected and base populations

The phenotypic trend for milk yield in the Lacaune nucleus and commercial flocks (base population in simplified recording) illustrates that the equilibrium was achieved at the beginning of the 1980s, 15 years after the start of the breeding programme (see Fig. 20.6). This result is in agreement with the studies carried out in the 1970s to design and optimize the breeding scheme (Elsen and Mocquot, 1974; Vallerand and Elsen, 1979). Now the phenotypic trend is the same in the nucleus and base populations, with a gap of about 40 l corresponding to 5–7 years of gain.

Under these conditions, the selection objective in dairy Lacaune sheep moved from milk yield to both milk yield and milk composition in order to maintain satisfactory cheese-making properties (Barillet, 1985). Therefore, the dairy selection criterion used from 1987 to 1992, U, enabled improvement in the genetic level for M while approximately maintaining F% and P%. The new goal of the end of the 1980s has actually been reached: for the 4342 AI Lacaune males born between 1984 and 1993, the genetic level for F% and P% was quite stable (see Fig. 20.5), whereas the genetic gain for M reached 56 l during the same time (Barillet *et al.*, 1994). From 1993, a new selection criterion will be used ($F + k_1P + k_2P\%$) in order to increase both M, F% and P% in the 1990s.

Secondary traits

The Lacaune breed has the oldest and most efficient breeding scheme, as it is based on a large use of AI and is able to select for both milk yield and milk composition. For this breed, the next objectives in the 1990s will focus on increasing the selection intensity on milk composition, eventually including genetic polymorphism of milk proteins based on studies of this topic in progress (Barillet *et al.*, 1993; Piredda *et al.*, 1993; Barbieri *et al.*, 1994). A direct selection for milking ability and udder traits will probably also be implemented, using an automatic milk meter, which has been designed and developed by INRA for dairy sheep and goats (Guillouet *et al.*, 1991). For other traits concerning dual purpose breeds, different studies are being carried out regarding the traits to be measured and the genetic improvement strategy (Barillet *et al.*, 1988). In particular, income from meat depends basically on prolificacy, because lambs are generally slaughtered just after weaning at 1 month old. Seeing that prolificacy has a very small heritability (~0.03–0.05) and that the genetic correlation with milk yield is slightly positive (~0.2–0.3), a very slow genetic increase in prolificacy (0.006 lamb born per year) is expected with or without including prolificacy in selection criteria of a dairy-sheep population.

More generally, genetic aspects of dairy-sheep selection based on breeding qualities (type traits, functional morphology, resistance to disease and mastitis) and lifespan are being studied, with the aim of making possible a decrease in production costs, particularly through less frequent undesired culling (Casu *et al.*, 1989; Baro *et al.*, 1994; Bergonier *et al.*, 1995; Fernandez *et al.*, 1995).

Conclusion

There are large differences between breeding programmes for the Mediterranean local breeds with respect to their levels of organization. In addition to the Lacaune breed, nowadays in southern Europe other breeding programmes are being carried out, using on-farm milk-recording, AI and an accurate genetic evaluation of the animals, particularly for the Churra, Karagouniko, Latxa, Manchega, Manech and Sarda breeds. The efficiency of the breeding programme depends directly on the rate of AI in the nucleus flocks. Thus, it would appear to be fundamentally important to increase the use of AI, since it is a powerful tool for genetic improvement. However, it is obvious that the local conditions of many Mediterranean areas reveal limiting factors for a wide use of on-farm milk-recording and AI.

Under these conditions, the question of the best breeding strategy has been an ongoing debate for Mediterranean areas (Mavrogenis, 1995). Different scenarios may be imagined for the future from the crossbreeding strategy to the implementation of local purebreed breeding programmes. Moreover, some authors propose an on-station closed nucleus flock if local conditions prohibit the development of an on-farm breeding scheme. This proposal requires a three-level pyramidal organization, the first level being the on-station nucleus unit, with a second multiplier level designed to provide breeding animals for the third level of commercial flocks. The use of modern reproduction techniques, such as embryo transfer or multiple ovulation and embryo transfer (MOET), have been proposed for the on-station level to reduce the generation intervals and increase family size (Rossi *et al.*, 1995). But the actual gain may be far from what is expected, especially if inbreeding and reduction of genetic variance are not checked. Moreover, past experience often showed that the transfer of improved animals to commercial flocks could be difficult, due to a lower adaptation of the improved material to traditional husbandry conditions and/or to the lack of simultaneous evolution of the breeding system along with the genetic improvement. Finally, regarding, on the one hand, the costs and the variability of technical results for embryo transfer, which limits its widespread use (Gabina, 1990; Zervas *et al.*, 1991; Cognié *et al.*, 1995), and on the other hand, the large size of the populations to be improved, the on-farm selection of local breeds using mainly AI may be recommended in case of purebreed selection. Nevertheless, it should be kept in mind that in the 1990s only some breeds show a real genetic trend for milk yield, which could modify the future situation for genetic improvement of dairy sheep in the Mediterranean basin, eventually promoting the use of some breeds in the basin.

For some countries where milking ewes is an exception, dairying with sheep appears as a new venture in the 1990s. From a genetic point of view, the first step in this situation is to choose which dairy-sheep germ-plasm resources to import, because the time to improve milk-yield level (at milking) of the local meat-sheep breeds would generally be too long. Furthermore, the specific abilities of each candidate breed for the local breeding system will be a main consideration to take into account.

References

Alps, H. and Averdunk, G. (1984) Genetische Parameter für die Merkmale der Milchleistung unter besonderer Berücksichtigung des Eiweisses beim Fleckvieh in Bayern. II. Summe aus Eiweiss und Fett. *Züchtungskunde* 56, 99–107.

Astruc, J.M., Barillet, F., Carta, A., Gabina, D., Manfredi, E., Moioli, B., Piacère, A., Pilla, A.M., Sanna, S.R., Sigwald J.P. and Ugarte E. (1995). Use of the animal model for genetic evaluation of dairy sheep and goats in several ICAR member countries. In: Lajoue, L., Lafontaine, S., Doyle, P. (eds) *Proceedings of the 29th Biennial Session of the International Committee for Animal Recording (ICAR), Ottawa, Ontario, Canada, 31 July–6 August 1994.* EAAP Publication No. 75, Agriculture and Agrifood Canada, Ottawa, pp. 271–275.

Barbieri, M.E., Manfredi, E., Rubino, R., Piredda, G., Fraghy, A., Sanna, S.R., Barillet, F. and Chianese, L. (1995) Genetic polymorphisms of goat and sheep milk proteins: allelic frequencies and relationships with dairy traits and cheesemaking properties of milk. In: Lajoue, L., Lafontaine, S., Doyle, P. (eds) *Proceedings of the 29th Biennial Session of the International Committee for Animal Recording (ICAR), Ottawa, Ontario, Canada, 31 July–6 August 1994.* EAAP Publication No. 75, Agriculture and Agrifood Canada, Ottawa, pp. 293–298.

Barillet, F. (1985) *Amélioration génétique de la composition du lait des brebis: l'exemple de la race Lacaune. Thèse de Docteur-Ingénieur, INA Paris-Grignon*, Paris, 144 pp.

Barillet, F. (1989) Expression de la production laitière à la traite des brebis en système allaitement × traite mecanique. In: Eitam, M. (ed.) *Proceedings of the 4th International Symposium on Machine Milking of Small Ruminants, Tel Aviv, Israel, 13–19 September 1989.* Ministry of Agriculture, Tel Aviv, pp. 463–495.

Barillet, F. (1990a) Les objectifs et les programmes d'amélioration génétique en brebis laitières. In: Options Méditerranéennes, séries A no. 12, *Les Petits Ruminants et lauers productions laitières dans la région méditerranéenne*. pp. 39–48.

Barillet, F. (1990b) Rapport du groupe du CICPLB sur la contrôle laitier des brebis: 'La simplification du contrôle laitier officiel de type A. In: Gaillon, P. and Chabert, Y. *Proceedings 27th Biennial Session of the International Committee for Animal Recording (ICAR), Paris, France, 2–6 July 1990.* EAAP Publication No. 50. pp. 119–125.

Barillet, F. and Astruc J.M. (1995) Survey of milk recording and genetic evaluation in dairy sheep in ICAR member countries. In: Lajoue, L., Lafontaire, S., Doyle, P. (eds) *Proceedings 29th Biennial Session of International Committee for Animal Recording (ICAR), Ottawa, Ontario, Canada, 31 July–6 August 1994.* EAAP Publication No. 75, Agriculture and Agrifood Canada, Ottawa, pp. 259–269.

Barillet, F. and Boichard, D. (1987) Studies on dairy production of milking ewes. I. Estimates of genetic parameters for total milk composition and yield. *Génétique, Sélection et Evolution* 19(3), 459–474.

Barillet, F. and Boichard, D. (1994) Use of first lactation test-day data for genetic evaluation of the Lacaune dairy sheep. In: *Proc. 5th World Congress on Genetics Applied to Livestock Production, Guelph, Ontario, Canada, 7–12 August 1994*, Vol. 18. pp. 111–114.

Barillet, F. and Elsen, J.M. (1979) Optimisation de l'utilisation de l'insémination artificielle dans les schémas de sélection des ovins. In: *Proceedings of the 5èmes Journées recherche ovine et caprine, Paris, 5–6 December 1979.* ITOVIC-SPEOC, Paris, pp. 186–204.

Barillet, F., Courot, M., Frebling, J. and Legault, C. (1984) Intérêts comparés, zootechniques et économiques, de la reproduction en insémination artificielle ou en monte naturelle selon l'espèce, le type de production, la taille du troupeau, le mode de conduite. In: Elsen, J.M. and Failleu, J.L. (eds) *Colloque INRA no. 29, Toulouse, 23–24 novembre 1983*. INRA, Versailles, France. pp. 97–111.

Barillet, F., Elsen, J.M. and Roussely, M. (1986) Optimization of a selection scheme for milk composition and yield in milking ewes: example of the Lacaune breed. In: Dickerson, G.E. and Johnson, R.K. (eds) *Proceedings of the 3rd World Congress on Genetics Applied to Livestock Production, Lincoln, 16–22 July*, Vol. 11. University of Nebraska, Lincoln, pp. 658–664.

Barillet, F., Elsen, J.M., Roussely, M., Belloc, J.P., Briois, M., Casu, S., Carta, R. and Poivey, J.P. (1988) Sélection lait–viande en brebis laitières. In: *3rd World Congress on Sheep and Beef Cattle Breeding, 19–23 June*, Vol. 2. INRA, Paris, pp. 469–490.

Barillet, F., Boichard, D., Barbat, A., Astruc, J.M. and Bonaïti, B. (1992) Use of an animal model for genetic evaluation of the Lacaune dairy sheep. *Livestock Production Science* 31, 287–299.

Barillet, F., Mahé, M.F., Pellegrini, O., Grosclaude, F. and Bernard S. (1993) Polymorphisme génétique des protéines du lait en race ovine de Lacaune. In: Kukovics, S. (ed.) *Proceedings of the 5th International Symposium on Machine Milking of Small Ruminants Budapest, Hungary, 14–10 May 1993*. Hungarian Journal of Animal Production 1993, supplement F1, pp. 199–207.

Barillet, F., Sanna, S.R., Boichard, D., Astruc, J.M., Carta A. and Casu S. (1993) Genetic evaluation of the Lacaune, Manech and Sarda dairy sheep with an animal model. In: Kukarics, S. (ed.) *Proc. 5th International Symposium on Machine Milking of Small Ruminants, Budapest, Hungary, 14–10 May 1993. Hungarian Journal of Animal Production* 1993, Supplement F2, pp. 289–304.

Baro, J.A., Carriedo, J.A. and San Primitivo, F. (1994) Genetic parameters of test day measures for somatic cell count, milk yield and protein percentage of milking ewes. *Journal of Dairy Science* 77, 2658–2662.

Bergonier, D., Lagriffoul, G., Concordet, D., Barillet, F. and Berthelot, X. (1995) Dynamique des infections mammaires subcliniques de la brebis laitière en relation avec les comptages de cellules somatiques. In: Institute l'Elevage (ed.), *2èmes Rencontres Recherches Ruminants, Paris, 13–14 December 1995*, Vol. 2, Paris. pp. 299–302.

Boichard, D. and Bonaïti, B. (1987) Etudes sur la production laitière des bovins: IV. Paramètres génétiques en première lactation. *Génétique, Sélection et Evolution* 19(3), 337–350.

Boyazoglu, J.G. and Flamant, J.C. (1990) Mediterranean systems of animal production. In: Galaty, J.G. and Johnson, D.L. (eds) *The World of Pastoralism*. pp. 353–393.

Boyazoglu, J., Casu, S. and Flamant, J.C. (1979) Crossbreeding the Sardinian and East Friesian breeds in Sardinia. *Annales de Génétique et Sélection Animale* 11(1), 23–51.

Carta, A., Sanna, S.R. and Casu, S. (1995) Estimating lactation curves and seasonal effects for milk, fat and protein in Sarda dairy sheep with a test day model. *Livestock Production Science* 44, 37–44.

Casu, S., Barillet, F., Carta, R. and Sanna, S. (1989) Amélioration génétique de la forme de la mamelle de la brebis Sarde en vue de la traite mécanique. In Eitam, M. (ed.) *Proceedings 4th International Symposium on Machine Milking of Small Ruminants, Tel Aviv, Israel, 13–19 September 1989*. Ministry of Agriculture. Tel Aviv, pp. 104–133.

Cognié, Y., Guerin, Y., Belloc, J.P., Boulenc, P., Briois, M., Girou, P., Lagriffoul, G.,

Solier, P., Barillet, F. and Brebion, P. (1995) Production d'embryons chez la brebis laitière: bilan des recherches conduites pour augmenter l'efficacité du transfert embryonnaire. In: Institut de l'Elevage (ed.), *2émes Rencontres Recherches Ruminants, Paris, 13–14 December 1995*, Vol. 2. Paris, pp. 429–434.

Cottier, M. (1972) Le controle laitier ovin simplifié dans la zone de Roquefort. In: *Proceedings of the Symposium on Milk Recording Practices in Sheep and Goats, Israel, 19–24 March 1972.*

de la Fuente, L.F., Baro, J.A. and San Primitivo, F. (1995) Breeding programme for the Spanish Churra sheep breed. In: Gabina, D. (ed.) *Strategies for Sheep and Goat Breeding: Proceedings of the Meeting of the Joint FAO/CIHEAM Network on Sheep and Goats, Subnetwork on Animal Resources, Sidi-Taheb, Tunisia, 26–28 March 1995.* Cahiers Options Mediterranéennes, no. 11, CIHEAM, Zaragoza pp. 165–172.

Elsen, J.M. and Mocquot, J.C. (1974) Recherches pour une rationalisation technique et économique des schémas de sélection des bovins et ovins. *Bulletin Technique du Département de Génétique Animale, INRA* 17, 76–97.

Epstein, H. (1985) The Awassi sheep with special reference to the improved dairy type. *FAO Animal Production and Health* 57.

Fernandez, G., Alvarez, P., San Primitivo, F. and de la Fuente, L.F. (1995) Factors affecting variation of udder traits of dairy ewes. *Journal of Dairy Science* 78, 842–849.

Flamant, J.C. and Barillet, F. (1982) Adaptation of the principles of selection for milk production to milking ewes: a review. *Livestock Production Science* 9, 549–559.

Flamant, J.C. and Poutous, M. (1970) Aspects quantitatifs de la production laitière des brebis. VII: Précision d'un contrôle laitier alterné (AT) et d'un contrôle laitier d'alternance quelconque corrigé pour les écarts moyens entre les performances du soir et du matin (AC). *Annales de Génétique et Sélection Animale* 2(1), 65–73.

Flamant, J.C. and Ricordeau, G. (1969) Croisements entre les races ovines Préalpes du Sud et Frisonne. I. La brebis laitière de Frise Orientale – elevage en race pure – utilisation en croisements. *Annales de Zootechnie* 18(2), 107–130.

Gabina, D. (1990) Les nouvelles techniques de reproduction et les programmes de sélection chez les ovins laitiers. In: Bouylez, J. and Tisserand, J.L. (eds) *Les Petits ruminants et leurs production laitières dans la région méditerranéenne*, Options Méditerranéennes, sér, A. no. 12, CIHEAM, Montpellier pp. 49–55.

Gabina, D. (1997) Current state of Mediterranean small ruminant germplasm and its characterization. In: *International Symposium on Mediterranean Animal Germplasm and Future Human Challenges, Benevento, Italy, 26–29 November 1995* (in press).

Gabina, D. and Barillet, F. (1991) Tendencias actuales en la seleccion del ovino de leche en la Europa communitaria. *Informacion Tecnica Economica Agraria* 87A(2–3), 227–234.

Georgoudis, A., Hatziminaoglou, I. and Pappas, V. (1995) The breeding scheme of the Karagouniko sheep in Greece. In: Gabina, D. (ed.) *Strategies for Sheep and Goat Breeding: Proceedings of the Meeting of the Joint FAO/CIHEAM Network on Sheep and Goats, Subnetwork on Animal Resources, Sidi-Taheb, Tunisia, 26–28 March 1995.* Cahiers Options Mediterranéennes, no. 11, CIHEAM, Zaragoza, pp. 61–65.

Goot, H. (1966) *Studies on the Native Awassi Sheep and Its Crosses with the Exotic East Friesian Milk Sheep.* Volcani Institute of Agricultural Research, Nat. Univ. Inst. Agric., Rehovot, pp. 115–168.

Guillouet, P., Ricard, E., Aurel, M.R., Jacquin, M., Astruc, J.M., Bibé, B. and Barillet, F. (1991) Conception d'un système de contrôle laitier automatisé pour les ovins et caprins laitiers. In: Guillon, P. and Chabet, Y. (eds) *Proc. 27th Biennial Session of International Committee for Animal Recording (ICAR), Paris, France, 2–6 July 1990.* EAAP Publication No. 50, Wageningen, Netherlands pp. 130–136.

Hanocq, E., Urarte, E., Ugarte, E., Arrese, F., Gabina, D., Arranz, J., Oregui, L., Bravo, M.V. and Beltran de Heredia, I. (1993) Situacion y problematica del programa de mejora genetica del ovino lechero en la Comunidad Autonoma Vasca. *Informacion Tecnica Economica Agraria* 89A(2), 43–161.

ICAR (Barillet, F., Astruc, J.M., de Brauwer, P., Casu, S., Fabbri, G., Feddersen, E., Frangos, K., Gabina, D., Gama, L.T., Ruiz Tena, J.L. and Sanna, S.R) (1992) *International Regulations for Milk Recording in Sheep*. Institut de l'Elevage, Paris, 15 pp. + appendix.

Jurado, J.J., Serrano, M., Perz-Guzman, M.D. and Montoro V. (1995) Improvements in the Manchega genetic breeding programme. In: Gabrina, D. (ed.) *Strategies for Sheep and Goat Breeding: Proceedings of the Meeting of the Joint FAO/CIHEAM Network on Sheep and Goats, Subnetwork on Animal Resources, Sidi-Taheb, Tunisia, 26–28 March 1995*. Cahiers Options Mediterranéennes, no. 11 CIHEAM, Zaragoza pp. 133–141.

Kalaissakis P.Papadimitriou, T., Flamant, J.C., Boyazoglou, J. and Zervas, N. (1977) Comparaison des races ovines Chios et Frisonne avec leurs croisements en Grèce continentale. II. Production laitière. *Annales de Génétique et Sélection Animale* 9(2), 181–201.

Maijala, K. and Hanna, M. (1974) Reliable phenotypic and genetic parameters in dairy cattle. In: *Proceedings 1st World Congress on Genetics Applied to Livestock Production, Madrid, 7–11 October 1974*, Vol. 1. Ed. Garsi, Madrid, pp. 541–563.

Maria, G. and Gabina, D. (1992) Simplification of milk recording scheme in Latxa milk sheep. *Livestock Production Science* 31, 313–320.

Marie, C., Bucquier, F. and Banillet, F. (1996) Influence du potential laitiez sur les composantes de l'efficacité alimentaire de brebis lacaune. In: Institute de l'Elevage (ed.) *3èimes Rencontres Recherches Ruminants, Paris, 4–5 Décembre 1996*, Paris, vol. 3, 297–300.

Mavrogenis, A.P. (1988) Genetic improvement of sheep in Cyprus by selection and/or crossbreeding. In: Thompson, E.F. and Thomson, F.S. (eds) *Proc. International Symposium on Increasing Small Animal Productivity in Semi-arid Areas, Aleppo, 1987*. Kluwer Academic Publishers pp. 189–195.

Mavrogenis, A.P. (1995) Breeding systems and selection strategies for sheep improvement in Cyprus. In: Gabrina, D. (ed.) *Strategies for Sheep and Goat Breeding: Proceedings of the Meeting of the Joint FAO/CIHEAM Network on Sheep and goats, Subnetwork on Animal Resources, Sid-Taheb, Tunisia, 26–28 March 1995.* Cahiers Options Mediterranéennes, no. 11, CIHEAM, Zaragoza, pp. 17–32.

Pagnacco, G., Moioli, B.M. and Pilla, A.M. (1991) Selezione della razza ovina delle Langhe: II. Valutazione genetica con un animal model. In: *Atti IX, Congr. Naz. ASPA Roma, 3–7 June 1991*. ISMEA, Rome, pp. 571–579.

Piredda, G., Papoff, C.M., Sanna, S.R. and Campus, R.L. (1993) Influenza del genotipo della αsl-caseina ovina sulle caratteristiche chimico-fisiche e lattodinamografiche del latte. *Sienza e Tecnica Lattiero-Casearia* 44(3), 135–143.

Ricordeau, G. and Flamant, J.C. (1969a) Croisements entre les races ovines Préalpes du Sud et Frisonne. II. Reproduction, viabilité, croissance, conformation. *Annales de Zootechnie* 18(2), 131–149.

Ricordeau, G. and Flamant, J.C. (1969b) Croisements entre les races ovines Préalpes du Sud et Frisonne (Ostfriesisches Milchschaf). III. Performances laitières. *Annales de Zootechnie* 8(2), 151–168.

Rossi, G., Cappio-Borlino, A., Macciotta, N.P.P. and Buttazzoni, L. (1995) The use of advanced reproductive techniques in a closed nucleus of dairy sheep: a deterministic study on genetic progress. In: Enne, G., Greppi, G.F., Lauria, A. (eds) *Proc. Atti XXX Simposio Internazionale di Zootecnica, Milano, 11–13 September 1995* Elsevier, Paris pp. 39–52.

Sanna, S.R., Carta, A., Casu, S., Pilla, A.M., and Pagnacco, G. (1994) Valutazione genetica della razza ovina Sarda. 3. Indici e andamenti genetici ed ambientali. *Zootecnica e Nutrizione Animale* 20, 313–318.

Sanna, S.R., Sanna, A., Carta, A. and Casu, S. (1995a) The Sarda dairy sheep production system: balance and perspectives of the breeding programme. In: Gabina, D. (ed.) *Strategies for Sheep and Goat Breeding: Proc. of the Meeting of the Joint FAO/ CIHEAM Network on Sheep and Goats, Subnetwork on Animal Resources, Sidi-Taheb, Tunisia, 26–28 March 1995.* Cahiers Options Mediterranéennes, no. 11, CIHEAM, Zaragoza, pp. 89–95.

Sanna, S.R, Carta, A. and Casu, S. (1995b) L'utilizzazione dell'inseminazione strumentale nella razza ovina Sarda. (The use of artificial insemination in the Sarda dairy sheep breed.) In: SIPAOC (ed) *Proc. Symp. 'Miglioramento genetico degli ovini e dei caprini: aspetti scientifici e problemi applicativi, 7 December 1995, Perugia, Italy,* Faculte d'Agrana de Purugia, pp. 55–77.

Sanna, S.R., Astruc, J.M., Carta, A., Rosati, A. and Barillet, F. (1995c) Simplification of the daily milk test-day recording in sheep. In: *Rajoue, L., Lafontaine, S. Doyle, P. (eds) Proc. 29th Biennial Session of International Committee for animal Recording (ICAR), Ottawa, Ontario, Canada, 31 July–6 August 1994.* EAAP Publication No. 75, Agriculture and Agrifood Canada, Ottawa, pp. 281–286.

Serrano, M., Jurado, J.J., Perez-Guzman, M.D. and Montoro, V. (1993) Estudio de la tendencia genetica y fenotipica de diversos caracteres en la raza ovina Manchega. *Informacion Tecnica Economica Agraria* 12(1), 233–235.

Ugarte, E., Urarte, E., Arrese, F., Arranz, J., Beltran de Heredia, I. and Gabina, D. (1995) Technical organization and economic needs of the breeding programme of Latxa and Carranzana dairy sheep in the Spanish Basque Country. In: Gabina, D. (ed.) *Strategies for Sheep and Goat Breeding: Proceedings of the Meeting of the Joint FAO/CIHEAM Network on Sheep and Goats, Subnetwork on Animal Resources, Sidi-Taheb, Tunisia, 26–28 March 1995.* Cahiers Options Mediterranéennes, no. 11, CIHEAM, Zaragoza, pp. 155–164.

Vallerand, F. and Elsen, J.M. (1979) La diffusion du progrès génétique: étude théorique et quelques applications. In: ITOVIC-SPEOCiea *Proc. 5emes Journées recherche ovine et caprine, Paris, 5–6 December 1979* Paris, pp. 76–98.

Zervas, N., Boyazoglu, J., Kalaissakis, P., Papadimitriou, T. and Flamant, J. (1975) Comparaison des races ovines Chios et Frisonne avec leurs croisements en Grèce continentale. I. Viabilité et reproduction. *Annales de Génétique et Sélection Animale* 7(3), 277–291.

Zervas, N.P., Hatziminaoglou, J., Boyazoglu, J. and Georgoudis, A. (1991) Production potential and breeding schemes of some Mediterranean dairy sheep breeds. *FAO Animal Genetic Resources Information* 8, 11–27.

Genetic Improvement of Sheep

21

B.P. KINGHORN
Department of Animal Science, University of New England, NSW 2351, Australia

Introduction	565
Breeding Objectives	566
The economically rational approach	566
The 'desired gains' approach	569
Use of breeding objectives	570
Genetic Evaluation	570
Evaluating breed resources	570
Evaluating individuals – estimated breeding values	571
Use of information from relatives – best linear unbiased prediction	572
Selection on multiple traits	578
Estimated breeding value distributions	579
Evaluation at quantitative trait loci	580
Design of Breeding Programmes	581
Within-flock design	581
Between-flock design	584
Assortative mating and mate selection	585
Crossbreeding	586
Development of a Unifying Approach	589
References	590

Introduction

The aim of sheep breeders is to bring about genetic change in their livestock, with a view to increasing profitability, sustainability and ease of management at the production level. This involves a large number of issues, which can be classified into three groups:

- Objectives to target: This relates to asking 'Where to go?', whereas tools and strategies below relate to 'How to get there?'. Breeding objectives describe the type of sheep we might want to develop or, more commonly, describe the utility of marginal genetic changes in each trait of commercial value.

- Tools to exploit: The sheep breeder's toolkit includes performance and pedigree recording systems, novel reproductive techniques, knowledge of population genetic parameters and genetic evaluation systems. Emerging tools include information from genetic markers and gene transfer.
- Strategies to adopt: This relates to the design of breeding programmes. The main components are deciding which sheep to use as parents and in what combination to mate them. These are the driving elements of both selection theory and crossbreeding theory. Other strategy decisions involve data recording (pedigree, trait performance and DNA test information) and the use of novel reproductive and gene transfer techniques.

This chapter provides an overview of progressive techniques in sheep breeding. Much of this is supported by material in other chapters, which deal with trait- or biology-specific topics.

Breeding Objectives

Breeding objectives declare a desired direction for genetic change. They should be constructed in a manner which allows their rational expression as part of a genetic evaluation system, in order to facilitate ranking of animals on genetic merit and implementation of effective breeding programme design.

Breeding objectives are generally expressed as economic weightings to be applied to each trait of commercial importance, these weightings describing the impact of a unit change in each trait. There are two approaches to calculating these weightings – the economically rational approach, and the desired gains approach.

The economically rational approach

The classic approach to calculating economic weightings is economically rational – it takes no account of genetic parameters. This makes sense in that the value of making a unit change in a given trait should not be influenced by how difficult it is to generate this change. These difficulties can be handled appropriately at the genetic-evaluation phase. In this setting, breeding objectives should reflect the costs and returns involved in a production system and should not consider costs and gains generated in a breeding programme. This section will give a critical overview of the economically rational approach.

A simple example

A very simple breeding objective is presented here. The reader is directed to Ponzoni (1988) for a comprehensive worked example. The key tactical objective of a selection programme is to choose animals of high breeding value to be used as parents. An animal's breeding value (BV) is the value of its genes to its progeny, as will be described more fully later. The breeding objective is simply a multitrait breeding value, with each trait weighted by a relative economic weight, for example:

Breeding objective:	6 × fleece weight	+	−1 × fibre diameter (FD)
Units = $:	$ kg^{-1} kg	+	$ μm^{-1} μm.

In order to combine different traits into such a single score, they have to be converted to a common scale. This is generally dollars or some other monetary unit. The economic weights in this simple example are taken to have been calculated from market prices of $6 kg^{-1} of wool and −$1 μm^{-1} for an average fleece. The units of expression are thus 'dollars (per unit (kg or μm)) per head shorn'. Note that these weights involve no consideration of genetic parameters.

Units of expression

All economic weightings in a breeding objective should have the same basis for units of expression, such as 'dollars per head shorn', as used above. Choice of this basis can have an important influence on the consequences of using the breeding objective.

A simple basis for unit of expression, such as 'dollars per head shorn', can be used for situations in which all traits are directly related to economic costs or returns and thus excludes reproductive traits, whose effect is at least partly manifested through progeny.

A less simple basis is 'dollars per breeding ewe per year', which accommodates both production and reproduction traits. In all cases, each trait should use this same basis, such that an objective might be, for example:

$7.20 kg^{-1} per breeding ewe per year × clean-fleece weight
+ −$1.20 μm^{-1} per breeding ewe per year × FD
+ $7.20 per lamb weaned per breeding ewe per year × number of lambs weaned

This means that, in a flock of 150 breeding ewes, a marginal increase of 1 kg in clean-fleece weight would increase profit by $6.00 × 150 = $900 per year. This accommodates wool shorn from all classes of stock, through the way in which the economic weight is calculated. The economic weight for an increase of one lamb weaned is more difficult to calculate, due to expression via progeny, but in this case it is the same as for clean-fleece weight. Delays in returns due to expression in progeny can be accommodated by considering the pattern of flow of genes through the population and discounting future returns to give current values (McClintock and Cunningham, 1974).

Economic weights calculated on a 'dollars per head shorn' or 'dollars per breeding ewe per year' basis suffer a potentially important drawback. They relate to dollars per livestock unit, rather than dollars per resource unit, such as 'dollars per hectare'. As an example, consider two breeds of meat sheep:

Breed	Value of weight at slaughter	Value of food consumed	Profit per head	Dollar efficiency
Small	\$200	\$100	\$100	2 : 1
Large	\$350	\$200	\$150	1.75 : 1

The large breed would be targeted by a breeding objective based on 'dollars per head'. However, a breeding objective based on 'dollars per hectare' would target the small breed. A breeding objective based on dollars per resource unit will usually be more appropriate, as long as proper account is taken of any fixed costs per head.

Matching environment to changing breeds

It is possible to derive inappropriate breeding objectives by not taking account of the fact that different breeds perform most economically under different environments or management policies. As an extreme example, consider the two breeds in Fig. 21.1. If food intake to a fixed slaughter weight of 25 kg is taken as a key objective trait, the small breed cannot compete, as it is very old when it reaches this weight. Also, if body weight achieved for a given level of accumulated food intake is taken as a key objective trait, the small breed cannot compete. These apparently convincing arguments have led to widespread selection for increased body size in meat sheep and other meat-producing species. However, if we consider two small sheep as a single large individual, they will compete equally well on both these objective traits. In fact, when we consider that smaller animals reach mature size more quickly than larger animals (Taylor, 1980), the small breeds come into favour under an objective that accounts for timely use of resources. But even this is an illusion: larger

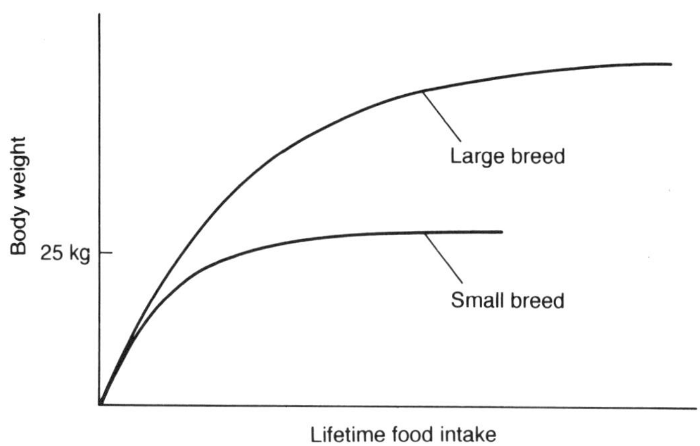

Fig. 21.1. Growth-efficiency curves for two breed sizes.

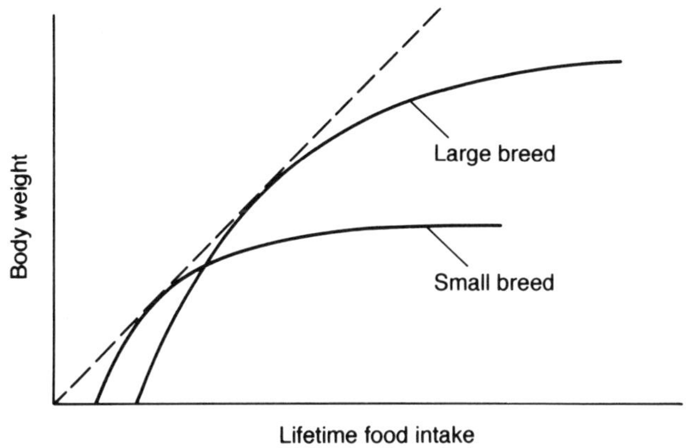

Fig. 21.2. Growth-efficiency curves accounting for maternal food costs.

animals consume less per kilogram of body weight (Taylor, 1980), such that fixed food resources can carry more kilograms of large animals, and the net result is that the two breed sizes are of equal value (Kinghorn, 1985).

This sort of debate can be avoided wherever it is possible to match the production system to the prevailing environment. For example, if the graphs in Fig. 21.2 were a close reflection of the patterns for dollar output plotted on dollar input, now accounting for maternal food costs, the production system should be to slaughter at a fixed proportion of mature size, about 60% in the current simplified example, and this should feature as the key breeding-objective trait.

However, such a simple solution is not always possible – for example, where:

- the market places a strong premium on a given slaughter weight or carcass weight; or
- the supply of grass or other food varies in magnitude or cost, such that slaughter before a given time has a large effect on profitability.

Under these conditions, the breeding objective calculations may have to accommodate non-linear value surfaces (Sivarajasingam *et al.*, 1984a) and possibly different management options (Sivarajasingam *et al.*, 1984b).

The 'desired gains' approach

An alternative approach to developing breeding objectives, the 'desired gains' approach, involves declaration of the relative magnitudes of genetic gain desired in the traits of importance. The breeding-objective calculations still result in relative economic weights, but these are now influenced by genetic parameters, with generally greater economic weightings for traits that are more difficult to change.

A simple subset of this approach is the restricted index, in which the objective is set up to give a predicted zero genetic change in one or more nominated traits. Examples are restrictions for no change in FD, back fat and milk fat.

Brascamp (1984) describes methods that can be used for restriction and desired gains. He also shows how to use a mixture of the economically rational and the desired gains approaches, with some traits constrained to prechosen levels of response and others influenced just by production economics. In all cases, relative economic weights are calculated, which is useful for demonstrating the 'effective economic weights' that nominated desired gains or restrictions imply.

Use of breeding objectives

A genetic evaluation system is used to rank animals on their estimated breeding value (EBV) for the breeding objective, using information from any number of traits, from candidate's relatives and from genetic parameters for all traits involved.

In many cases, such estimated breeding values are used only in conjunction with unscored traits, such as body conformation, leg weakness and pigmented fibres. This is not ideal. If these other traits have economic significance, they should be scored and included in the breeding objective and genetic evaluation steps. Not only will this give some 'economic balance', but it provides the framework to use information from correlated traits and information from candidates' relatives to speed up favourable genetic change for these traits.

The breeding objective can also be used to evaluate different breeds and crosses, as described in the next section.

Genetic Evaluation

Given a breeding objective, some form of genetic evaluation is required in order to provide information that can be acted on in a breeding programme.

Evaluating breed resources

A simple, but critically important type of information is the mean performance of available breeds and crosses and the predicted performance of untested crosses. This information provides an inventory of genetic resources, and the economic value of each breed genotype can be estimated by simply multiplying predicted performance for each trait by its corresponding economic weight and summing across traits. The feasibility and cost of characterizing breeds for complex or difficult-to-measure traits, such as food-conversion efficiency and disease resistance, are much more favourable than for similar-sized groups of individuals. We should make use of these opportunities when evaluating breed resources for the foundation of a breeding programme.

Breed evaluations can be extended to evaluation of different crossbreeding systems, with breeding objectives being calculated according to the specific role of each component breed or cross. For example, a breeding objective for a terminal-sire breed would involve little or no pressure on female fertility traits, as these will only be important within that breed, which will constitute only a small part of the total system.

Genetic evaluation of breed resources is relatively simple wherever good estimates of mean performance are available for the environment and production systems of interest. This is because the effects involved can be measured with high accuracy from many data and can be treated as fixed effects. However, when we come to evaluating the genetic merit of individual animals, there are much fewer data available per estimate and the random nature of BVs makes the process more complex.

Evaluating individuals – estimated breeding values

What causes an exceptional animal to be so much better than its contemporaries? There are two basic reasons:

1. The alleles it has inherited are more favourable, and/or they are more favourably configured, making the animal genetically superior.
2. It has probably experienced a better environment, through good luck.

In seeking genetic change we are not really interested in how much 'environmental luck' an animal has had – because that source of superiority cannot be transmitted to the next generation. Moreover, in selection programmes we are not generally interested in the configuration of alleles, as, in general, these configurations cannot be transmitted to the next generation (in the case of intralocus dominance) or are only weakly transmitted (in the case of interlocus epistasis).

We want to be able to choose the animals with alleles that will have the most beneficial effect on progeny, and we do this by selecting animals on the basis of their EBVs. The BV of an animal is a description of the value of that animal's genes to its progeny. In general, we do not know which genes an animal carries, so we can never fully know what an animal's BV is. However, we can estimate it from a wide range of information sources.

An important key concept is how to estimate an animal's BV, given information on just its phenotypic superiority (P, phenotype as a deviation from the contemporary mean). Ignoring epistatic interactions, for simplicity phenotypic superiority for animal i can be taken as:

$$P_i = G_i + E_i$$

where G is genetic value (the value of the animal's genes to itself) and E is environmental deviation.

This can be split further to:

$$P_i = A_i + D_i + E_i$$

where A is BV (the value of the animal's genes to its progeny) and D is dominance deviation.

Of the three components, A_i, D_i and E_i, only A_i can be transmitted to offspring; this is the sum of the average effects (in progeny) of the genes carried by animal i. Thus, we want to estimate A_i to give EBV_i. Given that P_i is positive (as animal i is superior), in the absence of further information each of A_i, D_i and E_i is more likely to be positive than negative. This means that P_i is an overestimate of A_i, and we must penalize it by the amount that D_i and E_i are likely, on average, to be positive.

The relative influence of A, D and E on P can be measured by their variance; if one of them does not vary, it does not contribute to differences in P. Thus, the estimate of breeding value is:

$$\hat{A}_i = \frac{V_A}{V_P} P_i = h^2 P_i$$

where ^ denotes 'estimate' and $h^2 = \frac{V_A}{V_P}$ is heritability of the trait concerned. Such an EBV would rank animals in the same order as phenotype or, indeed, in the same order as an objective version of the classical approach:

> the importance of the principle of selection in regard to MERINO SHEEP is so fully recognized, that men follow it as a trade: The sheep are placed on a Table and are studied, like a picture by a connoisseur; this is done three times at intervals of months, and the sheep are each time marked and classed, so that the very best may ultimately be selected for breeding.
>
> (Charles Darwin, *c.* 1859)

We are luckier than Darwin because we know about genes and how they are transmitted – and this has led to our ability to use information from relatives.

Use of information from relatives – best linear unbiased prediction

In selecting animals to act as parents, we are interested in choosing those with the most favourable alleles. An animal's own performance gives an indication of the value of its alleles to its progeny. However, some of this animal's alleles are also carried by each of its relatives, and so the performance of an animal's relatives can be used to give a more accurate assessment of the alleles it carries.

Thus progressive breeding programmes make use of information from all known relatives. This is of most value when heritability is low – when an animal's own performance is a poor indicator of BV. As heritability increases, there is a diminishing proportional value of information from relatives, until at a heritability of unity an animal's own performance is a perfect indicator of its

BV, with no room for improvement due to information on relatives.

Best linear unbiased prediction (BLUP) of EBV is a method by which appropriate use of phenotypic measurements on relatives can be made. It also accounts for differences between animals caused by fixed environmental effects, such as flock or year of birth, such that fair comparisons can be made between the EBVs of animals born or reared in different groups.

An overview of BLUP will be given here. Kennedy (1981) and Van Vleck *et al.* (1989) give more comprehensive, but digestible, treatments.

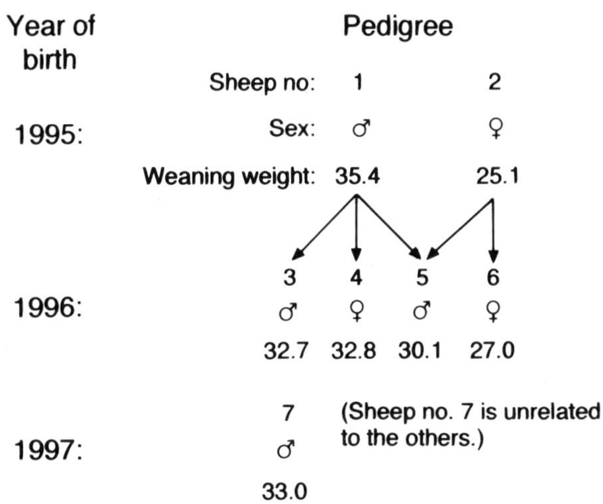

Fig. 21.3. Example data set for BLUP analysis.

Best linear unbiased prediction analysis by example

Figure 21.3 gives a hypothetical information set on seven animals, for which BLUP EBVs are to be calculated. According to convention, the BLUP analysis is set up using different terminology. The model is $\mathbf{Y} = \mathbf{Xb} + \mathbf{Zu} + \mathbf{e}$, where $\mathbf{Y}$ is the vector of observations, $\mathbf{b}$ is the vector of fixed effects fitted (such as year of birth and sex), and $\mathbf{u}$ is the vector of BVs. Matrices $\mathbf{X}$ and $\mathbf{Z}$ relate observations to $\mathbf{b}$ and $\mathbf{u}$, and $\mathbf{e}$ is the residual component, which contains non-additive genetic effects and environmental effects that are not accounted for. The following details the model for this example, assuming that the mean and year of birth are fitted as fixed effects:

$$\mathbf{Y} = \mathbf{X}\,\mathbf{b} + \mathbf{Z}\,\mathbf{u} + \mathbf{e}$$

$$\begin{pmatrix} 35.4 \\ 25.1 \\ 32.7 \\ 32.8 \\ 30.1 \\ 27.0 \\ 33.0 \end{pmatrix} = \overset{\text{mean}\ \ 1995\ \ 1996}{\begin{pmatrix} 1 & 1 & 0 \\ 1 & 1 & 0 \\ 1 & 0 & 1 \\ 1 & 0 & 1 \\ 1 & 0 & 1 \\ 1 & 0 & 1 \\ 1 & -1 & -1 \end{pmatrix}} \begin{pmatrix} b_{mean} \\ b_{1995} \\ b_{1996} \end{pmatrix} + \begin{pmatrix} 1 & 0 & 0 & 0 & 0 & 0 & 0 \\ 0 & 1 & 0 & 0 & 0 & 0 & 0 \\ 0 & 0 & 1 & 0 & 0 & 0 & 0 \\ 0 & 0 & 0 & 1 & 0 & 0 & 0 \\ 0 & 0 & 0 & 0 & 1 & 0 & 0 \\ 0 & 0 & 0 & 0 & 0 & 1 & 0 \\ 0 & 0 & 0 & 0 & 0 & 0 & 1 \end{pmatrix} \begin{pmatrix} u_1 \\ u_2 \\ u_3 \\ u_4 \\ u_5 \\ u_6 \\ u_7 \end{pmatrix} + \begin{pmatrix} e_1 \\ e_2 \\ e_3 \\ e_4 \\ e_5 \\ e_6 \\ e_7 \end{pmatrix}$$

What does this mean? For example, observation 1: The model says that the observed weaning weight of 35.4 kg is made up of 1 dose of b_{mean}, 1 dose of b_{1995}, no dose of b_{1996}, one dose of u_1, no dose of any of $u_2 \dots u_7$, plus whatever is left over undescribed by these effects (e_1). Notice that year 1997 is not fitted explicitly. Sheep no. 7, born in 1997, has its effect of b_{1997} equal to $-1\ b_{1995} + -1\ b_{1996}$. Thus the 3-year effects add to zero, and their value relates to deviation from an 'average' year. This sort of restriction is required in order to avoid confounding in the analysis.

The task is to get estimates of b_{mean}, b_{1995}, b_{1996}, and $u_1 \dots u_7$. Henderson (1975) showed that the mixed model can be solved using the following mixed-model equations:

$$\begin{pmatrix} \hat{\mathbf{b}} \\ \hat{\mathbf{u}} \end{pmatrix} = \begin{pmatrix} \mathbf{X'X} & \mathbf{X'Z} \\ \mathbf{Z'X} & \mathbf{Z'Z} + \mathbf{A}^{-1}\lambda \end{pmatrix}^{-1} \begin{pmatrix} \mathbf{X'Y} \\ \mathbf{Z'Y} \end{pmatrix} \qquad \text{Eqn 1}$$

where **A** is the numerator relationship matrix and $\lambda = \sigma_e^2/\sigma_u^2$. For non-inbred animals, the numerator-relationship matrix contains the coefficient of relationship for each pair of animals – the proportion of alleles they share in common by descent. For example, the coefficient of relationship between parent and offspring (e.g. sheep nos. 1 and 3) is ½. So $a_{1,3}$ equals ½ and so forth to give for our example:

$$\mathbf{A} = \begin{pmatrix} 1 & 0 & \tfrac{1}{2} & \tfrac{1}{2} & \tfrac{1}{2} & 0 & 0 \\ 0 & 1 & 0 & 0 & \tfrac{1}{2} & \tfrac{1}{2} & 0 \\ \tfrac{1}{2} & 0 & 1 & \tfrac{1}{4} & \tfrac{1}{4} & 0 & 0 \\ \tfrac{1}{2} & 0 & \tfrac{1}{4} & 1 & \tfrac{1}{4} & 0 & 0 \\ \tfrac{1}{2} & \tfrac{1}{2} & \tfrac{1}{4} & \tfrac{1}{4} & 1 & \tfrac{1}{4} & 0 \\ 0 & \tfrac{1}{2} & 0 & 0 & \tfrac{1}{4} & 1 & 0 \\ 0 & 0 & 0 & 0 & 0 & 0 & 1 \end{pmatrix}$$

Algorithms have been developed to derive the inverse of **A** directly, accommodating inbred animals, and this facilitates quick solution of the mixed-model equations.

These mixed-model equations can be related to the more digestible $\hat{A} = h^2 P$, described above, which is a single-animal EBV, with no fixed effects involved. Thus, we can ignore the fixed effects in equation 1 and present the multianimal estimation as:

$$\hat{\mathbf{u}} = \left(\mathbf{Z'}\,\mathbf{Z} + \mathbf{A}^{-1}\frac{\sigma_e^2}{\sigma_u^2}\right)^{-1} \mathbf{Z'}\,(\mathbf{Y} - \bar{\mathbf{Y}})$$

and, using the same terminology, the single-animal equivalent is:

$$\hat{\mathbf{u}} = \left(1 + \frac{\sigma_e^2}{\sigma_u^2}\right)^{-1} 1\,(\mathbf{Y} - \bar{\mathbf{Y}})$$

Changing to single-animal terminology gives:

$$\hat{A} = \frac{1}{1 + V_E/V_A} P = \frac{V_A}{V_A + V_E} P$$

$$= \frac{V_A}{V_P} P = h^2 P$$

which is the same as the conceptually simple result from the previous section.

To carry out the analysis, equation 1 must be solved. Here they are for our example, assuming heritability, $h^2 = 0.5$, giving $\lambda = \sigma_e^2/\sigma_u^2 = 0.5/0.5 = 1$:

$$\begin{pmatrix} \hat{\mathbf{b}} \\ \hat{\mathbf{u}} \end{pmatrix} = \begin{pmatrix} \mathbf{X'X} & \mathbf{X'Z} \\ \mathbf{Z'X} & \mathbf{Z'Z} + \mathbf{A}^{-1}\lambda \end{pmatrix}^{-1} \begin{pmatrix} \mathbf{X'Y} \\ \mathbf{Z'Y} \end{pmatrix}$$

$$\begin{pmatrix} \hat{b}_{mean} \\ \hat{b}_{1995} \\ \hat{b}_{1996} \\ \hat{u}_1 \\ \hat{u}_2 \\ \hat{u}_3 \\ \hat{u}_4 \\ \hat{u}_5 \\ \hat{u}_6 \\ \hat{u}_7 \end{pmatrix} = \begin{pmatrix} 7 & 1 & 3 & 1 & 1 & 1 & 1 & 1 & 1 & 1 \\ 1 & 3 & 1 & 1 & 1 & 0 & 0 & 0 & 0 & -1 \\ 3 & 1 & 5 & 0 & 0 & 1 & 1 & 1 & 1 & -1 \\ 1 & 1 & 0 & 19/6 & 1/2 & -2/3 & -2/3 & -1 & 0 & 0 \\ 1 & 1 & 0 & 1/2 & 17/6 & 0 & 0 & -1 & -2/3 & 0 \\ 1 & 0 & 1 & -2/3 & 0 & 7/3 & 0 & 0 & 0 & 0 \\ 1 & 0 & 1 & -2/3 & 0 & 0 & 7/3 & 0 & 0 & 0 \\ 1 & 0 & 1 & -1 & -1 & 0 & 0 & 3 & 0 & 0 \\ 1 & 0 & 1 & 0 & -2/3 & 0 & 0 & 0 & 7/3 & 0 \\ 1 & -1 & -1 & 0 & 0 & 0 & 0 & 0 & 0 & 2 \end{pmatrix}^{-1} \begin{pmatrix} 216.1 \\ 27.5 \\ 89.6 \\ 35.4 \\ 25.1 \\ 32.7 \\ 32.8 \\ 30.1 \\ 27.0 \\ 33.0 \end{pmatrix} = \begin{pmatrix} 31.2 \\ -0.92 \\ -0.89 \\ 2.83 \\ -2.89 \\ 1.83 \\ 1.88 \\ -0.09 \\ -2.24 \\ 0 \end{pmatrix}$$

The **X'Z** and **Z'X** blocks in the coefficient matrix (the square matrix to be inverted) provide a connection between the fixed and random effects. If they were full of zeros, the results would be the same as if two separate models had been fitted (one fixed and one random). The values in these blocks let the analysis account for the fact that, for example, progeny that are very good because of being born in a good year do not overly increase their parents' EBVs.

Once inverted, the coefficient matrix constitutes a set of custom index weights:

- The elements of the first three rows are multiplied by the elements of the (**X'Y** plus **Z'Y**) vector to give the estimates of the three fixed effects.
- The elements of the last seven rows are the index weights for estimating breeding values. Note that there are index weights in the '**Z'X** block', which are used to account for fixed effects when calculating EBVs.

Looking at the result vector (the last column), notice that:

- $b_{1997} = -(-0.92) - (-0.89) = 1.81$ kg and that the effects of year thus sum to zero.
- Animal 7 has a zero EBV. This is because it cannot be fairly compared with any other animal, as they are all born in different years. If animal 7 had had a relative born in a different year, it would have got a non-zero EBV, based on that relative's EBV.

Properties of best linear unbiased prediction estimated breeding values
This section lists the key properties of BLUP EBVs.

EBVs are generally additive For example, if a ram has an EBV of $\hat{A}_{ram} = + 1.0$ kg and a ewe has $\hat{A}_{ewe} = + 0.5$ kg for fleece weight, the prediction is that progeny will have a fleece-weight superiority of $(1.0 + 0.5)/2 = 0.75$ kg. This is actually a prediction of progeny genetic value, but, as progeny-dominance deviation and environmental deviation are unknown and thus have 'expectations' of zero, it is also a prediction of progeny BV and phenotype. Note also that the proportion of parental superiority in EBV which is transmitted to progeny is unity, after accounting for halving due to meiosis. Thus the heritability of EBVs is unity, as they have been preregressed.

BLUP makes full use of information from all relatives It does this by use of the numerator relationship matrix. BLUP does not have to give separate attention to sib testing, progeny testing, own performance, etc. Use of information from all relatives (even those long dead) is simultaneously handled. This gives more accurate EBVs and more selection response.

BLUP accounts for fixed environmental effects (management group, herd, season, year etc.) This means that animals can be compared across groups, giving wider scope for selection. For example, comparing across age-groups means that older animals have to prove their competitiveness at every round of selection. This property of BLUP usually accounts for most of its advantage over less powerful methods.

BLUP gives genetic trends The ability to compare the EBVs of animals born and measured in different years means that year-mean EBVs can be calculated and genetic trends reported.

BLUP can cater for non-random mating Thus, rams can be compared via their progeny even if some had been allocated better ewes. This can only be done where the ewes were allocated on the basis of their recorded performance, such that BLUP can account for their EBVs when evaluating the rams concerned.

BLUP can account for selection bias For example, consider ranking rams on the first two shearings of their daughters. The worse rams, who had worse daughters, will have benefited more from culling of daughters on first shearing performance. However, BLUP accounts for this, given that the information used to make selection decisions (first shearing results in this case) is included in the data set.

Accuracy and standard error of EBVs can be determined Accuracy is the correlation between EBV and true BV, $r_{\hat{A}A}$. It can be used to partition the variance between EBVs and the variance within individual EBVs – the latter being the square of the standard error of individual EBVs:

Variance between EBVs	$r^2_{\hat{A}A} \cdot V_A$
Variance within EBVs	$(1 - r^2_{\hat{A}A})\, V_A$
Total variance	V_A

Thus, with high accuracy of evaluation, animals will tend to differ more in their EBVs, and they will have lower standard errors of their EBVs. Where only one of these parameters – accuracy or standard error – has been presented in a genetic evaluation report, the other can be found quite simply:

$$SE_{EBV} = \sqrt{(1 - r^2_{\hat{A}A})\, V_A} \qquad r_{\hat{A}A} = \sqrt{\left(1 - \frac{SE^2_{EBV}}{V_A}\right)}$$

The standard error of the EBV of animal i can be calculated as $SE_{EBV_i} = \sqrt{C_i \sigma^2_e}$, where C_i is the diagonal element of the inverted coefficient matrix that corresponds to animal i and σ^2_e is the mean square error from the analysis, or, approximately, $\sigma^2_e = (1 - h^2)\, V_P$. When the analysis does not produce an inverted coefficient matrix, approximations to SE_{EBV} can be found by considering just close relatives of each animal.

Selection on multiple traits

In most cases, a number of traits will have been measured on animals, to be used as component selection criteria contributing to an overall index. The classical selection index has been used quite widely in sheep breeding. Brascamp (1984) gives an overview of selection-index methodology. However, BLUP of BVs is more flexible and powerful than the classical selection index, and software for its implementation has become widely available. Therefore, the selection index will be treated here only by overview.

Selection index

The selection index yields a multitrait EBV, in the same units as the breeding objective, in a manner conceptually similar to the single-trait version give above. As described above, for a single trait $\hat{A} = h^2\ P$ in trait units, or $a\hat{A} = ah^2\ P$ in dollar units, where a is the economic weight of the single trait. Thus, the overall weighting factor on phenotype is ah^2 and, by extension; for multiple traits, the selection index is $\hat{A}_\$ = a_1h^2{}_1\ P_1 + a_2h^2{}_2\ P_2 + a_2h^2{}_2\ P_2 + \ldots$ for traits 1, 2 etc. However, this is only true where traits are uncorrelated to each other, and the objective traits are the same as the measured criterion traits. These factors can be accommodated by calculating appropriate index weights b_1, b_2, etc., and this is illustrated here by analogy with the simple single-trait case:

For a single trait: $$b = h^2 \quad a$$

$$b = \frac{\mathrm{Cov}_{\mathrm{AP}}}{V_P} \quad a$$

$$b = \mathrm{Cov}_{PP}{}^{-1}\ \mathrm{Cov}_{PA} \quad a$$

For multitrait: $$\mathbf{b} = \mathbf{P}^{-1} \quad \mathbf{G} \quad a$$

where **b** is the vector of weights b_1, b_2 etc., which are used to calculate the selection index for animal i as follows: $\hat{A}_{\$_i} = b_1P_{1_i} + b_2P_{2_i} + \ldots$ **P** is the (co)variance of all criterion traits, **G** is the covariance matrix of criterion traits with BVs for objective traits and **a** is the vector of economic weights for objective traits.

This selection-index framework can be used to accommodate information from relatives, but this is more usefully done through BLUP genetic evaluation.

Multitrait best linear unbiased prediction

The extension from single-trait to multitrait BLUP is conceptually simple. The model is unchanged, **Y = Xb + Zu + e,** where **Y** now includes observations on all traits recorded for each individual, **b** includes fixed effects for all traits and **u** contains a BV for each trait for each individual. The mixed-model equations now involve genetic and environmental correlations. The result includes EBVs (or $\hat{A}$s) for the each of the traits fitted – which can include both measured criterion traits and breeding-objective traits, even if there is missing information on the latter. The breeder only needs to weight EBVs for the objective traits by their economic weights to provide an index which s/he can select on: Index = $a_1\hat{A}_1 + a_2\hat{A}_2 + a_3\hat{A}_3 + \ldots$. The index is itself an EBV – for economic merit.

Estimated breeding value distributions

Increased measurement (more traits, more relatives) results in higher accuracy, and this is reflected in the observed distribution of EBVs, or $EBV_{\$}$s, as shown in Fig. 21.4. If the only information available is animals' tag numbers, there is no power to identify superior (or inferior) animals and there is no variation in EBVs. If greasy-fleece weight is known, there is some power – and yet if FD is of key importance in the objective, this power is obviously limited. Animals of exceptional BV are difficult to identify, as the most important trait is not measured. As more information is gathered, there is more power to identify animals of low and high BV, and the EBV distribution widens. This is in agreement with the conclusion given above that the variance between EBVs is $r^2_{\hat{A}A} V_A$. Information from relatives also helps here, as in the distribution at the bottom of Fig. 21.4, which uses BLUP genetic evaluation. Of course, as with other forms of recording, pedigree recording involves expense and effort – more mating paddocks are needed and lambs need to be identified with their mothers.

This effect of accuracy on EBV distribution can be used to show why we tend to put more emphasis on measurement of rams (Fig. 21.5). A smaller

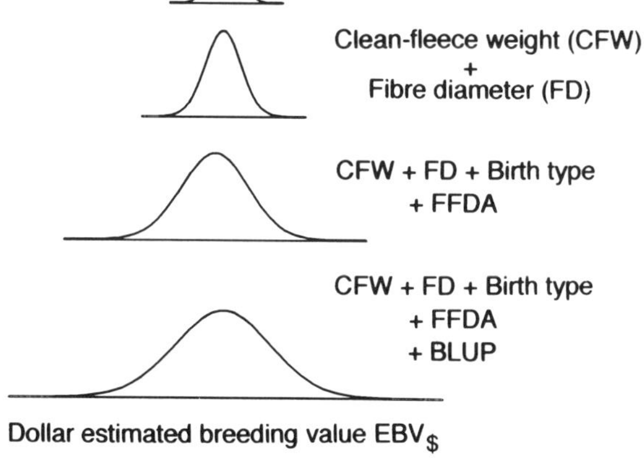

Fig. 21.4. More information gives more spread in dollar EBVs.

proportion of rams than ewes can be selected for breeding, due to their higher fecundity. This can be capitalized on by investing in more accurate EBVs for rams. If using multiple ovulation and embryo transfer, the value of taking more measurements in ewes (as well as rams) becomes higher, due to their increased fecundity.

Notice that the response is simply the average (or halfway point) between the selected rams' and ewes' mean EBV superiority. Response per generation is thus simply equal to $R_{gen} = (i_m \sigma_{\hat{A}_m} + i_f \sigma_{\hat{A}_f})/2$, where i_m is the selection intensity of males – the number of standard deviation units superior to contemporaries. The width of the EBV distribution of the progeny depends on how intensively they are measured.

Evaluation at quantitative trait loci

Some single loci of major effect, known as major genes or quantitative trait loci (QTL), have been identified and exploited directly. In sheep, these include the Booroola, Javanese and Inverdale genes for prolificacy (Piper and Bindon, 1982; Bradford *et al.*, 1986; Davis *et al.*, 1991) and the *callipyge* muscling gene

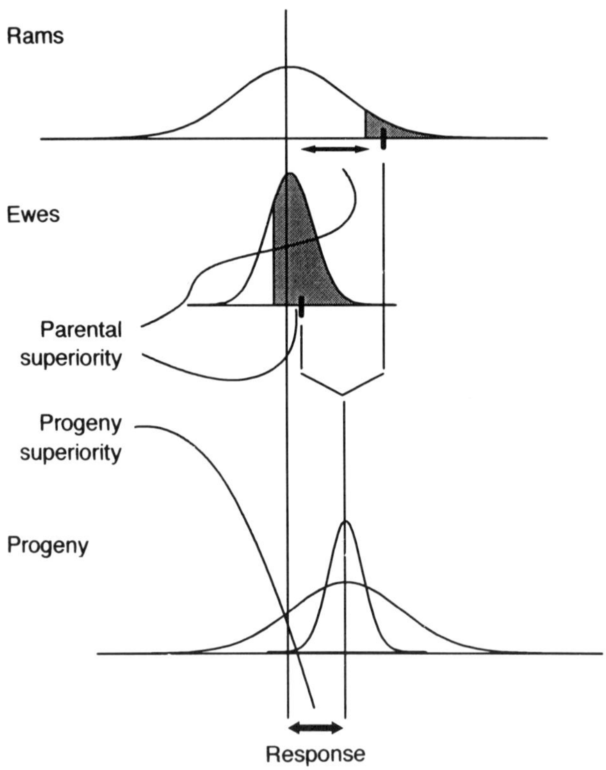

Fig. 21.5. Effect of EBV accuracy on selection response.

(Cockett *et al.*, 1994). Banks (Chapter 18) discusses some of these.

Most such QTL have been detected by inspection of data. However, systematic methods for computer screening have been developed. More importantly, with the recent development of genetic maps for sheep (see Chapter 11), genetic-marker plus trait performance data can be analysed to detect and locate other QTL of commercial value (reviewed by Kinghorn *et al.*, 1994).

Where QTL have been cloned and a DNA test developed to determine genotype for individual animals, genetic evaluation at QTL is relatively simple. The QTL genotypes can be treated as fixed effects, and these effects can be estimated very accurately, just as fixed effects of breed and cross means can be estimated accurately. The QTL effects may differ between genetic backgrounds (e.g. breeds) and between environments or production systems, and the power to estimate the range of effects involved constitutes a major advantage over evaluation of polygenes using, for example, BLUP. This advantage extends to the ability to market specific QTL-genotyped seed stock with a performance and product image, which is much more tangible than for competing 'high polygenic merit' seed stock.

Where QTL alleles can be inferred with imperfect accuracy through use of linked markers, marker-assisted selection (MAS) can be invoked. As with direct DNA tests for QTL, the value of MAS depends on a number of factors:

- Where heritability is low, the value of information on individual QTL tends to be higher.
- Where the trait(s) of interest cannot be measured on one sex, marker information gives a basis for ranking animals of that sex.
- If the trait is not measurable before sexual maturity, marker information can be used to select at a juvenile stage.
- If a trait is difficult to measure or requires sacrifice (as with many carcass traits), marker information can be used instead.

Marker-assisted selection is handicapped by the fact that, unless there is considerable linkage disequilibrium, no one marker allele is consistently associated with a favourable QTL allele, due to recombination events. This means that linkage phase in parents needs to be inferred – something which can be done readily with very large half-sib families, as in dairy cattle. However, sheep population structures lead to the need for MAS analysis methods which can operate on general pedigrees, and several appropriate methods have been put forward (for example, van Arendonk *et al.*, 1994).

Design of Breeding Programmes

Within-flock design

Breeding programmes within a closed flock use selection to target genetic change in the breeding objective. A classical approach to designing such breeding programmes is to set up prediction equations for genetic change and inbreeding, and find designs that maximize predicted genetic change while accommodating inbreeding at an acceptably low level. This approach results in a set of rules which the breeder can adopt.

A more powerful approach is to exploit the properties of a BLUP genetic evaluation system to implement design through tactical use of EBVs.

From prediction equations to breeding rules

This section will consider optimization of population age structure as an example. Predicted response per generation was given in a previous section as $R_{gen} = (i_m\sigma_{\hat{A}_m} + i_f\sigma_{\hat{A}_f})/2$. However, we are more interested in response per year R_{yr}, as this accounts for how quickly we can 'turn over a generation' and, more usefully, describes rate of genetic improvement. The generation interval is $(L_m + L_f)/2$, where L_m is the generation interval for males – the average age of rams when their progeny are born. Thus we have a general prediction of response per year:

$$R_{yr} + \frac{i_m\sigma_{\hat{A}_m} + i_f\sigma_{\hat{A}_f}}{L_m + L_f}$$

Noting that, for mass selection on individual phenotype, $\sigma_{\hat{A}_m} = \sigma_{\hat{A}_f} = \sigma_{h^2P} = h^2\sigma_P$, response per year to a policy of mass selection can be predicted by the equation:

$$R_{yr} = \frac{i_m + i_f}{L_m + L_f} h^2\sigma_P$$

This equation is used in Fig. 21.6. Consider a 1000-ewe flock with an age structure typified by the numbers in Fig. 21.6, which is taken from a computer modelling program. These numbers are given as real approximations to integers, and are driven by the parameters presented in the figure. It is assumed that selection takes place on fleece weight at time of entry to the breeding flock. Notice that, fairly typically, parents first drop progeny at 2 years, rams are kept for two matings and ewes are kept for seven matings. This structure gives a predicted response to selection of 80.6 g year^{-1}.

The curved lines on the graph within Fig 21.6 illustrate the relationships between generation interval and selection intensity for each sex. These are influenced by weaning rate, mating ratio and survival rate. The ticks on these curves represent the number of matings for which each sex is kept before culling. For the prevailing policy, these ticks are joined by a straight line whose slope is $(i_m + i_f)/(L_m + L_f)$. Maximizing the slope of this line maximizes the predicted response and thus optimizes the age structure.

Figure 21.7 shows the optimized structure, with an 8.4% increased predicted response to selection of 87.4 g year^{-1}. This is achieved by culling ewes 3 years earlier, after their fourth lambing opportunity. As the population is making genetic progress, younger ewes, born in later years, are expected to be genetically superior on average. In the prevailing case, earlier culling has benefited from this phenomenon, with an acceptable compromise in reduced selection intensity.

However, with selection on BLUP EBVs, this superiority in younger stock is 'automatically' accommodated, as BLUP can account for effects of year and can thus lead to fair comparison of animals across years.

From best linear unbiased prediction estimated breeding values to ad hoc *design*
Figure 21.8 illustrates hypothetical selection across years on BLUP EBV. For simplicity, this illustration assumes that all candidates are available for consideration after the 1998 measurement round. It can be seen that simple truncation selection sets up an age structure with appropriate emphasis on younger animals. Moreover, genetic gains are increased using this approach, as selection takes place not at entry to the breeding flock, but at each year of age. This is one example of a good genetic evaluation system driving good breeding-programme design – something that can work between flocks as well as within flocks.

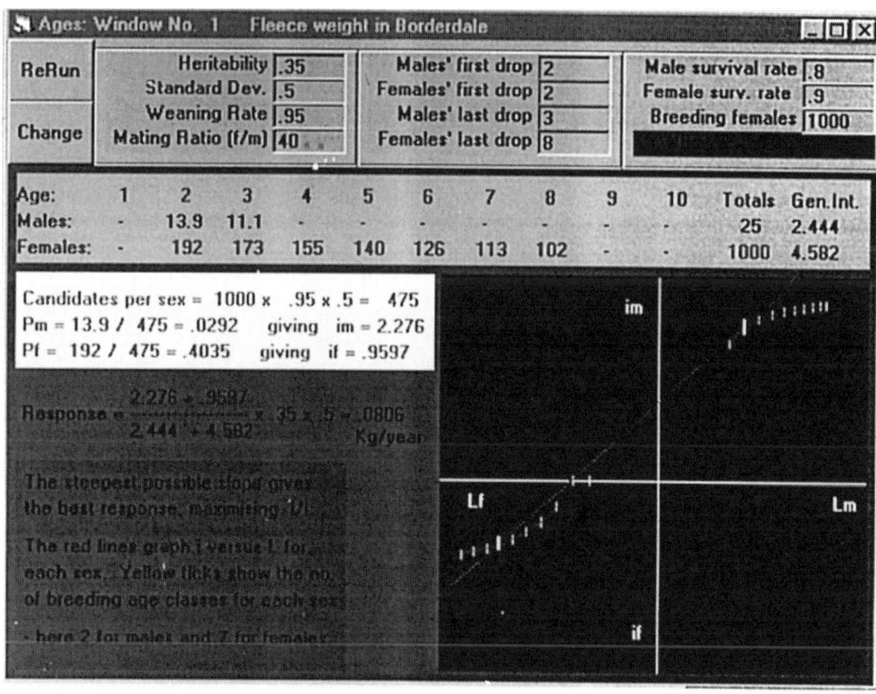

Fig. 21.6. Calculation of predicted response to mass selection on fleece weight.

Between-flock design

All animal industries have developed some form of pyramidal structure, with élite studs at the top delivering seed stock to the commercial level, sometimes through one or more multiplier levels (see Chapter 20, for an example). Due to the relatively low fecundity of sheep, there are usually one or two such multiplier levels, sometimes known as daughter and general studs. This structure evolves because it is usually not economically viable to record all animals for inclusion in a breeding programme.

Even among top-tier studs, sizeable genetic differences exist, and these can be exploited with migration of stock from better to poorer flocks, benefiting the population as a whole. A special case of this is where a group of breeders form a cooperative called an open-nucleus breeding scheme. Here, one flock is nominated as an élite nucleus, and all efforts are made to improve this flock as much as possible. At equilibrium, all flocks progress at the higher rate, although 'lower' flocks lag behind by about two generations.

Figure 21.9 shows the difference between such an open scheme and a traditional closed scheme. This is an example of a good genetic-evaluation system driving good breeding-programme design across flocks.

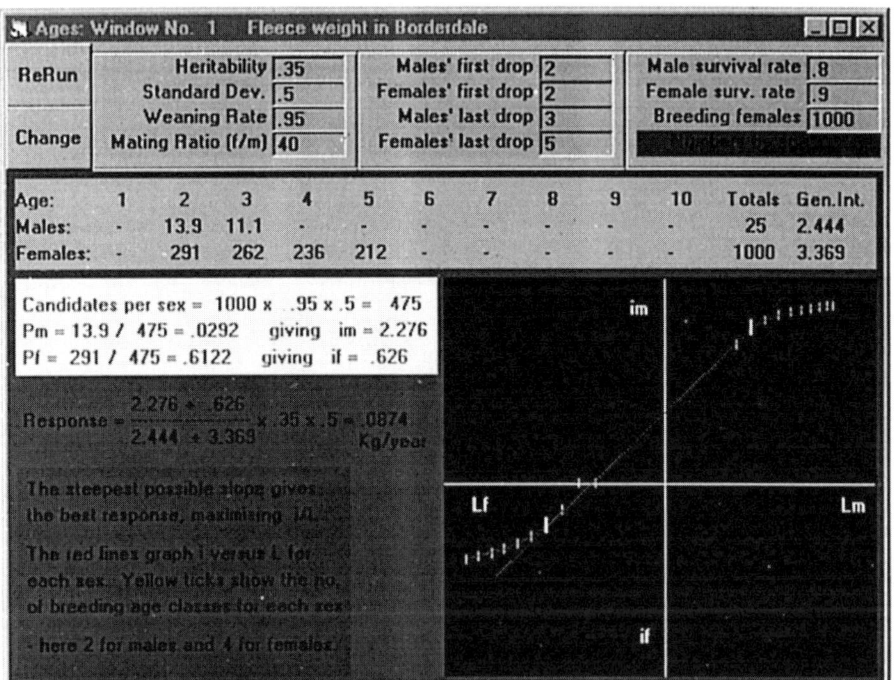

Fig. 21.7. Structure chosen to optimize response to selection on fleece weight.

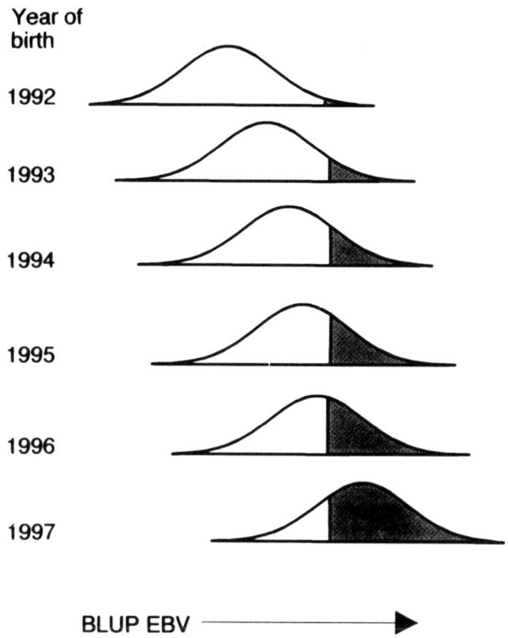

Fig. 21.8. Selection across years on BLUP EBV leads to appropriately young age structures.

Assortative mating and mate selection

The open-nucleus design exploits assortative mating by mating the élite animals in the nucleus. Where there is no information from relatives and selection is based on animal phenotypes, the open-nucleus design provides ancestral information, which helps drive the extra response enjoyed. This occurs because

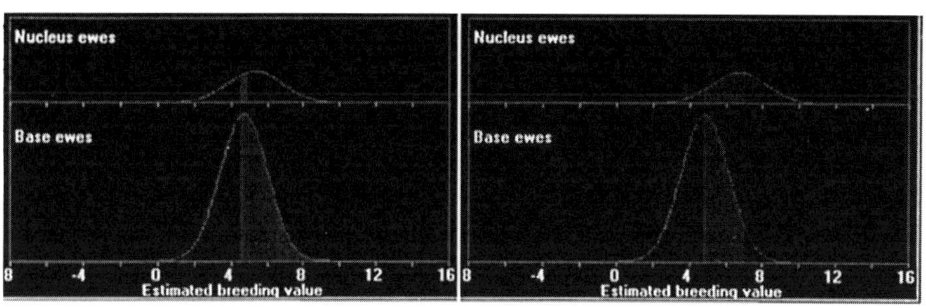

Fig. 21.9. Ewe migration in closed (left) and open (right) nucleus schemes at generation 4, taken from computer simulations. In the closed scheme, nucleus ewes are all born in the nucleus, despite availability of genetically superior ewes in the non-élite base flock(s). In the open scheme, nucleus ewes are selected as the best from both nucleus and base flocks, resulting in better gains in the nucleus. In this case, there is increased lag in the open scheme.

animal phenotypes are regressed by h^2 towards their own flock means – the P in $\hat{A} = h^2P$ is deviated from own-flock mean – boosting the EBVs of nucleus animals appropriately. However, where there is information from relatives and BLUP EBVs are calculated across flocks, full use of assortative mating can be made, with ranking of each sex across flocks on BLUP EBV and mating best to best and worst to worst across the whole population. This makes for even greater gains than an open nucleus scheme.

The processes of selection and mating can be conceptually combined to give mate selection. This is sometimes necessary, as the best animals to select may depend on how they are allocated for mating, and vice versa. The open-nucleus design provides an example of where this might be useful – with the need to balance genetic gains from making given matings with the cost of seed stock, transport, quarantine, etc. to achieve these matings.

Mate-selection methods have been developed to manage inbreeding in populations under selection (e.g. Klieve *et al.*, 1994), to target migration for more powerful comparison between flocks (Kinghorn and Shepherd, 1994) and to develop composite crossbred populations (Kinghorn and Atkins, 1987).

Crossbreeding

Crossbreeding is a powerful tool for the animal breeder. It is of relatively little value for specialized wool production, where the Merino breed is superior, but it is of critical importance to the meat-sheep industries.

The value of crossbreeding

The key reasons for crossbreeding are listed here:

- The averaging of breed effects: for example, to get an animal of intermediate size to fit a particular pasture cycle or market demand.
- Direct heterosis. Crossbred individuals often exhibit heterosis. Heterosis is measured as the extra performance of the crossbreds over the weighted average of their parent breeds. The percentage increase in performance ranges from about 0 to 10% for growth traits and 5 to 25% for fertility traits (e.g. Nitter, 1979). The effect of heterosis on the total production system can be even more than this, as effects accumulate over traits.
- Maternal heterosis. Crossbred ewes can exhibit considerable heterosis in their ability to raise many fast-growing, viable offspring.
- Sire–dam complementation. A good crossbreeding system aims to use breeding ewes that are small (but not so small for dystocia to be a problem) as well as prolific. When a large breed of ram is used the proportion of feed directed to growing animals is increased and the production system benefits accordingly.
- Possibly cheap source of breeding animals. This is evident in many sheep crossing systems – for example, in the Australian prime-lamb industry, where surplus Merino ewes are bought cheaply and crossed with Border Leicester rams to produce prime lamb dams.

The genetic basis of heterosis

We need to know the genetic basis of heterosis in order to predict the value of untested genotypes. There are two genetic mechanisms postulated as causing heterosis effects:

1. Dominance. Where the individual's parents come from two different breeds, the individual will carry a wider range of alleles, sampled from two breeds rather than just one. It is thought that this better equips the individual to perform well, especially under a varying or stressful environment. We would thus expect dominance to be a positive effect, and there is much evidence to support this.
2. Epistasis. When we cross breeds, alleles find themselves having to interact or 'cooperate' with alleles at other loci which they are 'not used to'. The crossbred animal may thus be out of harmony with itself, and we predict that epistasis, if important, is a negative effect.

The dominance model of heterosis is widely assumed and used, and so this model will be taken here. It should be borne in mind that epistatic loss could cause errors in prediction based on the dominance model alone.

Breed dominance is greatest when all loci consist of two alleles derived from different breeds, as in a first, or F_1, cross. Other crosses show a proportion of this heterosis equal to the proportion of loci that are heterozygous with respect to breed of origin. This can be seen in the column D_d (dominance for the direct subtrait) in Table 21.1.

Table 21.1 shows how to predict the merit of untested crossbred genotypes, given estimates of crossbreeding effects. These are additive (A) or 'purebreed' effects for each of the three breeds, and dominance (D) or heterotic effects, here assumed equal for each pair of breeds. Subscripts denote the direct subtrait (d) and the maternal subtrait (m) – both of these being of some importance for weaning weight in sheep. A least-squares analysis of the form $\hat{\beta} = (X'X)^{-1} X'Y$ can be used to estimate the crossbreeding effects (in vector $\hat{\beta}$) from Merit (Y), where X is the matrix formed by the body of Table 21.1. The section 'Evaluating breed resources', above, outlined the simple approach that can be used to consider all breeding-objective traits to help predict the economic merit of different crosses.

Choice of crossing system

The best crossing system to use depends to a large extent on the value of the breeds available and the amount of heterosis expressed in crossbred animals. This is illustrated in Table 21.2 by describing the conditions under which each crossbred genotype is worthy of choice.

Of course, care should be taken to consider factors other than the predicted genetic merit of candidate crosses for the traits of importance. The key factor here is the cost of maintaining structured crossing systems, where separate breeding units are required to give ongoing supply of purebred and/or crossbred parents. These costs often outweigh the genetic benefits of more structured crosses, especially in species of low fecundity, where the parental breeding units must be relatively large to supply the final cross. However,

Table 21.1. Example prediction of merit of weaning weight from estimated crossbreeding parameters. Multiply the coefficients shown in the body of the table by the Values of the corresponding Effects (see text). Adding the products gives the prediction of weaning weight, Merit, in the last column.

Effects: Values (kg):	Mean 28.0	A_{d1} +2.0	A_{d2} 0	A_{d3} −2.0	A_{m1} −0.6	A_{m2} −0.1	A_{m3} +0.7	D_d 2.0	D_m 1.0	Merit
Breed 1	1	1	0	0	1	0	0	0	0	29.40
Breed 2	1	0	1	0	0	1	0	0	0	27.90
Breed 3	1	0	0	1	0	0	1	0	0	26.70
Best F_1 (1 × 2)	1	0.5	0.5	0	0	1	0	1	0	30.90
Best three-breed cross (1 × 23)	1	0.5	0.25	0.25	0	0.5	0.5	1	1	31.80
Best backcross (1 × 12)	1	0.75	0.25	0	0.5	0.5	0	0.5	1	31.15
Balanced (1, 2)	1	0.5	0.5	0	0.5	0.5	0	0.5	0.5	30.15
Synthetics (1, 2, 3)	1	0.33	0.33	0.33	0.33	0.33	0.33	0.67	0.67	30.00
Optimum (1, 2)	1	0.63	0.37	0	0.63	0.37	0	0.47	0.47	30.24
Synthetics (1, 2, 3)	1	0.57	0.31	0.12	0.57	0.31	0.12	0.56	0.56	30.30
Rotations (1, 2)	1	0.5	0.5	0	0.5	0.5	0	0.67	0.67	30.65
(1, 2, 3)	1	0.33	0.33	0.33	0.33	0.33	0.33	0.86	0.86	30.57

Synthetics at equilibrium; rotations at equilibrium and averaged over years.

Table 21.2. General recommendations on use of crossbreeding.

Purebreed	When no cross is better: the Merino is a good example
F_1 cross	When direct heterosis is important
Three-breed cross	When both direct and maternal heterosis are important
Four-breed cross	When paternal heterosis is important as well
Backcross	When only two good parental breeds are available and/or when direct heterosis is not important
Rotational cross	When females are too expensive either to buy in or to produce in the same enterprise
Open or closed synthetic	When both males and females are too expensive: a few initial well-judged importations establish the synthetic, and it can then either be closed (which helps to establish a breed 'type') or be left open to occasional well-judged importations

crossbreeding systems can be driven largely by the supply/demand situation for parental stock, and this is quite prevalent in the meat sheep industries.

Development of a Unifying Approach

The tactical implementation of a breeding programme involves three classes of decision:

- What measurements to make, and on which animals? In its broadest sense, this question covers decisions on which animals to 'measure' via progeny test or sib test.
- Which animals to use as parents?
- How to allocate mates?

Although this tactical implementation is quite simple, it must accommodate a wide range of animal breeding issues, some of which are summarized in Table 21.3. Many of these issues have been addressed quite independently from others. A major task to be completed in animal breeding theory is the development of integrating frameworks that permit key issues to be simultaneously addressed in the implementation of animal-breeding programmes. The most powerful solutions here will involve tactical implementation, rather than rule development, as this will make full opportunistic use of the prevailing candidates for measurement and breeding.

Table 21.3. Some issues of importance in the design of breeding programmes.

Issue	Comment
Selection value	As indicated by estimated breeding value
Assortative mating	Giving increased genetic variation, as in open-nucleus schemes and Dairy 4 pathway structures
Pedigree structure	Information-rich structures lead to more accurate genetic evaluation
Connection	Giving better power to compare animals (or genetic groups) born and/or evaluated in different fixed-effect groups
Lifetime value	Of importance where selected animals continue to yield products
Parameter estimation	Where outside estimates are unreliable
Measurement strategies	Which animals to progeny-test or measure more intensively?
Crossbreeding value	Exploitation of breed differences and heterosis
Running costs	Accounting for the costs of selecting, migrating and mating
Others	Including factors related to inbreeding, reproductive manipulation, exploiting major genes and risk

References

Bradford, G.E., Quirke, J.F., Sitorus, P., Inounu, I., Tiesnamurti, B., Bell, F.L., Fletcher I.C. and Aan Torell, D.T. (1986) Reproduction in Javanese sheep: evidence for a gene with large effect on ovulation rate and litter size. *Journal of Animal Science* 63, 418–431.

Brascamp E.W. (1984) Selection indices with constraints. *Animal Breeding Abstracts* 52, 645–654.

Cockett, N.E., Jackson, S.P., Shay, T.L., Neilsen, D., Moore, S.S., Steele, M.R., Barendse, W., Green, R.D. and Georges, M. (1994) Chromosomal localization of the *callipyge* gene in sheep (*Ovis aries*) using bovine DNA markers. *Proceedings of the National Academy of Sciences of the USA* 91(8), 3019–3023.

Davis, G.H., McEwan, J.C., Fennessy, P.F., Dodds, K.G. and Farquhar, P.A. (1991) Evidence for the presence of a major gene influencing ovulation rate on the X chromosome of sheep. *Biology of Reproduction* 44, 620–624.

Kennedy, B.W. (1981) Variance component estimations and prediction of breeding values. *Canadian Journal of Genetics and Cytology.* 23, 565–578.

Kinghorn, B.P. (1985) Modelled relationships between animal size and production efficiency. *Journal of Animal Breeding and Genetics* 102, 241–255.

Kinghorn, B.P. and Atkins, K.D. (1987). Heterosis in crosses between Merino strains and bloodlines and its exploitation. In: McQuirk, B.J. (ed.) *Merino Improvement Programs in Australia.* Australian Wool Corporation, Leura, New South Wales, pp. 389–407.

Kinghorn, B.P. and Shepherd, R.K. (1994) A tactical approach to breeding for information-rich designs. In: Smioth, C. *et al.* (eds) *Proceedings of the 5th World Congress on Genetics Applied to Livestock Production, Guelph, 7–12 August 1994*, Vol. 18, pp. 255–261.

Kinghorn, B.P., van Arendonk, J.A.M. and Hetzel, D.J.S. (1994) Detection and use of major genes in animal breeding. *AgBiotech News and Information* 6, 297N–302N.

Klieve, H.M., Kinghorn, B.P. and Barwick, S.A. (1994) The joint regulation of genetic gain and inbreeding under mate selection. *Journal of Animal Breeding and Genetics* 111, 81–88.

McClintock, A.E. and Cunningham, E.P. (1974) Selection in dual purpose cattle populations: defining the breeding objective. *Animal Production* 18, 237–248

Nitter G. (1979) Breed utilisation for meat production in sheep. *Animal Breeding Abstracts* 46, 131–143.

Piper, L.R. and Bindon, B.M. (1982) Genetic segregation for fecundity in Booroola Merino sheep. In: Barton, R.A. and Smith, W.C. (eds) *Proceedings of the 1st World Congress on Sheep and Beef Cattle Breeding*, Vol. 1, World Congress on Sheep and Beef Cattle Society Inc., pp. 395–400.

Ponzoni, R.W. (1988) The derivation of economic values combining income and expense in different ways: an example with Australian Merino sheep. *Journal of Animal Breeding and Genetics* 105, 143–153.

Sivarajasingam, S., Burnside, E.B., Schaeffer, L.R. and Wilton, J.W. (1984a) Prediction of average daughter performance from sire proofs for use in linear programming sire profit models. *Journal of Dairy Science* 67, 3008–3014.

Sivarajasingam, S., Burnside, E.B., Wilton, J.W., Pfeiffer, W.C. and Grieve, D.G. (1984b) Ranking dairy sires by linear programming dairy farm model. *Journal of Dairy Science* 67, 3015–3024.

van Arendonk, J.A.M., Tier, B. and Kinghorn, B.P. (1994) Use of multiple genetic markers in prediction of breeding values. *Genetics* 137(1), 319–329.

Van Vleck, L.D., Pollack, E.J. and Oltenacu, E.A.B. (1989) *Genetics for the Animal Sciences*. W.H. Freeman, New York.

Standardized Genetic Nomenclature for Sheep

22

C.H.S. DOLLING

President of COGNOSAG, Box 74, McLaren Vale, South Australia 5171, Australia

Introduction 593
Development of Nomenclature 594
The first guidelines for sheep and goats 594
The 1991 guidelines for ruminants 594
The 1993 guidelines for ruminants 594
Categories of loci in sheep 595
The Guidelines for Genetic Nomenclature for Sheep – Locus Categories 1–4 595
Locus 595
Alleles 597
Genotype terminology 598
Phenotype terminology 598
Additional Comments for Categories 3 and 4 598
Listings of Loci by the Committee on Genetic Nomenclature of Sheep and Goats 599
Acknowledgements 600
References 600

Introduction

The Committee on Genetic Nomenclature of Sheep and Goats (COGNOSAG) was founded with the express purpose of drawing up guidelines for the nomenclature of loci and alleles in sheep and goats. It was launched at Palmerston North, New Zealand, in 1984 and is an association registered in France under the Loi de 1901 as the Comité de Nomenclature Génétique des Ovins et Caprins (COGOVICA).

The need for guidelines became manifest during the National Congress on Breeding Coloured Sheep and Using Coloured Wool in Adelaide, South Australia, in 1979, at which there were almost as many systems of nomenclature for coat colour and pattern as there were speakers from half a dozen countries.

Development of Nomenclature

The first guidelines for sheep and goats

A project for standardizing genetic nomenclature in sheep was developed by Lauvergne (1984) and presented to the COGNOSAG Workshop of 1986. At the same workshop, Searle (1988) reviewed the mouse nomenclature rules and gene nomenclature in the human and in other mammals. The rules proposed for sheep by the workshop were a compromise between nomenclature regulations for mice and those for humans.

The rules permitted the use of mainly human rules for loci defined by biochemical variants and tissue or blood groups, and retained useful rules from the mouse nomenclature for the loci where dominance and recessivity could be distinguished in the various genotypes (Lauvergne, 1988).

The 1986 Workshop guidelines for sheep and goats were reviewed at the COGNOSAG Workshop of 1987 (COGNOSAG, 1989).

The 1991 guidelines for ruminants

The guidelines from the 1987 Workshop were revised during the 1988 and 1989 Workshops. A further development was the revision at the 1991 Workshop to accommodate all ruminants and to facilitate the development of a comparative genome nomenclature. This development had been stimulated by the interest expressed by workers with cattle, at the 1990 Workshop, for the guidelines to be suitable for use with cattle as well as with sheep and goats. The ruminant guidelines were published by Andresen *et al.* (1991).

The 1993 guidelines for ruminants

Revised guidelines for gene nomenclature in ruminants were prepared by the COGNOSAG *ad hoc* committee at the 1993 Workshop (COGNOSAG *ad hoc* committee, 1995). The core of these revised guidelines consists of the 1991 proposals of Andresen *et al.* (1991). Their rewording was undertaken to reduce their length, and to increase their clarity. The earlier recommendations to limit the length of symbols of loci and alleles to a maximum of five and four characters, respectively, was relaxed. The designation of top dominant and codominant alleles by a capital initial letter was undertaken to assist in the recognition of alleles with visible effects. Thus, the changes made are intended to render the guidelines more permissive and user-friendly, while retaining consistency with the human and mouse systems of nomenclature. The new recommendations include the use of species prefixes – for example, *OOV* for sheep, *BBO* for cattle – and the adoption of the nomenclature for keratins and keratin-associated proteins proposed by Powell and Rogers (1993). An additional proposal for provisionally assigning symbols and for listing newly reported deoxyribonucleic acid (DNA) segments and proteins that have no known homologues, official names or symbols was also outlined.

As far as possible, COGNOSAG respects the names of loci and alleles proposed by authors, and COGNOSAG will propose new names only in the light of new knowledge or to maintain consistency with the existing nomenclature.

The use of names and symbols in italics for loci and their alleles is preferred. However, if it is not possible to comply with this, those names and symbols should be underlined. Authors are asked to be consistent in the underlining or use of italics for the names and symbols within a document and/or file. Locus and allele symbols need not be in italics or underlined in databases, but should be in italics in hard copy from these databases.

Categories of loci in sheep

In *Mendelian Inheritance in Sheep* (MIS96) (1996), COGNOSAG presents reports on loci that have been classified into one of four categories:

Category 1	Coat-colour loci
Category 2	Loci for visible traits other than coat colour
Category 3	Loci controlling blood and milk polymorphisms
Category 4	Mapped loci and other genetic systems

In the next section, the guidelines for genetic nomenclature for sheep loci in categories 1–4 are presented. Additional comments for categories 3 and 4 are given in the subsequent section.

The Guidelines for Genetic Nomenclature for Sheep – Locus Categories 1–4

Locus

Locus name

Choice of name

- The name in English should be as brief as possible but not consist of a single letter, and should convey as accurately as possible the character affected or the function by which the locus is recognized. The name may indicate a morphological character (*Ear Length*) or a disease character (*Photosensitivity*), a body system or body function (*Fecundity Java*), or a biochemical property (*Albumin*) or nucleotide segment (*DNA Segment: probe MAF 4*).
- As far as possible the locus name should reflect interspecies homology.
- All Greek symbols should be replaced by Roman letters and placed after the name, e.g. β *Haemoglobin* becomes *Haemoglobin beta*.
- If a newly described locus has an effect similar to one that has already been

named, it may be named according to the breed, geographical location or population of origin.

- Should a new locus be identified later as being the same as a locus already named, the name invoking breed, geographical location or population of origin should be abandoned.

Printing the name

- The locus name should be in Roman type or a combination of Roman type and Arabic numerals. Wherever possible, the locus name should be printed in italics; otherwise it should be underlined.
- The initial letter of the locus name should be a capital Roman character.
- If the locus name is two or more words, each word after the first word may begin with a capital or a lower case Roman character, e.g. *Ear Length*, *Haemoglobin beta*; Ear Length, Haemoglobin beta.

Locus symbol

Choice of symbol

- For newly reported loci, special care should be exercised in selecting an appropriate symbol to avoid duplication and confusion with existing nomenclature. Every effort should be made to ensure that the symbols selected conform with those in current use for homologous loci.
- The locus symbol should consist of as few Roman letters as possible, or a combination of Roman letters and Arabic numerals.
- The initial character should always be a capital Roman character which, if possible, should be the initial letter of the name of the locus.
- For loci other than those for coat colour and visible traits, upper-case Roman letters only, or upper-case letters combined with Arabic numerals, should be used.
- If the locus name is two or more words and the initial letters are used in the locus symbol, these letters should be in Roman capitals.
- All characters in a locus symbol should be written on the same line; no superscripts or subscripts should be used, nor should Roman numerals or Greek letters.
- Where appropriate, the symbol should indicate the biochemical property or designate a particular nucleotide segment.
- The rules of mammalian interspecific homology already used in the choice of the name of the locus should be applied to the choice of the symbol.
- The designation of prefixes denoting mammalian species of origin, when being used to distinguish between the species homologues of a locus (e.g. *OOV ALB* and *HSA ALB*), should follow that adopted by the Human Genome Nomenclature Committee.
- For unmapped DNA segments and proteins that have no known homologues or official names or symbols, the first character of the symbol should be \ (back-slash) followed by the letters and/or numbers designated by the authors (e.g. *\FCB11* for the dinucleotide repeat polymorphism at the FCB11 locus, Crawford and Buchanan, Genbank accession No LO1531).

- For the keratins and keratin-associated proteins, COGNOSAG recommends that authors adopt the symbols and names proposed by Powell and Rogers (1993).

Printing the symbol
Wherever possible the locus symbol should be in italics; otherwise it should be underlined, e.g. the symbol of the *Agouti* locus: *A* or A.

Alleles

Allele name
Choice of name

- The name should be as brief as possible, but should convey the variation associated with the allele. If not given names, alleles should be given symbols, as described in the 'Allele symbol' section below.
- If a newly described allele is similar to one that is already named, it should be named according to the breed, geographical location or population of origin. The names of new alleles at a recognized locus should conform to the nomenclature established for that locus.
- Should a new allele be identified later as being the same as an allele already named, the name invoking breed, geographical location or population of origin should be abandoned.

Printing the name
Wherever possible, the allele name should be in italics; otherwise it should be underlined. A lower-case initial letter of the allele name is preferred. This does not apply when a symbol is used instead of an allele name. For example, the allele for the polled condition at the *Horns* locus in sheep: *polled* or polled; an allele at the *Haemoglobin beta* locus: *B* or B.

Allele symbol
Choice of symbol

- The allele symbol should be as brief as possible, consisting of Roman letters and/or Arabic numerals.
- As far as possible, the allele symbol should be an abbreviation of the allele name, and should start with the same letter. In the loci detected by biochemical, serological or nucleotide methods, the allele name and symbol may be identical.
- Greek letters and Roman numerals should not be used.
- The symbol + can be used alone for identification of the standard allele ('wild type') for alleles having visible effects. Neither + nor – symbols should be used in alleles detected by biochemical, serological or nucleotide methods. Null alleles should be designated by the number zero.
- The initial letter of the symbol of the top dominant allele should be in capitals. The initial letter of all other alleles should be lower case. When

there are only codominant alleles, they should all have a capital initial letter.

Printing the symbol

- The allele symbol should always be written with the locus symbol. It may be written as a superscript following the locus symbol or following an asterisk on the same line as the locus symbol. The allele symbol should be printed immediately adjacent to the locus symbol, i.e. with no gaps.
- Wherever possible, the allele symbol should be in italics; otherwise it should be underlined. For example, the recessive allele *hornless* at the *Horns* locus in sheep will be printed in italics: Ho^{hl} or *Ho*hl*, or underlined, $\underline{\mathrm{Ho}^{\mathrm{hl}}}$, or <u>Ho*hl</u>. The dominant allele *polled* will have the symbol in italics, Ho^{P} or *Ho*P*, or underlined, $\underline{\mathrm{Ho}^{\mathrm{P}}}$ or <u>Ho*P</u>.

Genotype terminology

- The genotype of an individual should be shown by printing the relevant locus and allele symbols for the two homologous chromosomes concerned, separated by a solidus, e.g. Ho^{P}/Ho^{P} or $\underline{\mathrm{Ho}^{\mathrm{P}}}/\underline{\mathrm{Ho}^{\mathrm{P}}}$.
- Unlinked loci should be separated by semicolons.
- Linked or syntenic loci should be separated by a space and listed in alphabetical order when gene order and/or phase are not known.
- For X-linked loci, the hemizygous case should be designated by */Y* following the locus and allele symbol, e.g. *FecX*I/Y*.
- Y-linked loci should be designated by */X* following the locus and allele symbol.

Phenotype terminology

The phenotype symbol should be in the same characters as genotype and allele symbols. The difference is that the characters should not be underlined, should not be in italics and should be written with a space between locus characters and allele characters instead of with an asterisk. Square brackets [ ] may also be used. For example, the dominant genotype Ho^{P}/Ho^{+} is equivalent to the phenotype Ho P or [Ho^{P}].

Additional Comments for Categories 3 and 4

The group of COGNOSAG members working on category 3 – blood and milk polymorphisms – has given detailed consideration to the nomenclature for blood and milk polymorphisms in cattle, sheep and goats. To comply with the decision made at the 21st International Conference on Animal Genetics, Michigan, 1990, to change existing gene nomenclature closer to that used for humans, additional proposals were put forward for cattle, sheep and goats

(Larsen *et al.*, 1992). These were that the locus symbols would be written in capital letters but lower-case letters would be retained to designate recessive alleles. Lower-case letters were also retained for sheep and goat blood-group factors.

The guidelines for gene nomenclature for category 4 – mapped loci and other genetic systems – are as described in the previous section, but, in addition, COGNOSAG recommends that, if an unnamed, newly mapped sheep locus is identified, contact may be made with Dr Sue Povey, University College, London, Medical Research Council (MRC) Biochemical Genetics Unit, Wolfson House, 4 Stephenson Way, London NW1 2HE, phone +44–171–380 7410, fax +44–171–387 3496, email mpovey@hgmp.mrc.ac.uk, before the locus is named to ensure that the symbol proposed for sheep has not already been adopted for the human. Alternatively, researchers may contact their nearest available COGNOSAG member or the sheep-genome database (Sheep Base) for assistance.

Listings of Loci by the Committee on Genetic Nomenclature of Sheep and Goats

The following listings by COGNOSAG members document and record the loci and alleles of sheep reported in the scientific literature and illustrate the implementation of COGNOSAG's guidelines for gene nomenclature.

- Loci for category 2 – visible traits other than coat colour – have been published for sheep and goats (COGNOSAG, 1989).
- Loci for category 3 – blood and milk polymorphisms – have been published for sheep and goats (COGNOSAG, 1989) and for cattle, sheep and goats (Larsen *et al.*, 1992).
- Loci for category 1 – coat colour – have been published for sheep and goats (COGNOSAG, 1990).
- Listings of loci and alleles in categories 1–3 and of loci of category 4 – mapped loci and other genetic systems – are available on Internet at http ://www.angis. su.oz.au/ (contact person is Dr F.W. Nicholas, Department of Animal Science, University of Sydney, New South Wales 2006, Australia; fax: +61–2–692 2114; email: frankn@doolittle.vetsci.su.oz.au.
- In *Mendelian Inheritance in Sheep* (MIS96) (1996), being published by the University of Camezino, Italy, a full listing of loci and alleles in categories 1–3 is given, together with an abbreviated listing of loci in category 4.

Acknowledgements

This chapter has reported work on the genetic nomenclature of sheep which has been done by members of COGNOSAG from countries around the world, both during and between COGNOSAG Workshops. The chronology of COGNOSAG activities emphasizes the time and effort which members have put into the 10-year development of COGNOSAG guidelines for nomenclature and procedures for listing loci.

Year	Activities
1979	Original concept in Adelaide (South Australia)
1984	Foundation at Massey University (New Zealand)
1986	1st Workshop, Manosque (France)
1987	Registration in Antony Sub-Prefecture (France)
	2nd Workshop Manosque (France)
1988	3rd Workshop, Manosque (France)
1989	4th Workshop, Eugene (Oregon, USA)
1990	5th Workshop, Edinburgh (Scotland)
1991	6th Workshop, Manosque (France)
1992	7th Workshop, Manosque (France)
1993	8th Workshop, Turretfield (South Australia)
	Seminar, Australian Society of Animal Production/COGNOSAG (Adelaide)
1994	9th Workshop, Rome (Italy)
1995	10th Workshop, Matelica (Italy)
1996	11th Workshop, Nouzilly (France)

For the preparation of the chapter it is a pleasure to acknowledge the assistance of two members of COGNOSAG – Tom Broad, of New Zealand, for helpful comments on the manuscript, and Verle Wood, of South Australia, for the preparation of the typescript.

References

Andresen, E., Broad, T., di Stasio, L., Dolling, C.H.S., Hill, D., Huston, K., Larsen B., Lauvergne, J.J., Levéziel, H., Malher, X., Millar, P., Rae, A.L., Renieri, C. and Tucker, E.M. (1991) Guidelines for gene nomenclature in ruminants 1991. *Genetics, Selection, Evolution* 23, 461–466.

COGNOSAG (1989) Gene nomenclature in sheep and goats 1987. In: Lauvergne, J.J. (ed.) *Standardized Genetic Nomenclature for Sheep and Goats 1987. Loci for Visible Traits Other than Colour and Blood and Milk Polymorphisms. Proceedings of the COGNOSAG Workshop, 1987.* TEC and DOC-LAVOISIER Paris, pp. 17–21.

COGNOSAG (1990) Loci for coat colour of sheep and goats 1989. In: Lauvergne, J.J. (ed.) *Proceedings of the COGNOSAG Workshops of 1988 and 1989.* COGOVICA/ COGNOSAG, Clamart, France.

COGNOSAG *ad hoc* committee (1995) Revised guidelines for gene nomenclature in ruminants 1993. *Genetics, Selection, Evolution* 27, 89–93.

Larsen, B., di Stasio, L. and Tucker, E.M. (1992) List of alleles for blood and milk polymorphisms in cattle, sheep and goats. *Animal Genetics* 23, 188–192.

Lauvergne, J.J. (1984) A Project for Standardizing Genetic Nomenclature in Sheep. *Bulletin technique du Département de Génétique animale* No. 38, INRA Département de Génétique animale, CNRZ, Jouy-en-Josas, France, 59 pp.

Lauvergne, J.J. (1988) The project of genic nomenclature for the COGNOSAG. In: Lauvergne, J.J. (ed.) *Proceedings of the COGNOSAG Workshop, 1986.* TEC and DOC-LAVOISIER, Paris, pp. 21–25.

Mendelian Inheritance in Sheep (MIS96) (1996). Dolling, C.H.S., Rae, A.L., Lauvergne, J.J., Renieri, C. and Millar, P. (eds). COGOVICA/COGNOSAG, Clamart, France and University of Camerino, Camerino, Italy.

Powell, B.C. and Rogers, G.E. (1993) Differentiation in hard keratin tissues, hair and related structures. In: Leigh, I., Watt, F. and Lane, E.B. (eds) *Keratinocyte Handbook*. Cambridge University Press, Cambridge, UK.

Searle, A.G. (1988) The genetic nomenclatures of the mammals. In: Lauvergne, J.J. (ed.) *Proceedings of the COGNOSAG Workshop 1986.* TEC and DOC-LAVOISIER, Paris pp. 15–20.

Index

addax 5
additivity 576, 587
Aepycerotinae 5,6
albinism 96, 120
albumin 140, 247, 300
Alcelaphinae 5, 6
allele 586, 597
allozymes 3
American wild sheep 9
Ammotragus 1, 7, 8
aneuploidy
  autosomal 266
  sex chromosome 266
antibodies 134, 137, 186, 205, 216
Antilocapridae 4, 6
Antilopinae 4, 6
argali 9, 15, 55
Arylesterase 141
artificial insemenation 400, 552
Artiodactyla 1, 4, 10
assortative matings and mate selection 585

back-fat 531
Barbary sheep (*Ammotragus)* 1, 7, 8
behaviour 10
  feeding 355
  grazing 355, 356, 357
  maternal 363
  reproductive 362
  social 357
bharal *(Pseudois)* 7, 8
biochemical polymorphism 140
bison 5
blastocyst 415, 421
blastomere 413
blesbok 5, 6
blood and milk proteins 137
blood groups 3, 93, 94, 136, 137
BLUP 513, 514, 515, 576, 577, 579, 582, 583, 585, 586
body composition
  between-breed variation in composition 524
  between-breed variation in fat partitioning 526
  within-breed variation in composition 527
Booroola 29, 89, 98, 117, 146, 260, 299, 337, 460, 512, 580
Boorula locus 276, 299, 337, 384
Bovidae 1, 5, 6
Bovinae 4, 5
breed evaluations 509, 510, 552, 570–572
breeding objectives 495, 506, 507, 509, 543, 566, 570
breeding programmes 200, 204, 213, 472, 485, 493, 506, 508, 510–512, 517, 529, 531, 543, 546, 547, 550, 553, 556, 566, 572, 581, 589
breeding schemes 515, 550, 584
breeds
  Arab 27
  Awassi 27, 111, 361, 542, 543
  Baluchi 28
  Barbary 1, 28
  Bleu du Maine 19, 28, 29
  Booroola Merino 381, 444

breeds *continued*
Border Leicester 29, 31, 68, 106, 355, 356, 362, 364, 365, 367, 508, 533, 534
Charollais 530
Cheviot 30, 40, 42, 60, 228, 229–231, 234, 355, 362, 364, 366
Chios 30, 542, 543
Churra 30, 141, 552, 553, 559
Clun Forest 355, 361, 366
Columbia 31, 362
Comisana 553
Coopworth 31, 40, 207, 327, 360, 363, 533, 534
Cormo 31
Corriedale 31, 38, 105, 226, 356
Criollo 32, 442
Deccani 32, 264
Dorper 204, 212
Dorset Down 9, 32
Dorset Horn 33, 208, 355, 356, 362, 365
East Friesian 33, 543
Finnship 34, 111, 361, 362, 382, 444, 509
Fulani 34
German Mutton Merino 34, 362
Hair Sheep 35
Hampshire 35, 39, 92
Han 35
Ile-de-France 9, 36, 234, 368
Karakul 36, 67, 73, 78, 79, 112, 361
Kazakh Finewool 37
Kirgiz Finewool 37
Lacaune 232, 542, 552–554
Langhe 554
Latxa 552, 559
Leicester Longwool 9, 37, 106
Lincoln 9, 31, 38, 40, 43, 106, 362
Lincoln Longwool 38
Manchega 553, 555
Manech 552, 553, 559
Marwari 38
Merino 9, 10, 21, 38, 39, 68, 81, 112, 204, 207, 208, 212, 327, 337, 356, 360, 362–366, 508, 533, 534, 572
Merinolandschaf 39
Mongolian 39
Oxford down 39, 369, 509, 524, 533
Perendale 36, 40, 327, 360, 364
Polish Longwool 204
Polwarth 40
Poll Dorset 80, 533
Prècoce 40
Rambouillet 40, 362, 534
Red Maasai 204
Romanov 41, 204, 360, 368, 382, 444, 509
Romney 9, 41, 68, 81, 106, 204, 207, 263, 327, 337, 360, 362–365
Sarda 542, 552, 553, 559
Scottish Blackface 42, 204, 365–367
Soay 22, 23, 68, 208, 340, 366, 367, 524
Somali 42
Southdown 9, 42, 355, 356, 367, 524
Soviet Merino 43
Suffolk 9, 10, 30, 32, 33, 35, 36, 40, 42, 43, 66, 90, 226, 232, 234, 355, 382, 509, 526, 530, 533
Texel 10, 43, 226, 233, 327, 362, 509, 516, 525, 530, 533
Tibetian 44
Tsigai 44
Vendeen 45
Welsh Mountain 366, 367
Zackel 30, 44, 45, 462
breed evaluations 509, 510, 571
buffalo 5

Callipyge locus 430, 512, 513, 529, 533, 534, 580
Camelidae 4
camels 4, 5, 164, 165
Capra 1, 7, 8, 10
Caprinae 1, 5, 6, 7, 10
carcass leanness 29–33, 35, 36, 38, 40–43, 45, 526
CAT (computer aided tomography) 528, 529
cattle 4–6, 26, 90, 92, 101, 104, 106, 114, 116, 184, 256, 261, 278
cell culture 243–245, 259
centimorgan (cM) 327
centric fission 3, 7
centric fusion (Robertsonian translocation) 7, 256, 263, 264
Cephalophinae 5
Cervidae 4–6
chamois (*Rupicapra*) 7, 8
chevrotains 4
chiasmata 328
chromosome
acrocentric autosome 6, 260

aberrations 260, 262
arms 6, 7
banding 7, 151, 242, 244, 245, 260
C banding 246, 256
'dynamic' banding 257
elongated 244
GTG banding 151, 246
QFQ banding 245, 246
NOR staining 245, 246, 264
centric fusions 2, 3, 7, 256, 263, 264
fragile sites 259, 260
inversions 7
meiotic 246
metacentrics, 260
spreads 244
number 2, 3, 6, 7
cladistics, 2
cloning 409
coat 18, 22, 592
coat colour 95, 120
coat colour loci and alleles
afghan lethal 54, 74
Agouti 54, 56–64, 65, 597
Albino 54, 65
albino marrabel 54, 66
Australian Piebald 54, 70
badgerface 54, 61
bizet spotting 54, 73
black and tan 54, 61, 62
blue 54, 62
Brown 54, 67–69
Extension 54, 64, 65
eyepatch 54, 63
gotland grey 54
light blue 54, 62
moorit 67
non-agouti 54, 60
persian 54
Pigmented Head 54, 73-75
Roan 54, 75
spotted 54, 72
Spotting 54, 71–73
Sur 54, 70
Ticking 54, 76
white or tan 54, 57, 95, 119, 595
COGNOSAG 61, 65, 593–595, 597, 599, 600
conservation 437–467
justification 442–444
methods 449–453
corpus luteum 380
cow, 164, 165
crossbreeding 22, 30, 74, 438, 464, 526, 540, 543, 559, 566, 571, 586
cryoprotective agent 405, 406
cryptorchidism 95, 119
cysteine 53, 110, 159, 187
cysteine-rich proteins 158
cytokines 190

daily gain 29, 30, 34–37, 40, 43
dairy traits
breeding objectives and selection criteria 543–546
breeding programmes 546–550
deer 4, 5, 6
deletions 265
DNA
α-satellite 150, 151
fingerprinting 298
microsatellites 3, 298
mitochondrial 3, 5, 6
polymorphism 341
repeats 9, 151, 152, 158, 159, 298
replication 391, 401, 402
domestication 9, 13, 24, 80, 354, 357, 360
dominance 571
duikers 5
duplications 265

EBV (estimated breeding values) 579, 580
embryo production 396
in vitro production 397
laparoscopic techniques 397
non-surgical techniques 397
surgical recovery and transfer 396
embryo culture 398
embryo storage 405, 406
embryonic mortality 384
Eocene 4
epistasis 571
euchromatin 150
eumelanin 53, 67
evolution 3, 14, 18, 22, 46, 150, 151, 153, 154, 159, 161, 163, 167, 171, 256, 277, 375, 462
chromosomal 6, 7
genome 149, 340

fat 526, 534
backfat 531
intramuscular 530, 531
subcutaneous 526
partitioning 526
fat-tailed 27, 28, 30, 36, 39, 543
faecal worm egg count 203
fecundity 60, 89, 97, 276, 299, 337, 580
fertilization 398, 400, 413, 429
fetal development 415, 416
fibroblast cultures 243
fleece 1, 8, 22, 492–497
weight 480, 486–489
fleece-rot 207, 209, 210
fly strike 207, 210
follicle 378
follicle-stimulating hormone (FSH) 378, 379, 396, 400
foot-rot 201, 202, 207, 209, 216
Dichelobacter nodosus, 202, 208, 216
fossil record 3, 5
freemartins 263
foxes 19

gazelles 5
gene (genomic) mapping 242, 266, 270–272, 340, 341
gene (genetic) maps 7, 150, 272, 298, 328, 340, 341, 561
conserved type I loci 277, 278
highly polymorphic type II loci 277
interspecies cell hybrids 271
gene targeting 404
genetic conservation programmes 439, 440
genetic correlations 203, 205, 206, 209, 216, 368, 472, 479, 480, 482–485, 510, 527, 544, 558, 578
genetic distances 9, 448
genetic drift 18, 25, 65, 68, 438, 448, 455
genetic (genotype) evaluation 506, 510, 513–515, 528, 552, 554, 570–581
genetic information systems 514
genetic manipulation 403
genetic markers 9, 10, 88, 139, 143, 149, 191, 211–213, 242, 245, 271, 273, 277, 298, 328, 339, 448, 512, 566, 581
comparison with cattle gene map 277
comparison with human and mouse gene maps 278
genetic parameters 206, 215, 354, 362, 364, 368, 388, 389, 479–485, 510, 511, 513, 544, 566, 569
genome 6, 89, 150–152, 192, 272–278, 297, 328–336, 402, 421, 422
evolution, 338
mitochondrial 6
genomic libraries 270, 299
genotype 56, 59, 61, 70, 79, 137, 139, 143, 227, 229–236, 266, 327, 339, 342, 355, 365, 367, 376, 378, 444, 486, 506, 507, 509–511, 524, 531, 534, 543, 570, 581, 587, 594, 598
germ cells, 376, 377, 381, 387, 404, 428, 429
gestation 377, 379, 380, 381, 403, 418, 420, 422, 423, 426, 430
giraffes 4
Giraffidae 4, 6
globin α, β 142, 165–170, 251, 302
globin genes 165–170
glycine 159
goats 5, 6, 14, 26, 101, 164, 165, 256, 261, 593
gonadal development 376–378, 427
SRY sex determination 376, 377, 427
gonadotropin-releasing hormone (GnRH) 378, 379
goral (*Nemorhedus*) 7, 8, 10
grooming 367, 368

hair and wool structure loci 80, 81
Fibre Loss 80
Halo Hair 80
Lustrous Wool 80, 81
Silky Wool 80, 81
hair follicle 55, 57, 66, 79, 81
handling 359–361
hartebeest 5
haemoglobins 142, 152, 165, 211
Haemonchus contortus 203, 205, 211
heritability 143, 202, 204, 216, 354, 356, 357, 362, 368, 381–383, 386, 472, 479, 480, 482, 485, 527, 572, 575, 576, 581
carcass traits 528
coccidiosis 206
dairy traits 544, 550
embryonic mortality 385
fat depth 528
litter size 385
meat tenderness 533

resilience 203
testicular size 386
hermaphrodites 263, 266
heterosis 509, 587
Hippopotamidae 4
Hippotraginae 5, 6
horns 22, 26, 103
Horns locus 337
horses 26, 261
human genome 165, 278
hybridization 16, 268
fluorescence *in situ* hybridization (FISH) 2, 267, 279
in situ hybridization 191, 267–271, 337
Southern blot 192
hypothalamopituitary gonadal axis 378–380

immune responce 425
immunoglobulins 141, 184–186, 425
diversity, the primary repertoire 186
impala 5, 6
implantation 415, 420, 422
imprinted genes 429, 430
inbreeding 18, 20, 264, 450, 454, 455, 457, 463, 511, 559, 582, 586
incisors 4
inherited traits and disorders
adactyly 92
agnathia 92, 119
alopecia 92, 119
anencephaly 92
aprosopia 92
argininaemia 92, 119, 110
arthrogryposis 92
ataxia 92, 119
atresia ani 92
atresia ilei 93
atresia jejuni 93
birth coat 93
blindness 93
blood group system 93, 136, 137
blood group system A 93, 136
blood group system B 93, 136
blood group system C 93, 136
blood group system D 94, 137
blood group system F30 94
blood group system F41 94
blood group system I 94
blood group system M 94, 137
blood group system R 94, 137
blood group system X 94, 137
brachygnathia 94
cataract 94, 119
cerebellar cortical atrophy 94
ceroid lipofuscinosis 94–95, 119
chondrodysplasia 95
cleft palate 95, 119
cryptorchidism 95, 119
cystic renal dysplasia 95
Dandy-Walker syndrome 95
dwarfism, Ancon 95, 96
ears, short 97
ectrodactyly 97
Ehrers-Danlos syndrome 97, 119, 120
entropion 97
epidermolysis bullosa 97, 119, 120
epididymal aplasia 97
epitheliogenesis imperfecta 97
fecundity
Booroola, *see* Booroola
Cambridge 98, 383
Iceland 98
Inverdale 99, 580
Java 99, 578
gangliosidosis 100, 119, 120
Gaucher disease 100, 119, 120
glycogen storage disease 100, 119, 120
Goitre familial 100, 119, 120
golden ram effect 101
gonadal hypoplasia 101
haemolytic anaemia 102
haemophilia 102, 119, 120
halo hair 80, 102
heart defect 103
hemimelia 103
hepatic encephalopathy 103
hernia 103, 119
hydrocephalus 104
hydrops foetalis 104, 119
hyperbilirubinaemia 104, 119
hypomyelinogenesis 105
hypoplasia of mammary glands 105
hypospadias 105, 120
hypothyroidism 105, 120
hypotrichosis 105, 120
infertility 105
intersex 106
iris mottling 106
knobbed acrosome 106

inherited traits and disorders *continued*
Krabbe disease 106, 120
lethal grey 54, 106
lustrous wool, 106
mannosidosis, 106, 120
megaoesophagus 107
mesangiocapillary glomerulonephritis 107, 120
microchephaly 108
microphthalmia 108
muscular dystrophy 108
muscular hyperplasia 108
muscular hypertrophy 108
myofibrillar hypoplasia 108
nephropathy 108
neuroaxonal dystrophy 109
neurogenic atrophy 109
neuropathy 109
nucleoside transport effect 109
ocular squamous carcinoma 109
ophthalmia 109
osteohondrosis 110
osteogenesis imperfecta 110, 120
otognatia 110
patent urachus 110
polyceraty 110
potassium transport 110
progressive retinal atrophy 110
reduced glutathione deficiency 110, 111, 120
renal dysplasia 112
response to noradrenaline 112
roll curls 112
schistosomus reflexus 112
silky wool 112
spina bifida 112
split eyelid 113
spongiform encephalopathy 113, 120
swayback 114
syndactyly 115
tail abnormalities 115
tail, short 115
teat number 115
thread defect 115
twinning, conjoined 115
urolithiasis 115
ventricular septal defect 115
wattles 115
Wilms' tumour 116
Wilson disease 116
yellow fat 91
insulin-like growth factors 418–420, 423–425
interferon 416, 417
International Mapping Flock (IMF) 327, 328, 339
intersexes 263

karyotype 2, 6, 7, 150, 245, 261–263
standard sheep karyotype 261–263
keratin 149
keratin genes 153–162
kob 5

lactogen 416–418
lactoglobulin 138, 139, 143
leanness 516, 526, 529
Leidig cells 427
libido and serving capacity 360
LINEs 150–152
linkage 297
groups 298, 299
map 297–299, 327, 341
mapping 3, 277, 278, 297, 298, 339
list of physically mapped loci 247–256
litter size 385
llamas 4, 164
Lucilia cuprina 211
lutenizing hormone (LH) 378, 379, 382, 398
lymphocytes 242, 243
B-cells 184
T-cells 210, 211
T-cell receptors 185
lymphocyte cultures 242, 243
lysozyme genes 163–165

major genes
callypige 430, 512, 513, 529, 533, 534, 580
Fec^B (Booroola) 384, 580
Fec^I (Inverdale) 384, 580
mapping function (Kosambi) 327, 337
markers 9, 88, 139, 143, 149, 191, 211–213, 242, 245, 271, 272, 298, 299, 448, 512, 566, 581
marker-assisted selection (MAS) 581
meat quality 506, 523–535

genetic variation in visual meat quality 530
colour 530
fatty acids profile 530
intramuscular fat 530, 531
protein and amino acid concentrations 532
genetic variation in eating quality 562
flavour and odour 534
pH 534
tenderness 532, 534
response to single trait selection 527
MHC antigenes 188–192, 211, 212, 234, 339, 421, 426
meiosis 264, 400
diakinesis 328
metaphase I, 328
metaphase II 401
melanin 52, 53, 58
eumelanin 52, 53, 57, 62, 63, 65, 67–69, 79
phaeomelanin 52, 53, 55, 57, 62, 63, 65, 66, 79
melanoblast 60
melanocyte 52, 53, 55, 57, 64–66, 71, 76
migration 53, 60, 71, 75
melanocyte stimulating hormone (MSH) 53, 57, 64
melanogenesis 53, 65
melanosome 52, 53, 67
metallothionein genes 162, 163
microsatellites 9, 99, 151, 152, 271, 277, 278, 298, 299, 328, 340, 448
milk 9, 143, 539–560
milk recording 548–550
Miocene 4, 10
mitogens 242, 243
concavalin A 242
lectin mitogens 242
pokeweed mitogen 242
phytohaemagglutinin 242
mitotic index 243
mitosis 400
MOET (multiple ovulation and embryo transfer) 559, 580
molar 4
Moschidae 4
mouflon 8, 9, 15, 16, 46, 266
mountain goat (*Oreamnos*) 7, 8
mouse genome 165, 278
Müllerian ducts 378, 427
anti-Müllerian hormone 427, 428
musk deer 4
musk ox (*Ovibos*) 7, 10, 19
muscle depth 527, 530

nematode resistance 214
Nemorhedus 7
neural crest 52, 60, 71
nuclear transfer 400–403
coordination of cell cycle 400–401
stage of development of donor cells 401–403

oestradiol 378
oestrus 41, 363, 381, 382, 386, 387, 396, 407, 546
seasonality 381, 382
fertility 381, 382
okapi 4
ontogenesis 19
oryx 5
outbreeding 18, 21
ovary 376
Ovibos 7
Ovis 1, 3, 7, 8, 10, 354
O. ammon (argali) 15
O. aries, (sheep) 1, 9, 15, 260, 263
O. canadensis (bighorn) 15, 340
O. dalli (thinhorn) 15
O. musimon (European mouflon) 15
O. nivicola (snow ship) 15
O. orientalis (Asian mouflon) 15
O. vignei (urial) 9, 15
ovulation 19, 29, 98, 99, 337, 377, 380, 381, 383–385, 396, 397, 401, 402, 405, 415, 431, 438, 444, 512
major genes affecting ovulation rate 19, 98, 337, 383, 444, 512
polygenic control 383
oocyte 377, 397, 398, 400–402, 407, 413, 421
in vitro fertilization (IVF) 398, 400
in vitro maturation (IVM) 397, 398

Peccaries 4
Pecora 5, 10
Peleinae 5
pelt sheep 63, 70

peptide growth factors 418–420
Perissodactyla 4
phenetics 2
phenotype 59, 70, 78, 79, 384, 389, 430, 449, 571, 572, 574, 598
photophobia 66
phylogeny 1, 2, 6
physical maps 241, 274–277
pig 26, 261, 278
pigmentation 52–80
piramidal management of the population 546, 547
placenta 416
placental lactogens 417
plasminogen 422
Pleistocene 8
polled sheep 22, 27–45, 81, 103, 597
polledness 19
polygenic traits 338-340, 383
polymorphism 3, 134, 135, 137, 139–141, 143, 184, 189, 190, 211, 212, 230, 232, 261, 298, 361, 558
population size and inbreeding 455–457
pregnancy 381, 383, 416
pregnancy associated glycoproteins 420
preimplantation 413, 422
premolar 4
primordial follicles 377, 383, 400
prion 226–228
    PrP protein 227, 228
    scrapie diagnosis 227, 228
production systems and breeding objectives 507, 509
progesteron 378, 396
prolactin 417
prolificacy 19, 29, 34, 41, 98, 380, 381, 383, 384, 390, 540, 543, 548, 558
Pronghorn antelope 4
prostaglandin 378
puberty 380–382, 430
puku 5

QTL (quantitative trait loci) 212, 213, 338–340, 513, 580, 581

RAPD (Random Amplified Polymorphic DNA) 298
recombinants (crossover events) 298, 327, 338
recombination rate 328
    sex specific 328
red cell antigens 134–137
Reduncinae 5, 6
repeatability 354, 358, 362, 554
repetitive sequences 150–152, 256, 268
reproduction
    components of reproductive rate in the ewe 380–384
    male 386, 387
    modern technologies 395–408
resilience 202
resistance 203
to dermatophilosis and fly strike 206–208
    to disease, genetic aspects 216
    to internal parasites 10, 203, 204
    fleece-rot/fly strike 210, 211
    measurement 203–205
RFLP (Restriction Fragment Length Polymorphism) 157, 189, 298
rhebok 5
RNA 246
    12S rRNA 6
    16S rRNA 6
    tRNA 6, 152
RTUS (real-time ultrasound) 527–529
Ruminantia 4, 6
ruminants 2, 5, 15, 46, 164, 184, 187, 190, 205, 213, 594, 595
Rupicapra 7, 8, 10
Rupicaprini 8

saiga 5
scrapie 225–237
    clinical signs 226
    genetics of succeptibility 228–233
    transmission 234–237
selection
    carcase composition 527–530
    destabilizing 18
    disease resistance 209–211
    indexes 388, 492, 493, 495–498, 508, 512, 514, 516, 527, 558, 570, 576, 578
    mate selection 585, 586
    on multiple traits 578, 579
    prezygotic 16
    realized responses 3886, 485–492, 576, 582, 585
    direct responses 486–489, 527–529

correlated responses 385, 489, 496, 511, 524
reproductive performance 388, 389
Sertoli cells 380, 427
sex determination and differentiation 378, 427, 428
sexual maturity 29, 31, 33, 34, 36, 38–45, 381, 387, 581
Sheep Base 328, 336, 337
SINEs 150–152
skeleton 24
somatic cell hybrids 271–273
use for gene mapping 272
spermatogenesis 386
SRY gene 377, 427
SSCP (Single Strand Conformational Polymorphism) 298, 299
staple 26, 28, 29, 30, 32–38, 41–45, 476, 477, 484, 494
substitutions 3, 6, 115, 138, 158
Suidae 4
synteny 7
Systematics 1, 2
cladistics 2
classical evolutionary taxonomy 2
molecular taxonomy 2
phenetics 2

tahr (*Hemitragus*) 7, 8, 10
tail 24, 34, 35, 38, 41
fat-tailed 26–28, 30, 33, 35, 36, 39
takin (*Budorcas*) 7
taxonomy 2, 3, 8
Tayasuidae 4
T-cell receptor 185–188
textile 18, 22, 23, 472
testicular size 386
testosteron 377, 379, 380, 427
Tragulidae 4
Transferrin 140, 141
transgenics 160, 395, 403
translocations
reciprocal 265
Robertsonian 7, 256, 262, 263
transversions 6
Trichostrongylus colubriformis 203, 205
Tylopoda 46
tyrosinase 53, 65, 67, 120
tyrosinase-related protein (TRP) 53
tyrosine 53, 138, 159, 161, 339, 532

underwool 23
urial 8, 15, 22
uterus 378, 396, 415, 427

vaccination 215, 216
vicuna 4
Villafranchian 8

waterbuck 5
wildebeest 5
Wolfian ducts 378
wool
carpet 9, 23, 27, 28, 30, 32, 35, 37–39, 42, 44, 81, 90, 102, 161, 472
doggy 95
fibres 102
colour 78, 478, 485, 493
fibre diameter 28–30, 32–34, 36–38, 41–45, 475, 478, 483–485, 492–495
fleece weight 28, 29, 33, 34, 36–38, 41–45, 480–483, 486, 492–495
staple length 26, 28–30, 32-38, 41–45, 477, 484
staple strength 476, 477, 484, 494
style 477, 478, 485
yield 28–30, 34, 35, 44, 483, 493
hair mutations 80, 81, 161, 162
traits of importance 472–479
top-making 469
spinning 473
weaving/knitting 474
garment manifacture 474
vegetable matter 476
wool keratin genes 153–162
keratin IF genes 153–156
KAP genes 156-160

X chromosome inactivation 428, 429